The Physics of Submicron Semiconductor Devices

NATO ASI Series

Advanced Science Institutes Series

A series presenting the results of activities sponsored by the NATO Science Committee, which aims at the dissemination of advanced scientific and technological knowledge, with a view to strengthening links between scientific communities.

The series is published by an international board of publishers in conjunction with the NATO Scientific Affairs Division

A	**Life Sciences**	Plenum Publishing Corporation
B	**Physics**	New York and London
C	**Mathematical and Physical Sciences**	Kluwer Academic Publishers Dordrecht, Boston, and London
D	**Behavioral and Social Sciences**	
E	**Applied Sciences**	
F	**Computer and Systems Sciences**	Springer-Verlag
G	**Ecological Sciences**	Berlin, Heidelberg, New York, London,
H	**Cell Biology**	Paris, and Tokyo

Recent Volumes in this Series

Volume 175—Superstrings
edited by Peter G. O. Freund and K. T. Mahanthappa

Volume 176—Nonlinear Evolution and Chaotic Phenomena
edited by Giovanni Gallavotti and Paul F. Zweifel

Volume 177—Instabilities and Chaos in Quantum Optics II
edited by N. B. Abraham, F. T. Arecchi, and L. A. Lugiato

Volume 178—High-Brightness Accelerators
edited by Anthony K. Hyder, M. Franklin Rose, and Arthur H. Guenther

Volume 179—Interfaces, Quantum Wells, and Superlattices
edited by C. Richard Leavens and Roger Taylor

Volume 180—The Physics of Submicron Semiconductor Devices
edited by Harold L. Grubin, David K. Ferry, and C. Jacoboni

Volume 181—Fundamental Processes of Atomic Dynamics
edited by J. S. Briggs, H. Kleinpoppen, and H. O. Lutz

Volume 182—Physics, Fabrication, and Applications of Multilayered Structures
edited by P. Dhez and C. Weisbuch

Series B: Physics

The Physics of Submicron Semiconductor Devices

Edited by

Harold L. Grubin

Scientific Research Associates, Inc.
Glastonbury, Connecticut

David K. Ferry

Arizona State University
Tempe, Arizona

and

C. Jacoboni

Institute of Physics
University of Modena
Modena, Italy

Springer Science+Business Media, LLC

Based on a NATO Advanced Study Institute on
Physics of Submicron Semiconductor Devices,
held July 10-23, 1983,
in San Miniato, Italy

Library of Congress Cataloging in Publication Data

NATO Advanced Study Institute on Physics of Submicron Semiconductor Devices, (1983, in San Miniato, Italy)
The Physics of submicron semiconductor devices.

(NATO ASI series. Series B, Physics; v. 180)
Based on a NATO Advanced Study Institute on Physics of Submicron Semiconductor Devices, held July 10-23, 1983, in San Miniato, Italy.
Bibliography: p.
Includes Index.
1. Semiconductors—Congresses. 2. Electron transport—Congresses. 3. Microstructure—Congresses. I. Grubin, Harold L. II. Ferry, David K. III. Jacoboni, C. IV. Title. V. Series.
QC610.9.N37 1983 537.6′22 88-19713
ISBN 978-1-4899-2384-4

ISBN 978-1-4899-2384-4 ISBN 978-1-4899-2382-0 (eBook)
DOI 10.1007/978-1-4899-2382-0

Originally published by Plenum Press, New York in 1988
Softcover reprint of the hardcover 1st edition 1988

PREFACE

The papers contained in the volume represent lectures delivered as a 1983 NATO ASI, held at Urbino, Italy. The lecture series was designed to identify the key submicron and ultrasubmicron device physics, transport, materials and contact issues. Nonequilibrium transport, quantum transport, interfacial and size constraints issues were also highlighted. The ASI was supported by NATO and the European Research Office.

H. L. Grubin
D. K. Ferry
C. Jacoboni

CONTENTS

MODELLING OF SUB-MICRON DEVICES

Eugène Constant

Centre Hyperfrequences et Semiconducteurs
LA CNRS N° 287, GRECO Microonde N° 11
Université de Lille 1. Bât P3
59655 VILLENEUVE D'ASCQ Cedex, France

INTRODUCTION

Progress in the microelectronics industry is strongly coupled with the ability to make ever increasing numbers of smaller devices on a single chip. The advent of high-resolution electron and X-ray lithographic techniques is leading toward an era in which individual features sizes might well be fabricated on the scale of 10-20 nm. It will then become feasible to develop very small device structures where size and related effects may be as important as the bulk properties of the host semiconductor material. Moreover, it becomes obvious that we must now ask whether classical device modelling may be extrapolated down to the very small space and time scales usually encountered in sub-micron devices. It is the purpose of this paper to study in a first part what kind of new phenomena may occur in sub-micron devices, to suggest and to describe in a second part new methods of modelling which take them into account and to discuss in a third part what could be the future performance of submicronic logic or microwave devices.

WHAT NEW PHENOMENA OCCUR AND MUST BE TAKEN INTO ACCOUNT IN SUB-MICRONIC DEVICES

Background

The three main parameters characterizing the dynamics of the carrier in a semiconductor, the velocity, the effective mass and the energy depend on the band structure through the following equations:

$$\varepsilon = \varepsilon(\vec{k}), \ \vec{v} = \hbar^{-1} \left(\frac{\partial \varepsilon}{\partial \vec{k}}\right)_{\vec{k}} \quad \text{and} \ m^* = \hbar^2 \left(\frac{\partial^2 \varepsilon}{\partial k^2}\right)^{-1}_{\vec{k}} \tag{1}$$

As a result, these three quantities will be determined if the wave vector $\vec{k}$ is known. For this, we have to remember that $\vec{k}$ results from:

- the determinist effect of the electric field which modifies $\vec{k}$ every time step Δt through the equation:

$$\Delta\vec{k} = \frac{q\vec{E}\Delta t}{\hbar} \tag{2}$$

- the stochastic effect of the scattering mechanisms which may randomly change $\vec{k}$ with a probability

$$\lambda(\vec{k}) \ . \ \Delta t = \Delta t \int_{\vec{k}'} W(\vec{k}, \vec{k}') \, d\vec{k}' \tag{3}$$

In usual cases, the time and space variation of the electric field during the carrier mean free time and along the carrier mean free path can be neglected. Carriers can then be considered as in steady state conditions, since there is a balance between the effect of the electric field and the scattering mechanisms. As a result, if relatively broad energy and velocity distributions are observed due to the random nature of the scattering mechanisms, all the averages values of ε and of the drift velocity $\bar{v}$ depend only on the instantaneous local electric field regardless of the values of the electric field in the past and all around the point being studied in the semiconductor material. In this case, classical current equations and methods of modelling can be used assuming that all quantities only depend on the electric field.

"Non steady state carrier transport" will be achieved either when very high frequency or fast transient voltage are applied to large devices or when submicronic devices are used (even in steady state regime). In these cases, the variation of the electric field during either a mean free time between two collisions or along a mean free path cannot be neglected and new features will characterize electron transport and must be taken into account in the modelling of the devices.

In order to obtain a good understanding of all these phenomena which occur either in large device for a very high frequency operation or in submicronic device we intend to successively describe the main drift phenomena* which occur in a bulk semiconductor submitted to fast temporal variation of the electric field and the case of submicronic devices where the electric field applied is characterized by short spatial scales.

Time Dependent Phenomena in Bulk Semiconductor

Here, we intend to study the motion of an ensemble of electrons subjected in a uniform bulk semiconductor to various time configurations of the electric field. Ensemble average then allow to obtain the variation versus time t of all dynamic parameters such as the mean drift velocity and the mean carrier energy. Monte Carlo methods which have been described in previous lectures are well suited to carry out such study and consequently all the results which will be given will be obtained by this method using, in the case of GaAs, the standard 3 valleys model described in (2).

The main features of the transient drift phenomena which occur are summarized in figs. 1-3 where an ensemble of electrons are submitted to various time configuration of E:

a very short time pulse in the case of fig. 1

a time step in the case of fig. 2

a time pulse configuration and simultaneously a steady state field in the cases of fig. 3

In all these cases, it can be noted that very high values of the drift velocity (much higher than in steady states) can be achieved during very short times. However, it should be noted that in case 3, very low value of the velocity can also be achieved during a small while even for high values of the electric field. Obviously, all these features have to be taken into account in the modelling of the devices and consequently have to be studied in some more detail.

Most of the phenomena involved are due to the specific characteristics of electron dynamic in usual semiconductors, where in most cases, when the energy increases, the scattering rates and the effective mass also increase. As a result, low value of the drift velocity will be achieved for hot carriers characterized by high value of the energy since for this type of carriers, the effective mass is very high and the increase of the velocity due to the electric field:

$$\Delta\vec{v} = \frac{q\vec{E}}{m^*}\,dt \qquad (4)$$

*Diffusion phenomena can also be very different from those achieved in steady conditions (1) and obviously these features have also to be taken into account in the modelling of submicronic devices.

is very small while the randomization of the velocity due to the scattering mechanisms strongly occurs. On the contrary, cold carriers (characterized by low value of ε) will be able to reach very high value of the drift velocity since the electric field effect on the drift velocity is much higher (the effective mass is reduced) and the scattering rate is strongly reduced. Consequently average drift velocity which can be realized depend not only on the electric field but also on the instantaneous average energy of carriers $\bar{\varepsilon}$. Concerning this point, we have to remember that the variation of $\bar{\varepsilon}$ does not instantaneously follow the variation of E since increasing $\bar{\varepsilon}$ of $\overline{\Delta\varepsilon}$, need the work of the electric field acting on the carrier i.e. the travel $d\vec{r}$ of the carriers to be sufficient:

$$\vec{E}.d\vec{r} > \Delta\varepsilon$$

On the contrary, reducing $\bar{\varepsilon}$ needs suitable inelastic scattering mechanisms and a long time can be necessary for these scattering mechanisms to occur.

Based on this theoretical background, the results given in figs. 1-3 can be roughly explained. In case of fig. 1, which will be referred to as the ballistic case, a very short time pulse is applied to an ensemble of thermal equilibrium electrons. If the pulse duration is chosen in order to obtain a final energy slightly smaller than the intervalley gap $\Delta\varepsilon_{\Gamma L}$ (for $\varepsilon > \Delta\varepsilon_{\Gamma L}$, intervalley transfers strongly occur), scattering rate remains very low and for the very short pulse duration, scattering events do not occur. Only the effect of electric field is observed and as a result, very high drift velocity can be achieved without an important velocity distribution. When the electric field is cancelled, the initial very high velocity decreases very slowly since scattering mechanism are needed to randomize the drift velocity. The electric field vanishing to zero, a really ballistic motion of all the carriers is then achieved a little like a ballistic rocket where all the motion is carried out without driving field i.e. when the motor stopped and with this type of motion, quite long distances can be achieved over very short time.

In the case of fig. 2, which will be referred to as the overshoot case, a constant electric field is suddenly applied to an ensemble of thermal equilibrium electrons until the electron energy reaches a value close to $\Delta\varepsilon_{\Gamma L}$ (this is achieved in order to keep not too high values of the scattering rates). Since the time duration is short, a steady state is not obtained and due to the electric field effect high velocities can be achieved. However, it should be noted that, due to scattering, which has broadened the velocity distribution, the average velocity which can be achieved for an energy close to $\Delta\varepsilon_{\Gamma L}$ is much lower than the value which could be expected without distribution (about 10^8 cm/sec) from the classical equation:

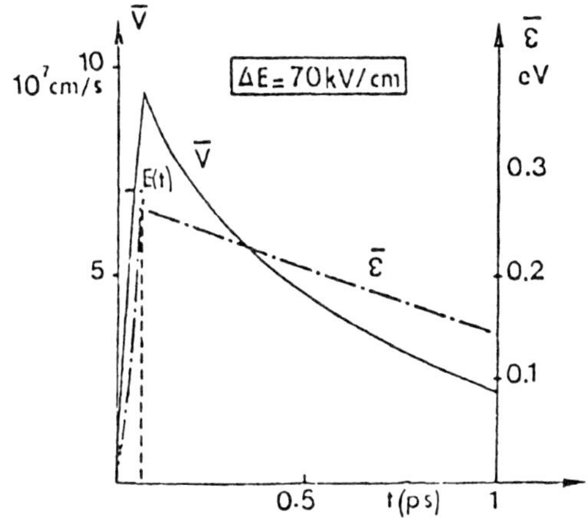

Figure 1. Ballistic case average velocity $\bar{v}$ and average ε functions of time of an ensemble of electrons submetted to various time configuration of the electric field (plotted on the same figure). All the results are obtained from Monte Carlo simulation. (GaAs N_D = 0 T = 77°K in the cases of fig. 1 and 2, T = 293°K in the case of fig. 3)

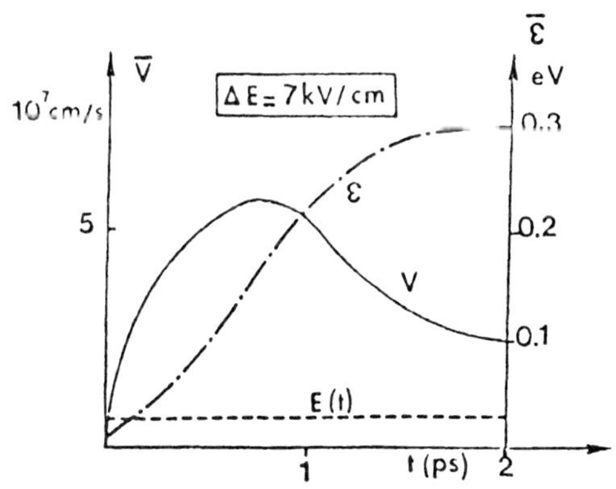

Figure 2. Overshoot case

$$\vec{v} = \frac{1}{\hbar} \left(\frac{\partial \varepsilon}{\partial \vec{k}}\right)_{\Delta\varepsilon_{\Gamma L}} \quad (5)$$

In addition, one can note that due to inertial effect, this overshoot velocity cannot be reached instantaneously. Just after the application of the electric field, scattering mechanisms have not yet occurred and this type of motion is also often called a ballistic one. However, one can remark that the velocity here achieved is not very high and consequently this type of ballistic behavior is not as advantageous as the motion achieved in the case of fig. 1.

In the result corresponding to case 3, not only the overshoot phenomena, but also that will be called the undershoot phenomena are illustrated. The GaAs semiconductor is, in this example, submitted to a time pulse configuration of E and simultaneously to a steady state field. Results obtained on the transient drift velocity can be roughly understood if we remember that this drift velocity depends not only on the electric field but also on the instantaneous value of the energy. State A is a few tenths of a picosecond after the application of the high electric field (20 kV/cm): the instantaneous energy has not yet increased and the scattering rate remains low despite the high value of E. A very high overshoot velocity can be obtained: in usual semiconductor, such phenomena will occur every time the instantaneous energy is lower than the steady one corresponding to the applied electric field. State B is 1 ps after the application of E; the energy has now reached its steady state value and a large number of carriers are now in the X valley. The scattering rates and the effective mass are very high and the average drift velocity is reduced although the electric field remains very high. State C is a few tenths of a picosecond after the application of the pulse, the instantaneous energy has not yet been reduced, the scattering rate remains very high and since the electric field is now very low, there is a drastic decrease in the drift velocity. Such phenomenon can be called the undershoot velocity effect; it will be achieved in usual semi-conductor every time the instantaneous energy is higher than the steady state value corresponding to the value of the applied electric field. However, with increasing time, the energy will decrease, carriers will return to the Γ valley, the scattering rate and the effective mass will strongly be reduced and the drift velocity will increase: a steady state identical to the initial state will be reached.

It has been shown that, due to overshoot or ballistic motion, average velocity much higher and in a few cases much lower than in steady state could be achieved. Such phenomena occurring in bulk semiconductor submitted to fast transient electric field might also occur in submicronic devices. However, in such devices, the electric field is characterized by short spatial and not always time scale.

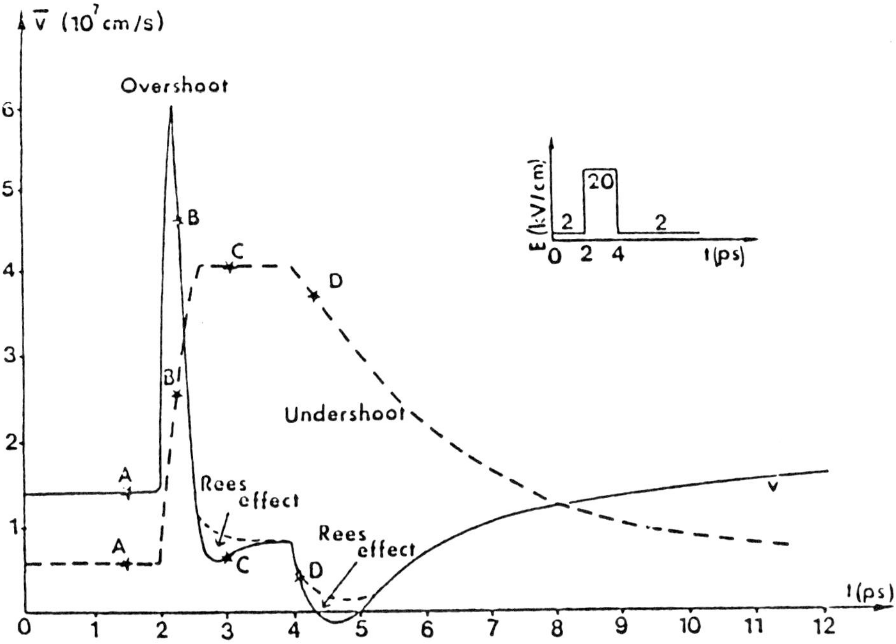

Figure 3. Overshoot and Undershoot Case

Due to this spatial nonuniformity, we now intend to show that additional phenomena occur which might considerably change the observed phenomena.

Space Dependent Phenomena

Let us now study the characteristics of nonsteady state electron dynamics in submicronic devices, i.e. in semiconductor subjected to space configuration of electric field characterized by short spatial scales.

In this case, some of the characteristics of the transport of carriers can be deduced from the results previously obtained when time configuration of E is applied to the whole semiconductor by using the usual relation

$$\overline{d\vec{r}} = \overline{\vec{v}}\ dt \tag{6}$$

However, it is very important to note that the spatial nonuniformity of the free carrier concentration, energy and of the distribution functions generate additional phenomena. The most important are:

Diffusion current due to the nonuniformity of n

Heat conduction due to nonuniformity of $\overline{\varepsilon}$

Thermoelectronic current due to the nonuniformity of $\overline{v^2}$

In order to illustrate such phenomena, Monte Carlo simulation can be carried out when a space pulse configuration of the electric field is applied to an ensemble of electrons in a semiconductor* and the results obtained in fig. 4 (when a stationary state is reached) can be compared with those obtained when a time pulse configuration of the electric field is applied [3]. One can note (and this appears to be the most striking feature) that the velocity of carriers is generally much lower than in the previous case corresponding to the study versus time. This difference can be partly explained by the diffusion current since the carrier concentration is no longer uniform: The highest values of the velocity which should have been observed at the end of the space pulse are not in fact obtained since the carrier concentration increases steeply and the diffusion velocity is opposite to the drift velocity. On the contrary, in front of the space pulse, the velocity of carriers is slightly higher than in the previous case since carrier concentration decreases and the diffusion velocity adds to the drift velocity. Moreover, one can note that the spatial nonuniformity of the average energy gives rise to heat conduction phenomena since the increase in the average energy is observed in front of the space step.

Let us now study other unusual features of carrier transport in submicron devices. In this type of devices, the active region thickness is often comparable to a few Debye lengths, and the transition region between a fully depleted layer and a quasi-neutral region cannot be neglected. Consequently, we have to know the characteristics of electron dynamics in this transition region where carriers are not only submitted to a driving field E_x along the axis of the conducting channel, but also to a transverse one (a value often com-

*For the sake of simplicity, we have not taken into account the effect of the space charge due to electron on the value of electric field so far (via Poisson's equation); this is to clearly put in light additional phenomena due to a space configuration of the electric field. Obviously, in the study of practical submicronic devices, space charge effect must be taken into account and this will be achieved in Section III.

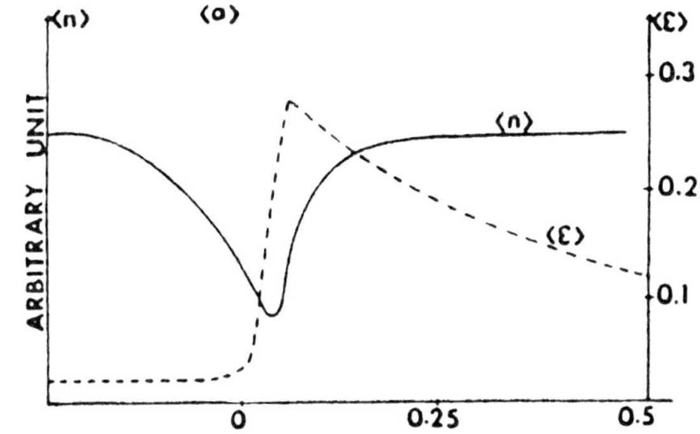

Figure 4a. Average electron concentration and average energy profile of electrons when a stationary state is reached in the structure, characterized by a space pulse configuration of the electric field (see ref. [4]).

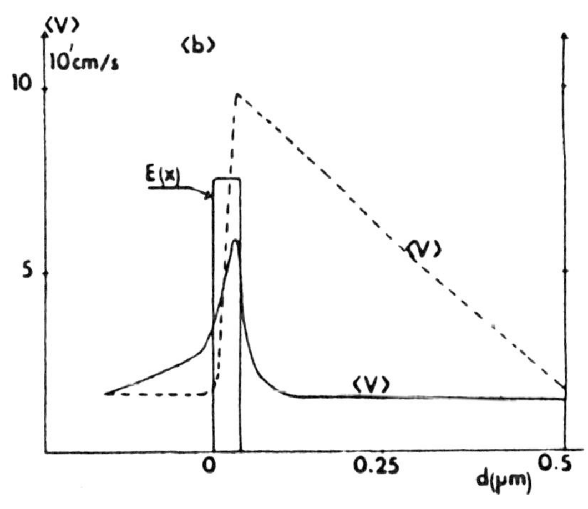

Figure 4b. Comparison between average velocity of electrons submitted to a space pulse (continuous curve) or time pulse (dashed curve) of electric field (see ref. [4]). The value of the electric field pulse is 70 kV/cm over a distance of 0.05 μm.

parable to E_x). This is particularly the case in nearly pinched-off GaAs FET operations and as an example, the characteristics of the carrier dynamic in an incremental section of a symmetrical FET will be shown for gate voltages close to the pinch-off value. Such a study can be easily achieved using MC procedures (5). Typical results obtained by this method are given in figure 5 in which are plotted along the O_y axis (perpendicular to the gate), variation in the transverse electric field E_y, the free-carrier concentration, the average energy and the average longitudinal drift velocity $\langle v_x \rangle$. It can be noted from figure 5(b) that most of the carriers are submitted not only to a longitudinal electric field (in the studied case E_x - 30 kV cm^{-1}), but also to a non-negligible transverse electric field. It is therefore necessary to take into account as exactly as possible the influence of this transverse electric field on the carrier dynamics in the active regions of the FET. In classical models, it is assumed that both the drift velocity and the average energy depend on the magnitude of the local electric field E_T acting on the carrier; in addition, the drift velocity is assumed to be isotropic (i.e. always to have the same direction as the local electric field). With these assumptions the longitudinal drift velocity along the channel is then given by

$$v_x = \left(\frac{E_x}{E_T}\right) \cdot v_{bulk}(E_T) \tag{7}$$

It is interesting to compare the values obtained using this classical equation with those obtained more exactly by MC procedures. This is achieved in figure (5b and c) for the drift velocity, and also for the average energy of the carriers. It can be seen that in both cases classical methods using relation (7) give results which are quite different from those obtained by MC procedures, and consequently their application to the modelling of FET's is inadequate and gives unreliable results.

From all the results previously described, it clearly appears that the main characteristics of carrier dynamics in submicronic devices but also in large devices submitted to fast transient electric field cannot be taken into account using classical approximations and electrokinetic equations. The modelling of such devices requires either the elaboration of new methods which will be more suitable for submicron devices, or the improvements of the methods presently used for large device and low frequency operation. In the second part of this paper, we will study these two types of methods.

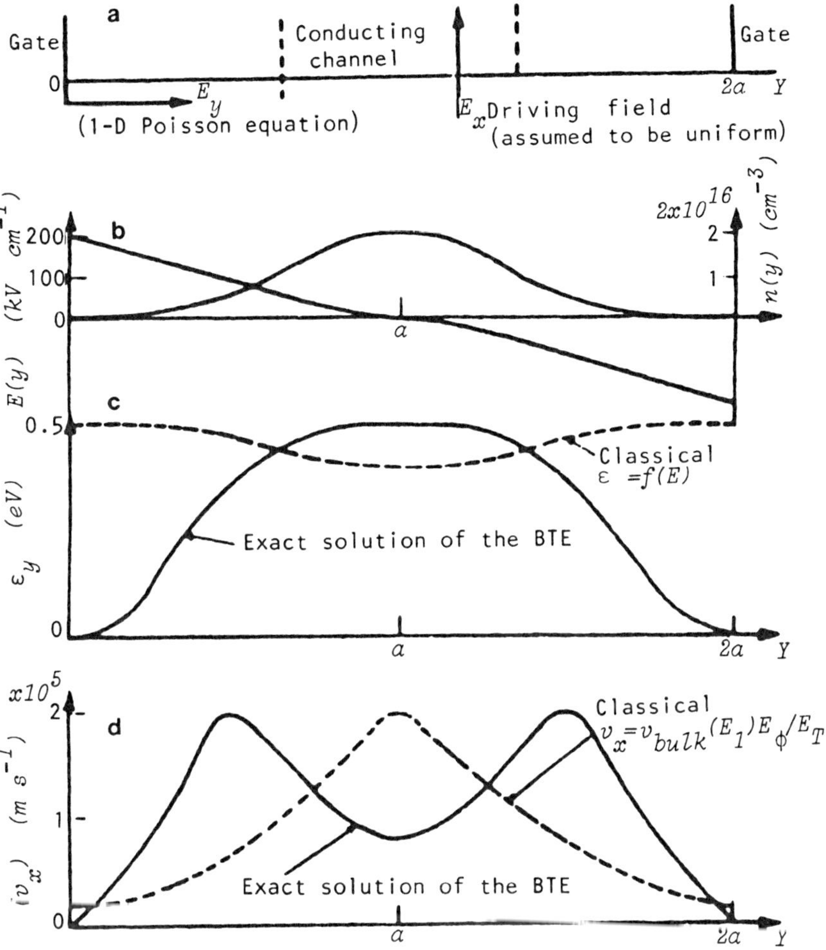

Figure 5a. Representation of a section of a symmetrical FET, (b) variation of the transverse electric field E_y and the free-carrier concentration along the O_y axis. (c) Variation of the average energy along O_y: full curve, calculated using MC procedures; broken curve, using classical bulk data. (d) Variation of the longitudinal average drift velocity vx along O_y; full curve, calculated using MC procedures; broken curve, using relation (6). MC simulation is achieved assuming a uniform driving field E_x, while E_y is obtained by integrating Poisson's equation along 0_y (E_x = 30 kV cm^{-1}, a = 0.13 μm impurity concentration of the active layer $N_D = 10^{17}$ cm^{-3} and $(V_G + V_B)/W_{oo}$ = 0.93).

HOW TO MODEL SUBMICRONIC DEVICES

Monte Carlo Particle Models: the Process and its Basic Features

Monte Carlo particle models appear to be the most promising methods for modelling and will therefore be described first. In such a model the motion of particles representative of the carriers in the device is studied simultaneously in k and r-space using (i) MC simulation of the scattering process in three-dimensional k-space and (ii) a space (one, two or three-dimensional) description of the electric field of the device (obtained from Poisson's equation).

The MC simulation is first used to obtain the drift velocity of all the simulated carriers from which the instantaneous positions of carriers in the device are deduced. The carrier density can then be obtained and used to solve Poisson's equation and to determine the electric field required at each point of the device to carry out further MC simulations.

Two main methods can be employed to determine the carrier density when the positions of the particles have been obtained. In the single-carrier method, the motion of only one particle is studied and the carrier density is determined progressively from the time ΔT spent by the particle in an elementary volume $\Delta^3 r$ of the device using the relation

$$n(r) = \frac{K\Delta T}{T\Delta^3 r} \tag{8}$$

where T is the observation time during which the motion has been studied, and K is a normalisation constant determined from the total number of carriers in the device. Such a method does not allow one to obtain the time dependence of n because, by using relation (8) the integration of n over time is carried out automatically. This method can therefore only be used to obtain stationary values of the various parameters of the device. In the multi-carrier method, the carrier density is determined from the number of simulated carriers observed at an instant t in an elementary volume $\Delta^3 r$. The time dependence of the carrier density can be obtained, and the high frequency behavior of the device can then be studied.

The main application of MC methods were to FET's [6], [7] and an example of results obtained are given in fig. 6 [8]. In view of the importance of this method of modelling, it is interesting to highlight the main features of MC methods in relation to other more classical methods. Its main advantages appear to be:

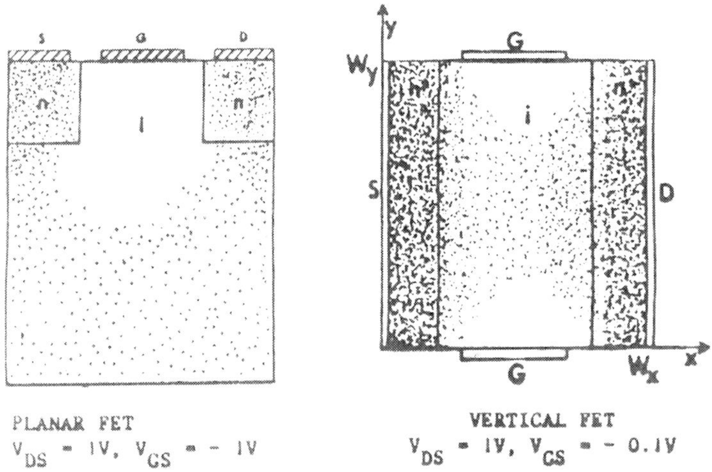

Figure 6. Monte Carlo simulation of space charge injection FET's i.e. FET's using a nondoped active layer. Instantaneous locations of the electrons in the devices.

1. The continuity and current equations (including the diffusivity effect) are automatically taken into account when carrying out the simulation process. Consequently the computer has to solve only one differential equation (Poisson's), and this remains true even if complex devices (such as bipolars) are studied.

2. All the characteristics of carrier transport (for example, nonstationary features) which occur in submicron devices are taken into account exactly. The only condition is that the BTE and the band structure approximation must be valid.

3. Surface and interface scattering effect due to heterostructure and heterojunction can be taken into account without too many complications. An example of results obtained for ID GaAlAs GaAs heterostructure is given in fig. 7.

4. Additional phenomena (such as IMPATT ionization, electron-electron interactions) can be described, and preliminary results have already been obtained.

5. The MC method can be used to determine all the device properties, including noise.

6. Bipolar devices can be studied without greatly increasing the computation time and the fig. 8 is related to the study of a planar photo-resistance where the motion of both the holes and the electrons created by the laser excitation has to be taken into account and studied [9] in the 3D band structure (conduction and valence bands) and in the 2D $\vec{r}$ space.

On the other hand, a few serious problems are generally encountered using MC particle methods; the main ones are as follows:

1. The method requires very long computation times and consequently needs a very powerful computer.

2. The numerical results obtained by this method cannot be very accurate due to the stochastic motion of the simulated carriers and to the resulting random fluctuation of the calculated parameters (the physical meaning of these fluctuation resulting from thermal noise can be pointed out). Concerning this point, it should be noted that the accuracy achieved is inversely proportional to the square root of the number of simulated carriers and/or the observation time; very expensive calculations are therefore needed if highly accurate results are to be obtained.

3. Taking into account that the computation time is roughly proportional to the observation time, the study of phenomena over a very long period will be partucularly expensive and so some physical phenomena such as trapping or generation and recombination effects are very difficult to investigate.

The first two problems appear to be the most difficult features of MC particle methods, but it should be noted that by using single-carrier methods some improvement can be achieved. Since only one carrier has to be studied, computer time and memory occupation can be reduced. In addition due to the time integration of all quantities, thermal noise and random fluctuations of the calculated parameters are greatly reduced. Unfortunately only the stationary state can then be studied, and effects such as those due to microwave or optical fields cannot be investigated.

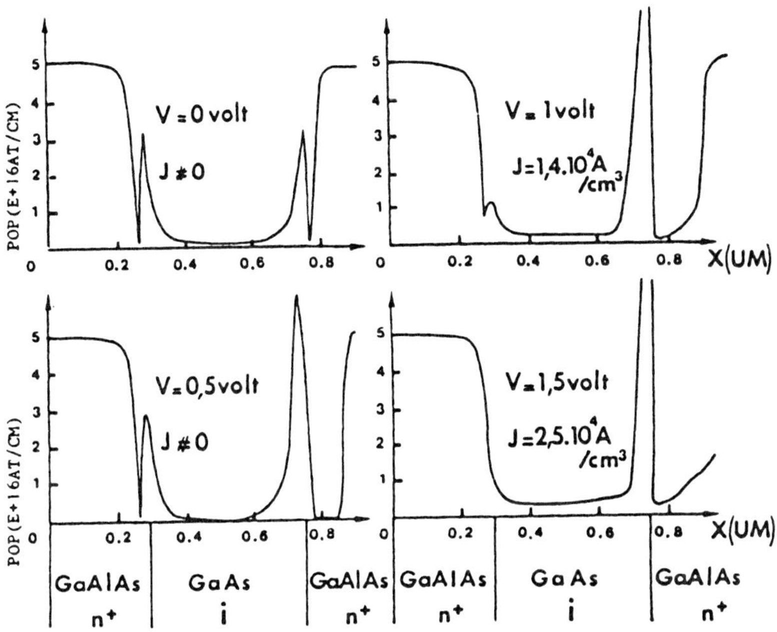

Figure 7. Electron concentration profile in 1D GaAlAs/GaAs/GaAlAs heterostructure for various applied voltages. The alloy concentration of GaAlAs is 0.3 and the doping concentration $5.10^{16}/cm^3$.

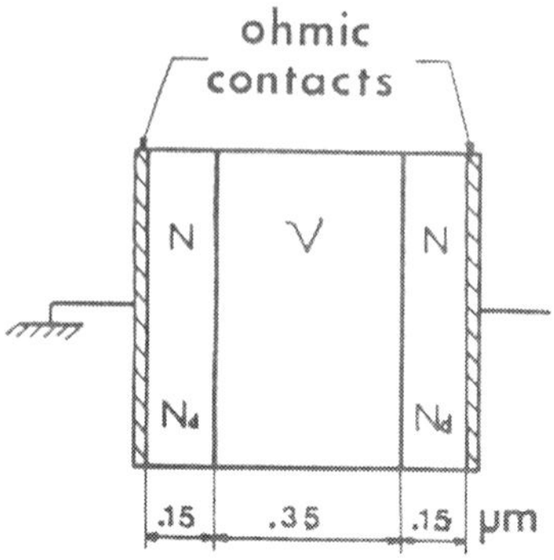

a) Structure of the device
($N_d = 3.10^{16}/cm^3$, V=0)

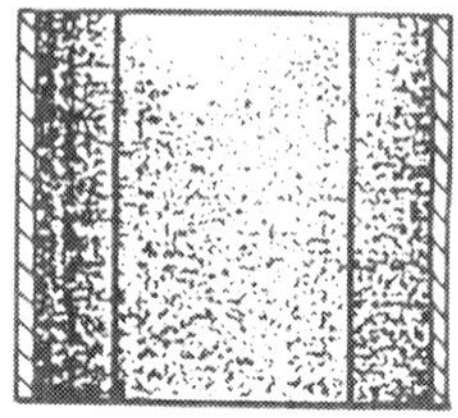

b) Distribution of electrons in darkness

c) Distribution of electrons with continuous illumination

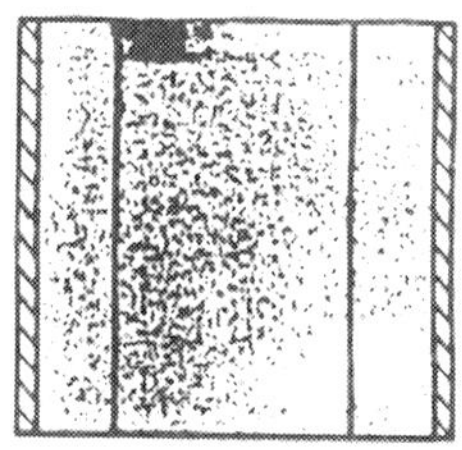

d) Distribution of holes continuous illumination

Figure 8. Monte Carlo Simulation of a Planar Photoconductor a,b,c,d captions.

In a more general point of view, it should be noted that obviously the results obtained will be exact only if all the parameters introduced in the model are exactly known. Consequently, if we want accurate results, we have to know with a sufficient accuracy:

- the band structure and all the scattering mechanisms characterizing electrons not only in bulk semiconductor but also near to surface or interfaces.

- the boundary condition characterizing the studied submicronic device.

In many cases our knowledge about all these parameters is not sufficient and additional theoretical or experimental investigations are needed to achieve accurate modelings of most submicronic devices.

IMPROVEMENT OF CLASSICAL METHODS: THE RELAXATION TIME APPROXIMATION: BASIC FEATURES

It has been shown that it is very difficult to obtain by MC methods very accurate numerical results due to the stochastic motion of the simulated carrier. If when studying static properties of a device, these random fluctuation can be almost cancelled by a suitable time integration this is not the case when time dependent properties of the device have to be studied. In this case more macroscopic methods which don't take into account thermal noise seem more useful. Consequently, it appears interesting to consider whether suitable improvements can be made to the classical macroscopic methods of modeling large devices in order for them to become useable for submicron devices. The basic equations usually employed to calculate the density current in large unipolar devices are as follows:

$$\vec{J} = q\, n\, \overline{\vec{v}}(E) + q\, D_n(E)\, \vec{\nabla} n \tag{9}$$

It can be noted that the least accurate and also the main assumptions employed in these equations are that the average drift velocity (and also the diffusion coefficient D) are instantaneous functions of the electric field since it has been shown that the drift velocity also depends on the average energy of the carrier. However, a considerable improvement could be achieved by using the relaxation time approximations in order to obtain the energy of the carriers and then to determine the drift velocity.

Two cases will be successively studied. In the first one, only the effect due to time variation of the electric field will be considered. The two basics equations which can then be employed are as follows:

$$\frac{d m^*(\overline{\varepsilon})\overline{v}}{dt} = qE \quad \frac{m^*(\overline{\varepsilon})\ \overline{v}}{\tau_m(\overline{\varepsilon})} \tag{10}$$

$$\frac{d\bar{\varepsilon}}{dt} = qE\bar{v} - \frac{\bar{\varepsilon} - \varepsilon_o}{\tau_\varepsilon(\bar{\varepsilon})} \tag{11}$$

where ε_o is the thermal energy 3/2 kT_o of the lattice.

These relations can be roughly justified taking into account the conservation of momentum and energy, but can also be deduced more rigorously from a suitable integration of BTE under certain conditions as discussed by NOUGIER and al [10]. In these two relations, scattering mechanisms are taken into account by introducing the momentum and energy relaxation times τ_m and τ_ε which will be assumed to depend on the energy $\bar{\varepsilon}$ of the carrier. As a result, it should be noted that the transient motion of the carriers which will be determined by using equations (10) and (11) will depend only on the instantaneous values of the electric field and of the average energy $\bar{\varepsilon}$ of the carriers. This means that, in equations (10) and (11) it is implicitly assumed that the instantaneous distribution function depends only on the average energy of the carriers and of course in a few cases, such an assumption can be very rough. One of the main problems to be solved when we want to use equations (10) and (11) is to obtain the dependence of τ_m, τ_ε and m^* against $\bar{\varepsilon}$. This can be achieved practically by using the static characteristics $v_s(E_s)$, $\varepsilon_s(E_s)$ and $m^*(E_s)$ relating stationary average drift velocity, energy and effective mass to the electric field (obtained for example by MC methods in steady state conditions). In this case very simple relations between drift velocity, average energy, momentum and energy relaxation times are obtained. For example, equating $dm\bar{v}/dt$ and $d\bar{\varepsilon}/dt$ to zero in equation (10) and (11) gives:

$$\tau_p(\varepsilon_s) = \frac{m^* v_s}{qE_s} \tag{12}$$

$$\tau_\varepsilon(\varepsilon_s) = \frac{(\varepsilon_s - \varepsilon_o)}{qE_s v_s} \tag{13}$$

Typical variation of τ_p, τ_ε and also $\mu(\varepsilon) = q\tau_p/m^*$ for Si and GaAs have been given in [11]. It can be noted that when the energy increases, scattering rate generally increases and as a result τ_m is reduced; in addition due to the nonparabolicity of valley, the effective mass increases. Consequently $\mu(\varepsilon)$ decreases strongly particularly in the case of GaAs where, due to intervalley scattering for energy greater than the intervalley gap, the effective mass can strongly increase. The conservation equation (10),(11) have already been employed successfully by Shur [12], Carnez and Cappy et al. [13] to determine the variation versus time of the average velocity and the average energy when a time configuration of the electric field is applied to the semiconductor and it can be noted that good agreement with a more exact calculation is generally observed. This is clearly shown in Figure 9 where a comparison between the relaxation time description and the transient response obtained by Monte Carlo or by iterative methods has been carried out. It can be seen that for the five semiconductor materials studied, results obtained on the drift velocity by the relaxation time method agree very well with the more exact calculation. Consequently, the relaxation time approximation appears to be very useful for the study of time dependent phenomena.

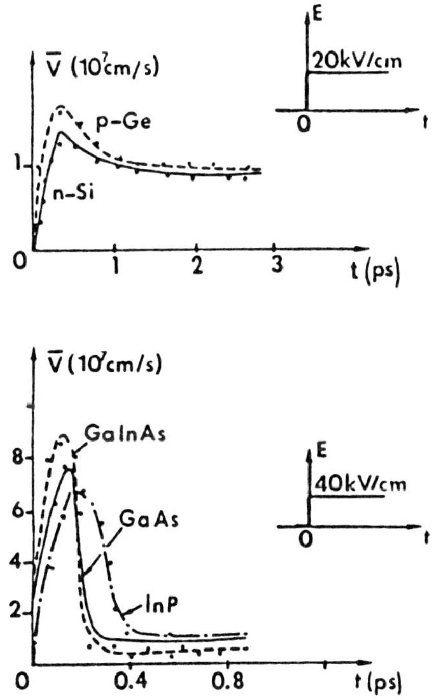

Figure 9. Transient drift velocity against time when a time step of the electric field is applied to a semiconductor. (a) p-Ge and n-Si Curves correspond to the relaxation time description, and dots to MC calculations (n-Si), or results obtained from iterative methods (P-Ge). (b) n-GaInAs, n-GaAs and n-InP. Curves correspond to the relaxation time description, and dots to MC calculations.

The problem is now to find another way to process to study space dependent phenomena and to take into account the space variation of the electric field usually met in submicronic devices.

The simplest method to employ is to deduce the characteristics of carrier transport versus space from the results obtained versus time by using the relation:

$$d\vec{r} = \bar{\vec{v}}\, dt, \text{ i.e (for 1D problem) } dx = \bar{v}\, dt$$

This is generally a very rough assumption since the additional phenomena previously described caused by the spatial nonuniformity are not taken into account. Nevertheless, this method can be a useful approproximation when the space variation of the carrier density along the active region of the submicronic device are not too strong and FET models based on this approximation have been successively developed [14].

However, balance equations obtained by integration of the Boltzmann equation over the $\vec{k}$ space, taking into account the space variation of all quantities appear to be much more rigorous, consequently, they can be very useful in the modeling of submicronic devices.

In the single valley case, the integration of Boltzmann equation leads to three fundamental macroscopic equations which express respectively the conservation of particles, momentum and energy. In the case of 1D problems, we have:

$$\frac{\partial n}{\partial t} + \frac{\partial nv}{\partial x} = 0 \tag{14}$$

$$\frac{\partial}{\partial t}(n\, m^* \bar{v}) = qnE - \frac{\partial}{\partial x}(nkT) - \frac{\partial}{\partial x} n\, m^* \overline{v^2} - \frac{n m^* \bar{v}}{\tau_m(\varepsilon)} \tag{15}$$

$$\frac{\partial}{\partial t}(n\bar{\varepsilon}) = q\, n\, v\, E - \frac{\partial}{\partial x}(n\bar{v}(\bar{\varepsilon} + kT)) - n\,\frac{\bar{\varepsilon} - \varepsilon_0}{\tau_\varepsilon(\bar{\varepsilon})} \tag{16}$$

where T is the electronic temperature, $\bar{\varepsilon}$ the average instantaneous energy and τ_m and τ_ε are the momentum and energy relaxation time. Such equations can be employed instead of the usual continuity and currents equation and together with the Poisson equation for the modeling of submicron device.

However, in the case of multivalley semiconductor, these equations must be written in each valley together with the coupling continuity equation since an additional term $(\partial n/\partial t)_c$ has to be taken into account in the second member of equation (13). A rough but useful assumption [15] is to average all the relevant quantities over the all valleys of the considered band structure. By doing so, equations 13-15 can still be applied to multivalley semiconductor. This procedure means again that we assume that the distribution function is fully determined by the knowledge of the instantaneous energy $\bar{\varepsilon}$ and obviously such assumptions can be in a few cases, very rough. Nevertheless, following Shur [16] the energy and momentum relaxation times, the effective mass and electronic temperature can be determined from Monte Carlo simulations under static steady state condition and one of the great interests of this method is that the steady state results are valid whatever the shape of the distribution function since all the operating parameters are analytically fitted from Monte Carlo results.

As a result, good agreement between Monte Carlo and modeling based on equations (14)-(16) are generally obtained. This appears in Fig. 10 which illustrates the spatial evolution of the average drift velocity obtained by the two aforementioned methods for a GaAs N+ N N+ structure under static conditions at liquid nitrogen temperature.

Such methods can also be used for the modeling of 2D submicron devices such as FET[17]. Equations (14)-(16) then become for this 2D problem

$$\frac{\partial n}{\partial t} + \vec{\nabla}\,\vec{J} = 0 \tag{17}$$

$$\frac{\partial (n\bar{\varepsilon})}{\partial t} + \vec{\nabla}.\vec{J}\,\varepsilon = \vec{J}\,\vec{E} - \vec{\nabla}\,(\vec{J}\,k\,T\,(\varepsilon)) - \frac{n(\bar{\varepsilon} - \varepsilon_o)}{\tau_\varepsilon(\bar{\varepsilon})} \tag{18}$$

$$\vec{J} = \frac{q\tau_m(\bar{\varepsilon})}{m^*(\bar{\varepsilon})}\left[n\,\vec{E} - \vec{\nabla}\,(n\,kT(\vec{\varepsilon}))\right] \tag{19}$$

and have to be solved together with the Poisson's equation. An example of results which can be obtained is illustrated in Figure 11.

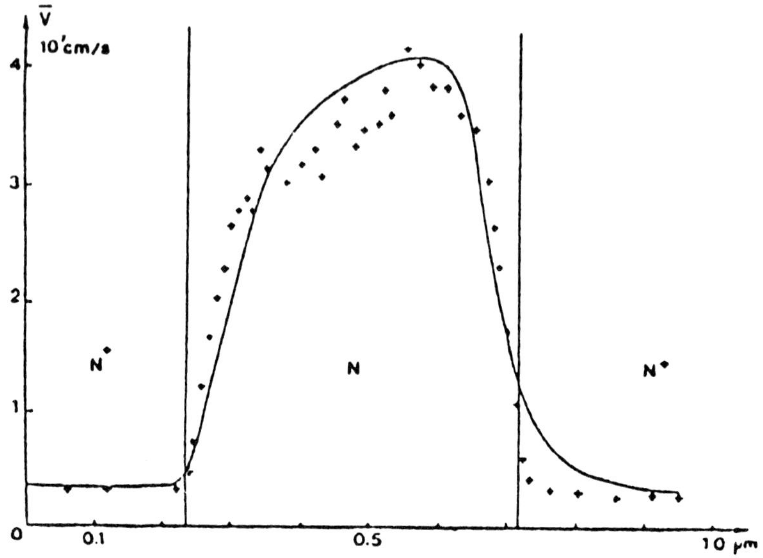

Figure 10. Spatial evolutions of electron drift velocity for a $N^+ N N^+$ GaAs structure at 77°K, obtained from Monte Carlo simulation and analytical formulation, under static conditions.
$N^+ = 10^{17}/cm^3$; $N = 5\ 10^{15}/cm^3$; $V_o = 0.2$ V

——— analytical formulation
\+ Monte Carlo simulation

WHAT IMPROVEMENT IN PERFORMANCE CAN BE EXPECTED FROM SUBMICRONIC DEVICES?

In the previous section, it has been shown that due to nonsteady state transport, velocities much higher than the steady values could be realized and suitable methods of modeling taking into account these phenomena have been described. In the last section of this paper we intend to discuss the improvement in performances which can be expected from the future submicronics and VLSI devices.

When we want to reduce the switching delay of logic devices, when we want to increase the operating frequency of microwave devices, one of the main conditions to meet is to reduce the transit time τ of the carriers in the active region of the device.

This must be realized in analogic operation, in order to increase the gain bandwith product f_c. This quantity is given by the following equation (where parasitics capacitance are neglected)

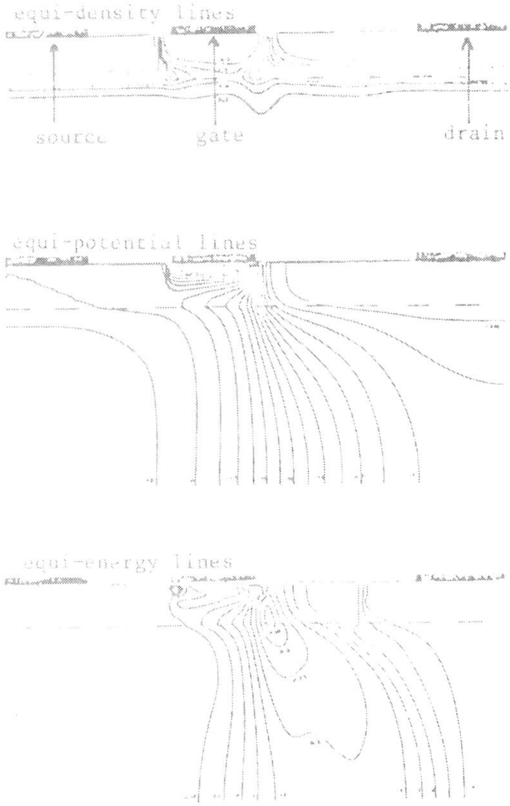

Figure 11. Simulation of FET using equations (16)-(18). The doping concentration of the active layer is: $N_D = 2.10^{17}$

The drain voltage is $V_D = +\ 2$ volt

The gate voltage is $V_G = 0$ volt

$$f_c = \frac{g_m}{2\pi C_{input}} = \frac{1}{2\pi}\left(\frac{dI_D}{dV_g}\right)\left(\frac{dV_g}{dQ}\right) = \frac{1}{2\pi}\,\frac{dI_D}{dQ} = \frac{1}{2\pi\tau} \qquad (20)$$

and it can be noted that to increase $\frac{g_m}{2\pi c}$, the transit time τ has to be reduced.

This condition must also be realized in digital operation in order to reduce the propagation delay t_{pd}. This delay t_{pd} corresponds to the time required to charge the output capacitance C_S to an additional voltage ΔV with a current ΔI (assumed to be constant)

$$t_{pd} = \frac{C_S \cdot \Delta V}{\Delta I_{DS}} \tag{21}$$

and it can be noted that, in the simplest and the very favourable case, where C_S reduce to the input capacitance of the transistors (assumed to be identicals) used in the logic circuit, the propagation delay reduces to the transit time in the active region:

$$t_{pd} = \frac{C_{input}}{g_m} = \tau \tag{22}$$

This transit time τ is given by the following equation

$$\tau = \frac{d}{\bar{v}} \tag{23}$$

where d is the active region length and $\bar{v}$ the average velocity of the carriers. Consequently, if we want to reduce τ, we have to reduce d, i.e. to use submicronic devices. However, due to overshoot or ballistic phenomena, $\bar{v}$ could be strongly increased and as a result, one can hope to further reduce the transit time of these carriers.

The maximum average velocity of the carriers which can be achieved over a distance d can be determined in a bulk semiconductor submitted to various time configuration of the electric field [18]. The results obtained are given in Figures 12 and 13 for two operating temperatures and two values of the doping concentration. Two cases have been studied:

- the overshoot case where an ensemble of thermal equilibrium electrons are subjected to a time step configuration of the electric field: this roughly simulates the electric field acting on the carriers when it travels through the active layer of a common submicronic device.

- the ballistic case where the ensemble of thermal equilibrium electrons are subjected to a very short time pulse of electric and can then travel through the semiconductor with a high initial energy, this roughly simulated the case of electron injected in the base of a ballistic transistor GaAlAs/GaAs (the initial energy is then equal to the "spike" in the conduction band at the interface GaAlAs/GaAs).

It can be noted that the average drift velocity can be increased by using very short submicronic distances, low impurity concentration,

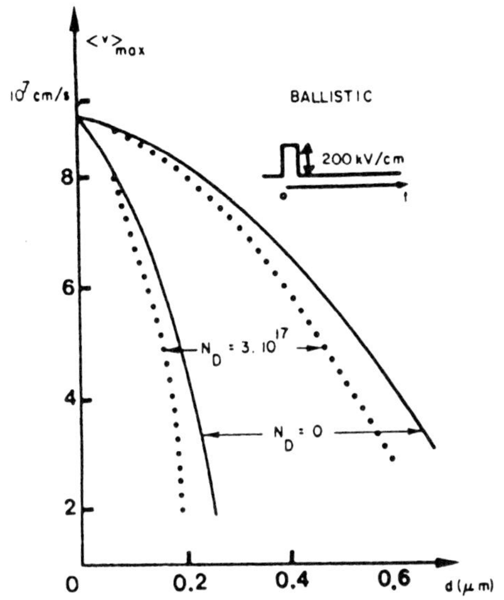

FIGURE 12

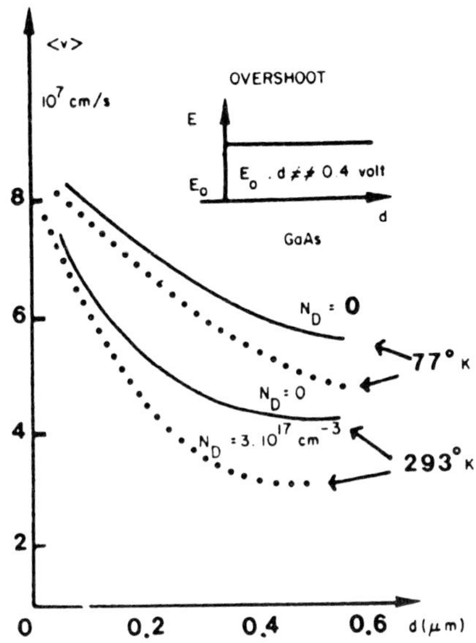

FIGURE 13

Figure 12 and 13. Maximum average velocity of the carriers over a distance d in GaAs

low temperature operations and ballistic motions rather than the overshoot motion. Ballistic motion does not strongly depend on the impurity concentration and this can be explained since due to the initial high value of the energy of the carriers, impurity scattering is very weak. In addition, for very small values of d, ballistic and overshoot phenomena progressively occur regardless of the impurity concentration and the temperature. This can be easily understood since for such distances scattering has no time to occur and the average velocity which can be achieved only depends on the characteristics of the band structure.

If the average velocity which can be achieved over a distance d is known, it is possible to roughly determine the transit time τ characterizing a device versus its active region width and then to obtain a rough evaluation of its limit performance i.e. of the minimum value of the propagation delay and the maximum value of the cutoff frequency. The results appear in Figure 14 and 15 and Table I where one can remark that rather impressive performance can be expected.

TABLE I

LIMIT PERFORMANCES FOR A DEVICE CHARACTERIZED BY 0.2μ ACTIVE REGION WIDTH

The performances are given for room temperature operation. The performances for nitrogen temperature are written in the brackets.

	OPERATING CONDITION	CUTOFF FREQUENCY (GHz)	PROPAGATION DELAY (ps)	TYPE OF DEVICE WHICH COULD BE USED
Overshoot	$N_D = 3\ 10^{17}/cm^3$	278 (466)	0,57 (0,34)	very submicronic gate FET
	$N_D = 0$	360 (512)	0,44 (0,31)	HEMT Space charge injection FET [8]
Ballistic	$N_D = 3\ 10^{17}/cm^3$	317 (635)	0,5 (0,25)	Ballistic transistor GaAlAs/GaAs
	$N_D = 0$	396 (661)	0,4 (0,24)	Space charge injection HETERO FET

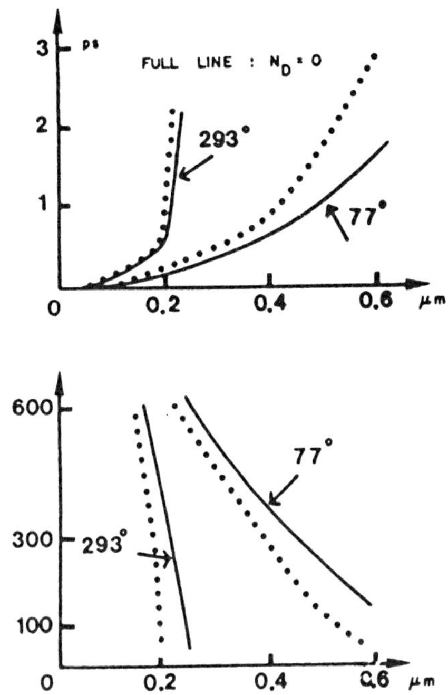

FIGURE 14

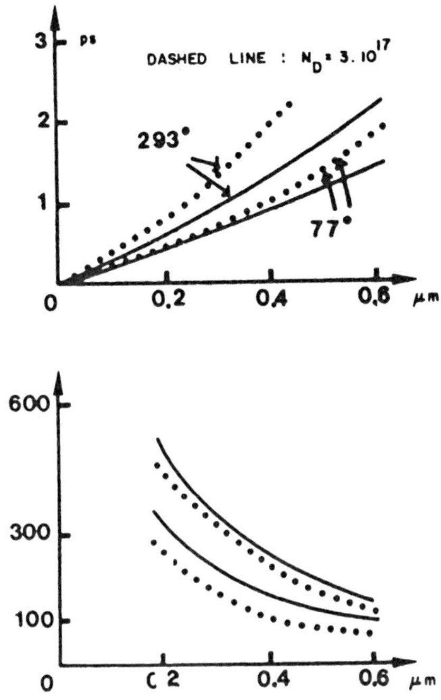

FIGURE 15

Figure 14 and 15. Minimum values of the propagation delay (ps) and maximum values of the gain bandwidth product (GHz) versus the active region length.

However, such submicronic devices are now beginning to be realized in laboratories and one can wonder why, so far, such performances are not generally obtained. This discrepancy can be easily explained using the following arguments:

EDGE EFFECTS

For submicronic devices, the edge effects strongly increase. Due to this effect, for example on each side of FET gate, the effective active region length L_e is always greater than the gate length L. Very roughly $L_e = L + 2a$ where a is the thickness of the active epilayer. For example, FET with 0.2μ gate length and 0.15μ epilayer thickness is characterized by an active region length of about 0.5μ and not (as it could be expected) of 0.2μ. For vertical structures such as used in bipolar transistor, edge effects are generally reduced but it should be noted that in the determination of the active region thickness, not only the base thickness has to be included, but also the length of the collector base depleted region.

PARASITIC CAPACITANCES

To increase the impedance level in analogic operation, to reduce the power consumption in logic circuits, very small area devices are generally used. For these very small devices, the fringing part of the input or output capacitance can no more be neglected and the parasitics capacitance C_p (generally independent of the gate width) has to be taken into account. Gain bandwidth product and propagation delay are then given by the following equations:

$$f_c = \frac{1}{2\pi\tau}\left(\frac{1}{1+\frac{C_p}{C_{in}}}\right) \tag{24}$$

$$\tau_{pd} = \bar{\tau}\left(1+\frac{C_p}{C_{in}}\right) \tag{25}$$

Based on these two relations, the following remarks can be done:

1) When, for very small devices, C_p becomes greater than C_{in}, the performance can be strongly reduced

2) The effect of C_p can be reduced by using transistor characterized by high value of C_{in} i.e. by using very thin and high doping concentration epilayer or HEMT rather than FET.

3) The performances remain proportional to the transit time τ or to $1/\tau$. As a result, if we are able to increase the average velocity of the carriers and to reduce τ, we should be able to achieve an increase of the performance of the same order of magnitude, whatever the C_p value of parasitics capacitance.

PARASITIC RESISTANCES

When we decrease the active region width, the impedance is often reduced and as a result, an increasing influence of parasitic resistances can be observed. The effect of the source drain resistance can be the most important. In common source configuration, the transconductance value can be strongly reduced, when the source resistance R_s increases. Using the classical equation:

$$g_{me} = \frac{g_m}{1 + R_s g_m} \tag{26}$$

gain bandwidth product and propagation delay values can be obtained from the following equation

$$f_c = \frac{1}{2\pi(\tau + R_s C_{in})\left(1 + \frac{C_p}{C_{in}}\right)} \qquad (27)$$

$$\tau_{pd} = (\tau + R_s C_{in})\left(1 + \frac{C_p}{C_{in}}\right) \qquad (28)$$

It can be noted that the time constant $R_s C_{in}$ directly adds to the transit time τ.

Consequently the effect of R_s can be very drastic and may determine the limit performances of very submicronic devices. This could be particularly the case of vertical transistors where recess structure cannot be used to reduce R_s and the effect of contact resistances.

The effect of the gate resistance R_g in logic circuit is not drastic since by employing active load, constant current is generally used to charge capacitance and the only main effect of R_s is to increase the power consumptions. This is not the case in analogic operation where an additional time constant $R_g C_{in}$ has to be taken into account and can determine the limit in performances when $R_g C_{in} >> \tau$. Such effect can occur when the gate length is reduced since the gate metallization resistance generally increases; consequently, a great number of unit gate widths must be used to achieve reasonable value of R_g and $R_g C_{in}$.

SPACIAL NONUNIFORMITY

The overshoot or ballistic velocity value which have been determined and used in Table 1 and in Figure 12-15 to obtain the transit time values have been obtained with a bulk semiconductor submitted to a uniform time configuration of the electric field. This is not the case of submicronic devices which are always characterized by short spatial scale of the electric field. Due to this spatial nonuniformity, we have shown that additional phenomena occur which might considerably change and often reduce the velocity of the carriers. Big mistakes can be made by not taking into account phenomena such as diffusion current, heat conduction, and thermoelectronic current. Consequently, the methods of modelisation described in this paper have to be used to obtain the exact performance of a specific device.

CONCLUSION

It seems possible to achieve very high performance submicronic devices. However, when designing and optimizing the devices, we have to use new methods of modeling and to be careful about additional phenomena due to strong space uniformity and also about the increasing influence of edge effects and parasitic elements.

If these various problems can be solved, picosecond and millimetric devices should become possible to achieve. This appears to be an interesting goal for the future.

ACKNOWLEDGEMENT

I would like to thank professor R. Castagne, J.P. Nougier, L. Reggiani, G. Salmer, Dr. B. Boittiaux, A. Cappy, R. Fauquemberque, P.A Rolland, J. Zimmermann for many useful discussions.

REFERENCES

1. Boittiaux, B., Constant, E., Reggiani, L., Brunetti, R., Jacoboni, C., 1982, Appl. Phys. Lett 40:5, 407.

2. Littlejohn, M.A., Hauser, J.R., Glisson, T.H., J. Appl. Phys.,1977 48 N° 11 pp 4587-4590.

3. Ghis, A., Constant, E., Boittiaux, B., January 1983, J. Appl. Phys. 54:1.

4. The results obtained by the study versus time are plotted versus the distance d travelled by the electron by using the relation:

$$d = \int_{o}^{t} \langle v(t')\rangle dt'$$

Monte Carlo simulation is carried out with a nonhomogeneous semiconductor where the space pulse of electric field is obtained from the study versus time. Ohmic contacts are assumed on each side of the nonhomogeneous semiconductor and consequently, the number of studies electrons remains constant.

5. Deblock, M., Fauquemberque, R., Constant, E., Boittiaux, B., 1980, Appl. Phys. Lett. 36, 756.

6. Warriner, R.A., 1977, Solid St. Electron Devices, 14:97.

7. Rees, H., Sanghera, G.S., Warriner, R.A., 1977, Electron Lett. 1:56.

8. Fauquemberque, R., Pernisek, M., Constant, E., July, 1982, Electronics Letters. Vol. 18, N° 15, p. 670.

9. Vaterkowski, J.L., Constant, E., Pernisek, M, Berdai, Solid State Electronics, to be published.

10. Nougier, J.P., Vaissiere, J.C., Gasquet, D., Zimmermann, J., Constant, E., February, 1981, J. Appl. Phys. Vol. 52, N° 2, pp. 825-832.

11. Constant, E., Nonsteady State carrier Transport in Semiconductor Submicron Devices, Chapter 7, Springer Verlag, to be published.

12. Shur, M., 1976, Electron Lett. 12:615.

13. Cappy, A. Carnez, B., Fauquemberque, R., Salmer, G., Constant, E. 1980, IEEE Trans. Electron Devices ED.

14. Carnez, B., Cappy, A., Kaszynski, A., Constant, E., Salmer, G., 1980, J. Appl. Phys. 51:1.

15. Friscourt, M.R., Rolland, P.A., Cappy, A., Constant, E., Salmer, G. March 1983, IEEE ED, Vol. ED 30, N° 3, pp. 223-229.

16. Shur, S., 1980, IEEE Trans. Electron Devices ED.

17. Salmer, G., May 1983, Workshop on Active Microwave Semiconductor Devices, Maidenhead (U.K.)

18. Ghis, A. Constant, E., Boittiaux, B., 1982, J. Appl. Phys. 52.

BOLTZMANN TRANSPORT EQUATION

Karl Hess

Department of Electrical Engineering & Coordinated Science Laboratory
University of Illinois at Urbana-Champaign
Urbana, IL 61801

The theory of nonlinear electronic transport in semiconductors is mostly based on the Boltzmann transport equation (BE_T) and "natural" extensions of it. Consequently there exists an enormous amount of original papers and reviews on that subject {1}. From a general point of view, there is probably no need for reviewing the Boltzmann equation once more. The following treatment is meant to provide an introduction and reference guide for the participants of this summer school who are new in the area.

The BTE describes the distribution $f(\vec{v},\vec{r},\vec{t})$ of electrons as a function of velocity $\vec{v}$, space coordinate $\vec{r}$ and time t. $f(\vec{v},\vec{r},\vec{t})\, d\vec{v}\, d\vec{r}$ defines the probability of finding an electron (hole) in a velocity range between $\vec{v}$ and $\vec{v} + d\vec{v}$ having a space coordinate between $\vec{r}$ and $\vec{r} + d\vec{r}$. Therefore, the BTE is not a quantum mechanical equation as it specifies $\vec{v}$ and $\vec{r}$ simultaneously, ignoring Heisenberg's uncertainty relations.

The BTE is most easily derived by considering the rate of change of particles in a cube in 6-dim phase space which can be factored in a cube in real space and one in velocity space (see Fig. 1). We first calculate how many electrons come from the left, enter the cube in the left dydz plane and how many leave at the corresponding plane to the right in a time dt. Since the traveling distance in x-direction of electrons with velocity $\vec{v}$ is $v_x dt$ we have:

$$\text{incoming:} \quad f(\vec{v},\vec{r},t)\, d\vec{v}\, dydzv_x dt \tag{1}$$

$$\text{outgoing:} \quad f(\vec{v},\, x + dx/y/z/,\, t)\, d\vec{v}\, dydzv_x dt \tag{2}$$

and the net particle gain is therefore:

$$-v_x \{ f(\vec{v}, x + dx/y/z, t) - f(\vec{v},\vec{r},t)\} \, dydz \, d\vec{v} \, dt$$

$$= -v_x \frac{\partial f}{\partial z} dxdydzd\vec{v} \, dt \tag{3}$$

or $\quad = -\vec{v}.\vec{\nabla}_r \, fd\vec{v} \, d\vec{r} \, dt$ in three dimensions.

In a totally analogous manner we obtain the change of electrons at $\vec{r}$ because of accelerations (v changes in velocity space):

$$-\dot{\vec{v}}.\nabla_v \, f \, d\vec{v} \, d\vec{r} \, dt \tag{4}$$

where $\dot{\vec{v}} = \frac{1}{m} e\vec{F}$ and $\vec{F}$ is the electric field. There is still another possibility to change the number of electrons with velocity $\vec{v}$ at $\vec{r}$. The electrons can be scattered and change their velocity from $\vec{v}$ to $\vec{v}'$ at a given point $\vec{r}$ in space.

Fig. 2 shows the two infinitesimal volumes in $\vec{v}$-space to illustrate the scattering events. The outgoing (out of state $\vec{v}$) electrons are:

$$\text{out} = - \sum_{\vec{v}'} S(\vec{v},\vec{v}')f(\vec{v},\vec{r},t) \tag{5}$$

The factor $f(\vec{v})$ is necessary because an electron has to be in the $\vec{v}$ state to be scattered out. In degenerate systems a factor $1 - f(\vec{v}')$ arises from the Pauli principle. Notice that this treatment implies that the collision processes are irreversible and <u>completed locally in space and time</u>. The incoming (into the $\vec{v}$ state) electrons are:

$$\text{in} = \sum_{\vec{v}'} S(\vec{v}',\vec{v}) \, f(\vec{v}',\vec{r},t) \tag{6}$$

for obvious reasons. The Boltzmann equation is obtained by balancing the particle numbers and the change in f given by the net change of incoming and outgoing electrons, therefore we have:

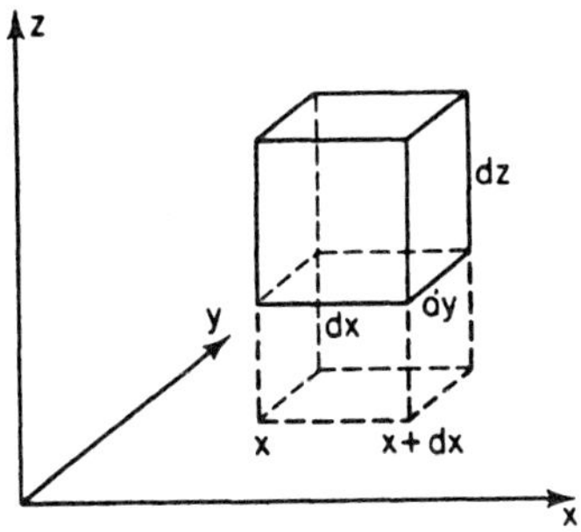

Figure 1. Cube in real space with electrons incoming and leaving.

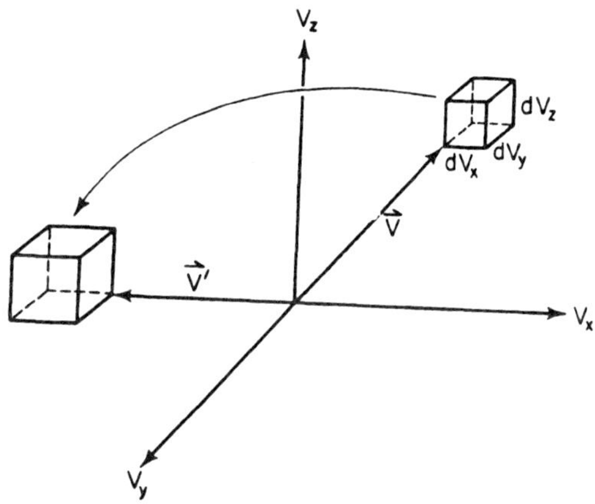

Figure 2. Transition in velocity space because of scattering.

$$\frac{\partial f(v,r,t)}{\partial t} = -\vec{v}.\vec{\nabla}_{\vec{r}} \; f(\vec{v},\vec{r},t) - \frac{1}{m}\vec{F} \cdot \vec{\nabla}_{\vec{v}} \; f(\vec{v},\vec{r},t) + \quad \vec{F} \cdot \vec{V}$$

$$\int d\vec{v}' \; \{ f(\vec{v}',\vec{r},t) \; S(\vec{v}',\vec{v}) - f(\vec{v},\vec{r},t) \; S(\vec{v},\vec{v}')\} \quad (7)$$

This equation is now more than 150 years old but because of the complicated integral form of the scattering term explicit solutions are only possible in very special cases. Numerous approaches are known from the transport theory in gases. Some of them can be directly applied to semiconductors. It is usually demonstrated that the complicated collision integral can be replaced by a relaxation time if the collisions are elastic.

For most practical cases in semiconductors, however, the following approach is more appropriate. We partition the distribution functions into a part f_o which is even in the velocity and a contribution f_1 which is odd, i.e.,

$$f_o(\vec{v}) = f_o(-\vec{v}), \quad f_1(-\vec{v}) = -f_1(\vec{v}) \tag{8}$$

Furthermore, we assume:

$$S(\vec{v},\vec{v}') = S(\vec{v},-\vec{v}') = S(-\vec{v},\vec{v}') = S(-\vec{v},-\vec{v}') \tag{9}$$

This is exact for optical deformation potential scattering and a good approximation for acoustic phonon scattering. Polar optical scattering cannot be treated this way and the reader is referred to {2} for a physical description of the reasons. With the above assumptions, the collision term in the Boltzmann Equation reduces to:

$$\int d\vec{v}' \{ f_o(\vec{v}',\vec{r},t)S(\vec{v}',\vec{v}) - f_o(\vec{v},\vec{r},t)S(\vec{v},\vec{v}') - f_1(v,r,t)S(v,v')\} \tag{10}$$

The term containing $f_1(\vec{v}',\vec{r}'t)$ vanishes because f_1 is odd and S is even in $\vec{v}'$. In equilibrium the terms containing f_o cancel each other since f_o is the Maxwell-Boltzmann distribution and we have detailed balance, i.e., an equal number of particles is scattered in and out. In very high fields this is not true anymore and f_o is different from the Maxwell- Boltzmann distribution. Then we denote the terms containing f_o symbolically by:

$$\left.\frac{\partial f_o}{\partial t}\right]_{\text{collision}}$$

the Boltzmann equation is then written as:

$$\frac{\partial f(v,r,t)}{\partial t} = -\vec{v}.\vec{\nabla}_{\vec{r}}\, f(v,r,t) - \frac{1}{m}\vec{F}\cdot\vec{\nabla}_{\vec{v}}\, f(\vec{v},\vec{r},t) +$$

$$\left.\frac{\partial f_o}{\partial t}\right]_{coll} - f_1(\vec{v},\vec{r},t)\int d\vec{v}'\; S(\vec{v},\vec{v}') \qquad (11)$$

We now define

$$\frac{1}{\tau_m(\vec{k})} = \frac{V}{8\pi^3}\int d\vec{k}'\; S(\vec{k},\vec{k}') \qquad (12)$$

with $\vec{k}$ being the wavevector of the electrons. $1/\tau_m$ represents the total scattering rate and, in this case, also the momentum relaxation rate since the scattering has been assumed to be randomizing (assumptions for $S(\vec{v},\vec{v}')$).

As mentioned, for the case of small electric fields, the term $\partial f_o/\partial t$ collision vanishes because f_o is the Maxwellian distribution function and the Boltzmann equation reduces to a partial differential equation of first order which is easy to solve, especially if the coefficients are constant. For the case in which the term $\partial f_o/\partial t$ collision is not negligible, explicit solutions can still be reached as long as one can separate the BTE in separated equations for f_o and f_1 (one usually has to assume that $f_1 << f_o$). A number of such solutions have been discussed by Conwell {3} and by Hess and Sah {4}. The case of polar optical scattering requires special treatments but explicit solutions are also available for this case {5}. In other words, the BTE can be solved for a free electron gas weakly interacting with phonons (impurities) quite easily.

In a semiconductor, the equations of motion are close to the free electron equation only at the minimum of the conduction band (maximum of the valence band) where the mass m of the electron has to be replaced by an effective mass m*. The energy-momentum relation (or $E(\vec{k})$ relation with $\hbar\vec{k}$ being the crystal momentum) usually takes a complicated form. Then we have for the acceleration in an electric field F

$$\hbar\dot{\vec{k}} = eF \qquad (13)$$

Magnetic fields need special treatment and are not considered in this paper (transport in magnetic fields is discussed, e.g., by J.R. Barker {6}). Furthermore,

$$\vec{v} = \frac{1}{\hbar} \vec{\nabla}_{\vec{k}} E(\vec{k}) \qquad (14)$$

To take this into account, we have to rewrite the Boltzmann equation as

$$\frac{\partial f(\vec{k},\vec{r},t)}{\partial t} = -\frac{1}{\hbar} \vec{\nabla}_k E(\vec{k}) \cdot \vec{\nabla}_r f(\vec{k},\vec{r},t) - \frac{\vec{F}}{\hbar} \cdot \vec{\nabla}_k f(\vec{k},\vec{r},t)$$
$$= f_1 \frac{V}{8\pi^3} \int dk' \, S(k,k') - \left.\frac{\partial f_o}{\partial t}\right|_{coll} \qquad (15)$$

This equation is sometimes called the Boltzmann-Bloch Equation.

Figures 3 - 5 show details of the $E(\vec{k})$ relation of silicon as obtained from the empirical pseudopotential method. It is clear that these complicated geometrical forms make a solution of the Boltzmann equation very difficult and explicit solutions impossible without very crude approximations. Only close to the band minima can some explicit solutions be found, e.g., by including the non-parabolicity as obtained from the k.p method. If details of the band structure are important, Monte Carlo computations are more powerful and conceptually simpler than a direct solution of the BTE.

Although Eq. (15) is more appropriate for semiconductors than Eq. (11) it is still not a quantum transport equation for the following reasons. First, Eq. (15) only describes the acceleration in an electric field of relatively low frequency and moderate strength and

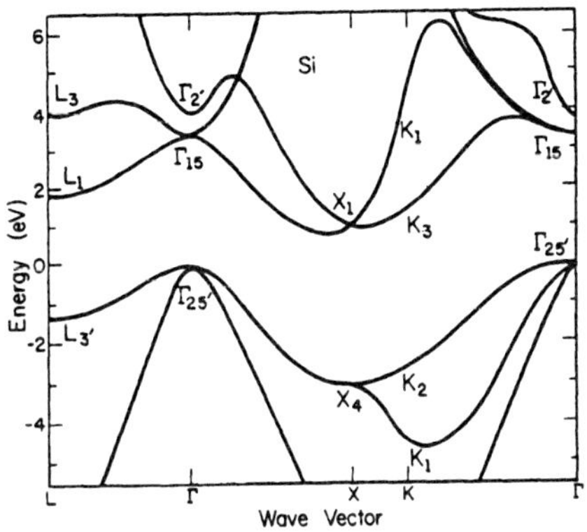

Figure 3. The band structure of silicon calculated by the empirical pseudopotential method.

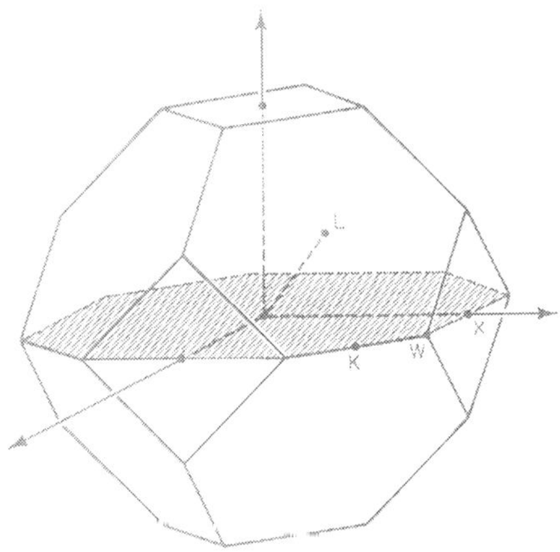

Figure 4. A cross-section of the Brillouin zone.

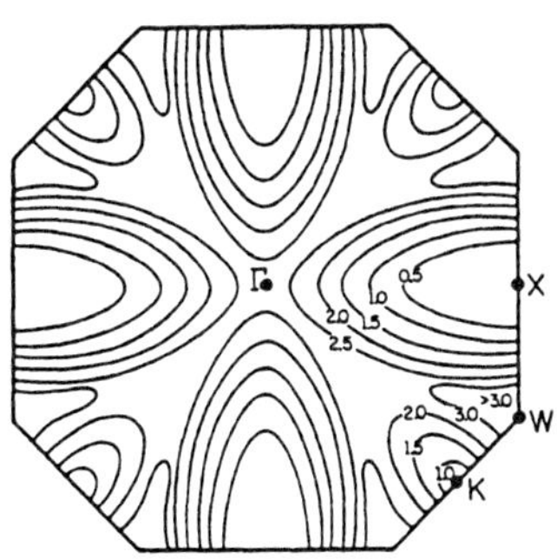

Figure 5. Equal energy surfaces for the first conduction band of silicon corresponding to the cross-section of Figure 2.

is therefore unsuited for, e.g., the calculation of optical processes and fields strong enough to make the Stark effect important. In order to apply Eq. (15) the natural broadening of the energy levels $\hbar/\tau$ must be larger than the photon energy $\hbar\omega_{phot}$ or the Stark level separation ΔE_{st}. This gives

$$\omega_{phot}\,\tau < 1 \qquad (16a)$$

and

$$\frac{eFa}{\hbar}\,\tau < 1 \qquad (16b)$$

where a is the lattice constant. Eq. (16a) restricts the validity range of the modified BTE typically to photon energies of $\sim$ 7 meV and corresponding wavelength of 30 μm. The limitations of the second equation are not as easy to find because τ is a function of energy and the average energy is a function of the electric field F. Typically for high fields $\tau \gtrsim 1.5 \times 10^{-15}$ s, which gives a limiting electric field of about $F \gtrsim 10^5$ V/cm. This is the field at which impact ionization phenomena become very strong in GaAs, Si, InP, and many other semiconductors.

The short relaxation times τ at high energies bring us to the next limitation of the BTE, which is contained in the scattering integral. The basic implication in this term is that the scattering is local in space and <u>time</u> and in addition, that the interaction is weak. Although the electron-phonon coupling is weak in most of the important semiconductors, the sum over all possible final states in Eq. (15) can become large (and so does the total scattering rate), if the density of states is high. A large peak in the density of states occurs in many semiconductors 1 - 3 eV above the conduction band edge. This is the reason why $1/\tau$ can reach values of several 10^{14} s^{-1} in high electric fields. As a consequence, the electrons are in continuous interactions with phonons and the scattering events can hardly be separated and assumed to occur at certain instances of time. It follows that the scattering rate has to be renormalized and collision broadening plays a major role.

The effect of collision broadening can easily be understood. For a noninteracting system, the wave function of a particle in momentum state $\vec{k}$ evolves as

$$\Psi(t) \sim \exp(-iE(\vec{k})t/\hbar) \tag{17}$$

where $E(\vec{k})$ is the energy of the free-particle state. The Fourier transform of Eq.(17) is the spectral density function $A(\omega,E) \sim \delta(\omega-E(\vec{k}))/\hbar$. For an interacting system (in our case electron-phonon and impact ionization) we have

$$\Psi(t) \sim \exp(-i(E(\vec{k})+\Sigma(\vec{k}))\ t\ /\ \hbar \tag{18}$$

The self energy $\Sigma(\vec{k})$ is a complex number ($\Sigma(\vec{k}) = \Delta(\vec{k}) + i\Gamma(\vec{k})$) and the spectral density has a Lorentzian shape

$$A(\omega,E) \simeq \frac{1}{\pi}\ \frac{\Gamma(\vec{k})}{(\hbar\omega - (E(\vec{k})+\Delta(\vec{k}\ )^2 + (\Gamma(\vec{k}))^2} \tag{19}$$

which determines the broadening (line width) of the state.

Δ and Γ have been calculated for the electron-phonon interaction as shown in Figure 6. The details of this calculation are given in Ref. {7}. It therefore appears to be natural to extend the Boltzmann Equation within a quasiparticle picture and to replace the scattering terms S(k,k') by expressions which include the above effects. This can be accomplished by replacing the δ-function appearing in the Golden rule result for the scattering rate by {8}

$$\frac{\overline{\Gamma}}{((E(\vec{k})-E(\vec{k})')-\hbar\omega_{ph})^2 + \overline{\Gamma}^2} \tag{20}$$

where $\overline{\Gamma}$ is a suitable average of the initial and final state value of Γ.

The quasi-particle BTE as described above should be valid as long as

$$E \gtrsim \overline{\Gamma} \tag{21}$$

Since Γ is high only at high energies E, this is normally the case.

The quasi-particle concept is also useful when size quantization effects are important. At interfaces or in heterojunction quantum wells, the electrons are not free to move in one direction. When

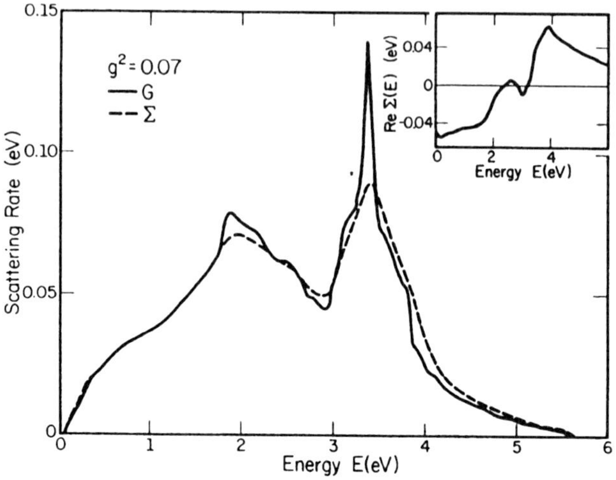

Figure 6. Real and imaginary part of the self energy Σ of silicon.

the quantization is significant (which can easily be estimated using the square well model), i.e., when the electrons occupy one "subband", they can be regarded as quasi-two-dimensional electrons and the Boltzmann Equation is valid as discussed above if all three-dimensional vectors are replaced by two-dimensional ones. Similar arguments can be applied to one-dimensional conductors, if the electron density is relatively low (compared to the density of atoms) and Peierl's distortions {9} therefore do not occur.

There is still one more effect which needs to be included in more exact transport treatments. This effect is commonly called the intracollisional field effect. Again, this effect is not properly accounted for by using the BTE and Golden rule because of the implication that the scattering interactions are local in time. It is clear that the emission and absorption of a phonon, for example, takes some period of time which we do not want to specify here. During this time the electron can gain (or lose) energy from (to) the electric field. Therefore, energy conservation in the collision process has to include the effect of the electric field. This effect can be handled very elegantly within the Feynman path integral formalism {10}. In this formalism the scattering rate is proportional to

$$\int_0^\infty dt \; \exp(-i\omega_{ph} t + \frac{i\hbar^2 k^2}{2m^*} t - \frac{ikpt}{m^*} + i\vec{k} \cdot \vec{F} \frac{t^2}{2m}) \tag{22}$$

which gives for F = 0 the energy conserving δ-function appearing in the Golden rule. For large electric fields the δ-function is broadened and skewed. Notice that for $\vec{k} \perp \vec{F}$ the intracollisional field effect vanishes. It also is negligible if the collision broadening discussed before is large.

In summary, I have described necessary modifications to the BTE in semiconductors. These modifications whould be sufficient to describe electronic transport in a wide range of material and external parameters as they arise in modern transport problems. A rigorous quantum treatment of transport in solids will be discussed in this issue by G.J. Iafrate.

REFERENCES

1. See, e.g., D.K. Ferry, J.R. Barker, and C. Jacoboni, eds., Physics of Nonlinear Transport in Semiconductors (New York: Plenum, 1980), and references therein.
2. See, e.g., the paper by K. Hess in Ref. 1.
3. E.M. Conwell, Solid State Physics, Suppl. 9 (New York: Academic Press, 1967), F. Seitz, D. Turnbull, and H. Ehrenreich, eds.
4. K. Hess and C.T. Sah, Solid State Electronics 18,667 (1975).
5. See, e.g., J.P. Leburton and K. Hess, Phys. Rev. B 26, 5623 (1982) and references therein.
6. See, e.g., K. Seeger, Semiconductor Physics (New York: Springer, 1973).
7. Y.C. Chang, D. Z.-Y. Ting, J.Y. Tang, and K. Hess, Appl. Phys. Lett. 42, 76 (1983).
8. J.R. Barker, J. Phys. C 6 2663 (1973).
9. R. Peierls, Quantum Theory of Solids (London: Oxford University Press, 1974), pp. 140-141.
10. K.K. Thornber, Solid State Electronics 21, 259 (1978).

TRANSPORT AND MATERIAL CONSIDERATIONS FOR SUBMICRON DEVICES

H.L. Grubin

Scientific Research Associates Inc.

Glastonbury, Connecticut USA

INTRODUCTION

The purpose of this lecture series is to identify the important areas of device physics and the questions related to them. In addressing these problems we will emphasize an understanding of transport within these devices and the role of the circuit, the boundaries and the material variations. We will also scrutinize some of the formal underpinnings of device physics. Thus, the lectures will be primarily theoretical. With regard to experimental device physics studies, direct evidence of high speed, submicron scale effects is sparse, and is likely to remain so for the near future. Although, here, recent activities by workers at Bell[1], Los Alamos[2], and the University of Paris--Orsay[3], on submicron, high speed device phenomenology, may provide some important keys to device related effects.

In the absence of a broad experimental program in the area of submicron devices, we are likely to see an increased reliance on numerical simulation as the principle vehicle for determining the physics of device operation. This, of course, places a considerable burden on those who perform device simulations. But the general availability of numerical programs to those who do not develop them will widen the use of numerical studies, particularly in a direct coupling to experiment. In any case, the accumulated studies of the past half century has tended to provide a natural classification of device physics into three groupings: (1) classical device physics, (2) submicron device physics, and (3) ultrasubmicron device physics, where quantum effects may dominate.

The domain of classical semiconductor device physics is one where the temporal scales of interest are generally large enough that accelerative effects are often ignored. These approximations find their most common application in the drift and diffusion equations, where the relation between carrier velocity and applied field is in terms of a local mobility.

The first departure from classical device physics introduces the idea of non-locality where the response of a particle at a point "x" and time "t" is dependent on disturbances at earlier times $t'<t$ and at points $x'<x$. This is the critical feature of the Boltzmann transport equation (BTE), and is one that has found expression in Chambers[4] path integral solutions as well as some initial work of Pippard[5]. While non-local effects have appeared only recently in semiconductor device studies, they have been the object of serious study for the past twenty-five years. Such phenomena as the anomalous skin effect[6], magneto-acoustic attenuation[7] are based on these effects. They are also critical to hydrodynamic studies.

Other departures arise when spatial scales are of the order of the deBroglie wavelength associated with a quasi-particle, and arise when scattering events can no longer be regarded as occurring instantaneously. Approximate spatial and temporal scales associated with these three device divisions are identified in the first two tables and figure 1. Further, each of the above device groupings are identified by a specific set of generic device equations. The key point that must be kept in mind, is that the development of these equations is keyed to the development of high frequency, high speed electronic devices. This will be the common thread through our entire discussion, and also forms the basis for this NATO ASI.

TABLE 1. Approximate spatial scales (GaAs)

Term	Value
Active region length	10^{-5}cm, 10^{-4}cm
Impurity separation ($N_0^{-1/3}$)	10^{-5}cm @ 10^{15}/cm^3
10^{-6}cm @ 10^{18}/cm^3	
Mean free path	10^{-5}cm
Thermal deBroglie wavelength	2.6×10^{-6}cm

TABLE 2. Approximate temporal scales (GaAs)

Term	Value
Transit time (for $v=2\times10^{7}$/cm/s)	5×10^{-13}s @ $L=10^{-5}$cm 5×10^{-12}s @ $L=10^{-4}$cm
Momentum relaxation	3×10^{-13}s @ 300K
Collision duration	2×10^{-14}s

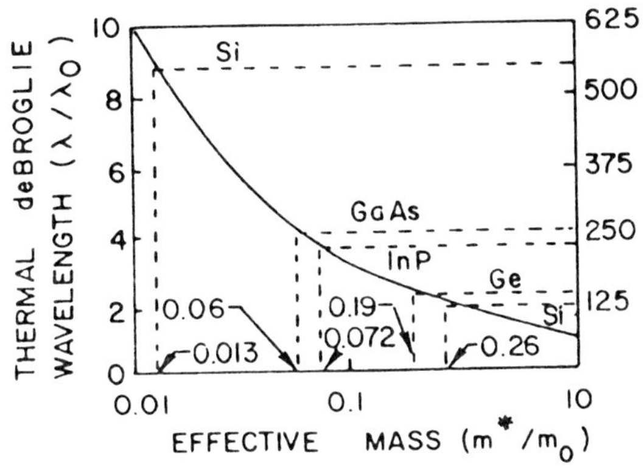

$$\lambda/\lambda_0 = 1/\sqrt{m^*/m_0}$$

$$\lambda_0 = \hbar/\sqrt{m_0{}^3 k T_0}$$

FIGURE 1 . Thermal deBroglie wavelength vs effective mass.

We begin the discussion of submicron device physics with a simple description of the Boltzmann transport equation and its approximate solutions, as well as those areas of such common interest as overshoot phenomena.

THE BOLTZMANN TRANSPORT EQUATION

The conceptual picture most frequently used to describe conduction in semiconductors is that of a gas of carriers interacting with external forces. The gas is characterized by the fact that the state of any of its constituents, e.g., electrons, are independent of all others except at the instant of collision. The state of each constituent is described by specifying its position and momentum at each instant of time. Rather than treating the transport of individual particles, transport within the semiconductor is discussed statistically. Volume elements within phase space are identified as

$$d^3r = dx_1 dx_2 dx_3 \tag{1}$$

$$d^3p = dp_1 dp_2 dp_3 \tag{2}$$

The differentials above are chosen large enough to contain a large number of carriers and small enough to permit the neglect of density variations within the differential elements.

The number of particles at a point (**r**, **p**) is taken as

$$dN = f(\mathbf{r},\mathbf{p},t)d^3r d^3p \tag{3}$$

which may be taken as the definition of the distribution function. Several other definitions are in order. The total number of particles is given by

$$N = \int f(\mathbf{r},\mathbf{p},t)d^3r d^3p \tag{4}$$

The mean value of a function $G(\mathbf{r},\mathbf{p})$ is

$$\langle\langle G\rangle\rangle \equiv \int G(\mathbf{r},\mathbf{p})f(\mathbf{r},\mathbf{p},t)d^3r d^3p/N \tag{5}$$

In much of the discussion below, we will be interested in local mean values.

If $\phi(\mathbf{p})$ denotes any function of momentum, the local mean value of $\phi(\mathbf{p})$ is given by

$$\langle\phi\rangle = \int \phi(\mathbf{p})f(\mathbf{r},\mathbf{p},t)d^3p/n \tag{6}$$

where

$$n = \int f(\mathbf{r},\mathbf{p},t)d^3p \tag{7}$$

We note that $\langle\ \rangle$ is a function of position and time through the distribution function $f(\mathbf{r},\mathbf{p},t)$.

To determine the variation of the distribution function with time, we assume the presence of external and self-consistent forces, as well as inter-particle forces. We consider time intervals which are large, compared to the duration of a collision, modifications of which have been considered by Ferry and Barker[8]. We also assume that the time interval Δt is small compared to the mean time interval between collision.

If no collisions occur during the time interval Δt, the following transformation applies

$$\mathbf{r} \rightarrow \mathbf{r}' = \mathbf{r} + (\mathbf{p}/m)\Delta t \tag{8}$$

$$\mathbf{p} \rightarrow \mathbf{p}' = \mathbf{p} + \mathbf{F}\Delta t \tag{9}$$

and

$$f(\mathbf{r},\mathbf{p},t)d^3rd^3p \rightarrow f(\mathbf{r}',\mathbf{p}',t + \Delta t)d^3r'd^3p' \tag{10}$$

Application of Louville's theorem concerning elements of volume in phase space

$$d^3rd^3p = d^3r'd^3p' \tag{11}$$

leads to the expansion

$$f(\mathbf{r},\mathbf{p},t) \cong f(\mathbf{r},\mathbf{p},t) + \left(\frac{\partial f}{\partial \mathbf{r}} \cdot \frac{\mathbf{p}}{m} + \frac{\partial f}{\partial \mathbf{p}} \cdot \mathbf{F} + \frac{\partial f}{\partial t}\right)\Delta t + \cdots (\quad)(\Delta t)^{\ell} \tag{12}$$

In examining transport in semiconductors we generally neglect terms of order $(\Delta t)^{\ell}$ with $\ell>1$; the assumption being that the distribution function does not change appreciably during the interval Δt. According to Sommerfeld[9], this assumption is compatible with considerable changes in $f(\mathbf{r},\mathbf{p},t)$ within one mean free path, because $\Delta t<\tau$, where τ is the mean time between collisions. This latter result is extremely important when examining overshoot phenomena in devices. Thus within this approximation, we obtain the collisionless Boltzmann transport equation

$$\frac{\partial f}{\partial \mathbf{r}} \cdot \frac{\mathbf{p}}{m} + \frac{\partial f}{\partial \mathbf{p}} \cdot \mathbf{F} + \frac{\partial f}{\partial t} = 0 \tag{13}$$

Collisions cause carriers to leave different elements of phase space with the following balance holding:

$$f(\mathbf{r},\mathbf{p},t)d^3pd^3r - f(\mathbf{r}',\mathbf{p}',t' + \Delta t)d^3p'd^3r' = \left(\frac{\partial f}{\partial t}\right)_{coll} d^3pd^3r, \qquad (14)$$

from which we obtain the BTE with collisions,

$$\frac{\partial f}{\partial \mathbf{r}} \cdot \frac{\mathbf{p}}{m} + \frac{\partial f}{\partial \mathbf{p}} \cdot \mathbf{F} + \frac{\partial f}{\partial t} = \left(\frac{\partial f}{\partial t}\right)_{coll} \qquad (15)$$

Among the various problems in semiconductor device physics, one involves the proper formulation of the collision integral. We will not concern ourselves with this during the following lectures. Rather, we will borrow the necessary terms for establishing a suitable submicron device physics.

Before proceeding to an actual description of submicron device transport, which will necessarily involve solving a complicated set of equations derived from the BTE, it is worthwhile presenting a formal solution to the BTE within the framework of the relaxation time approximation. This formal solution will highlight the significant differences between the more standard approach to examining device physics, and what is currently required.

Within the framework of the relaxation time approximation, the collision integral takes the form

$$\left(\frac{\partial f}{\partial t}\right)_{coll} = -\left(\frac{f-f_o}{\tau}\right) \qquad (16)$$

in which we assume that τ may be a function of energy. The formal solution to the above equation basically involves transforming the BTE, as a partial differential equation into an ordinary differential equation

$$\frac{df}{ds} + \frac{f}{\tau} = \frac{f_o}{\tau} \qquad (17)$$

where "s" denotes a "characteristic" path variable. There is, of course, considerable history associated with this, see e.g., Chambers[4], and Thornber[10]. Basically we consider the variables t(s), r(s) and p(s) such that

$$\frac{df}{ds} = \frac{\partial f}{\partial t} \cdot \frac{dt}{ds} + \frac{\partial f}{\partial \mathbf{r}} \cdot \frac{d\mathbf{r}}{ds} + \frac{\partial f}{\partial \mathbf{p}} \cdot \frac{d\mathbf{p}}{ds} \tag{18}$$

where we require

$$\frac{dt}{ds} = 1 \quad , \quad \frac{d\mathbf{r}}{ds} = \frac{\mathbf{p}}{m} \quad , \quad \frac{d\mathbf{p}}{ds} = \mathbf{F} \tag{19}$$

For the simple case[10] where **F** is taken to be independent of t, **r** and **p**

$$\begin{aligned} t(s) &= s-s_o + t_o \\ \mathbf{r}(s) &= \frac{\mathbf{p}}{m}(s-s_o) + \mathbf{r}_o \\ \mathbf{p}(s) &= \mathbf{F}(s-s_o) + \mathbf{p}_o \end{aligned} \tag{20}$$

Equation (18) has the formal solution

$$f\left[\mathbf{r}(s),\mathbf{p}(s),t(s)\right] = f\left[\mathbf{r}(s_o),\mathbf{p}(s_o),t(s_o),\right]\exp - \int_{s_o}^{s}\frac{ds'}{\tau[\mathbf{p}(s')]}$$

$$+ \int_{s_o}^{s}\frac{ds' f_o[\mathbf{r}(s'),\mathbf{p}(s'),t(s')]}{\tau[\mathbf{p}(s')]}\exp - \int_{s'}^{s}\frac{ds''}{\tau[\mathbf{p}(s'')]} \tag{21}$$

When the following identification is made

$$s_o = t_o \ , \qquad t(s) = s \tag{22}$$

The formal solution takes the form

$$f(\mathbf{r},\mathbf{p},t) = f\left[\mathbf{r} - \frac{\mathbf{p}}{m}(t-t_o),\mathbf{p}-\mathbf{F}(t-t_o),t_o\right] \exp\left[-\int_{t_o}^{t}dt''/\tau[\mathbf{p}-\mathbf{F}(t-t'')]\right]$$

$$+ \int dt' \, \frac{f_o[\mathbf{r} - \frac{\mathbf{p}}{m}(t-t'),\mathbf{p}-\mathbf{F}(t-t'),t']}{\tau[\mathbf{p}-\mathbf{F}(t-t')]} \exp\left[-\int_{t'}^{t}\frac{dt''}{\tau[\mathbf{p}-\mathbf{F}(t-t')]}\right] \tag{23}$$

The physical significance of the above solution is contained in the exponential contributions. Here the probability that an electron initially at

$$\mathbf{r} - \frac{\mathbf{p}}{m}(t-t_o), \quad \mathbf{p}-\mathbf{F}(t-t_o), \quad t_o \tag{24}$$

has arrived at

$$\mathbf{r}, \quad \mathbf{p}, \quad t \tag{25}$$

without scattering is

$$\exp - \int_{t'}^{t} \frac{dt''}{\tau[\mathbf{p}-\mathbf{F}(t-t'')]} \tag{26}$$

Thus, all transport is governed by non-local events. This is the point we wish to emphasize. We now apply this to a very specific and familiar problem.

We consider the situation where a system of carriers is subjected to an applied external field and scattering centers, and that the field represents a small departure from equilibrium:

$$f = f_o + f_1 \tag{27}$$

where f_o represents an equilibrium time independent distribution function. Under the typical small perturbation approximation, the BTE becomes

$$\frac{\partial f_1}{\partial t} + \frac{\mathbf{p}}{m} \cdot \frac{\partial f_1}{\partial \mathbf{r}} + \frac{f_1}{\tau} = - \left(\mathbf{F}\cdot\frac{\partial f_0}{\partial \mathbf{p}} + \frac{\mathbf{p}}{m}\cdot\frac{\partial f_0}{\partial \mathbf{r}}\right) = g \tag{28}$$

If we now solve this equation using the method of characteristics and concentrate only on the particular solutions

$$f_{1,p} = \int_{t_o}^{t} dt' g[\mathbf{r} - \frac{\mathbf{p}}{m}(t-t'), t'] \exp - \left(\int_{t'}^{t} \frac{t-t''}{\tau(\mathbf{p})} \right) \tag{29}$$

we see an explicit dependence of the perturbed distribution function on past history. Now while this will be important for submicron devices, there exists an entire class of devices whose spatial scales are longer than a mean free path and whose temporal scales are longer than the mean time between collisions. On these longer spaces and time dependent scales the spatial and temporal variation of the integral in equation (29) is not important, and can be taken out of the integral. Under these circumstances the perturbed distribution function, for times longer than the mean time between collisions, becomes

$$f_{1,p} \approx -\tau \left(\mathbf{F} \cdot \frac{\partial f_0}{\partial p} + \frac{p}{m} \cdot \frac{\partial f_0}{\partial r} \right) \tag{30}$$

Equation (30) is the typical starting point for most semiconductor device transport descriptions. It requires for its validity that only insignificant variations occur over a mean free path, or over a mean time between collisions. It has nevertheless been used for situations where there are strong field gradients over very short distances, such as that within the vicinity of a strongly asymmetrical PN junction. It may be expected that many of the more common device calculations such as those associated with bipolar silicon devices, MOS devices will require significant scrutiny to determine if the more commonly used assumptions are valid in design on a submicron scale.

The above description of transport as governed by the non-local feature of the BTE will now be applied to examining submicron high field transport. The requirements for this discussion have at their basis, the need for a set of governing differential equations. The equations that we use to describe submicron and near micron transport are based upon the moments of the BTE. Typically, these moments are often truncated after the first three moments although in principle, an infinite number of moment equations will provide an exact solution to the BTE. In the discussion that follows we will apply the use of the displaced Maxwellian distribution function to obtain a set of device equations.

THE DISPLACED MAXWELLIAN DISTRIBUTION AND THE MOMENT EQUATIONS

As indicated in the above discussion, within the relaxation time approximation, a momentum relaxation time may be defined, where $\tau(p)$ depends on energy through momentum. As discussed by Keyes[11], we can add a term to equation (16) which describes the effects of electron-electron scattering by observing that if the external fields and scattering mechanisms (exclusive of e-e scattering) were suddenly removed, the e-e collisions would provide a mechanism for transferring energy and momentum between different parts of the distribution function. This would bring f to some other equilibrium function, f_o^*. f_o^* must be different from f_o, since with f_o the electrons have zero total momentum. Since e-e collisions cannot alter the total electron momentum, f_o^* must have the same total momentum as the nonequilibrium distribution function f, at the time the field and scattering centers were removed.

The determination of f_o^*, has been carried out by the methods of statistical mechanics, and it has been shown that f_o^* is a Boltzmann distribution moving with velocity $\mathbf{v}=\mathbf{p}^*/m$, i.e.,

$$f_0^*(\mathbf{r},\mathbf{p},t) = ae^{-|\mathbf{p}-\mathbf{p}^*|^2} \tag{31}$$

The prominence of the displaced Maxwellian lies in its successful application to examining transport in III-V materials, particularly gallium arsenide (GaAs). Its use in transport is primarily through the moments of the BTE.

More generally, the moments of the BTE are obtained by multiplying equation (15) by successive powers of $\mathbf{p}$, i.e.,

$$\phi^{(\ell)}(\mathbf{p}) = \mathbf{p}^{\ell}, \ell = 0,1,2, \cdots \tag{32}$$

The result is

$$\frac{\partial}{\partial t}(n\langle\phi^{(\ell)}\rangle) + \frac{\partial}{\partial \mathbf{r}}\cdot\left[\frac{n}{m}\langle \mathbf{p}\phi^{(\ell)}\rangle\right] = \mathbf{F}n\cdot\left\langle \frac{\partial\phi^{(\ell)}}{\partial \mathbf{p}}\right\rangle + \frac{\partial}{\partial t}(n\langle\phi^{(\ell)}\rangle)_{coll} \tag{33}$$

where

$$\frac{\partial}{\partial t}(n\langle\phi^{(\ell)}\rangle)_{coll} = \int\phi^{(\ell)}\left(\frac{\partial f}{\partial t}\right)_{coll} d^3p \tag{34}$$

In the absence of collisions, an exact solution to the BTE can be obtained by solving the infinite set of coupled equations. This is not satisfactory, particularly when the effects of collisions must by accounted for. The situation then calls for an estimate of the distribution function. The often used assumption is that of a displaced Maxwellian which cuts off the moment equations at the number "3". The degree to which the distribution function is well represented by the displaced Maxwellian has been examined by a number of workers with the general conclusion that as far as bulk transport is concerned, it is likely to be a reasonable assumption at high carrier densities[12]. It is likely to be seriously unrealistic to assume that the distribution function is spherical in momentum space. Such a situation would tend to ignore important thermal gradient effects as well as viscous effects, which are known to appear in fluid flow and in plasma physics. The situation is such that the displaced Maxwellian needs generalization. One such approach, which we consider below was discussed by Sommerfeld[9]. Here the distribution function of interest is expressed as

$$f(\mathbf{r},\mathbf{p},t) = \left[1 + A_k \frac{\partial}{\partial p_k} + A_{k\ell} \frac{\partial^2}{\partial p_k \partial p_\ell} + A_{k\ell m} \frac{\partial^3}{\partial p_k \partial p_\ell \partial p_m} + \cdots \right] f_0^*(\mathbf{r},\mathbf{p},t) \tag{35}$$

with the coefficients to be determined below. Additionally, the following convention is assumed:

$$A_k \frac{\partial}{\partial p_k} = A_x \frac{\partial}{\partial p_x} + A_y \frac{\partial}{\partial p_y} + A_z \frac{\partial}{\partial p_z} \tag{36}$$

The constraints on this expansion are that the mean carrier density

$$n = \int f(\mathbf{r},\mathbf{p},t) d^3p = \int f_0^*(\mathbf{r},\mathbf{p},t) d^3p \tag{37}$$

and that the mean momenta

$$n\langle p\rangle = \int f(\mathbf{r},\mathbf{p},t) p d^3p = \int f_0^*(\mathbf{r},\mathbf{p},t) p d^3p \tag{38}$$

which yields

$$\langle p\rangle = p^* \tag{39}$$

It is further required that the mean, isotropic, thermal kinetic pressure satisfy the following requirement

$$\mathcal{P} = \frac{2}{3m} \int (p-p^*)\cdot(p-p^*) f d^3p = \frac{2}{3m} \int (p-p^*)\cdot(p-p^*) f_0^* d^3p = \frac{n}{2\gamma m} \tag{40}$$

The identification of a suitable electron temperature model is through an assumed gas law, which is taken as

$$\mathcal{P} = nk_BT \tag{41}$$

Thus,

$$\gamma = \frac{1}{2mk_BT} \tag{42}$$

And along with a suitable normalization that includes spin the displaced Maxwellian is of the following form

$$f_0^* = \frac{n}{(2\pi mk_BT)^{3/2}} \exp\frac{-|p-p^*|^2}{2mkT} \tag{43}$$

The expansion coefficients, which identify the important nonspherical contributions of the distribution function are identified as follows: (1) the linear expansion coefficient is zero; (2) the expansion coefficient A_{kl} is identified as a stress tensor; and (3) the expansion coefficient A_{klm} is associated with a heat flux. The incorporation of these terms leads to the following set of balance/moment equations, from which the numerical simulation of the submicron devices is obtained.

Carrier Balance:

$$\frac{\partial n_r}{\partial t} + \frac{\partial}{\partial x_i} v_r^i n_r = - n_r \Lambda_r + n_s \Lambda_s \tag{44}$$

Momentum Balance:

$$\frac{\partial p_r^j}{\partial t} + \frac{\partial}{\partial x_i} v_r^i p_r^j = -en_r F^j - \frac{\partial}{\partial x_i} n_r k_B T_r + \mu_r \underline{\underline{\frac{\partial^2}{\partial x_i \partial x_i}}} v_r^i - p_r^j \Pi_r \tag{45}$$

Energy Balance:

$$\frac{\partial W_r}{\partial t} + \frac{\partial}{\partial x_i} v_r^i W_r = -en_r v_r^i F^i - \frac{\partial}{\partial x_i} v_r^i k_B T_r + \underline{\underline{\frac{\partial}{\partial x_i} \kappa_r \frac{\partial}{\partial x_i} T_r}}$$

$$- \frac{3}{2} n_r k_B T_r \Sigma_r + \frac{3}{2} n_s k_B T_s \Sigma_s \tag{46}$$

Equations (44) through (46) are the governing equations used below to discuss submicron transport. In these equations

$$P_r \equiv n_r p_r \equiv n_r m_r v_r \tag{47}$$

and

$$W_r = n_r\left(\frac{1}{2}m_r v_r^i v_r^i + \frac{3}{2}k_B T_r\right) \tag{48}$$

the subscripts are used to emphasize the fact that we are considering multi-valley species transport. The underlined terms in these equations identify the stress term contributions (equation (45)) and the thermal heat transport contribution (equation (46)). Stress contributions in equation (46) have been ignored. The coefficient in the momentum balance equation has dimensions of viscosity, and this term behaves as a viscous contribution.

We examine the above equations in detail below. In order to do this effectively, we will demonstrate how, for the case of the dipole transit, the results obtained from these equations differ from those obtained using the standard drift and diffusion

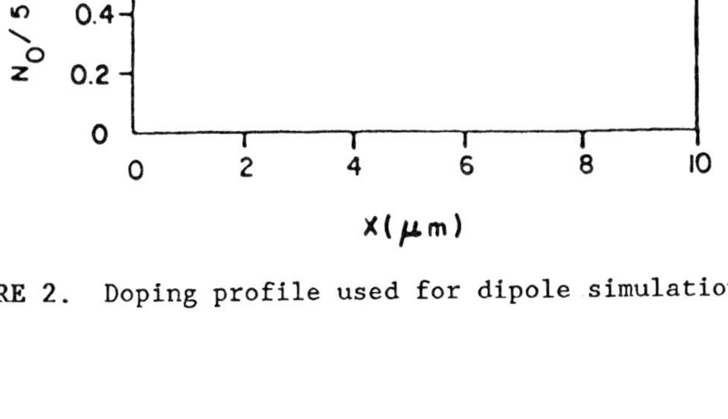

FIGURE 2. Doping profile used for dipole simulations.

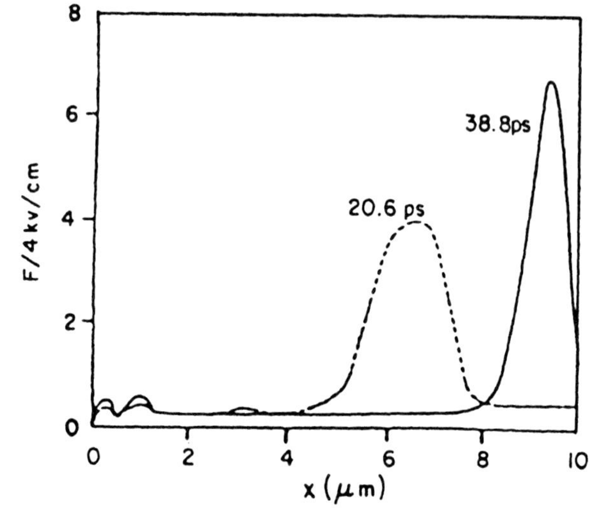

FIGURE 3a. Electric field vs distance vs time, for a propagating layer.

equations. To illustrate this, we consider a 10μm-long GaAs element with a doping level of $5x10^{15}/cm^3$ and a doping notch near the cathode boundary, as shown in figure 2. The device is clamped at 4 volts (there is no load line). At this voltage level domain oscillations occur. The oscillations are displayed in figures 3 and 4. (We note: In this simulation Poisson's equation is solved, along with the moment equations.)

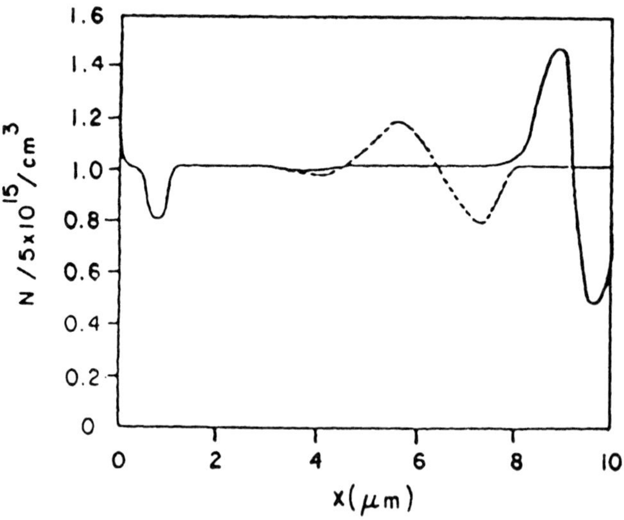

FIGURE 3b. Mobile carrier density at the time of figure 3a.

Figures 3a through 3d display, respectively, the time dependent behavior of electric field, total carrier density, gamma valley carrier density and gamma valley velocity, for a propagating domain (additional development is shown in figure 4). The domain is first shown approximately halfway down the element at a time of 20.6ps; and later at the anode at 38.8ps.

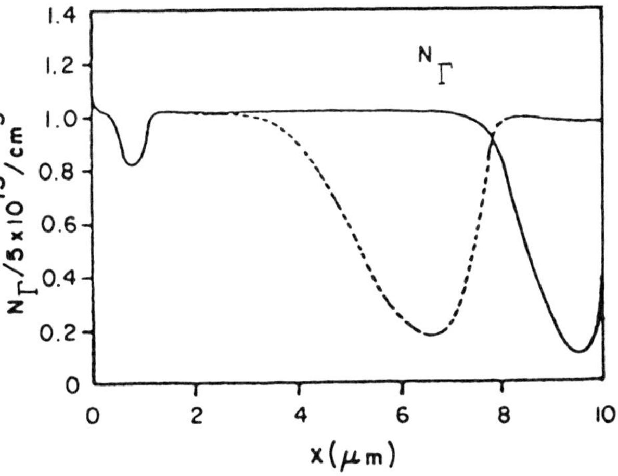

FIGURE 3c. Γ-valley carrier density at the time of figure 3c.

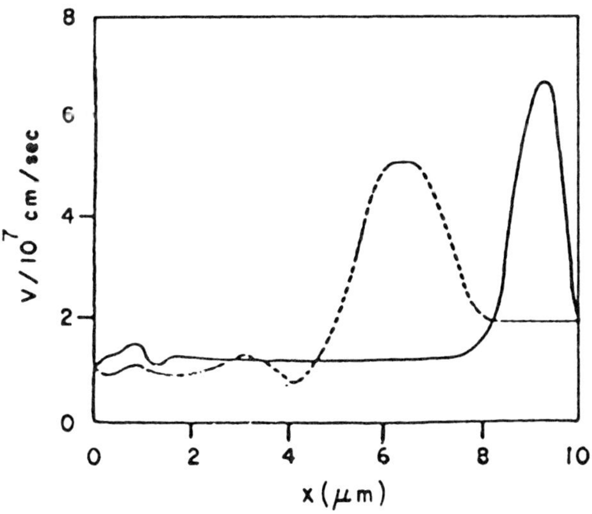

FIGURE 3d. Γ-valley carrier density at the time of figure 3a.

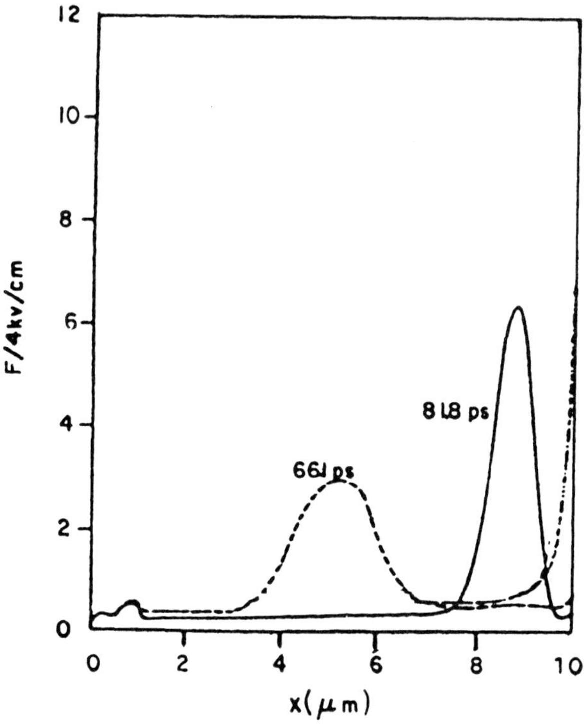

FIGURE 4a. As in figure 3a but at later times.

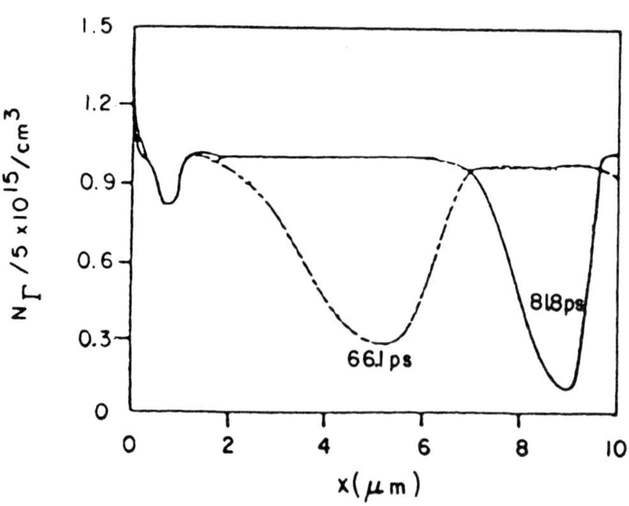

FIGURE 4b. As in figure 3c but at later times.

The travelling domain has all the appearance of a classical transit time oscillation. This, however, is deceptive because immediately upstream from the domain where the field has decreased to below the threshold for negative differential resistance (NDR), and where semiconductor drift and diffusion would indicate that relaxation to quiescent values of carrier density and near ambient electron temperatures is complete, the moment equations indicate otherwise. The moment equations show a substantial satellite valley population even after the high field region has passed and only low values of field remain. This significant relaxation effect was first discussed by Rees[13]. Its numerical origin arises from the first order spatial and temporal derivatives in the left hand side of the moment equations (45) and (46). In the absence of these terms, we, of course, recover the drift and diffusion equations. The results of this calculation are significant and cannot be described in any reasonable way by the drift and diffusion concepts. Further differences arise here, because at these moderate values of electric field, the satellite valleys carry only a negligible amount of current. It is as though in an approximate sense, once the carriers have been transferred to a satellite valley, they are effectively out of the conduction process. (This is clearly not the case at high bias levels). Thus is a real sense, stable domain theory must be reexamined. Of interest here is that at the peak field where the total carrier density is equal to the background, i.e., $N=N_0$, the Γ valley population is less than $N/4$, yet the velocity of these carriers, figure 3d, is greater than twice the mean velocity, obtained from the relation

$$\langle v \rangle = J/n_0 e \qquad (49)$$

Nonequilibrium transport as depicted above is of more than academic interest. Classical modeling of a 10μm-long device is predicated on the notion that proximity effects are negligible. Based upon the size of the high field domain as depicted in figure 3a, this would appear to be a valid assumption. On the other hand, transport in the wake of these domains, at least for doping levels, near $5\times10^{15}/\text{cm}^3$ extends over a distance of nearly twice the width of the domain, figures 3c and 4d. This factor of two should be cause of considerable concern when transit time devices below 5μm in active region length are designed.

PARAMETRIC DEPENDENCE OF THE MOMENT EQUATIONS

At this point, it is necessary to examine several parametric aspects of the moments of the BTE. To facilitate this parametric study we introduce the following dimensionless primed variables:

$$n' = n/n_{ref} \quad v' = v/v_{ref}$$
$$t' = t/t_{ref} \quad \Psi' = \Psi/\Psi_{ref}$$
$$x' = x/x_{ref} \quad t' = t/t_{ref}$$

with similar descriptions for other variables, and where

$$x_{ref} = v_{ref}\, t_{ref}$$

Additionally, Ψ is obtained from a solution to Poisson's equation

$$\frac{\partial^2 \Psi}{\partial x_i \partial x_i} = + \frac{e}{\epsilon}(n - n_0) \tag{50}$$

for a permittivity ϵ, and we are observing the summation convention. The moment equations and Poisson's equation in the primed variables become

Carrier Balance:

$$n'_r = n'_s \frac{\Lambda'_s}{\Lambda'_r} - \frac{1}{\Lambda'_r}\left\{\frac{\partial n'_r}{\partial t'} + \frac{\partial}{\partial x'_i} v_r^{i'} n'_r\right\} \tag{51}$$

Momentum Balance:

$$n'_r m'_r v_r^{j'} =$$

$$a n'_r \frac{\partial \Psi'}{\partial x'_j} - \frac{1}{\Pi'_r}\left\{b\frac{\partial n_r T_r}{\partial x'_j} - c\frac{\partial^2 \mu v_r^i}{\partial x'_i \partial x'_i} + \frac{\partial}{\partial t'} n'_r m'_r v_r^{i'} + \frac{\partial}{\partial x'_i} n'_r m'_r v_r^{i'} v_r^{j'}\right\} \tag{52}$$

Energy Balance:

$$n'_r T'_r = n'_s T'_s \frac{\Sigma'_s}{\Sigma'_r} + d n'_r v_r^i \frac{\partial \Psi'}{\partial x'_i} \cdot \left(\frac{2}{3b}\right)$$

$$- \frac{1}{\Sigma'_r}\left\{b\frac{\partial}{\partial x'_i} n'_r v_r^{i'} T'_r + e\frac{\partial}{\partial x'_i}\kappa'\frac{\partial}{\partial x'_i} T_r + \frac{\partial}{\partial t'} W'_r + \frac{\partial}{\partial x'_i} v_r^{i'} W'_r\right\} \cdot \left(\frac{2}{3b}\right) \tag{53}$$

Poisson's equation:

$$\frac{\partial^2\Psi'}{\partial x_i'\partial x_i'} = -\frac{\partial F_i'}{\partial x_i'} = \mathbf{f}(n'-N_0') \tag{54}$$

In the above

$$W_r' = \frac{1}{2}n_r'm_r'v_r^{i'}v_r^{i'} + \frac{3}{2}\mathbf{b}n_r'T_r' \tag{55}$$

The primed scattering rates in the above equations are, $\Lambda'_r = \Lambda_r \cdot t_{ref}$ with similar expressions for the other scattering rates. For dimensional reasons, as well as to establish a dialogue concerning the dissipative terms and the thermal conductivity contribution, the viscosity has been expressed via the Maxwell relation

$$\mu = \frac{1}{3}n_{ref}\, m_{ref}\, \langle v\rangle \ell \tag{56}$$

where $\langle v\rangle$ is a characteristic mean velocity and ℓ an effective mean free path. The thermal conductivity has been expressed in terms of a Widemann-Franz relation

$$\kappa = \frac{\mathcal{L}\sigma T k_B^2}{e^2} \tag{57}$$

where

$$\sigma = \frac{n_{ref}\, e^2\langle\tau\rangle}{m_{ref}}$$

and $\langle\tau\rangle = \ell/\langle v\rangle$. In these terms the bold coefficients in equations (51) through (54) are

$$\mathbf{a} = \frac{e t_{ref}}{m_{ref}\Pi'_r}\frac{F_{ref}}{v_{ref}} \tag{58}$$

where

$$F_{ref} = \frac{\Psi_{ref}}{x_{ref}}, \tag{59}$$

$$\mathbf{b} = \frac{k_B T_{ref}}{m_r v_{ref}^2} \tag{60}$$

$$\mathbf{c} = \frac{1}{3} \frac{\langle v \rangle \ell}{v_{ref} x_{ref}} \tag{61}$$

$$\mathbf{d} = \mathbf{a} \frac{\Pi_r}{\Sigma_r} \tag{62}$$

$$\mathbf{e} = \mathbf{b}^2 \frac{\ell \langle \tau \rangle}{t_{ref}} \tag{63}$$

$$\mathbf{f} = \frac{e}{\epsilon} \frac{x_{ref} n_{ref}}{F_{ref}} \tag{64}$$

The connection between these bold letter coefficients and dimensionless hydrodynamic quantities is

$$\mathbf{b} = \frac{1}{M^2 \gamma} \qquad (\gamma = \frac{3}{5}) \tag{65}$$

where M is the Mach number, and

$$\mathbf{c} = \frac{1}{\text{Rey}} \tag{66}$$

where Rey is the Reynolds number. We consider the broad consequences of the normalization.

The first point that should be noticed is that the effects of the nonspherical distribution, as represented by thermal conductivity and viscosity are likely to be more pronounced at the shorter device lengths. The situation with regard to the thermal conductivity is illustrated in figures 5 and 6.

In figure 5, we display the distribution of field and mobile carrier density for a 2μm-long element with the indicated structure and a high thermal conductivity consistent with Eq. (57)

[14]. We note the presence of highly nonuniform fields, figure 5b, marginal electron transfer, figure 5c, and local Γ-valley carrier velocities in excess of $5x10^7$ cm/sec, figure 5d.

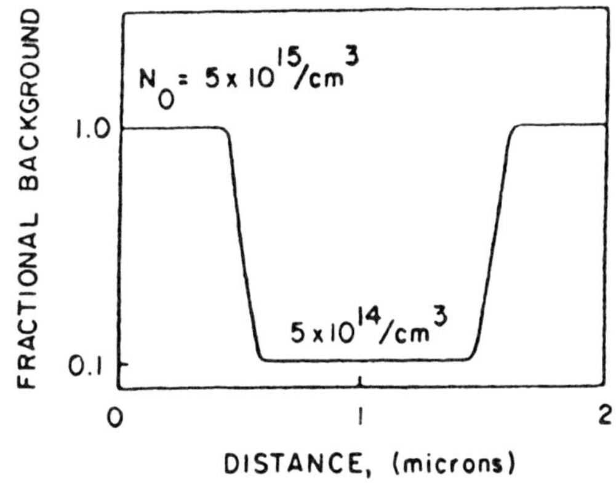

FIGURE 5a. Doping profile for high thermal conductivity calculations.

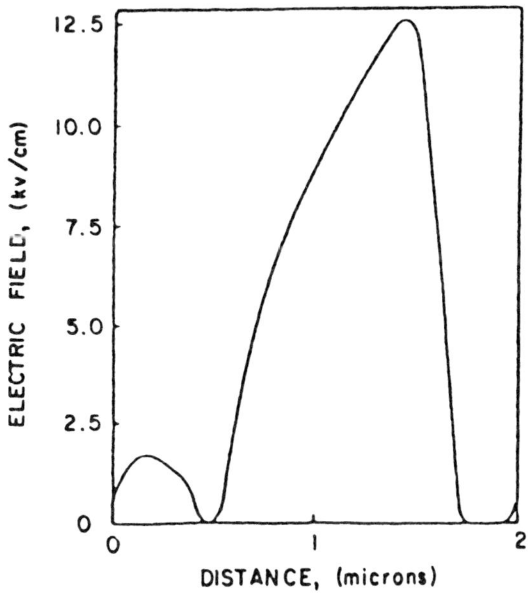

FIGURE 5b. Field profile for high thermal conductivity calculations.

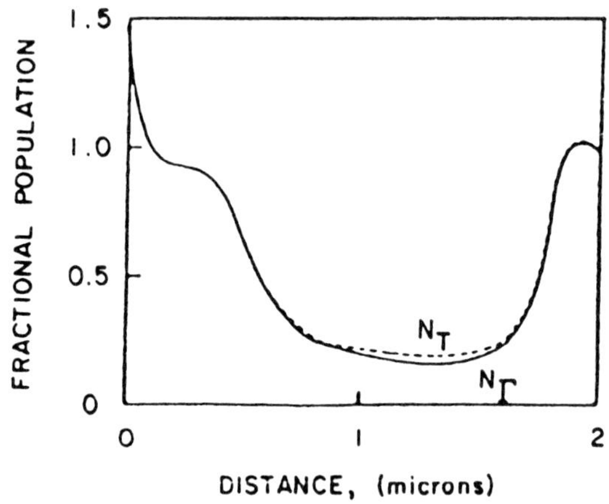

FIGURE 5c. Carrier distribution for high thermal conductivity calculations.

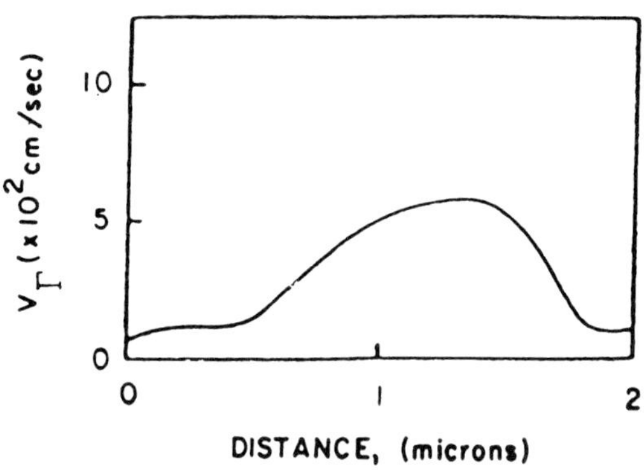

FIGURE 5d. Velocity distribution for high thermal conductivity calculations.

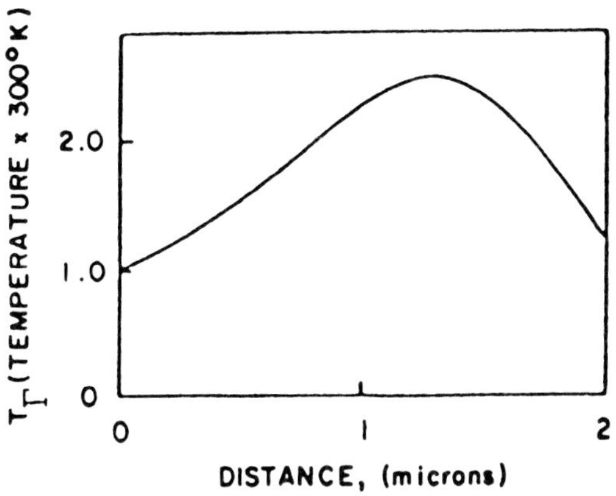

FIGURE 5e. Temperature distribution for high thermal conductivity calculations.

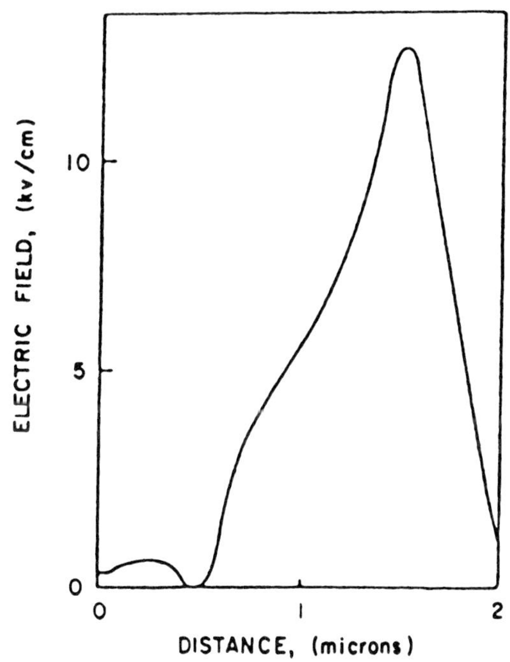

FIGURE 6a. Field profiles for a lower thermal conductivity of.

Figure 6 is for a lower value of the thermal conductivity [14]. A decrease in the thermal conductivity by an order of magnitude results in an expected increase in the overall carrier temperature and an increase in the number of carriers that have transferred to subsidiary valleys, figure 6b. The results suggest that further study is required with respect to the importance of the thermal conductivity with regard to device design.

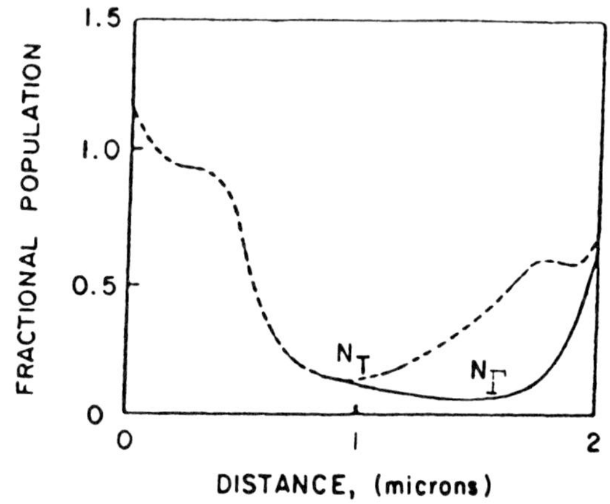

FIGURE 6b. Free carrier density for a low thermal conductivity.

The situation with regard to the viscous contribution is considered next. First, viscous contributions are basically dissipative. This is displayed in figures 7a and 7b, where for the structure of figure 2 and a bias level of two volts, the steady

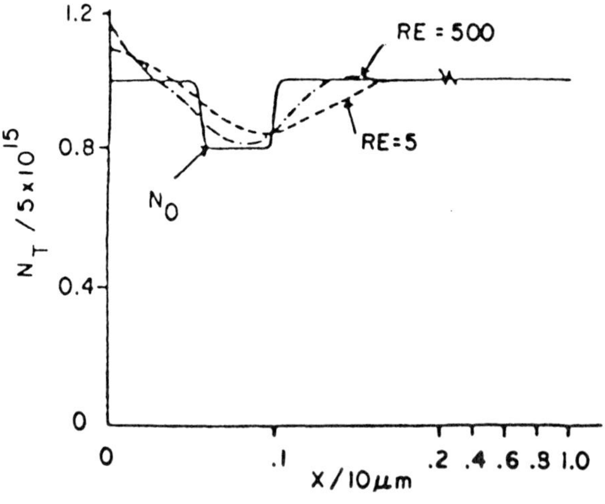

FIGURE 7a. Dependence of charge density on Reynolds number. (Ψ = 2 volts)

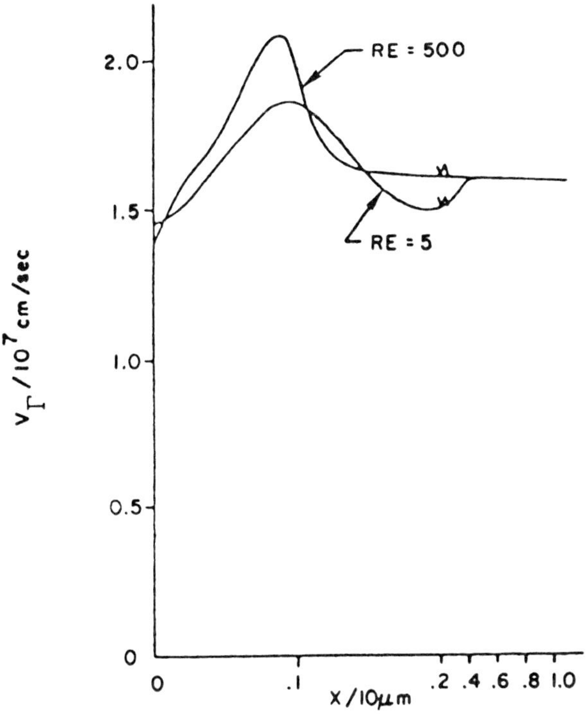

FIGURE 7b. Dependence of Γ-valley velocity on Reynolds number. (Ψ = 2 volts).

state carrier density and Γ valley velocity are sketched for two different values of the Reynolds number - the viscous contributions are different. It is seen from these figures that the smaller the Reynolds number, the smoother the carrier density and velocity profiles. This of course, is expected. The smooth profile, however, tends to result in a larger variation in electric field, as seen from figure 7c as well as from Poisson's equation. Thus, simple predictions about the effect of viscosity on the detailed performance of a device cannot yet be made. More study is clearly in order. We will come back to these equations and their scaling implications later. Meanwhile, we consider the detailed transient behavior of submicron devices.

SPATIALLY INDEPENDENT TRANSIENT TRANSPORT

We have emphasized in these lectures that one of the more important features of transient micron and submicron transport in

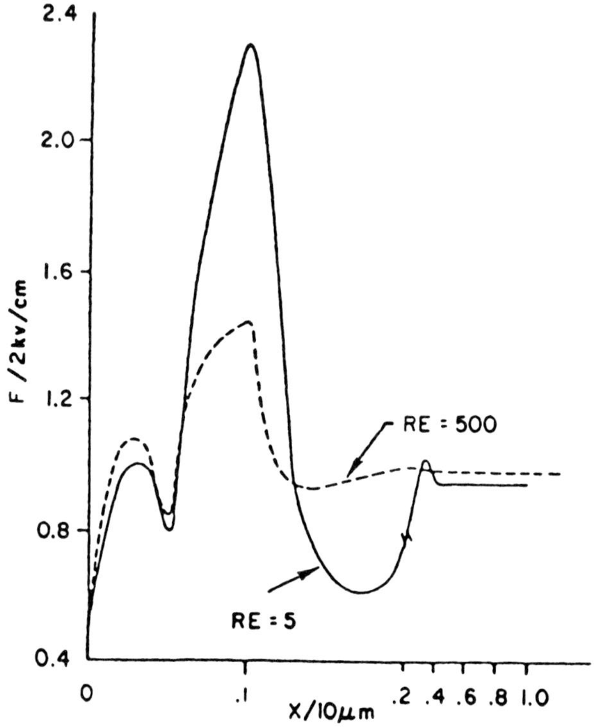

FIGURE 7c. As in Figure 7a, but for electric field distribution.

semiconductors is the non-local contribution. Thus, events at time t depend on an earlier time history, and transient phenomena at a point x depend on disturbances at a point x′ and on the propagation characteristics from x′ to x. In addition, as dimensions shrink and proximity effects become pronounced, reflections and, hence, time dependent contributions from the downstream, as well as the upstream boundary comes into importance. While these effects are implicit in the discussion of dipole transport through solution to the moments of the Boltzmann transport equation, a fuller discussion is given below. We begin with a discussion of nonequilibrium high field transient effects.

Transient nonequilibrium transport, or velocity overshoot as we have come to know it, is prominently identified with the studies of Ruch[15], although historically one of the earliest contributions is due to Butcher et al[16]. The phenomena, as we begin to understand it better, arises from at least two effects: (1) the incorporation of acceleration in the governing equations, and (2) energy dependent relaxation rates. Additionally, in the III-V there is also intervalley transfer. Let us first identify the overshoot and then go through a description of its origin.

Figure 8 is a plot of velocity versus time for a GaAs uniform field device subject to a sudden change in electric field. In this curve the long time asymptotic value is that appropriate to steady state values. The interest in this curve lies in it's short time transient which has a peak velocity in excess of it's steady state value. In the case of Si, the peak velocities are only marginally higher than their steady state values, and nothing truly dramatic occurs. The situation as represented by figure 8 is for GaAs. Indeed, it is this difference that is, in good part, responsible for the high activity in the III-V compounds.

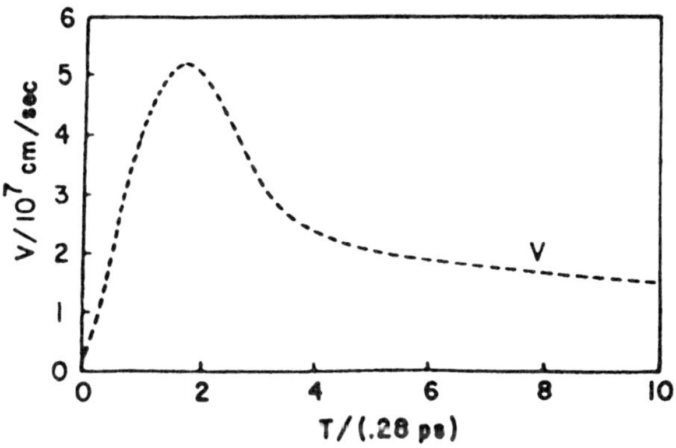

FIGURE 8. Velocity transient for GaAs, from the moment equations. The applied field is 9.7kv/cm.

While much of the early work on velocity overshoot was obtained within the framework of Monte Carlo calculation, the discussion below will be through application of the Moment/Balance equations. These have been discussed in fairly general terms earlier. For spatially uniform fields these equations reduce to a set of simpler coupled ordinary differential equations, which will be used to identify significant nonequilibrium phenomena. For uniform fields, the continuity equation, as well as that for momentum and energy balance reduce to:

$$\frac{\partial n_r}{\partial t} = - n_r\Lambda_r + n_s\Lambda_s \tag{67}$$

$$\frac{\partial p_r}{\partial t} = - en_rF - p_r\Pi_r \tag{68}$$

$$\frac{\partial W_r}{\partial t} = - en_rv_rF - \frac{3}{2}n_rk_BT_r\Sigma_r + \frac{3}{2}n_sk_BT_s \tag{69}$$

The first equation states that the time rate of decrease of the population of the "r"th conduction (or valence) band is proportional to the rate at which the particles leave, less the rate at which they are replenished. The second equation is, of course, Newton's law with a driving force, the electric field, and dissipative term (momentum relaxation). The third equation describes energy balance. We note that with regard to the acceleration equation, nothing other than the driving term introduces gain. With regard to the energy balance equations, the carriers gain energy from the field when

$$\mathrm{sgn}v_i = -\mathrm{sgn}F \tag{70}$$

and return energy when the signs are the same. There is also the possibility of either an energy loss through intravalley electron-phonon interaction and energy gain or loss through intervalley transfer.

For uniform fields the entire description of the transient transport through the semiconductor is governed by a balance between the driving field and the scattering rates. These rates identify all of the relevant physics associated with electron-impurity scattering, electron-hole and hole-impurity scattering, electron-phonon and hole-phonon scattering, etc. The scattering rates fold-in the matrix elements and are obtained through several layers of approximation. One approximation is the use of the BTE for situations where inelastic scattering occurs. Another approximation is associated with the fact that we do not generally obtain matrix elements from first principle arguments, and that there does not exist a set of corresponding experiments that allow us to confidently use the results of these matrix elements. Generally, the strength of a specific scattering matrix element is identified by a coupling coefficient, e.g., the deformation potential, and a scattering rate is computed. Similar procedures are applied to different scattering mechanisms. The resulting scattering rates are then summed. The assigned scattering rates are deemed satisfactory when the calculated steady state field-dependent velocity curve agrees with experiment. The difficulty with this approach is that (1) it is not unique, a variety of different coefficients are likely to give similar results, (2) steady field dependent velocity curves are not obtained experimentally under steady state conditions, (3) consistency does not exist between alternative methods of calculation. Here Monte Carlo and displaced Maxwellian calculations yield somewhat different results (although they are qualitatively similar). The above statements are not meant to suggest that all transport studies should be delayed until all of

TABLE 3. GaAs Parameters Used in Calculations

Parameter	Γ	*L*	X	Common
Number of equivalent valleys	1	4	3	
Effective mass (m_e)	0.063	0.222	0.58	
Γ-*L* separation (eV)		0.33	0.522	
Polar optical scattering				
Static dielectric constant				12.90
High-frequency dielectric constant				10.92
LO phonon (eV)				0.0354
Γ-*L*, X scattering				
L-X scattering				
Coupling constant (eV/cm)		0.800×10^9	0.200×10^9	
Phonon energy (eV)		0.100×10^9		
Γ-*L*, X		0.0278	0.0299	
L-X		0.0293		
L-*L*, X-X scattering				
Coupling constant (eV/cm)		2.0×10^9	2.0×10^9	
Phonon energy (eV)		0.0299	0.0290	
Acoustic scattering				
Deformation potential (eV)	7.0	9.2	9.2	
Nonpolar scattering (*L*)				
Coupling constant (eV/cm)		0.300×10^9	0.300×10^9	
Phonon energy (eV)		0.0343	0.0343	

the above problems are overcome. They are mentioned to indicate that caution is in order and that an over reliance on available parameters for describing scattering rates is not acceptable.

As indicated earlier, all scattering rates will be obtained within the framework of the displaced Maxwellian. The integrals as used in the discussion below are summarized in Table 3.We note that the scattering matrix elements for electron-phonon interaction were applied, within the early fifties to calculate the scattering rates via the displaced Maxwellian, by Froelich and coworkers[12]. They have since been repeated by a variety of other workers, in some cases with important generalizations. One of the earliest applications of the moment equation and the above scattering integrals was to the semiconductor GaAs. These calculations were performed under the assumption of transport within two portions of the GaAs conduction band, one at Γ, and the second at X. The parameters for this calculation are displayed in Table 4[16].

TABLE 4. GaAs Parameters Used in Calculations

Parameter	Γ	X	Common
Number of equivalent valleys	1	3	
Effective mass (m_e)	0.067	0.40	
Γ-X separation (eV)			0.36
Polar optical scattering			
Static dielectric constant			12.53
High-frequency dielectric constant			10.82
LO phonon (eV)			0.0354
Γ-X scattering			
Coupling constant (eV/cm)			0.621
Phonon energy (eV)			0.0300
X-X scattering			
Coupling constant (eV/cm)		1.064	
Phonon energy (eV)		0.0300	
Acoustic scattering			
Deformation potential (eV)	7.0	7.0	

In the late seventies experimental evidence pointed to the L valley as the lowest satellite valley, but with X sufficiently close to L to provide enough coupling at high fields. Thus GaAs was regarded as a three-level transfer semiconductor. Transient and steady state calculations with three levels of transfer have been considered by many workers. Results within the framework of the displaced Maxwellian will be illustrated below. Additionally, several two-level spatially dependent electron transfer calculations, with Γ-L ordering, using the parameters of Table 5.

For three-level transfer with the parameters of Table 3, the scattering integrals are shown in figure 9, where scattering rates are plotted as a function of electron temperature. For example,

TABLE 5. GaAs Parameters Used in Calculations

Parameter	Γ	*L*	Common
Number of equivalent valleys	1	4	
Effective mass (m_e)	0.067	0.222	
Γ-*L* separation (eV)			0.33
Polar optical scattering			
Static dielectric constant			12.90
High-frequency dielectric constant			10.92
LO phonon (eV)			0.0354
Γ-*L* scattering			
Coupling constant (eV/cm)			0.800×10^9
Phonon energy (eV)			0.0278
L-*L* scattering			
Coupling constant (eV/cm)		2.0×10^9	
Phonon energy (eV)		0.0354	
Acoustic scattering			
Deformation potential (eV)	7.0	9.2	
Nonpolar scattering (*L*)			
Coupling constant (eV/cm)		0.300×10^9	
Phonon energy (eV)		0.0343	

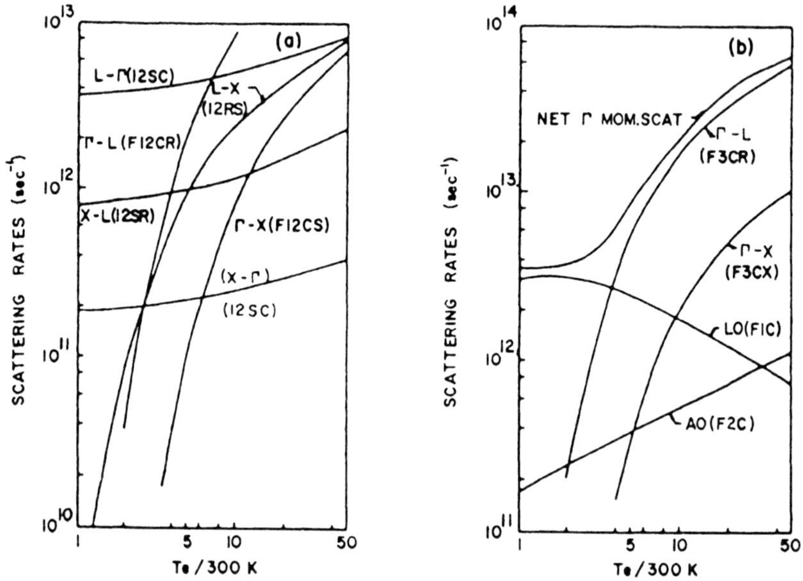

FIGURE 9. Three-level **(a)** carrier scattering rates for use in equation (44); and **(b)** Γ valley momentum scattering rates for use in equation (45).

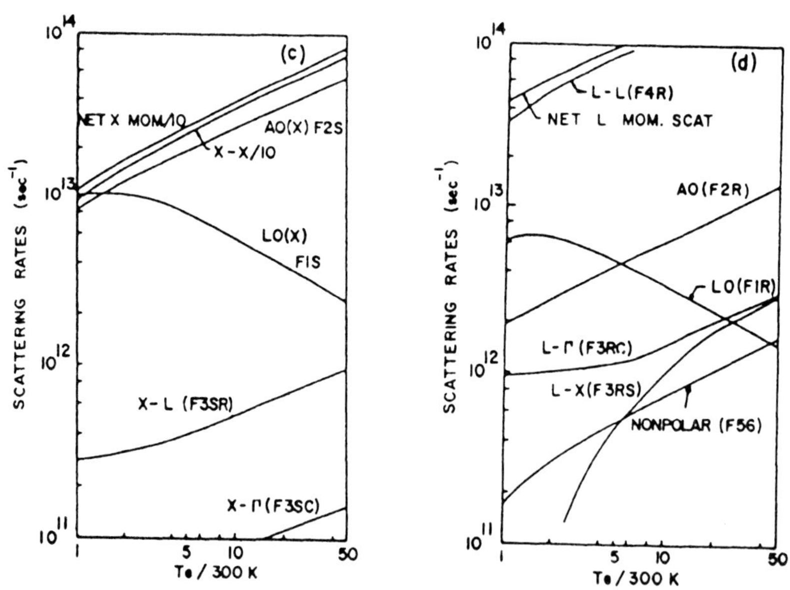

FIGURE 9. Three-level **(c)** L valley and **(d)** X valley momentum scattering rates, for use in equation 45.

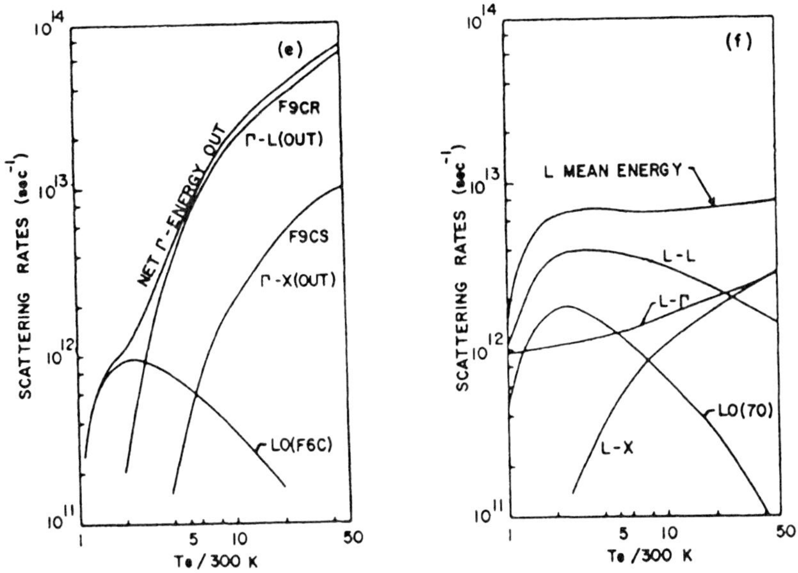

FIGURE 9. Three-level **(e)** Γ valley energy scattering, and, f) L valley energy scattering, for use in equation (46).

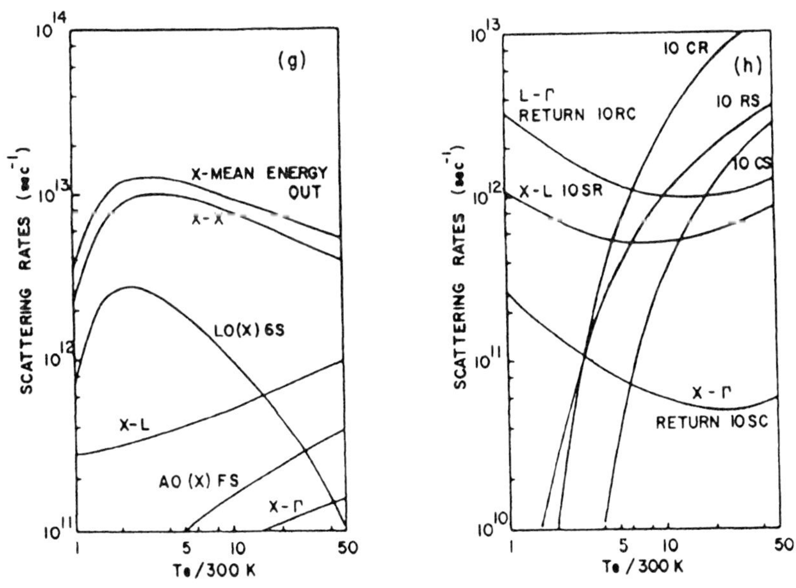

FIGURE 9. Three-level **(g)** X valley energy and **(h)** return energy scattering for use in equation (46).

figure 9a displays the carrier density scattering rates. Note that the ratio, at T_e=300K, of the LΓ carrier scattering to ΓL scattering is governed by the equilibrium distribution of the Γ and L valley carriers, as given by equation 71. The scattering contributions to the Γ valley momenta are shown in figure 9b. At low energies scattering is dominated by the LO phonon, whereas at high energies by the intervalley phonon. The scattering contributions to the L and X valleys momentum are shown in figures 9c and 9d, respectively. The dominant contribution is equivalent intervalley scattering. The Γ, L and X valley energy scattering rates are shown in figures 9e, 9f, and 9g, respectively. Note the increased scattering rate at high energies for the Γ valley carriers, and the decreased scattering rate for the X-valley carriers. The energy loss for the Γ valley carriers is dominated by transfer to valleys with higher density of states. The L valley energy loss is intermediate between Γ and X. The return energy scattering is shown in figure 9i.

Note that there are 42 scattering integrals, exclusive of ionized impurity scattering, to contend with. After summing, there are 18 scattering rates of interest. The steady state calculation arising from these integrals is shown in figures 10 and 11.

Figure 10 displays the field dependent mean velocity for Γ-L-X ordering. Figure 11a is the distribution of carriers as a function of applied field. The electric field versus Γ-valley electron temperature is displayed in figure 11b. In figure 11c and 11d, we sketch (e/mΠ)F for select Γ and L valley integrals, as well as the mean velocity for Γ and L-valley electrons. Note that the Γ valley velocity, whose contributions are dominated by LO phonons at low energies, and intervalley phonons at high energies, is at least an order of magnitude greater than the L-valley velocity.

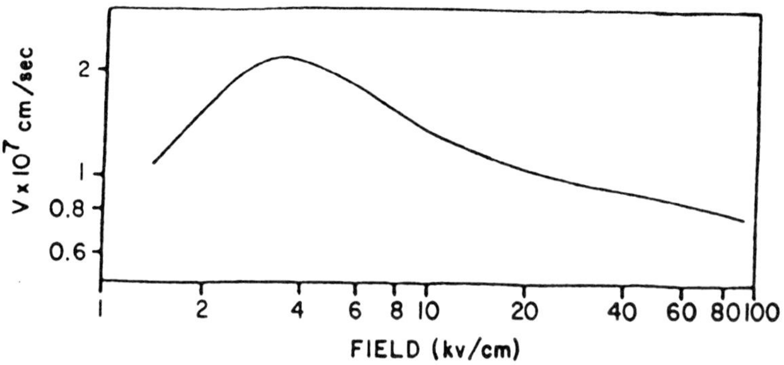

FIGURE 10. Steady state velocity vs field for Γ-L-X ordering in GaAs.

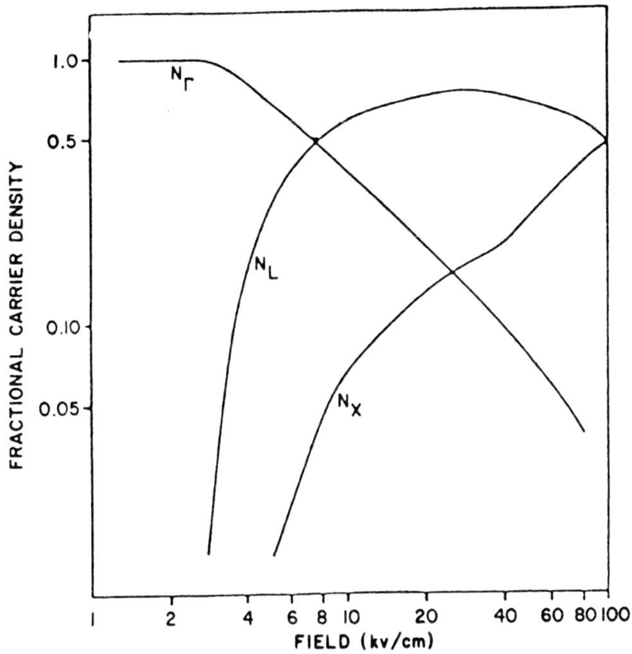

FIGURE 11a. Carrier density of Γ-L-X valleys for the calculation of figure 10.

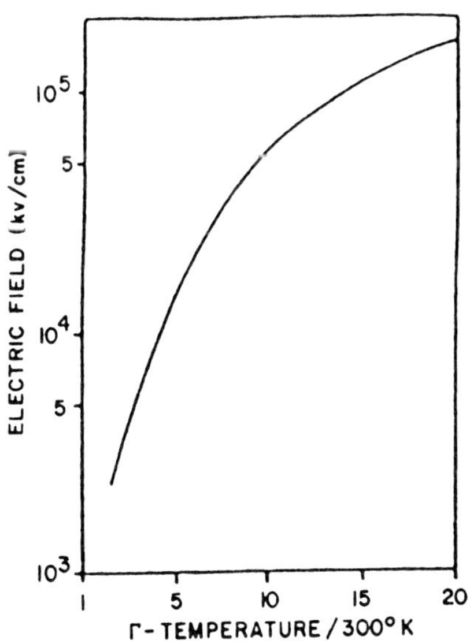

FIGURE 11b. Field vs Γ-valley electron temperature.

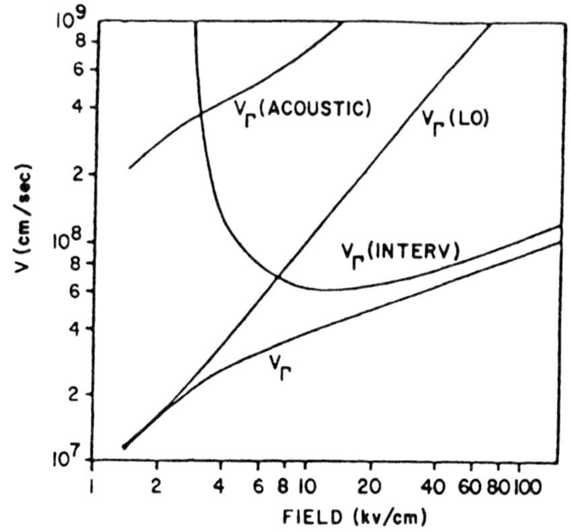

FIGURE 11c. Contributions to Γ-valley velocity.

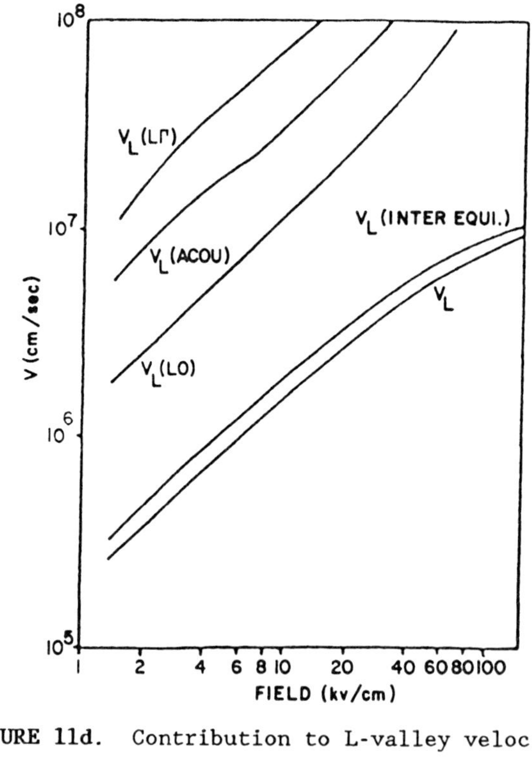

FIGURE 11d. Contribution to L-valley velocity.

There are several points to be made about these integrals. The first is that the energy scattering rates, particularly those associated with the central valley are several orders of magnitude less than those of the corresponding momentum scattering rates. Thus, while initially in an applied field the momentum increase without any significant relaxation, subsequent momentum relaxation is energy dependent and dominated by the longer energy relaxation times. Prior to significant energy relaxation, the momentum and carrier density scattering rates are being altered by the increasing electron temperature as well as the increasing carrier velocity. Thus in this case of momentum balance, there is a decrease in carrier momentum, and in the case of carrier balance, there is a decrease in central valley population. The consequences of this is a net decrease in carrier velocity. These results are born out numerically and are displayed in figure 12 for carriers

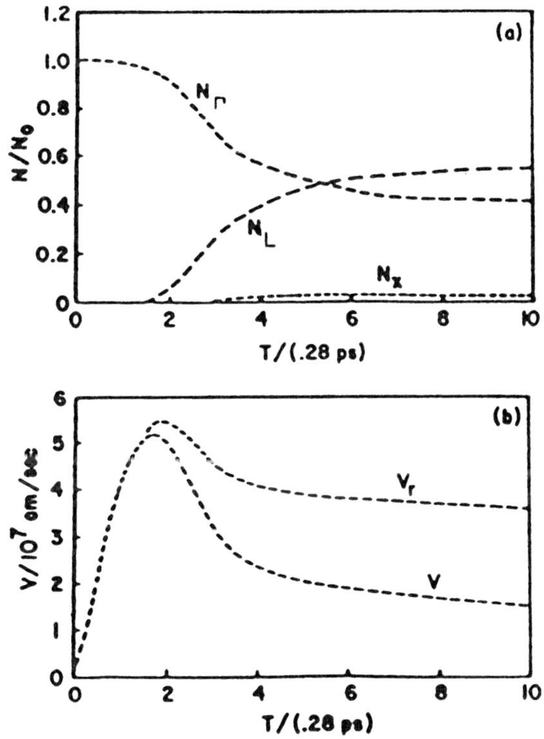

FIGURE 12. Transient uniform field response, <F> = 9.7Kv/cm μ.

with a zero initial velocity, an initial electron temperature in the central and satellite valley equal to the ambient temperature, and an initial equilibrium distribution of carriers:

$$\frac{n_\Gamma}{n_L} = \left(\frac{m_\Gamma}{m_L}\right)^{\frac{3}{2}} \exp \Delta_{\Gamma L}/kT \tag{71}$$

$$\frac{n_\Gamma}{n_X} = \left(\frac{m_\Gamma}{m_X}\right)^{\frac{3}{2}} \exp \Delta_{\Gamma X}/kT \tag{72}$$

Note: There is overshoot in the Γ-valley carrier velocity, as well as the mean velocity.

As the discussion above indicates, the initial transients are dependent in a detailed way on the way the energy relaxes. Thus the details of transient transport are likely to depend significantly on at least the following items: (1) rise time of the external source, (2) initial conditions, (3) boundary conditions, and (4) device structure. The first two items are taken up next.

The study of the rise time is basically a matter of matching time constants. For example, if the time it takes electrons to reach steady state equilibrium is of the order of one picosecond, then rise times of the order of one picosecond or less will introduce strong nonequilibrium effects. These effects will decrease as the rise time increases. This is illustrated in figure 13 where we plot peak velocity versus rise time for a device subject to a specific value of bias field.

The significance of initial conditions for uniform field calculations is that it is thought to identify the properties of the entrance contact. In the strictest sense the properties of the entrance contact are determined by a set of boundary conditions, with the initial conditions on the boundary playing a less prominent role. Calculations with both will be illustrated below with more emphasis on the contact as a boundary conditions.

The first set of calculations is for uniform field and GaAs with three level transfer. The bias field for the case is 17.6 kv/cm and the initial carrier velocity is zero, figure 14, dashed line. Strong velocity overshoot is apparent from this calculation, and there is overshoot in the temperature. (Please note: that for this calculation capacitive effects have not been included. They are incorporated implicitly in the spatially dependent transient calculations). The situation when the entrance carrier velocity and temperature are varied is indicated by the solid curves of figure 14.

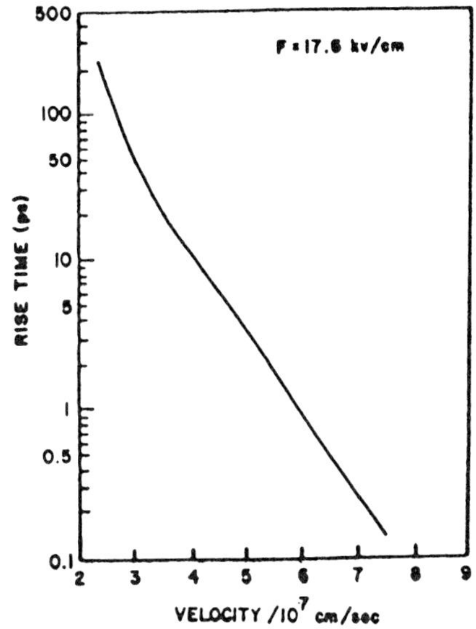

FIGURE 13. Velocity vs rise time.

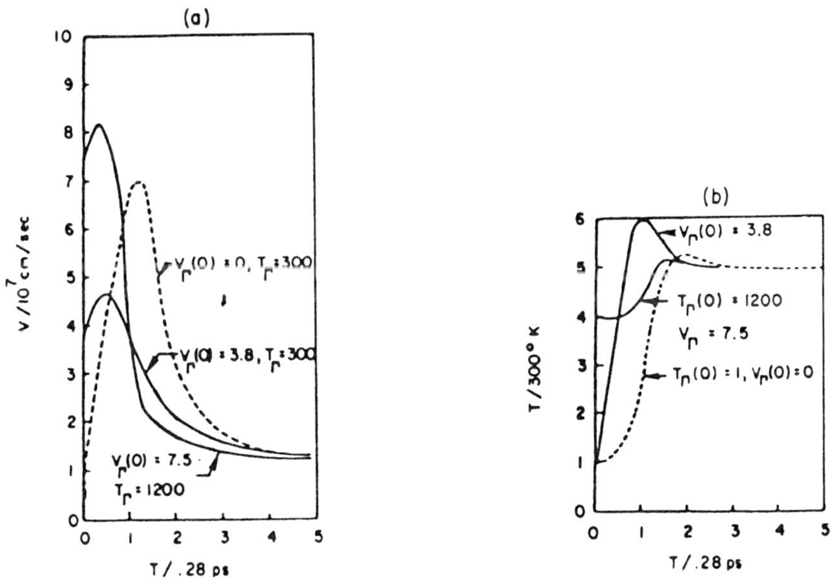

FIGURE 14. (a) Transient response as a function of initial velocity and temperature. (F=17.6kv/cm). Load line included; (b) As in 14a, but for Γ-valley temperature.

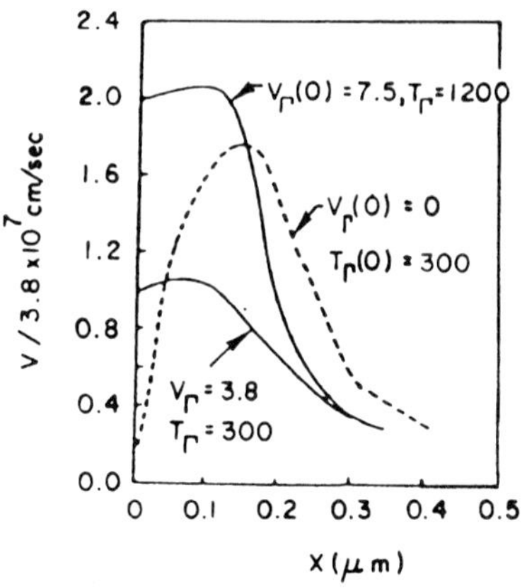

FIGURE 15. As in 14a, but for velocity vs distance.

Several things are apparent from these results. First, for an initial entrance temperature equal to the ambient temperature, an increase in the entrance velocity results in an increase in the peak velocity and the rate at which electron temperature increases. As a consequence of the latter, the momentum and density relaxation rates increase. Thus, the time to steady state is reduced. We can project that a tradeoff exists for an optimum set of initial conditions in which an electron will traverse a given distance in the shortest time interval.

A very similar set of results was presented by Iafrate et al.,[17] using Monte Carlo techniques. There are some very important differences in the assumptions associated with the Monte Carlo technique and the displaced Maxwellian, particularly with regard to the assumption concerning the distribution of the entering carriers. Nevertheless, the qualitative features of the two results are similar. The Monte Carlo studies are displayed in figures 16 and 17. Note, the F=20 curve of figure 16 is qualitatively similar to that of the $v_\Gamma(0)=7.5$ curve of figure 15, which is the same as figure 14a but with distance as the abscissa. Figure 17 indicates that the rate of carrier velocity decrease increases with increasing energy. Again, in qualitative agreement with figure 15a.

INFLUENCE OF SPATIAL NONUNIFORMITIES ON TRANSIENT TRANSPORT

The above discussion was concerned with uniform fields, a situation that is not expected to occur. Indeed if it did occur, we would necessarily conclude that submicron two-terminal GaAs devices would possess a region of dc negative differential conductivity. Further, the implication of the spatially uniform velocity overshoot calculations is that they are representative of

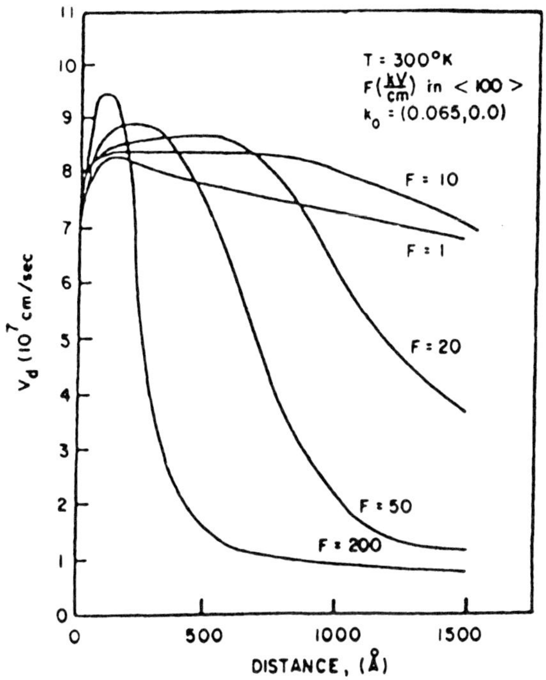

FIGURE 16. Velocity vs distance vs applied field.

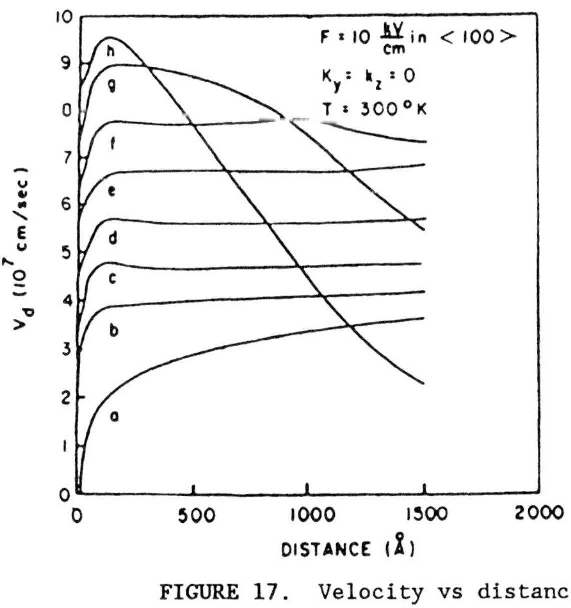

	k_x $\frac{2\pi}{a}$	E_0(eV)
a	0	0
b	0.02	0.04
c	0.03	0.07
d	0.04	0.11
e	0.05	0.16
f	0.06	0.21
g	0.07	0.27
h	0.08	0.34

FIGURE 17. Velocity vs distance vs energy.

measured current transients profiles. As discussed below, this is not likely to be the case.

To illustrate the role of spatial nonuniformities on transient transport, a limited number of numerical calculations involving space and time dependent solutions to the first three moments of the BTE. These calculations have been performed as a function of device length, boundary conditions, doping level and material variations. None of these calculations display a region of dc NDC. Indeed, all calculations show considerable departure from the familiar homogeneous uniform field studies. All calculations are performed using a model for GaAs that contains two levels of transfer, and assumes a conductivity effective mass equal to the density of state effective mass. In addition, the boundary conditions are chosen to represent the effects of injecting contacts. It is possible to argue against the use of a specific set of boundary conditions to represent the properties of the physical contact. These arguments would, however, tend to remove us from the essential features of the problem, which are to

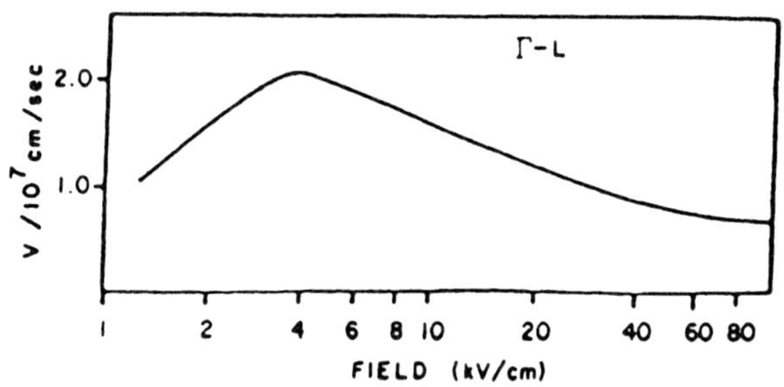

FIGURE 18a. Field dependent velocity for Γ-L two-level transfer.

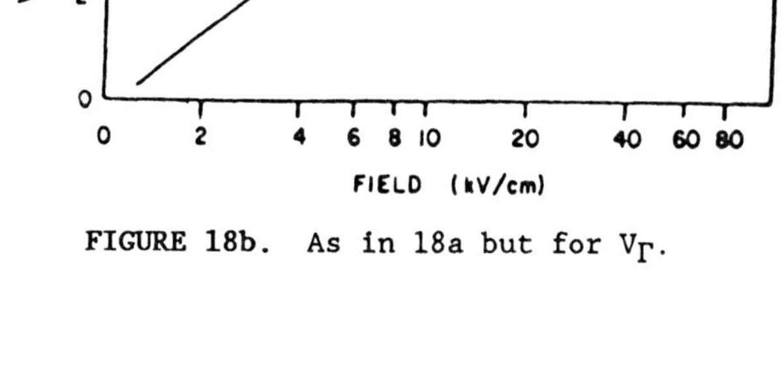

FIGURE 18b. As in 18a but for V_Γ.

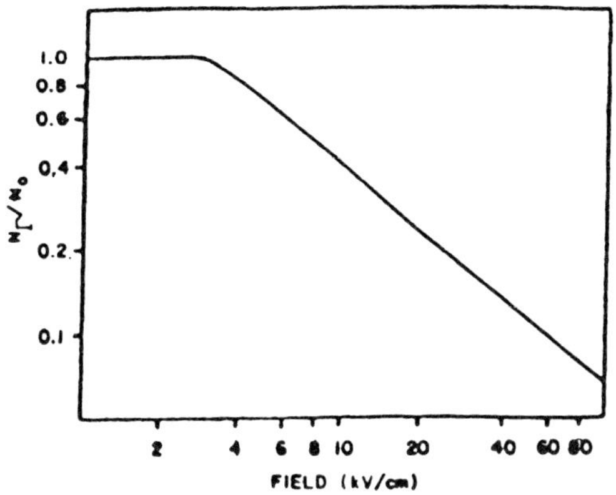

FIGURE 18c. As in 18a but for NΓ.

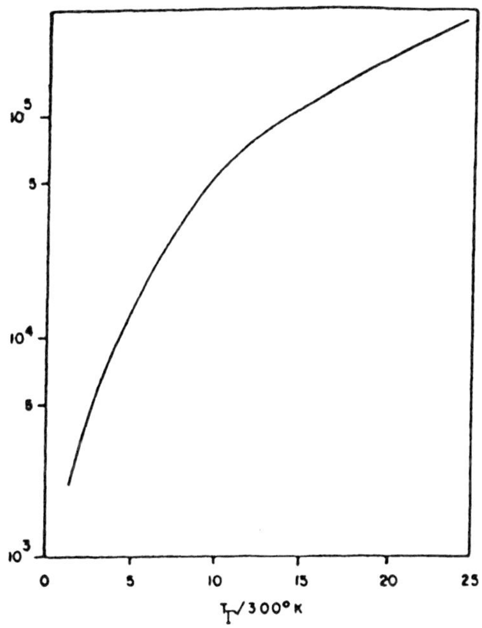

FIGURE 18d. Field vs T_Γ.

determine the role of the boundary and of nonuniform space charge effects on device performance. The parameters used for the calculations are displayed in Table 5. For reference, the steady state field dependent transport results are displayed in figure 18.

The first set of calculations shown is for a uniform field. These calculations are performed for reference only. **For the uniform field calculations,** the device is in series with a load resistor and an inductor. The device is also in parallel with a capacitor. For the first set of calculations, the inductor is set to zero, and the load resistor is set to $0.01R_o$, where R_o is the low field resistance of the element. The capacitor was set to $C_o = \tau_R/R_o$, where τ_R is the dielectric relaxation time for the semiconductor element. **For the space charge dominated device,** the device was subjected to a specified time dependent voltage source and all capacitive effects were included self-consistently through application of the continuity equation. (An earlier study incorporated the load resistor). The two space charge dependent calculations were for two terminal devices, one with an active region length of 0.25μm, the second with a device length of 1.0μm. Both had doping levels of $5x10^{15}/cm^3$ and were subject to injecting boundary. Further in these calculations, ionized impurity scattering was ignored.

The uniform field calculations were performed: (1) at a bias level of 6kv/cm, and (2) at 10kv/cm. The familiar overshoot results discussed earlier are displayed here in figure 19.

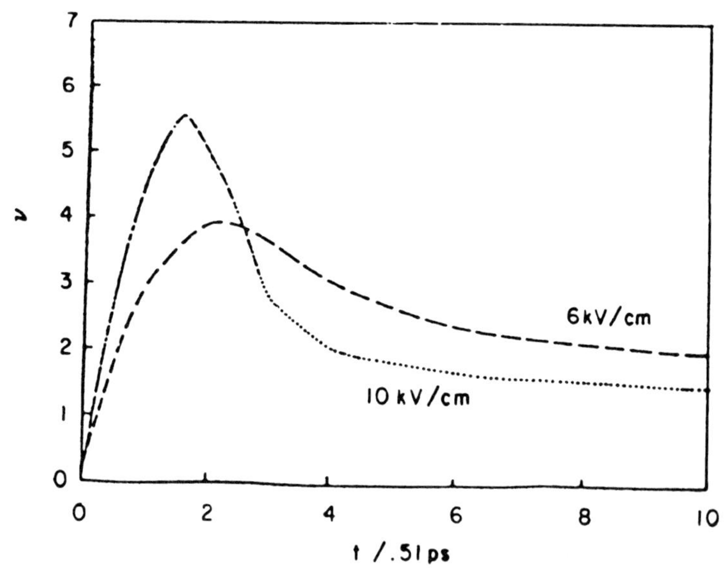

FIGURE 19. Transient transport uniform field for two-level Γ-L transport.

The situation with regard to the space charge dependent calculations with injecting boundary conditions is shown in the next set of figures. Figure 20 displays the current time transients for the 0.25 and 1.0μm devices at two different bias levels. The use of current rather than velocity is significant

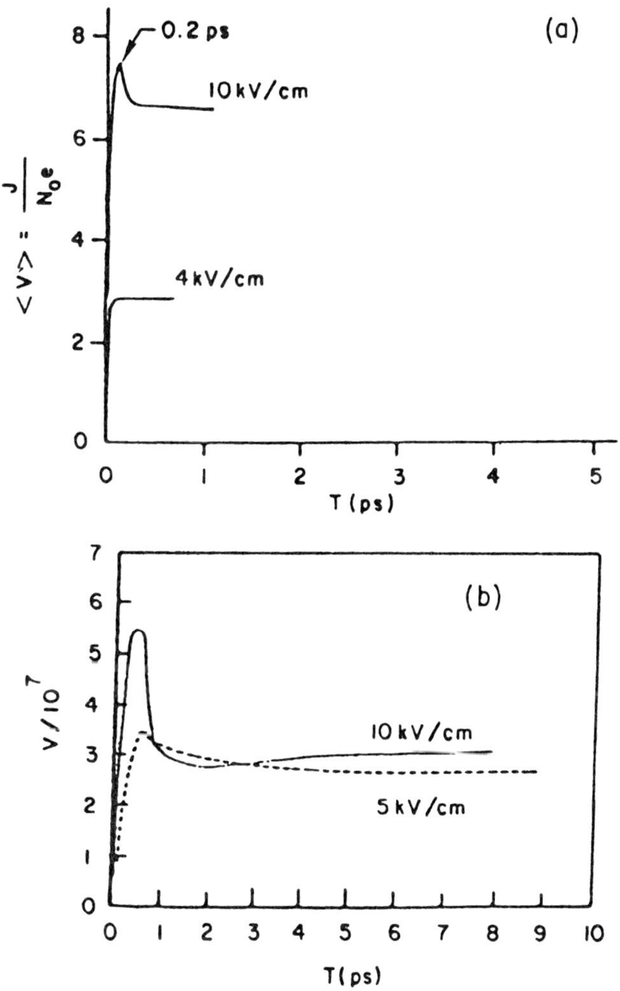

FIGURE 20. (a) <v> vs time for .25μm device with nonuniform fields; (b) <v> vs time for 1.0μm device with nonuniform fields. Each transient was taken from a non-zero initial state.

here - it is current that is conserved - not velocity. In one dimension the total current is independent of position. Note the significant result that at the higher bias levels the steady current level increases - a result opposite to that for the uniform field calculations.

The steady state distribution of electric field in these devices is nonuniform and an understanding of them is crucial for understanding device operation. The field profiles for the 0.25μm device are displayed in figures 21a and 22a, and show a

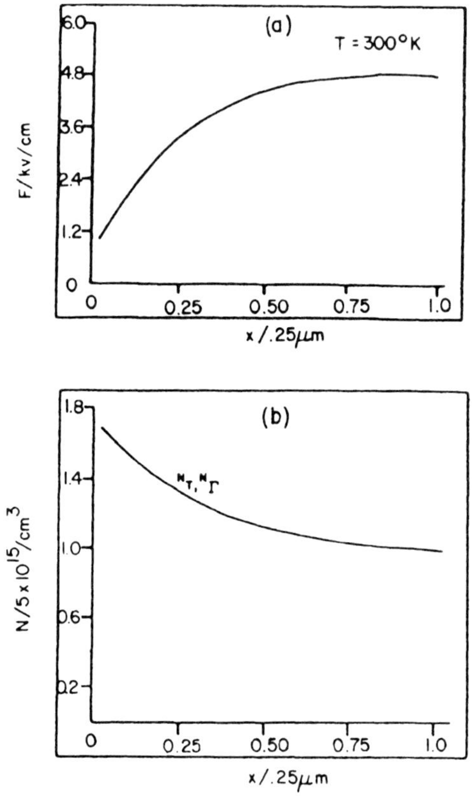

FIGURE 21. (a) Electric field vs distance for 0.25μm device with injecting cathode and an applied bias of 0.1 volts; (b) N-total and N_Γ carrier density.

characteristic increase in field as the downstream boundary is approached. In both cases the initial increase from the cathode is a consequence of the injecting boundary condition. In the 0.25μm

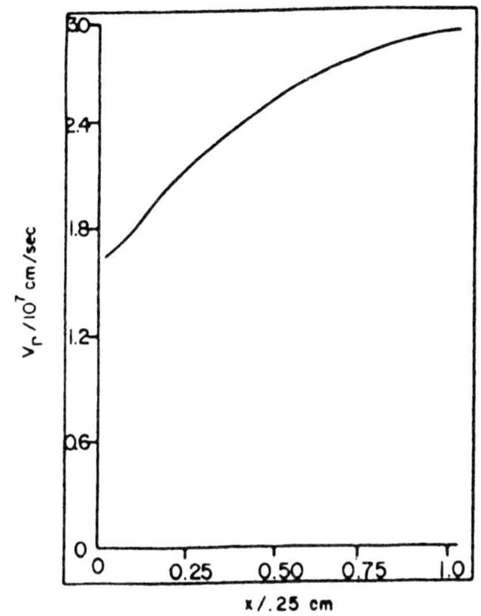

FIGURE 21c. Γ-valley velocity distribution.

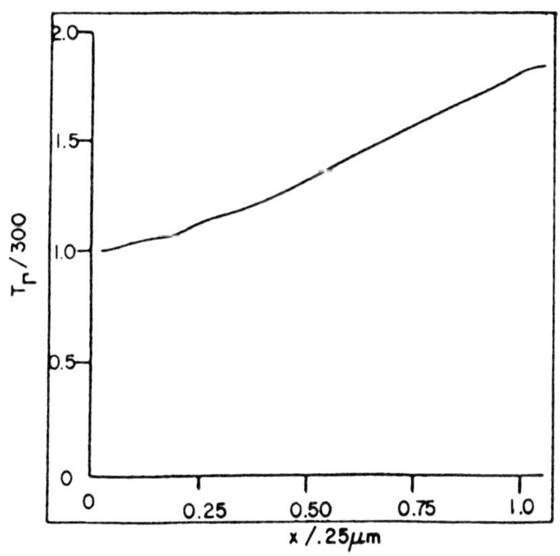

FIGURE 21d. Γ-valley temperature distribution.

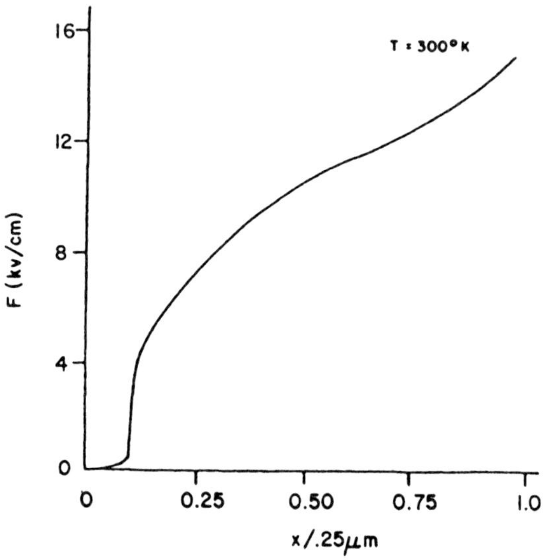

FIGURE 22(a). Electric field vs distance for 0.25μm device with injecting cathode and a bias of 0.25 volts.

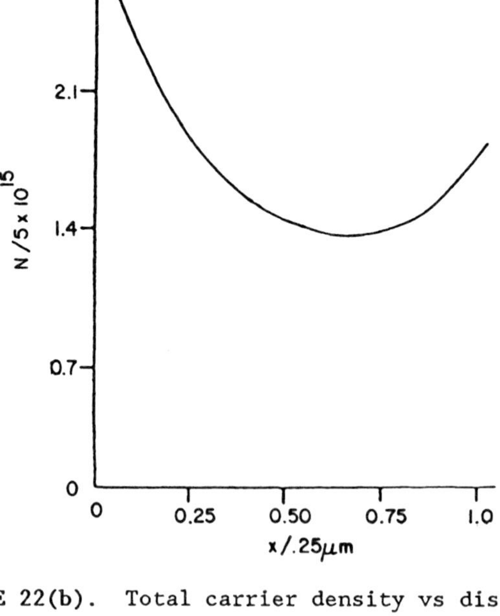

FIGURE 22(b). Total carrier density vs distance.

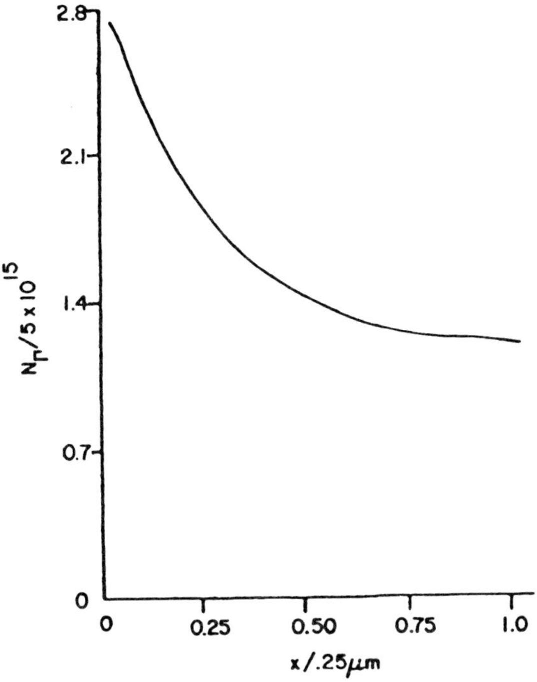

FIGURE 22(c). Γ-valley density vs distance.

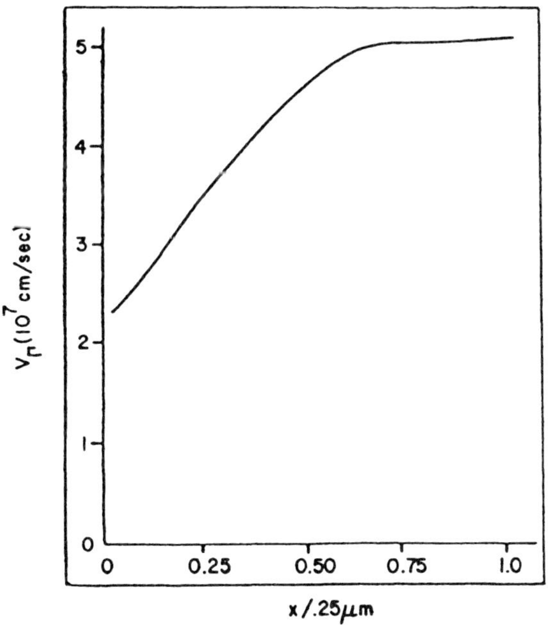

FIGURE 22 (d). Γ-valley velocity vs distance.

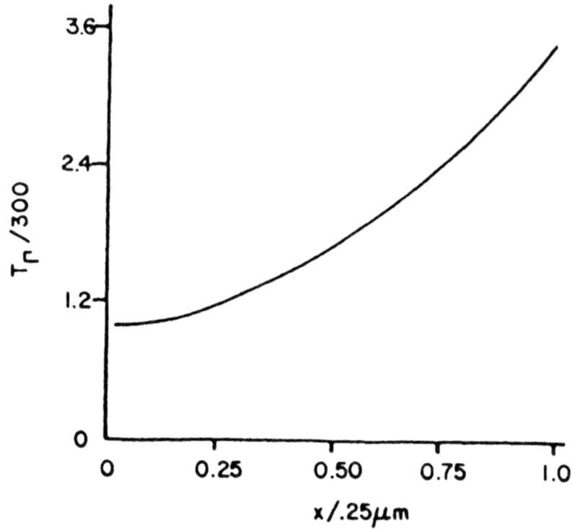

FIGURE 22(e). Γ-valley temperature.

structure and a bias of 0.1v the anode field is flat, while that for the 0.25v case the field continues to increase. There is little electron transfer at the lower bias level, as indicated in figure 21b, but a substantial number for the higher bias level, as indicated in figures 22b and 22c. It is anticipated that increasing the applied bias will cause an increase in the numbers of carriers scattered to the satellite valleys where the carrier velocity is reduced. Experimentally, however, in the absence of device heating an increase in current density with an increase bias field is expected, at least under dc conditions. The most likely reason for this to occur is through net carrier accumulation within the region of electron transfer. This, indeed, is what is occurring and is seen in figure 22b. We note that the decreasing carrier concentration for the short device and low bias requires, through current continuity an increase in carrier velocity, as seen in figure 21c. Further, the Γ-valley electron temperature increases only to 600°K, not enough to result in carrier transfer.

The increasing electric field at high bias levels in the vicinity of the anode is a feature of materials with a region of negative differential mobility and does not appear to be dependent upon the specific boundary conditions applied. It was predicted by Shockley[18], and its consequences have been detected experimentally[19]. We note that, as in the low bias calculation, the Γ-valley carrier velocity increases substantially toward the anode (figure 22d) and the electron temperature exceeds 900°K (figure 22e), forcing electron transfer.

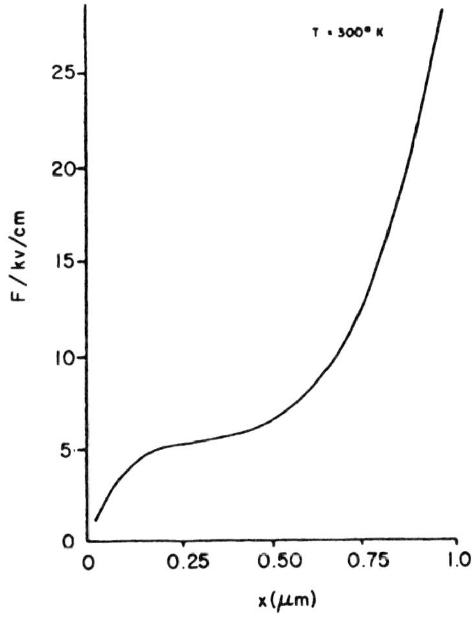

FIGURE 23. Electric field vs distance of 1μm device with an injecting cathode and a bias of 1.0 v.

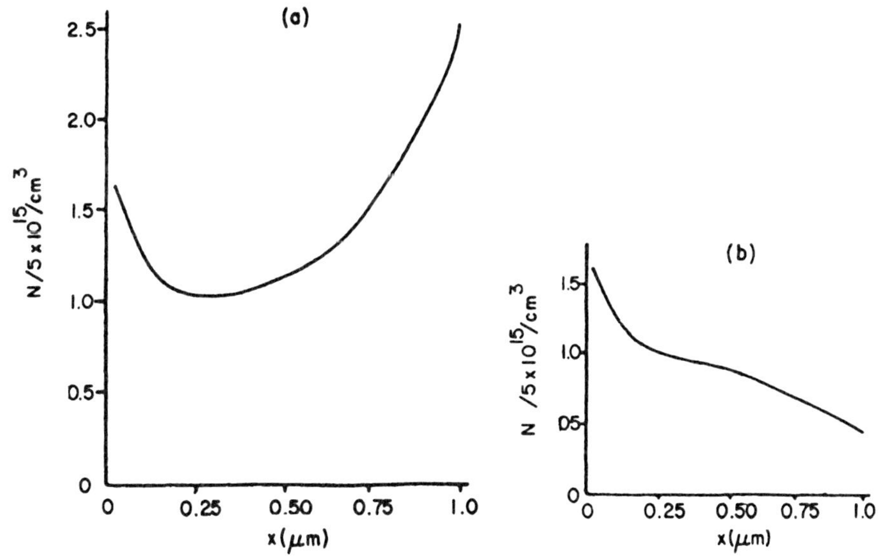

FIGURE 24(a). Total carrier density corresponding to the field distribution; (b) Γ-valley carrier density.

The features of the 0.25μm calculation are exaggerated for the longer 1.0μm calculation of figure 23. Note the high anode field near 30kv/cm, and the substantial electron transfer (figures 24a and b). The Γ-valley carrier velocity for this case is **larger** than that of the 0.25μm case as required by current continuity.

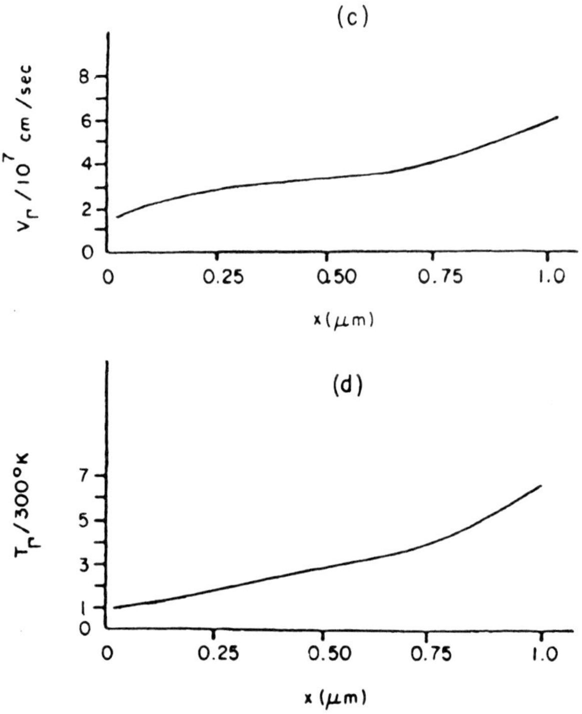

FIGURE 24(c,d). As in figure 24a, b, but for Γ-valley carrier velocity and electron temperature.

It is important to reiterate the significance of these results. It was originally anticipated that the uniform field calculations and the consequent predictions were relevant to high frequency and, therefore, submicron devices. With regard to the uniform field calculations, it must be noted that these are applicable to devices of any length, as long as the field is uniform. With regard to submicron devices with nonuniform fields, the time dependent uniform field results are not at all adequate. They do not predict the peak or asymptotic currents correctly.

The numerical results are somewhat model dependent. Certainly the details are dependent upon boundary conditions, device structure, length, doping, and etc. But there are several interesting features that should be highlighted as they appear in a number of different studies. The first point to be noted is that the carriers are entering the device at high velocities of $1.5x10^7$ cm/sec. Within the device and near the downstream boundary, as indicated above, a large fraction of the carriers transfer to the satellite valley. These low mobility carriers contribute only an insignificant amount to the net current through the device. The high current drive is a consequence of the high level of injection throughout the active region of the device. (In longer "active region" devices, the advantages of charge injection are reduced and intervalley transfer leads to lower current levels).

The importance of space charge injection is not limited to the structure just discussed. An extremely interesting device that appears to exhibit many of the properties of the 0.25μm device is a one dimensional version of the space charge injection field effect transistor (Constant et al[20]). The geometry of this structure is shown in figure 25, and the space charge dependent

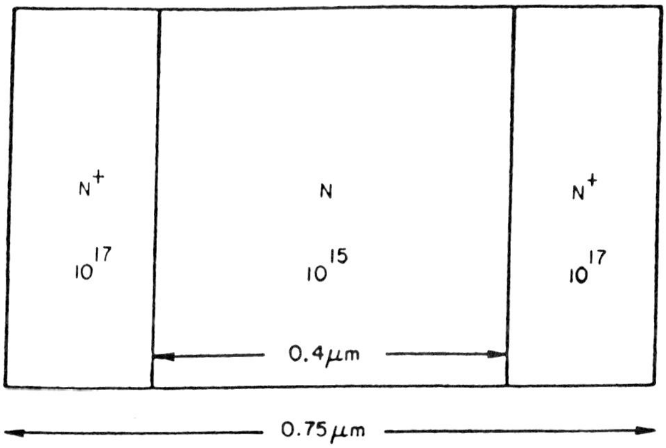

FIGURE 25. Structure for simulation of submicron space charge limited transport.

profiles for a one-volt bias across the device are displayed in figures 26a through 26c.

For the injection device depicted in figures 25c and 26, the space charge dependent profiles display considerable similarity to the results of the space charge limited diode. There is a potential minimum some distance away from the physical cathode, resulting in a retarding electric field. The consequence of this

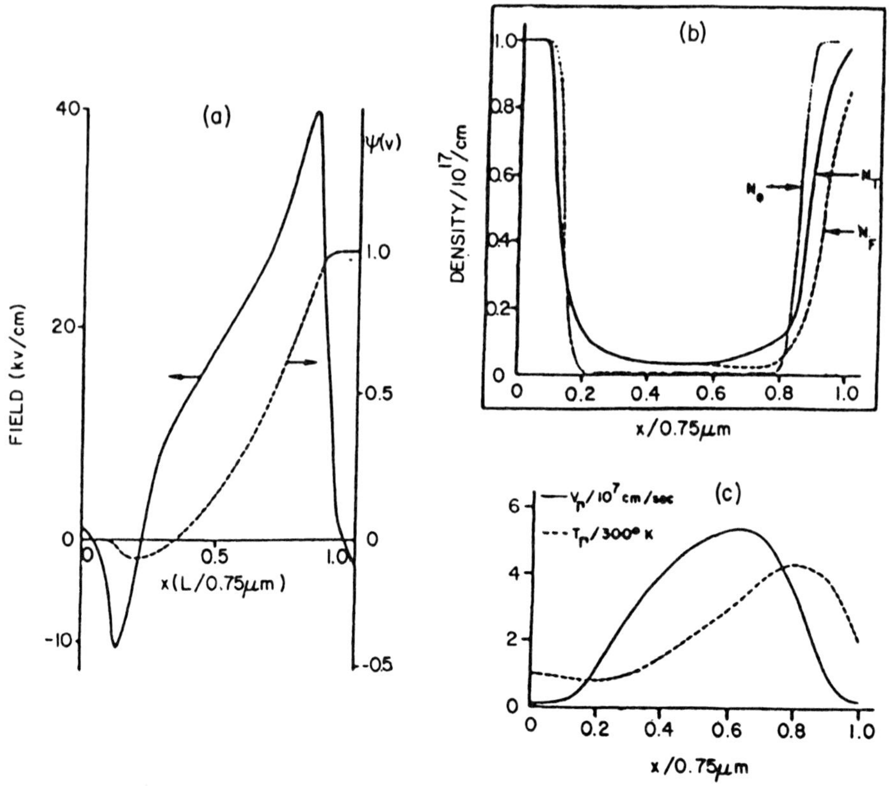

FIGURE 26. (a) Field and potential vs distance for the structure of figure 25; (b) carrier density; (c) carrier velocity and electron temperature.

retarding field is that the electrons are cooled as they approach the minimum[21]. (Note: we are not accurately representing the scattering at temperatures below the ambient. The results are therefore only qualitatively significant). The carriers are entering the active region at relatively high values of velocity and are exiting at even higher values. The field is nonuniform with high values near the downstream contact. There is considerable transfer which is compensated by an excess of injected carriers.

ASSESSING THE UPPER FREQUENCY LIMIT OF SMALL SIGNAL OPERATION

The next item of interest in connection with any of these devices, and certainly with regard to assessing material properties is the upper frequency limit of operation. The determination of the upper frequency limit depends on the structure of interest and the mode of operation. If we confine ourselves to the transferred electron effect, then there are at least five items of interest: (1) large versus small signal operation, (2) device length, (3) boundary conditions, (4) doping variations, and (5) circuit. There has been only a limited amount of work done with regard to the last four items and here only with long devices. We will consider these later. At this point, we will confine our activity to large and small signal operation.

Consider first small signal high frequency operation, within the context of the following type of "experiment" (see figure 27). We assume a device is in a quiescent state at a voltage level Ψ_0 and a current level I_0. The device is then subjected to a sudden perturbation in the form of a square wave pulse. The questions to be asked are: what is the output response, and what is the upper frequency limit for small signal negative conductance?

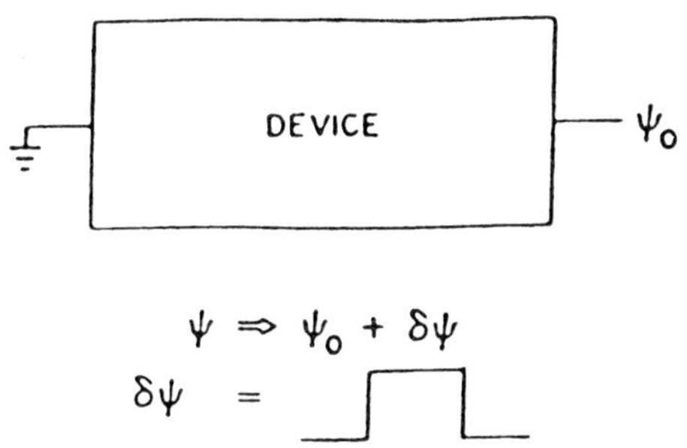

FIGURE 27. Small signal configuration.

Under uniform field conditions, NDR occurs when carriers transfer to the low mobility satellite valleys. Since the number of electrons transferring to subsidiary values requires a finite time to be completed, the extent of negative conductance is frequency dependent. This frequency dependence is complicated by the presence of nonuniform fields which completely eliminates dc NDR. Indeed for nonuniform fields, the most significant contribution to small signal NDR is the non-local transit time effect. (Note: this does not preclude the possibility of a transient "uniform-field"-like response to an external source). We first examine the way a material like GaAs responds to such an external stimulus. Consider figure 28, which displays the voltage pulse and a qualitative sketch of a current response (based on the overshoot calculations). The current response includes an apparent overshoot and a relaxation to a current level below that of the unperturbed system. We are clearly considering uniform fields here. There could, additionally, be some nonequilibrium undershoot as represented by the dashed lines. When the pulse is turned off the system returns to its steady state value. Hereto, there may be some residual nonequilibrium overshoot.

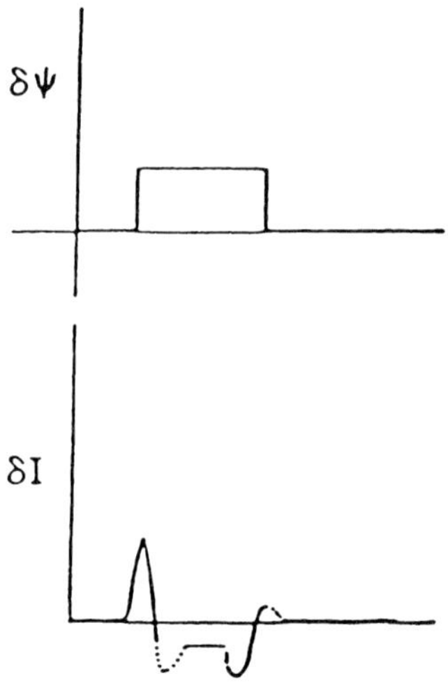

FIGURE 28. Small signal current response.

With reference to figure 28, the product $\delta\Psi\delta I$ represents the time rate of change of energy across the device. When it is positive, the system is lossy, when it is negative, the device is returning power. For uniform fields where an increase in the bias field results in a decrease in current, there is necessarily a region of small signal frequency dependent NDR[22]. The frequency dependence is due to several sources. First, there is over and undershoot prior to steady state (dashed lines in figure 28). Second, the extent of the frequency dependence is sensitive to the time duration of the disturbance. For example, if the excitation is turned off before the system had a chance to reach the NDR region, no power would be delivered to the load. Thus, the limitation to small signal negative resistance is principally at high frequencies.

The small signal NDR is dependent upon the scattering rates. These in turn are dependent upon the bias field. Thus, the magnitude and extent of the small signal NDR is bias dependent. Specifically, at sufficiently high bias levels a substantial amount of transfer occurs and the small signal NDR will decrease. There should be an optimum bias level for small signal NDR, and this is displayed in figure 29a. The data summarized in 29a is obtained from dispersion relationships. The field dependence of the real part of the dispersion, yields the frequency dependent mobility, as is shown in figure 29b. Here, as the quiescent field increases,

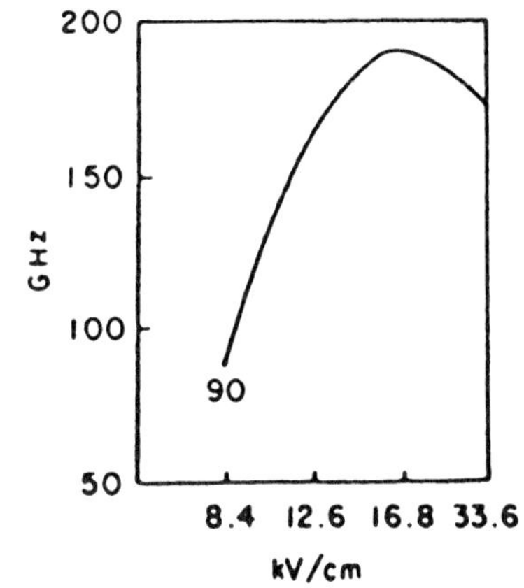

FIGURE 29a. Maximum frequency for small-signal oscillations (from uniform field MBTE).

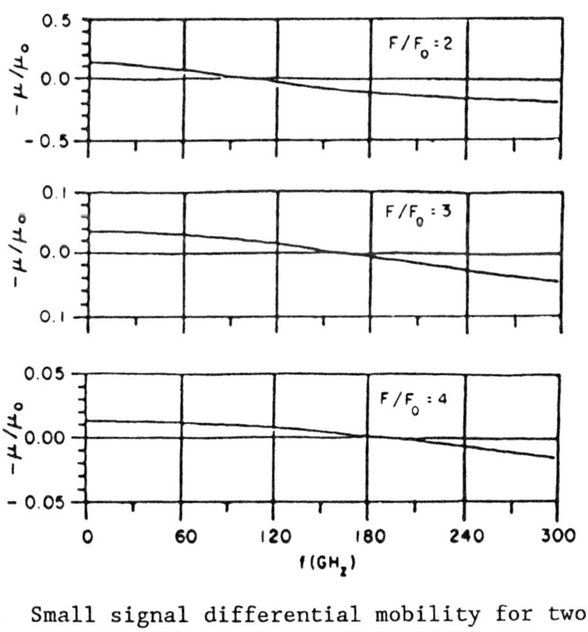

FIGURE 29b. Small signal differential mobility for two-level transfer (from uniform field MBTE). Here F_0 = kv/cm.

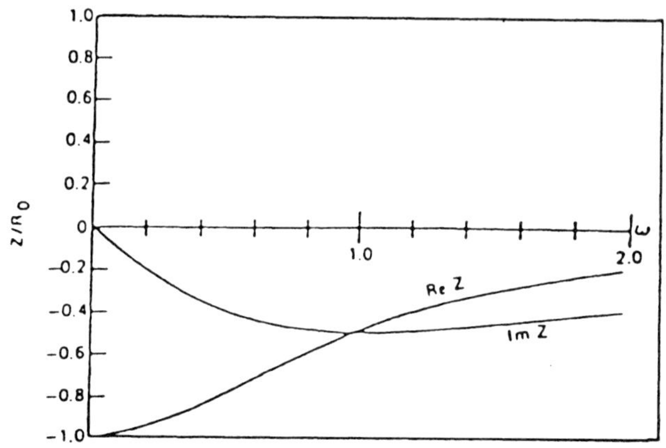

FIGURE 30. Small signal impedence as a function of normalized frequency from the drift and diffusion equations[19].

the magnitude of the small signal mobility decreases, while the maximum frequency increases.

It is important to note that the dispersion relations, obtained from the drift and diffusion equation, will not cross over to positive values at high frequencies. The absence of this cross over is shown in figure 30c[19].

The above small signal discussion has been for uniform fields. The question, of course, is what happens for nonuniform fields. For nonuniform fields and structures several microns in length, transit time contributions introduce significant phase delays. For submicron devices, the extent to which these transit time delays enter is not yet clear. What is apparent, however, is that if Shockley's positive conductance theorem is valid for submicron devices, then a residual pre-relaxation under- or overshoot is necessary. This was not apparent in the 0.25μm calculation of figure 20a, but it was present in the 1.0μm result (figure 20b).

THE UPPER FREQUENCY LIMIT OF SELF-EXCITED OSCILLATIONS

The above discussion was concerned with small signal NDR. The other extremes are the large signal oscillations, often identified as relaxation oscillations. For transferred electron devices, these oscillations received prominent attention when LSA devices were being studied. These self-excited oscillations require the presence of an inductor in the circuit. When dealing with transferred electron devices, they are a fundamental phenomena to be dealt with, and the question, of course, is how to describe them.

The simplest description of the self-excited oscillator for uniform fields is through the coupled circuit equations (see figure 31)

$$\Psi_B = IR + L\frac{dI}{dt} + \Psi \tag{73}$$

$$I = I_c + C\frac{d\Psi}{dt} \tag{74}$$

where Ψ is the voltage across the device and I_c is the conduction current and is equal to $I_c = N_o ev$. These equations can, in turn, be represented by a second order ordinary differential equation:

$$\Psi_B = \left(1 + \frac{I_0}{\Psi}R\right)\Psi + \frac{Z_0}{R_0}\left(R_0\frac{dI_c}{d\Psi} + \frac{R_0 C}{L/R}\right)\frac{d\Psi}{dt'} + \frac{d^2\Psi}{dt'^2}$$

$$Z_0 = \left(\frac{L}{C}\right)^{\frac{1}{2}} \qquad t' = \frac{t}{(LC)^{\frac{1}{2}}} \tag{75}$$

For the situation where the relation between I_c and Ψ includes that of a region of NDR, we have essentially a generic form of a Van der Pol oscillator. Typically the NDR region displays no hysterisis and is not frequency dependent. The normalization identifies the quantity Z_0/R_0 as the strength of the nonlinear element. Its effect on long devices is that it will determine when space charge contributions will dominate an oscillation.

The relevance of the dynamic properties of high frequency devices is that the origin of the self-excited oscillation lies in the presence of a region of NDR. Thus the upper frequency limit of oscillation is essentially defined by the intervalley transfer rates. Further, because the scattering rates are sampled differently from that of the small signal oscillation, the upper frequency limit of the two will be different. Figure 31 summarizes the dependence of the upper frequency limit of the self-excited oscillations. Note the significantly lower maximum frequencies.

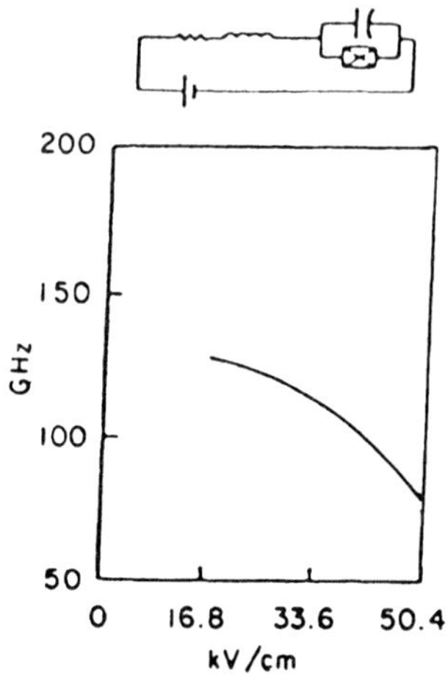

FIGURE 31. Maximum frequency for self-excited oscillations.

Figures 32a-32c illustrate a dynamic large signal self-excited oscillation at 78GHz as computed from the moments of the Boltzmann

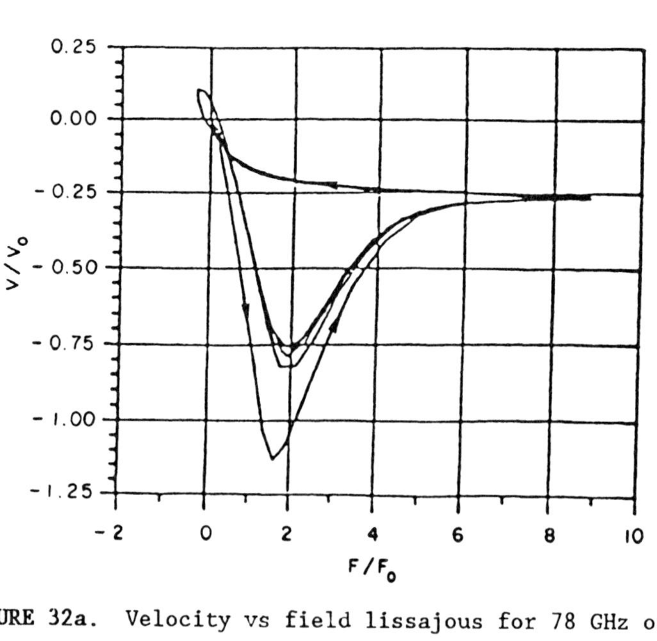

FIGURE 32a. Velocity vs field lissajous for 78 GHz osc.

transport equation. Figure 31a displays the velocity, not the magnitude of the velocity, as computed self-consistently from the MBTE and the circuit equation. Note that the velocity field figure arises from eliminating time between the velocity-time profile and the field-time profile. In figure 32a, there is an initial very high velocity arising from the initial overshoot transient. Under steady state, the time dependent voltage, or field, as displayed in figure 32b sustains a softer rise time and overshoot effects are reduced. Note the hysterisis in velocity arises because the response of the carrier density, velocity and temperature are not instantaneous. In the case of the carrier density, there is a delay following a field increase, before the electrons transfer to the satellite valley, as shown in figure 32c, and a delay following

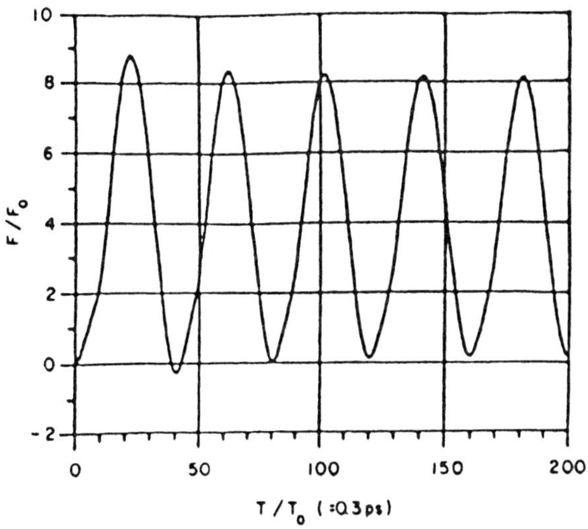

FIGURE 32b. Field vs time for the 78 GHz oscillation.

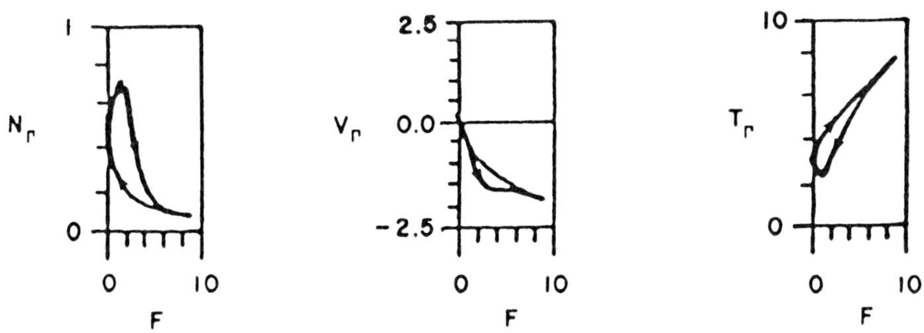

FIGURE 32c. Dynamic lissajous figures for Γ-valley carrier density, velocity and temperature.

a field decrease, before the electrons return to the Γ-valley. At higher frequencies, obtained by reducing the inductance, the finite intervalley relaxation times prevent sustained oscillations.

The question, of course, is how are space charge dependent effects going to enter the picture. For submicron devices, this has not yet been determined. Let us, however, go through some discussion to see if we cannot pick up a thread of prediction. To do this we rely on some analysis done earlier. Basically, we are concerned with three time constants: (1) an intervalley relaxation time, (2) a circuit frequency, (3) the propagation time of a space charge wave. In the case of low frequency operation, space charge control requires that the circuit frequency be sufficiently high to suppress residual space charge nonuniformities. (This will also require a sufficiently high Z_0/R_0 circuit. See Eq. 75) The above arguments are also required for short submicron length devices. But if we look at the transit time associated with the smaller devices, particularly the 0.25μm device, the transit time is of the order of the LO phonon scattering time. It is unrealistic to suggest self-excited oscillations on this time scale for GaAs devices.

In addition to both the self-excited oscillation and the small signal oscillation, there is the driven oscillator, where a sinusoidal driving potential is imposed on the device as shown in figure 33. The dynamic situation associated with the above figure

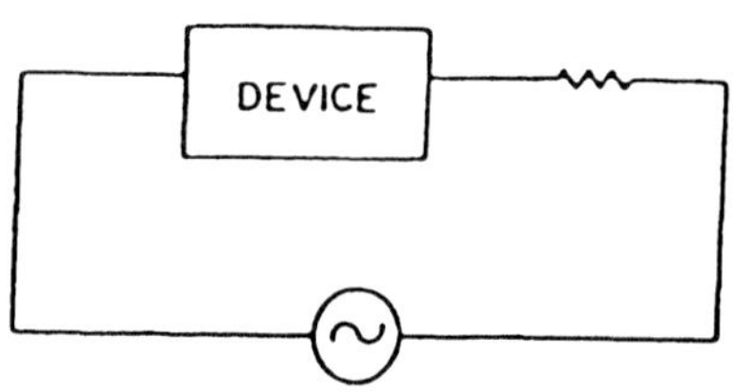

FIGURE 33. Driven oscillator configuration.

is as follows. When the power is delivered to the device-plus-load, power is dissipated to the load in an amount equal to

$$R\langle I^2\rangle \tag{76}$$

where the brackets denote a time average over one period. The power delivered to the device is

$$\langle I\Psi\rangle \tag{77}$$

When the above quantity is positive, the device is lossy. Interest in the transferred electron device arises from the fact that Eq. (77) can be negative and the device can deliver power to the load. The situation is clearly limited by the frequency of operation, and one particular set of results is displayed in figure 34. Which shows a Lissajous between velocity and applied field.

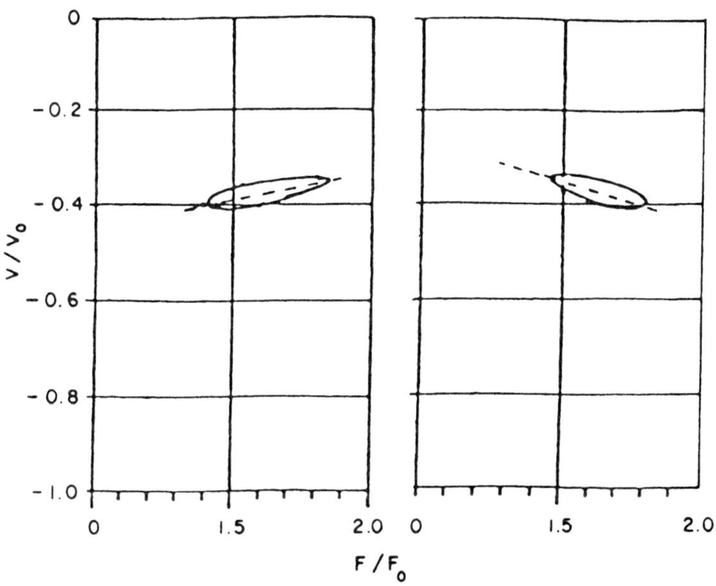

FIGURE 34. Velocity vs driving field; positive slope denotes gain, negative slope denotes loss.

When the net slope of the major axis is positive power is delivered by device. When the net slope is negative, power is dissipated. Both results are consistent with the phenomena of negative differential mobility. For nonuniform fields the question again arises as to how high in frequency we can go and still obtain amplification of the imposed signal. The situation is likely to be similar to that obtained from the small signal and large signal self-excited oscillation studies, and to depend critically upon device design. While the upper frequency limit for a particular device has not yet been obtained, steps have already been taken to couple solutions to the BTE with a driving sinusoidal source to design transferred electron amplifiers. These transferred electron amplifiers were designed using indium phosphide (InP)[23].

THREE-TERMINAL DEVICES: TWO-DIMENSIONAL DRIFT AND DIFFUSION STUDIES

The situation with three terminal elements is somewhat different. In this case, unless we specifically are interested in active oscillations as will occur in transferred electron logic elements, or with switching devices the relevant figures of merit are associated with the frequency at which a device amplifies the power of an incoming signal. For example, the frequency at which the current gain of an FET drops to unity is referred to as the cutoff frequency. In an approximate sense, the cutoff frequency is obtained from

$$f_T \sim \frac{g_m}{2\pi Cgs} \tag{78}$$

and is usually reduced by the presence of unwanted parasitic elements. The quantity f_T, when it is sufficiently high is usually the driving force behind the development of new device structures. The general expressions for f_T which are used in the design of a device depend in large part on the mobility model used. This is clearly unsatisfactory, and the situation requires assessment, which is difficult at this time. Frey, et al., using a highly approximate approach to examining submicron transport in two dimensions estimated f_T for a 2-micron long, 0.25-micron gate length device at approximately 60GHz.

More commonly, In its simplest description, f_T is expressed in terms of specific material quantities. For example, in the Shockley model, it is proportional to the carrier mobility and inversely proportional to the gate length.

$$f_T \approx \frac{\mu}{Lg^2} \tag{79a}$$

Under conditions of saturation in velocity, this expression has been estimated as

$$f_T = \frac{V_s}{2\pi Lg} \tag{79b}$$

and for Vs = 10^7cm/sec, Lg = 0.25μm, $f_T \approx 66$ GHz. Note: velocity overshoot contributions should improve f_T, as well as the transconductance from which these expressions are obtained.

The experimental situation on the gate-length dependence of FETs is also somewhat uncertain on this matter. For example, a recent study by Chao et al.[25], showed (1) an increased drain current with shorter gates, (2) significantly larger pinch-off voltage for shorter gate devices, (3) a slight increase in the transconductance of the shorter gate devices, and (4) the output conductance of the short gate device was not necessarily larger than that of the long gate device. The general conclusion of the study was that by keeping other device material parameters constant, no improvement would necessarily occur by going to short gate length devices.

Whether the conclusions of the Chao et al., study are going to be confirmed by a number of other studies remains uncertain. But diagnostic techniques are required for determining the effect of the space charge distribution on device performance and, in particular, when the speed enhancement can be utilized. One such set of measurements, which appears to be sensitive to the distribution of space charge within the device are the small signal circuit parameters. This was first demonstrated by Engelmann et al.[26], for longer gate length InP and GaAs devices. We review this below for long devices and suggest this as a method of examining the space charge structure of short gate length devices. The method involves obtaining the self-consistent solutions for the small signal microwave parameters of a device and identifying their frequency dependence with specific space charge profiles within the device. The calculation does not stem from the moments of the BTE, which, of course, should be used. Rather, it is derived from the standard drift and diffusion equations. In this case using parameters appropriate to GaAs.

CALCULATION OF THE SMALL SIGNAL "Y" PARAMETERS

This calculation is a numerical one. The equations solved are the two-dimensional continuity equation

$$- e \frac{\partial n(x,t)}{\partial t} + \mathrm{div} \cdot J = 0 \tag{80}$$

the drift and diffusion equation

$$J = - e \left\{ n(x,t)\ v(F) - D(F) \nabla n(x,t) \right\} \tag{81}$$

Poisson's equation

$$\nabla^2\Psi(\mathbf{x},t) = +\frac{e}{\epsilon}\left[n(\mathbf{x},t) - n_0(\mathbf{x})\right] \tag{82}$$

and the relevant circuit equations. In the above **X** denotes the two dimensional position vector. The FET configuration is shown in figure 35, for a structure with a doping of $10^{15}/cm^3$, the low doping was chosen to reduce computational time.

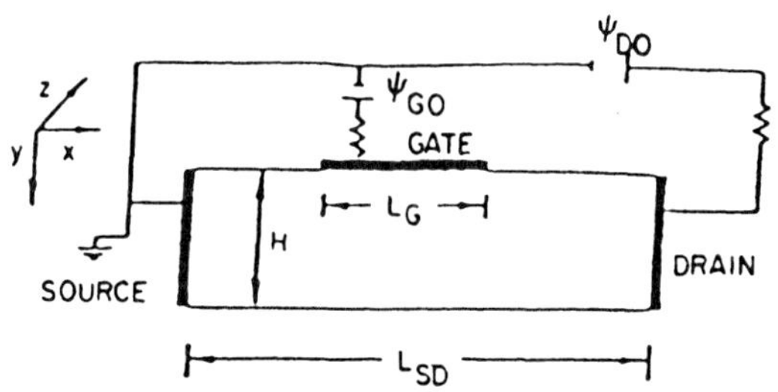

FIGURE 35. Two-dimensional device circuit configuration for the small signal calculations. Relevant dimensions are $L_G = 1.2\mu m$, $L_{SD} = 10\mu m$ and $H = 1.95\mu m$.

The small signal calculations are obtained as small time dependent perturbations of time independent steady state solutions. Since we are treating GaAs, the possibility of large signal domain transit time oscillations must be considered. Since this would only serve to complicate matters and render the concept of small signal parameters specious, attention was concentrated on thin "ND" product FET's that do not sustain instabilities [Grubin et al., 1980][27].

The steady state or dc characteristics of the device were self-consistently computed and are displayed in figure 36. Typical carrier density profiles are represented in figures 37 and 38. Figure 37 displays results for a bias sufficiently high to generate a high field domain under the gate contact.

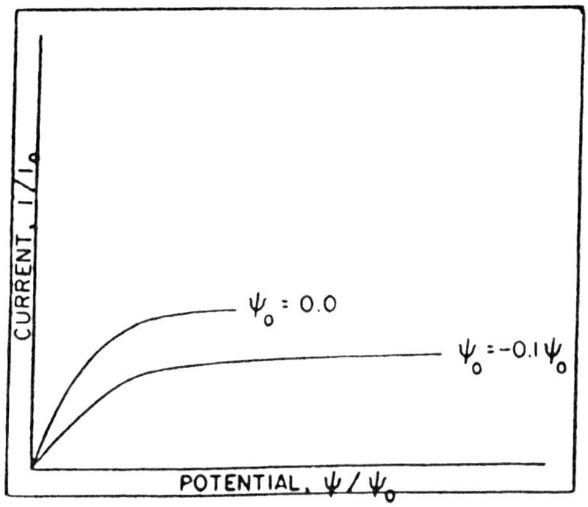

FIGURE 36. Steady state drain current vs drain potential for two values of gate bias. Here $I_o = N_o evpA$ and $\Psi_o = FpL$. Where v_p is the peak gallium arsenide carrier velocity and F_p is the threshold field for negative differential mobility.

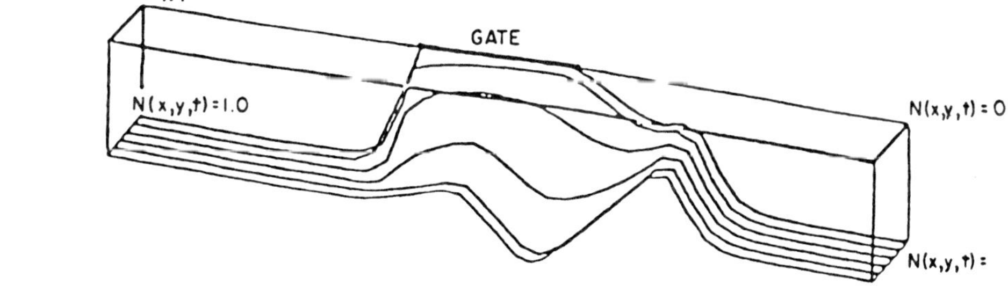

FIGURE 37. Projection of electron density within the FET. Channel height increases toward the bottom of the channel. Channel length is along a direction of increasing x. Density, which is in normalized units, increases in the downward direction. For this calculation $\Psi_G = 0.0$. and $\Psi_D = 1.0\Psi_o$.

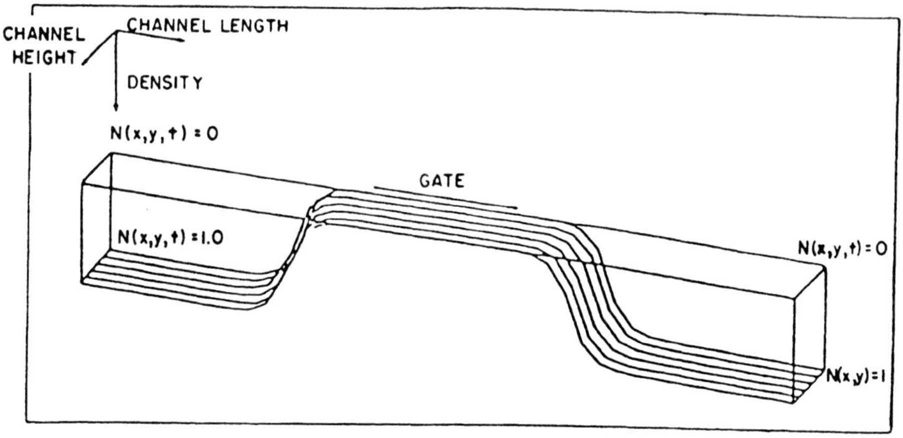

FIGURE 38. Carrier density for $\Psi_G = -0.6\ \Psi_o$ and $\Psi_D = 0.4\ \Psi_o$. The current density for this case is $I_D \approx 0.0$.

Figures 37 and 38, and their respective current, and voltage values represent the starting point of the calculation. Subsequently, a small signal square wave voltage (see e.g., figure 39) was superimposed on the drain contract potential, with the gate contact potential fixed at its steady state value. The resulting

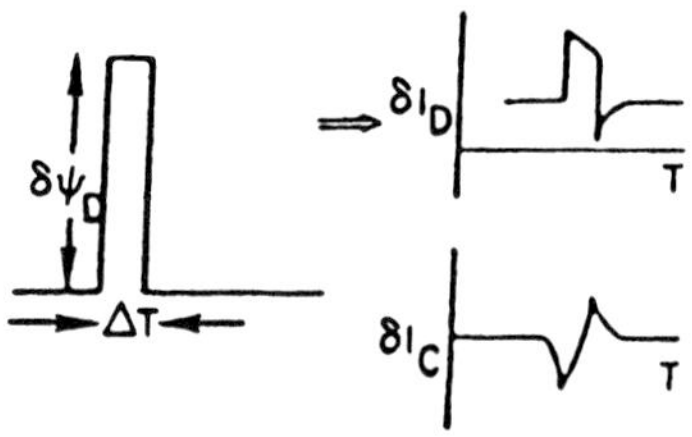

FIGURE 39. Representation of the small signal disturbance. A square wave pulse is applied to the drain contact, altering the drain contact boundary condition: $\Psi_D = \Psi_D + \delta\Psi_D$. There is a subsequent change in the drain and gate currents, δI_D and δI_C.

change in source, gate and drain currents are then computed subject to the constraint

$$\delta I_s(T) = \delta I_d(T) + \delta I_g(T) \tag{83}$$

The gate and drain current, and the drain potential are then Fourier analyzed. The ratios for a fixed gate potential are identified as the admittance parameters:

$$Y_{12}(\Omega) = \frac{\delta I_G(\Omega)}{\delta \Psi_D(\Omega)} \tag{84}$$

$$Y_{22}(\Omega) = \frac{\delta I_D(\Omega)}{\delta \Psi_D(\Omega)} \tag{85}$$

A similar exercise is performed for a perturbation on the gate contact, with the identification of two additional admittance parameters

$$Y_{11}(\Omega) = \frac{\delta I_G(\Omega)}{\delta \Psi_G(\Omega)} \tag{86}$$

$$Y_{21}(\Omega) = \frac{\delta I_D(\Omega)}{\delta \Psi_G(\Omega)} \tag{87}$$

The "Y" or admittance parameters are dependent on the space charge and potential profiles and are thus bias dependent. In addition, it is generally assumed that, about a given bias point, the small signal currents add in a linear way. Then:

$$\delta I_G = (Y_{11} + Y_{12})\ \delta\Psi_G - Y_{12}(\delta\Psi_G - \delta\Psi_D) \tag{88}$$

$$\delta I_D = (Y_{21} - Y_{12})\ \delta\Psi_G + Y_{12}(\delta\Psi_G - \delta\Psi_D) + (Y_{22} + Y_{12})\ \delta\Psi_D \tag{89}$$

with the equivalent circuit as shown in figure 40. The terms

$Y_{11} + Y_{12}$ are generally identified as source-gate admittance parameters, while $-Y_{12}$ is referred to as the gate-drain admittance.

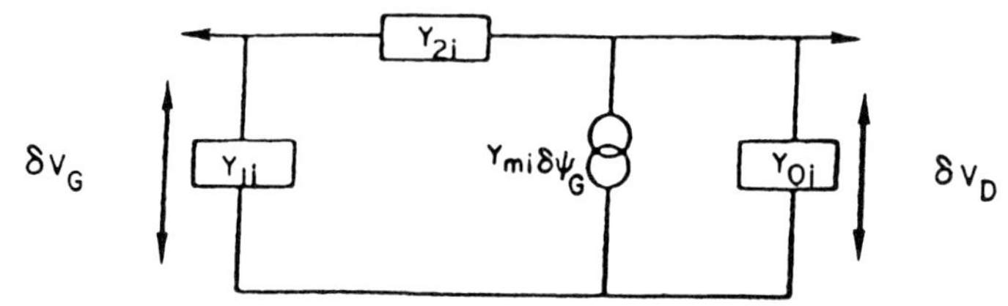

FIGURE 40. General form of the FET intrinsic circuit model with reference to equations 88 and 89:
$Y_{0i} = Y_{22} + Y_{12} \cdot Y_{11} - +Y_{12} \cdot Y_{21} - \nabla^{12}$
and $Y_{mi} - \nabla_{21} - \nabla_{12}$. See also Cobbold (1970).[28] Figure 5.3.

The advantage of this type of representation is that a heuristic connection is often made to circuit elements as shown and intuitive design models may be expanded upon as shown in figure 41, which represents the first order equivalent circuit for small signal calculations. This circuit, while commonly used, does not adequately express the presence of high field dipole layers; as we show below.

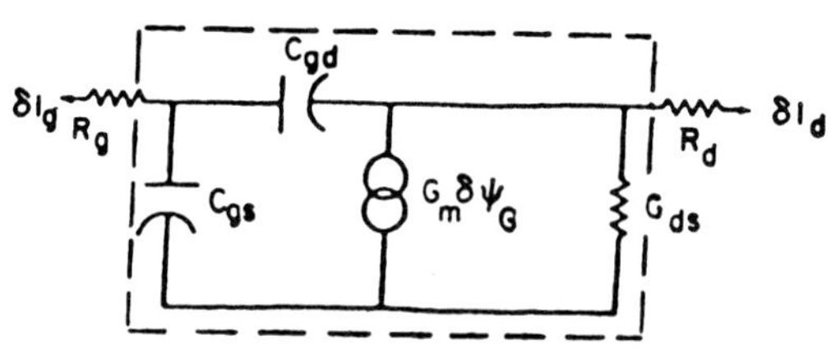

FIGURE 41. First order equivalent circuit terms within dashed border represent zeroth order circuit.

Selective "Y" parameters versus frequency are displayed in figures 42 through 44 for a range of drain bias values and a moderate value of gate bias[29].

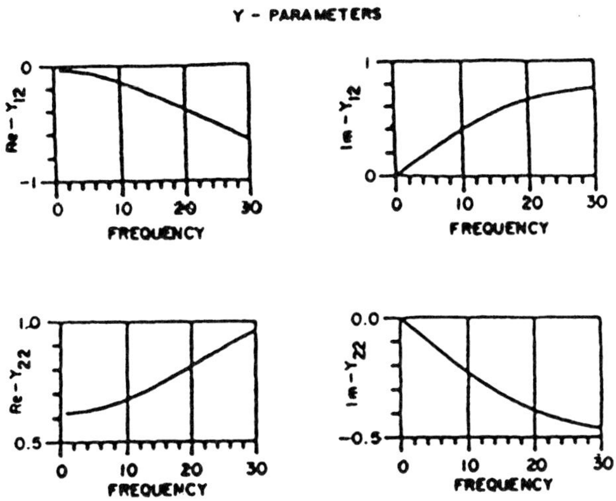

FIGURE 42a. Frequency dependent admittance parameters Y_{12} and Y_{22} obtained by Fourier analysis. The dc bias level for this calculation are $\Psi_{Do} = 0.1\Psi_o$. $\Psi_{Go} = -0.1\Psi_o$.

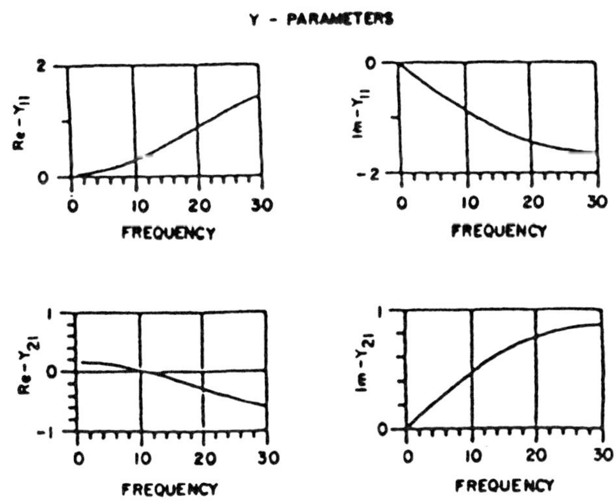

FIGURE 42b. Frequency dependent admittance parameters Y_{11} and Y_{21}.

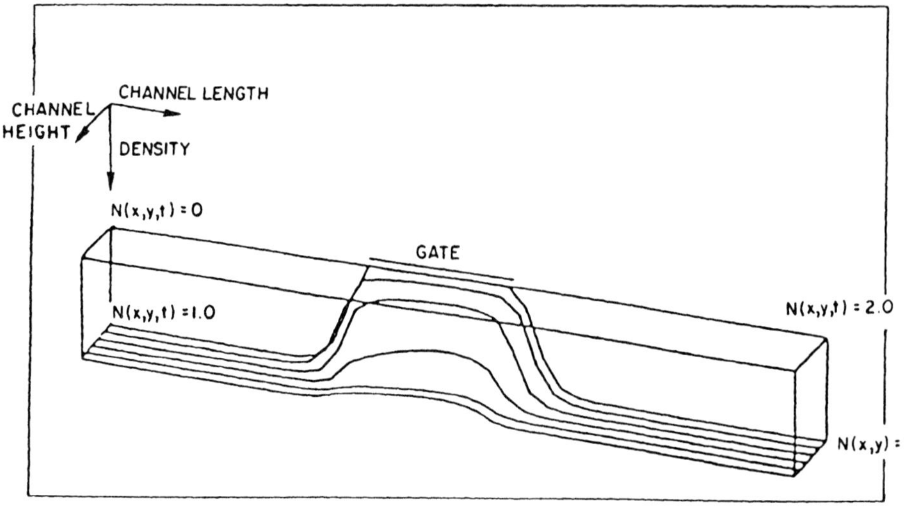

FIGURE 42c. Projection of steady state electron density for this bias level at which the admittance parameters were obtained.

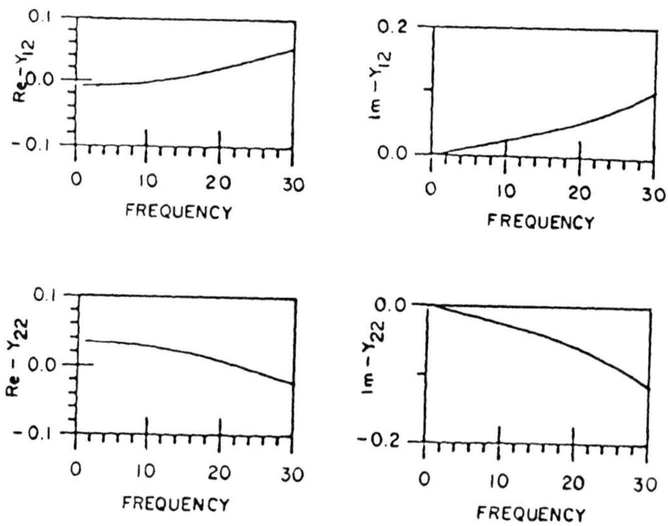

FIGURE 43a. As in 43a but for $\Psi_{Do} = 1.0\Psi_o$, $\Psi_{Go} = -0.1\Psi_o$.

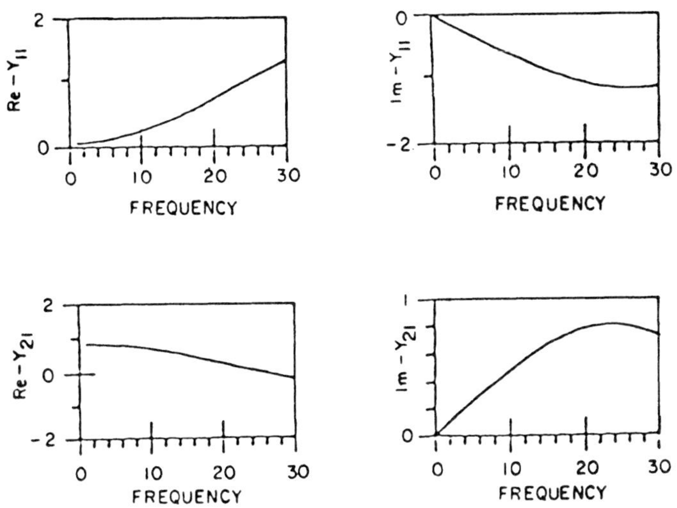

FIGURE 43b. As in 42a but for $\Psi_{Do} = 1.0\Psi_o$, $\Psi_{Go} = -0.1\Psi_o$.

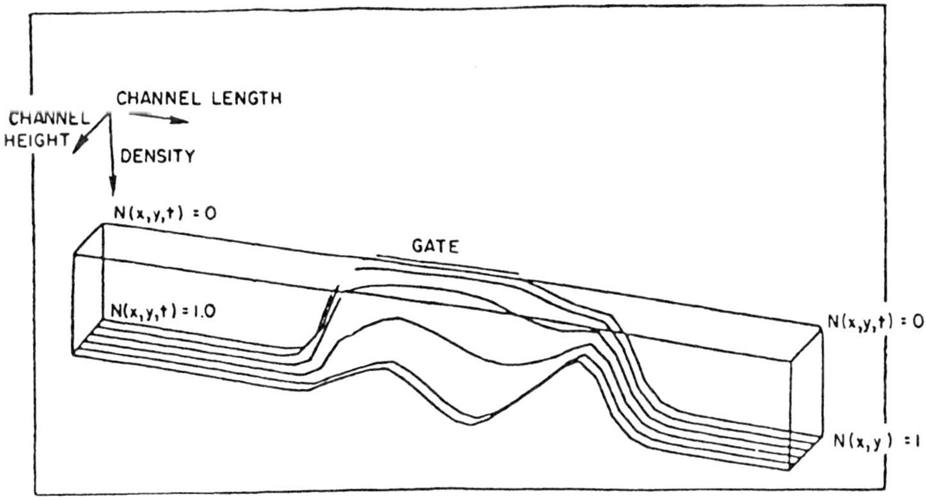

FIGURE 43c. Carrier density projections.

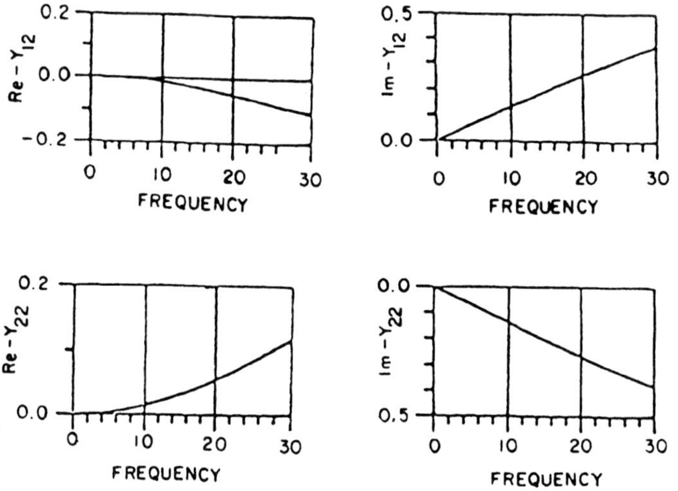

FIGURE 44a. As in figure 42, but for $\Psi_{Ho} = -1.0\Psi_o$.

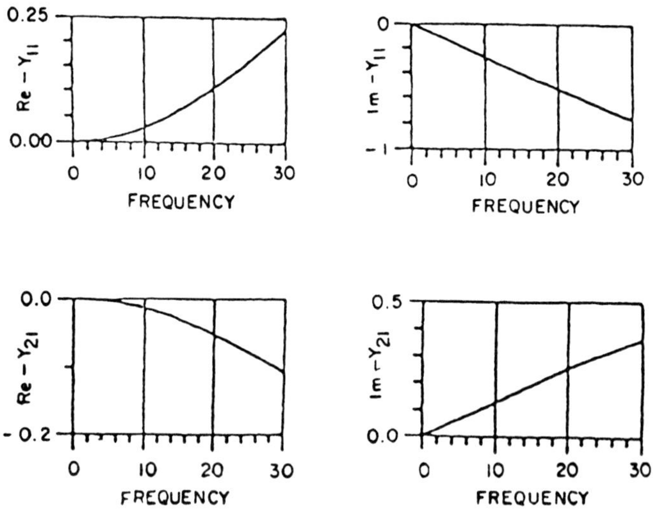

FIGURE 44b. As in figure 42, but for $\Psi_{Ho} = -1.0\Psi_o$.

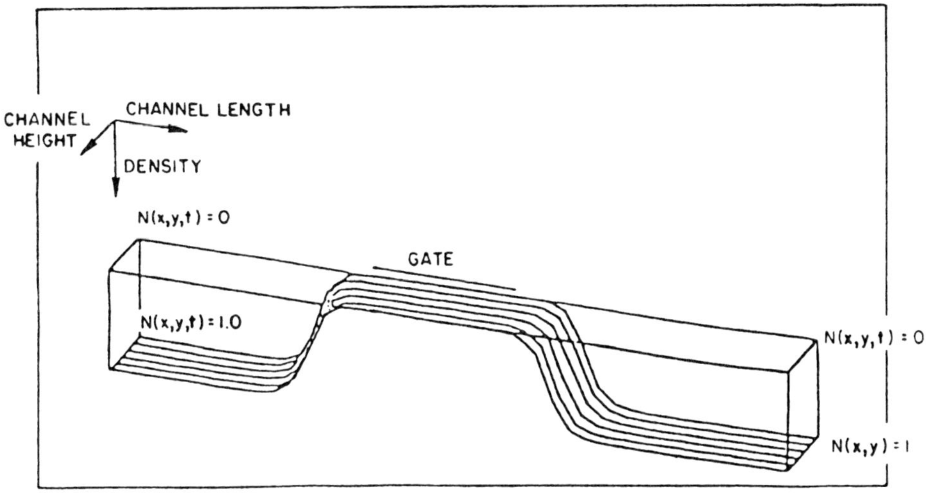

FIGURE 44c. As in figure 42, but for $\Psi_{Ho} = -1.0\ \Psi_o$.

In these figures, the admittance parameters are expressed units of G_o, where

$$G_o = \frac{N_o e \mu_o HW}{L} \tag{90}$$

the frequency is in multiples of f_o

$$f_o = 0.542\text{GHz} \tag{91}$$

and the bias levels are in multiples of Ψ_0=3.2 kv/cmx10^{-4}cm=.32v. Some broad general features are clear and these are expressed in terms of the zeroth order circuit parameters of figure 41.

$$Y_{110} = -\ j\Omega C_{gs} - j\Omega C_{gd} \tag{92}$$

$$Y_{120} = j\Omega C_{gd} \tag{93}$$

$$Y_{210} = -G_m + j\Omega C_{gd} \tag{94}$$

$$Y_{220} = -G_{ds} + j\Omega C_{gd} \tag{95}$$

We begin a discussion of the "Y" parameter with an examination of drain components Y_{12} and Y_{22}. For a fixed gate bias of -0.1 normalized units and at low values of drain bias, $|\text{Im } Y_{22}| < |\text{Im } Y_{12}|$. This situation remains until a high field domain forms within the channel and $|\text{Im } Y_{22}| \geq |\text{Im } Y_{12}|$. The situation in which the drain bias is fixed and the gate bias is increased, results in movement of the depletion layer toward the bottom of the channel, with $|\text{Im } Y_{22}| \geq |\text{Im } Y_{12}|$. Insofar as the zeroth order element requires $|\text{Im } Y_{22}| = |\text{Im } Y_{12}|$, at the very least, the

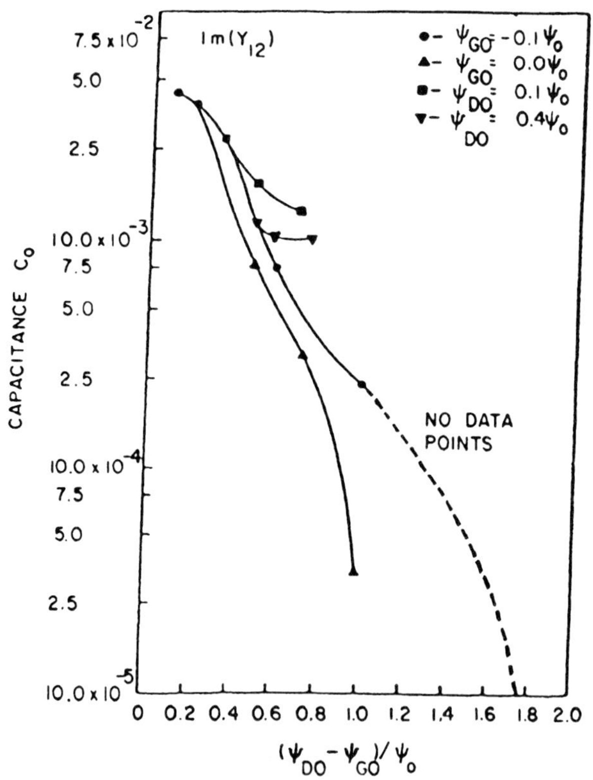

FIGURE 45. Gate to drain capacitance, as obtained from Y_{12}, vs $(\Psi_{DO} - \Psi_{GO})/\Psi_o$. Circular and triangular results are for a fixed gate bias. Square and inverted triangular data are for a fixed drain bias. For this calculation $C_o = 3.25 \times 10^2 \epsilon HW/L$.

drain element Y_o appears to require the inclusion of a bias dependent capacitor.

Additional study of the imaginary part of Y_{12} is shown next. As seen in figure 45, the capacitance of the system undergoes a precipitous drop. A feature also seen experimentally (Engelmann et al., 1977)[26]. The broad features of this result were discussed by Engelmann, et al, (1977)[26], and is represented in figure 46. Essentially, the presence of the high field domain is cutting off the gate drain coupling.

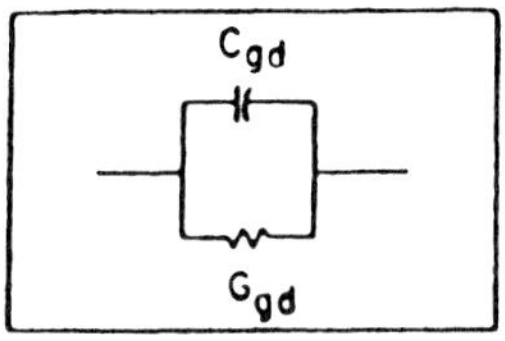

FIGURE 46. Schematic representation of the gate-drain admittance in the presence of a high field domain. C_{gd} represents a small signal conductance which exhibits a frequency dependent region of negative differential conductivity when a high field domain is present.

We recall that C_{gd} represents the gate-to-channel capacitance on the drain side of the channel as indicated in figure 47, and is a measure of the change in channel depletion charge as a result of changes in drain bias. Now, while a capacitance change is expected

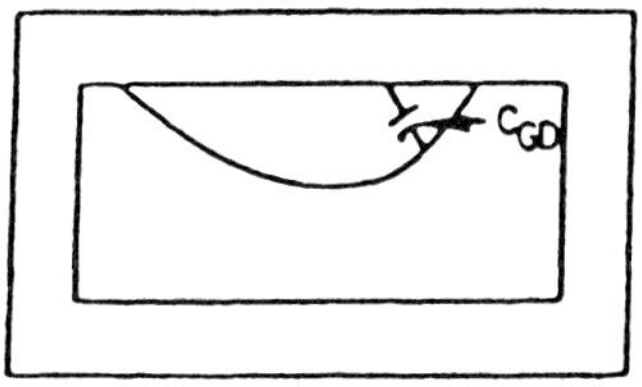

FIGURE 47. Schematic of the gate-to-drain capacitance as a measure of the change in channel depletion charge resulting from changes in drain bias.

as the edge of the depletion layer moves toward the bottom of the channel, the enhanced drop in capacitance suggests that most of the modulation is across the dipole layer. This is illustrated in figure 45.

There are basically two sets of data in figure 45. The bold line is primarily for a moderate value of gate bias. The most significant drop in capacitance occurs when a domain forms. The situation when the increase in gate to drain voltage is a result of an increased gate bias does not result in the dramatic decrease in capacitance. For the latter case, dipole domain do not form. We note that the qualitative features of this calculation are in agreement with the experimental results of Engelmann, et al., (1977)[26].

For the parameter Y_{22}, we simply note that at low bias levels $G_{ds}>0$, whereas at high values of bias, the results strongly suggest a frequency dependent region of small signal negative conductance, a feature consistent with the presence of high field domains.

We next consider Y_{11} and note again that to account for the high frequency high bias behavior of Y_{11}, it is necessary that C_{ds} exhibit a frequency dependence which allows for a range of small signal negative conductance. With regard to Im Y_{11}, which we write as

$$\text{Im } Y_{11} = - \Omega\, C_{11}(\Omega) \tag{96}$$

we note from figures 42 and 44 that at low bias levels C_{11} is somewhat greater than twice C_{gd}. At high bias levels when domains form $C_{11}(\Omega)$ is at least an order of magnitude greater than C_{gd}. Under low or moderate gate bias levels $C_{11}(\Omega)$ does not exhibit a precipitous drop in value. Rather, at first the capacitance decrease corresponding to a movement of the depletion layer toward the bottom of the channel. This was also the initial behavior of the gate to drain capacitance. Further increases in drain bias result in domain formation and space charge injection into the depleted zone. The effective capacitance shows a corresponding increase corresponding to an increase in the stored charge. This is displayed in figure 48. For the situation where the net voltage increase is due to an increase in gate bias, where no domains form there is the expected drop in capacitance, as also reflected in the gate drain capacitance without domains. This is also shown in figure 48. The experimental situation shows broad agreement with the numerical results.

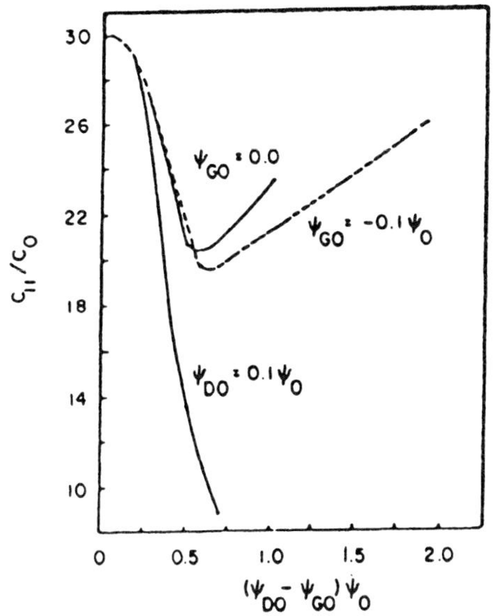

FIGURE 48. Capacitance C_{11}, obtained from Y_{11}, vs $(\Psi_{DO}-\Psi_{GO})/\Psi_o$. For this calculation $C_o = cHW/L$. Note $C_{11} = C_{gs} + C_{gd}$.

The two remaining items of interest here are theoretical/experimental comparison of the transconductance, and current-gain cutoff frequency. Figure 49 displays the transconductance, Re Y_{21}, vs $V_{DS} - V_{GS}$, for two values of gate bias. We see

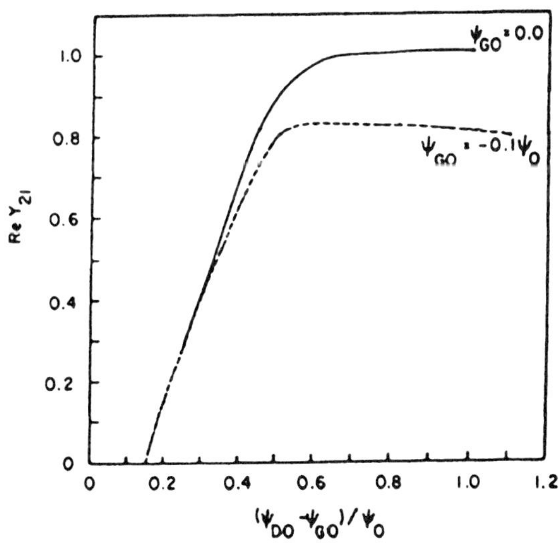

FIGURE 49. Re Y_{21} (in multiples of C_o) vs $(\Psi_{DO} - \Psi_{GO})\Psi_o$. There is saturation at high bias levels.

the presence of near saturation in both sets of data. This is also seen experimentally. Our data does not extend to high enough drain bias levels to determine whether a corresponding decrease in transconductance occurs. The current-gain cutoff frequency is obtained from the expression

$$f_t = \frac{\Omega \text{ Re } Y_{21}}{2\pi |\text{Im } (Y_{11} + Y_{12})|} \tag{97}$$

and is shown in figure 50. It is seen that saturation in f_T occurs under the presence of domain formation. The decrease in f_T at the higher drain bias levels appears to be associated with an increase in the source-gate capacitance. Experimental observations again show qualitative agreement with theory.

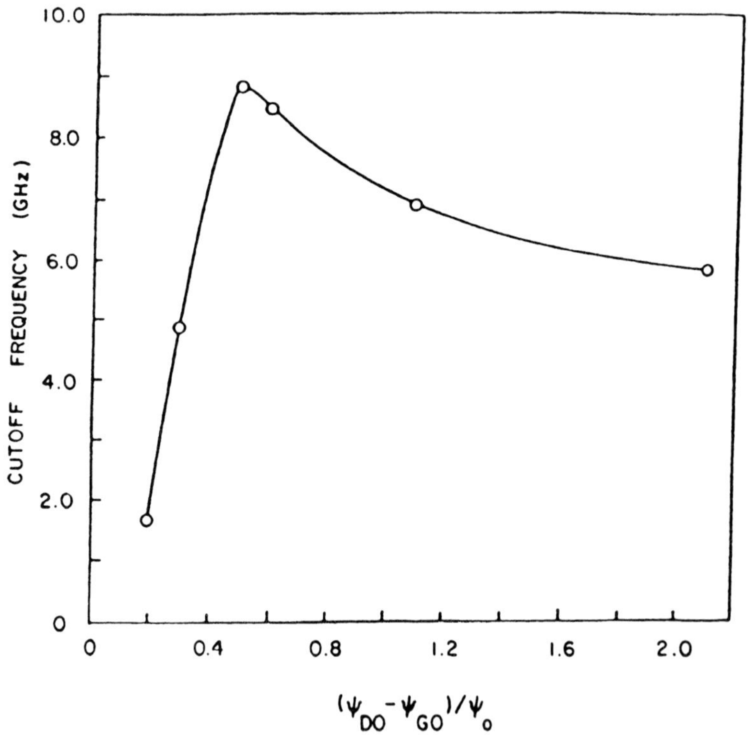

FIGURE 50. Cutoff frequency vs $(\Psi_{DO} - \Psi_{GO})/\Psi_o$ for $\Psi_{GO} = -0.1\Psi_o$.

THREE-TERMINAL DEVICES: TWO-DIMENSIONAL MOMENTS OF THE BOLTZMANN TRANSPORT EQUATION

The two-dimensional simulations are new and as a result there has not been much experience to rely upon. The boundary conditions are more complicated insofar as parallel components of velocity are also required to characterize the boundary and the contact. The result is that the familiar "simple" one-dimensional contact descriptions need rethinking. The next few figures will illustrate some of the present observed features. The simulations have been performed for a two-micron long device, with a gate length of 0.5μm with structure of figure 51. Because most of the

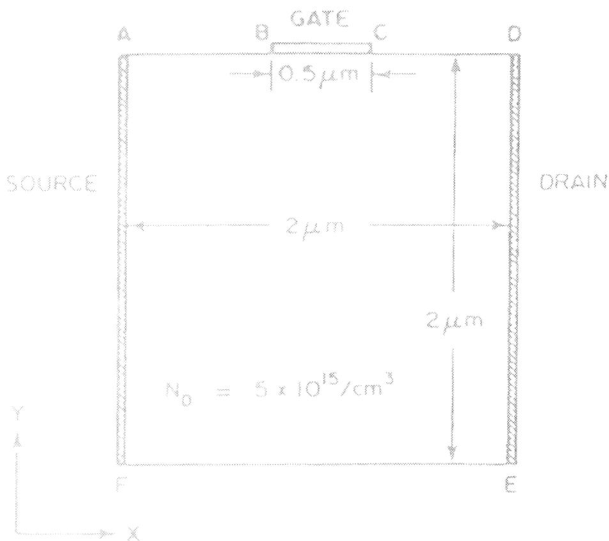

FIGURE 51. Schematic of "horizontally" placed submicron gate MESFET used in submicron dimensioned BTE "moment" simulations

one-dimensional calculations were performed for doping levels of $5x10^{15}/cm^3$, the doping levels of the two-dimensional calculation were taken at this value. The boundary conditions are listed in Table 6.

Figure 52a displays the distribution of total charge within the two-micron long device with a gate potential kept at ground, and a drain potential at 1 volt. The source contact was an injecting contact. The normal component boundary conditions were similar to those obtained from the one-dimensional case. For the Schottky

TABLE 6. Boundary Conditions

AB, CD, EF	$V_1^y = V_2^y = 0$, otherwise derivatives are zero
BC	$V_1^y = V_2^y = 0, \quad \frac{\partial V_1^x}{\partial Y} = \frac{\partial V_2^x}{\partial Y} = 0, \quad \frac{\partial T_2}{\partial Y} = 0, \quad T_1 = \text{constant}$ $\frac{\partial}{\partial Y} N_1 N_2 = \frac{\partial}{\partial T} N_2 T_2 = 0$
DE	$\frac{\partial^2(\text{all})}{\partial X^2} = 0, \quad \Psi = \text{constant}$
FA	$V_1^x = \text{constant} \cdot \frac{\partial \Psi}{\partial X}, \quad \Psi = 0, \quad \frac{\partial V_2^x}{\partial X} = 0, \quad T_1 = 1, \quad \frac{\partial T_2}{\partial X} = 0$ $V_1^y = V_2^y = 0, \quad \frac{\partial^2 V_1}{\partial X^2} = 0, \quad \frac{\partial^2 N_2}{\partial X^2} = 0, \quad T_1 = \text{constant}$

gate, the normal component of velocity was set to zero, and the Γ-valley component of electron temperature was set to 900°K. Thus some transfer was effected at the gate. The parallel component of velocity was set to zero. What is apparent from these calculations is the space charge accumulation at the drain boundary, a feature already observed in the one-dimensional simulations. This downstream charge accumulation is a consequence of electron transfer and is likely to lead to reduced current drives. An increase in drain bias serves to enhance the level of space charge accumulation without any real improvement in current level. The Γ-valley charge density is displayed in figure 52b, and for this bias level, there is a certain degree of electron transfer. Within the channel near the bottom of the device the fractional component of $N_\Gamma \approx 0.9N_0$. Again, most of the current is carried by the Γ-valley electrons. The distribution of satellite valley carriers is shown in figure 52c. It tends to emphasize that a larger fraction of L-valley carriers are partaking in transport immediately downstream from the gate contact than within the channel, but their net contribution to the total current is negligible. (There are also fewer total carriers in this

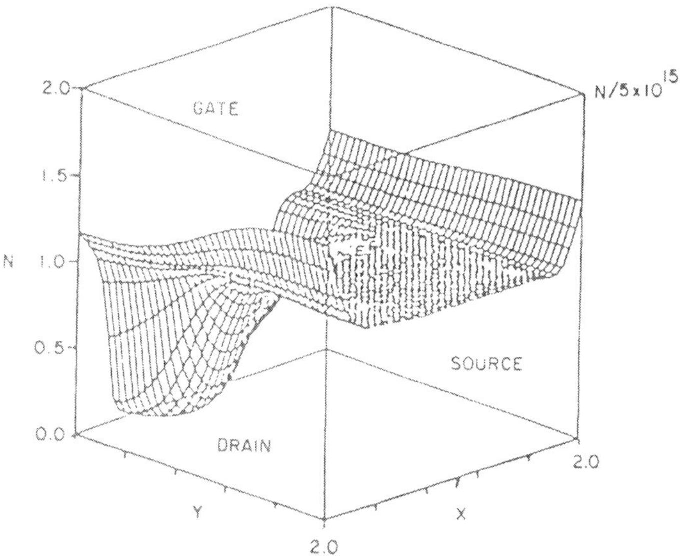

FIGURE 52a. Distribution of total charge within a three-terminal device (distance scales are nonlinear, the depletion region is approximately 1000A deep), for a potential of 1 volt on the drain.

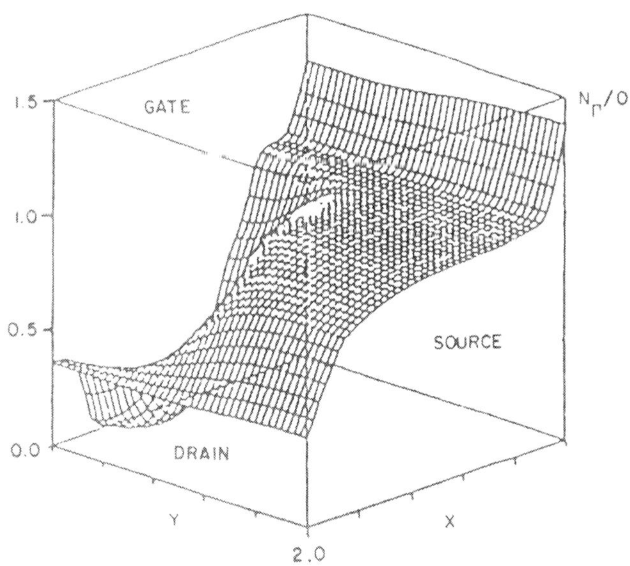

FIGURE 52b. As in (a), but for Γ-valley.

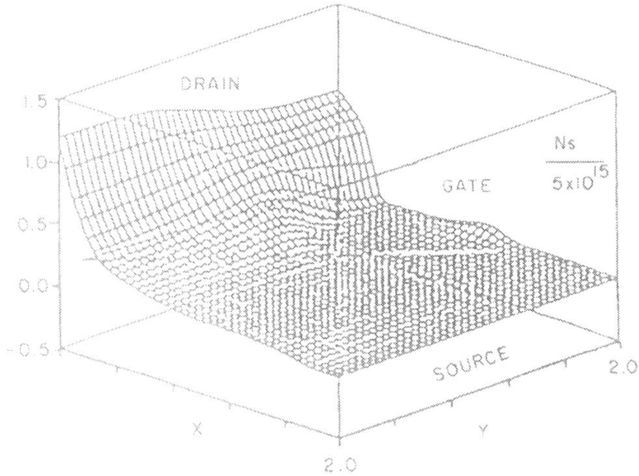

FIGURE 52c. As in (a), but for satellite valley.

region.) In figure 53 we have included the result of a calculation in which the boundary condition on the gate electron temperature was increased to 1500°K to show how this alters the charge distribution. At these bias levels no significant alteration in current level was observed. The total charge distribution for 1.25v on the case is also displayed in figure 54a. The observation is made of increased carrier accumulation at the drain boundary and enhanced electron transfer (figures 54b and 54c).

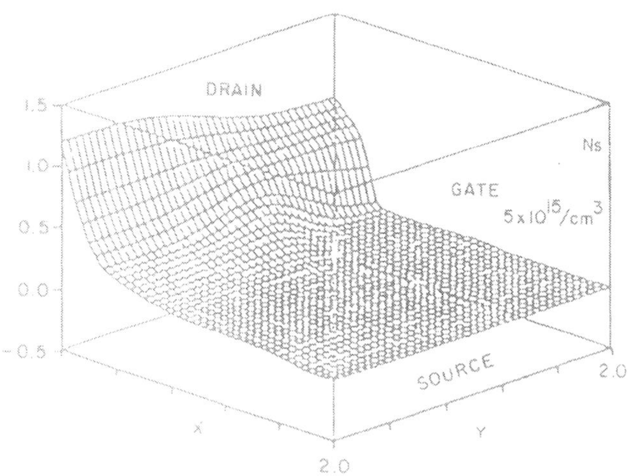

FIGURE 53. As in 52c. But for Tg = 1500°K.

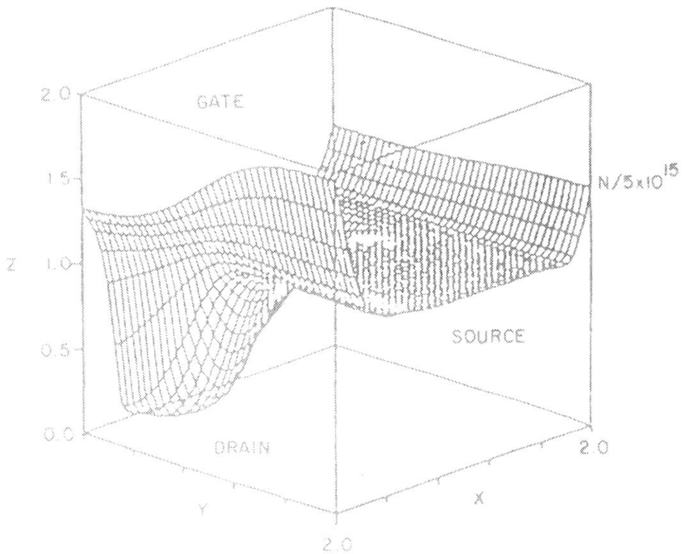

FIGURE 54a. As in figure 52a, but for a drain potential of 1.25V.

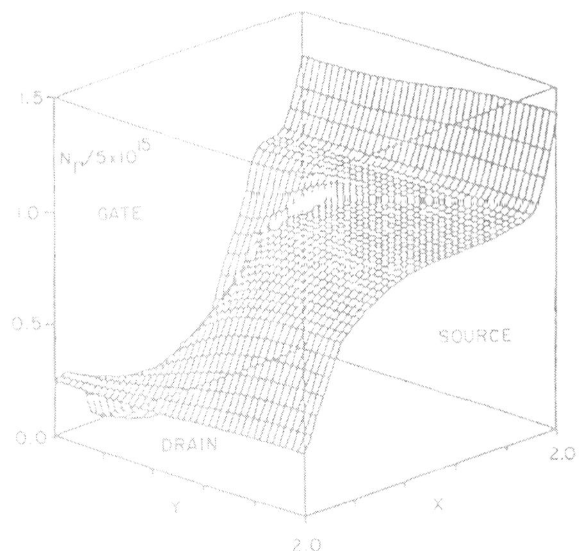

FIGURE 54b. Γ-valley carrier density.

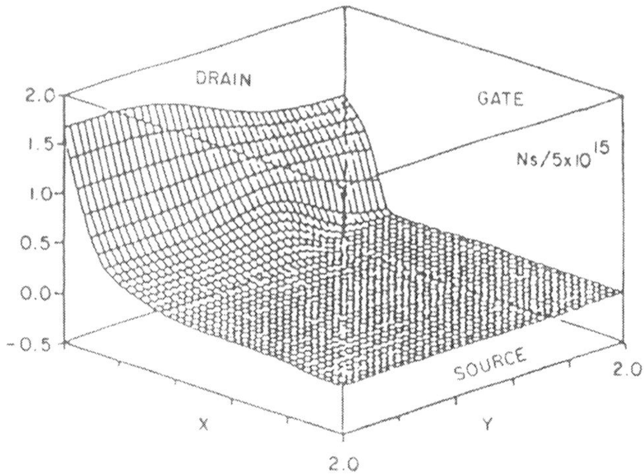

FIGURE 54c. Satellite valley population.

SCALING AND MATERIAL CONSIDERATIONS

There are two questions I want to address, and they have to do with the kinds of materials we should be looking for. The guide to this answer will come from the specific device goals. The question of interest concerns the significance of super velocity materials and how do we balance high speeds with sometimes reduced mobilities. To examine these issues we return to the moment equations as written in equations (51), (52) and (53). The first thing we will consider is what we will refer to as intrinsic scaling. This type of scaling is not dependent on device length explicitly, although it enters the discussion through the relation

$$x = \int v(t)dt \tag{98}$$

We turn to equations (51) through (53), assume a constant field across the device and set all other spatial derivatives to zero. Then

$$n'_r = n'_s \frac{\Lambda'_s}{\Lambda'_r} - \frac{1}{\Lambda'_r}\frac{\partial n'_r}{\partial t'} \tag{99}$$

$$n'_r m'_r v_r^{i'} = a n'_r \frac{\partial \Psi'}{\partial x'_i} - \frac{1}{\Pi'_r}\frac{\partial n'_r m'_r v_r^{i'}}{\partial t'} \tag{100}$$

energy:

$$n_r' T_r' = n_s' T_s' \frac{\Sigma_s'}{\Sigma_r'} + \left(d n_r' v_r' \frac{\partial \Psi'}{\partial x_i'} - \frac{1}{\Sigma_r'} \frac{\partial w_r'}{\partial t} \right) \cdot \frac{2}{3b} \tag{101}$$

We consider the situation as in Ref. 30 where all of the scattering rates are altered by the multiplicative constant λ, e.g.,

$$\Lambda_\Gamma \Rightarrow \lambda \Lambda_\Gamma \tag{102}$$

while the primed scattering rates are kept constant. Thus, e.g.

$$t_{ref} \Rightarrow t_{ref}/\lambda \tag{103}$$

We note that such materials as GaAs and InP are identified by their scattering rates, and constant scaling is directly relevant to a comparison of the two. We next consider the case where the coefficients **a** and **d** are invariant. For a constant v_{ref}, this is achieved by assuming the product

$$t_{ref}\ F_{ref} \tag{104}$$

to be constant. This requires that

$$F_{ref} \Rightarrow \lambda F_{ref} \tag{105}$$

The first consequence of this scaling is associated with the steady state velocity field curves. If

$$v_o(F)$$

represents the steady state field dependent velocity, taken as reference; and

$$v_\lambda(F)$$

represents the scaled velocity, then

$$v_\lambda(f) = v_o(F/\lambda) \tag{106}$$

In simplest terms this equations states that the magnitudes of the steady state and saturated drift velocities are unchanged by this scaling. The low field mobilities differ by the factor λ. In a qualitative sense, many of the III-V NDR materials are subject to this type of scaling. It would not stand up in a quantitative sense because of variations in the phonon frequencies etc. In figure 55 we sketch the select scattering rates for InP and GaAs at room temperature. In figure 56 we display the scaled GaAs and InP

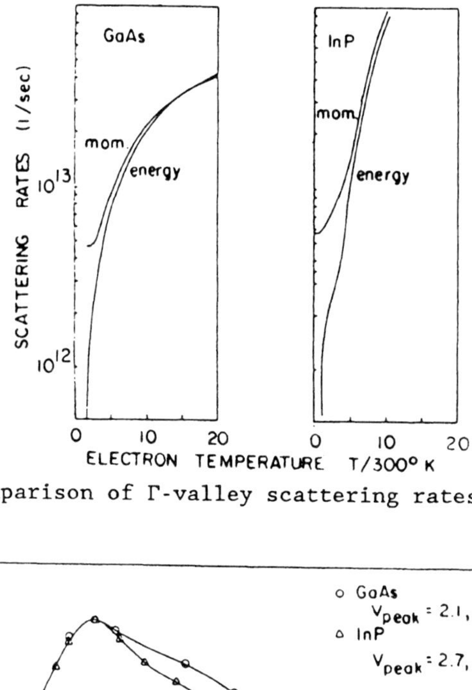

FIGURE 55. Comparison of Γ-valley scattering rates for GaAs and InP.

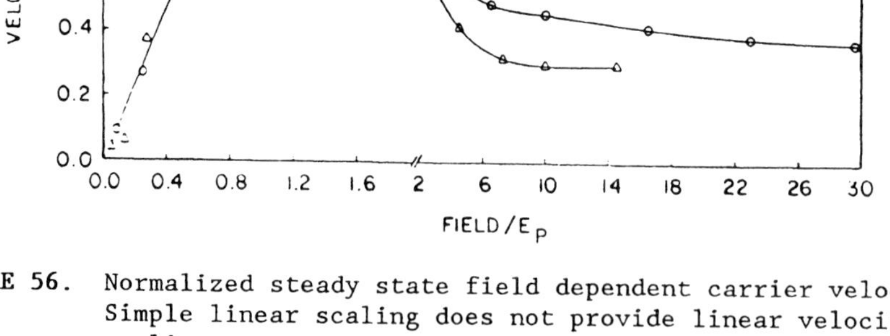

FIGURE 56. Normalized steady state field dependent carrier velocity. Simple linear scaling does not provide linear velocity scaling.

steady state velocity field relations. It is seen that linear scattering rate scaling does not yield linear velocity scaling.

The next item of interest concerns the transient response. Here with reference to equations (99) through (101) it is clear that constant λ scaling with

$$t' = t/t_{ref}$$

implies

$$v_\Gamma(t,F) = v_o(\lambda t, F/\lambda) \tag{107}$$

or in another form

$$v_o(t,F) = v_\Gamma(t/\lambda, \lambda F) \tag{108}$$

Consider the situation where the material under consideration is undergoing self-excited oscillations. Then the scaling law states that if GaAs has an upper frequency limit of 150GHz, and it were possible to design a scaled semiconductor material with λ = 2, then the latter would have an upper frequency limit of 300GHz. Similar remarks can be made with respect to small signal and driven oscillators. Simulations with InP self-excited oscillations have been performed and frequencies in excess of those found in gallium arsenide were obtained.

With regard to the more familiar transient transport calculations, as extrapolation from equations 107 and 108 yields the results shown in figure 57. Thus, transport in materials with enhanced scattering rates appears to approach equilibrium sooner.

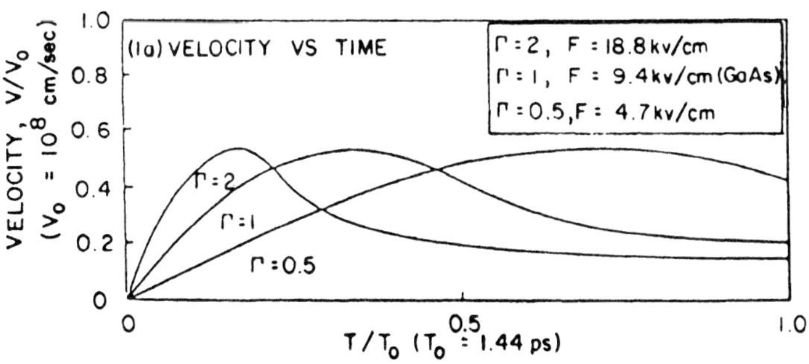

FIGURE 57. Scattering rate scaling as applied to velocity transients.

These conclusions, of course, must be folded into the fact that at higher values of field the carriers are driven to higher values of velocity and the scattering rates which are field dependent also increase. In terms of a comparison to real materials, we are not interested in scaled values, but rather in the response of two dissimilar materials to the same value of field. A simple sketch based upon previous comments is displayed in figure 58, which displays an estimate of the velocity for the Γ=1 element at a field of 18.8 kv/cm (dashed line). Note, the higher peak carrier velocity. Also shown in figure 58, is the velocity versus distance relation obtained through application of equation (98). The

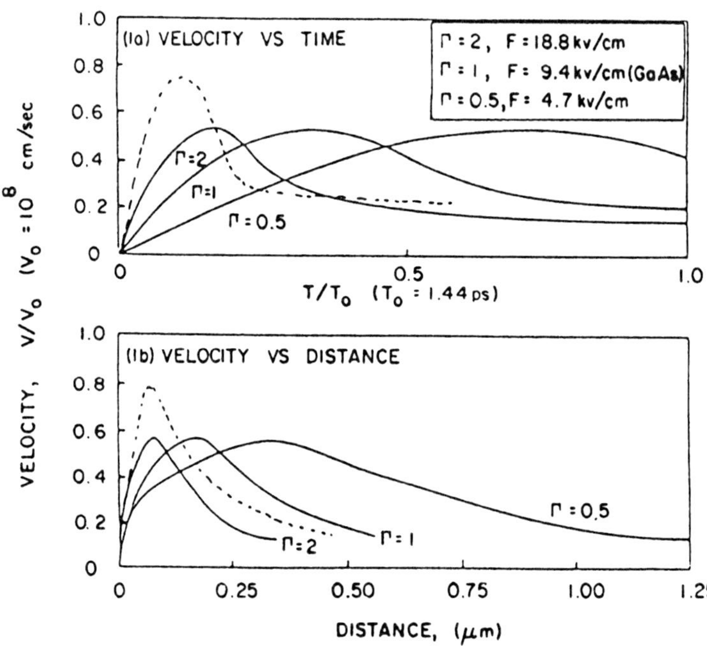

FIGURE 58. Scattering rate scaling as applied to velocity transients.

question is, how well do these very general comments, which can be obtained from some very general scaling of the Boltzmann transport equation, stand up against detailed numerical calculation. The answer is displayed in figures 59 and 60, where the general scaling principles appear to hold up fairly well.

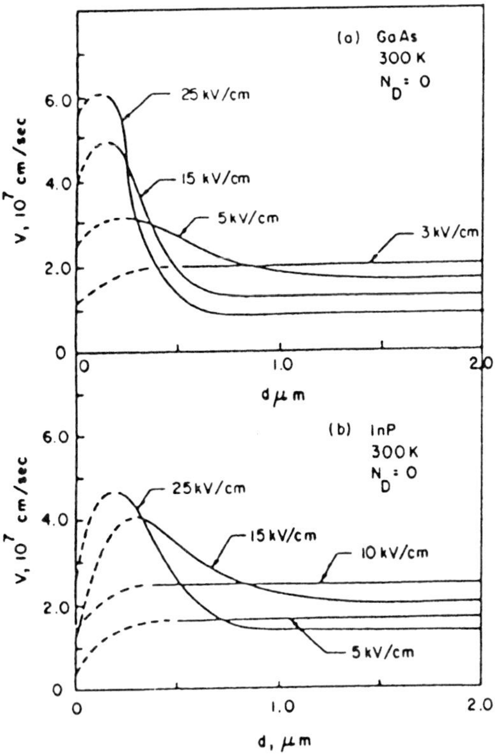

FIGURE 59. Drift velocity vs position in GaAs and InP for electrons release at $\chi = 0$ into various uniform fields: (a) GaAs, from Ruch [15]; (b) GaAs and InP, from Maloney and Frey[31].

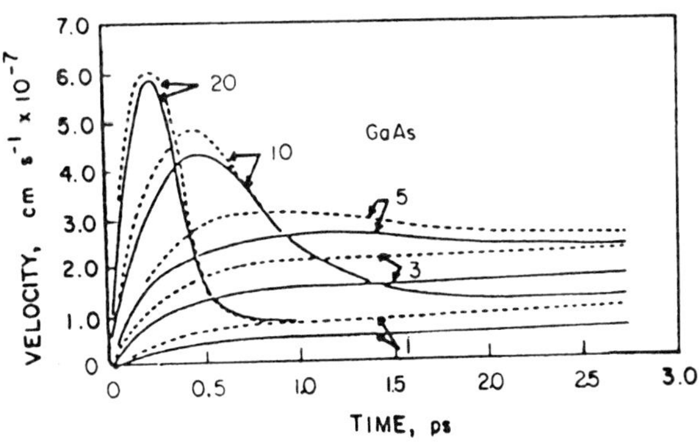

FIGURE 60a. GaAs transients. Solid curves denote impurity scattering; $N_p = 10^{17} cm^{-3}$.[32]

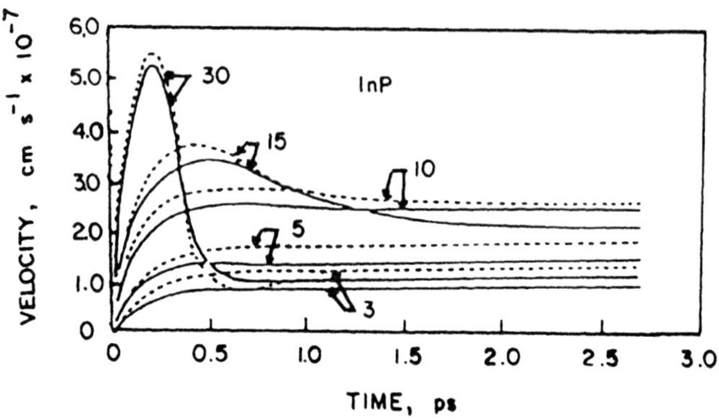

FIGURE 60b. As in (a), but for InP.[32]

The results of the above figure appear to suggest that for a fixed submicron length the transit time versus distance should show advantages for GaAs against InP. This is borne out in figure 61. Indeed, these results should not be taken as noncontroversial. There appears to be some differences between these results and those of Maloney and Frey who conclude that InP will always show some speed advantage. Indeed, based upon the Maloney and Frey results, we concluded recently, the results shown in figure 62. Based on figure 62, a calculation of

$$f_T = \frac{1}{(2\pi \text{ transit time})} \tag{109}$$

for channel length of 0.4μm f_T yields 114GHz for GaAs. It is

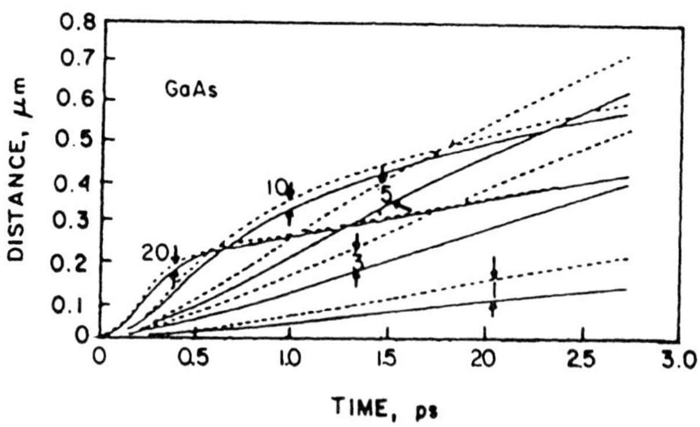

FIGURE 61a. Variation of drift distance with time (GaAs). N_p=10^{17}cm^{-3} ---N_p 0. Fields in kV/cm.[32]

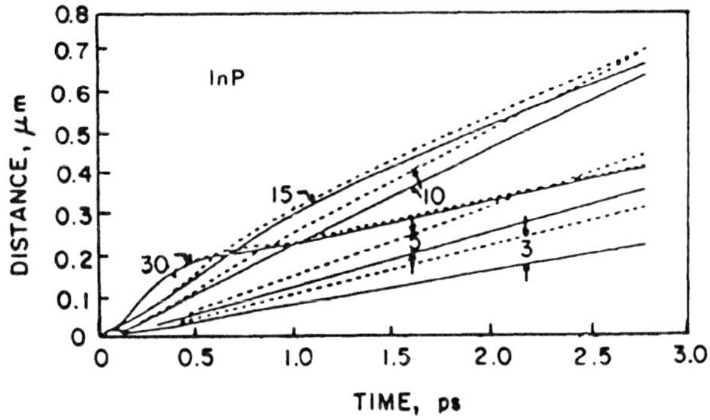

FIGURE 61b. As in (a), but for InP.[32]

slightly higher for InP. At 2500 A, it is approximately 265GHz for GaAs. For InP it is approximately 227GHz. But the numbers derived form figure 61 would indicate that the initial transient must be included in the material assessment. Thus, clear predictions based upon uniform field calculations cannot be given.

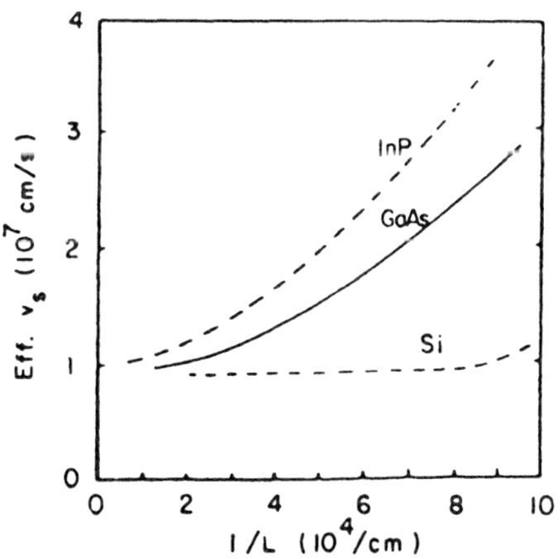

FIGURE 62. The effective, or time-of-flight, saturated velocity as a function of the inverse channel length. The InP curve is estimated from the data of ref. 31.[33]

The next situation of interest concerns itself with extrinsic scaling, where we are concerned with the effects of device length for nonuniform field situations. To deal with this situation we return to equation 52 and concentrate on the coefficients **a** and the normalized scattering rate. (A similar analysis holds for the energy balance equation). For extrinsic scaling.

$$t_{ref} = \frac{\text{Device Length}}{v_{ref}} \tag{110}$$

Thus

$$\mathbf{a} = \frac{eF_{ref}}{m_{ref}\ \Pi_r\ v_{ref}} \tag{111}$$

and

$$\Pi_r = \frac{v_{ref}}{x_{ref}}\ \Pi_r' \tag{112}$$

With regard to the coefficient **a**, if the scattering rates are constant, and F_{ref} is constant, then **a** is independent of device length. Insofar as **b** is independent of device length, as the device length begins to decrease, the "non-drift and diffusion" terms

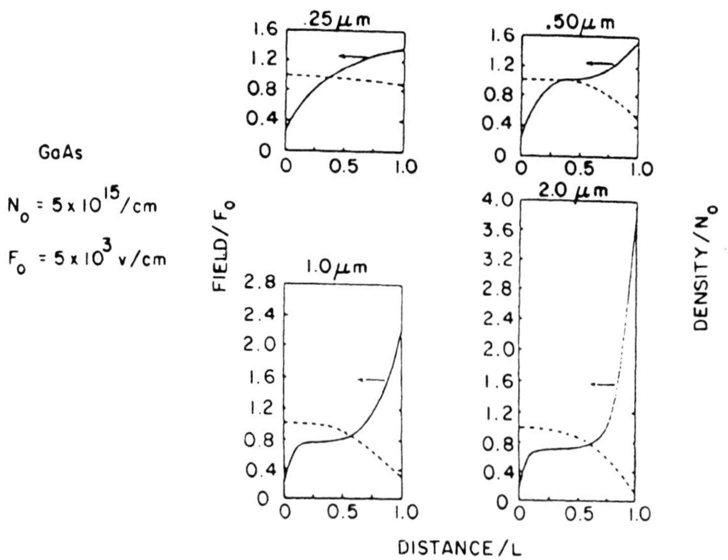

FIGURE 63. Effect of device length on electron transfer, for a constant average field.

begin to dominate. Additionally, the coefficient f in Poisson's equation is length dependent and as x_{ref} increases, the gradients in F increase. These contributions, when coupled to the fact that electrons need a specific device length to traverse before they undergo electron transfer, leads to the following key results. Figure 63 shows the calculated electric field distribution for GaAs subject to an average field of 5kv/cm. It is seen that the highest nonuniform field distribution occurs as the device length increases.

We note that electron transfer tends to be synonymous with low current levels. This is displayed in figure 64. Additionally, two other important features arise. First, there is the increased drive current for the submicron dimensions. Second, there is the absence of negative differential conductivity. A related question of importance at this time concerns the choice of material. If a material with an increased scattering rate were chosen, then simple scaling theory suggests that to achieve similar current levels, shorter device lengths are needed, and higher fields. If this

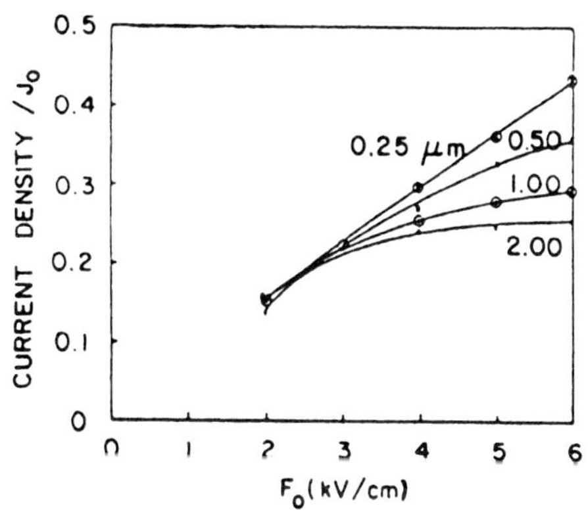

FIGURE 64. Current density versus average field versus device length.

conclusion holds true then the achievement of the high InP velocities may place extreme demands on available technology, particularly, if very small gate lengths are required, as appears to be the case.

The next type of scaling we consider is that of extrinsic carrier density scaling. For this case, the important equation to examine is Poisson's equation, and the relevant scaled quantity for this case is the coefficient f of equation (64). This coefficient tends to indicate that for a given value of average electric field all results are unchanged if the product of carrier density and device length are kept constant. 'NL' products are common in examining transport in long GaAs diodes. They are, however, irrelevant for submicron length devices in that they ignore the fact the intervalley transfer is length dependent, as the results in figure 65 indicate.

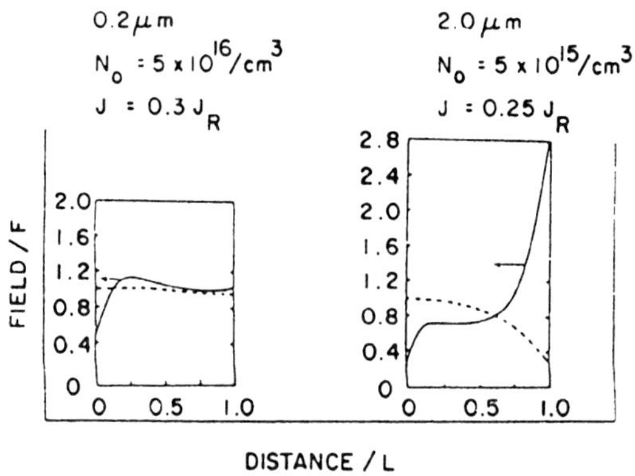

FIGURE 65. Field and Γ-valley carrier density for an average field of 5kv/cm. For this calculation $J_R = N_0 e v_R$, where $v_R = 1.0 \times 10^8$ cm/sec.

One very general conclusion can be drawn from the above scaling arguments, it is that the achievement of high speed submicron length devices does not necessarily require materials with high steady state velocities. High steady state velocities are irrelevant for submicron devices. Rather what is needed, are high mobility materials. For this case, it is clear that of the three materials of interest, GaAs, InP and Si, GaAs has the clear advantage. But there has been considerable interest in a variety of other materials.

MATERIALS CHOICE

The situation with material choice is best illustrated by several examples, but again the material choice is based upon device application. If the device conceived is such that submicron effects do not enter, then the steady state field dependent velocity curves are the ones of interest. as illustrated in figures 66 through 69. Note, particularly high values of field and velocity associated with the InGaAs alloys (figure 66), in addition to very high values of mobility. Also note the reduced mobility for the AlGaAs ternary, as Al is introduced; for InPAs as arsenic is introduced; and GaAsP, as phosphorous is introduced. The method

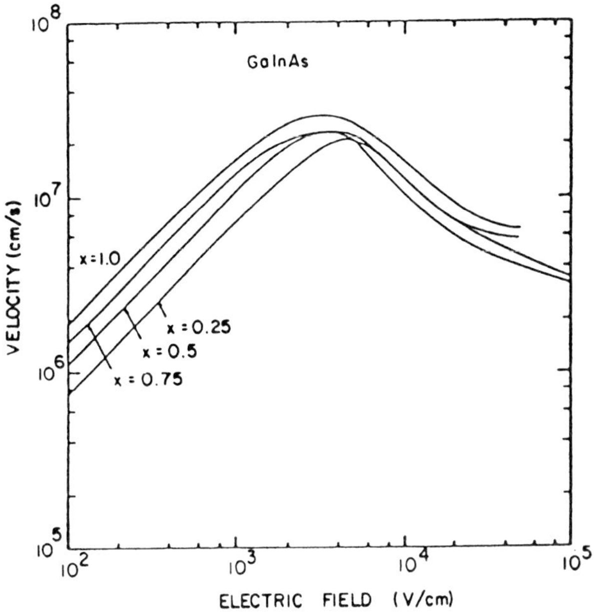

FIGURE 66. Velocity-Field Curves for $Ga_{1-x}In_xAs$ at 300°K with a doping level of 10^{16} cm^{-3} (From Ref. 31).

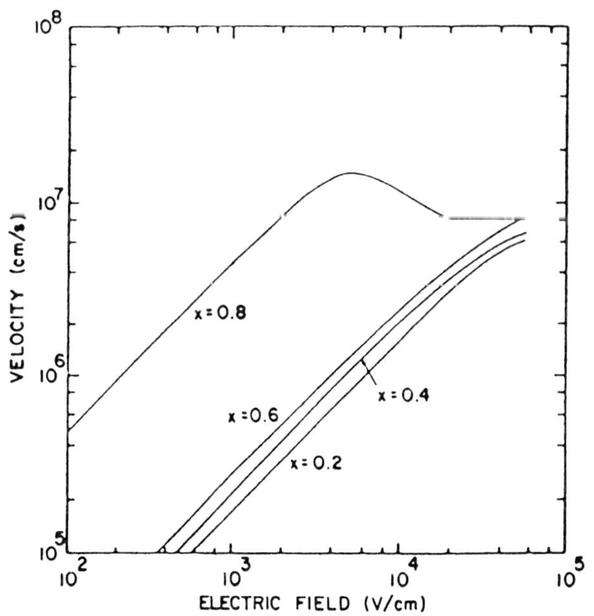

FIGURE 67. Velocity-Field Curves for $Al_{1-x}Ga_xAs$ at 300°K with a doping level of 10^{16} cm^{-3} (Ref. 31).

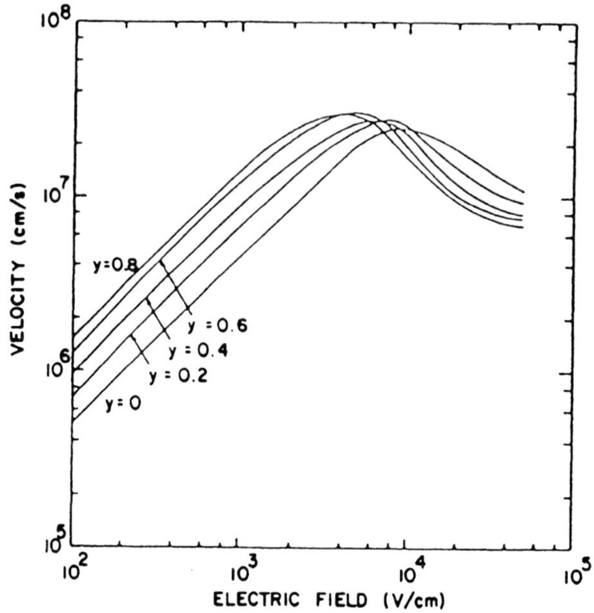

FIGURE 68. Velocity-Field Curves for $InP_{1-y}As_y$ at 300°K with a doping level of 10^{16} cm^{-3}. (Ref. 31)

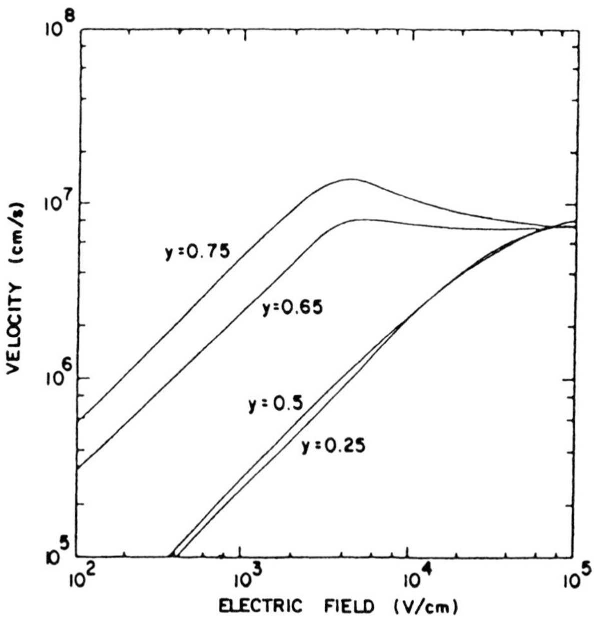

FIGURE 69. Velocity-Field Curves for $GaP_{1-y}As_y$ at 300°K with a doping level of 10^{16} cm^{-3} (Ref. 31).

for calculating these field dependent curves is essentially similar to the technique used for calculating the field dependent curves for GaAs, InP, etc.

Because of the narrow band gap of some of these materials, nonparabolic effects enter and additional care must be exercised in performing the calculations, but basically a set of material constants is needed for each binary component in the ternary element. We illustrate this in figure 70 for the ternary $Ga_xIn_{1-x}A_s$, where x represents the mole fraction of the GaAs element.

Figure 70 displays the band gap variation of the three principle portions of the conduction band for GaInAs. We note that InAs has the same Γ-L-X ordering as GaAs. Such is not the case with AlAs. The band gap variation of these curves is represented by (photoluminesence studier)[39]

$$E_\Gamma = 0.422 + .7x + 0.4x^2 \quad \text{at} \quad 2°K \tag{113a}$$

$$E_\Gamma = 0.324 + 0.7x + 0.4x^2 \quad \text{at} \quad 300°K \tag{113b}$$

where at room temperature for GaAs (x=1), E_Γ = 1.43ev and for InAs

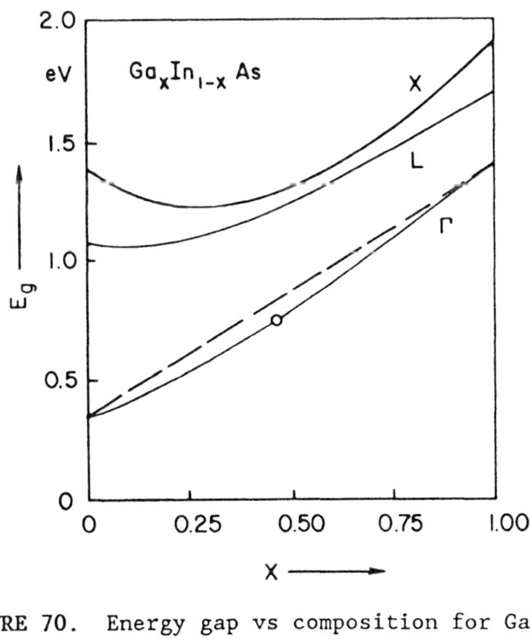

FIGURE 70. Energy gap vs composition for GaInAs.

(x=0), E_Γ = 0.35ev. There is a clear presence of bowing in the band gap variation with composition, but linear approximations for the L and X band for X>.47 are useful (note: $Ga_{.47}In_{.53}As$ has the same lattice parameter as InP)

$$E_L = 0.38 + 1.73x \tag{114}$$

$$E_X = 0.3929 + 1.93x \tag{115}$$

The above band gap variations is specific to GaInAs; there are, however, fairly general rules[40] often used for calculating other terms. The lattice constant **a** is computed from

$$\mathbf{a} = x\mathbf{a}_{GaAs} + (1-x)\mathbf{a}_{InAs} \tag{116}$$

The effective mass is:

$$\frac{1}{m^*} = \frac{x}{m^*_{GaAs}} + \frac{(1-x)}{m^*_{InAs}} \tag{117}$$

for each section of the conduction band. For the dielectric constants[40]

$$\frac{\epsilon_{o,\infty}-1}{\epsilon_{o,\infty}+2} = x\cdot\left(\frac{\epsilon_{o,\infty}-1}{\epsilon_{o,\infty}+2}\right)_{GaAs} + (1-x)\left(\frac{\epsilon_{o,\infty}-1}{\epsilon_{o,\infty}+2}\right) \tag{118}$$

For the transverse and longitudinal optical frequencies

$$\omega_T = (x\omega_T{}^2{}_{GaAs} + [1-x]\omega_T{}^2{}_{InAs})^{\frac{1}{2}} \tag{119a}$$

$$\omega_L = \omega_T\sqrt{\frac{\epsilon_o}{\epsilon_\infty}} \tag{119b}$$

To calculate the acoustic velocity U[40]

$$\left(\frac{U}{a}\right)^2 \omega_T^2 = x^2\omega^2_{AGaAs}\omega^2_{TGaAs} + (1-x)^2\omega^2_{AInAs}\omega^2_{TInAs}$$

$$+ \frac{x(1-x)\ M_C(M_A+M_B)}{(M_A+M_C)(M_B+M_C)}\omega^2_{TGaAs}\omega^2_{TInAs} \tag{120}$$

where

$$\omega_A = \frac{U_{GaAs}}{a_{GaAs}}, \qquad \omega_A = \frac{U_{InAs}}{a_{InAs}} \tag{121}$$

and M_A, M_P, M_C are the masses of the constituent atoms. For example, in GaInAs::

$M_A = M_{GA} =$ 69.72 gm/mole
$M_B = M_{IN} =$ 114.82 gm/mole
$M_C = M_{AS} =$ 74.92 gm/mole

Intervalley phonons are obtained from a linear extrapolation of the intervalley phonons of each of the constituent elements which in turn are obtained by application of a set of selection rules[41]. The selections rules were not obtained under hot carrier conditions, and so it is unrealistic to suggest that these be rigorously applied. Nevertheless, consider the intervalley phonon from Γ(00) → X(100). The selection rule is that an LO phonon at X is involved when $M_{III}<M_V$ (e.g., GaAs). An LA phonon is involved when $M_{III}>M_V$ (e.g., InAs). The phonon frequency for the ternary is taken as:

Intervalley Γ-X phonon frequency

$$\hbar\omega_{\Gamma X} = x\hbar\omega^{GaAs}_{\Gamma X} + (1-x)\hbar\omega^{InAs}_{\Gamma X} \tag{122}$$

The intervalley phonon from Γ(000) to L(111) is an average of LO and LA at L. Again,

$$\hbar\omega_{\Gamma L} = x\hbar\omega^{GaAs}_{\Gamma L} + (1-x)\hbar\omega^{InAs}_{\Gamma L} \tag{123}$$

For the intervalley phonon from L(111) to X(100), an average of the LA, TO and LO at L is used. Again,

$$\hbar\omega_{LX} = x\hbar\omega_{\substack{GaAs\\LX}} + (1-x)\hbar\omega_{\substack{InAs\\LX}} \tag{124}$$

From X(100) to X(100), from L(111) to L(111) for equivalent intervalley scattering, a linear extrapolation is used. Similarly for the nonpolar intervalley phonon (LO+TO) a linear extrapolation is used. The other remaining quantities of interest are the deformation coupling coefficients all of which undergo extrapolation.

Another feature of importance here concerns material choice associated with nonparabolic bands. In all of the calculations we have assumed a parabolic dispersion relation, namely

$$E = \frac{\hbar^2 k^2}{2m^*} \tag{125}$$

A simple extension to nonparabolic is generally given as

$$E + \frac{E^2}{Eg} = \frac{\hbar^2 k^2}{2m^*} \tag{126}$$

where Eg is an effective energy gap. The significance of this is

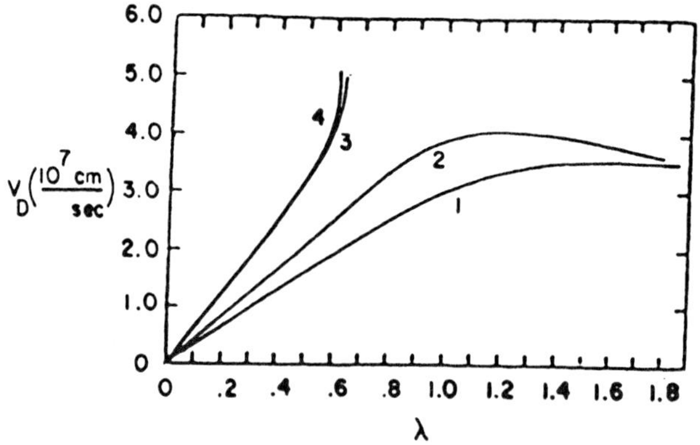

FIGURE 71. Drift velocity as a function of normalized field, λ: (1) InSb at 290°K, nonparabolic calculation; (2) InAs at 300°K, nonparabolic calculation; (3) InSb at 290°K, electron, parabolic band; (4) InAs at 300°K, parabolic band[42].

that the effective mass increases with increasing energy. The consequences of this is that negative differential conductivity in the absence of electron transfer can occur. On the basis of the above expression it is clear that nonparabolic effects, while important for GaAs, will be even more pronounced for InAs. The effects of nonparabolicity on the polar phonon scattering were discussed at a very early stage by Matz[42], and his results are shown in figure 71, where we see a clear contribution to NDR.

A general discussion of NDR due to nonparabolic energy bands was presented by Harris and Ridley[43] who applied their results to PbTe at 77 K. The general applicability of their conclusions are discussed below. Harris, et al[43] used the displaced Maxwellian approximation, and examined scattering due to acoustic and polar optical scattering. The results of their study are summarized in Figure 72. The symbols in their paper have the following significance. W is the ratio of the low field acoustic to polar optical mobilities:

$$W = \frac{\mu_a{}^\circ}{\mu_p{}^\circ} \tag{127}$$

when W = ∞ scattering is purely polar optical, when W = 0 scattering is purely acoustic. When W = 1 there is a mixture of acoustic and polar scattering. The bold lines signify calculations with nonparabolic contributions. The dashed lines are for purely parabolic bands. The nonparabolic contributions indicate an avoidance of runaway.

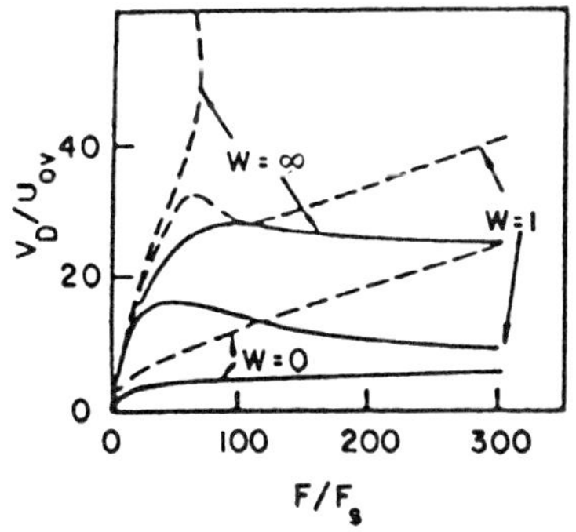

FIGURE 72. Field dependent variation of velocity for different combinations of acoustic and polar optical scattering. Dashed line denotes parabolic model, solid line denotes nonparabolic model[43].

Another important contribution to transport is the alloy scattering. Hauser, et al[41] treated a completely random array of allow scatters whose scattering rate increased with increasing electron energy. The results indicate a decrease in peak velocity and in the magnitude of the negative differential mobility.

The velocity - field curves of a variety of ternary compounds were shown in figures 66 through 69. We briefly reconsider AlGaAs. The interesting feature of this material is that for pure AlAs, the lowest portions of the conduction band is at X (see figure 73) Of particular interest, is the crossover, as reflected in figure 73. Figure 74 displays a normalized conductivity versus aluminum concentration where it is seen that (1) the lowest conductivities occur for highest aluminum concentration, and (2) the precipitous

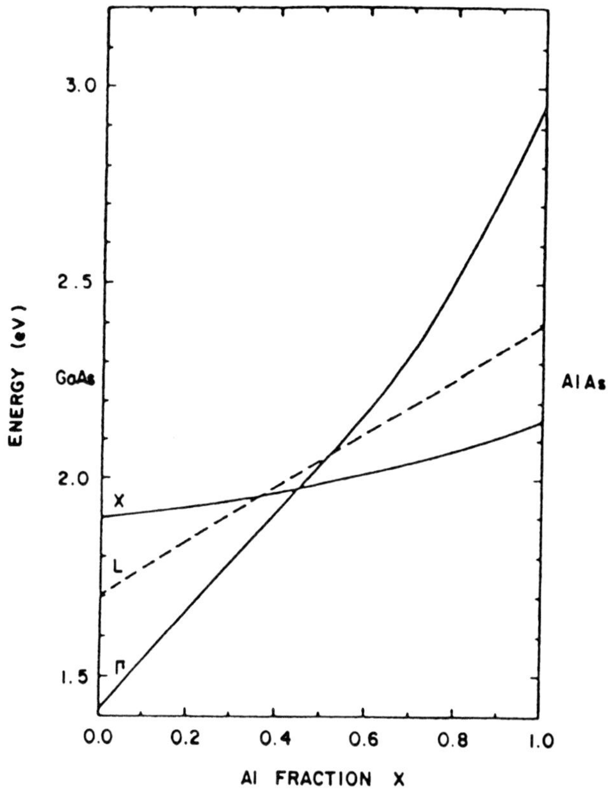

FIGURE 73. Energy band gaps (Γ,X, and L) of $Ga_{1-x}Al_xAs$ (at 300° K) calculated from Eqs. (6)-(8), as a function of Al concentration x. Dashed line represents the L band[45].

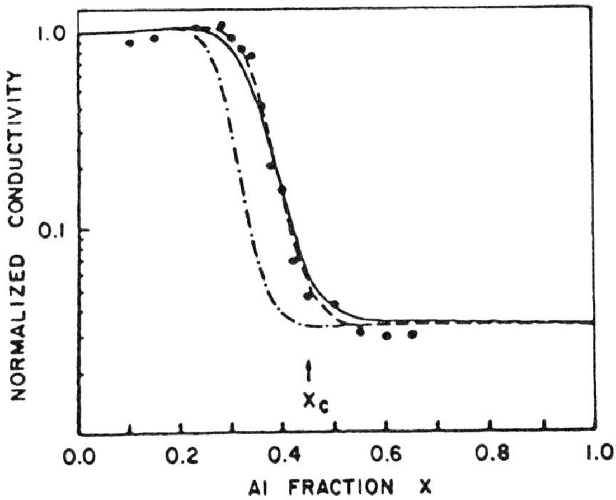

FIGURE 74. The data points represents normalized conductivity of $Ga_{1-x}Al_xAs$ vs Al concentration. A direct-to-indirect bandgap crossover occurs at $x_c = 0.45$. The solid line corresponds to a three-band (Γ, X, and L) calculating, only Γ and X are considered in the calculation resulting in a dashed line. The dot-dash line corresponds to $x_c = 0.37$, as explained in Ref. 45.

drop in conductivity occurs near the crossover. These results are also reflected in the field dependent velocity relation (figure 67).

At this junction we reiterate a point made earlier for parabolic bands, namely that central valley transport appeared to dominate device behavior. In figure 75a the calculated field

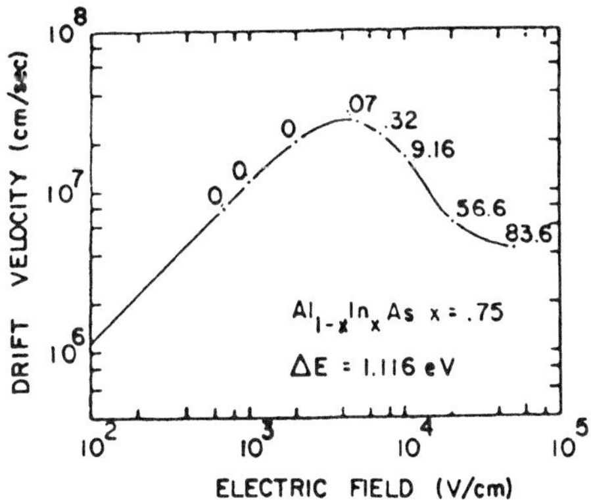

FIGURE 75a. Calculated velocity-field curve for $Al_{0.25}In_{0.75}As$ values along the curve show the percentages of electrons in the upper (111 and 100) valleys[47].

dependent velocity for $Al_{0.25}In_{0.75}As$ is shown along with the percentage of carriers in the subsidiary valleys. In Figure 75b the contribution from the Γ-valley velocity relative to the total velocity is shown. It is seen that transport is dominated by the central valley. Note: the calculated energy gaps for AlInAs are displayed in figure 75c. There is a discrepancy between the L-valley AlAs results of this calculation and that of figure 73. The calculations for gallium indium arsenide show a similar behavior, as displayed in Fig. 76.

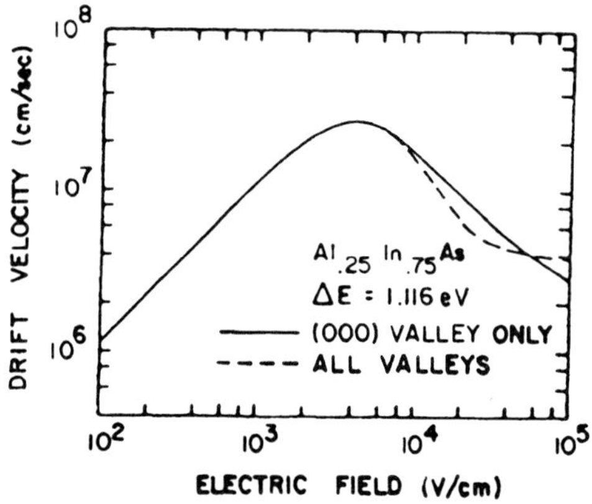

FIGURE 75b. Comparison of velocity-field curves for $Al_0l_{.25}In_{0.75}As$ use g central valley only and using all valleys[47].

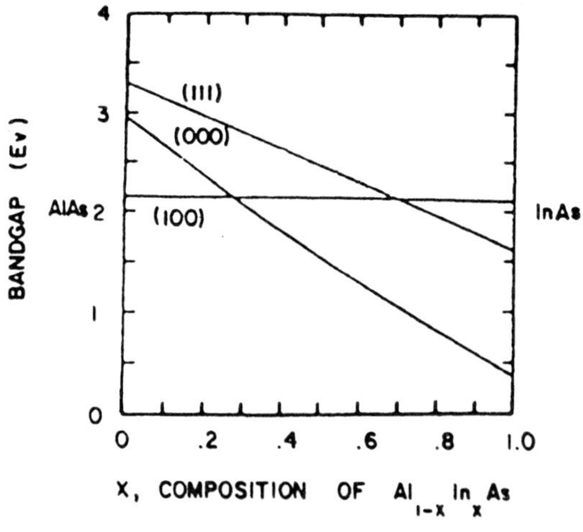

FIGURE 75c. Calculated energy gaps vs composition for $Al_{1-x}In_xAs$[47].

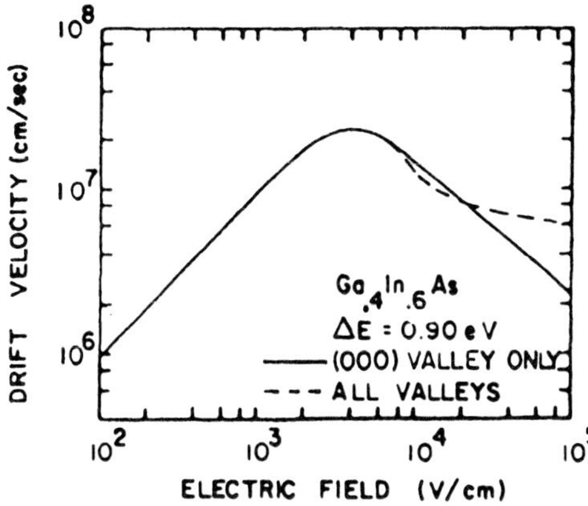

FIGURE 76. As in figure 75b, but for $Ga_{0.4}In_{0.6}As$[47].

To tie the varieties of field dependent transport coefficients together, several figures of merit have been proposed. One, put forth by Hauser, classifies materials in terms of the peak drift velocity obtained from polar phonon scattering including nonparabolic effects. This was estimated by Hauser, et al[42] as

$$V_p \leq V_{MAX} = \left(\frac{\hbar\omega_o}{m^2} + \tan\left(\frac{\hbar\omega_o}{2kT}\right)\right)^{\frac{1}{2}} \qquad (128)$$

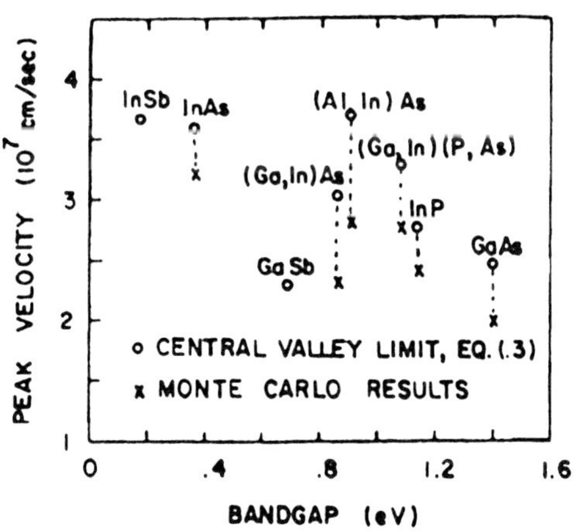

FIGURE 77. Calculated upper limits to peak velocity from equation (128)[42].

The results of the above expression for a variety of different materials are summarized in Figure 77, where we note that the highest peak drift velocity is for the material InSb.

In another study (Ferry[43]) summarized in Figure 78 the high field saturated drift velocity was plotted as a function of $8\hbar\omega/3\pi m^*$. The curve in figure 78 follows the relation:

$$V = (8\pi\omega_0/3\pi m^*)^{\frac{1}{2}} \tag{129}$$

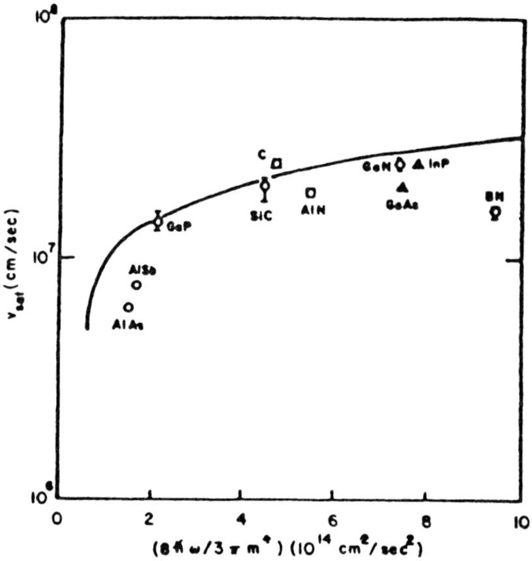

FIGURE 78. Saturated velocities calculated here are shown as functions of the parameter relating to energy relaxation.

BOUNDARY CONDITIONS TO DEVICES

We close this lecture series with a more detailed discussion of field nonuniformities.

While most of the previous discussion has tended to focus on the importance of field nonuniformities for realistically assessing device performance only a small fraction of the discussion was devoted to the single most important contribution to these field nonuniformities - boundary conditions. Here we provide a brief review of its influence on the behavior of two-terminal NDC devices. Figure 79 displays typical boundary-dependent data from three different gallium arsenide two-terminal devices. The lower portion of each diagram displays current versus voltage characteristics, while the upper portion shows voltage versus distance at one bias point. Figure 79a shows measurements for a device in which the metal contacts are far removed from the active region of the device. The current-voltage relation is relatively linear until a point where current oscillations occur. The field profile just prior to the oscillation is relatively uniform within the active region of the device, and is near zero at the ends of the active region. Figure 79b represents a set of measurements in which the metal contact abuts the active region of the device. The current-voltage characteristics remain linear to threshold which again is manifested by a current oscillation. Notably different here is the lower average field prior to the instability and the enhanced voltage drop at the cathode. Figure 79c displays results for another device with a metal contact abutting the active region. For this case there is a sublinear current voltage characteristic and no instability. The probed voltage versus distance shows a large voltage drop at the vicinity of the cathode.

The electrical characteristics associated with Figs. 79a

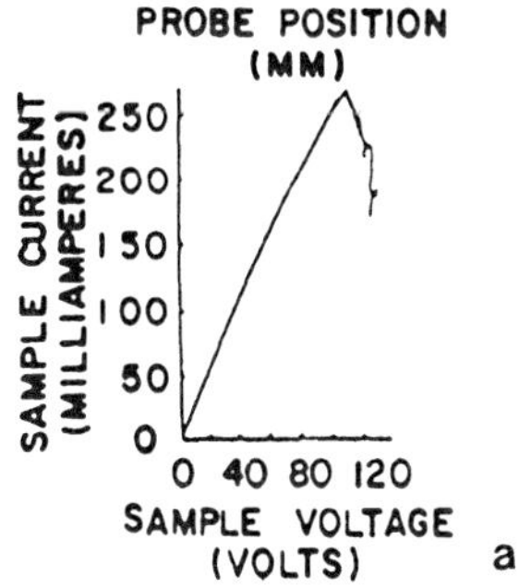

FIGURE 79a. Probed voltage versus distance, and current versus voltage for a two-terminal GaAs device with low boundary fields. From Ref. 48.

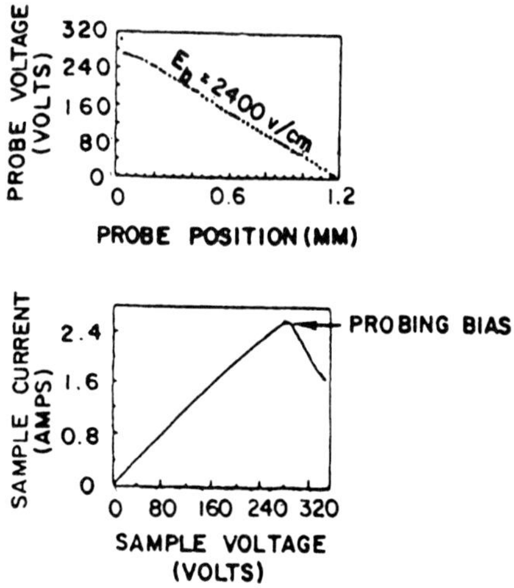

FIGURE 79b. As in Fig. 79a but for a metal contact abutting the cathode.

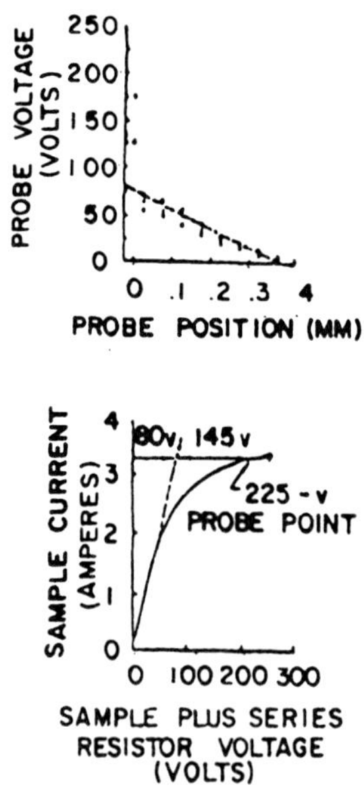

FIGURE 79c. As in Fig. 79a but for a metal contact abutting the cathode.

through c have been described as representative of "ohmic contacts", "slightly blocking contacts" and "strongly blocking" contacts, respectively. One of the earliest models employed for explaining these results assumed a "pinned" value of cathode electric field. Other models in which the cathode conductivity or doping profile was varied have also been suggested with varying degrees of success. The "pinned" cathode field model, discussed below, developed partially as a consequence of the way the equation's current instabilities was written. Here the continuity equation is rewritten in one dimension,

$$J(t) = q\left(nv - \frac{D\partial n}{\partial x}\right) + \frac{\epsilon\partial F}{\partial t} \tag{130}$$

where J(t) is the total current, and rewritten, again using Poisson's equation, as

$$J(t) = qN_o v + \epsilon\left(nv - \frac{v\partial F}{\partial x} + \frac{D\partial^2 F}{\partial x^2} + \frac{\partial F}{\partial t}\right) \tag{131}$$

This is a second order nonlinear partial differential equation on E requiring two boundary conditions and one initial condition.

Solutions to equation (131) have been used successfully to simulate device results similar to those of figure 79 for gallium arsenide. Qualitatively similar results occur for solutions with cathode fields F(=0,t) falling within any of the following three groups:

$< 0 \leq F(x=0,t) \leq F_{TH}$, $F_{TH} \leq F(x=0,t) < 4F_{TH}$, $F(x=0,t) > 4F_{TH}$.

Where F_{TH} is the threshold field for negative differential mobility. The simulations with pinned fields falling in either group 1, 2 or 3 yield electrical characteristics similar to those of figures 79a, b and c, respectively. The crucial feature of this model is that the cathode field is pinned, necessitating that any instabilities in current occur at a critical value of current density. The field profiles associated with cathode fields in the range

$$< 0 \leq F(x=0,t) \leq F_{TH}, \text{ and } F_{TH} \leq 4F(x=0,t) < 4F_{TH}$$

are sketched in figure 80. For reference, a velocity field curve with velocity scaled to current as $qN_o v(F)=J_n(F)$, and with a region of negative differential mobility is also included. Figure 80 is understood as follows: The second column of each section shows the electric field versus distance profiles. F(x) begins with a value F_c at the cathode and extends downstream to a value F_b. By current continuity, the current everywhere within the device is given by $J = q N_o v(F_b)$. For $J < J_n(F_c)$, first row, a region of charge depletion forms near the cathode for F_c below and within the region of negative differential mobility (NDM). Increasing the current until $J \simeq J_n(F_c)$, second row, introduces charge neutrality everywhere for $F_c < F_{TH}$. However, because F is a double valued function of V, for F_c within the NDM region; approximate charge neutrality exists near the cathode for $J \simeq J_n(F_c)$, and for regions sufficiently far downstream from the cathode. Charge neutrality breaks down between these two regions. Finally, for $J > J_n(F_c)$, third row, an accumulation layer forms near the cathode. For figure 80a, the accumulation layer is stable until the bulk field exceeds F_{TH}. For Fig. 80b,

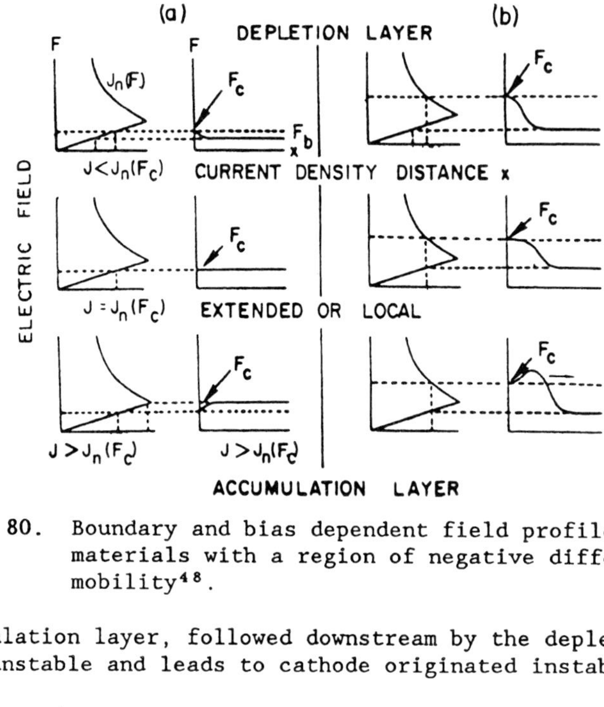

FIGURE 80. Boundary and bias dependent field profiles for materials with a region of negative differential mobility[48].

the accumulation layer, followed downstream by the depletion layer is often unstable and leads to cathode originated instabilities.

The situation corresponding to Fig. 79c is often represented by very high cathode fields. The field profiles are those appropriate to a wide region of charge depletion near the cathode. The field profiles are stable.

The pinning of the cathode field is not necessarily common, however, to all semiconductor devices. For example, while it was also applied to InP devices, where it worked for a significant number of cases, a broad class of InP device behavior could not be accounted for through it's use[13,41]. The latter showed anonymously high efficiency and low dc current levels. Spontaneous Gunn type oscillations did not occur. Rather, device operation required a tuned circuit. InP device operation was also thought to depend critically on the cathode boundary condition, and the experiments were explained through a fixed cathode conduction condition. The distinction between "pinned" cathode field and "pinned" cathode conduction current is placed in perspective in figure 81 and in the following equation

$$J(t) = J_c(F_c) + \frac{\epsilon dF_c}{dt} \tag{132}$$

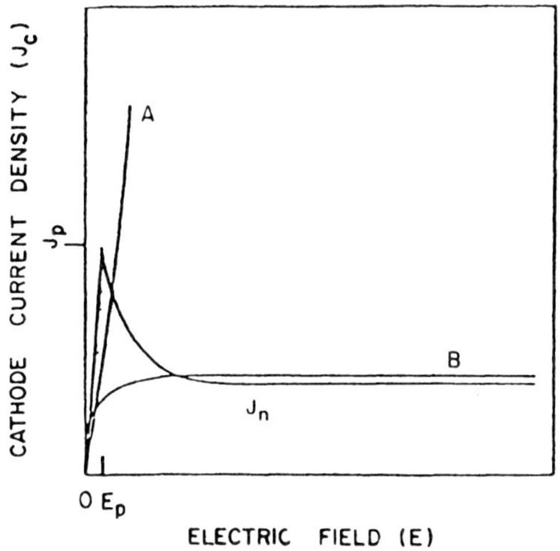

FIGURE 81. Cathode current density, from Equation (133).

Equation 132 is the equation for total current through the boundary to the device. $J_c(F_c)$ represents the current-field relation at the cathode, also referred to as a control characteristic which may be expected to differ from that of the semiconductor device. Two such types of curves are represented by curve A and curve B of figure 88. Curve A is closely related to the pinned cathode field model while curve B is associated with the pinned cathode current model. The similarity in "form" of curves A and B to, respectively, moderate barrier height tunneling and thermionic emission dominated contacts is deliberate, and the equation used to arrive at these curves is shown below.

$$J_c(F_c) = -J_R \left(\exp[-qF_cL_c/nkT] - \exp[-(1/n - 1)(qF_cL_c/kT]\right) \tag{133}$$

which was adapted from studies on the unalloyed metal/semiconductor contact[50]. It's use here presumes a similar description. For the unalloyed contact "n" is the ideality factor and describes the contact as dominated by thermionic emission ($n \geq 1$) or by tunneling ($n >> 1$). J_R is the reverse current flux and may be related to the barrier height phenomenologically through the Richardson equation.

Detecting a particular contact effect on a device is a difficult procedure. For long devices current voltage characteristics as represented by figure 79 are often signatures of a contact classification. For short devices proximity effects introduce an additional complication and current-voltage measurements are less valuable. One type of measurement which may serve to provide information about the boundary is a noise measurement.

Here the situation to envision is that if the field is pinned within the negative differential mobility region, increasing the bias will result in an amplification and the noise will increase. If an increasing the bias results in carrier injection into the device, the field at the cathode is likely to decrease and the noise is expected to decrease. While these results should be folded in with the field dependence of velocity and diffusion, a simple analytical noise calculation assuming a three piece approximation to represent GaAs has been performed.

In this calculation, the "impedence field method"[51] is applied to calculating noise due to thermal velocity fluctuations amplified within the device. The mean squared noise voltage per unit band width is computed,

$$\left\langle \frac{\delta V_N^2}{\Delta f} \right\rangle = 4q^2 \int_{vol} |\nabla Z|^2 \, ND \, d(vol), \tag{134}$$

where ∇Z is the impedence field vector. The calculation was performed for a ten-micron long element with a doping of $10^{15}/cm^3$. The element sustains the field profile with a cathode depletion where it is seen that the NDM region increases with increasing bias. The calculations, which are discussed in detail elsewhere, [Ref. 29] are expressed in terms of the noise

$$NF = 1 + \left\langle \frac{\delta V_N^2}{\Delta f} \right\rangle \cdot \frac{1}{4k_o T |R|} \tag{135}$$

figure 51 where R is the real part of the device impedence. The results of the calculation are displayed in figure 82 where the noise figure is sketched as a function of bias current and transit angle. The

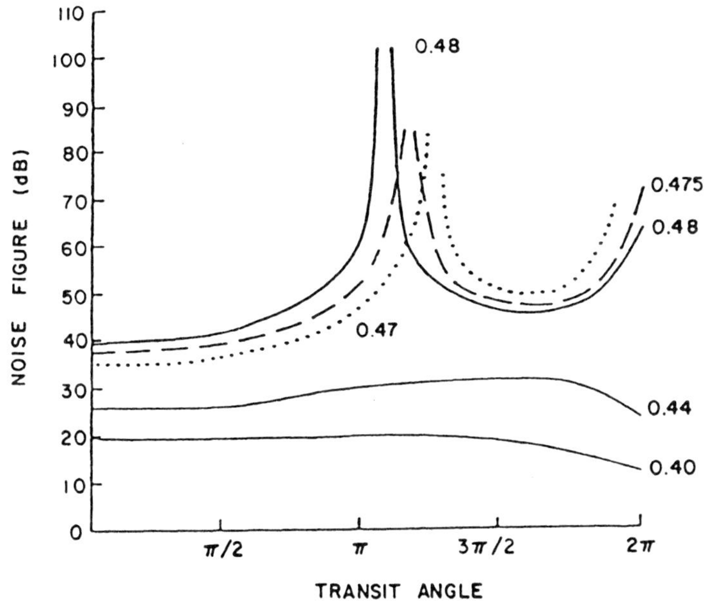

FIGURE 82. Noise figure versus transit angle and current.

results appear as a signature of the effects of the cathode boundary. First, at low values of transit angle $\vartheta = \omega T(\Delta)$, where $T(\Delta)$ is the transit time across the negative differential mobility region, the noise figure increases with increasing bias. This corresponds to an increase in the length of the negative differential mobility region and enhances amplification of any fluctuation originating there. More interesting structure is present at higher frequencies and higher bias where the noise figure increases and then shows a singularity. On the other side of the singularity there is a "U"-shaped region ending again at a singularity. The strong increase in noise figure represents the approach of $|R| \to 0$. Here at low frequencies the real part of the impedence is positive, and becomes negative at frequencies somewhere between $\pi \leq \vartheta < \omega T$. In going from positive to negative values it passes through zero, hence the singularity. The frequency range for small signal negative resistance increases with increasing bias, reflecting the broadening of the negative differential mobility regions - a broad "U"-shaped region appears. Both the increasing noise figure at low frequencies, and the "U"-shaped region at high frequencies are characteristics of an increasing depletion layer width. Note that increasing the bias still further will result in an electrical instability.

The discussion of the above sections dealt with devices whose lengths were typically 10μm-long or longer and the analysis was through the drift and diffusion equation. For near and submicron length devices the Boltzmann transport equation is required. A set of one dimensional steady state calculations are displayed below, using the moment equations discussed earlier. The boundary conditions for this problem are stated below[14]

$$\frac{\partial}{\partial X} \log N_1 = -A \tag{136}$$

$$V_1 = -\mu_c F \tag{137}$$

$$T_1 = B \tag{138}$$

$$\frac{\partial N_2}{\partial X} = \frac{\partial V_2}{\partial X} = \frac{\partial T_2}{\partial X} = 0 \tag{139}$$

The boundary condition represented by Eq. (136) dictates the slope of the Γ valley carrier density downstream from the cathode. Thus for "A" positive (negative) local charge accumulation (depletion) occurs at the cathode boundary. The carrier velocity within the interior of the device is governed by the assigned scattering rates. The velocity of the entering carriers will, in general, differ from that within the bulk. The assigned entrance velocity is governed by Eq. (137). The temperature of the entering carriers is assigned a constant value, represented by Eq. (138). "B" is generally greater than, or equal to 300 K and is a measure of the mean thermal energy of the central valley carriers. As in the discussion of Fig. 80, the results are placed into three categories, "ohmic", "slightly" and "strongly" blocking contacts. The "ohmic" results are shown in Fig. 83, for a 2.0μm-long structure with a doping of $5\text{x}10^{15}/\text{cm}^3$.

For "ohmic" boundaries an appropriate set of constants are: $A=0.2, \mu_c=12{,}000\text{cm}^2/\text{v}_{\text{sec}}$, and $T_1 = 300°\text{K}$. For this case carriers enter the device with speeds greater than that associated with the central valley mobility. Additionally there is an accumulation of carriers at the boundaries, resulting in low values of cathode field. The field starts off at nearly 1kv/cm and approximately 2500 A must be traversed before significant transfer occurs. Increased transfer results in a lowering of the mean carrier velocity, necessitating an increase in mobile charge as the mode is approached. The average velocity across this device obtained from the relation

$$\langle v\rangle = \frac{J}{qN_o} \tag{140}$$

is $\langle v\rangle=1.78\text{x}10^7$ cm/sec, with $N_o=5\text{x}10^{15}/\text{cm}^3$.

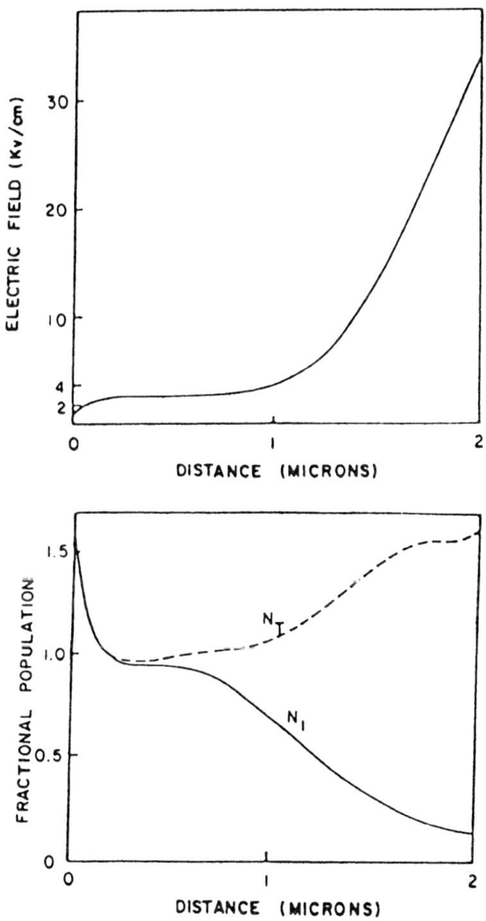

FIGURE 83. Distribution of field and carriers for ohmic boundaries, with an applied bias of 2 volts. From [14]

Lowering the cathode mobility to a value below that associated with the Γ-valley results in a more rapid dispersal of carriers, and cathode adjacent charge depletion, associated with slightly blocking contacts, occurs. This is seen in figure 84 for A=-0.11, μ_c=6000 cm^2/V$_{sec}$, and T=300°K. It is noticed that the cathode field for this case is approximately 4kv/cm, which is higher than that associated with the "ohmic" contact condition of Fig. 83. There are, however, important similarities between figures 83 and 84. In both cases the carriers adjacent to the cathode are, for all practical cases, Γ-valley electrons. Very little transfer, which is determined by carrier energy

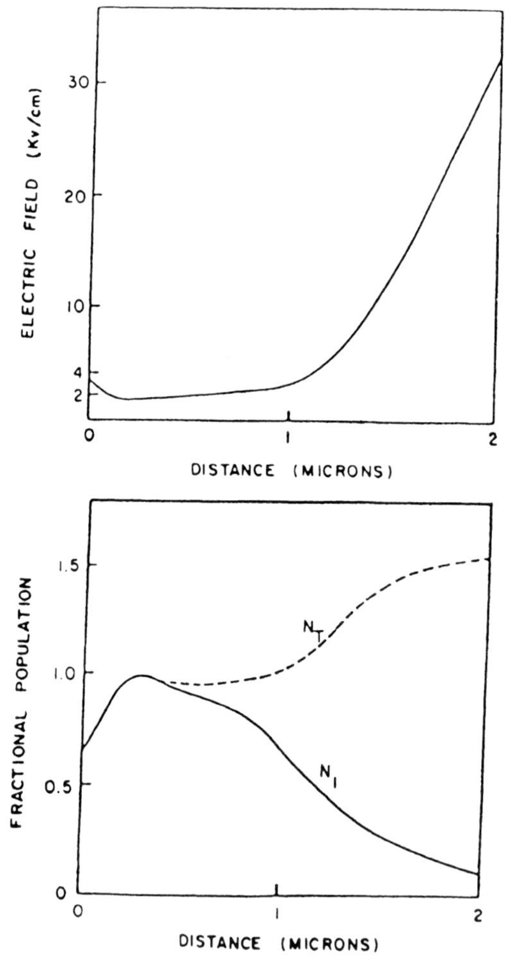

FIGURE 84. As in Fig. 83, but for a partially depleted cathode boundary. [From 14]

(temperature) has occurred. In addition, sufficiently downstream from the cathode the carriers appear to be ignoring the cathode condition and are dominated by the downstream voltage drop, whose spatial distribution is about the same for both. The average velocity for this case is $\langle V\rangle = 1.75\times10^7$ cm/sec, slightly below that of figure 83.

A significant change occurs when the mean energy of the Γ-valley entering carriers is elevated. For the parameters of figure 84, but with $T_1 = 1200$ k, a substantial amount of transfer occurs at the cathode, resulting in a lowering of the current through the device. The cathode is approximately 7kv/cm, higher than that associated with figure 84, and the downstream field is lower (see figure 85). The average velocity of the carriers in this case is lowered to 1.28×10^7cm/sec, even though the central valley carriers are traveling at higher speeds.

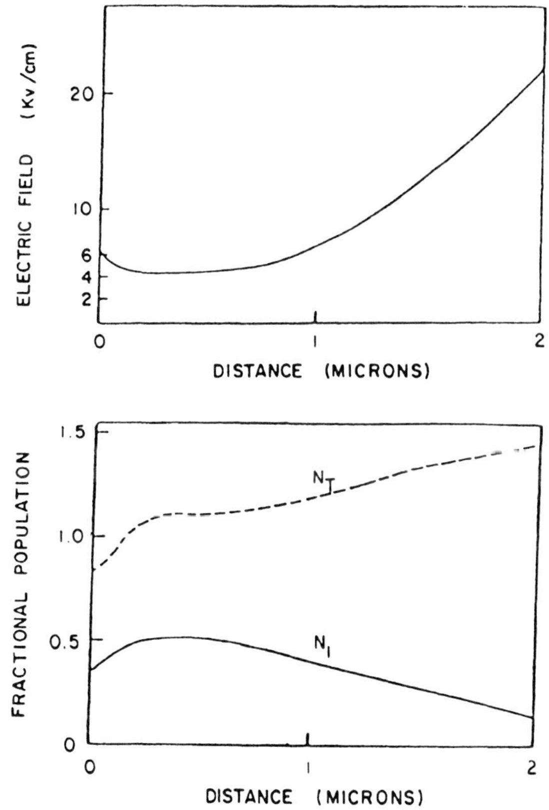

FIGURE 85. As in Fig. 84 but with an elevated cathode electron temperature. From [14]

The presence of moderately high cathode fields is attractive if a sufficient number of carriers can be retained in the central valley where they can sustain high transit velocities. While this case is discussed in more detail below, the simple ruse of injecting excess carriers into a device with the contact conditions of figure 85 does not always yield the sought after current levels. This is illustrated in figure 86, where now A=0.2. Here, the excess charge serves to lower the cathode field, which does not change the downstream characteristics in any significant way. The average velocity for this case is virtually unchanged when compared to figure 85. In this case v_{AV}=1.30x10^7/sec.

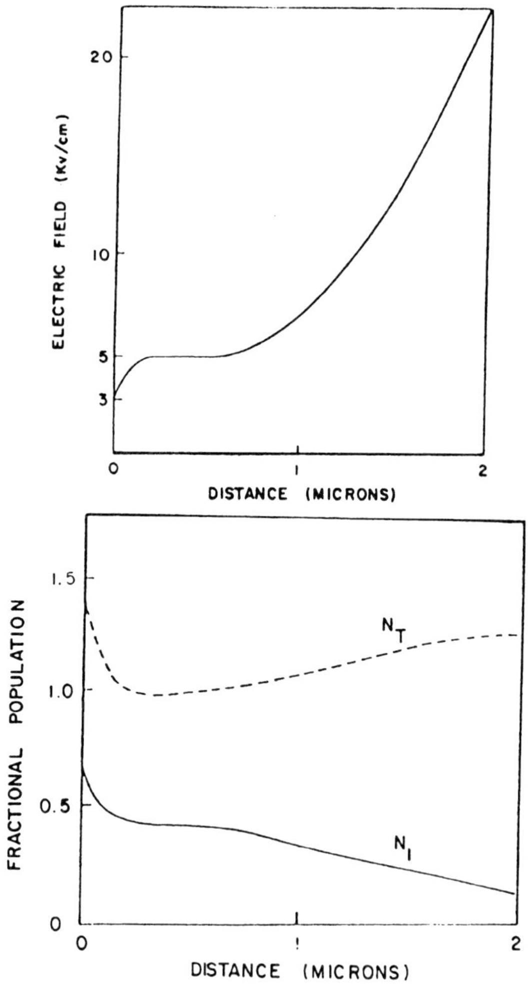

FIGURE 86. As in Fig. 85, but for injecting boundaries. [From 14]

The situation of a strongly blocking contact is illustrated in figure 87 for A=9.0, μ_c = 1000 cm/V_{sec}, and T=3000 K. For this case virtually all carriers are swept away from the cathode, with the satellite valley carriers accounting for most of the transport in the transition region. The cathode field is extremely high, approaching 60kv/cm. The average velocity is approximately zero.

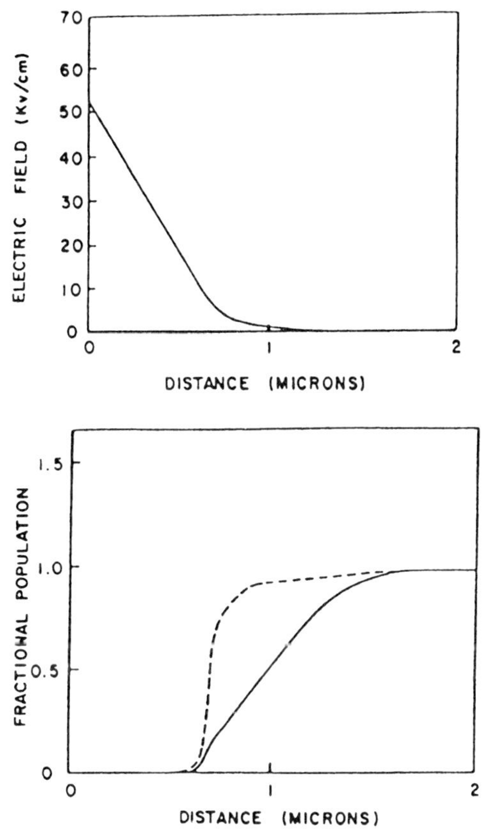

FIGURE 87. As in Fig. 86, but for a strongly blocking contact. [From 14]

In addition to determining the distribution of carriers, etc, boundaries through their influence on the field distribution, also affect current transients. This was shown for an accumulation boundary in Fig. 20. It is shown for a depleted boundary below, where a calculation was performed for a five-micron element with a high cathode temperature, T_1=1200. The transient results which are displayed in figure 88, and are expressed in terms of a mean velocity. The detailed field profiles are shown in figures 88 and 89. We note, approximate saturation in current, as displayed in Fig. 88. A significant voltage drop in the vicinity of the cathode, as displayed in Figs. 89a and 90a. Note that as the

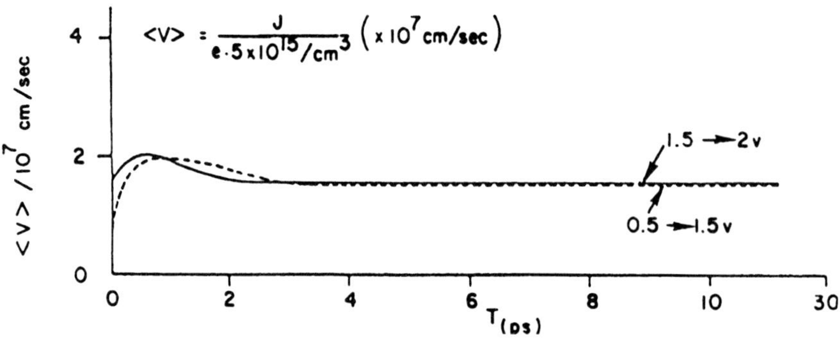

FIGURE 88. Transient current versus time for the indicated voltage change.

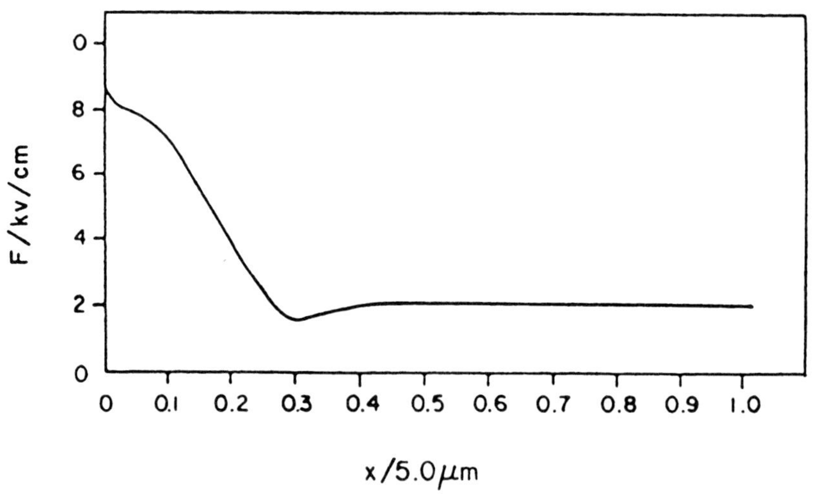

FIGURE 89a. For a bias of 1.5v, steady state distribution of: field.

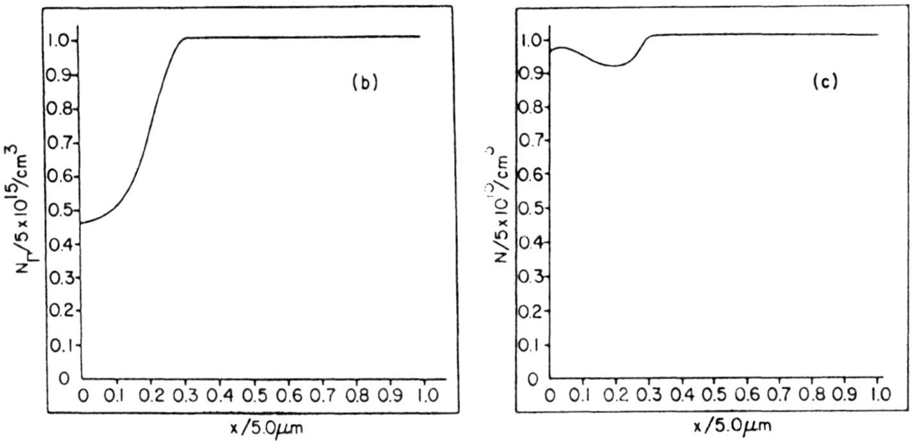

FIGURE 89b, c. For a bias of 1.5v, steady state distribution of: (b) Γ valley carriers; (c) total carriers.

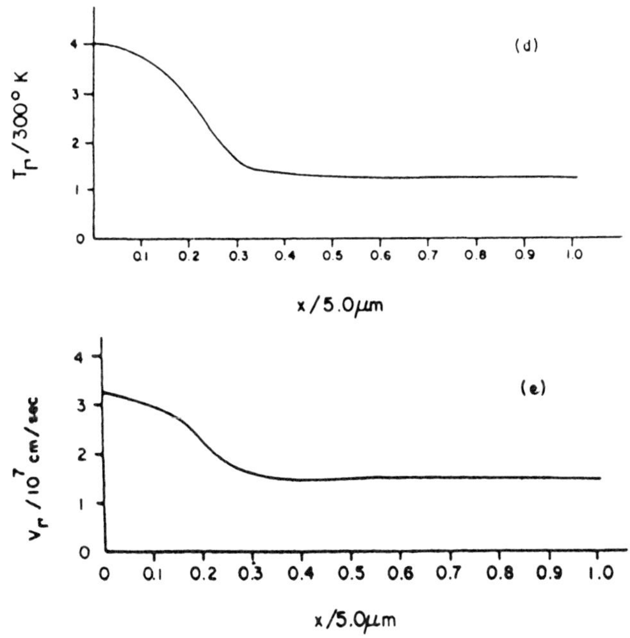

FIGURE 89d, e. For a bias of 1.5v, steady state distribution of: (d) Γ valley temperature, (e) Γ valley velocity.

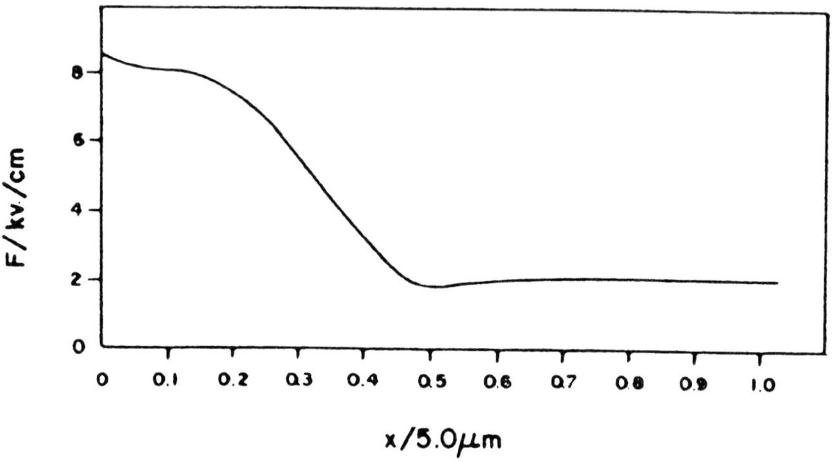

FIGURE 90a. As in Fig. 89, but for a bias of 2 volts.

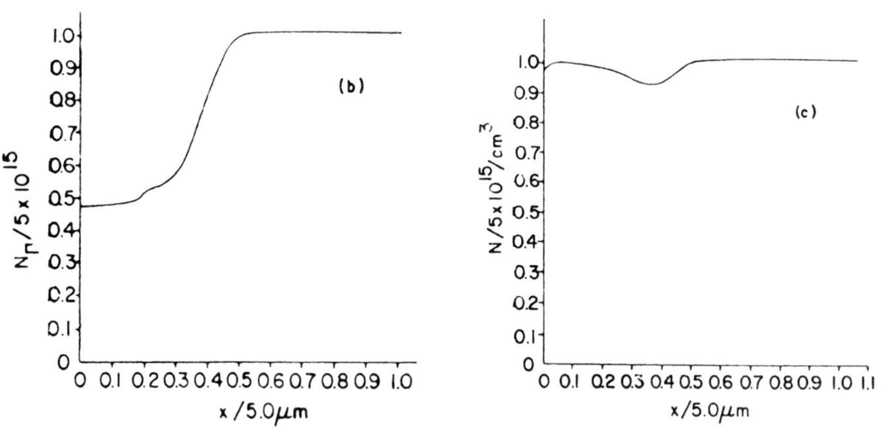

FIGURE 90b, c. As in Fig. 89 but for a bias of 2 volts.

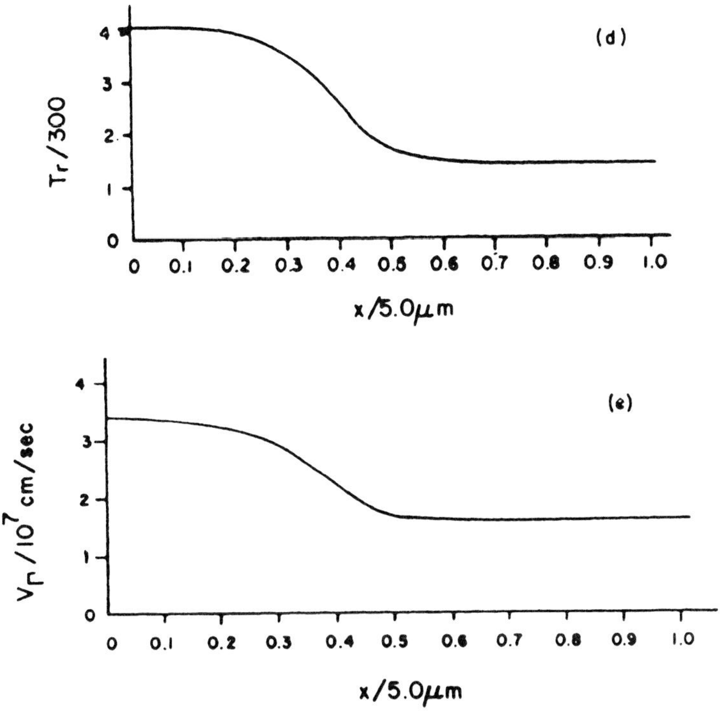

FIGURE 90d, e. As in Fig. 89 but for a bias of 2 volts.

voltage increases a region of local charge neutrality is forming at the cathode. Insofar as the Γ valley carriers are significantly depleted, the cathode region is rich in satellite valley carriers. But the distribution of velocity, as displayed in Figs. 89e and 90e, indicate again that most of the current is carried by the Γ valley carriers. What is clear from these figures is the apparent migration of the depletion layer toward the anode boundary, a situation similar to that obtained from the drift and diffusion equations.

ALTERNATIVE SOURCES OF FIELD NONUNIFORMITIES

All of the above calculations have tended to focus on the role of the cathode on device performance. Several calculations shown below illustrate the influence of selective reductions of "notches" in doping. One calculation is for a 2μm-long device with an 8000 A notch at the anode boundary (figure 91). The second was for a device with a narrow notch (figure 92). In both cases the electric

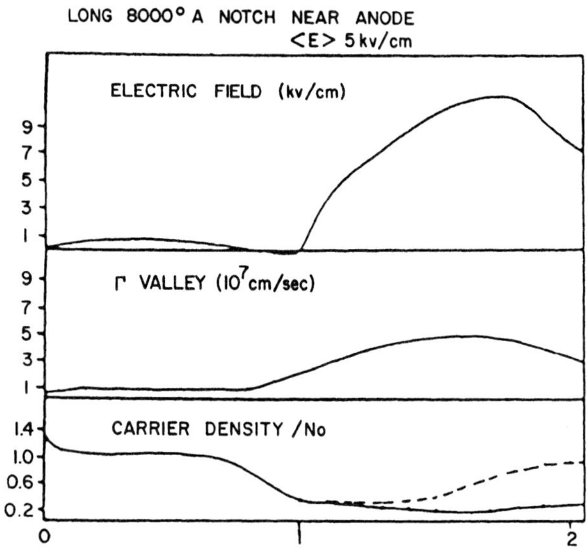

FIGURE 91. Long 8000°A notch anode. Ψ = 1v.

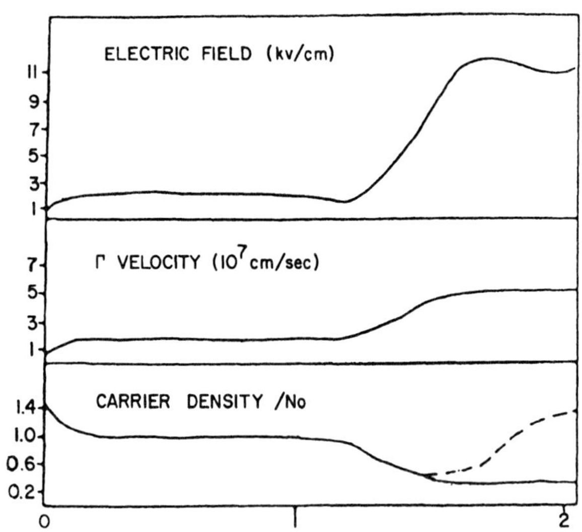

FIGURE 92. Narrow 4000°A notch is near anode Ψ = 1 v.

field showed an excess voltage over the notch along with significant transfer. The current level was higher for the narrow notch device and again most of the current level was carried by the Γ-valley carriers.

The situation when a section of periodic notches is introduced is shown in figures 93 through 95. In figure 93, the device is 2μm-long with an applied bias of 1.0v. Two notches were used here. Comparing the results of Fig. 93 with that of 94, it is seen that the Γ valley velocity is higher in the former case, as is

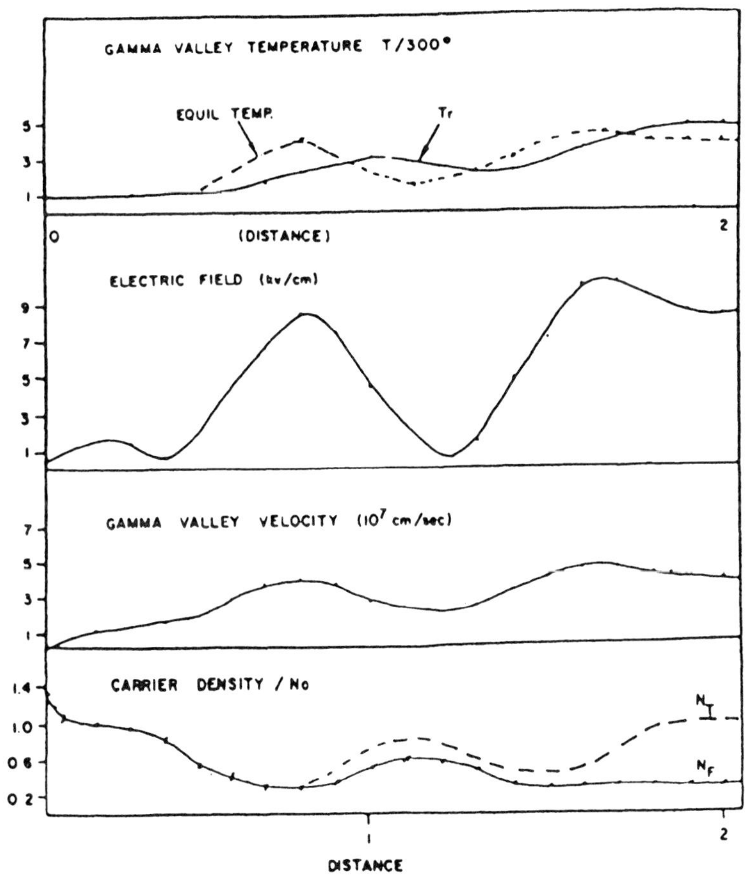

FIGURE 93. Double notch 2μm-long device. Ψ = 1 v.

the current level. Figures 94 and 95 are for 3 and 4 notches within a one-micron long device. Each with an applied bias of .8v. In each case as the number of notches increases, the modulation of carrier density decreases, although not the field, as indicated in Fig. 95b.

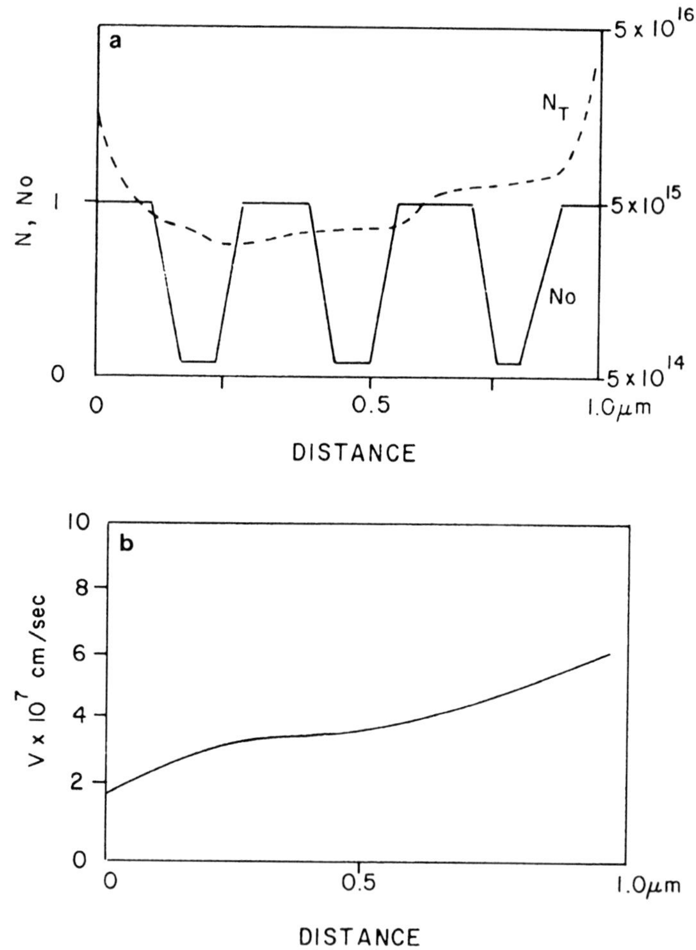

FIGURE 94. Repeated overshoot , **(a)** carrier density and doping profile, **(b)** gamma valley carrier velocity. <F> = 8Kv/cm.

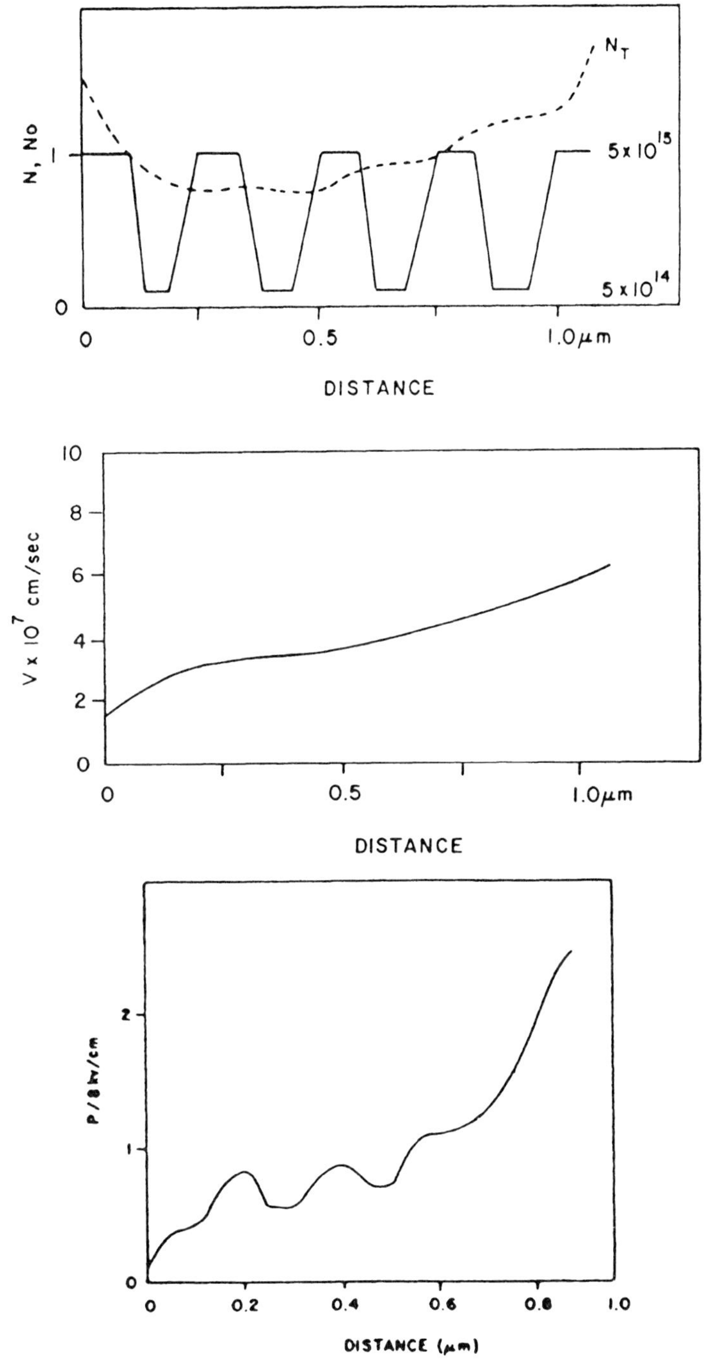

FIGURE 95. As in figure 94, but for four notches.

SUMMARY

The experimental situation is such that, with the exception of long compound semiconductor devices, there is very little data on the role of boundaries and contacts to submicron devices. The reason for the paucity of data lies in the fact that most submicron devices are three terminal devices designs and the third terminal tends to mask the role of the contact boundaries. This is unfortunate in that it is likely that two terminal device measurements will indicate what can be achieved in controlling the entrance dynamics of the carriers. To date, most two terminal device measurements on simple device structures have concentrated on the role of transport within the device, and raise the question of whether "ballistic" motion is possible. Based on the history of vacuum tube dynamics it should be recalled that, if transport is ballistic, the electrical characteristics will be controlled by the contacts.

The situation in submicron devices is further complicated by communication between the up- and downstream contacts. Thus is may be expected that the influence of a blocking contact on the electrical characteristics of long and submicron devices will be different. For submicron devices simple current voltage measurements may be rendered useless as a diagnostic tool. This is certainly not the case in long devices.

The role of numerical simulations in these boundary and device studies has been to act as surrogates for measurements that are not feasible. In one case, obtaining cathode boundary fields from measurements was not possible. Thus for long devices the sensitivity of the numerical results to numerical changes in the boundary conditions, when coupled to experiments, provided the key to the role of contacts on device behavior. For submicron devices, the difficulties of direct correlation of experiment with specific transport phenomena are apparent and simulation through parametric studies will provide a key to the role of boundaries. But the description of transport on a submicron scale is still inadequate and the descriptive role of boundaries is correspondingly weak. For example, most space charge dependent problems still treat the background as a "jellium" distribution. The discrete nature of impurities is ignored, as are structural variations in the contacts. The extent to which this affects such measurements as current-versus-voltage is yet to be determined. Notwithstanding these uncertainties, a considerable amount of information can be obtained by extrapolation from the ideal cases which can provide bounds on the limits of transport through both the boundary and active region of the device.

ACKNOWLEDGEMENTS

The author is grateful for the support of the Office of Naval Research and the Army Research Office.

REFERENCES

1. See, e.g. D.H. Austin, A.M. Johnson, P.R. Smith, and J.C. Bean, 1980, Appl. Phys. Lett. 37:371; and C.V. Shank, R.L. Fork, B.I. Greene, F.K. Reinhart and R.A. Logan, 1981, Appl. Phys. Lett. 38:104.

2. R. Hammond, This Lecture Series.

3. S. Laval, C. Bru, C. Arnodo and R. Castagne, Proc. 1980 IEDM, 626.

4. R. G. Chambers, 1952, Proc. Phys. Soc. (London) A65:458.

5. See, e.g., A.B. Pippard, 1962, "The Dynamics of Conduction Electrons" in "Low Temperature Physics", Eds: C. DeWitt, B. Dreyfus and P.G. DeGennes, Gordon and Breach, NY.

6. See, e.g., G.E.H. Reuter and E.H. Sondheimer, 1949, Proc. Roy. Soc., A195:336.

7. See, e.g. T. Kjeldaas, Jr. and T. Holstein, 1959, Phys. Rev. Lett., 2:340.

8. See, e.g, J.R. Barker, 1980, in " Physics of Nonlinear Transport in Semiconductors" (D.K.Ferry, R. Barker and C. Jacoboni, eds), 126, Plenum Press and D.K. Ferry, ibid.

9. A. Sommerfeld, 1956, "Thermodynamics and Statistical Mechanics", Academic Press, NY.

10. K. Thornber, 1978, Solid State Electronics, 21:259.

11. R. W. Keyes, 1958, J. Phys. Chem. Solids, 6:1.

12. See, e.g., H. Frohlich and B.V. Paranjape, 1956, Proc. Phys. Soc. B, 69:21; and C.J. Hearn, 1979, in "Physics of Nonlinear Transport in Semiconductors", (D.K. Ferry, J.R. Barker, and C. Jacoboni, eds.) Plenum Press.

13. H. D. Rees, see, e.g., Institute of Physics Conf. Series # 22:105; and K.W. Gray, J.E. Pattison, H.D. Rees, B.A. Prew, R.C. Clarke and L.D. Irving, 1975, Proc. Fifth Biennial Cornell Electrical Engineering Conf., 215, Cornell University Press, Ithaca, NY.

14. H.L. Grubin and J.P. Kreskovsky, 1983, Surface Science 132:594.

15. J.G. Ruch, 1972, IEEE Trans. Electron Devices, ED-19:652.

16. P.N. Butcher and C.J. Hearn, 1968, Electron Lett. 4:459.

17. G.J. Iafrate, R. Malik, K. Hess and J. Tang (to be published).

18. W.F. Shockley, 1954, The Bell System Telephone Journal, 33:799.

19. M.P. Shaw, P.R. Solomon and H.L. Grubin, 1969, IBM J. Res. Devel. 13:587.

20. R. Fauquembergue, M. Pernisek and E. Constant, 1984, "The Physics of Submicron Structures" (H.L. Grubin, K. Hess, G.J. Iafrate and D.K. Ferry, eds.), 171, Plenum Press NY.

21. R.K. Cook and J. Frey, 1971, IEEE Trans. Electron Devices, ED-28:951.

22. H.D. Rees, 1969, IBM J. Res. Devel. 12:537.

23. H. Kroemer, R.E. Hayes, and R.M. Raymond, 1978, IEEE Trans. Electron Devices, ED-35.

24. R.F. Cooke and J. Frey, 1982, COMPEL, 1,2.

25. P.C. Chao, P.M. Smith, S. Wanuga, W.H. Perkins and E.D. Wolf, 1983, IEEE Electron Dev. Lett., EDL-4:326.

26. R.W.H. Englemann and C. Liechti, 1977, IEEE Trans. Electron Devices, ED-24.

27. H.L. Grubin, D.K. Ferry and K.R. Gleason, 1980, State Electronics, 23:157.

28. R.S.C. Cobbold, 1970, "Theory and Application of Field Effect Transistors", Wiley-Interscience, NY.

29. H.L. Grubin and J.P. Kreskovsky, 1983, SRA Report R930007-F.

30. K.K. Thornber, 1980, J. Appl. Phys. 52:279 and 52:2127.

31. T.J. Maloney and J. Frey, 1977, J. Appl. Phys. 48:781.

32. G. Hill, P. N. Robson, A. Majerfeld and W. Fawcett, 1977, Electronics Lett.

33. D.K. Ferry and H.L. Grubin, 1981, Microelectronics Journal, 12:5.

34. J.W. Harrison, J.R. Hauser, T.H. Glisson and M.A. Littlejohn, 1977, Report No. AFAL-TR-77-129, figure 22.

35. Ibid, figure 26.

36. Ibid, figure 23.

37. Ibid, figure 25.

38. W. Porod and D.K. Ferry, 1983, Phys. Rev. B27:2587.

39. Y.F. Biryulin, S.P. Vul, V.V. Chaldychev, Y.V. Shmartsev, 1983, Phys. Semicond., 17:65.

40. J.W. Harrison, 1972, "Electron Mobility in Alloys of Direct Band Gap III-V Compounds", Ph.D Thesis, North Carolina State University.

41. See, e.g., J.L. Birman, 1962, Phys. Rev 127:1093; and 1963, Phys. Rev. 131:1489.

42. D. Matz, 1965, Phys. Rev. 168:843.

43. J.J. Harris and B.K. Ridley, 1973, J. Phys. and Chem. Solids, 34:197.

44. J.R. Hauser, M.A. Littlejohn and T. Glisson, 1976, Appl. Phys. Lett., 28:458.

45. H. Temkin and V.G. Keramidas, 1980, J. Appl. Phys. 51:3268.

46. J.R. Hauser, T.H. Glisson and M.A. Littlejohn, 1979, Solid State Electronics, 22:487.

47. D.K.Ferry, 1972, Phys.Rev.12, 2360.

48. M.P. Shaw, H.L. Grubin and P.R. Solomon, 1979, "The Gunn-Hilsum Effect", Academic Press, NY.

49. H.L. Grubin, 1976, IEEE Trans. Electron Devices, ED-23:1012.

50. See, e.g., V.L. Rideout, 1975, Solid State Electronics 18:541.

51. W. Shockley, J.A. Copeland and R. P. James, 1966, "Quantum Theory of Atoms, Molecules and the Solid State", (P.O. Lowdin, ed.) 537, Academic Press, NY.

EPITAXIAL GROWTH FOR SUB MICRON STRUCTURES

Colin E.C. Wood

Gec. Hirst Research Centre
East Lane
Wembley, England

INTRODUCTION

From realization of the first electronic quality semiconductor films back in 1970, MBE has developed into a powerful tool for both research into crystal growth processess, doping incorporation phenomena and multilayer structure behavior as well as the most sophisticated technique for preparing all forms of electronic and opto-electronic devices with a precision, control and reproducibility otherwise unobtainable.

This paper attempts to overview very briefly the impact of the development of understanding and capability of MBE on conventional device design and fabrication, and on the evolution of a new series of devices which capitalize on MBE's unique dimensional control. The present limitations both in understanding and capabilities in certain structures and materials are also briefly treated. Suggestions for emphasis of future device and materials research utilizing MBE are included.

BACKGROUND

GaAs was the first semiconductor to be grown by the technique which is now known as molecular beam epitaxy (MBE) followed shortly by GaP and GaAlAs. Improvements of equipment, basic furnace design, vacuum systems, diffusion cells, etc.) (see **figure 1)** has allowed the quality of films to improve with time. Parallel developments in surface analytical techniques: reflection electron diffraction (RED) (see **figure 2)** and Auger electron spectroscopy (AES) (see figure 3) allow substrate surface-cleaning techniques

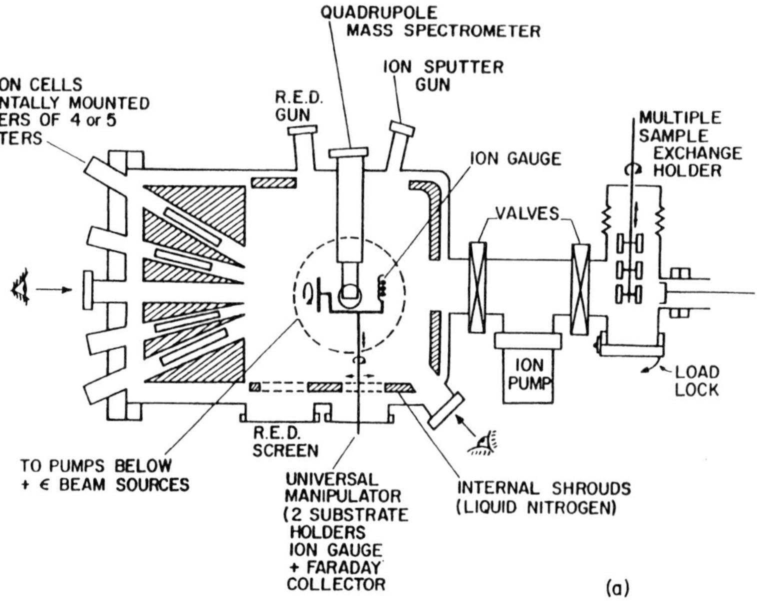

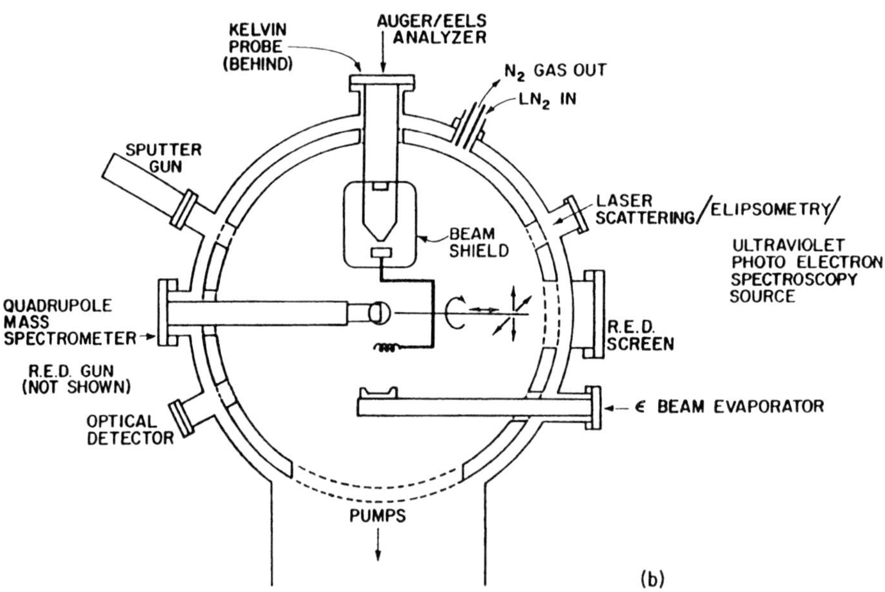

Figure 1. Diagrammatic representation of a typical modern MBE system; (a) plan cross section; (b) vertical cross section.

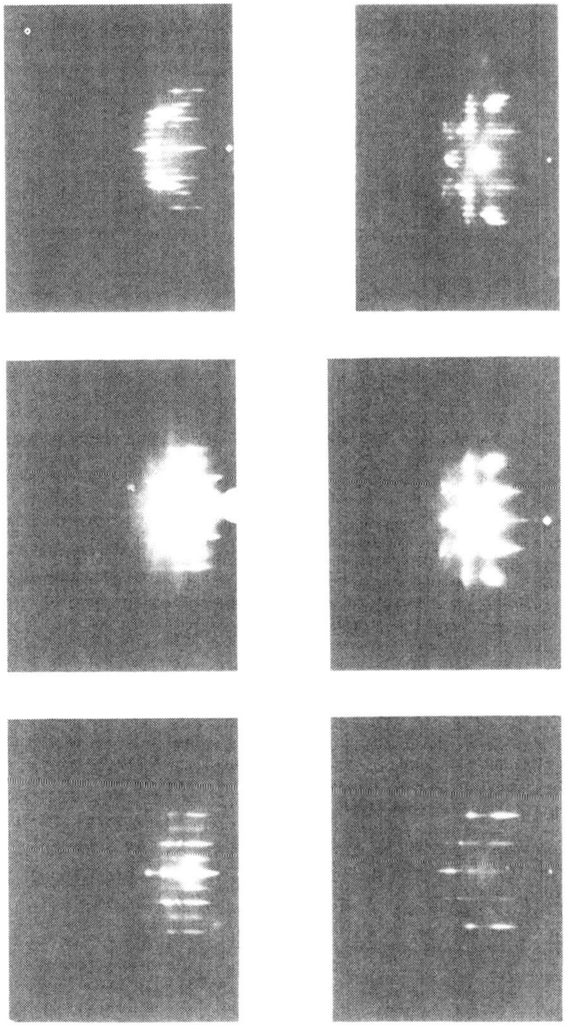

Figure 2. 10 KV reflection electron diffraction patterns of (100) MBE GaAs. Stabilization [100] As-Ga As_4Ga [110] As-Ga As_4Ga.

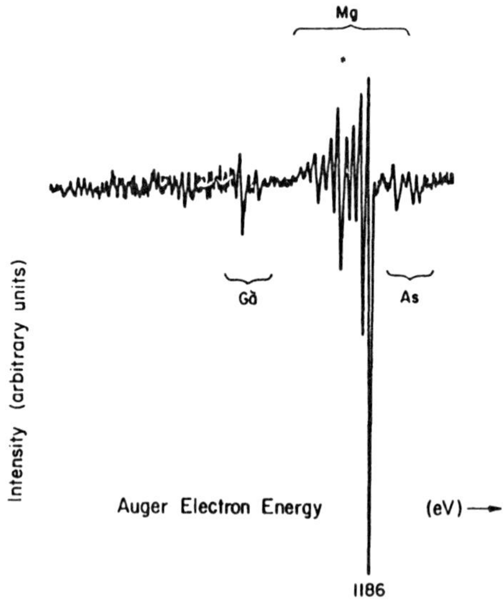

Figure 3. Auger electron spectrum of magnesium deposited on GaAs.

and nucleation, growth and doping conditions to be determined. Other techniques such as laser reflection, raman spectroscopy, electron-enery-loss-spectro-scopy, and secondary-ion-mass-spectrometry have all been used to understand fundamental parameters for epitaxy of alloy semiconductors by MBE.

Similar techniques were later used for II-VI and IV-VI compounds amd more recently for silicon.

Group IV elemental semiconductors (Si and Ge) require electron-beam-evaporation and benefit from ionized doping and is, therefore, not strictly MBE. As an ultra-high vacuum evaporation and growth technique, (See **figure 4)** it is considered a close MBE relative, and as such, is included in this comparative discussion.

ADVANTAGES

Early in the evolution of MBE it was shown that 0.5-2.0 micron-per-hour rates were convenient for study of expitaxial processes, and are now normal for practical device layer growth. At these rates, shuttering effusion-cells allows independent changes in doping level or composition within monolayer depths. It was apparent that improvements could be made in heterojunction abruptness and control of doping profiles over conventional liquid-phase (LPE) or vapor-phase-epitaxy (VPE).

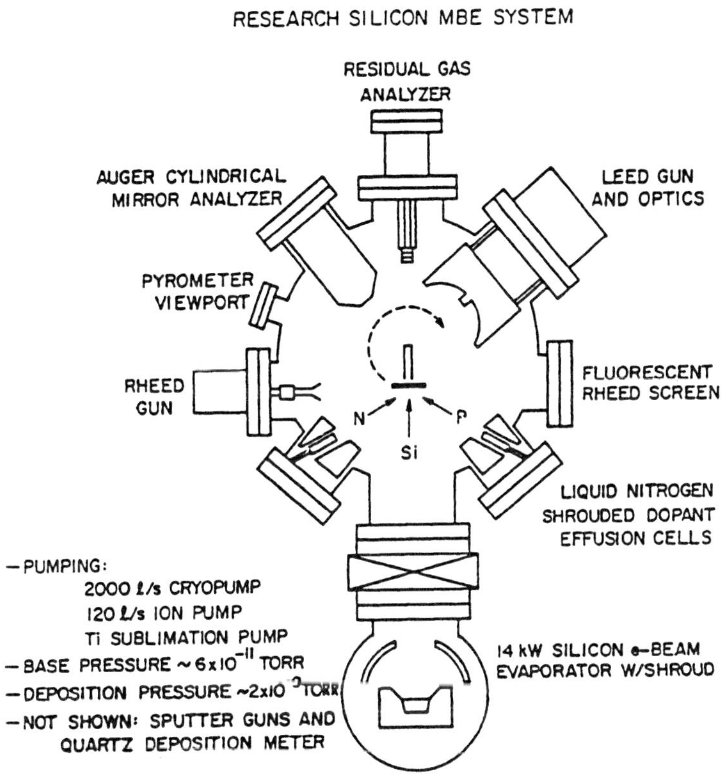

Figure 4. Ultra-high vacuum evaporation and growth technique (from J.C. Bean, Proc. IEDM 1981, p. 7).

More recently the technique of metal-organic chemical-vapor-deposition has been used to demonstrate profile and composition changes within several tens of angstroms, however this technique is still very operator and equipment dependent and relies for its purity and control on the supply of unstable organo-metallic materials.

Using ion-gauges to monitor neutral-beam-fluxes, in conjunction with thickness, doping-level and free carrier profile measurements, growth of films by MBE is now a predictable, controllable and reproducible technique for most thermally stable semiconductor materials.

As device sizes shrink to increase integrated circuit packing densities, thickness and doping control have to be improved. As processing techniques allow the limits of elemental and compound semiconductors to be approached, further improvements require alloy semiconductors and multilayer structures, in which heterojunction discontinuities can be expoited for electron and optical confinement and selectively doped multilayer heterojunction systems.

As vacuum technology develops then features such as continuous azimuthal-substrate-rotation allow increased uniformity of the three important parameters, thickness, carrier concentration, and alloy composition.

Electron beam evaporation of refractory metals and high-temperature-melting or low volatility species has instigated great interest in silicon molecular beam epitaxy. The relatively high temperatures used for silicon deposition has meant that conventional neutral doping-species from effusion cells have relatively low surface-life- times, are thus not incorporated efficiently and severe temperature dependence of doping levels are observed for specific dopants. Low energy dopant-ion-beam implantation during growth has recently yielded highly controlled and reproducible epitaxial films for mixer, tunnet and hyperabrupt varactor diodes.

In all materials studied to date, growth-temperatures are significantly lower than in other techniques, so diffusion of major components across heterojunctions and of impurities is virtually absent.

IMPROVEMENTS IN CONVENTIONAL DEVICES

MICROWAVE AND HIGH SPEED DIGITAL DEVICES

Transferred-electron, field-effect and avalanche devices with highest performance are typically fabricated from III-V compounds, and most commonly from GaAs.

Having demonstrated early that it was possible to grow FETs with reasonable performance and having understood certain dopant in corporation mechanisms, exponential channel doping-levels **(figure 5)** were prepared for transconductance linearity.

Troublesome undoped GaAs buffer layers have been replaced with high-resistivity AlGaAs to exploit the electron confining conduction band discontinuity (ΔE_C) between this ternary alloy and GaAs active layers.

Single-crystal metal Schottky-barrier gates can now also be deposited to avoid troublesome metal-semiconductor interface problems and low-resistance non-alloyed 'superdoped' and heterojunction ohmic contacts have been demonstrated.

Gunn diodes for millimeter wave applications have been prepared by exploiting the abrupt n+n-n+ doping capability of MBE and double drift impatt diodes with unparallelled uniformity and performance are now made on a semi-commercial scale by Varian.

OPTICAL DEVICES

Long-lifetime AlGaAs double-heterojunction lasers (DHLS) with GaAs and AlGaAs active layers are now made by MBE on a semi-routine basis.

High efficiency GaAsSb and AlGaAs light-emitting diodes, negative-electron-affinity photocathodes and PIN diode optical detectors have also been demonstrated.

In the area of optical sources MBE PbSnTe alloys have been successfully fabricated into D.H. lasers and recently IV-VI compounds Cd Te and ZnSe have received much attention in Japanese laboratories and HgCdTe has recently been prepared in France.

NOVEL DEVICES

Using the novel concepts of selectively doped, heterojunctions, **(figure 6)** planar-doped barriers **(figure 7)** and heterojunction superlattices a new generation of devices is now possible. Selectively Doped or Single Dingle 77k FET logic structures are already capable of switching times less than 12pS.

Planar-doped barriers, either simple or associated with heterojunctions, are currently very interesting for a variety of hot electron injection cathodes and optical barrier detection devices.

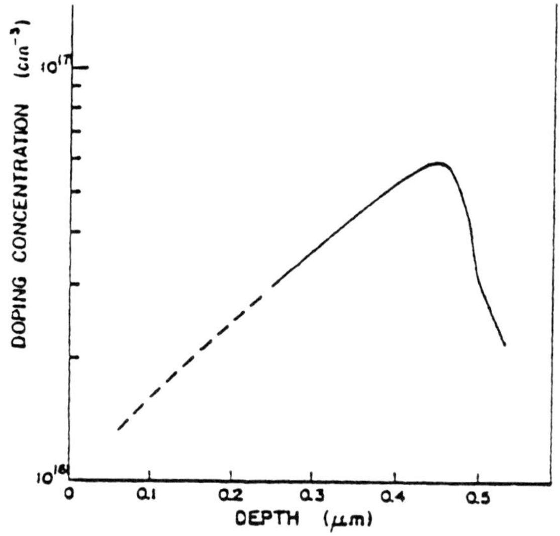

Figure 5a. Predeposition doping profile of a layer used for power FET fabrication; and I_D versus V_G for a power FET,

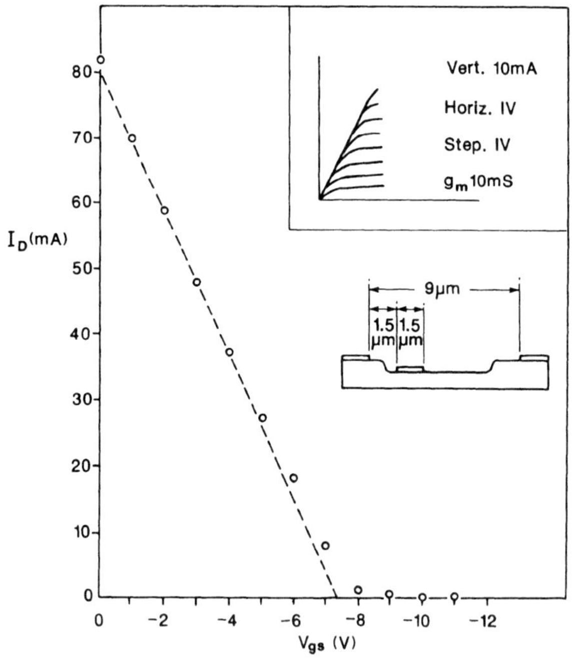

Figure 5b. I_O versus V_G for a power FET made from the layer whose profile shown in figure 7. The deviation at higher values of V_G are caused by a space charge-limited current in the interfacial substrate region. The corresponding I_O-V_O characteristics are shown inserted.

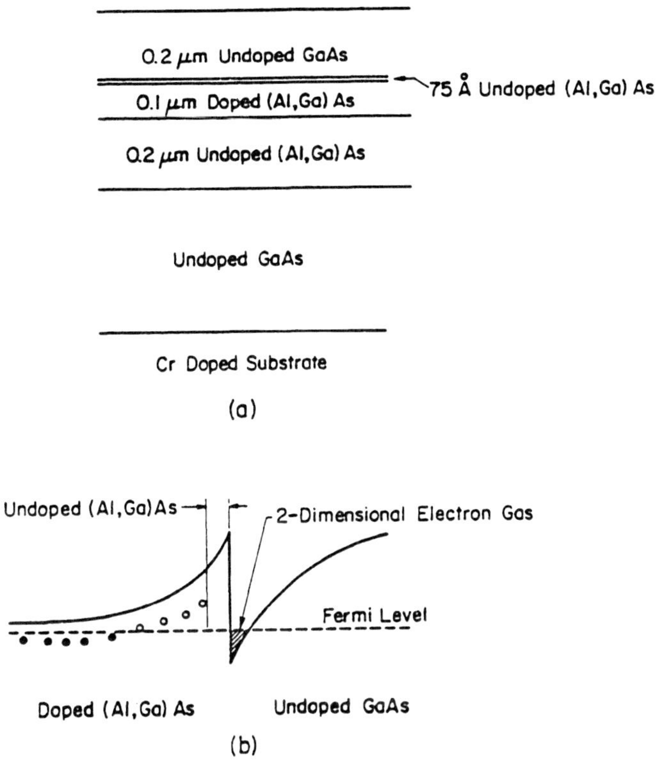

Figure 6. Selectively doped heterojuctions (from T.J. Drummon, H. Morkoc, S.L. Su, R. Fischer, and A.Y. Cho, Electronics Lett., **17**, 870 (1981).

Periodic multilayer structures provided the unpredictable advantages of reducing DH laser degradation mechanisms, reducing threshold currents and increasing conversion efficiency.

When multi-layer heterojunction periods approach 100Å thickness (or less) then sub-bands are formed in the quantum wells and luminescent emission wavelengths can be much shorter than bulk band-to band recombinations **(figure 8).**

The ability to produce abrupt heterojunction interfaces, and to position dopants within a few atomic layers is thus imperative for most of the future generation of III-V microwave and optical devices using hese concepts.

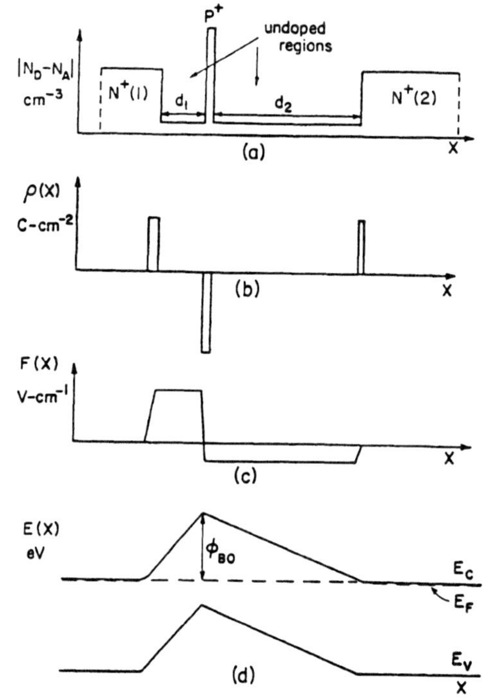

Figure 7a. doping, space cahrage, electric field, and band-energy diagrams of a PDB at zero bias.

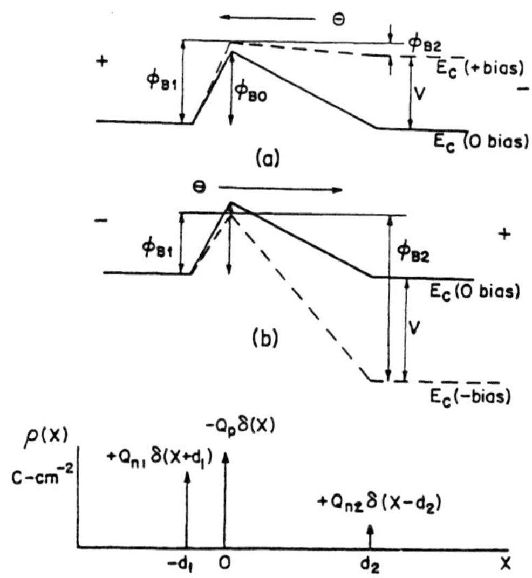

Figure 7b. Barrier height response of positive and negative bias; simiplified atomic plane doping model of a PDB.

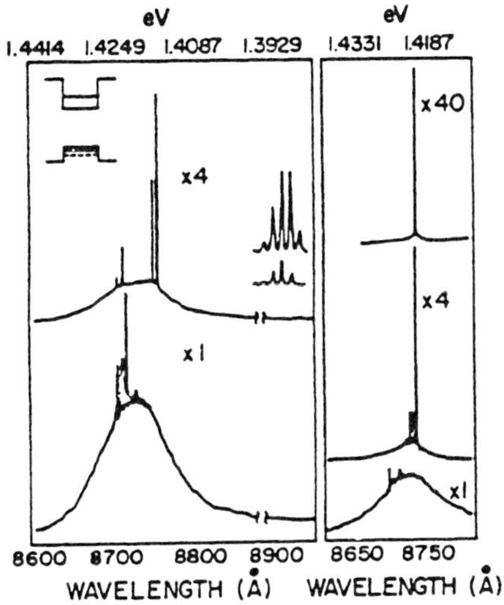

Figure 8a. Lasing spectra for a MQW lser 375 μ m long and 200 μm wide at two current injection levels. **8b.** Single-longitudinal operation of a broad-area MQW laser. (From W.T. Tsang, C. Weisbuch, R.C. Miller, R. Dingle, Appl. Phys. Lett. 35, 673)).

LIMITATIONS AND AREAS FOR PROFITABLE FUTURE RESEARCH

Silicon

As mentioned above silicon MBE is still in its infancy. The effects of secondary electrons, X-rays, ionization and incidential sputtering on the quality of epitaxial films needs to be studied. The number of available dopants, simpler, more robust and inexpensive ion-sources together with larger volume electron-beam evaporation hearths need to be developed.

III-V Compounds

The competition between different group V tetrameric molecules for group V sites during epitaxy of III-V, V' alloys is dependent upon temperature and absolute flux and is little understood. At present, composition-control of these alloys is not good and improvement will depend upon new ideas.

The present quality of alloys containing both Al and In is limited, as the optimum growth conditions (congruent sublimation temperatures etc.) for the binary compounds (eg. AlAs and InAs) are mutually exclusive. Sources that produce group V dimers (crackers) will probably help reduce this problem although device quality problems are still expected.

II-VI Compounds

These componds usually demonstrate a preference for one residual free carrier type. It is therefore very difficult to change the carrier type or form p-n junctions etc. The origins or reasons for this unitype behavior needs study and methods to overcome it are developed.

HgCdTe

As mentioned above, the successful growth of this type of strategically important material demands that techniques be developed to prepare very (volatile) thermally unstable compounds.

Substrates

There is a limited number of compatible (stable, lattice matching, insulating substrates (especially for Si epitaxy) available for epitaxy. One of the most obvious problems is the thermal instability of InP as a fundamental material for epigrowth of InP and GaInAs and AlInAs and AlGaInAs alloys.

Thermal Stability

Thermal decomposition is also a problem in the growth of HgCdTe alloys and other II-VI compounds. Growth system modification (closed) will probably necessary if MBE is expected to prepare this strategic material with control.

Of the above problems the dominant ones are specific to the particular material system that is being considered, however, there are future considerations that are common to all MBE growth systems.

Throughput

It is necessary to increase both wafer size (4" ϕ_{minimum}) and if possible introduce multiple-substrate MBE systems for future industry-compatible production volume.

Cost

System cost is at present too high for production of many consumer device layers. It is necessary that a new generation of stripped down, robust, fast throughput, single purpose machine should be designed for commercial viability.

In conclusion much has been learned about fundamentals and practices of compound and elemental semiconductor epitaxy. Much of the new expertise has led to new ideas for devices not possible with other material growth processes. There are many other discoveries and advantages that are not covered here and doubtless many more to be made. The scope of this paper does not allow detailed or complete coverage. There are also many problems that can at present be identified but no fundamental limitations are yet apparent for all currently projected devices.

For details and further reading on this subject attendees are referred to the following articles:

MBE of III-V Compounds

C.E.C Wood, "Progress, Problems, and Applications of Molecular-Beam Epitaxy" Chapter in Physics of Thin Films, Ed. G. Hass M. Fancome, Academic Press, 1981.

A.Y. Cho and J.R. Arthur, "Molecular Beam Epitaxy", Prog. Solid State Chem. 10, 157 (1975).

K. Ploog, Chapter in Crystals, Properties and Applications.

III-V Alloy

C.E.C. Wood III-V Alloy Growth by MBE; Chapter in GaInAsP Alloy Semiconductors, John Wiley, (1982).

MBE of II-VI Compounds

D.L. Smith and V.Y. Pickardt, J.Appl. Phys. 46, 2366 (1975).

T. Yao, S. Amano, Y. Makwa and S. Maekawa, Jap. J. Appl. Phys. 16, 369, (1977).

J.P. Faurie and A. Million, J. Crystal Growth, 54, 577, (1981) and 54, 582, (1981).

MBE of III-VI Compounds

J.N. Walpole, A.R. Calawa, S.R. Chinn, S.H. Groves and T.C. Harman, Appl. Phys. Lett. **29**, 307 (1976).

Doping of III-V Compounds

C.E.C. Wood, Chapter in Technology and Physics of MBE, Eds. E.H.C., Parker and M.G. Dowsett (Plenum, NY, 1982).

MBE Silicon

J.C.Bean, Proc. Int. Electron Device Meeting, Washington, DC (1981), P.6.

Power FETs

A.Y. Cho, J.V. DiLorenzo, B.S. Hewitt, W.C. Niehaus, W. Schlosser and C. Radice, J. Appl. Physics, **48**, 336, (1977).

Linear Power FETs

C.E.C. Wood, D. DeSimone and S. Judap!rawira, J. Appl. Phys. **51**, 2074 (1980).

Low Noise FETs

S.G.Bandy, D.M. Collins, and C.K. Nishimoto, Electron. Lett. **15**, 218, (1979).

Low Noise Varactor and Mixer Diodes

J.S. Hammis and J.M. Woodcock, Electron Lett., 16, **16**, 319, (1980).

Impatt

W.S. Knodle and P.E. Luscher, Semiconductor International, **3**, (10), 39, (1980).

Heterojunction Buffer Layer

W.I. Wang, S. Judaprawira, C.E.C. Wood, L.F. Eastman, Appl. Phys. Lett. **38**, 708 (1981).

Metal Epitaxy

R. Ludeke, L.L. Chang and L. Esaki, Appl. Phys. Lett. **24**, 417 (1974).

Heterojunction Ohmic Contacts

R.A. Stall, C.E.C. Wood, K. Board and L.F. Eastman, Electron. Lett. **15**, 800 (1979).

mmWave Gunn Diodes

W. Haydl, R. Smith and R. Bosch, Appl. Phys. Lett. **37**, 556, (1980).

D.H. Lasers

W.T. Tsang and J.A. Ditzenberger, e.g., Appl. Phys. Lett., **39**, 193 (1981).

LEDs

A.Y. Cho, H.C. Casey and P.W.Foy, Appl. Phys. Lett. **30**, 397 (1977).

Solar Cells

D.L. Miller and J.S. Harris, Appl. Phys. Lett. **37**, 1104, (1980).

Planar Doped Barriers

R.J. Malik, T.R. AuCoin, R.L. Ross, K. Board, C.E.C. Wood, L.F. Eastman, Electron. Lett. **16**, 836, (1980).

Modulation (selectively) Doped Structures

e.g., S. Hiyamizu, T. Fuji, T. Mimura, K. Nanbu, J. Saito, H. Hashimoto, Jap. J. Appl. Phys., **20**, L455 (1981).

Quantum Well Lasers

P.M. Petroff, Weisbush, et al. Proc. 2nd.Int. MBE Workshop, Cornell Univ. 1980.

Planar Doped Barrier Transistors

See L.F. Eastman in this workshop.

INSULATOR/SEMICONDUCTOR INTERFACES

Carl W. Wilmsen

Department of Electrical Engineering
Colorado State University
Fort Collins, Colorado 80523

INTRODUCTION

Often the insulator-semiconductor interface is modelled as an abrupt change from a crystalline semiconductor to an amorphorous insulator. For some applications this idealized model is satisfactory but for others it leads to incorrect results. Realistic modeling of the insulator-semiconductor interface requires an understanding of the chemical and structural properties of the interface and how it was formed. In this series of lectures, these properties will be discussed for the Si/SiO_2, InP and GaAs/native oxide and the InP/deposited insulator interfaces. These interfaces are both technologically important and typical of the wide variety of possible insulator-semiconductor couples.

The formation and properties of the interfaces formed with Si are less complex than those with InP and GaAs. Even so, the relationship between the chemical, structural and electrical properties of the Si/SiO_2 interface are not fully understood. However, as we shall see from the following, there is sufficient knowledge to provide a fairly detailed picture of the Si/SiO_2 interface while those of InP and GaAs are just beginning to emerge. For both Si and the III-V's the correlation of interface chemistry and structure with interface states is better understood than it is for inversion layer transport. This is largely due to the disproportionate effort applied to the former.

GENERAL OXIDE/SEMICONDUCTOR INTERFACE CONCEPTS

Oxide/semiconductor interface formation begins with the interaction of a gaseous or liquid oxidant with the clean semiconductor surface. The oxidant may first physisorb or chemisorb on the surface before breaking back bonds to form the initial oxide layer. When the semiconductor is composed of more than one element, there may be preferential oxidation of one element over the other and there may also be selective evaporation or dissolution of the initial or subsequent oxide layers. In this regard, the column V elements and their oxides are much more volatile than the column III elements, and the column V oxides are usually more water soluble. However, solution pH and the type of anion can change the solubility significantly. In addition, the oxidation may proceed through a nucleation and island growth process. Thus, the initial oxide layer and possibly the resultant thick oxide/III-V semiconductor interface may be nonuniform in both chemical composition and thickness.

After the initial oxide layer is formed, diffusion of either the oxidant or the substrate species (or possibly both) must take place in order to grow additional oxide. This growth may change the chemistry and structure of the initial interface or the interface may remain essentially unperturbed. The amount of change which occurs in the interface is determined by kinetic and thermodynamic factors. Equilibrium thermodynamics indicates which is the favored final state for the interface composition, while the kinetic factors determine if this state will be reached during a given set of growth conditions. Understanding these factors provides the key to understanding oxide/semiconductor interfaces and how they evolve with increasing oxide thickness, growth method and postoxidation treatment.

Si-SiO_2 Interface

It is instructive to first examine the physical structure of the Si-SiO_2 interface as seen by high resolution transmission electron microscopy (HRTEM), Figure 1, Goodnick et al. (1982). The sharpness of the boundary between the amorphous SiO_2 and the crystalline Si substrate is striking even though further examination reveals some interface roughness on the order of atomic steps with very little, if any fuzzy boundary. This observation leads to a number of important questions and possible conclusions.

First, the abruptness of the interface suggests that bonding changes from crystalline Si to amorphous SiO_2 over one bond length. Using a chemical stick model, Pantelides and Lang (1978) showed that this was possible with only small changes in bond angle, at least for a planar (100) interface. The appearance of atomic steps does not alter the fact that the structure can change phase over a single bond length but it does, however, introduce the possibility of defects such as surface states and/or highly strained bonds. From the HRTEM photomicrographs the atomic steps are seen to occur every few lattice sites. If a surface state were associated with each, then a surface state density in the 10^{13} to 10^{14} cm^{-2} range would result. This is definitely not the case and thus one must conclude that the Si-SiO_2 bonding at the atomic steps seen in the HRTEM micrographs are passivated in some way or the atoms are rearranged in such a manner as to

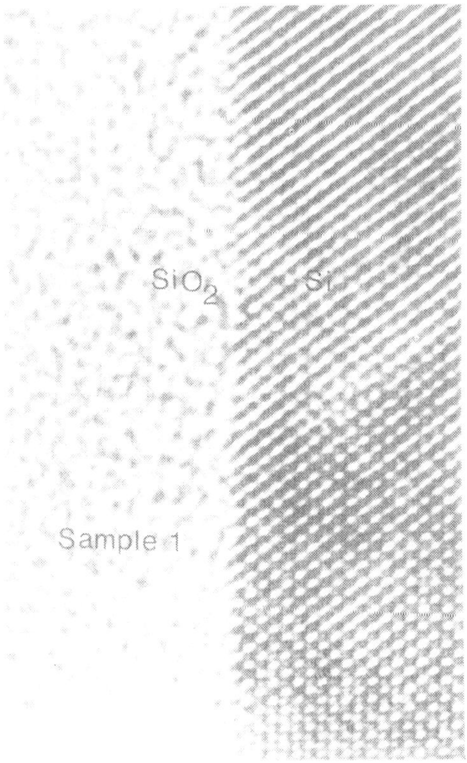

Figure 1. High resolution transmission electron micrograph of the Si/SiO_2 Interface. (Goodnick et al. 1982).

prevent dangling bonds. It is known that a cleaned Si surface has many dangling bonds (Tamm states); each of which contributes an electronic state in the gap. Oxidation of the surface passivates a large number of the states and the surface state density is greatly reduced but not eliminated (Many et al. 1965).

The concept of relating a defect at the Si/SC_2 interface with a surface state was modeled by Poindexter et al. (1981) based on extensive electron spin resonance (ESR) measurements. Surface states are seen in Figure 2 to occur at the equivalent of an atomic step, i.e. one missing Si atom. If the Si/SiO_2 samples are properly annealed, the density of interface states is reduced, Johnson et al. (1980). HRTEM photos of both annealed and unannealed samples are not available, however, the density of atomic steps as seen in these photos probably would not be significantly different! Thus, it could be argued that the atomic steps seen in the HRTEM photo are either an artifact of the technique or have their electrical properties altered by its chemical surroundings, e.g. H on N.

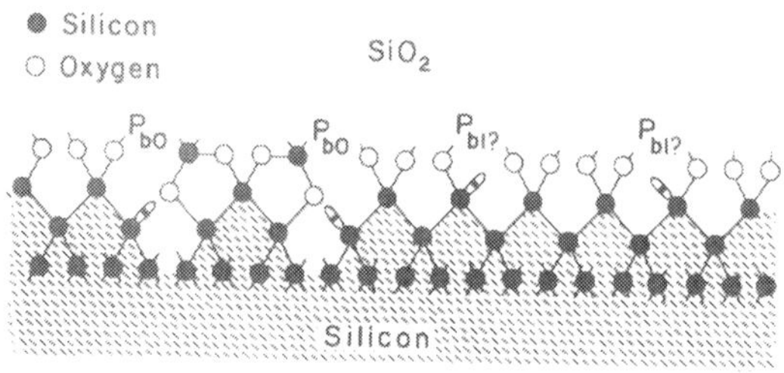

Figure 2. Possible defects at the Si/SiO_2 interface (Poindexter et al. 1981).

In addition to surface states, the atomic steps cause an apparent roughening which is thought to reduce the mobility of inversion layer electrons. There are, however, a number of other factors that can reduce the inversion layer mobility (Moore and Ferry, 1980), including interfacial charge scattering and interface phonons. The average rate of surface roughness scattering has been modeled by a number of researchers who assume a Gaussian distribution for the roughness, Hartstein et al. (1976) and Ferry (1976). This results in an expression containing the rms height, Δ and the correlation length L. Curve fitting allows determination of Δ and L which should be independent of temperature. This, however, is not born out by experiment even though the Δ and L obtained from the HRTEM micrographs are close to that obtained by curve fitting the electrical data, Goodnick et al. (1982). This seems to suggest that the steps seen with HRTEM are indeed scattering sites but that a large number of them are passivated by the attachment of an H atom, for example, which may add a weak temperature dependence to the scattering. This could be due to the vibration of the passivating H, a perturbation of the charge scattering or a change in the phonon coupling to the inversion layer electrons.

While the exact meaning of the HRTEM photos can be questioned, the abruptness of the Si/SiO_2 interface has been verified by a variety of techniques. Chemical profiles of the Si/SiO_2 interface also indicate an abrupt change in bonding. This was first shown by Johannersen and his co-workers (1976a,b) using Auger sputter profiles of thick $Si)_2$ samples. Later, Wager and Wilmsen (1979) applied these techniques to thermally grown oxides in the thickness range 18 to 45A. Their work showed an apparent broadening of the interface with increasing oxide thickness. Since the local interface is seen from the above to be abrupt, the data can be interpreted in terms of a nonuniform growth model, Figure 3, with surface undulations. However, more recent measurements by Taubenbott and Helms (1983) indicate that the sputtering introduces an antifact which confuses the profiles.

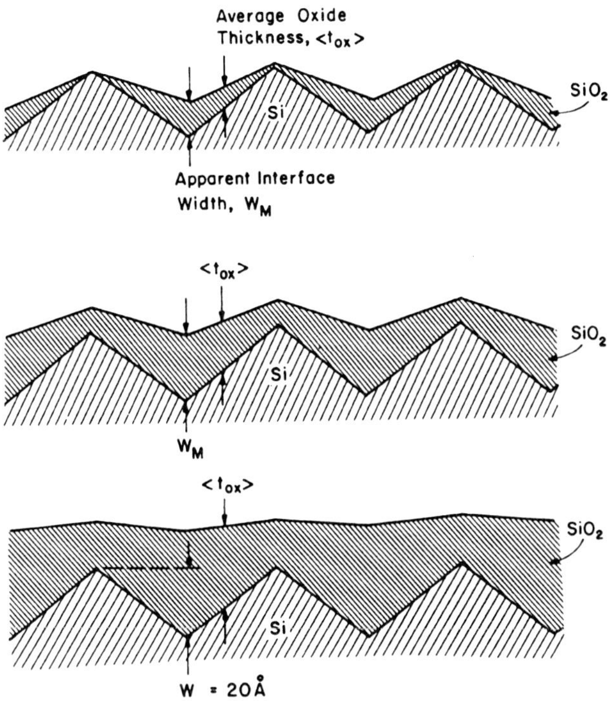

Figure 3. Proposed nonuniform SiO_2 growth on Si (Wager and Wilmsen, 1979).

By comparing the measured Auger lines with standard Si and SiO_2 lines, it was shown that these are the primary bonding states independent of oxide thickness. A possible third state was also isolated by this technique. Grunthaner et al. (1979) also investigated the Si/SiO_2 interface but used a chemical etching method to obtain their XPS profiles. Three bonding configurations, Si^{+3}, Si^{+2}, Si^{+1}, within 0.8 monolayer of the interface were identified.

Figure 4a illustrates a typical energy diagram for the Si/SiO_2 interface indicating an abrupt junction. From the chemical and structural evidence presented thus far, this diagram appears justified. Figure 4b expands the horizontal scale to dramatize this point. The Si/SiO_2 interface is shown linearly graded over 5A while the depletion layer is spread over 1000A corresponding to X_{dmax} for Si doped $N_A - N_D = 10^{17}$ cm^{-3}. The interfacial barrier height is 3.2eV as determined by Williams (1965) with internal photoemission. The barrier rounding was measured by DiStefano (1978), also with internal photoemission. His measurements indicated no change in the SiO_2 band gap to within 4A of the interface. From the previous discussion, this energy band diagram should hold for an SiO_2 thickness of as little as 20A. If a metal is deposited on such a thin oxide, then metal-SiO_2

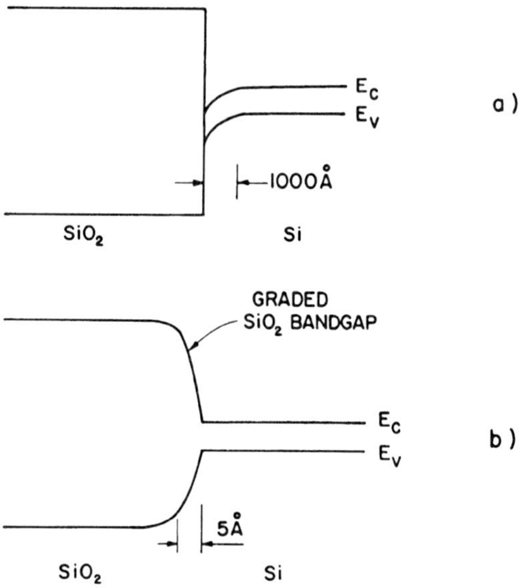

Figure 4. Energy diagram for the Si/SiO_2 interface (Di Stefano, 1978), a) normal scale, b) expanded scale.

reactions can alter this diagram.

In summary, the structure and chemical bonding of the Si/SiO_2 interface changeover is approximately 1 monolayer. Within this monolayer there are possibly three Si-O bonding configurations and atomic steps which cause gap states and a degradation of inversion mobility. Within this monolayer the band gap of the SiO_2 changes from 8 eV to 1.1 eV.

III-V/NATIVE OXIDE INTERFACES

The topic of III-V compound semiconductor/native oxide interface formation and characteristics is more difficult than the previous discussion of Si because there are 2 substrate elements instead of 1. The two elements and their oxides usually have different diffusion, oxidation, evaporation, dissolution and thermodynamic characteristics and there seems to be a constant competition between the elemental oxides that adds to the complexity of these interfaces. Even though it is difficult to generalize about the III-V/native oxide interfaces, the rules which determine the physical and chemical characteristics are the same for all of the substrates.

In this section, these rules are discussed using GaAs and InP as examples. These materials are well suited for this task since their oxides have different thermodynamic properties. In addition, GaAs and InP are the more important device substrate materials.

Thermodynamic Properties

The oxidation of an element normally yields one stable oxide. A good example of this is the oxidation of Si which yields SiO_2 except at the interface where several Si-O bond coordinations have been observed. These interfacial bonding states result from interfacial strain and bond coordination problems and should not be reviewed as stable bulk oxides. For other elements, such as Fe and Cu there are two stable bulk oxides. When there is another element present, as with GaAs and InP, then the equilibrium chemical composition of mixed oxides depends upon possible reactions between the oxides. At the interface, continued reaction with substrate is also possible. Equilibrium thermodynamics indicates which set of compounds is more stable for a given situation. For the In-P-O and Ga-As systems the thermodynamics can be most easily summarized in a phase diagram, Schwartz (1983), such as those shown in Figure 5. Solid tie lines connect two compounds which do not react with each other. Unconnected compounds can react if the kinetic barrier can be overcome. From the phase diagrams it is determined that under equilibrium conditions, InP will oxidize to form $InPO_4$ and GaAs will form Ga_2O_3 plus elemental As, As_2O_3 and $GaAsO_4$. As seen later, kinetic factors such as the chemical barrier and diffusion rate often prevent the system from reaching the equilibrium state. The kinetic factors affect both the interfacial and the bulk oxides and while the interface is of primary concern to the present discussion, the bulk oxide must also be considered.

Thermal Oxide InP and GaAs Interfaces

InP. The initial oxide layer on InP is the equilibrium product, $InPO_4$. However, as the $InPO_4$ layer thickens, the diffusion of P becomes increasingly difficult, Wager et al. (1983). As a result, elemental P remains behind at the oxide/InP interface and the bulk oxide becomes P deficient, Wilmsen (1976), resulting in a bulk oxide composed of a mixture of $InPO_4$ and In_2O_3. Evidence for the collection of elemental P at the interface is shown in the XPS profile of Figure 6. The elemental P may affect the physical and electrical properties of the interface in several ways. First, the P could change the shape of the interfacial oxide barrier. The P can also coalesce into islands and act as inversion layer scattering sites and induce nonuniform barrier height change. At high temperatures, T - 625^oC, the P vaporizes and physically lifts the oxide from the surface and since the oxide begins to soften near this temperature, bubbles such as those shown in Figure 7 are formed, McLaren et al. (1983).

The In_2O_3 that forms in the oxide layer as a result of the slow diffusion of P appears to create a tunneling site which traps inversion layer electrons. This trapping reduces the inversion density

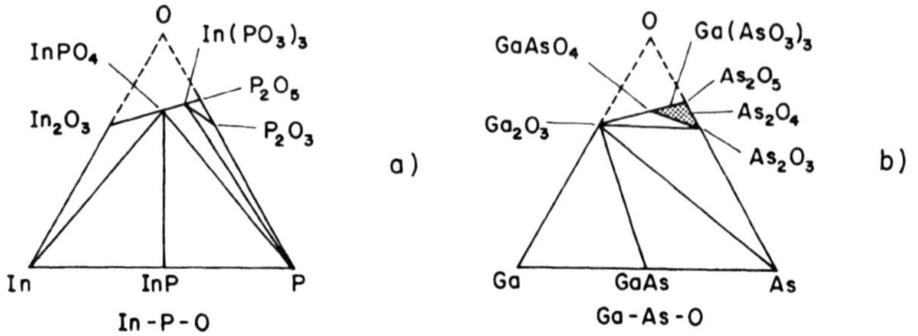

Figure 5. Ternary phase diagrams for a) In-P-O, b) Ga-As-O (Schwartz, 1983).

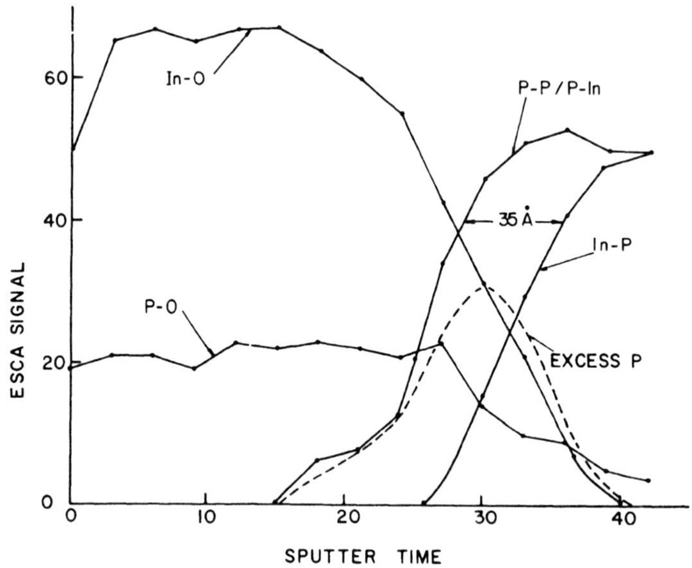

Figure 6. ESCA profile of a thermal oxide grown on InP at 400°C.

over a period of time and thus the MOSFET characteristics are unstable, Lile and Taylor (1983).

A full understanding of the effects of a thermal oxide at the deposited insulator/substrate interface are not known; however, from this discussion it would seem that an interfacial native oxide is not desirable, and indeed this is the case if the oxide is too thick or is formed at too high a temperature. However, the oxide does seal the surface and prevents substrate evaporation.

GaAs. The thermal oxidation of GaAs leads to the collection of elemental As at the interface, however, the cause is more thermodynamic than kinetic. As seen from the phase diagram in Fig. 5, both

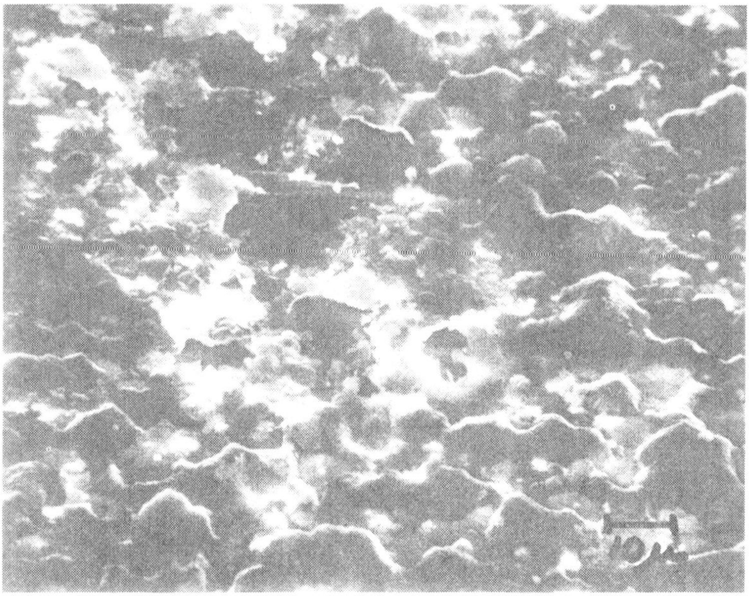

Figure 7. Scanning electron micrograph of thermal oxide of InP grown at 650°C. (McLaren, et al, 1983).

As_2O_3 and $GaAsO_4$ are possible oxidation products, Schwartz (1983). Due to a reaction with the GaAs substrate, these two compounds decompose to form Ga_2O_3 and As. The interfacial layer is thus composed of Ga_2O_3 and elemental As. The outer layers can contain Ga_2O_3, As_2O_3 and/or $GaAsO_4$ since they are not in physical contact with the GaAs and therefore cannot react. However, the As_2O_3 is highly volatile and most of it evaporates from the oxide film. The elemental As at the interface behaves similarly to the elemental P in that it can form islands and cause nonuniformities. The effect on inversion layer transport is much less important because interface states prevent inversion layer formation except under vary fast pulsed operation.

Anodic Oxide InP and GaAs Interfaces

InP. In the anodic oxidation process, a very strong electric field appears across the oxide and the kinetic barriers to diffusion are greatly reduced. Elemental P can now easily move through the oxide and does not collect at the interface. However, there is not sufficient thermal energy to overcome the chemical barrier to the formation of $InPO_4$ and the anodic oxide is composed of In_2O_3

and P_2O_5, not $InPO_4$ and In_2O_3 as is the case with thermal oxides, Wilmsen (1981). Both In_2O_3 and P_2O_5 are soluble in aqueous solutions and significant amounts of the growing film are dissolved. The solubility of In_2O_3 is strongly pH dependent, Laughlin and Wilmsen (1980), which provides a means of varying the P_2O_5/In_2O_3 ratio of the oxide film as shown in Figure 8. However, note that the P_2O_5/In_2O_3 ratio is constant throughout the interfacial region and is independent of the oxide thickness.

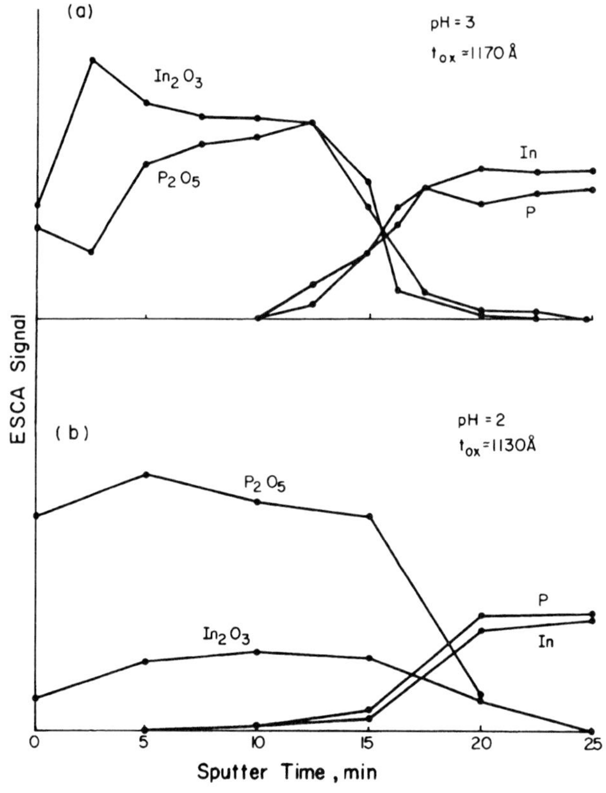

Figure 8. ESCA profiles of anodic oxides grown on InP a) pH = 3, b) pH = 2 (Laughlin and Wilmsen, 1980).

The anodic oxide on InP is initiated by a nucleation and island growth stage, Makky et al. (1983), as illustrated by the TEM photomicrographs of Figure 9. These islands are seen to be smooth and droplet-like. Their coalescence does not appear to result from surface diffusion but rather from accelerated growth in the neck region surrounding the touching islands. These islands show no sign of significant dissolution and this may be the reason for the uniformity of the anodic oxide/InP interface. Even though dissolution does not strongly effect this interface, the incorporation of water and impurities from the electrolyte occurs. The water causes the resistivity of the oxide to decrease considerably and the hysteresis, i.e., trap density to increase. The water can be removed by annealing.

The bulk oxide and probably the interface too is a phase separated structure composed of In_2O_3 and P_2O_5 islands. This strongly effects the insulating quality of the oxide and may introduce a large lateral variation in the interface potential. Annealing the anodic oxide partially converts the layer to $InPO_4$ but no elemental P is formed. There are, however, many changes in topography, Fathipour et al. (1983).

GaAs. The interface that forms between GaAs and its anodic oxide is notibly different than with InP. This is clear from the islands which characterize the initial growth (Figure 10), Makky et al. (1982). These islands are seen to be very rough and pitted when compared to those of InP. The pitting may be an indication of oxide dissolution. There is also a larger variation in size and a sharper boundary between coalescing islands.

Figure 9. Anodic oxide islands on InP (Makky and Wilmsen, 1983)

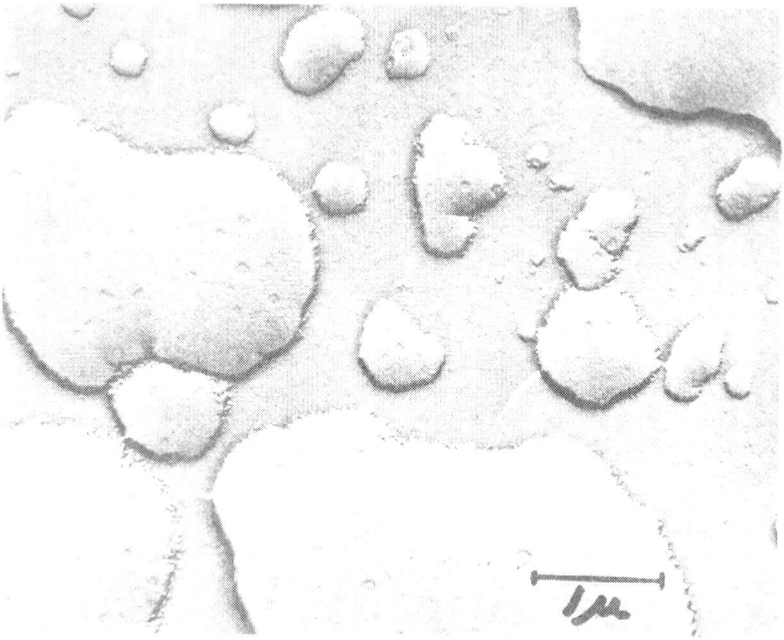

Figure 10. Anodic oxide islands of GaAs (Makky et al, 1982).

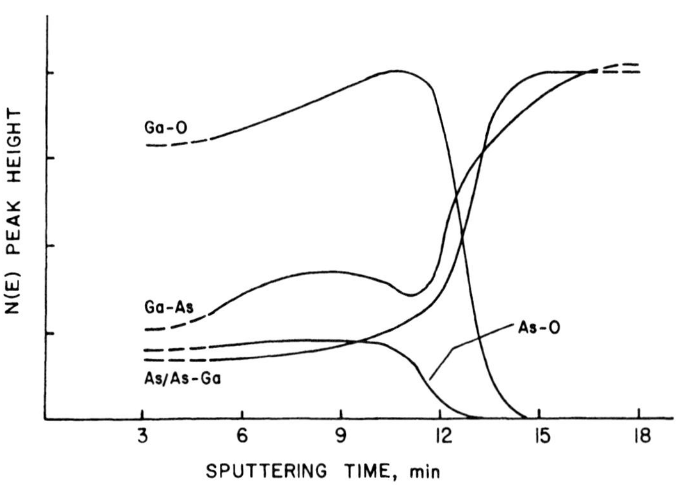

Figure 11. ESCA profiles of a thick GaAs anodic oxide (Wilmsen and Kee, 1978).

The composition of the interface as seen in the XPS profiles of Wilmsen and Kee (1978), in Figure 11, also indicates a departure from that observed on InP. On GaAs, an inner layer of Ga_2O_3 is found without the presence of As_2O_3. The XPS profiles of very thin layers, Figure 12, Geib and Wilmsen (1980), show that this inner Ga_2O_3 layer forms during the island stage. There is also some evidence for elemental As at the interface of these thin layers but if it exits, it is probably removed by the high electrical field once the islands form a continuous oxide layer. The bulk oxide is a one-to-one mixture of Ga_2O_3 and As_2O_3 which contains some water and probably traces of the electrolyte. While the structure of the oxide has not been extensively investigated, the oxide in known to be amorphous and is thought to be a homogeneous mixture of the two oxide components. This later characteristic is in contrast with the phase separated structure of the InP anodic oxide.

Annealing of the anodic oxide has little effect until water begins to evaporate at 300°C. At 450°C, As_2O_3 begins to evaporate and initiates the conversion of the amorphous mixed oxide into crystalline β Ga_2O_3. At this temperature, the As_2O_3 also decomposed at the interface to form elemental As which forms a granular, Ishii and Jeppsen (1977).

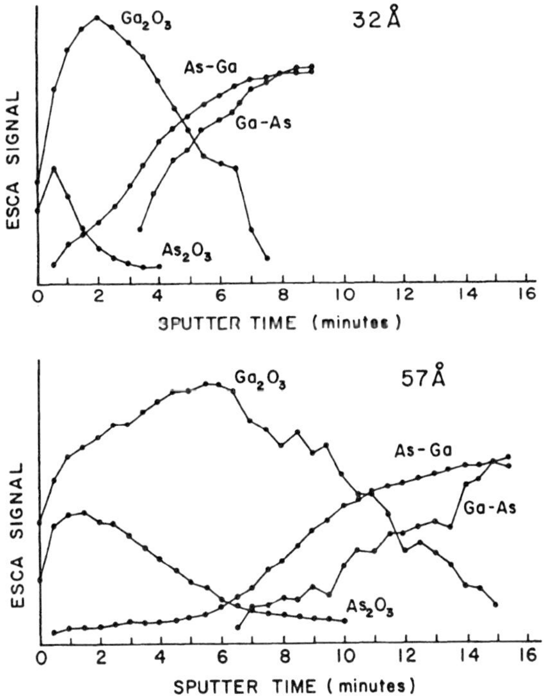

Figure 12. ESCA profiles of thin GaAs anodic oxides a) t_{ox} = 32A, b) t_{ox} = 50A. (Geib and Wilmsen, 1980).

Plasma Oxide

InP

On InP, a plasma oxide will not grow thicker than a few tens of angstroms unless the substrate is heated, Clark and Fok (1981) and Wager et al. (1982b). The heating, however, causes the growth of a thermal oxide and is not a true plasma oxide. The plasma oxide itself is composed of $InPO_4$ but the addition of heat causes the formation of In_2O_3 which mixes with the $InPO_4$ in a similar fashion as the thermal oxide. While the presence of elemental P at the interface has not been detected, it is presumed to exist for oxides grown at T - 300°C.

GaAs

Deposited Insulator/InP Interfaces

While native oxides can be easily grown on the III-V compounds, most of them are mechanically soft, absorb water from the atmosphere and have only marginally acceptable insulating properties. Thus, technologically viable MOSFET devices are fabricated by depositing an insulating film, usually SiO_2 or Al_2O_3. This immediately suggests difficulty since it is known that devices with deposited insulators on Si have a much higher interface state density than do those with a thermally grown oxide. The difficulties are even greater for the III-V compounds as indicated by the following list, Wilmsen et al. (1979, 1980):

1. Surface decomposition and evaporation at elevated temperature.

2. Oxidation in the deposition chamber.

3. Interdiffusion of the insulator and substrate atoms.

4. Reactions between the substrate and the deposited atoms.

Each of these will be discussed later.

Cleaning of the III-V surfaces is also a major problem of concern. In the manufacture of Si MOSFET devices, industrial processing requires very elaborate surface cleaning procedures which are designed to remove organic contaminates and heavy and alkali metals. The sequence of cleaning steps takes about an hour to complete. The very severity of the Si cleaning process suggests the importance of contamination in determining the interfacial properties. Since such a procedure has not been worked out for the III-V's, it is not surprising that improvement in device performance has been slow.

In most cases, the III-V surface is chemo-mechanically polished with Br-methanol and stored in air until ready for use. The wafers are then dipped in a cleaning agent such as Br-methanol or HF and rinsed in H_2O just prior to deposition. In some processes in situ cleaning such as etching with HCl vapor is also performed. With this knowledge of the initial state of the surface, the above 4 factors are discussed for the SiO_2/InP interface.

Surface decomposition -- the works of Bayliss and Kirk (1976) and Clawson et al (1979) have shown that, when heated in vacuum, the bare surface of InP decomposes and loses phosphorous which causes the surface to become very rough and In rich. On the other hand, when heated in an oxygen atmosphere, the surface readily oxidizes if the temperature is sufficiently high. These results are summarized in Fig. 13. Note that evaporation from the bare surface and thermal oxidation have the same temperature dependence. From a device fabrication stand point, this curve indicates that the deposition temperature should not be greater than 350° unless a thermal oxide at the interface is deemed advantageous.

Silicon wafers are also known to degrade due to surface evaporation when heated in a nonoxidizing atmosphere. When the Si surface is oxidized, the SiO_2 seals the surface and prevents thermal etching. Similarly, an oxide can help seal the InP surface, however, thermal or plasma grown oxides can cause interface instabilities and thus they do not provide a suitable MOS interface. The thin air grown oxide does not appear to introduce interface traps. With an air grown oxide, the InP surface does not evaporate below 550^oC and thus this oxide provides a seal for the surface, Wager et al. (1981).

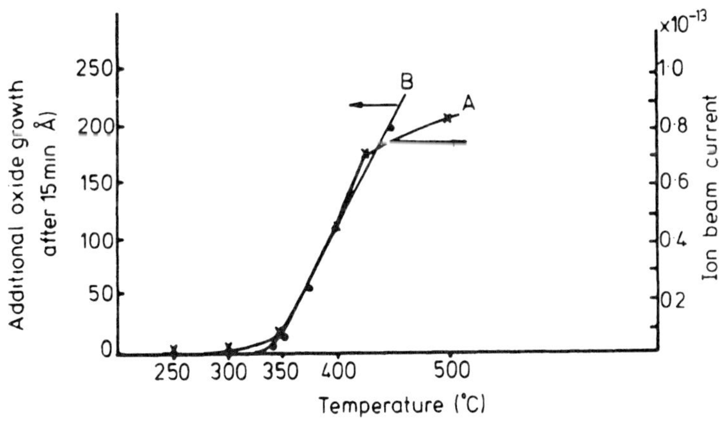

Figure 13. Thermal evaporation of P from the InP surface (Bayliss and Kirk 1976; cuve A, right-hand ordinate) and the thickness of a thermal oxide layer after 15 min. (Wager and Wilmsen 1979· cuve B, left-hand ordinate) plotted against temperature.

Oxidation -- Oxidation of InP has previously been discussed in detail, however, several points need to be emphasized with regard to SiO_2 deposition on InP. First, the plasma oxide grows faster at all temperatures than does thermal oxides, Wager et al. (1983), however, the plasma oxide contains less In_2O_3. In addition, the interfacial oxide caused by thermal or plasma oxidation is always present independent of pre or in situ cleaning procedures. Thus, the oxygen in the deposition chamber unavoidably reacts with the InP.

Interdiffusion -- From the growth of InP crystals in quartz crucibles, it is known that elemental silicon enters the InP and resides on In sites, Mullin et al. (1974), Popov (1976), Baumann et al. (1976). This reaction does not appear to have been investigated in as much detail as has the GaAs--SiO_2 reaction. Cochran and Foster (1962a,b) reported that the following reaction for GaAs proceeds with a value of ΔH 170 kcal mol^{-1}:

$$4Ga_{(from\ GaAs)} + SiO_2 = Si_{(into\ GaAs)} + 2Ga_2O.$$

A similar reaction is expected to occur between InP and SiO_2. This reaction does not involve any In-Si or In-SiO_2 compound formation but instead leaves the In or In oxide free to diffuse through the SiO_2. Indeed Grove et al. (1964) reported that In diffuses very rapidly through thermally grown SiO_2. It has recently been reported by Jain et al. (1978) that Ga diffuses through SiO_2 Ga_2O; thus it may be expected that In will be found in the CVD SiO_2 in an oxidized state.

An XPS profile for a sample fabricated by depositing SiO_2 onto InP at 450° is shown in Fig. 14a. The profile is spread out because of the nonuniformity of the SiO_2 and the large (about 2 mm) area of the x-ray beam. Nonetheless, certain points are clear; there is a thick native oxide layer at the interface and there is In_2O_3 distributed throughout the SiO_2 layer. There is, however, no detectable P in the SiO_2. This establishes that a thermal oxide can grow readily in a CVD reactor and that In can diffuse through the SiO_2 layer (apparently as an oxide) while the P does not diffuse significantly.

The XPS profile (Fig. 14b) of a sample fabricated at 340°C reveals a similar result. The profile indicates that In is distributed throughout the SiO_2 layer (including the surface) at a concentration of about 1%, the In concentration increases sharply approximately 50A before the first detectable P signal and there is no detectable P in the bulk of the SiO_2. Not shown explicitly in Fig. 14b are the bonding states of In and P. No P oxide was observed in the SiO_2 or at the interface; however, the major portion of the In in the SiO_2 layer and at the interface is bonded to oxygen. These data further emphasize the fact that

In diffuses readily through SiO_2 whereas P does not. The lack of a native oxide layer at the interface indicates that there was no significant growth of thermal oxide. This is in keeping with the thermal oxidation rates of InP at 340°C.

Reactions -- There are several techniques for depositing SiO_2; thermal evaporation, sputtering, chemical vapor deposition (CVD) and spin on liquid suspensions. All evaporation and sputtering and some CVD methods result in a partial pressure of SiO_2 and/or Si. These often impinge on the surface where they can react with the InP or the native oxide. The thermodynamic calculations summarized in Table I indicate the possible results, Wager (1982c).

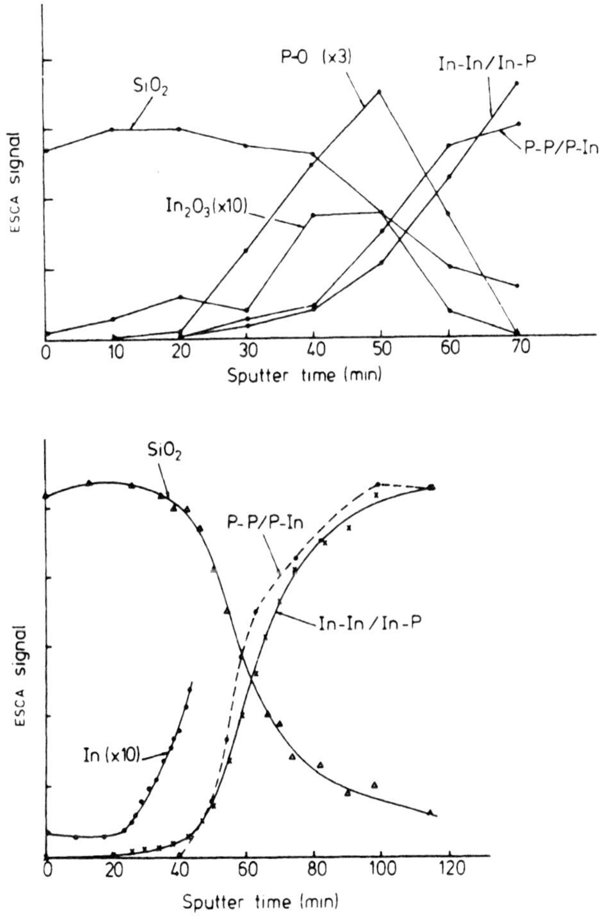

Figure 14. ESCA profile of 200A SiO_2 deposited on InP at a deposition temperature of a) 450°C, b) 340°C, (Wilmsen et al, 1980).

Table I. A Summary of the Most Stable Reaction Product for a Given Set of Reactants

Reactants	Most Stable Products
$InP + SiO_2$	stable
$InP + SiO + O_2$	$SiO_2 + InPO_4$
$InP + SiO$	stable
$InP + Si + O_2$	$SiO_2 + InPO_4$
$InP + Si$	$SiP + In$ or $Si + InP$
$InPO_4 + SiO_2$	stable
$InPO_4 + SiO$	$SiO_2 + InP$
$InPO_4 + Si$	$SiO_2 + InP$
$In_2O_3 + SiO_2$	stable
$In_2O_3 + SiO$	$SiO_2 + InO$
$In_2O_3 + Si$	$SiO_2 + In$
$P_2O_5 + SiO_2$	stable
$P_2O_5 + SiO$	$SiO_2 + P$
$P_2O_5 + Si$	$SiO_2 + P$

Note that while there are often possible reactions, the resulting products do not always appear to cause difficulty, e.g., $Si+InPO_4$ SiO_2+InP. Those that form In_2O_3 or elemental P or In are probably the most damaging to the interface.

Interface Traps

Deposited insulators have been successfully used by a number of laboratories to fabricate MISFETs and MISFET integrated circuits on InP. While these devices can operate at high speed, the drain current, I_D, has been found to decrease with time after a voltage step has been applied to the gate. In some devices I_D decreases very slowly while others have a very rapid decay as shown in Fig. 15. The cause of the current decay has been attributed by Lile (1983) and Okamura (1980b) and their co-workers to a trap slightly above the InP conduction band edge indicated in the band diagram of Fig. 16.

Lile and Taylor (1983) have found that their experimental results on the current drift can be fitted to the following two empirical equations:

$$I(t) = K_1 \ln(t) + C_1 \quad (1)$$

$$\ln (K_1) = K_2/T \ . \quad (2)$$

These combine to yield:

$$I(t) = Ke^{K_2/T} \ln(t) + C_1 \quad (3)$$

where K, K_1, K_2, and C_1 are parameters obtained from plots of the experimental data. The data suggest a single trap level located in the dielectric layer and above the conduction band minimum, E_c. When a step voltage is applied to the gate of the MISFET, some of these electrons are trapped by a thermally assisted tunneling process. The trap density was calculated to be of the order of $10^{17}/cm^3$ and to lie approximately 16 meV above E_c.

Okamura and Kobayashi (1980a,b) had previously adapted the two trap models of Koelmans and DeGroff (1967) to fit their Al_2O_3 MISFET results. One trap is placed below E_c and the other above. The resulting equation for the drain current is:

$$I(t) = 1 - \frac{N_t}{N_o^a} e^{-2E_t/kT} \ln(\frac{t}{t_o})$$

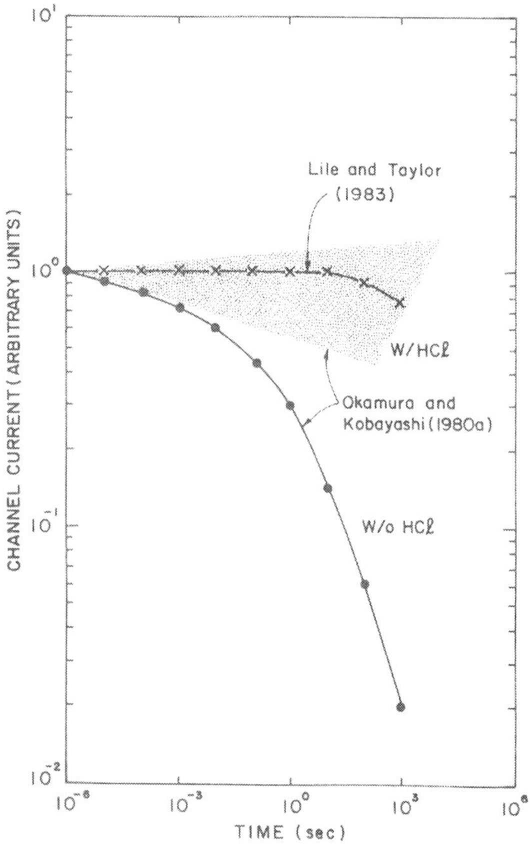

Figure 15. Inversion channel current versus time after application of a gate voltage (Lile and Tayler, 1983).

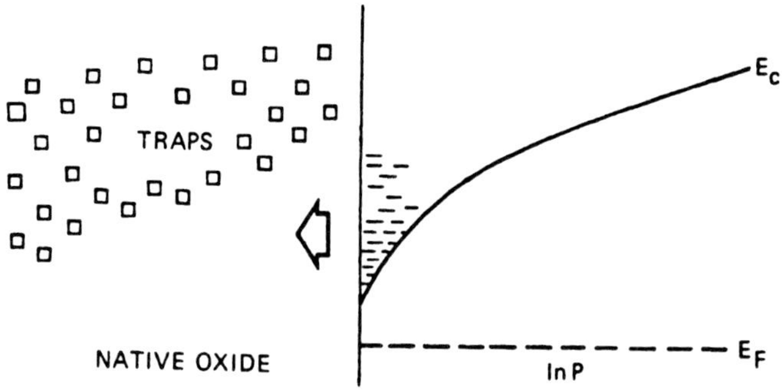

Figure 16. Energy diagram showing traps in the gate insulator (Wager et al, 1983).

There are problems with both of the above formations since they do not describe the current over the full range of time and temperature, nor do they fit samples tested, e.g., in same samples, I_D increases with time. These problems notwithstanding, there appears to be a thermally activated process which causes a loss of inversion layer elections.

It is desired to formulate a physical model for the cause of the interfacial traps. It should be made clear that these states are not located in the InP but rather in the insulator or the transition region between the insulator and the InP. From previous discussions on interface formation and additional information on the insulating films there are several possible causes for the traps:

*The interfacial native oxide

*In diffused into the insultor

*Defects in the insulator

*Elemental P

Table II summarizes, in a simplified form, the electrical stability of InP MIS structures which were fabricated with a number of insulator materials, primarily SiO_2 and Al_2O_3, Wilmsen et al. (1983). The simplified classification of the interface stability as 'high' or 'low' is chosen because of the variety of ways in which the various authors have presented their data. Obviously, the classification is subjective but hopefully we have not biased the data to fit a preconceived model. Also included in the table are the insulator formation temperature and some statement concerning the possible interface chemistry. For the latter, the main concern is the presence and composition of a native oxide and how it was formed. We are also interested in any inter-diffusion and possible reactions between the insulator and the oxides. Again for some of the structures there is uncertainty about the nature of the interfacial chemistry and an intelligent guess must be used.

From Table II it is seen that:

1. There is no correlation between the existence of an interfacial oxide and the electrical stability of the completed MIS structure.

2. Insulator deposition at 340°C or above leads to a high trap density interface, although low deposition temperature does not assure a low trap density interface.

The observation that an interfacial oxide in and of itself is not the direct cause of the electrical instability is very important since there has been speculation to the contrary. The best example of a MISFET structure which has an interfacial oxide but which does not show rapid I_D decay, is the plasma enhanced CVD SiO_2 layers on InP fabricated by Meiners (1981) (line 8 of the table). We have extensively investigated the interface of this type of device and have shown that a plasma oxide exists at the interface for all surface preparations and deposition conditions, Wager and Wilmsen (1982a). The thickness of the interfacial oxide has been shown to decrease when the plasma is remote from the deposition chamber, but it is always present. The Langmuir-Blodgett films (line 1) of Vincett and Roberts (1980) and the prethermally oxidized Al_2O_3 deposited samples of Okamura and Kobayashi (1981) (line 4) also emphasize the fact that the native oxide is not the direct cause of the traps.

The inert Langmuir-Blodgett (L-B) films were deposited in a very gentle way at low temperature. Thus, the oxide which grows in air and in the water bath before deposition is not disturbed by induffusion or strong interfacial reactions. Unfortunately, measurement of the electrical stability has not bee reported in the same manner and detail as has been done by Lile and Taylor (1983) on the PECVD SiO_2 samples and by Okamura and Kobayashi (1980a) on the CVD Al_2O_3 devices. Thus, a true comparison cannot be made even though a low surface state density appears to result from the L-B technique.

From the preceeding discussion it is clear that the presence of an interfacial oxide does not necessarily mean a high interfacial trap density and that In_2O_3 could be the cause of the trap. It would appear that the cause of the traps is complex and is not simply due to the presence or absence of a native oxide at the interface. To date only the $In_2O_3/InOP_4$ native oxide levels have been measured, Wager et al., (1983). As discussed below, these indicate that the In_2O_3 can create a state suitable for tunneling from the conduction bond. Thermal and plasma grown oxides have been shown to have In_2O_3 incorporation into their outer surface.

Figure 17 is an approximate energy band diagram for a native oxide of InP which has In_2O_3 at its outer surface. The In_2O_3 bandgap and energy separation between the vacuum level and the valence band maximum ($E_{vac}-E_{vbm}$) are those obtained in recent work. It is seen that the energy separation between the conduction band minima of In_2O_3 and InP is very small (0.1 eV). Thus, tunneling of electrons from InP to In_2O_3 appears to be a likely mechanism responsible for electrical interface instabilities.

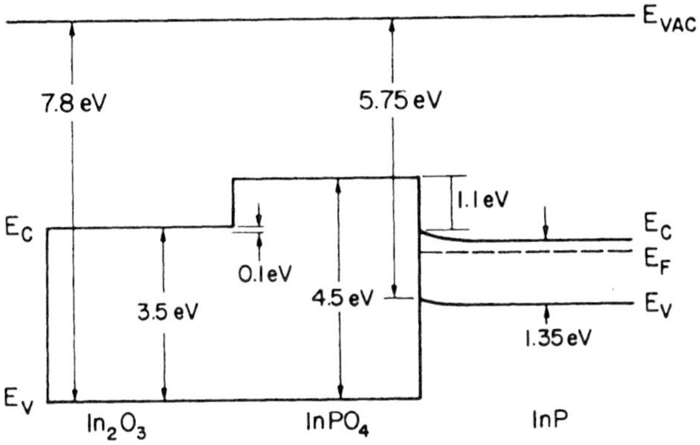

Figure 17. Energy diagram for a native oxide on InP (Wager et al., 1983).

Table II. Interfacial Properties of MIS Structures on InP

Technique for Insulator Deposition/ References	Temperature of Deposition	Qualitative Trap Density	Type of Interfacial Oxide	Remarks
1. Langmuir-Blodgett/ Vincett & Roberts, 1980	20°C	Low	*AGO	The oxide would not be disturbed
2. CVD Al_2O_3/ Okamura and Kobayashi, 1980a	330°C	High	AGO	Possible Reaction
3. CVD Al_2O_3 within situ vapor etch/ Okamura and Kobayashi, 1980a	330°C	Low	*NO	It is possible that some native oxide exists
4. CVD Al_2O_3 on thermally oxided InP/ Okamura and Kobayashi, 1981	330°C	Low	*TO	Probably a layer of elemental at the interface
5. CVD SiO_2/ Fritzsche, 1980	340°C	High	TO	
6. CVD SiO_2/ Stannard, 1979	340-370°C	High	TO	
7. CVD SiO_2/ Messick Personal Comm.	300-325°C	Low	TO	
8. Plasma enhanced CVD SiO_2/ Meiners, 1981	300°	Low	*PO	
9. Plasma enhanced CVD SiO_2/ Meiners, Personal Comm.	370°	High	PO	
10. Plasma enhanced CVD SiO_2/Woodward et al. 1981	300°	Low	PO	

*AGO - Air Grown Oxides
NO - No Oxide
TO - Thermal Oxide
PO - Plasma Oxide

Acknowledgements. I wish to thank my many sutdents for their research and help in preparing this manuscript. The work was supported in part by ARO and ONR.

References

Baumann, G. G., Benz, K. W. and Pilkuln, M. H., 1976, J. Electrochem. Soc. 123:1232.

Bayliss, C. R. and Kird, D. L., 1976, J. Phys. D.: Appl. Phys. 9:233.

Clark, D. T. and Fok, T., 1981, Thin Solid Films 78:271.

Clawson, A. R., Lum, W. Y. and McWilliams, G. E., 1979, J. Crystal Growth 46:300.

Cockran, C. N. and Foster, L. M., 1962a, J. Electrochem. Soc. 109:144. 1962b, J. Electrochem, Soc. 109:149.

DiStefano, T. H., 1978, The Physics of SiO_2 and Its Interfaces, ed. S. T. Pantelides, Pergamon Press, New York.

Fathipour, M., Makky, W. H., McLaren, J., Geib, K. M. and Wilmsen, C. W., 1983, J. Voc. Sci. Technol.

Ferry, D. K., 1976, Phys. Rev. B 14, 5364.

Fritzsche, D., 1980, Inst. Phys. Conf. Sev. 50:258.

Geib, K. M. and Wilmsen, C. W., 1980, J. Voc. Sci. Technol.

Goodnick, S. M., Gann, R. G. Ferry, D. K., Wilmsen, C. W. and Krivanek, O. L., 1982, Sur. Sci. 114:238.

Grove, A. S., Leistiko, O. and Sah, C. T., 1964, Phys. Chem. Solids 25:985.

Grunthaner, F. J., Grunthaner, P. J., Vasquez, R. P., Lewis, B. F. and Maserjuan, J., 1979, J. Vac. Sci. Technol. 16:1443.

Hartstein, A., Ning, J. H. and Fowler, A. B., 1976, Sur. Sci. 58:178.

Ishii, T. and Jeppson, B., 1977, J. Electrochem. Soc. 124:1784.

Jain, G. C., Prasod, A. and Chakravary, D. C. 1978, Phys. Stat. Solidi A46, K151.

Johannessen, J. S., Spicer, W. E. and Strausser, Y. E., 1976a, J. Vac. Sci. Technol. 13:849.

Johannessen, J. S., Spicer, W. E. and Strausser, Y. E., 1976b, J. Applied Phys. 47:3028.

Johnson, N. M., Biegelsen, D. K. and M. D. Moyer, 1980, The Physics of MOS Insulators, Pergamon Press, New York.

Koelmans, H. and DeGraff, H. C., 1967, Sólid State Electronics 10:997.

Laughlin, D. H. and Wilmsen, C. W., 1980, Appl. Phys. Lett. 37:915.

Lile, D. and Taylor, J., 1983, J. Appl. Phys.

Makky, W. H., Cabrera, F., Geib, K. M. and Wilmsen, C. W., 1982, J. Vac. Sci. Technol. 21:417.

Makky, W. H. and Wilmsen, C. W., 1983, J. Electrochem. Soc. 130:659.

McLaren, J., Nelson, A., Geib, K. M., Gann, R., and Wilmsen, C. W., 1983, J. Vac. Sci. Technol.

Meiners, L. G., 1981, J. Vac. Sci. Technol. 19:373.

Many, A., Goldstein, Y. and Grover, N. B., 1965, Semiconductor Surfaces, North-Holland, New York.

Moore, B. T. and Ferry, D. K., 1980, J. Vac. Sci. Technol. 17:1037.

Mullin, J. B., Royle, A., Straushan, B. W., Tuftan, P. J. and Williams, E. W., 1974, J. Cryst. Growth 13:640.

Okamura, M. and Kobayashi, T., 1980a, Jpn. J. Appl. Phys. 19:2143.

Okamura, M. and Kobayashi, T., 1980b, Jpn. J. Appl. Phys. 19:2151.

Okamura, M. and Kobayashi, T., 1981, Electronics Lett., 17:941.

Pantelides, S. T. and Lang, M., 1978, The Physics of SiO_2 and Its Interfaces, ed. S. T. Pantelides, Pergamon Press, New York.

Poindexter, E. H., Caplan, P. J., Deal, B. E. and Razouk, R. R., 1981, J. Appl, Phys. 52:879.

Popov, A. S., 1976, Phys. Stat. Solidi A37:K53.

Schwartz, G. P., 1983, Thin Solid Films.

Stannard, J., 1979, J. Voc. Sci. Technol. 16:1462.

Taubenott, M. A. and Helms, C. R., 1983, J. Appl. Phys. 54:2667.

Vincett and Roberts 1980.

Wager, J. F. and Wilmsen, C. W., 1979, J. Appl. Phys. 50:874.

Wager, J. F. Ellsworth, D. L., Goodnick, S. M. and Wilmsen, C. W., 1981 J. Vac. Sci. Technol. 19:513.

Wager, J. F. and Wilmsen, C. W., 1982a, J. Appl. Phys. 53:5789.

Wager, J. F., Makky, W. H. and Wilmsen, C. W., 1982b, Thin Solid Films 95:343.

Wager, J. F., 1982c, Thesis, Colorado State University, Fort Collins, Colorado, USA.

Wager, J. F., Geib, K. M., Wilmsen, C. W. and Kazmerski, L., 1983, J. Vac. Soc. Technol.

Williams, R., 1965, Phys. Rev., 140:509.

Wilmsen, C. W., 1976, Thin Solid Films, 39:105.

Wilmsen, C. W. and Kee, R. W., 1978, J. Vac. Sci. Technol. 15:1513.

Wilmsen, C. W., Kee, R. W., Wager, J. F., Stannard, J. and Messick, L., 1979, Thin Solid Films, 64:49.

Wilmsen, C. W., Wager, J. F. and Stannard, J., 1980, Inst. Phys. Conf. Ser. 50:251.

Wilmsen, C. W., 1981, J. Vac. Sci. Technol. 19:279.

Wilmsen, D. W., Wager, J. F., Geib, K. M., Hwang, T. and Fathipour, M., 1983, Thin Solid Films, 103:

Woodward, J., Cameron, D. C., Irving, L. D. and Jones, G. R., 1981, Thin Solid Films, 85:61.

THEORY OF THE ELECTRONIC STRUCTURE OF SEMICONDUCTOR SURFACES AND INTERFACES

C. Calandra

University of Modena, Italy

I. INTRODUCTION

In these lectures I will survey some of the theoretical results, which have been obtained in the study of the electronic structure of semiconductor surfaces and interfaces in the last few years. After the first pioneering works at the beginning of the seventies (Applebaum and Hamann 1973, Bortolani et al., 1973, Jones 1975) our description of the electronic properties of these systems has become more and more accurate and for some specific systems a remarkable agreement between theory and experiments has been achieved.

To a large extent this huge amount of theoretical work has been made possible by two major factors: i) the availability of large and fast computer facilities without which surface calculations, specially those based on first principles self-consistent approaches, would not be feasible; ii) the diffusion of surface sensitive spectroscopical techniques (Auger Electron Spectroscopy, Electron Energy Loss Spectroscopy, Photoemission Spectroscopy, etc.) and of experimental methods for the structural characterization of the interfaces (Low Energy Electron Diffraction, Ion Scattering, etc.). As in any other field of solid state physics, the interplay between theory and experiments has been particularly fruitful.

For the clean surfaces experimental and theoretical investigations aim at the characterization of the surface phase i.e., at specifying: i) the chemical identity of the atoms in the surface and near surface region; (ii) their geometrical or structural arrangements; (iii) the energy level structure of the valence electrons in the outermost planes; (iv) the role of surface defects, such as steps, vacancies or dislocations, in determining surface properties.

The need of such characterization stems ultimately from the fact that surfaces show properties which depart considerably from those of the bulk crystal and often from those of a simple ideal termination of the solid lattice.

Work on interfaces ranges from the investigation of thin film diffusion and solid state reactions (Tu and Mayer 1978, Ottaviani 1981) to the spectroscopical study of the electronic and vibrational properties of a metal or semiconductor overlayer deposited on a semiconductor substrate (Brillson 1982). For the purpose of achieving a theoretical understanding of the microscopic processes underlying the electronic behaviour of an interface, the early stages of its formation are particularly interesting. It is, in fact, commonly accepted that some of the most important properties of an interface are already established at very low coverages i.e. when a few atoms are deposited on a semiconductor substrate (Margaritondo et al., 1976). Such a belief has been amply supported by the experimental findings of the last few years. For this reason, in these lectures, I will mainly discuss interfaces formed after low atomic depositions on semiconductor surfaces. Both for clean surfaces and for interfaces, I will consider only results concerning the intrinsic or adsorption induced electronic properties. Defect induced states are discussed in Professor Monch lectures.

II. THEORETICAL METHODS AND APPROXIMATIONS

Two approaches have been adopted in the study of the electronic properties of clean surfaces. The first one is semiempirical in nature and is based on a tight binding description of the electron states. Its application to surface physics is an aspect of the so called renaissance of the tight binding, which has taken place during the seventies. To a large extent, the reclamation of this method has been motivated by the necessity of understanding the electronic behavior of systems, as amorphous materials or crystals with defects, where translational periodicity is not present and more sophisticated theoretical approaches may not be applicable. Since the main difficulty in the theoretical treatment of the electronic structure of clean surfaces is the lacking of periodicity normal to the surface plane, tight binding methods have become popular among theoreticians interested in surface problems (Bortolani et al., 1973, Joannopoulos and Cohen 1974, Pandey and Phillips 1974, Calandra e Santoro 1975).

In the most common version of this approach the surface electronic structure is derived through a calculation of the electron states of a semiconductor slab. Suppose we have a finite crystal made of a number of regularly stacked atomic planes with two dimensional periodicity and one atom per surface unit cell. At a given wave vector $\underline{k}_{\parallel}$ of the two dimensional Brillouin zone (2DBZ), the one electron wave functions can be written as

$$\Psi_{n\underline{k}_{\parallel}}(\underline{r}) = \sum_{m}\sum_{\alpha} C_{m\alpha}(\underline{k}_{\parallel}n)\Phi_{\alpha\underline{k}_{\parallel}}(\underline{r}-m\underline{\tau}) \tag{1}$$

where n is a band index, α labels the atomic orbitals of a given atom, m is a plane index and the three dimensional vectors $\underline{\tau}$ give the origin of the two dimensional lattice on each layer. Vectors $\underline{R}_{\ell}$ describe the position of the atomic sites on the two dimensional lattice. To satisfy the Block conditions the planar orbitals $\Phi_{\alpha\underline{k}_{\parallel}}$ are given by

$$\Phi_{\alpha\underline{k}_{\parallel}}(\underline{r}-m\underline{\tau}) = \frac{1}{\sqrt{N}}\sum_{\ell}\exp\left[i\underline{k}_{\parallel}\cdot\underline{R}_{\ell}\right]\phi_{\alpha}(\underline{r}-\underline{R}_{\ell}-m\underline{\tau}) \tag{2}$$

With the assumption of vanishing overlapping integrals, the coefficients of the wave functions are obtained by solving the secular problem,

$$\sum_{m'}\sum_{\beta}\left\{H_{m\alpha,m'\beta}(\underline{k}_{\parallel}) - E_{n}(\underline{k}_{\parallel})\delta_{mm'}\delta_{\alpha\beta}\right\}C_{m'\beta}(\underline{k}_{\parallel}n) = 0 \tag{3}$$

where the matrix elements of the Hamiltonian are,

$$H_{m\alpha,m'\beta}(\underline{k}_{\parallel}) = \sum_{\ell'}\exp\left[i\underline{k}_{\parallel}\cdot R_{\ell'}\right]\int\phi^{*}_{\alpha}(\underline{r}-m\underline{\tau})\hat{H}\phi_{\beta}(\underline{r}-\underline{R}_{\ell'}-m'\underline{\tau})d\underline{r} \tag{4}$$

In order to obtain information about the electronic structure of the surface by this method two basic requirements must be satisfied. First, the two surfaces of our slab must not interact; Second, for a plane well inside the slab, the basic features of the bulk band structure are to be reproduced. In practice, these requirements are fulfilled by performing the calculation with a sufficiently high number of planes.

The behavior of the surface states and the modifications induced by the surface on the bulk states can be visualized at different energies by giving the local density of states,

$$\rho(E,m) = \sum_{n}\sum_{\alpha}\sum_{\underline{k}_{\parallel}}\left|C_{m\alpha}(\underline{k}_{\parallel}n)\right|^{2}\delta\left[E-E_{n}(\underline{k}_{\parallel})\right] \tag{5}$$

This quantity tells us how many electron states are available at a given energy in a given atomic plane. Its modifications on passing from the surface to the inner layers reflect the changes in the energy level structure of the valence electrons. Indeed, in many circumstances it is this quantity which can directly be compared with the experimental data. This is specifically true of angle integrated photoemisson spectra for low escape depth of the photoelectron. Theoretical techniques have been proposed which directly determine the local density of states without solving the secular problem (Haydock 1980). These approaches have been used to study transition metal surfaces, but their application to semiconductors has been limited (Bortolani et.al., 1973).

The previous expressions are quite general and indeed they could be used in a first principle LCAO approach. However, it is the semi-empirical version of this method, which is commonly used in the calculations. Rather than attempting to evaluate the interaction integrals starting from a given set of atomic orbitals and a given form of the one electron hamiltonian, the method relies upon a semi-empirical parametrization of these quantities, usually obtained by a more or less accurate fit to the known electronic properties of the bulk of the material under consideration. This sort of fitting procedure is quite familiar to band theoretists, since the tight binding method is commonly used as an interpolation scheme in the study of the electronic structure of solids (Slater and Koster 1954). Usually the additional approximation is made that only two center integrals are retained in the calculation. For semiconductors a good fit to the bulk bands can be obtained with ten or twelve parameters, which essentially amounts to retain interaction integrals up to the second neighbors, assuming all the interactions with more distant cells to be zero. Once these parameters have been obtained, the secular problem can be easily solved for a slab.

This procedure is simple, but not exempt of ambiguities. There is not a unique set of parameters which reproduce a given band bulk structure with a prefixed accuracy. Besides a given set of parameters cannot reproduce with the same accuracy both the valence and the conduction band. Some, but not all, of these ambiguities can be eliminated either by using a number of orbitals higher than the valence states (Hjalmarson et al., 1980) or by imposing some restriction to the parameterization (Pandey, 1976).

There are, however, serious objections to the application of semi-empirical tight binding approaches to the determination of the surface electronic structure. In the first place, the set of parameters is appropriate to reproduce the bulk Hamiltonian and not the Hamiltonian of the crystal in the presence of a surface. Second, there is no explicit form of the lattice potential and of the atomic wavefunctions corresponding to the parameters obtained by the fitting procedure, so that the method cannot be carried to self-consistency. Since the distribution of the valence charge is different at the surface with respect to the bulk, self-consistency is often crucial to obtain realistic results. Third, the effect of relaxation or reconstruction which frequently occur at surfaces, lead to changes in the inter-atomic distances, which imply modifications in the interaction integrals. They are often introduced empirically (Pandey 1976), but the lacking of any explicit form of the Hamiltonian does not allow to do this unambiguously.

In spite of these limitations, tight binding calculations have been remarkably successful in providing a description of the electronic structure of low index surfaces of elemental and partially ionic semiconductors. This is partly due to the fact that some important concepts as those of dangling or back bond surface states are easily formulated in a tight binding scheme.

A more fundamental and rigorous approach is based on the pseudopotential description of the electronic structure of the valence electrons (Schluter et al. 1975, Chelikowsky and Cohen 1979). It has the advantage of being exempt from empirical parameterizations and fully self-consistent. As usual in the pseudopotential theory, the basic idea is to replace the potential of the ion core with a model potential that takes account of all the interactions between the core and the valence electrons. The replacement allows to carry out the calculations for the valence electrons only.

Rather than evaluating the electronic properties of a semi-infinite crystal, the common version of this method works with a slab geometry, periodically repeated in the direction normal to the surface (Schluter et al., 1975). Each slab is composed by regularly stacked atomic planes of the semiconductor under consideration and it is separated from the neighboring slabs by a region of vacuum. The introduction of the periodicity normal to the surface allows to calculate the electronic structure using essentially the same techniques adopted in bulk band structure calculations. The size of the slab depends upon the system at hand and determines the volume of the orthohombic unit cell of the periodic structure. The basic requirements to be satisfied is that two consecutive surfaces do not interact appreciably either through the bulk or through the vacuum. In the study of ordered chemisorbed overlayers, the structure to be repeated is made by a slab of atomic planes of the substrate with an overlayer on both sides and by a region of vacuum (Bertoni et al., 1983).

One electron wavefunctions are then expanded in plane waves and the secular equation

$$\det\left|\left\{|\underline{k}+\underline{G}|^2 - E_n(\underline{k})\right\}\delta_{\underline{G}\underline{G}'} + V_r(\underline{G}-\underline{G}')\right| = 0 \tag{6}$$

is solved, leading to the full band structure of the system. In this equation $V_T(\underline{G})$ is the Fourier transform of the crystal pseudo-potential given by:

$$V_T(\underline{G}) = V_b(\underline{G}) + V_H(\underline{G}) + V_{xc}(\underline{G}) \tag{7}$$

Here $V_H(\underline{G})$ is the Hartree potential, connected to the valence charge density by the Poisson equation

$$V_H(\underline{G}) = 8\pi\rho(\underline{G})/G^2 \tag{8}$$

V_{xc} is the exchange and correlation potential. This is taken as a local functional of the pseudo-charge density

$$V_{xc}(\underline{r}) = -\alpha\frac{3}{\pi}\left[3\pi^2\rho(\underline{r})\right]^{1/3} \tag{9}$$

The factor α ranges between 0.7 and 1 (Manghi et al., 1981). $V_b(\underline{G})$ is the Fourier transform of the bare crystal pseudopotential, which can be written as

$$V_b(\underline{G}) = \sum_i v_i(\underline{G})S_i(\underline{G}) \tag{10}$$

where $v_i(\underline{G})$ is the bare pseudopotential of the i-th ion and $S_i(\underline{G})$ is the static structure factor. $S_i(\underline{G})$ specifies the geometry of the system, since it depends upon the location of the ions in the unit cell.

The most serious limitation of the method is the use of the previously given local form of the exchange and correlation potential. When applied to bulk band structure calculations with the recommended value α=0.7-0.8, this approximation leads to gaps which are considerably smaller than the experimental ones (Zunger 1980). For a spectroscopical investigation a better choice is α=1, which gives wider gaps in agreement with the experiments. On the other hand, for total energy calculations the value 2/3 is the more appropriate.

Another shortcoming of the form of the exchange and correlation potential is that it does not approach the classical image potential at large distances from the surface, but it decays exponentially. Such a behaviour is not relevant in the calculation of surface states with binding energy higher than few eV, but it has a great importance in the determination of the states lying in the proximity of the vacuum level.

Self-consistent pseudopotential calculations have been made for a great number of systems. On the whole it seems that the results are more reliable than those obtained by tight binding calculations, specially for what concerns the location of the first excited states and the electronic properties of interface systems.

III. SURFACE STATES IN CLEAN SURFACES

The presence of a surface introduces a number of new features in the spectrum of the eigenvalues of the valence electrons of a solid, which are called surface states or surface resonances. Unlike the bulk states, whose $\Psi(\underline{r})^2$ is periodical with the same periodicity of the lattice, surface states show a pronounced maximum of the probability amplitude at the surface and decay exponentially into the solid. Surface resonances have a somewhat intermediate behaviour, since they have a maximum at the surface, but behave as a bulk periodic state far from it.

Surface states are found inside gaps of the projected bulk bank structure. These are energy regions forbidden for bulk states, which exist throughout the two dimensional Brillouin zone or in some portion of it. These gaps can be obtained by projecting the bulk electronic structure onto the two dimensional Brillouin zone, i.e., considering for every two dimensional lattice wave-vector $\underline{k}$ the continuum of bulk states E_n ($\underline{k}$, $\underline{k}$) having the same planar component of the three dimensional wave-vector.

To illustrate these concepts we consider the case of the cleavage (110) surfaces of III-V compounds. The geometry of the surface and the two dimensional Brillouin zone are given in **figure 1.** There are two atoms per unit cell belonging to the first and second sublattice, respectively. Each surface atoms has two nearest neighbors on the surface and one neighbor in the first sublayer. Compared to the configuration of a bulk atom, at the surface one neighbor is missing. If we assume an ideal termination of the crystal, so that the atoms are not allowed to relax near the surface, this is the only effect of the creation of the surface.

The projected bulk band structure at high symmetry points of the two dimensional Brillouin zone for GaP (110) is given in **figure**

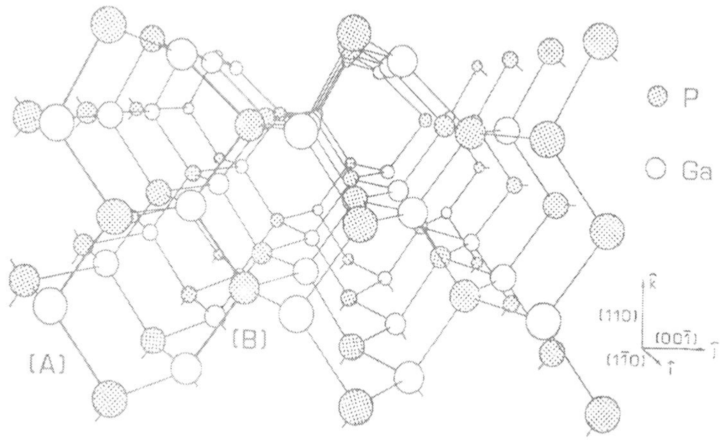

Figure 1. Side view of a slab of five GaP (110) atomic layers. The topmost layer is relaxed according to the rotation-relaxation model (Tong et al., 1978) with a surface bond rotation angle = 27.5. The lowest is unrelaxed.

2, where bulk continua with different symmetries are shown separately (Manghi et al., 1981). The same figure shows the results of a theoretical calculation carried out self-consistently using a pseudopotential approach and the repeated slab geometry. Surface states are indicated by arrows and labelled by A_i and G_i depending on whether they are localized on the anions and cations respectively. They always occur in couples since there are two surfaces in the semiconductor slab.

Two bands of surface states denoted by A_5 and G_4 are found to lie in the main gap between the valence and the conduction band. The origin of these bands is clearly understood if we determine the charge density distribution associated to these valence states. This is shown in **figure 3a and 3b** giving a plot of the square modulus of the wavefunction along (110) planes normal to the surface and passing through either the metal or the non-metal atom. It is seen that both states are surface states localized in the first atomic layer with a pronounced charge maximum pointing out of the surface. This is the charge density distribution expected for a state arising from the unsaturated hybrids of P and Ga atoms at the surface, so that A_5 and G_4 are called dangling bond states. The presence of these states in the band gap region can be easily understood, if we recall the fact that the binding properties in tetrahedrally coordinated materials can be approximately described by sp^3 hybrid orbitals. On each atom four equivalent sp^3 hybrids are formed from the s and p valence states which point toward the four neighboring

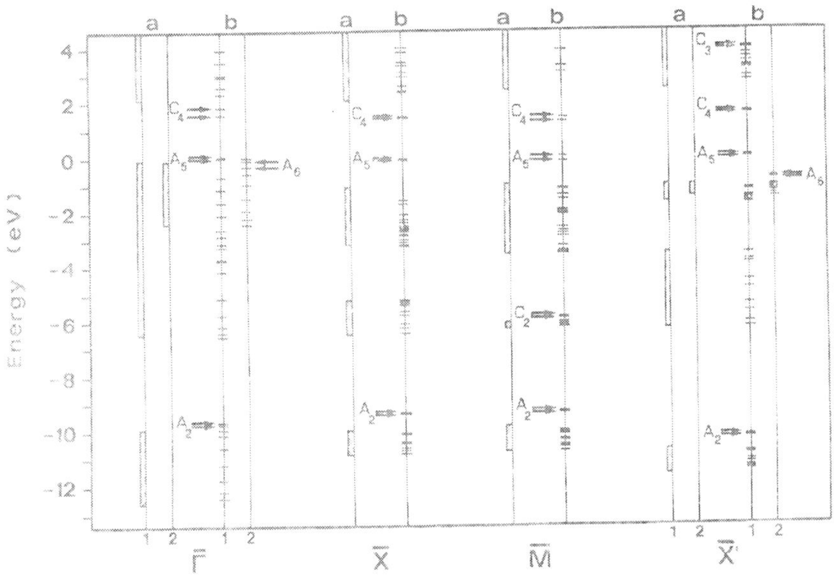

Figure 2. Energy spectrum for GaP (110) ideal surface. Electron states are divided according to the representations of the point group of the bulk directions corresponding to the high symmetry points of the 2DBZ. In column **(a)** the projected bulk band structure is displayed. Column **(b)** shows the eigenvalues resulting from the slab calculation. Energies are referred to the top of the valence band.

which point toward the four neighboring atoms. Bonding combinations of these hybrids form the completely filled valence band, while anti-bonding states give rise to the empty conduction band. When the crystal is cut to create the (110) surface, a dangling bond is created on each atom. Since the two atoms in the surface unit cell are different, we are left with two bands, a filled one located on group V element and an empty one located on group III atoms. The energy of these bands is expected to be close to the energy of the corresponding hybrid, so that they lie in the band gap.

Beside dangling bond states, a number of other surface features are found in the calculation. The most important are indicated by A_2 and G_2. The analysis of the charge density distribution shows that they are essentially s-like P and Ga surface states. The existence of these surface features is a direct consequence of the modifications of the one electron potential, which take place on passing from the bulk to the surface. A_2 lies above the border of the P s-like bulk band near the lower edge of the ionic gap. State G_2 is found in a gap of the bulk band structure which exists along XM and MX directions but is closed at $\bar{\Gamma}$. Therefore, it is seen clearly only near $\bar{M}$. On going toward $\bar{X}$ or $\bar{X}'$ mixes with bulk states and loses its surface character.

Perhaps one of the most significant achievements of the application of surface sensitive techniques to the study of semiconductor surfaces has been the experimental evidence that the cleavage surfaces III-V compounds are not in the ideal form, but undergo a significant relaxation (Duke et al., 1976, Kan et al., 1978, Tone et al., 1978, Tear et al., 1980). The occurrence of relaxation can be explained using the hybrid orbital model. Hybridization of the atomic orbitals of an atom depends on the number of neighbors surrounding it. At the surface, atoms are in a threefold coordination and dehybridization takes place. The group III atoms tend to go toward a planar sp^2 configuration with three equivalent planar bonds and an empty p orbital perpendicular to the surface. This configuration can be obtained by an inward displacement. The group V atom undergoes an outward atomic displacement, which leads to an electron configuration with three bonding orbitals of p-character, each containing one electron and a filled non-bonding orbital of s character. Such configuration is typical of compounds as NH_3 or PH_3 having bond angles close to 90°. The situation is shown in figure 1.

Two important consequences stem from this distortion of the atomic geometry at the surface. First the "so called" dangling bond states are removed from the gap: the band of anion states is lowered, whereas the empty cation state is rised in energy (Calandra et al., 1977). This usually leads to a gap free of surface states a fact which has important consequences on the electronic properties of the free surface, since no Fermi level pinning can occur in these conditions, unless extrinsic states are present (van Laar an Scheer 1967, van Laar and Huijser (1976). Second, a number of new surface features appear in the electronic spectrum which are caused by the modifications of the one electron potential induced by the relaxation.

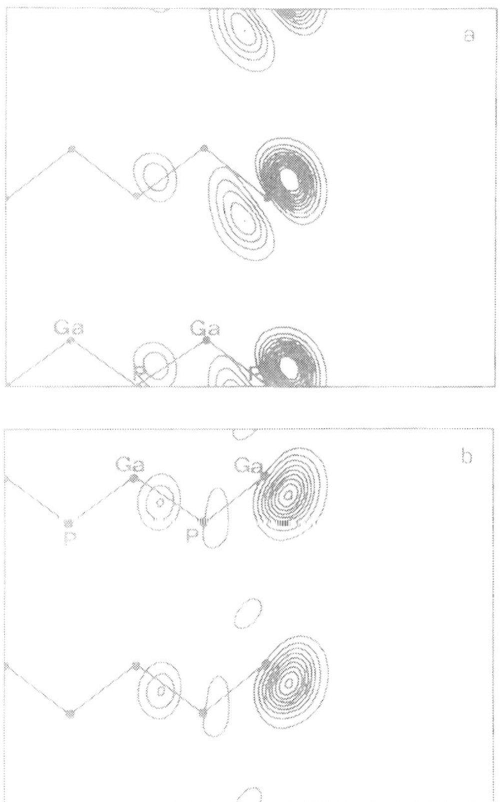

Figure 3. Charge density distribution for **(a)** filled surface state A_5; **(b)** empty surface state C_4 at the $\bar{M}$ point of the 2DBZ.

To illustrate these effects we give in **figure 4,** the electronic structure of InP (110) surface along the high symmetry directions of the two dimensional Brillouin zone (Manghi et al., (1982). It is seen that the dangling bond bands are removed from the gap and partially overlap the bulk continua near the Brillouin zone center. Two new surface states appear in the near gap region of the bulk bands, which are denoted by A_4 and G_3. Their nature is better illustrated by the charge densities of **figures 5 and 6.** It appears that these states are localized at the surface with a pronounced charge maximum between the first and the second layer. They are usually referred as back bond surface states. Notice that the metal derived back bond state G_3 crosses the G_4 band and around $\bar{M}$ it lies at lower energies. Interesting modifications occur also at higher binding energies in the s-derived states. Two features indicated by A_1 and G_1 are, in fact, s-states localized on the second atomic plane. It is clear from this example that the real surface is considerably richer of surface states than one would expect on the basis of simple dangling bond arguments.

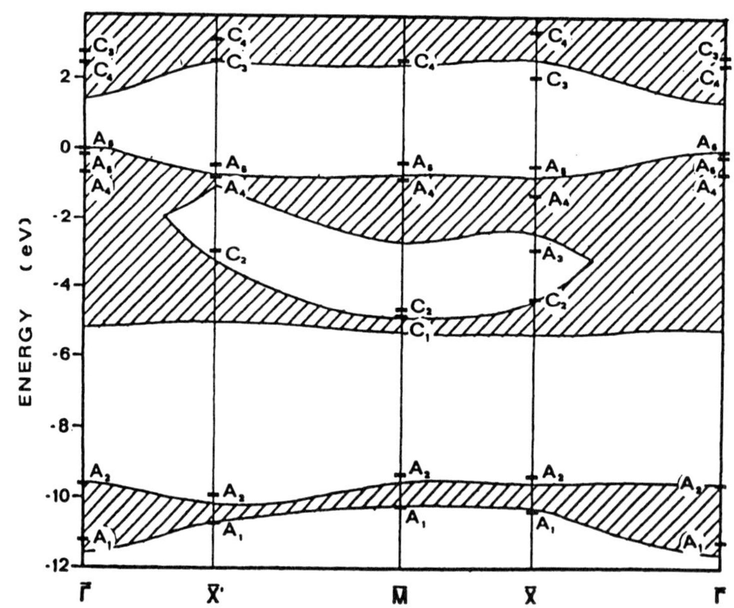

Figure 4. Projected bulk band structure and surface state energies for the relaxed InP (110) surface. The shaded regions correspond to bulk states. Energies are referred to the top of the valence band.

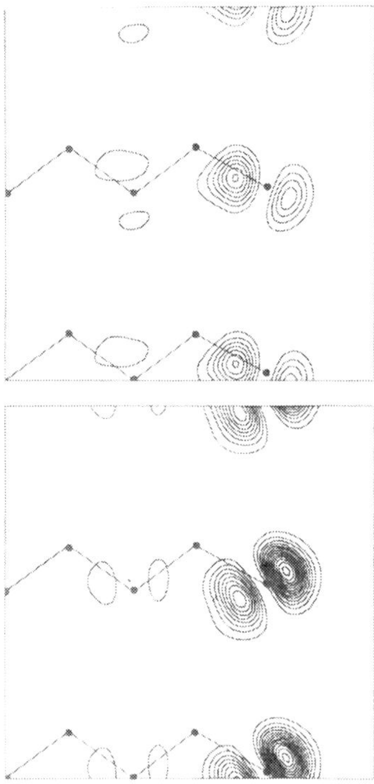

Figure 5. Charge density distribution for state A_4 at $\bar{X}$ (upper panel) and A_5 at $\bar{X}$ (lower panel) of the InP (110) relaxed surface.

This surface relaxation is expected to occur in the cleavage surface of all III-V compounds. It is now quite well established that this is the reason why no Fermi level pinning is found in these surfaces. The only exception is GaP (110), where Fermi level pinning for n-type samples has been found, in spite of the fact that LEED data indicate a significant relaxation. The explanation of this anomalous behavior has been provided by theoretical calculations (Manghi et al., 1981). Unlike what was found in the other III-V's the theory shows that dangling bond surface states are not completely removed by relaxation in this case. This is shown in **figure 7**, where the surface states at the high symmetry points of the two dimensional Brillouin zone are given for GaP (110) in the relaxed form. The possibility of explaining the experimental findings relative to the Fermi level pinning is a significant achievement of the first principle pseudopotential theory of the surface electronic structure.

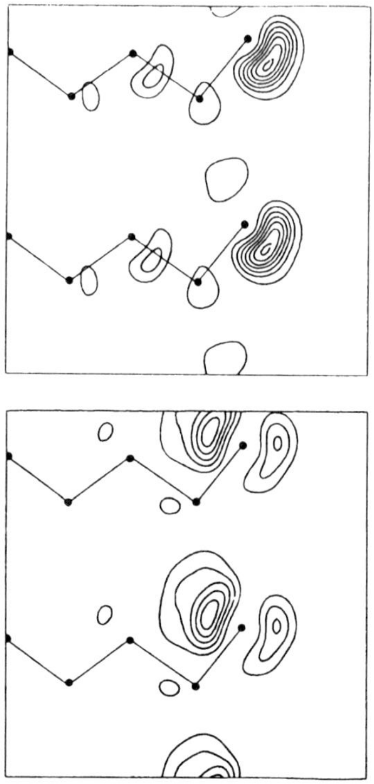

Figure 6. Charge density distribution for state G_3 at $\bar{M}$ (upper panel and G_4 at $\bar{X}$ (lower panel) of InP (110).

One aspect of the previous results deserves some comments. The surface bands obtained by the calculations are very narrow. For example the dangling bond bands have a width of approximately 1 eV. One could ask whether under such condition the band description is appropriate since correlation effects are expected to be important in narrow bands. Although a detailed investigation of the importance of correlation effects in surface bands has not yet been performed, there is some evidence that a single particle description may not be completely appropriate in the case of surface states.

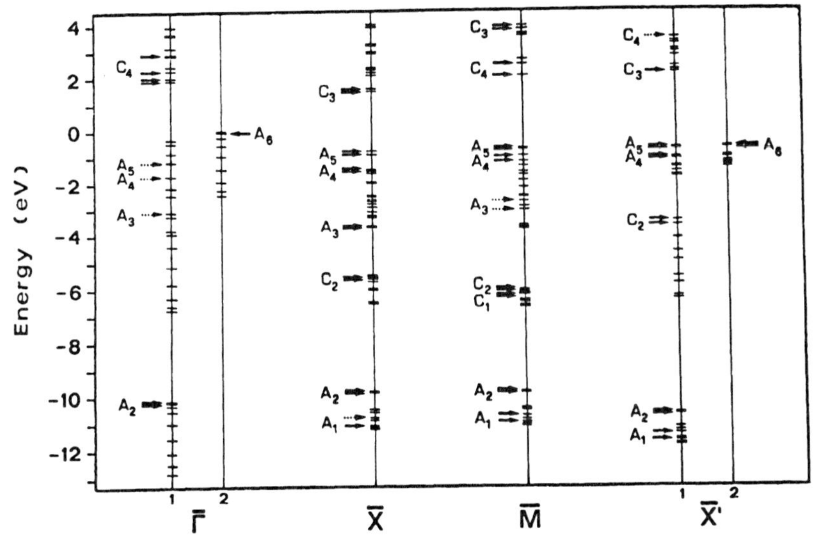

Figure 7. Energy spectrum of the relaxed surface of GaP (110) at high symmetry points of the 2DBZ.

To illustrate this point, we consider some results about the Si (111) surface. This is the cleavage surface of Si and usually shows a 2x1 reconstruction, i.e., a two dimensional periodicity different from an ideal termination of the bulk crystal (Lander 1965). The ideal unreconstructed surface should have one atom per two dimensional unit cell and one dangling bond per atom. In this condition one would expect a single half-filled surface band in the main gap. Theoretical calculations performed with different methods confirm this conclusion (Bortolani et al., 1973, Appelbaum and Hamann 1973, Pandey and Phillips 1974, Schluter et al., 1975, Ciraci et al., 1975). Recently, a metastable 1x1 unreconstructed surface has been prepared by laser annealing (Bedair and Smith 1967, Zehner et al., 1980). Detailed studies of its structural properties (Zehner et al., 1981) have been performed showing that a significant relaxation takes place in the first atomic planes. The electronic structure has been studied by angular resolved photoemission spectroscopy (White et al.,1981). The results indicate the absence of filled surface states in the band gap region, in disagreement with the outcomes of theoretical calculations (see figure 8).

A possible explanation of this discrepancy has been suggested by Duke and Ford (1981), who assumed the surface state band to be described by a two dimensional version of the Hubbard Hamiltonian,

$$\hat{H} = \epsilon_0 \sum_{i,\sigma} \hat{n}_{i,\sigma} + t_1 \sum_{\substack{i \neq j \\ nn}} \sum_{\sigma} \hat{a}^+_{i,\sigma} \hat{a}_{j,\sigma} + U \sum_{i,\sigma} \hat{n}_{i,\sigma} \hat{n}_{i,-\sigma} \quad (11)$$

Here is the creation operator of an electron at site i with spin and U is the intra-atomic coulomb interaction. If we set U=0, the surface state eigenvalue spectrum is given by the simple tight-binding formula

$$\begin{aligned} \mathcal{E}(\underline{k}_{\parallel}) &= \mathcal{E}_0 + 2t_1 \left[\cos k_x b + 2\cos(k_x b/2)\cos(k_y b\sqrt{3}/2)\right] \\ &= \mathcal{E}_0 + 2t(\underline{k}_{\parallel}) \end{aligned} \quad (12)$$

where b is the distance between surface neighbors. This formula reproduces the results of theoretical calculations with good accuracy (figure 8). On the other hand, in the limit $U >> t_1$ the Hamiltonian has two solutions (Doniach 1969),

$$\mathcal{E}^+(\underline{k}_{\parallel}) = \mathcal{E}_0 + U + t(\underline{k}_{\parallel})/2. \quad (13a)$$

$$\mathcal{E}^-(\underline{k}_{\parallel}) = \mathcal{E}_0 + t(\underline{k}_{\parallel})/2. \quad (13b)$$

which are shown in **figure 8.** This is the situation typical of the Mott-Hubbard insulator. If we assume U>~1eV, with the value of t_1 of the order of 0.1eV, which is appropriate to reproduce the results of the pseudopotential calculations, we can explain the discrepancy between single particle theory and photoemission experiments. As indicated in figure 8, under such conditions the filled surface band can be removed from the main gap, so that no significant emission takes place from filled surface states in the gap region.

This example indicates that the description of the surface states may be considerably more complicated than expected on the basis of single particle calculations. A more detailed investigation of this point for the low index surfaces of semiconductors is likely to lead to a better understanding of the surfaces electronic spectra.

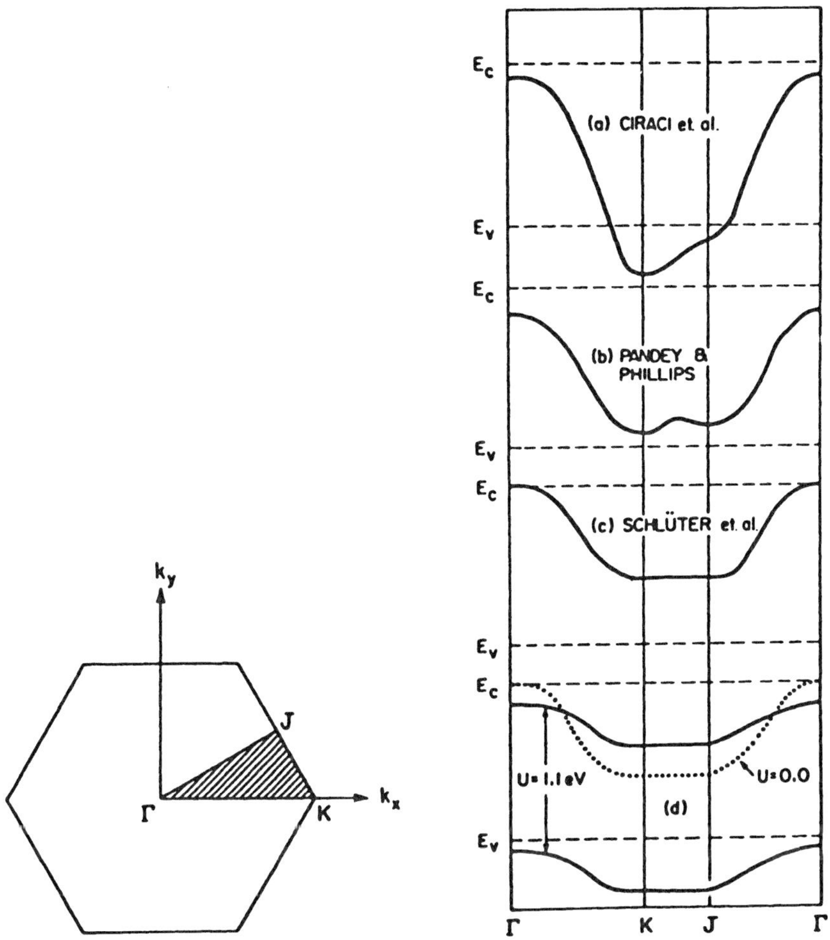

Figure 8. Comparison of the eigenvalue spectra along the high symmetry lines of the 2DBZ of Si (111) 1x1 surface predicted by: **(a)** Ciraci et al., (1975); **(b)** Pandey and Phillips (1974); (c) Schulter et al., (1975); **(d)** the Hubbard Hamiltonian with U=0 (dotted) and with U-1.leV (solid). Also shown is the two dimensional Brillouin zone (after Duke and Ford 1981).

IV. INTERFACE ELECTRONIC STRUCTURE

Both metal-semiconductor and semiconductor-semiconductor interfaces have been subject of much theoretical and experimental work (Brillson 1982, Harrison 1977, Pickett et al., 1977, Bauer and McMenamin 1978, Monch et al., 1982, Zurcher et al., 1982). In both cases the interest has been primarily addressed to the occurrence of states localized at the interfaces in their early stages of formation. Localized states of this sort can have a fundamental role in the optical and transport properties of the interface by influencing important interface parameters, e.g. the band bending and the built-in potential. In particular, there is a correlation between the appearance of interface states and the shift of the pinning position of the Fermi level (Milnes and Feucht 1972).

For metal semiconductor systems we can distinguish between nonreactive and reactive interfaces. The first ones are formed when deposition of metal overlayers on a given semiconductor substrate is not followed by any chemical solid state reaction or by atomic interdiffusion. Under such circumstances, the interface is abrupt and localized states arise from the coupling of the surface states of the free surface with the continuum of metal states. Reactive interfaces are produced when compound formation takes place as a consequence of metal deposition or after thermal treatment. The first one is typical of Si-transition metal interfaces (Rubloff and Ho 1982 and references therein); the second, has been found, for example, in the study of noble metals-Si or Al-GaAs systems (Braicovich et al., 1980, Perfetti et al., 1982, Duke et al., 1981, Huijser et al., (1981). The distinction between abrupt and reactive interfaces is not sharp. There are systems which behave as non-reactive at room temperature, but show strong diffusion and significant reactions when heated at convenient temperatures.

As an example of a non-reactive interface we shall discuss the one formed by monolayer deposition of Cs onto GaAs (110). Combined LEED and Auger studies have shown that Cs is absorbed in a C (4x4) superstructure for coverages ranging from 0.8 to 1.2 monolayers (van Bommel and Crombeen 1976), Derrien and Arnaud dAvitaya 1977). This geometry is schematically shown in **figure 9.** The overlayer can be seen as a compressd (110) plane of metallic Cs with the Cs [100] axis parallel to the GaAs [110] direction. The compression deduced from experimental LEED data is about 14%. Experiments indicate also that (i) the Fermi level is pinned around 0.4eV above the valence band maximum (Spicer et al., 1975); (ii) the As-dangling band states are removed by the deposition (Huijser, 1979); and (iii) transitions having Ga surface states as final states are neither removed nor significantly shifted in energy (Derrien et al., 1976, Huijser 1979).

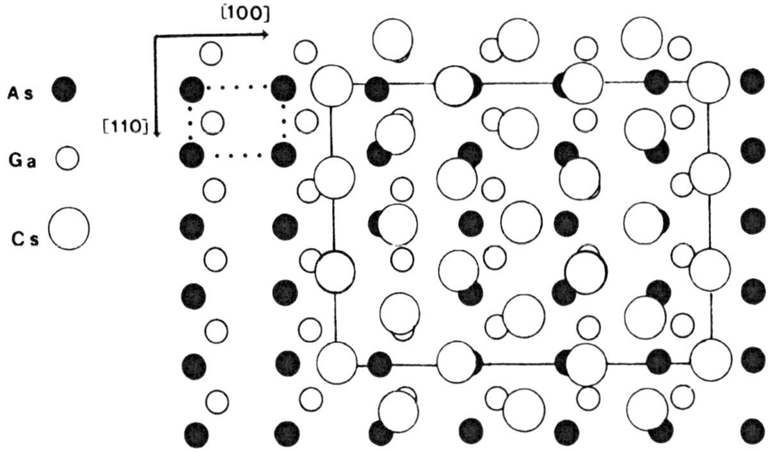

Figure 9. Top view of the GaAs (110)-Cs C (4x) interface.

Self-consistent pseudopotential calculations for this interface have been carried-out by Manghi et al., (1983). Since no information is available on the modifications of the substrate geometry induced by Cs absorption, the calculations have been performed assuming both the ideal and the relaxed configuration of the surface. The question about which model is appropriate to describe the system has been answered by comparing the outcomes of the calculations with the experimental findings. The slab band structures obtained for the two absorption models are shown in **figure 10 and 11.** The results for the ideal geometry show two main bands called S_3 and S_4, which couple with the bulk continuum and with a more dispersive surface band, called S_5, ongoing toward the center of the 2DBZ. This behavior can be qualitatively understood by considering a model Hamiltonian where a parabolic free electron band of Cs states hybridizes with the two dangling bond states of the free surface. Compared with the results of the clean surface, the As dangling bond moves slightly down in energy, while the Ga state lies in the upper part of the gap. It is this surface band that is responsible of the Fermi level position, which is found at 0.7eV above the valence band maximum. These which results are quite in agreement with those found by Cohen and co-workers in their study of the electronic properties of metal-semiconductor interfaces carried-out using a periodic repetition of a semiconductor slab alternated by thick jellium layers to simulate the metal (Schluter 1982 and references therein). In

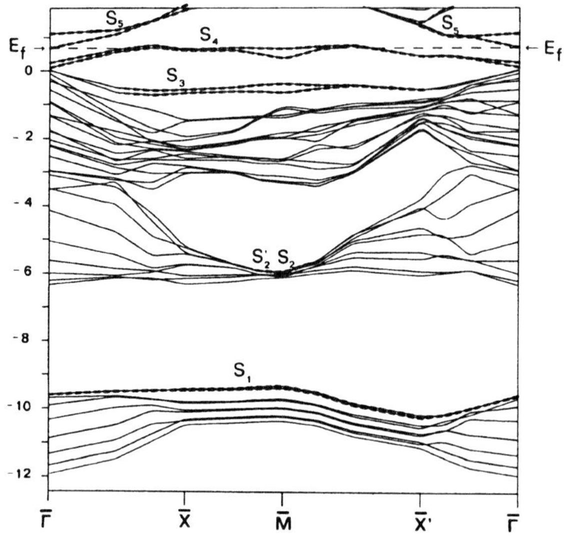

Figure 10. Two dimensional band structure of a 11 layer GaAs (110) film in the presence of a thin Cs overlayer. The topmost substrate layer is in the ideal configuration. Surface states are indicated by dotted lines and labelled by S_i. Energies are referred to the valence band maximum.

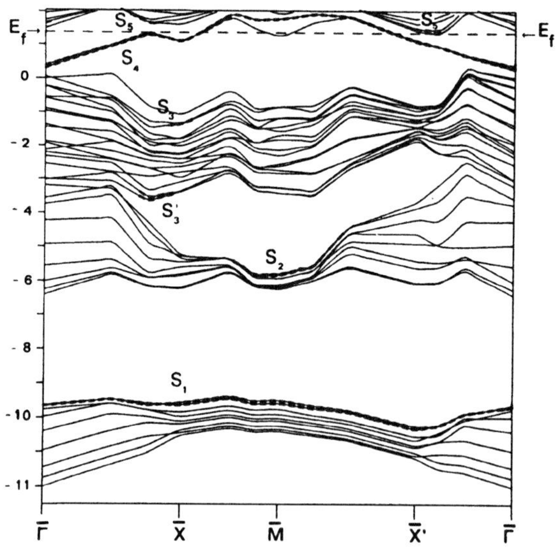

Figure 11. Energy spectrum of a thin Cs slab deposited onto the relaxed GaAs (110) surface.

those calculations gap states are found which behave as free electron states at the metal side and decay exponentially into the semiconductor, but right at the interface they retain the characteristic behavior of the dangling bond states of the free unrelaxed surface.

The band structure obtained from the calculation for the overlayer deposited on the relaxed substrate is shown in figure 11. It is seen that the presence of the overlayer leads to important changes in the distribution of the surface states. The most significant is the absence of any surface feature with As dangling bond character, except near $\bar{X}$, where a resonance denoted as S_3 with As character is found. The same is true for the As back bond state, which is present in the theoretical spectrum for the clean surface along the upper edge of the projected valence band, but is completely absent in the theoretical spectrum of the interface. The band gap accomodates one band starting at $\bar{\Gamma}$ with parabolic dispersion. This is clearly an overlayer Cs state, which in the proximity of the upper edge of the gap gets hybridized with Ga surface states. For this reason S_4 has a Ga back bond character on going from $\bar{X}$ to $\bar{M}$ and from $\bar{X}'$ to $\bar{M}$ after hybridization with state S_5, which disperses upward around $\bar{X}$ and $\bar{X}'$. Near $\bar{M}$, S_4 behaves much like a Ga dangling bond. Due to the absence of high density surface bands in the gap, the Fermi level is located at the valence band maximum nearly leV higher than in the previous case.

Comparison with the experimental findings seems to indicate that the model with a relaxed substrate is in agreement with the data. In particular, it explains the absence of As induced surface states in the experimental photoemission spectra and the insensitivity to cesiation of the transitions having Ga empty surface states as final state. Both features cannot be explained on the basis of the results for the unrelaxed substrate, which show a high density of filled surface states near the valence band maximum and a significant modification of the Ga empty states compared with the results for the free relaxed surface. However, the model with the relaxed substrate disagree with the experimental findings in that it locates the Fermi energy leV above the measured value. Such a discrepancy in a model that shows qualitative agreement with the spectroscopical data is somewhat disturbing. The only possible explanation of it seems to be the possibility that the surface Fermi level is pinned by some extrinsic surface states lying in the lower part of the gap. This conclusion would be in agreement with some of the models recently proposed to explain the origin of the Schottky barriers in metal semiconductor interfaces (Spicer et al. 1980, Monch and Gant 1982).

The main problem with the reactive interfaces is the identification of the surface phases, i.e., the compound or solid solution formed in the interface region (Ottaviani et al., 1981). Among the reactive systems, the transition metal-silicon interfaces have been studied intensively in the last few years. Interdiffufusion at room temperature with compound formation has been monitored spectroscopically by photoemission for near noble Ni, Pd and Pt and refractory metals deposited on Si (111) surface (Branicovich et al. 1982). The identification of the surface phases can be obtained by systematic comparison with the photoemission spectra for bulk silicides and theoretical calculations of the electron density of states for these compounds. The results indicate a graded concentration between the silicides and Si. Metal silicides with high formation enthalpy have been found to exhibit a higher concentration gradient than those with low heat of formation and therefore to have narrower interface region.

To see how theoretical results for bulk silicides can be useful in interpeting experimental data for the interfaces, we consider the case of Ni deposited on Si. **Figure 12** shows a comparison between valence band photoemission spectra at hv=50eV for bulk silicides and theoretical calculations of the density of states (Bisi and Calandra 1981, Franciosi et al., 1982). The maximum of intensity at these photon energies comes from emission caused by d-metal states, since the cross section for photoionzation of d-states is considerably larger than from s or p states. The experiments indicate a narrowing of the metal d-band and a shift of its maximum at higher binding energies upon increasing the Si content. There is a considerable agreement between the experimental energy distribution curves and the theoretical densities of states. Both the trend and the absolute location of the d-metal band with respect to the Fermi energy are well reproduced.

If we now consider the experimental results for interfaces obtained by depositing Ni on freshly cleaved Si surfaces **(figure 13)**, we find a somewhat similar behavior of the d-band on passing from low to high metal coverages. This behavior provides evidence of the existence of a mixed Ni-Si phase even in the early stages of formation of the interface. **Figure 14** shows the effects of thermal treatments on the interface electronic structure for different metal coverages. As expected the main effect of heating is to promove the reaction through an increased out-diffusion of Si and to lower the metal d-band with respect to the Fermi energy. Stationary phases found during the annealing correspond to the formation of stable Ni silicides at the interface. Similar results have been found for Pt-Si systems. In all cases, the comparison with bulk calculations allows to identify the nature of the phases formed at the interface with reasonable accuracy. The question whether intrinsic or extrinsic states exist at these reactive interfaces, which are responsible of the Fermi level location and of the height of the Schottky barrier has not yet received an answer (Ottavani et al. 1981, Brillouin 1982, Grunthaner et al., 1981). The microscopical characterization is not yet detailed enough to draw conclusions on the origin of the Schottky barrier in these interfaces.

V. CONCLUSIONS

We have examined the contribution of the modern theory of solids to our understanding of the electronic properties of clean surfaces and metal-semiconductor interfaces. Although a number of important issues and results have been considered, it is clear from the examples presented in these lectures that the description of the electronic properties of these systems on a microscopic basis is still far from being complete. Fundamental problems, as the origin of reconstruction, the formation of the Schottky barrier, the role of extrinsic and intrinsic states in interfaces, remain to be solved. This will be the task of surface theoreticians in the next few years.

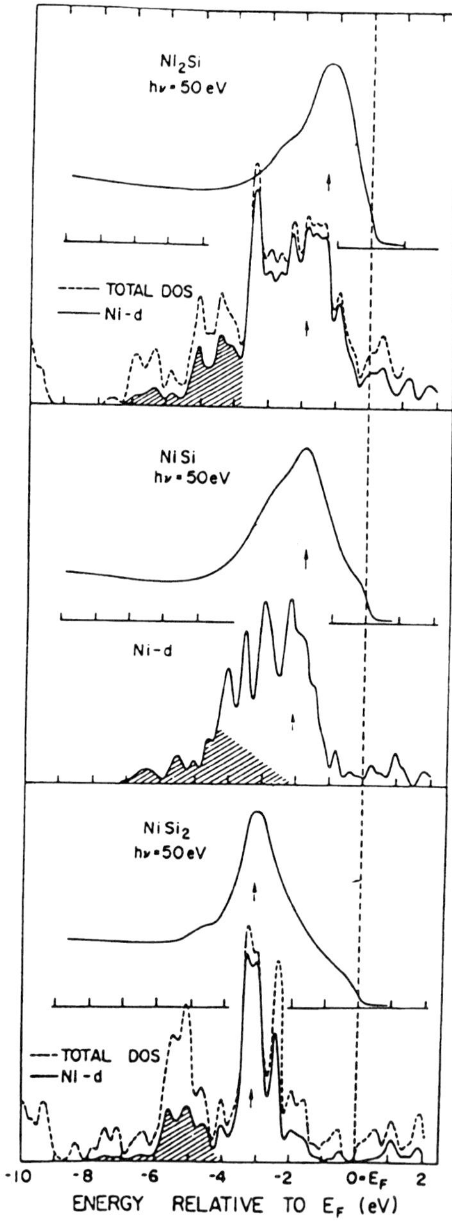

Figure 12. Comparison of experimental photoemission spectra for bulk crystals of NiSi, NiSi and NiSi with calculated densities of states. The shaded areas indicate Si-p/Ni-d hybridized states in bonding combinations. The unshaded areas are primarily non-bonding d-states. Total density of states (dashed line), Ni-d (solid line) and energy distribution curve at hν=50 (upper solid line).

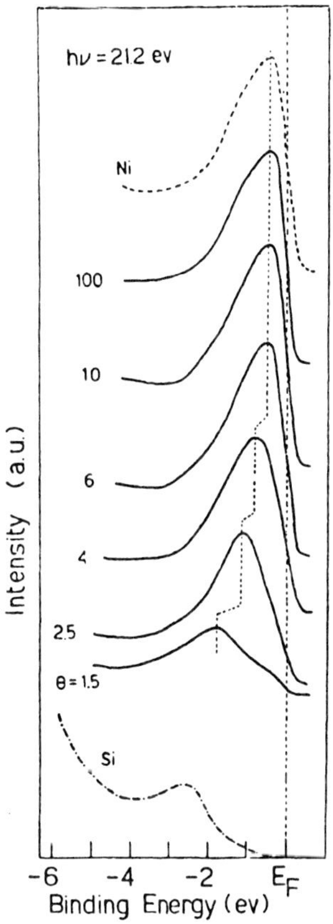

Figure 13. Photoemission curves for different Ni coverages onto Si (111) deposited at liquid nitrogen temperature (after Abbati et al., 1982).

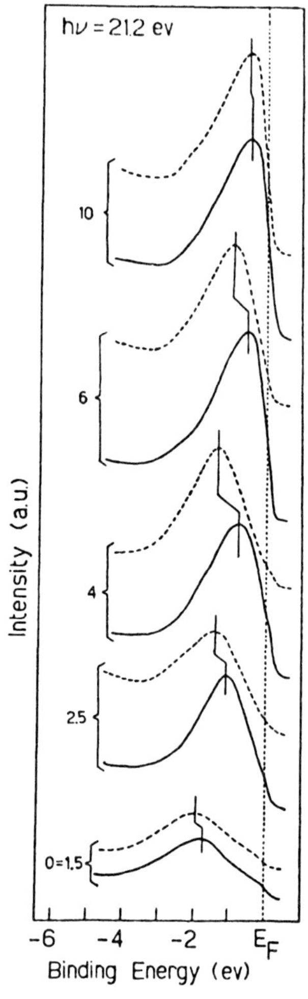

Figure 14. Effect of heating the samples of figure 13 from liquid nitrogen to room temperature (after Abbati et al., 1982).

REFERENCES

1. Abbati, I., Braicovich, L., DeMichelis, B., del Pennino, U., and Valeri, S., 1982 Solid St. Commun. 43; 199.
2. Appelbaum, J.A., and Hamann, D.R., 1973, Phys.Rev.Letters 31; 106.
3. Bauer, R.S., and McMenamin, J.C., 1978, J. Vac. Sci. Tech., 15; 1444.
4. Bedair, S.M., and Smith, H.P. 1967, J. Appl. Phys. 40; 4776.
5. Bertoni, C.M., Calandra, C., Manghi, F., and Molinari, E., 1983, Phys. Rev. B27; 1251.
6. Bisi, O., and Calandra, C., Braicovich, L., Abbati, I., Rossi, G., Lindau, I., and Spicer, W.E., 1982, J. Phys. C 15; 4707.
7. Bortolani, V., Calandra, C., and Kelly, M., 1973, J. Phys. C 6; L349.
8. Braicovich, L., Garner, G.M., Skeath, P.R., Su, C.Y., Chye, P.W., Lindau, I., and Spicer, W.E., 1979, Phys. Rev. B20; 5131.
9. Braicovich, L., Abbati, I., Miller, J.N., Lindau, I., Schwartz, S., Skeath, P.R., Su, C.Y., and Spicer, W.E., 1980, J.Vac.Sci.Technol. 17; 1005.
10. Brillson, L.J., 1982, Surface Sci. Reports 2.
11. Calandra, C., and Santoro, G., 1975, J. Phys. C 8; L86.
12. Calandra, C., Manghi, F., and Bertoni, C.M., 1977, J.Phys.C. 10; 1911.
13. Chelikowsky, J.R., and Cohen, M.L., 1979, Phys. Rev. B20; 4150.
14. Ciraci, P., Batra, I., and Tiller, W.A., 1975, Phys.Rev. B12; 5811.
15. Derrien, J., Arnaud d'Avitava, F., and Bienfait, M., 1976, Solid State Commun. 20; 557.
16. Derrien, J., Arnaud d'Avitava, F., 1977, Surface Sci. 65; 668.
17. Doniach, S., 1969, Advances in Phys. 18; 819.
18. Duke, C.B., Lubinsky, A.R., Lee, B.W., and Park, P., 1976, J.Vac. Sci. Technol. 13; 761.
19. Duke, C.B., Paton, A., Meyer, R.J., Brillson, L.J., Kahn, A., Kanani, K., Carrelli, J., Yeh, L.J., Margaritondo, G., and Katnani, A.D., 1981, Phys.Rev. Letters 46; 440.
20. Duke, C.B., and Ford, W.K., 1981, Surface Sci. 111; L685.
21. Francisosi, A., Weaver, J.H., O'Neill, D.G., Chabal, Y., Rowe, J.E., Poate, J.M., Bisi, O., and Calandra, C., 1982, J.Vac.Sci.Technol., 21; 624.
22. Grunthaner, P.J., Grunthaner, F.J., Madhukar, A., and Mayer, J.W., 1981, J. Vac. Sci, Technol., 19; 649 and references therein.
23. Grunthaner, P.J., Grunthaner, F.J., Madhukar, A., 1983, Physica 117-118B; 831.
24. Harrison, W.A., 1977, J. Vac. Sci, Technol., 14; 1016.
25. Haydock, R., 1980, in "Solid State Physics" Eds. Ehrenreich, H., Seitz, F., and Turnbull, D., 35; 215.

26. Hjalmarson, H.P., Vogl, P., Wolford, D.J., and Dow, J.D., 1980, Phys. Rev. Letters 44; 810.
27. Huijser, A., 1979, Thesis.
28. Huijser, A., van Laar, J., and van Rooy, T.L., 1979, Surface Sci. 102; 264.
29. Kahn, A., Cisneros, G., Bonn, M., Mark, P., and Duke, C.B., 1978, Surface Sci. 71; 387.
30. Joannopoulos, J.P., and Cohen, M.L., 1974, Phys.Rev. B10; 5075.
31. Jones, R.O., 1975 in "Surface Physics of Semiconductors and Phosphors"
32. Margaritondo, G., Stoffel, N.G., Kathnani, A.D., Edelmann, H.S., and Bertoni, C.M., 1981, J. Vac. Sci. Technol. 18;784.
33. McKinley, A., Williams, R.H., and Park, A.W., 1979, J.Phys.C, 12; 2447.
34. Milnes, A.G., and Feucht, D.L., 1972, "Heterojunctions and Metal-Semiconductor Interfaces", Academic Press.
35. Monch, W., and Gant, H., Phys. Rev. Letters 48; 1982.
36. Monch, W., Bauer, R.S., Gant, H., and Murshall, R., 1982, J. Vac. Sci. Technol. 21; 498.
37. Ottaviani, G., Tu, K.N., and Mayer, J.W., 1981, Phys. Rev. B24; 3354.
38. Ottaviani, G., 1981, J.Vac.Sci. Technol. 18; 924.
39. Pandey, K.C., and Phillips, J.C., 1974, Phys. Rev. Letters 32; 1433.
40. Pandey, K.C., 1976, Phys. Rev. B14; 1557.
41. Perfetti, P., Nannarone, S., Patella, F., Quaresima, C., Capozi, M., Savoia, A., and Ottaviani, G., 1982, Phys. Rev. B26; 1125.
42. Pickett, W.E., Louie, S.G., and Cohen, M.L., 1977, Phys. Rev. B17; 815.
43. Rubloff, G.W., and Ho, P.S., 1982 in "Thin Film and Interfaces" Eds. Ho, P.S., and Tu, K.N., North-Holland, 21, and references therein.
44. Slater, J.C., and Koster, G.F., 1954, Phys. Rev. 94; 1498.
45. Schluter, M., Chelikowsky, J.R., Louie, S., and Cohen, M.L., 1975, Phys. Rev. B12; 4200.
46. Schluter, M., 1982, in "Thin Film and Interfaces" Eds. Ho, P.S., and Tu, K.N., North-Holland, 3.
47. Spicer, W.E., Gregory, P.E., Chye, P.W., Babalola, I.A., and Sukegawa, T., 1975, Appl.Phys. Letters 27; 617.
48. Tear, S.P., Walton-Cook, M.R., Prutton, M., and Walter, J.A., 1980; Surface Sci. 99; 598.
49. Tong, S.Y., Lubinsky, A.R., Mrstik, B.J., and van Hove, M.A., 1978, Phys.Rev. B17; 3303.
50. Tu, K.N., and Mayer, J.W., 1978, in "Thin Film Interdiffusion and Reactions" Eds, Poate, J.M., Tu, K.N., and Mayer, J.W., J. Wiley, Ch.10.

51. van Bommel, A.J., and Crombeen, J.E., 1976, Surface Sci. 57; 109.
52. van Laar, J., and Scheer, J.J. 1967, Surface Sci. 8; 342.
53. van Laar, J., and Huijser, A., 1976, J. Vac. Sci. Technol. 13; 769.
54. White, C.W., Zehner, D.M., Eastman, D.E., Heimann, P., Himpsel, F.J., and Reighl, B., 1981, Bull. Am. Phys. Soc. 26; 350.
55. Zehner, D.M., White, C.W., and Ownky, G., 1980, Appl. Phys. Letters 37; 456.
56. Zehner, D.M., Noonan, J.R., Davis, H.L., and White, C.W., 1981, J.Vac.Sci.Techno. 18; 852.
57. Zunger, A., 1980, Phys. Rev. B22; 959.
58. Zurcher, P., Lapeyre, G.J., Anderson, J., and Frankel, D., 1982, J.Vac.Sci.Technol 21; 476.

DEEP LEVELS AT COMPOUND-SEMICONDUCTOR INTERFACES

Winfried Mönch

Laboratorium für Festkörperphysik
Universität Duisburg
D 4100
Duisburg, Germany

INTRODUCTION

One of the main features of semiconductors is that their electronic properties may be drastically modified by the introduction of small numbers of foreign atoms or lattice defects. Impurities and imperfections are determining the type and the density of the majority carriers as well as the lifetime of injected minority carriers. Broadly speaking, the energy levels of conductivity dopants are close to the conduction or the valence band edges while they are nearer to midgap position for effective lifetime dopants. Therefore, such levels are traditionally referred to as shallow or deep, respectively.

In III-V compound semiconductors, deep impurities are not only important as recombination centers, but also as conductivity dopants. If, for example, shallow acceptors and deep donors are simultaneously present in a crystal the deep donors provide electrons compensating the acceptors. The material then has high resistivity and is called to be semi-insulating. At present, undoped semi-insulating GaAs, for example, is receiving much attention since it is applied in integrated circuits fabricated by direction implantation and as substrates for doped epitaxial films. The main problems in the growth of such material still are the control and the identification of those deep impurities which cause the undoped crystals to become semi-insulating.

In optoelectronic devices, such as light-emitting diodes and above all, lasers, deep impurities are very undesirable since they cause recombination of injected carriers. By the associated radiation-less transitions not only the quantum efficiency drops, but heat is generated that is considered to contribute to the degradation of the device via the formation of defects.

Deep levels may also be localized at surfaces or interfaces. Surface and interface states may be associated with the perfect arrangement of the surface and interface atoms, respectively, or with defects and impurities. Accordingly, they are labeled as intrinsic and extrinsic surface states, respectively. A perfect surface does not mean a truncated bulk lattice neither with respect to atomic positions nor to chemical composition but rather a clean surface exhibiting long-range order. Surface cleanliness does thus not only refer to foreign atoms but also comprises an excess or a deficiency of one or the other type of atoms at the surfaces of compound semiconductors. Bulk stoichiometry cannot always be regarded as reference level since, for example, polar surfaces may exhibit several different surface structures each correlated with a definite composition.

Electrically, surface states behave donor- or acceptor-like and depending on their energetic position relative to the Fermi level they are either neutral or charged positively and negatively, respectively. In thermal equilibrium, the surface altogether has to be electrically neutral. This is achieved by an adjustment of the bulk bands and thereby of the surface states with respect to the Fermi level. The resulting excess charge in the surface states cancels the one in the surface space-charge layer in the substrate. The band bending observed is thus determined by the energetic distribution of the surface states and the bulk doping. Rather low densities of surface states fix the band bending with regard to changes in bulk doping or temperature, for example. This behavior is traditionally referred to as pinning of the Fermi level at the surface. A detailed discussion of surface space-charge layers and the occupation statistics of surface states may be found in the books by Many et al. (1965) and Frankl (1967).

As deep levels in the bulk, surface states may also act as recombination and trapping centers for injected carriers reaching the surface. The actual surface recombination velocity critically depends on the population of the surface states and thus on the band banding under the specific experimental conditions (see Many et al. (1965) and Frankl (1967)). This subject will not be pursued in the present paper.

Surface states on clean surfaces are of basic interest, and the investigation of their properties is important for the understanding of surfaces in general. Devices, on the other hand, contain no bare semiconductor surfaces but many different interfaces. These may be semiconductor homo- or heterojunctions, semiconductor-metal contacts and semiconductor/insulator as well as metal/insulator boundaries. States localized at such interfaces may not only influence but determine the performance of semiconductor devices and they are thus very carefully considered in device modeling. This shall be illustrated by two examples.

The electrical properties of metal-insulator-semiconductor (MIS) devices sensitively depend on the density of electronic states at the semiconductor-to-insulator interface. In particular, the threshold voltage of a MIS-transistor is directly related to the flat-band voltage and their high-frequency operation can be also limited by states at the IS-boundary. Therefore, the fabrication processes are designed such as to minimize the density of states at the semiconductor-to-insulator interface. On the other hand, a high density of interface states may be desirable in other cases. For example, a strong surface depletion layer between adjacent integrated devices minimizes leakage currents. This objective is accomplished by an interface between the semiconductor and an insulator that exhibits a high density of interface states.

The existence of interface states at semiconductor-insulator boundaries as well as in semiconductor heterojunctions and their influence on device properties are common knowledge. Metal-semiconductor contacts have been studied since the discovery of the crystal rectifier more than 100 years ago and modern semiconductor-device design and technology are inconceivable without the use of such Schottky barriers. However, interface states in metal-semiconductor junctions and their importance for the understanding of such contacts are still debated. The recent development of the discussions may be found in the Proceedings of the Annual Conferences "Physics and Chemistry of Semiconductor Interfaces."

In this paper we will consider deep levels in the bulk as well as at interfaces of III-V compound semiconductors only. It is even out of the scope of this contribution to attempt reviews of these two restricted fields. Therefore, only simple models and some of the very many experimental results will be discussed. We have tried to select such topics that might provide some physical insight into problems of current interest and approaches to solve them.

DEEP LEVELS IN THE BULK

As pointed out, deep impurities in semiconductors are of great technological importance. However, the large gap between the ever growing pool of data and the qualitative as well as quantitative understanding of the physics of deep levels has been bridged only recently. In the following neither the present theoretical approaches will be outlined in detail nor a compilation of the data will be attempted. For this we should like to refer to some of the recent and excellent review articles and books on this subject (see e.g. Queisser (1971), Kaufmann & Schneider (1980), Vogl (1981, 1983), Scheffler (1982), Jaros (1982)). In the following two sections we will rather try to sketch some of the basic concepts on which our present understanding of deep impurities is based. As a specific example we will then discuss the identification of the most important deep donor, EL2, in GaAs as an As_{Ga} antisite defect. Such antisite defects are native and most characteristic defects in III-V semiconductors(van Vechten (1975(F)).The As_{Ga} antisite defect has been chosen since in the discussion of Fermi-level pinning at interfaces of III-V semiconductors we will have to consider such defects again.

SHORT-VERSUS LONG-RANGE POTENTIAL OF IMPURITIES

Shallow impurities in semiconductors are understood quite well. The behavior of substitutional donors and acceptors is adequately described in terms of the hydrogenic model which leads to the well-known effective-mass approximation (see, for example, Altarelli & Bassani (1982)). This approach assumes that the impurity electron or hole moves in the long-range Coulomb potential $e_0^2/4\pi\varepsilon_s\varepsilon_0 r$ of the singly charged impurity ion which is embedded in a uniform dielectric. The semiconductor is thus modeled by its static dielectric constant ε_s and the effective mass m* of an electron or a hole in the unperturbed periodic potential. The ionization energy of such a hydrogenic impurity measures $E_1^H m^*/m_0\varepsilon_s^2$ where E_1^H=13.6 eV is the ionization energy of atomic hydrogen. Since the static dielectric constants of semiconductors are typically of the order of 10 the ionization energies of typical donors or acceptors are quite small, i.e., they are shallow impurities.

In this approach, the ever present short-range part of the impurity potential is completely neglected. Let us, however, consider nitrogen atoms being incorporated in GaP on phosphorous sites. N and P have the same number of valence electrons; they are "isoelectronic". There is no long-range Coulomb tail and the electron is rather bound through the short-range potential. The associated level exhibits the properties of a deep trap although its activation energy is only 10 meV(Bergh & Dean (1976), Hsu et al. (1977), Wolford et al. (1979)).

An inclusion of short-range corrections in the hydrogenic model results in slight chemical shifts. If, however, the short-range component dominates the long-range Coulomb part of the potential then the impurity can no longer be described by the hydrogenic model and its levels become deep.

This can be illustrated in terms of a simple model(Vogl (1981)). Let us consider a donor. Its electron shall be subject to a screened Coulomb potential and a short-range model potential which is assumed as an attractive square well. In Figure 1 the energy of the lowest bound-state is shown as a function of the depth of the well. For small short-range potentials the long-range Coulomb part controls the

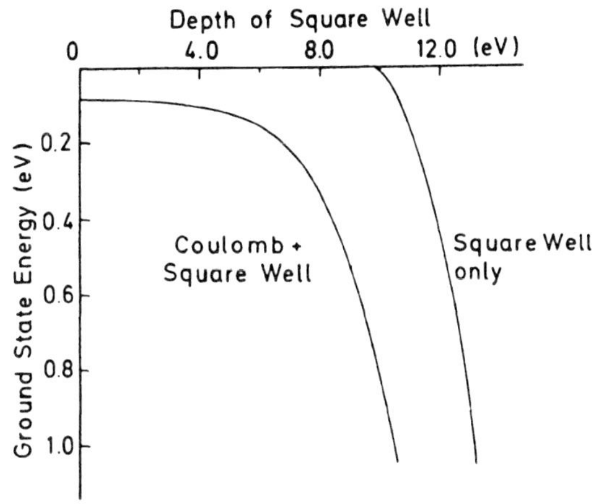

Figure 1. Schematic illustration of the effect of the short-range part of the impurity potential on the ionization energy of a donor electron. The potential has been taken to be the sum of a screened Coulomb potential $e_0^2/4\pi\varepsilon_s\varepsilon_o r$, and an attractive square well, with ε_s=10, m^*=0.1 m_o and a well-width of 5 Å. From Vogl (1981).

bound states and the energy of the first one of these levels measures several 10 meV as predicted by the effective-mass approximation. With increasing depth of the well, finally the short-range component alone is able to bind the electron and the ionization energy of the donor then rapidly increases. The impurity turns from shallow to deep with increasing strength of the short-range potential.

Therefore, an impurity in a semiconductor is defined to be a deep trap theoretically if its short-range potential alone is sufficiently strong to bind a state within the band gap of the host (Hjalmarson et al. (1980)). In contrast to a shallow level, such a deep impurity is not exclusively derived from the nearest band edge. Experimentally a trap may thus be defined as deep if it does not follow a nearby band edge when that edge is perturbed by alloying or pressure (Jantsch et al. (1982)).

THE DEFECT-MOLECULE APPROACH

The short-range or central-cell component of the potential arises from the difference in the atomic term energies ε_i and ε_h of the impurity atom and the replaced host atom, respectively. This is qualitatively modeled in a molecular LCAO picture (Harrison (1980)). Following Hjalmarson et al. (1980) we will examine the isoelectronic impurity N in the host GaP. This example has been already mentioned above. Fig. 2 explains the model used. The chemical bonds of the Ga atoms surrounding a P atom, or an N atom replacing it, are described by the interaction of s-like atomic orbitals. The Ga atomic level lies above the energy of the P s-orbital. Bringing Ga and P

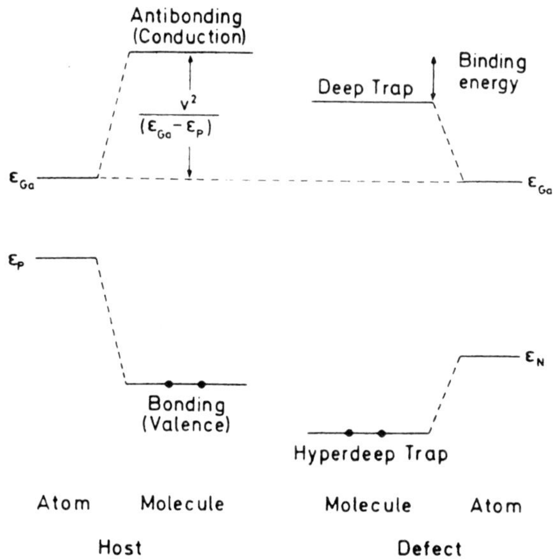

Fig. 2. The two-state model schematically shows the host GaP antibonding and bonding states, derived from atomic energies, compared with the same model with N replacing P. From Hjalmarson et al. (1980).

together in a molecule leads to a bonding-antibonding splitting of $2\sqrt{V_2^2+\frac{1}{2}(\varepsilon_{Ga}-\varepsilon_P)^2}$. V_2 is the anion-cation interaction matrix-element, and it is known to depend on the bond length only (Harrison (1980)) . The antibonding level is the parent of the conduction band in the solid. Let us now consider a defect molecule formed by the impurity N and its neighbors Ga. The energy of the N s-orbital lies by approximately 7 eV below the P level. The bonding-antibonding splitting is proportional to $V_2^2/(\varepsilon_{Ga}-\varepsilon_N)$. If we neglect relaxations around the impurity the matrix element V_2 is the same for the host and at the defect. Since the denominator $(\varepsilon_{Ga}-\varepsilon_N)$ is by approximately 7 eV larger than $(\varepsilon_{Ga}-\varepsilon_P)$ the antibonding and also the bonding states lie lower for the defect than for the host. The net result is the creation of two impurity levels:

- One level is a hyperdeep bonding state within or below the valence band. It is impurity-like and is always filled and electrically inactive.

- The deep level is a Ga-like antibonding state slightly below the GaP conduction state. It is unoccupied and therefore electrically active. A change of the impurity s-like level by a large amount results in only a small change of the deep level. This is, qualitatively, the reason why impurity atoms introduce trap states within the energy gap. Increasing ε_N towards $-\infty$ the Ga-like antibonding level, i.e., the deep level, asymptotically approaches ε_{Ga}, the Ga dangling-bond energy.

Figure 3 contains experimental binding energies of donor electrons. They are measured from the conduction band edge and are plotted as a function of the difference of the atomic-orbital energies of the impurity and the replaced host atom. This difference reflects the strength of the short-range central-cell potential. Chemical trends are indicated by the dashed lines connecting data points of similar impurities in a particular semiconductor. The pronounced trends may be easily understood in terms of the simple model discussed in connection with Figure 1. With increasing strength of the short-range potential the bonding energy of the shallow donors slightly increases which trend is commonly referred to as a chemical shift. Finally, the short-range component dominates the coulombic part and the ionization energy of the impurities increases dramatically; they now become deep traps.

THE DEEP DONOR EL2 IN GaAs

The most important deep donor in GaAs is the so-called EL2 trap. In recent years it has been the subject of extensive studies with various experimental techniques. This level was considered by Henry and Lang (1977) and Martin et.al. (1980) and finally shown by Holmes et al. (1982) to control the high resistivity of undoped GaAs. However, the chemical nature of this most important deep donor has turned out only recently. At present, there is strong evidence that the deep donor EL2 is an As_{Ga} antisite defect. In the following some of the investigations supporting this assignment shall be discussed.

Holmes et al. (1982) grew GaAs crystals by the liquid encapsulation Czochralski technique (LEC). They varied the arsenic to gallium ratio in the melt by a few percent below and above the stoichiometric composition. The concentration of EL2 was determined by optical absorption measurements in the 1 to 1.4 μm range and found to increase by almost an order of magnitude with increasing As content in the melt. A similar study was undertaken by Lagowski et al. (1982). They grew GaAs crystals by the Horizontal Bridgman Method (HBM) and determined the density of electron traps by standard Deep Level Transient Spectroscopy (DLTS). The measurements at room temperature revealed EL2 to be the major trap and its density to increase with increasing As pressure during growth. They concluded that the deep donor EL2 forms during postgrowth cooling of the crystal by the reaction

$$As^{o}_{As} + V^{-}_{Ga} \rightleftarrows As^{++}_{Ga} + V^{+}_{As} + 4\,e^{-}. \qquad (1)$$

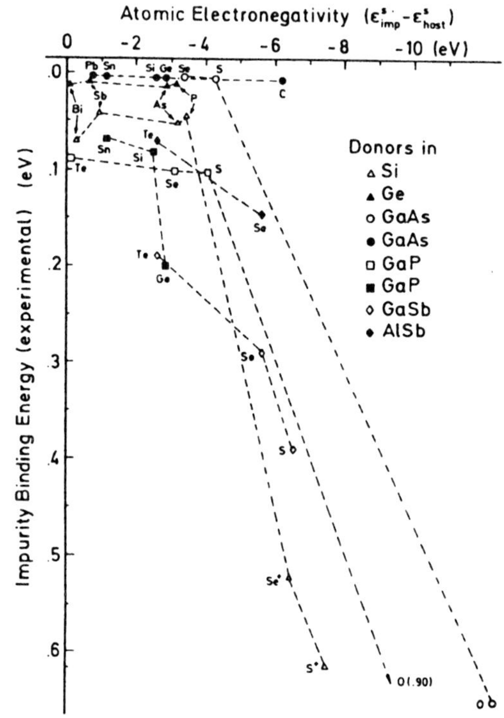

Fig. 3. Experimental ionization energies of donors in several semiconductors versus the central-cell impurity-potential strength, which is measured by the difference between the s-orbital energies of the impurity atom and the replaced host atom. From Vogl (1981).

As^{o}_{As} is a neutral As on its lattice site, V^{-}_{Ga} and V^{+}_{As} are charged cation and anion vacancies, respectively. As_{Ga}, finally, is an antisite defect and the EL2 level is attributed to this defect.

It should also be mentioned that no deep impurities and specifically no EL2 levels have been detected in GaAs layers grown by Liquid Phase or Molecular Beam Epitaxy (LPE, MBE). Tegude (1983), for example, gives the detection limit in his measurements using the photocapacitance technique as approximately 10^{12} traps per cm^3. Bulk HBM or LEC samples as well as layers grown by Vapor Phase Epitaxy (VPE) always contain large concentrations of EL2 levels. Since in these techniques the growth temperatures are consideraly higher than those employed in LPE and MBE, it has been concluded that the absence of EL2 traps in these layers is due to the lower growth temperature.

Using Electron Paramagnetic Resonance (EPR) Weber et al. (1982) identified As_{Ga} antisite defects in plastically deformed GaAs.

Wosinski et al., (1983), on the other hand, studied the influence of plastic deformation on the spectrum of deep traps in GaAs by using DLTS. They could correlate the increase of the EL2 concentration and of the EPR signal from As_{Ga} antisite defects as measured by Weber et al. (1982) under the same experimental conditions.

In conclusion, the identification of the most important deep donor EL2 in GaAs as an As_{Ga} antisite defect seems to be well established.

By using EPR, the As_{Ga} antisite defect in GaAs, as previously already the P_{Ga} defect in GaP, has been identified as a double donor. The energy levels associated with As_{Ga} antisite defects have been determined by Weber et al. (1982). Their samples, as grown as well as plastically deformed, were illuminated with light in the spectral range between 0.2 and 1.3 eV while they recorded the EPR spectra of the As_{Ga} defects. From photo enhancement and quenching of the EPR signals they determined the two energy levels

$$E\ (0/+) = E_c - 0.77\ \text{eV}, \qquad (2)$$

$$E\ (+/{++}) = E_v + 0.52\ \text{eV} \qquad (3)$$

for the change of the occupancy from neutral to positively charged and from singly to doubly positively charged, respectively.

During plastic deformation of GaAs the complementary Ga_{As} antisite defect might be expected to form since the formation energies of both antisite defects have been estimated by van Vechten (1975) to be equal.

However, till now the Ga_{As} defect has not been identified positively. Schneider (1982) has tentatively assigned the two native hole traps, commonly labelled A and B to the double acceptor Ga_{As}. The levels of these deep acceptors are $E(A) = E_v + 0.41$ eV and $E(B) = E_V + 0.71$ eV.

METAL-SEMICONDUCTOR JUNCTIONS

Chemical Trends of the Barrier Height

The rectification of metal-semiconductor contacts was discovered by Braun (1874) more than 100 years ago. Any successful theory to describe this effect had to wait until quantum mechanics was applied to solids and as a consequence, the theory of the electronic band structure and electronic conduction in semiconductors was developed. Schottky (1939) was the first to present a theory which explains the properties of metal-semiconductor contacts by a space-charge layer on the semiconductor side of the junction. Rectifying junctions then exhibit a depletion layer. Figure 4 shows the energy-band diagram of such a Schottky barrier at the interface between a metal and an n-type semiconductor. The contact is characterized by the barrier height $\phi_{Bn} = E_{ci} - E_F$, i.e. the distance from the Fermi level to the bottom of the conduction band edge at the interface.

The main theoretical problem is to predict the barrier heights for any combination of a metal and a semiconductor. The first proposal in this regard again goes back to Schottky (1942). He assumed that in forming an intimate contact between the two materials their vacuum levels adjust. Since in thermal equilibrium the Fermi energy is constant across the interface the barrier height then results as

$$\phi_{Bn} = \phi_m - \phi_s \qquad (4)$$

where ϕ_m is the work function of the metal and ϕ_s is the electron affinity of the semiconductor. This model tries to predict properties of an interface from properties of the bare surfaces of both materials in contact.

Later on, Anderson (1962) employed the same approach in estimating the conduction-band discontinuity across a semiconductor heterojunction as the difference of the electron affinities of the two semiconductors in contact. The band diagram at the interface of a semiconductor heterostructure is also sketched in Figure 4. Kroemer (1975) has strongly criticized Anderson's and thereby also Schottky's approach and has pointed out that bulk instead of surface quantities should be used to predict interface properties. And, indeed, the experimentally determined band discontinuities are not resembled by Anderson's values. Harrison (1977, 1980) on the other hand, calculated the absolute energies of the valence-band maxima relative to the vacuum level in a LCAO approach and obtained the discontinuities in the valence bands across a heterojunction by subtracting the energies of the valence bands in the bulk of the two semiconductors in contact. Katnani and Margaritondo (1983) and Kroemer (1983) have shown these predictions to be in good agreement with the experimental results for quite a number of heterojunctions.

Schottky's rule seems to be only obeyed by some of the more ionic semiconductors (Kurtin et al. (1970)). Motivated by Schottky's model the data points ϕ_B/ϕ_m measured with various metals on a particular semiconductor were then fitted by a straight line. The slopes of these lines varied between 0 and 1 and they were thought to correlate with the ionizities of the semiconductors. However, more recent experimental results obtained with contacts prepared by evaporation of metals on clean and well-characterized surfaces cleaved from n-

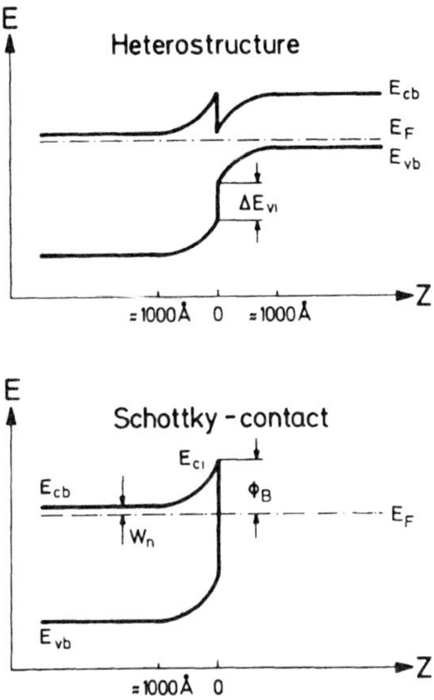

Fig. 4. Band scheme in intimate and abrupt metal and semiconductor-semiconductor junctions.

and p-type samples question such linear models. As an example, the barrier heights of metal-InP(110) junctions are plotted versus the metal work functions of the metals used in the upper panel of Figure 5. The dashed line gives Schottky's prediction based on an electron affinity of 4.5 eV as determined by van Laar et al. (1977) for InP. These experimental data reported by Williams et al. (1979) neither follow Schottky's model nor can they be fitted by any straight line. However, they reveal two distinct values of the barrier height, namely ϕ_B=0 and ϕ_B=0.5 eV.

The diagram in the lower part of Figure 5 offers the correlation of the observed barrier heights with another quantity. The motivation for this choice shall be explained briefly. It has been known since Poganski's work (1952) that metal overlayers may interact with semiconductor substrates. Technical selenium rectifiers consisted of a Se layer covered with a Cd-Sn film. As has been shown by Poganski (1953) the rectification, however, does not occur at a metal-semiconductor junction but rather at a Se-CdSe heterojunction which forms during the fabrication of the device. The importance of such metal-semiconductor reaction is meanwhile also realized for silicon and the III-V compound semiconductors. The barrier heights at interfaces between silicon and transition-metal silicides vary by many tenths of an eV. Andrew and Phillips (1975) found a linear correlation between the barrier heights measured with these interfaces and the formation energy of the respective silicides. Their plot is given in Figure 6.

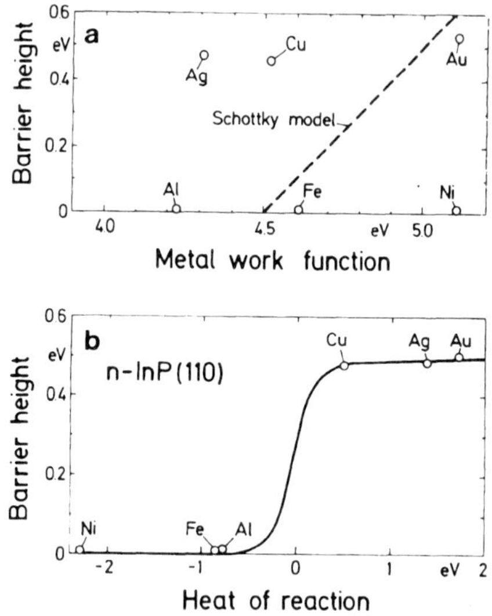

Fig. 5. a) Plot of the barrier height versus the metal work function for metal-InP(110) junctions. The dashed line is the prediction by Schottky's model.

b) Barrier height of metal-InP(110) junctions as function of the heat of reaction at the interface as defined in equation 6. After Williams et al. (1979).

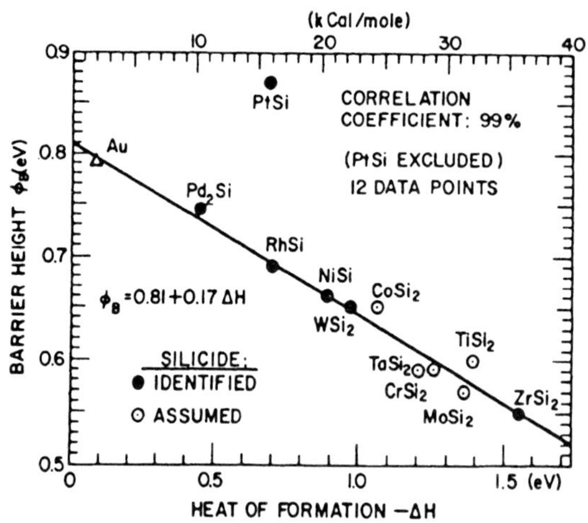

Fig. 6. Barrier heights of transition metal-silicon junctions plotted against the heat of formation of transition-metal silicides. From Andrew and Phillips (1975).

Since III-V and II-VI compounds are known to interact with some metals Brillson (1978) proposed a correlation between the measured barrier heights with a heat of reaction. He postulated reactions of the type

$$M + \frac{1}{x} CA \rightarrow \frac{1}{x}(M_xA + C), \tag{5}$$

to occur at the interface between a compound semiconductor CA and a metal M. The heat of reaction is then given by

$$\Delta H_r = \frac{1}{x}(H_f(CA) - H_f(M_xA)), \tag{6}$$

where H_f are the heats of formation of the compounds given in brackets. Reactive and nonreactive metals are distinguished by negative and positive heats of reactions.

In the lower diagram of Figure 5 the barrier heights given in the upper panel are now plotted versus the heats of reaction calculated according to Equation 6 from tabulated heats of formation (Wagman et al. (1968-71), Kubaschewski & Alcock (1979)) of the compounds listed. All data points can now be connected by a S-shaped curve. The same correlation has been obtained for metal-semiconductor contacts on n-type ZnO, ZuS, CdS and GaP by Brillson (1978) as well as on CdTe and GaSe by Mönch (1983) using the data reported by Williams & Patterson (1982) and Hughes et al. (1982). Figure 7 shows the InP and CdTe data together with those published for GaAs by Spicer et al. (1976) and Skeath et al. (1981). The S-shaped curves identify two different barrier heights on each of the n-type semiconductors. GaAs does not follow that rule; it exhibits the same barrier height independent of the metal overlayer. Au on GaAs and CdTe does not follow the common trends contrary to Au:InP(110). These observations will be discussed separately. Unfortunately not many metal-semiconductor contacts have been investigated on surfaces cleaved from p-type substrates. Only some metals on p-type GaAs(110) and Al on p-type InP(001) have been studied (Spicer et al. (1976), Waldrop (1983)). The GaAs as well as the InP data show that the positions of the Fermi level on p- and n-type substrates are not the same but differ by 0.27 eV for both semiconductors. It is thus not generally true that the barrier heights $\phi_{Bn}=E_{ci}-E_F$ on the n-type and $\phi_{Bp}= E_F- E_{vi}$ on p-type substrates of the same semiconductor add up to the width of the band gap as frequently was postulated. Au on GaAs again is an exception.

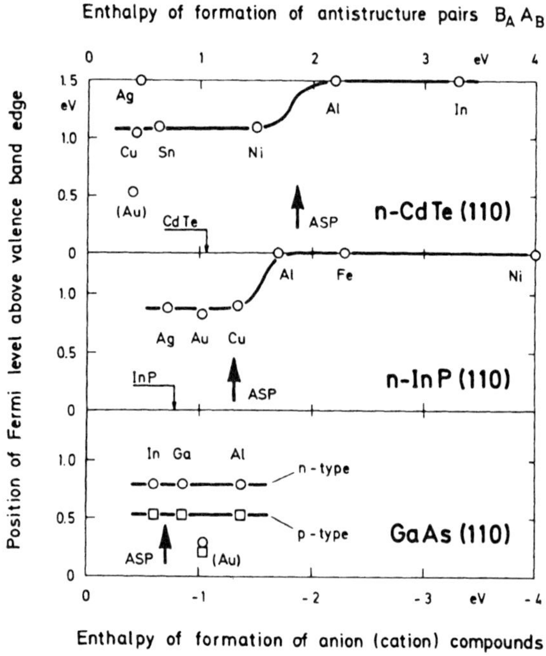

Fig. 7. Positions of the Fermi level in the band gap for Schottky contacts on surfaces cleaved from various semiconductors versus the heat of formation of metal-anion compounds and additional data for some other interfaces.

Barrier heights for GaAs(110) from Spicer et al. (1976) and Skeath et al. (1981), for InP(110) from Williams et al. (1979), and for CdTe from Williams and Patterson (1982).

SCHOTTKY CONTACTS OF Au AND COMPOUND SEMICONDUCTORS

As Figures 7 and 8 show, the pinning positions of the Au contacts on GaAs, CdTe and GaSe considerably deviate from the other data points while Au:InP(110) fits the general trend. In the case of GaAs, no Au-As anion-compound is known but the cation compounds AuGa and $AuGa_2$ exist (Hansen (1958)). The heat of formation of $AuGa_2$ was estimated from the chemical trend observed with $AuAl_2$ and $AuIn_2$ as compiled by Wagman et al. (1968-71) and Kubaschewski and Alcock (1979). For CdTe the heat of formation of the anion compound $AuTe_2$ is by a factor of two smaller than the one of the cation compound AuCd which was considered in Fig. 7. The compounds AuSe and $AuSe_2$ are less stable than $AuTe_2$ (see e.g. Puddephat (1978)) and should thus exhibit an even lower heat of formation. Therefore, in Figure 8 again $AuGa_2$, i.e., the cation compound, was considered. On the other hand, in the

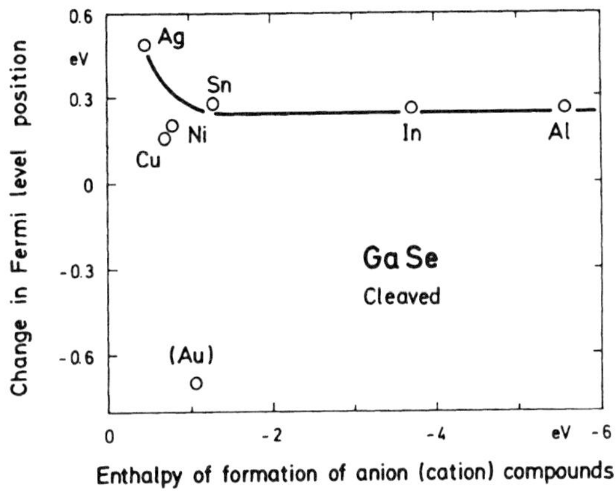

Fig. 8. Change in position of the Fermi level at the interface of metal-GaSe junctions Hughes et al. (1982) versus the heat of formation of metal-anion compounds.

case of Au-InP contacts, the anion compound Au_2P_3 has the larger enthalpy of formation of 1.01 eV compared to 0.78 eV for $AuIn_2$, the cation compound and the barrier height fits the general trend. The same behaviour is observed with Au:GaP (see Fig. 7 in Brillson (1978)). These findings support the interpretation that the barrier height of a metal-semiconductor junction depends on whether the formation of binary compounds is more likely to occur between the overlayer atoms and the anion or the cation atoms of the compound semiconductor substrate.

ON Al:GaAs(001) CONTACTS

The interaction of Al with GaAs(001) surfaces has been studied very extensively by Ludeke et al. (1973), Cho & Dernier (1978), Marra et al. (1979), Massies et al. (1979, 1982), Petroff et al. (1981), Landgren et al. (1982 a, b), Andersson et al. (1982), Barret et al. (1983), Svensson et al. (1983). This wide-spread attention is at least partly motivated by the potential use of this particular metal-semiconductor contact in high speed-logic, optoelectronic and microwave devices. However, the Al-GaAs system is also quite interesting from a merely scientific point of view.

Al and GaAs are both cubic materials, and they crystalize in the face-centered and the zinc-blende structure, respectively. The lattice constant of Al is smaller than that of GaAs by almost exactly a factor of 2. The ideal (001) surfaces of both materials thus match if the Al is rotated azimuthally by 45° relative to the GaAs. Therefore, epitaxial growth of Al films on GaAs(001) surfaces may be expected. Indeed, growth of single-crystalline Al films on GaAs(001) substrates has been observed. In addition to the expected (001) orientation, however, films oriented along (110) were found to grow (Ludeke et al. (1973), Cho and Dernier (1978), Marra et al. (1979)). The nucleation and thus the growth of the Al films was found to depend critically on the surface composition of the GaAs substrate (Massies et al. (1979 a, b), Petroff et al. (1981). Landgren et al. (1982 a)). Bulk truncated GaAs single crystals oriented along (001) would be either Ga- or As-terminated. Real GaAs(001) surface exhibit a number of different reconstructions which are each correlated with a particular chemical surface-composition that varies from almost purely Ga-terminated to As-terminated with an excess of chemisorbed As (Drathen et al. (1978), Bachrach et al. (1981)). This correlation has been critically reviewed by Mönch (1983). Surfaces exhibiting different structures and thus compositions are usually prepared by the technique of Molecular Beam Epitaxy (MBE). A crucial difficulty in these experiments is to control the amount of disordered and loosely bound arsenic on the surface, which may be either adsorbed from the As-background in the residual gas of the growth chamber during the cooling procedure of the sample or which may out diffuse from the bulk (Landgren et al. (1982 a)). Inclusions of As in single crystals prepared by the Horizontal Bridgman technique as well as by Liquid Encapsulation Czochralski growth have been detected by Bartels et al. (1983) on surfaces prepared by cleavage in ultra-high vacuum. These difficulties in surface preparation are also reflected by the scatter in the data reported for the relative positions of the valence band maximum and the Fermi level at (001) surfaces exhibiting the same reconstruction as observed by Rheed.

For the As-rich c(4 x 4) and 2 x 4) structures van der Veen et al. (1983) determined $E_F-E_{vs}=(0.4\pm0.1)$ eV. Svensson et al. (1983). on the other hand, reported $E_F - E_{vs} = (0.73 \pm 0.05)$ eV for both the As-rich c(4 x 4)- and the Ga-rich (4 x 6)-reconstructed surfaces. Despite these difficulties in the characterization of the bare GaAs(001) surfaces good agreement exists with regard to the barrier heights published for Al:GaAs(001) contacts. Barret et al. (1983) as well as Svensson et al. (1983) reported barrier heights of 0.8 and 0.75 eV, respectively. These values are independent of the initial reconstruction and thus the chemical composition of the GaAs(001) surfaces prior to the condensation of Al. The surface compositions were varied from the very As-rich c(4 x 4)- to the most Ga-rich c(8 x 2)-structure. Furthermore, Barret et al. (1983) measured the same barrier height with (001)- and (110)-oriented single-crystalline Al films. The work functions of both Al surfaces differ by 0.35 eV (Eastman & Mee (1973), Grepstad et al. (1976)), and thus this experimental result is another argument against Schottky's model.

DEEP LEVELS AT COMPOUND SEMICONDUCTOR INTERFACES

The Defect Model

In the past many attempts have been made to correlate the barrier heights in metal-semiconductor junctions with other properties of the associated two materials. They were motivated by the intentions to elucidate the physical mechanism that determines the position of the band edges with respect to the Fermi level at such interfaces. For III-V, II-VI and III-VI compound semiconductors, the correlation with the heat of formation of metal-anion compounds has been most successful. Examples of such plots are contained in Figures 7 and 8. The main observations to be explained are:

- The positions of the Fermi level with respect to the band edges are generally not the same at interfaces with n- or p-type semiconductors and they differ by approximately 0.3 eV.

- For both types of doping, one or two particular barrier heights are observed with a specific semiconductor.

Among the III-V compounds, GaAs has been most widely studied, and as Figure 9 reveals, the same positions of the Fermi level relative to the band edges are not only observed with various metals (Spicer et al. (1976)) but also in heterostructures (Mönch et al. (1982), G. Margaritondo (1982)) and on oxidized (Spicer et al. (1979), Grant et al. (1981)) and hydrogenated surfaces (Bartels et al. (1983)). Starting from these observations, which were not that complete at that time, Spicer et al. (1979, 1980) concluded that the mechanism pinning the Fermi level at the same position independent of the type of overlayer atoms has to be related with properties of the substrate rather than the overlayer atoms and they proposed native defects and their associated energy levels to determine the band bending at the interface. Similar conclusions have been reached by Wieder (1978) and by Williams et al. (1979). These defects are considered to be formed during the early stages of adsorption of the overlayer atoms.

Already in 1947, Bardeen suggested that surface states buried under a metal overlayer may play an essential role in fixing the position of the Fermi level at metal-semiconductor interfaces. Spicer's model now proposes states that originate from chemisorption-induced defects. In the following section we will discuss additional experimental results which strongly support Spicer's defect model and then review some of the speculations on the nature of these defects.

CHEMISORPTION-INDUCED INTERFACE STATES

From many recent studies it became evident that the electrical characteristics of metal-semiconductor contacts as well as semiconductor heterojunctions are essentially determined within an inter-

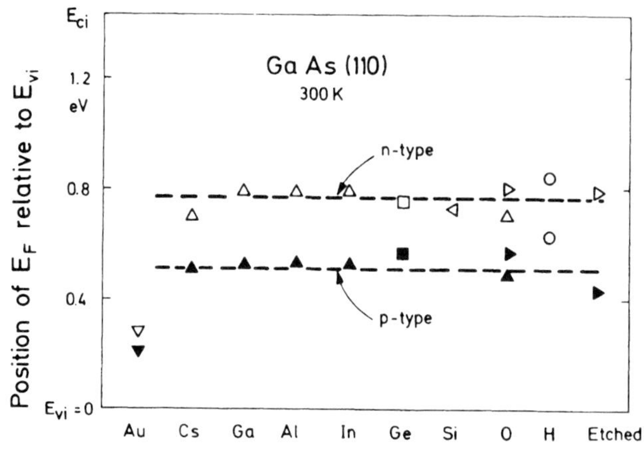

Fig. 9. Fermi-level pinning-position at various interfaces on cleaved GaAs (110) at 300k. (Data from Monch et al. (1982) (□), Bartels et al. (1983) (○), Margaritondo (1982) (◁), Spicer et al. (1976) (△), Skeath et al. (1981) (▽), and Grant et al. (1981) (▷)).

facial region only a few layers thick. Already much earlier it had been pointed out by Mönch (1970) that the electrical properties of metal-semiconductor junctions are fully developed after the deposition of the first monolayer of the metal. This conclusion which was based on studies of the chemisorption of Cs on Si(111) cleaved surfaces is again substantiated by the data plotted in Figure 10 for Al, Ga, In, and Ge evaporated on GaAs(110) surfaces. These data will be analyzed in the following. Although the GaAs(001) surfaces are the ones that are most important technologically, the (110) surfaces are considered since they can be prepared most easily and reproducibly by cleavage and are well characterized meanwhile.

Clean and well cleaved GaAs(110) surfaces exhibit no intrinsic surface states in the bulk band gap(van Laar & Scheer (1967) Huijser & van Laar (1975), Gudat et al. (1976), Spicer et al. (1976), Mönch & Clemens (1979)), and the bands are thus flat up to the surface. Band bendings observed immediately after cleavage are caused by extrinsic surface states. They are associated with cleavage imperfections which may be steps or point defects. With the bulk doping levels used the surface band-bending always measures less than 50 meV on p-type samples while it may amount to up to 0.6 eV depending on the quality of the surfaces cleaved from n-type samples. This is clearly visible in Figure 10. To our experience, only less than 10% of the cleavages from low-doped p-type samples exhibit a band bending larger than 50 meV.

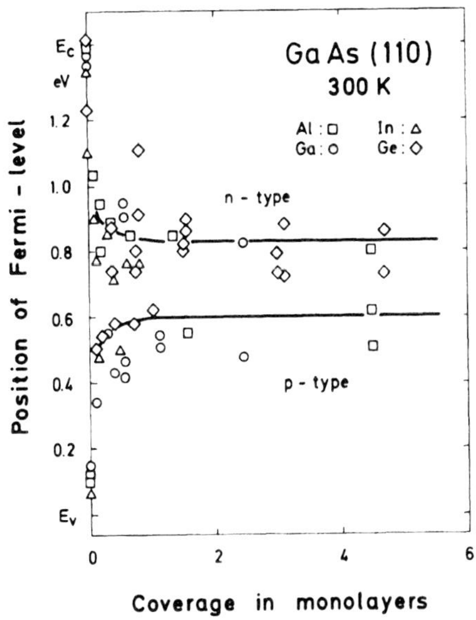

Fig. 10. Fermi level position at GaAs (110) surfaces as a function of coverage with various metals (Skeath et al. (1979)) and germanium (Mönch et al. (1982), Gant (1983)).

The deposition of the metal atoms Al, Ga and In as well as of the semiconductor atoms Ge on cleaved GaAs surfaces creates depletion layers on both n- and p-type samples. The changes of surface band-bending as a function of coverage are the same for these four different atoms and also for oxygen as has been recently shown by Mönch (1983). To achieve charge neutrality at the surface, the space charge Q_{sc} of the depletion layers has be be compensated by an excess charge Q_{ss} in surface states,

$$Q_{sc} + Q_{ss} = 0. \quad (7)$$

The deposition of metal and semiconductor atoms as well as the adsorption of oxygen thus creates extrinsic surface states. The space charge in depletion layers on p-type and n-type samples are of opposite sign since they are carried by negatively charged acceptors and positively charged donors, respectively. Together with Eq. (7) the data plotted in Figure 10 immediately indicate that acceptor- and donor-like surface states dominate on the n- and p-type doped samples, respectively.

The changes of the surface band-bending are already completed after deposition of a few tenths of a monolayer and the Fermi level then remains "pinned" up to the highest coverages. These positions of the Fermi level relative to the valence-band top at the interface are contained in Figure 9. As will be discussed in detail these values depend on the doping level of the samples used. This has to be kept in mind in comparing results published by different groups.

The thickness of the deposited metal or semiconductor films is restricted by the technique used to determine the band bending at the interface. A comparison of different methods has been given by Mönch (1983). For the data displayed in Figure 10 photoemission spectroscopy was employed. The electrons photon-excited at the interface have to penetrate the evaporated film to reach the surface. The escape depth of these electrons, which describes the exponential attenuation, measures in the order of 10A for the electron energies involved (Seah & Dench (1979), Gant & Monch (1981)) and is thus the limiting factor.

SURFACE SPACE CHARGE AND OCCUPATION OF SURFACE STATES

We will now discuss the evaluation of the energy levels and the densities of the surface states that are created by the chemisorption of metal and semiconductor atoms as well as of oxygen on GaAs (110) surfaces.

In a space charge layer the potential energy $e_o V(z)=E_{cb}-E_c(z)$, where $E_c(z)$ is the conduction band edge and E_{cb} its position in the bulk, is related to the charge density $\rho(z)$ by Poisson's equation

$$\frac{d^2V}{dz^2} = - \frac{\rho(z)}{\varepsilon_s \varepsilon_o} \tag{8}$$

As usual, ε_s is the dielectric constant of the semiconductor and ε_o the permittivity of vacuum. In strong depletion layers, as to be discussed in the present context, the space charge is carried by the charged impurities and the mobile carriers at the edge of the layer may be neglected, i.e. $\rho(z) = e_o N_D$ in an n-type semiconductor containing shallow donors only. Poisson's equation can then be integrated directly and yields the result

$$V = V_s(1 - z/w)^2 \tag{9}$$

where $w = \sqrt{2\varepsilon_s \varepsilon_o |V_s| / e_o N_D}$ is the width of the depletion layer.

For the space charge we obtain

$$Q^n_{sc} = e_o N_D w = (2\varepsilon_s \varepsilon_o N_D e_o |V_s|)^{1/2} \qquad (10)$$

The charge density Q_{ss} in surface states is determined by their energetic distribution and the Fermi Dirac distribution function f. The simplest model-distribution is a set of discrete surface states. The charge density in acceptor-like surface states, for example, is then given by

$$Q^A_{ss} = -e_o N^A_{ss} f(E^A_{ss}-E_F)$$
$$= -e_o N^A_{ss} f([E^A_{ss}-E_{cs}]-e|V_s|+[E_{cb}-E_F]). \qquad (11)$$

For the simple case of Eq. (10) and (11) the charge neutrality at the surface, Eq. (7), is illustrated in Figure 11. The intersections between the curve for the space charge and the set of curves for increasing densities of surface states are marked with circles and give those surface band-bendings that fulfill the condidion of charge neutrality. Starting from flat bands, a stepwise increase in the density of surface states first results in rather large changes of the band bending. For higher densities of surface states the band bending seems to saturate and the Fermi level is seemingly pinned at or, to be more precise, near to the energetic position of the surface states. However, an increase of the density of surface states to 2 x 10^{13} cm^{-2}, which still corresponds to only a few hundredths of a monolayer on semiconductors, increases the band bending at the surface. By this, the acceptor-type surface states move to above the Fermi level and are now only partly occupied. This is more clearly illustrated in Figure 12. The equilibrium band-bendings at the surface, as marked in Figure 11 by the circles, are plotted as a function of the density of surface states. This plot again "suggests" the pinning of the Fermi level at the energetic position of the surface states, but the insert shows that the band bending does not truly saturate with increasing density of surface states.

Band bending at the surface and thus the position of the Fermi level relative to the band edges also depends on the density of the bulk doping atoms.

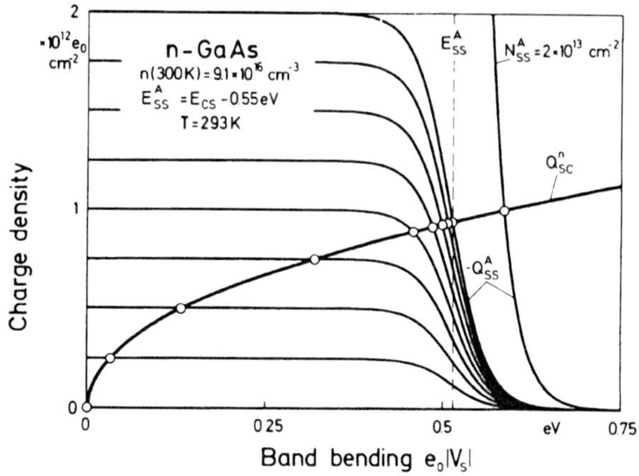

Fig. 11. Space charge in a depletion layer on n-GaAs and occupancy of acceptor-type surface states as a function of surface band-bending.

OCCUPANCY LEVEL OF CHEMISORPTION-INDUCED DEFECTS ON GaAs(110)

Figure 10 contains the Fermi level position measured from the top of the valence band as a function of the coverage of Al, Ga, In(Skeath et al. (1979)) and Ge (Mönch et al. (1982), Gant (1983) on initially clean (110) surfaces cleaved in situ from n- and p-type GaAs. These data were obtained from photoemission spectroscopy using soft X-rays (SXPS) as well as ultraviolet light (UPS). These techniques measure the photoemission from the valence band or from core levels which follow the band bending in the space charge layer. The experimental results clearly show that the Fermi level is pinned at approximately 0.75 and 0.5 eV above the top of the valence band on n- and p-type substrates, respectively, after covering the surfaces by about a tenth of a monolayer. Some of the data have to be corrected by approximately 60 meV since photoemission not only probes the very surface layer and its potential but also the curvature of the bands within the space charge layer up to a depth of some escape lengths of the photoelectrons (Mönch et al. (1982)). From the saturation of the band bending above coverages of about one tenth of a monolayer the density of interface states can be estimated to be some percent of a monolayer by considering the results of Figure 12. In SXPS and UPS - as presently used - the limits of experimental error amount to 0.08 eV for the determination of band bendings and its changes, Thus, from such data, the energetic position and the density of the interface states can only be approximately evaluated. Therefore, additional experimental techniques exhibiting smaller limits of experimental error should be used.

Another quantity containing the band bending at the surface is

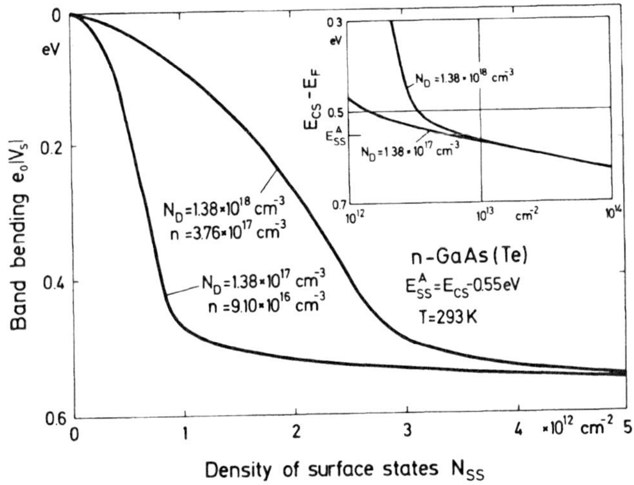

Fig. 12. Surface band-bending of a depletion layer on n-GaAs as a function of the density of acceptor-type surface states for two different bulk doping-concentrations.

the work function

$$\Phi_s=E_{vac}-E_F=I-e_o|V_s| +(E_{cb}-E_F). \tag{12}$$

With a Kelvin probe, for example, changes of the contact potential difference between the probe and the sample, and thus of the work function of a sample

$$\Delta CPD =\Delta\Phi_s = \Delta I- e_o\Delta|V_s| \tag{13}$$

can be recorded to an accuracy of 0.1 meV. However, the reproducibility after resetting the probe is limited to 10 meV. Provided the ionization energy remains constant, one obtains

$$\Delta CPD =\Delta\Phi_s = e_o\Delta|V_s| \tag{14}$$

Within the limits of experimental error this behaviour was found with Ge chemisorbed on clean, cleaved GaAs(110) surfaces at 300°C by Gant & Mönch (1982, 1983). Figure 13 compares the change in work function measured with a Kelvin probe and the band bending determined from ultraviolet photoemission spectra as a function of coverage. The

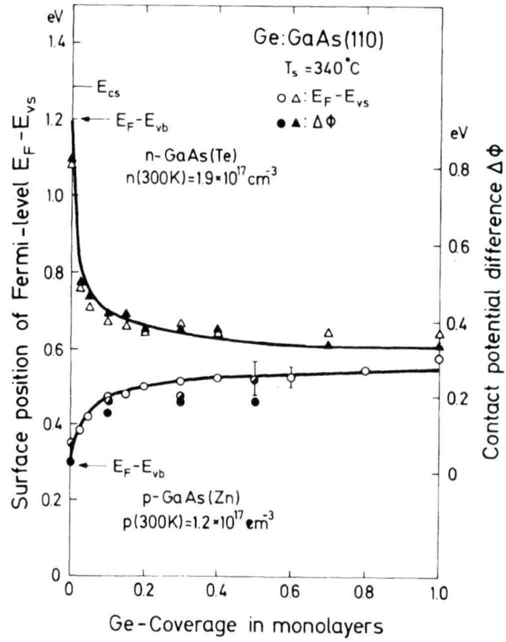

Fig. 13. Surface band-bending and work function as a function of Ge-coverage on (110 surfaces cleaved from p- and n-type GaAs (Gant and Mönch (1982, 1983)). The curves are at least-square fits to the experimental data with the parameters listed.

coverage $\Theta = 1$ measures the total number σ_{110} of sites in a Ge and GaAs(110) plane, i.e. $\sigma_{110} = 8.85 \times 10^{14}$ cm^{-2}. Both sets of measurements agree and therefore the CPD data can be read as the sought-for but now more precisely determined, changes in surface band-bending. The insensitivity of the ionization energy to the coverage with Ge is plausible. Ga, Ge, and As follow each other in the periodic table of the elements and their electronegativities are nearly equal. According to Pauling's concept (1940), the ionic character of Ge-Ga and Ge-As covalent bonds is estimated to be only a few percent. For the chemisorption of Ge on GaAs(110) the ionization energy is thus expected to change by less than 50 meV which value is smaller than the experimental error and thus escapes detection.

Ge chemisorption on GaAs(110) surfaces at room temperature, however, not only introduces a band bending but also decreases the ionization energy by (70 $\pm$ 20) meV (Gant (1983)). This indicates that the atomic arrangement and thus the charge transfer is different for Ge atoms condensed at room temperature and above 300°C where thicker films are growing amorphously and epitaxially, respectively.

The Kelvin method measures the difference in work function between probe and surface under investigation and cannot be used for the study of interfaces. Above half a monolayer of coverage the formation of the overlayer tends to dominate the chemisorption at the substrate (Monch & Gant (1980)). Therefore, the CPD-versus-Θ curves measured with Ge on GaAs(110) are analyzed for the parameters of the surface states up to only Θ = 0.5. The space charges in the depletion layers formed on n- as well as on p-type substrates are of opposite sign and acceptor- and donor-type surface states have to be present - or at least predominantly charged- on n- and p-type samples, respectively. Since the surface states are induced by chemisorption we assume their density to increase proportional to the coverage with Ge.

As figure 13 clearly shows good fits for n- as well as for p-type substrates are achieved with acceptor- and donor-type surface states being simultaneously present at

$$E_{ss}^{A} = E_{vs} + (0.73 \pm 0.05)\,eV \qquad (15)$$

$$E_{ss}^{D} = E_{vs} + (0.42 \pm 0.42\ eV \qquad (16)$$

with slightly different density-factors

$$q^{A} = 0.06 \pm 0.03 \qquad (17)$$

$$q^{D} = 0.04 \pm 0.02 \qquad (18)$$

The energies of these levels are temperature dependent, and at room temperature their positions are

$$E_{ss}^{A} = E_{vs} + (0.78 \pm 0.05)\,eV, \qquad (19)$$

$$E_{ss}^{D} = E_{vs} + (0.51 \pm 0.05)\,eV \qquad (20)$$

The already mentioned data measured with metal-GaAs (110) contacts by Spicer and coworkers cannot be directly compared with the above given results obtained with Ge:GaAs (110) heterojunctions since the doping levels of the samples used in both studies were quite different. However, with the parameters of the chemisorption-induced interface states the band bending at room temperature may be calculated for samples exhibiting the doping level used by the Stanford group and it is plotted as a function of coverage in Figure 14. The experimental points are very well fitted by the calculated curves. This strongly indicates that the barrier heights in metal: GaAs (110) as well as the band bending on the GaAs side of Ge: GaAs (110) junctions can be described by the same set of chemisorption-induced interface states. An additional feature is that the density of these states is independent of whether they are created by condensation of metal or Ge atoms at room temperature or by epitaxial growth of Ge at 300°C.

ON THE NATURE OF THE CHEMISORPTION-INDUCED DEFECTS

From the observation that in some metal: GaAs (110) junctions the positions of the band edges relative to the Fermi level are insensitive to the metals and differ by approximately 0.25 eV on n- and p-type substrates Spicer et al. (1979, 1980) have inferred that the interface states responsible for that pinning of the Fermi level originate in defects created at or near to the metallurgical interface during the condensation of the metal atoms. Gant and Mönch (1982) then proposed a single defect being chemisorption-induced since they observed both donor- and acceptor-type levels on p- as well as on n-type samples. A single defect model for GaAs(110) was also favored by Grant et al. (1981).

The nature of these chemisorption-induced defects could not be identified by experiments till now and speculations based on theoretical models dominate (Daw & Smith (1979, 1980, 1981), Nishida (1980 a, b), Allen & Dow (1981, 1982 a,b), Dow & Allen (1982)). The most widely discussed defects are vacancies and antisite defects.

The formation of chemisorption-induced vacancies in metal- and semiconductor-GaAs(110) junctions may be excluded. Studies of radiation damage have shown vacancies to be stable up to only 220°C (Lang (1977)) while the chemisorption-induced interface states are observed at room temperature and at 300° C with the same density as shown by the good agreement of the data plotted in Figure 14.

The As_{Ga} antisite defect is the candidate most favored as the chemisorption-induced defect in GaAs(110). Such defects have been identified by electron paramagnetic resonance (EPR) by Wagner et al. (1980) in as-grown, by Wörner et al. (1982) in neutron-irradiated and by Weber et al. (1982) in plastically deformed GaAs. The As_{Ga} defects in GaAs are double donors and their occupation levels have been determined by Weber et al. (1982) as

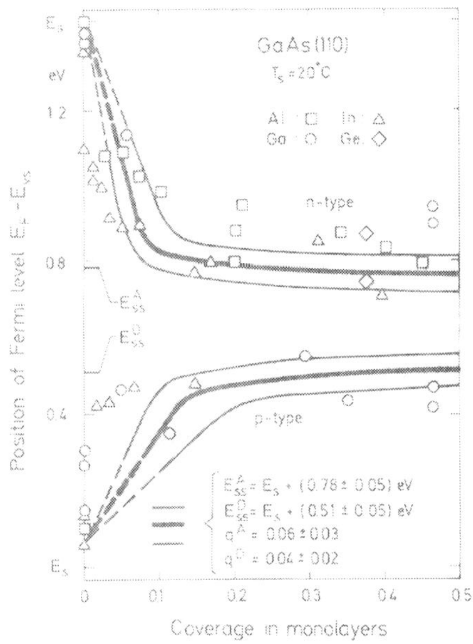

Fig. 14. Position of the Fermi level with respect to the top of the valence band on GaAs(110) surfaces as a function of coverage with metals (Skeath et al. (1979)) and Ge (Mönch et al. (1982)) at room temperature and as calculated from the parameters of the chemisorption-induced surface states as measured with Ge coverage (Gant (1983)).

$$E(o/+) = E_v + 0.75 \text{ eV and,} \qquad (21)$$

$$E(+/{++}) = E_v + 0.52 \text{ eV.} \qquad (22)$$

They are quite close to the energy levels induced by chemisorption of Ge on GaAs (110) surfaces.

As they are double donors, As_{Ga} antisite defects alone cannot explain the Fermi-level pinning in metal- and semiconductor-GaAs (110) interfaces. As_{Ga} antisite defects can be only part of then more complicated chemisorption-induced defects which are characterized by the net charge-states C^-, C^o, and C^+. Most simply, these net charge-states can be achieved, for example, by adding

- shallow acceptors A_s^- or
- double-acceptors with the charge states A_d^{2-}, A_d^{4-} and A_d^0 to the As_{Ga} double-donors with D_d^{2+}, D_d^{1+} and D_d^0.

For a more detailed discussion we should like to refer to a paper by Mönch (1982).

The As_{Ga} antisite defects in neutron-irradiated, plastically deformed and as-grown semi-insulating GaAs samples are not thermally stable but decay during isochronal anneals at temperatures above 400° C. Results published by Schneider and coworkers are plotted in the upper panel of Figure 15.

Motivated by these observations Gant and Mönch (1983) investigated the annealing behavior of the band bending created by the chemisorption of 0.3 of a monolayer of Ge on (110) surfaces cleaved from n- as well as p-type GaAs samples. They studied photoemission spectroscopy with 21.2 eV photons and determined the band bending and its changes during isochronal anneals of 10 min. at temperatures up to 590° C from the recorded energy distributions of the electrons emitted. Their data are plotted in the lower part of Figure 15.

Figure 15 reveals the Fermi-level position to remain unchanged by anneals up to 400° C. On the p-type sample the distance of the valence band edge to the Fermi level is decreased by 0.3 eV after anneals above 400° C and at the surface the positions of E_F relative to E_v is now quite close to the one in the bulk. On the n-type sample, however, a strong band bending still exists after the anneal at 590° C and it even increases by approximately 100 meV during the heat treatments between 400 and 500° C. Additional experimental data, which will be discussed elsewhere, prove that these results cannot be explained by an indiffusion of Ge or a decomposition of the surfaces. The observed change of the band bending during the anneals rather demonstrate that the chemisorption-induced surface levels of donor-type, which carry the surface charge to compensate the charge in the depletion layer on the p-doped sample, decay above 400° C while the acceptor-type states prevail up to 600° C.

As Figure 15 illustrates, the annealing behaviour of As_{Ga} antisite defects, which are double donors, and of chemisorption-induced surface states of donor-type is evidently the same. Together with the already mentioned close agreement of the energy levels this finding represents another experimental support for As_{Ga} antisite defects to play a major role in the pinning of the Fermi level at metal-, insulator and semiconductor-GaAs interfaces. As already inferred from the double-donor character of As_{Ga} antisite defects, the results presented in Figure 15 establish that the chemisorption-induced defects cannot be simple As_{Ga} antisite defects but have to exhibit a more complicated structure. Possibly, antisite defects of Ga_{As} which are double acceptors are additionally formed or As_{Ga}-Ga_{As} antistructure pairs are created.

While with metal-semiconductor contacts prepared on clean surfaces cleaved from n-type GaAs the barrier height is insensitive of

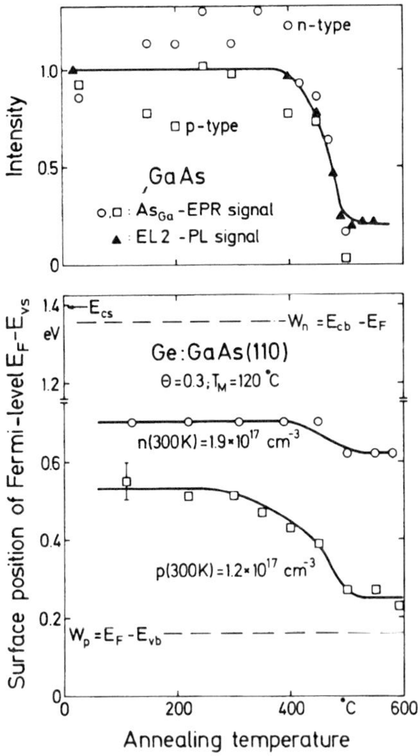

Fig. 15. Annealing behavious of the EPR signal of As_{Ge} antisite defects produced by neutron irradiation (Wörner et al. (1982)), of the 0.8 eV photoluminescence (PL) signal observed with LEC-GaAs Windscheif et al. (1983), and of the band bending induced by the deposition of 0.3 monolayers on surfaces cleaved from p- and n-type GaAs (Gant and Mönch (1983)).

the metal used two distinctly different barrier heights are observed with n-type InP, GaP, GaSe, CdS, CdTe, ZnS and ZnO (see e.g. Figs. 5, 7 and 8 and Figure 2 in the paper by Brillson (1978)). Following the defect model, two different defects are responsible for the two distinct pinning positions of the Fermi level at the interface.

It has been first noted by Brillson (1978) that these two distinct values of the barrier height correlate well with positive and negative heats of reaction as defined by Eq. (4). However, the turning points of the S-shaped ϕ_B - versus-ΔH_r curves deviates by up to $\pm$ 1 eV from $\Delta H_r = 0$. In all cases, with the exception of ZnO, a much better correlation is obtained by using the enthalpies of formation of antistructure pairs, as estimated by van Vechten (1975), rather than the heat of formation of the specific compound semiconductor. This correlation between the change from one value of the barrier height to the other one and the heat of formation of the antistructure pair may be accidental. Nevertheless, this observation is suggestive since it means that for heats of formation of metal-anion compounds exceeding the enthalpy of formation of the antistructure pair such defects are created. For lower $H_f(MA)$ the energy is not sufficient and another defect with lower enthalpy of formation and different energy levels is induced. These might be, for example, isolated antisite defects since as has been shown by van Vechten (1975) their heats of formation are only half of that needed for the production of antistructure pairs. Even in the case of GaAs where no -shaped ϕ_B-versus- H_f curve is observed different defects might be formed, too, above and below a specific H_f threshold. However, they might not be detectable from the pinning position of the Fermi level in Schottky barriers with different metals if their energy levels are at nearly the same positions in the energy gap. In any case, the defect model offers a plausible explanation for the S-shaped ϕ_B-versus- H_f curves even if the nature of the defects possibly involved cannot be concluded from those curves.

FINAL REMARKS

In the present paper we have tried to outline some physical concepts on deep levels in the bulk and at interfaces of compound semiconductors. We have limited our attention mostly to GaAs. Both types of deep levels are essential for device performance, and with decreasing size and increasing density of devices in Very Large Scale Integration (VLSI) especially interface states are becoming more and more important. Even if the bulk of GaAs layers, for example, grown by the techniques of Liquid Phase or Molecular Beam Epitaxy (LPE and MBE, respectively) contains no deep impurities this is not necessarily true for the interface between the substrate used and the overlayer grown on it. In any case, however, the preparation of metal or insulator junctions on top of such high-quality GaAs layers introduces deep interface states. In this regard other semiconductors, as for example InP, might prove to be more favorable but further studies are needed to clarify this. Not all aspects of this subject could be treated in the present paper. For further details we should like to refer to some recent review papers by Wieder (1982) and Mönch (1982).

The physical insight in deep bulk-impurities has made considerable progress during recent years. With respect to interfaces the model of chemisorption-induced defects has proven to be extremely fruitful. Even if no defect has been unambiguously identified until now, many experimental results strongly point to As_{Ga} antisite defects to play a major but not the only role in metal-, insulator- and also some semiconductor-GaAs interfaces. Further studies are needed in this field, too.

This article was completed in August 1983. Since then, much progress has been achieved in the understanding of semiconductor surfaces and interfaces. With regard to the topics dealt with in the present paper, the most important development has been that the barrier heights in metal-semiconductor contacts may described by the tailing of the metal wave-functions into the virtual gap states of the complex band structure of the semiconductor. In a simple picture, the virtual gap states of the semiconductor are accupied according to the difference in electronegativity of the metal and the semiconductor (Mönch (1986 a)). The chemical trends in barrier heights of Schottky contacts may be completely understood by knowing the charge-neutrality levels of the virtual gap states (Tejedor et al. (1977), Tersoff (1984 a, 1986)) and the electronic polarizibilities of the semiconductors (Mönch (1986 a)). This picture is based on early considerations by Heine (1965) and was shown to explain the line-up of the bands in semiconductor heterostructure, too (Tejedor and Flores (1978), Tersoff (1984 b)). By now, defects are assumed to play a less important role in determining the band alignments in semiconductor interfaces (Mönch (1986 b)).

Acknowledgements

The manuscript was typed by Sabine Reif and the drawings were prepared by J. Krusenbaum.

REFERENCES

1. Allen, R.E., and Dow, J.D., 1981, J.Vac. Sci. Technol. 19:383.
2. Allen, R.E., and Dow, J.D., 1982 (a), Applic. Surface Sci. 11/12/:362.
3. Allen, R.E. and Dow, J.D., 1982 (b), Phys. Rev. B 25:1423.
4. Altarelli, M., and Bassani, R., 1982, in Handbook on Semiconductors, Ed. W. Paul, North-Holland, New York, Vol. 1, p. 269.
5. Andersen, R.L., 1962, Solid State Electronics 5:341.
6. Andersson, T.G., Svensson, S.P., and Landgren, G., 1982, J. Phys. C: Solid State Phys. 15:6673.
7. Andrew, J.M. and Phillips, J.C., 1975, CRC Crit. Rev. Solid State Sci. 5:405.
8. Bachrach, R.Z., Bauer, R.S., Chiaradia, P. and Hansson, G.G., 1981, J. Vac. Sci. Technol. 19:335.
9. Bardeen J., 1947, Phys, Rev. 71:717
10. Barret, C., Chekir, F., Nefatti, T., Vapaike, A., and Massies, J., 1983, Physica 117 B & 118 B:851.
11. Bartels, F., Clemens, H.J. and Monch, W., 1983, Physica 117 B & 118 B:801
12. Bartels, F., Surkamp, L., Clemens, H.J., und Mönch. W., 1983, J. Vac. Sci. Technol. B1 : 756

13. Bergh, A.A., and Dean, P.J., 1976, "Light-emitting Diodes", Oxford, London.
14. Braun, F., 1874, Pogg. Ann. 153:556; 1877, Wied. Ann. 1:95, 1878, ibid. 4:476, 1883, ibid. 19:340.
15. Brillson, L.J., 1978, Phys, Rev. Letters 40:1378.
16. Cho, A.Y. and Dernier, P.D., 1978, J. Appl. Phys. 49:3328.
17. Daw, M.S. and Smith, D.L., 1979, Phys. Rev. B 20:5150.
18. Daw, M.S. and Smith, D.L., 1980, Appl. Phys. Lett. 36:690.
19. Daw, M.S. and Smith, D.L., 1981, Solid State Commun. 37:205.
20. Dow, J.D. and Allen, R.E., 1982, J. Vac. Sci. Technol. 20:659
21. Drathen, P., Ranke, W. and Jacobi, K., 1978, Surface Sci. 77:L162.
22. Eastman, R.M. and Mee, C.H.B., 1973, J. Phys. F:Metal Phys. 3:1738.
23. Frankl,D.1967, "Electrical Properties of Semiconductor Surfaces", Pergamon, Oxford.
24. Gant, H. and Mönch, W., 1981, Surface Sci. 105:217.
25. Gant, H. and Mönch, W., 1982, Applic. Surface Sci. 11/12:332
26. Gant, H., 1983, Thesis Universität Duisburg.
27. Grant, R.W., Waldrop, J.R., Kowalsczyk, S.P. and Kraut, E.A., 1981, J. Vac. Sci. Technol. 19:477.
28. Grepstad, J.K., Gartland, P.O. and Slagsvold, 1976, Surface Sci. 57:348.
29. Gudat, W., Eastman, D.E. and Freeouf J.L., 1976, J.Vac. Sci. Technol. 13:250.
30. Hansen, M., 1958/1965/1969, "Constitution of Binary Alloys" and Supplements by Elliott, R.P. and Shunk, F.A., McGraw Hill, New York.
31. Harrison, W.A., 1977, J.Vac. Sci. Technol. 14:1016.
32. Harrison, W.A., 1980, "Electronic Structure and the Properties of Solids", Freeman, San Francisco.
33. Henry, C.H. and Lang, D.V., 1977, Phys, Rev. B 15:989.
34. Hjalmarson, H.P., Vogl, P., Wolford, D.J. and Dow, J.D.. 1980, Phys, Rev. Lett. 44:810.
35. Holmes, D.E., Chen, R.T., Elliott, K.R. and Kirkpatrick, C.G., 1982, Appl. Phys. Lett. 40:46.
36. Hsu, W.J., Dow, J.D., Wolford, D.J. and Streetman, B.G., 1977, Phys. Rev. B 16:1597.
37. Hughes, G.J., McKinley, A., Williams, R.H. and McGovern, I.T., 1982, J. Phys. C: Solid State Phys. 15:L159.
38. Huijser, A. and Laar, van, J., 1975, Surface Sci. 52:202.
39. Jantsch, W., Wunstel, K., Kumagi, O. and Vogl, P., 1982, Phys. Rev. B 25:5515.
40. Jaros, M., 1982, "Deep levels in Semiconductors", Hilger, Bristol.
41. Katnani, A.D. and Margaritondo, G., 1983, Phys. Rev. B in press.
42. Kaufmann, U. and Schneider, J., 1980, Festkörperprobleme XX:87, Ed. by J. Treusch, Vieweg, Braunschweig.
43. Kaufmann, U., Schneider, J., Worner, R., Kennedy, T.A., and Wilsey, N.D., 1981, J. Phys. C: Solid State Phys. 14:L951.
44. Kroemer, H., 1975, CRC Crit. Rev. Solid State Sci. 5:555.

45. Kroemer, H., 1983, Proc. 2nd Course of the Intl. School on Solid-State Device Research, Erice "NATO ASI-Series" Series E, No. 87 Dordrecht 1985, p.331.
46. Kubaschewski, O., and Alcock, C.B., 1979, "Metallurgical Thermochemistry", Pergamon, Oxford.
47. Kurtin, S., McGill, T.C. and Mead, C.A., 1970, Phys. Rev. Lett. 22:1433.
48. Laar, van, J. and Scheer, J.J., 1967, Surface Sci. 8:343.
49. Laar, van, J., Huijser, A., and Rooy, van, T.L., 1977, J. Vac. Sci. Technol. 14:894.
50. Lagowski, J., Gatos, H.C., Parsey, J.M., Wada, K., Kaminska, M., and Walukiewicz, 1982, Appl. Phys. Lett. 40:342.
51. Landgren, G., Ludeke, R. and Serrano, C., 1982, J. Crystal Growth 60:393.
52. Landgren, G., Svensson, S.P. and Andersson, T.G., 1982, Surface Sci. 122:55.
53. Lang, D.V., 1977, Inst. Phys. Conf. Ser. No. 31:70.
54. Ludeke, R., Chang, L.L., and Esaki, L., 1973, Appl. Phys. Lett. 23:801.
55. Many, A., Goldstein, Y., and Grover 1965, "Semiconductor Surfaces", North-Holland, Amsterdam.
56. Margaritondo, G., 1983, Surface Ser. 132:469
57. Marra, W.C., Eisenberger, P. and Cho, A.J., 1979, J. Appl. Phys. 50:6927.
58. Martin, G.M., Farges, J.P., Jacob, G., Hellais, J.P. and Poibland, G., 1980, J. Appl. Phys. 51:2840.
59. Massies, J., Etienne, P. and Linh, N.T., 1979 (a), Surface Sci. 80:550.
60. Massies, J., Chaplart, J. and Linh, N.T., 1979 (b), Solid State Commun. 32:707.
61. Massies, J. and Linh, N.T., 1982, Surface Sci. 114:147.
62. Mönch, W., 1970, Surface Sci. 21:443.
63. Mönch, W. and Clemens, H.J., 1979, J.Vac. Sci. Technol. 16:1238.
64. Mönch, W., and Gant, H., 1980, J. Vac. Sci. Technol. 17:1094.
65. Mönch, W., and Gant, H., 1982 (a), Phys. Rev. Lett. 48:512.
66. Mönch, W., Bauer, R.S., Gant, H., and Murschall, R., 1982 (b) J. Vac. Sci. Technol. 21:498.
67. Mönch, W., 1982, Proc. 2nd IUPAP-UNESCO Semiconductor Symp., Trieste, Surface Sci. in press.
68. Mönch, W., 1983, Proc. 2nd Course of the Intl. School on Solid-State Device Research, Erice, "NATO ASI-Series", Series E No. 87 Dordrecht, 1985, p.105
69. Mönch, W., 1983, Thin Solid Films 104:285.
70. Nishida, M., 1980 (a), Surface Sci. 99:L 384.
71. Nishida, M., 1980 (b), J. Phys, Soc. Japan 49 Suppl. A:1093.
72. Pauling, L., 1940, "The Nature of the Chemical Bond", Cornell, Ithaca, N.J.
73. Petroff, P.M., Feldman, L.C., Cho, A.J. and Williams, R.S., 1981, J. Appl. Phys. 52:7313.

74. Poganski, S., 1952, Z. Elektrochemie, 56:193.
75. Poganski, S., 1953, Z. Physik, 134:469.
76. Puddephat, R.J., 1978, "The Chemistry of Gold", Elsevier, Amsterdam.
77. Queisser, H.J., 1971, Festkörperprobleme XI:45, Ed. by O. Madelung, Vieweg, Braunschweig.
78. Scheffler, M., 1982, Festkörperprobleme XXII:115, Ed. by. P. Grosse, Vieweg, Braunschweig.
79. Schneider, J., 1982, Proc. 2nd Conference on Semi-Insulating III-V materials, Evian (France) in press.
80. Schottky, W., 1939, Z. Physik 113:367.
81. Schottky, W., 1942, Z. Physik 118:539.
82. Seah, M.P., and Dench, W.E., 1979, Surface Interface Analysis 1:2.
83. Skeath, P., Lindau, I., Chye, P.W., Su, C.Y. and Spicer, W.E. 1979, J. Vac. Sci. Technol. 16:1143.
84. Skeath, P., Su, C.Y., Hino, I., Lindau, I., and Spicer, W.E., 1981, Appl. Phys. Letters 39:349.
85. Spicer, W.E., Lindau, I., Gregory, P.E., Garner, C.M. Pianetta, P. and Chye, P.W., 1976, J. Vac. Sci. Technol. 13:780.
86. Spicer, W.E., Chye, P.W., Skeath, P.R., Su, C.Y. and Lindau, I., 1979, J. Vac. Sci. Technol. 16:1422.
87. Spicer, W.E., Lindau, I., Pianetta, P., Chye, P.W. and Garner, C.M., 1979, Thin Solid Films 56:1.
88. Spicer, W.E., Chye, P.W., Skeath, P.R., Su, C.J. and Lindau, J., 1979, J. Vac. Sci. Technol. 16:1422.
89. Spicer, W.E., Lindau, J., Skeath, P. and Su, S.Y., 1980, J. Vac. Sci. Technol. 17:1019.
90. Svensson, S.P., Kanski, J., Andersson, T.G. and Nilsson, P.O., 1983, Surface Sci. 124:L31.
91. Tegude, F.-J., 1983, Thesis Universität Duisburg.
92. Vechten, van, J.A., 1975, J. Electrochem. Soc. 122:423.
93. Veen, van der, J.F., Smit, L., Larsen, P.K. and Neave, J.H., 1983, Physica 117 B & 118 B:822.
94. Vogl, P., 1981, Festkörperprobleme XXI:191, Ed. by. J. Treusch, Vieweg, Braunschweig.
95. Vogl, P., 1983, J. Electronics and Electron Physics, to be published.
96. Wagman, D.D., Evans, W.H. Parker, V.B., Halow, I., Bailey S.M. and Schumm, R.S., 1968 - 1971.
NBS Tech. Notes 270:3 - 270:5 (National Bureau of Standards, U.S. Department of Commerce).
97. Wagner, R.J., Krebs, J.J., Stauss, G.H. and White, A.M., 1980, Solid State Commun. 36:15.
98. Waldrop, J.R., Kowalczyk, S.P. and Grant R.W., 1983, Appl. Phys. Letters 42:453.
99. Weber, E.R., Ennen, H., Kaufmann, U., Windscheif, J., Schneider, J. and Wosinski, T., 1982, J. Appl. Phys. 53:6140.
100. Wieder, H., 1978, J. Vac. Sci. Technol. 15:1498.
101. Wieder, H., 1983, Surface Sci. 132:390

102. Williams, R.H., Varma, R.R. and Montgomery, V., 1979, J. Vac. Sci. Technol. 16:1418.
103. Williams, R.H. and Patterson, M.H., 1982, Appl. Phys. Lett. 40:484.
104. Windscheif, J., Ennen, H., Kaufmann, U., Schneider, J. and Kimura T., 1983, Appl. Phys. A 30:47.
105. Wolford, D.J., Hsu, W.J., Dow, J.D. and Streetman, B.G., 1979, J. Lum. 18/19:863.
106. Wosinski, T., Morawski, A. and Figielski, T., 1983, Appl. Phys. A 30:233.
107. Wörner, R., Kaufmann, U. und Schneider, J., 1982, Appl. Phys. Lett. 40:141.

ENSEMBLE MONTE CARLO TECHNIQUES

Carlo Jacoboni

Instituto di Fisica dell'Università di Modena
Via Campi 213/A
Modena, Italy

INTRODUCTION

Charge transport in semiconductors is in general a difficult problem, both from the mathematical and the physical point of view. In fact, the integro-differential equation (The Boltzmann equation) which describes the problem does not offer simple, or even complicated, analytical solutions except for very few cases, usually not applicable to real systems. Furthermore, since transport quantities derive from average procedures over many physical processes whose relative importance is not known a priori, the formulation of reliable microscopic models for the physical system under investigation is difficult. When one moves from linear to nonlinear response conditions, difficulties become even greater: the analytical solution of the Boltzmann equation without linearization with respect to the external force is a formidable mathematical problem which has resisted many attacks in the last decades. In order to get any result, it is necessary to perform such drastic approximations that it is no longer clear whether the features of interest in the results are due to the microscopic model or to mathematic approximations.

From the foregoing it is understandable why, when two new numerical approaches to this problem, i.e. the Monte Carlo (MC) technique and the iterative technique [2] were presented at the Kyoto Semiconductor Conference in 1966, hot-electron physicists received the new proposals with great enthusiasm. It was in fact clear that, with the aid of modern large and fast computers, it would become possible to obtain exact numerical solutions of the Boltzmann equation for microscopic physical models of a high complexity. These two techniques were soon developed to a high degree of refinement, [3,5] and since then they have been widely used to obtain results for various situations in practically all materials of interest.

The MC method is by far the most popular of the two techniques mentioned above because it is easier to use and more directly interpretable from the physical point of view. MC is in fact a statistical numerical means used, in general, for solving mathematical problems; as such, it was born well before its application to transport in semiconductors [6] and has been applied to a large number of scientific fields. In the case of charge transport, however, the statistical numerical approach to the solution of the Boltzmann equation proves to be a direct simulation of the dynamics of charge carriers inside the crystal, so that, while the solution of the equation is being built up, any physical information required can be easily extracted. In this respect it is perhaps useful to stress that, once a numerical solution of a given problem is obtained, its subsequent physical interpretation is still a very important state to be overcome, in order to achieve a further step in the knowledge of the phenomenon under investigation. The MC method shows itself to be a very useful tool for this purpose, since it enables simulations of particular physical situations unattainable in experiments, or even investigations of nonexisting materials in order to emphasize special features of the phenomenon under study. This use of the MC makes it similar to an experimental technique: the simulated experiment can in fact be compared with analytically formulated theories.

In these lectures we shall first summarize the principles of the MC method and shall then move to discuss its generalization to the many-particle simulation, which is necessary when space or time dependent phenomena are to be investigated. Particular attention will be paid to the problem of diffusion, of the possible inclusion of particle-particle interaction and of device simulation.

THE BASIC MONTE CARLO METHOD

Fundamentals

The MC method, [5,7,8] as applied to charge transport in semiconductors, consists of a simulation of the motion of one or more electrons (for brevity we shall often write "electrons" meaning in fact "charge carriers", that is electrons or holes, indifferently) inside the crystal, subject to the action of external forces due to applied electric and magnetic fields and of given scattering mechanisms. The duration of each carrier flight, i.e. the time between two successive collisions, and the scattering events involved in the simulation are selected stochastically in accordance with some given probabilities describing the microscopic processes. As a consequence, the foundation of any MC method relies on the generation of a sequence of random numbers with given probability distributions. A discussion of the techniques used to generate such numbers in connection is given in [7,8]

When the purpose of the analysis is the investigation of a steady-state, homogeneous phenomenon, then it is sufficient in general to simulate the motion of one single electron; from ergodicity we may assume that a sufficiently long path of this sample electron will give information on the behavior of the entire electron gas. When, on the contrary, the process under investigation is not homogeneous or not stationary, then it is necessary to simulate a large number of electrons and follow them in their dynamic histories in order to obtain the desired information on the process of interest.

A Typical Monte Carlo Program

Let us summarize here the major structure of a typical MC program suited to the simulation of a stationary, homogeneous, transport process. More details on each step of the procedure will be given in the following. For the sake of simplicity we shall refer to the case of electrons in a simple semiconductor(*) subject to an external electric field $\vec{E}$.

The simulation starts with an electron in given initial conditions with wave vector $\vec{k}_o$; then the duration of the first free flight is chosen with a probability distribution determined by the scattering probabilities. During the free flight the external forces are made to act according to the relation

$$\hbar \dot{\vec{k}} = e\vec{E} \tag{1}$$

where $\vec{k}$ is the carrier wave vector, e its charge, and $\hbar$ the Planck constant. In this part of the simulation all quantities of interest, velocity, energy, etc..., are recorded.

Then a scattering mechanism is chosen as responsible for the end of the free flight, according to the relative probabilities of all possible scattering mechanisms. From the differential cross section of this mechanism, a new $\vec{k}$ state after scattering is randomly chosen as initial state of the new free flight, and the entire process is iteratively repeated. The results of the calculation become more and more precise as the simulation goes on, and the simulation ends when the quantities of interest are known with the desired precision.

A simple way to determine the precision, that is the statistical uncertainty of transport quantities, may consist of splitting the entire history into a number of successive subhistories of equal time duration, and making a determination of a quantity of interest for each of them. We then determine the average value of each quantity and take its standard deviation as an estimate of its statistical uncertainty. Figure 1 shows a flowchart of a typical MC program for the

*A semiconductor with a single, spherical and parabolic energy band.

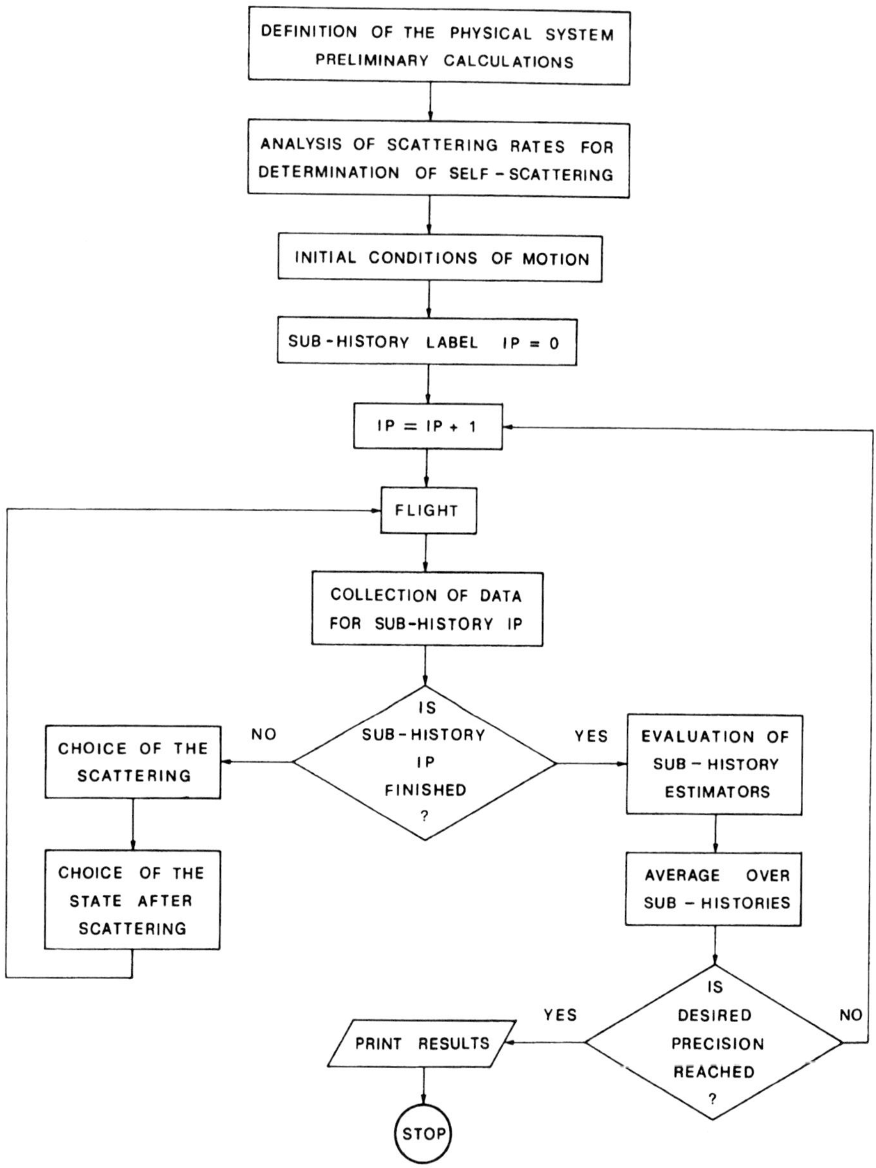

Figure 1. Flowchart of a typical Monte Carlo program.

simulation of a stationary, homogeneous, transport process.

Figure 2 illustrates the principles of the method by showing the simulation in $\vec{k}$-space and real space and the effect of collecting statistics in the determination of the drift velocity.

Definition of the Physical System

The starting point of the program consists in defining the physical system of interest as regards the parameters of the material and the values of physical quantities such as lattice temperature T_o and electric field. It is worth noting that among the parameters to be entered in order to characterize the material, the least known ones are usually the coupling strengths describing the interactions of the electron with the lattice and/or with defects inside the crystal.

At this level we also define the parameters which control the simulation, such as the duration of the simulation or the desired precision of the results.

The Initial Conditions of Motion

In the simulation of a steady-state situation, the initial conditions of the electron motion will not influence the final results when the simulation is sufficiently long. If, however, a highly improbable initial value for the electron wave vector is chosen, the first part of the simulation can be strongly influenced by this inappropriate choice. In order to avoid this undesired effect and to obtain a better convergence, the elimination of the first part of the simulation from the collection of the results may be conveient. When the history is split into many subhistories, the initial state of each new subhistory is taken as equal to the final state of the previous one. Thus, only the initial condition of the first subhistory will influence the final results in a biased way.

On the other hand, when a simulation is made to study a transient phenomenon or a transport process in a nonhomogeneous system, then it is necessary to simulate many electrons separately; in this case the distribution of the initial electron states must be taken according to the particular physical situation under investigation and the initial part of the simulation becomes an essential part of the desired results.

Flight Duration - Self Scattering

The electron wave vector $\vec{k}$ changes continuously during a free flight because of the applied field according to Equation (1). Thus, if $P(\vec{k}(t))$ is the probability that an electron in the state $\vec{k}$ suffers a collision during the time dt, the probability that an electron which suffered a collision at time t = 0 has not yet suffered another collision after a time t is

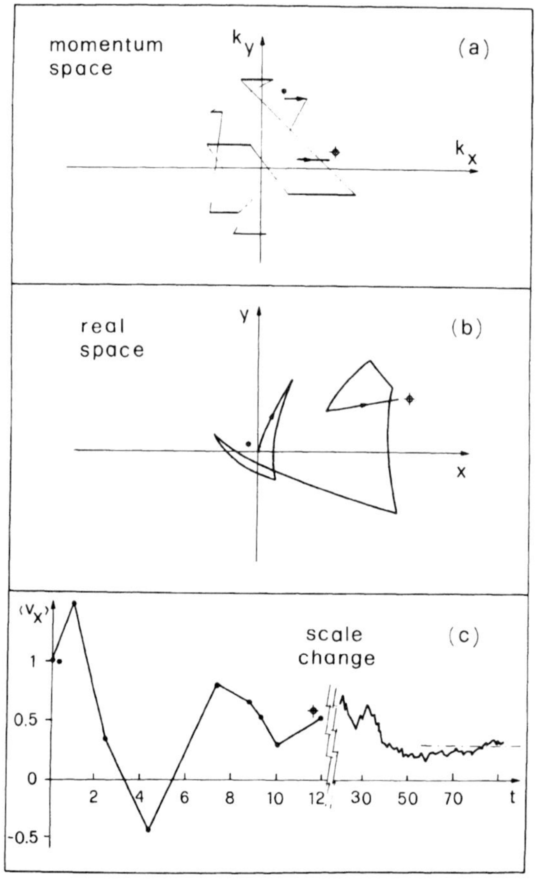

Figure 2. The principles of the Monte Carlo method. Part (a) shows the simulation of the sampling particle, in the wavevector space, subject to an accelerating force oriented along the positive x direction. Part (b) shows the path of the particle in the real space. Part (c) shows the average velocity of the particle obtained as a function of the simulation time. The horizontal dot-dashed curve represents the "exact" drift velocity obtained with a very long simulation time. Arbitrary units.

$$\exp\left(-\int P\left[\vec{k}(t')\right] dt'\right) \tag{2}$$

Consequently, the probability $\rho(t)$ that the electron will undergo its next collision during dt around t is given by

$$\rho(t)\, dt = P\left[\vec{k}(t)\right] e^{-\int_0^t P\left[\vec{k}(t')\right] dt'} dt \tag{3}$$

Because of the complexity of the integral at the exponent, it is very impractical to generate stochastic free flights with the distribution in Equation (3) starting from evenly distributed random numbers r: an integral equation should be solved for each scattering event [1]. Rees [4,9] has devised a very simple method to overcome this difficulty. If $\Gamma = 1/\tau_0$ is the maximum value of $P(\vec{k})$ in the region of $\vec{k}$-space of interest, a new fictitious "self scattering" is introduced such that the total scattering probability, including this self scattering, is constant and equal to Γ. If the carrier undergoes such a self scattering, its state $\vec{k}$ after the collision is taken to be equal to its state $\vec{k}$ before the collision, so that in practice the electron motion continues unperturbed as if no scattering at all occurred. Now, with a constant $P(\vec{k}) = \tau_0^{-1}$, Equation (3) reduces to

$$\rho(t) = \frac{1}{\tau_0} e^{-t/\tau_0} \tag{4}$$

and random numbers r evenly distributed between 0 and 1 can be used very simply to generate stochastic free flights t_r. They will be given by

$$t_r = -\tau_0 \ln(r) \tag{5}$$

The computer time "wasted" in taking care of self scatterings is more than compensated by the simplification of the calculation necessary to evaluate the free-flight durations.

As regards the choice of the constant Γ, we may note that in general $P(\vec{k})$ is simply a function of the electron energy ϵ; a suitable choice for Γ is then the maximum value of $P(\epsilon)$ in the region of energies which are expected to be sampled during the simulation. When $P(\epsilon)$ is not a monotonic function of ϵ, its maximum value must be evaluated in some way, for example with a tabulation at the beginning of the computer program. When $P(\epsilon)$ is an increasing function of ϵ, as is often the case, this is accomplished by taking $\Gamma = P(\epsilon_M)$, where ϵ_m is a maximum electron energy with small probability of being reached by the carrier during the simulation. It must be observed, however, that the range of energies "visited" by the electron during the simulation is not known at the beginning, when Γ is to be chosen. Therefore, an estimate must be made for keeping in mind that ϵ_M cannot be taken too large, otherwise a corresponding unnecessarily large value of Γ would result in a waste of computer time for self-scattering events. It is therefore necessary to decide what action to take when, during the simulation, the electron energy ϵ happens to exceed the maximum value ϵ_M set up at the beginning of the computer run. Some authors in such circumstances arbitrarily reduce the electron energy to a value below ϵ_M and simply check at the end of the run that such a situation occurred a limited number of times. This procedure may be dangerous since it may obscure the occurrence of indefinite increase of electron energy (runaway) when this happens at energies above a critical value close to ϵ_M. A safer method is to increase the value of ϵ_M and correspondingly of Γ for the simulation to come, without intervening in the electron energy. Since Γ must be independent of the simulated flights, it is always necessary to check that Γ has not been changed too many times.

For a semiconductor model that contains several valleys, a different value of ϵ_M may be appropriately taken in each type of valley.

Sometimes the total scattering probability $P(\epsilon)$ has a large variation around some threshold value due to a strong scattering mechanism with a given activation energy. A typical case is intervalley scattering from central to upper valleys in polar materials. In this case a step-shaped scattering rate $\Gamma(\epsilon)$ may be more convenient [10].

The Choice of the Scattering Mechanism

During the free flight the electron dynamics is governed by Equation (1) so that at its end the electron wave vector and energy are known, and all scattering probabilities $P_i(\epsilon)$ can be evaluated, where i indicated the i-th scattering mechanism. The probability of a self scattering will be the complement to Γ of the sum of the P_i's. A mechanism must then be chosen among all possible ones: given a random number r, the product $r\Gamma$ is compared with the successive sums of the P_i's and a mechanism is selected as described in Figure 3.

If all scatterings have been tried and none of them has been selected, it means that $r\Gamma > P(\epsilon)$, and a self scattering occurs. Thus, in such a procedure, a self scattering is the most time consuming, since all P_i's must be explicitely calculated. However, an expedient can be used to overcome this shortcoming, which can be called "fast self scattering." It consists of setting up a mesh of the energy range under consideration at the beginning of the simulation and then recording in a vector the maximum total scattering probability $P^{(m)}$ in each energy interval. At the end of the flight, if the electron energy falls in the n-th interval, before trying all

P_i's separately, $r\Gamma$ is compared with $P^{(n)}$; at this stage, if $r\Gamma > P^{(n)}$, then a self scattering certainly occurs, otherwise all P_i's will be successively evaluated. Consequently, the improvement is obtained that only when $P(\epsilon) < r\Gamma < P^{(n)}$ does a self scattering occur which requires the evaluation of all P_i's.

The Choice of the State After Scattering

Once the scattering mechanism which caused the end of the electron free flight has been determined, the new state after scattering $\vec{k}_a$ must be chosen. If the free flight ended with a self scattering, $\vec{k}_a$ must be taken as equal to $\vec{k}_b$, the state before scattering.

When, on the contrary, a true scattering occurred, then $\vec{k}_a$ must be generated stochastically, according to the differential cross section of the particular mechanism that caused the end of the free flight.

Collection of Results for Steady-State Phenomena

The data collected at each free flight will form the base for the determination of the quantities of interest.

Time Averages

When a steady state is investigated, we may obtain the average value of a quantity $A(\vec{k}(t))$ (e.g. the drift velocity or the mean energy) during a simulation of duration T as

$$\langle A \rangle_T = \frac{1}{T}\int_0^T A(\vec{k}(t))\,dt = \frac{1}{T}\sum_i \int_0^{t_i} A(\vec{k}(t'))\,dt' \tag{6}$$

where the integral over the whole simulation time T has been separated into the sum of integrals over all free flights of durations t_i.

In a similar way we may obtain the electron distribution function: a mesh of $\vec{k}$ space (or of energy) is set up at the beginning of the computer run; during the simulation the time spent by the sample electron in each cell of the mesh is recorded. For large T this time, conveniently normalized, will represent the electron distribution function, that is the solution of the Boltzmann equation [5]. Such an evaluation of the distribution function can be considered a special case of Equation (6) in which we choose for A the functions $n_j(\vec{k})$ defined as 1 if $\vec{k}$ lies inside the j-th cell of the mesh and 0 otherwise.

Synchronous Ensemble

Another possible method for obtaining an average quantity <A> from the simulation of a steady-state phenomenon has been introduced by Price [3] and it is called the synchronous-ensemble method. Let us consider Figure 4 in which for each of the N electrons of the physical system a time axis is considered. Circles in these axes indicate scattering events. In principle, each of the axes can represent a simulated electron and self scattering can be included, if wanted. In general, the average value of a quantity A is defined

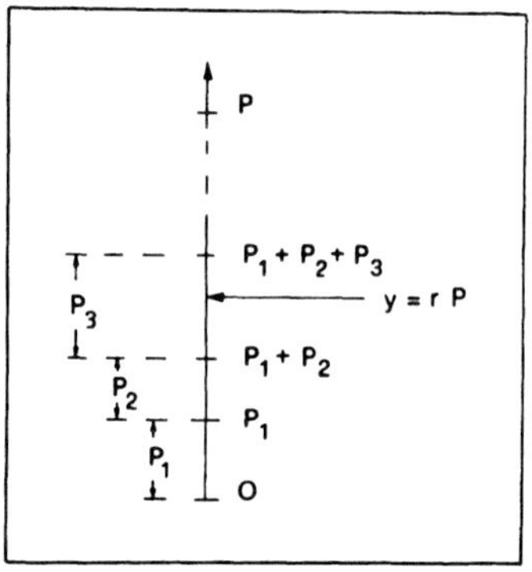

Figure 3. Stochastic determination of an event in the discrete case. P_i with i=1,2,3,... gives the probability of occurrence of the i-th event. P_j is selected if $r\Gamma$, evenly distributed between 0 and Γ falls in the corresponding interval.

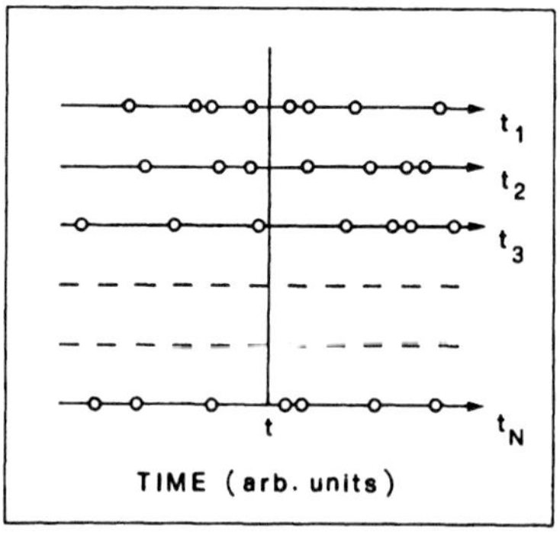

Figure 4. Sketch illustrating the syncronous ensemble method. Index i (1,2,3,...,N) labels the time axis of the i-th particle; open circles indicate scattering events; t is the generic time of a steady-state condition.

as the ensemble average at time t over the N electrons:

$$\langle A \rangle = \frac{1}{N}\sum_i A_i(t_i = t) \tag{7}$$

In particular, the steady-state distribution function is proportional to the number of electrons $n(\vec{k})$. Equation 7 can also be evaluated as

$$\langle A \rangle = C\sum_{\vec{k}} n(\vec{k})A(\vec{k}) \tag{8}$$

where C is an appropriate normalization constant, and $n(\vec{k})$ can be considered as proportional to the probability of finding any given electron in the state $\vec{k}$. If we now consider the before-scattering distribution function $n_b(\vec{k})$, proportional to the probability that an electron is found in $\vec{k}$ immediately before a scattering event, it can be shown that

$$n(\vec{k}) = \frac{1}{\tau_0} n_b(\vec{k})\tau(\vec{k}) \tag{9}$$

where τ_0 is an appropriate normalization constant. If, by including self scattering, we use a constant $\tau(\vec{k})$ then the steady-state distribution function becomes equal to the before-scattering distribution*:

$$n(\vec{k}) = n_b(\vec{k}) \tag{10}$$

and Equation (8) can be used in a MC simulation in the following form:

$$\langle A \rangle = \frac{1}{M}\sum_i A_{b_i} \tag{11}$$

where the sum covers all M electron free flights and A_{bi} indicates the value of the quantity A evaluated at the end of the i-th free flight.

If a self-scattering with a step-shaped total scattering rate is used, then the various terms in Eq. (11) must be weighted with a factor $\Gamma^{-1}(\vec{k})$.

It may seem strange that with a constant $\tau(\vec{k})$ the before-scattering distribution $n_b(\vec{k})$ is equal to the steady-state distribution since the states just before the scattering events seem to be influenced "at the maximum" by the applied field. However, one should consider that, while $n_b(\vec{k})$ weights equally all free flights (short and long ones) with average duration τ, when an instantaneous picture of the electron gas is taken at any time, longer free flights are more probably caught. In other words, in the latter case, the vertical line in Figure 4 crosses free flights whose mean duration is longer than average over all free flights; in fact, the distribution of the half-flights on the right and on the left of the line t reproduces the distribution of flight durations, so that the average length of the flights crossed by t is 2τ.

THE ENSEMBLE MONTE CARLO SIMULATION (EMC)

For time or space dependent problems the analytical solution of the Boltzmann equation is even more difficult than for homogeneous and stationary problems, while for MC programs little work needs be added to attack such problems. This confirms again the usefulness of the MC method in particular for the analysis of small devices, where it is often necessary to consider both the transient dynamic response to voltage changes and the electronic behavior at different points of the device with different field or material properties [11].

As previously mentioned, for such problems we cannot rely on the ergodic property of the system, and an ensemble of particles must be explicitely simulated [12, 13]. One exception to this rule is that of phenomena which are periodic in space and/or time. In this case the different situations that the electron will experience in equivalent position or times will take into account the many possibilities of various particles, and again the simulation of a single electron may yield the necessary information about the entire electron gas [14, 15].

Time Dependent Phenomena - Transients

We shall here consider the simulation of a homogeneous electron gas with a time-dependent behavior. In particular, it is of interest to study the case of a sudden change in the value of an applied field and to investigate the transient dynamic response to such a change. Provided the number of simulated particles is sufficiently large, the average value of a quantity of interest, obtained on this sample ensemble as a function of time, will be representative for the average on the entire gas. It must be stressed that the estimators given by Equations (6) and (11) are based on the hypothesis of steady-state conditions and cannot be used when a time dependent phenomenon is analyzed: the ensemble average of a quantity A must actually be estimated according to its definition given by Equation (7).

To determine the precision of the results obtained, the entire ensemble can be separated into a certain number of subensembles; for each of them an estimate of the quantity of interest A will be performed. Then, their average value and its standard deviation can be taken as the most probable value and the statistical uncertainty of A.

The duration of the transient response is not known a priori and will be of the order of the largest of the characteristic times of the electron system. This time may be called the "transport-transient time" and in general depends upon the values of the applied field and temperature; in our case of high fields in semiconductors it may roughly correspond to the energy relaxation time or to the time

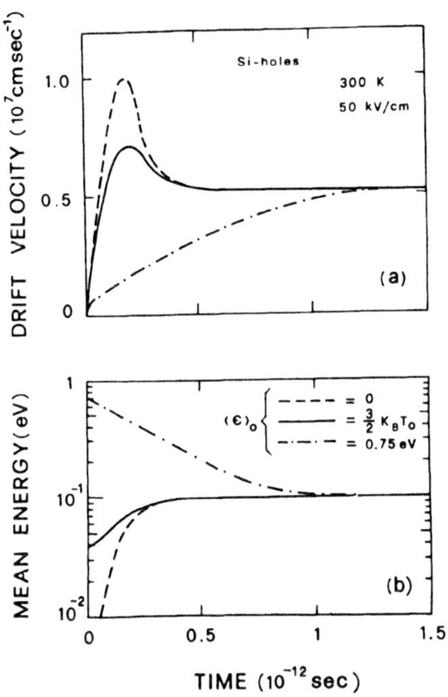

Figure 5. Drift velocity (a) and mean energy (b) as a function of time for the case of holes in Si. Different curves refer to the reported different values of the initial mean energy with velocities randomly distributed.

necessary for repopulation of different valleys.

The transient dynamic response which will be obtained by means of the simulation, will of course depend upon the initial conditions of the carriers, and these must be assumed according to the situation which is to be explored. Initial velocity distributions that may be of interest are, for example, a Maxwellian distribution at the lattice temperature and no drift, or a Maxwellian with an electron temperature higher than that of the lattice with or without drift.

Figures 5 and 6 show some examples of transient transport quantities in Si and in GaAs. Figure 5 reports EMC calculations for the case of holes in Si and T_o = 300K and E = 50 kV/cm [16]. Here the transient behavior of the drift velocity and the mean energy is investigated at different initial mean energies. An overshoot in the drift velocity can be seen that is larger for colder initial conditions and disappears for the hottest case. This overshoot phenomenon is now well known and is related to the energy dependence of the momentum relaxation time: as long as the electrons are cold, their average momentum relaxation time is longer and yields a larger drift velocity; then the carriers become hot, their momentum relaxation time

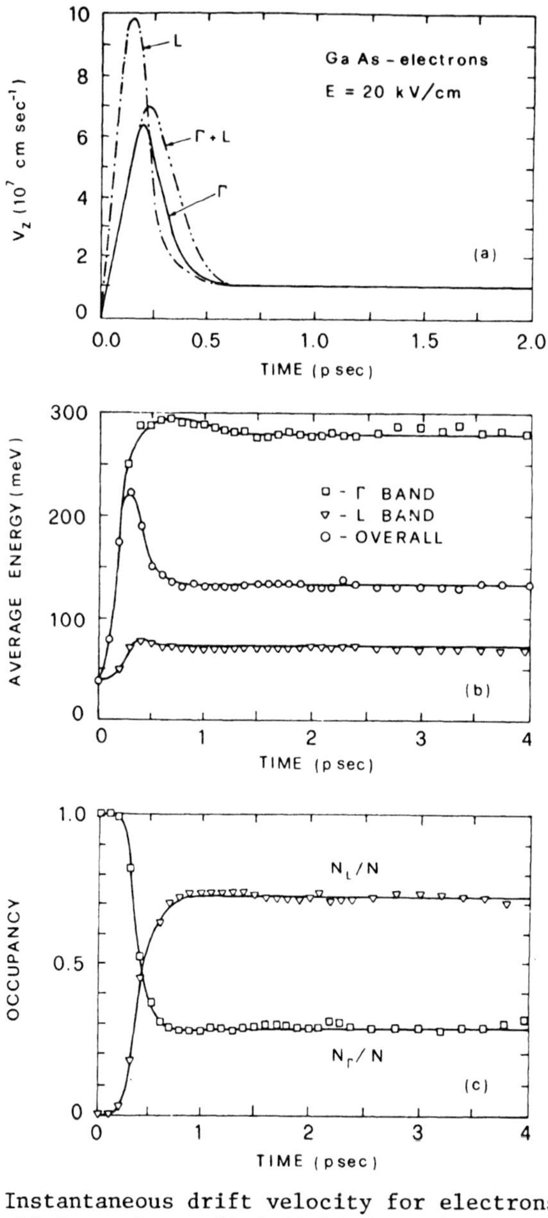

Figure 6. a) Instantaneous drift velocity for electrons in GaAs parallel to the applied field for Γ band, L band and both Γ +L bands; b) average mean energy for electrons in GaAs versus time for Γ band, L band, and both bands; c) fractional occupancies $N_Γ/N$ and N_L/N for electrons in GaAs. All the calculations are made with E = 20kV/cm.

decreases, since the scatterings are more effective, and the drift velocity settles down to its steady-state value. When the electron energy is higher in the initial conditions than in the steady state, we have the opposite situation, and the drift velocity is at first lowered by a higher scattering and reaches its stationary value more slowly.

The above interpretation of the velocity overshoot is confirmed by Figure 5b, where the transient behavior of energy is shown.

A similar phenomenology is found for electrons in elemental semiconductors. In some cases the velocity overshoot involves an excess absorbed power, so that an energy overshoot is also present that, in turn, lowers the drift velocity below its steady-state value. An oscillatory behavior of v_d and $< \epsilon >$ may then result. When we consider transient transport overshoots of electron quentities in compound materials that present Gunn effect, we have to take into account also the populations of the valleys. Figure 6 shows the drift velocity and the mean energy of electrons in the central valley and the valley populations for the case of GaAs at T_o = 300K with = 20 kV/cm [17].

Space Dependent Phenomena

The simulation of a steady-state phenomenon in a physical system where the electron transport depends upon the position in space is of particular interest for the analysis and modelling of devices. For this case an ensemble of independent particles must again be used, and averages must be performed over particles at given positions.

In a steady-state situation particles will enter the region of interest continuously, and in the simulation some initial wave vector must be assumed when a new particle is considered, according to the momentum distribution of the particles in the physical system under investigation. For example, a cold hemi-Maxwellian distribution, that is with only positive components of the velocities along a given direction, may be convenient in order to simulate a metallic contact of a device. If the simulated electron entering the device goes back to the contact, a new electron must be generated, but the information relative to the "lost" electron must be accounted for in the final results in a way that depends on the particular analysis that is being carried out.

Space and time dependent phenomena may present similar characteristics and they have sometimes been confused in simulation problems. For instance, we have seen that, if a field is suddenly switched on from zero to a large value, during the initial transient the drift velocity reaches values larger than in steady-state conditions, and this is sometimes used in discussing the behavior of an electron stream coming from a low-field region and entering a region with a high applied field.

If this reasoning is correct from a qualitative point of view, in rigorous calculations the kind of average that must be done is different whether time or space dependent phenomena are considered. In a steady-state phenomenon with a space dependence of the applied field it is in general necessary to consider the electron properties at given positions, so that the simulation must perform recording and averages for given points at any times, while when the object of the investigation is the evolution in time of a homogeneous system, average quantities must be evaluated at given times independent of the particle positions.

Results of EMC calculations for space dependent phenomena will be presented in the section devoted to device simulation.

Diffusion, Velocity Autocorrelation, and Noise

Diffusion may be considered a special, important case of space dependent phenomenon, that is, in general, also time dependent and a large amount of work has been done in this field with M techniques.

In the linear response regime, diffusivity D and mobility μ are related by the Einstein relation

$$D = \frac{1}{e} \mu K_B T_0 \tag{12}$$

Thus, an independent determination of the diffusion coefficient D does not add any particular information about the transport properties of a material with respect to the knowledge obtained from a determination of the Ohmic mobility*.

At high fields, however, the Einstein relation fails, and the study of hot-electron diffusion becomes useful for two reasons: it provides an independent check of theoretical models for the material under investigation and yields useful information for the analysis of most solid-state devices. It must be added that deviations of the diffusion coefficient from its equilibrium value can occur also at fields lower than those at which hot-electron phenomena are usually present because of the possibility of intervalley diffusion [19].

Diffusion is described at a phenomenological level by the first Fick equation

$$j_i = -eD_{ij} \frac{\partial n(\vec{x})}{\partial x_j} \tag{13}$$

*In fact, MC can be used more conveniently the other way around: the ohmic mobility can be obtained from the Einstein relation, once D has been calculated with a MC simulation at zero field [18].

where $\vec{j}$ is the current density, D the diffusivity tensor or diffusion coefficient, $\vec{x}$ the position in space, $n(\vec{x})$ the particle density; sum over repeated indexes is implied. If an electric field is also present that would produce, in absence of diffusion, a drift velocity v_d, and if diffusion and drift do not influence each other* then Equation (13) is substituted by

$$j_i = \left[n(\vec{x}) v_{d_i}(E) - D_{ij} \frac{\partial n(\vec{x})}{\partial x_j} \right] \tag{14}$$

By combining this equation with the continuity equation, we obtain the diffusion equation

$$\frac{\partial n}{\partial t} = - v_{d_i} \frac{\partial n}{\partial x_i} + D_{ij} \frac{\partial^2 n}{\partial x_i \partial x_j} \tag{15}$$

where v_d and D have been supposed space independent.

Under linear response conditions, v_d depends linearly on field and D is field independent. The generalization at high fields of Equation (15) is usually performed by assuming $D = D(\vec{E})$. However Equation (15) even modified holds only at times longer than both the transient transport time, defined in Time Dependent Phenomena - Transients, and and the time necessary for setting up the correct space-velocity correlations at the basis of the diffusion phenomenon (correlation time).

If, for simplicity, n is assumed to be a function only of the coordinate z parallel to the direction of v_d, from Equation (15) we obtain

$$D_\ell = \frac{1}{2} \frac{d}{dt} \langle (z - \langle z \rangle)^2 \rangle = \langle (z - \langle z \rangle)(v_z - \langle v_z \rangle) \rangle \tag{16}$$

where D_ℓ is the diagonal component of D along the direction of v_d. This expression can be used for an EMC determination of the diffusion coefficient: a number of particles is independently simulated and their positions are recorded at fixed times. For large enough simulation times, the second central moment shows the linearity predicted by Equation (16) and from its slope D_ℓ is obtained. The last expression in Equation (16) may be more convenient for numerical computations.

Often instead of Equation (16) the following expression has been used:

$$D_\ell = \frac{\langle (z - \langle z \rangle)^2 \rangle}{2t} \tag{17}$$

*This is not a trivial point; for example let us consider an open-circuit situation with a high applied electric field. Here, diffusion and drift currents cancel each other to produce a vanishing net current. The electron distribution function is not the usual hot-electron distribution because power does not flow through the particle gas, and yet electrons experience a high field between successive collisions.

D_ℓ as obtained from it will coincide with the correct D_ℓ only for $t \to \infty$, since Equation (17) neglects the effect of the initial transient, when Equation (15) and therefore Equation (17) do not hold. Much longer simulation times are then necessary in order to obtain a good estimate of D_ℓ, than by using Equation (16), that yields the correct answer as soon as the diffusion equation holds.

Other components of D can be obtained in a similar way from

$$D_{xy} = \frac{1}{2}\frac{d}{dt} <(x - <x>)(y - <y>)> \tag{18}$$

The diffusion coefficient D can also be determined from a MC simulation through the evaluation of the steady-state autocorrelation function of velocity fluctuations [20, 22].

$$C_{ij}(t) = \overline{<\delta v_i(t')\delta v_j(t'+t)>} \tag{19}$$

where the bar indicates time average. In fact the following relation holds:

$$D = \int_0^\infty C(t)\,dt \tag{20}$$

The calculation of C(t) may be of interest in itself and can be performed as follows: Let T be the interval of time where the autocorrelation function is to be sampled. In general T must be taken longer than the autocorrelation time, i.e. such that $C(t) \simeq 0$ for $t > T$. Then T is divided into a number N of intervals of duration $\Delta T = T/N$ in order to determine C(t) at times $0, \Delta t, 2\Delta T, \ldots N\Delta T = T$. During the simulation, the velocity of the sample particle is recorded at the time values $i\Delta T$, $i = 0, 1 \ldots$. When i becomes $\geq N$ the products

$$v(i\Delta T) \cdot v\left[(i-j)\Delta T\right], \qquad j = 0, 1, 2, \ldots, N$$

are evaluated for each i. Products corresponding to the same value of j are averaged over the simulation, thus obtaining

$$\overline{v(t)\,v(t+j\Delta T)} = C(j\Delta T) + v_d^2 \tag{21}$$

since in a steady state the ensemble average is included in the time average.

The knowledge of the autocorrelation function of velocity fluctuations allows also a determination of the power spectrum of velocity fluctuations, defined as

$$S_v(\omega) = \lim_{T \to \infty} \frac{1}{T} < \left| \int_0^T \delta v(t) e^{i\omega t} dt \right|^2 > \tag{22}$$

In fact the Wiener-Khintchine theorem connects the autocorrelation function C(t) and S_v through

$$S_v(\omega) = 2\int_0^\infty C(t) e^{i\omega t} dt = 2D(\omega) \tag{23}$$

Thus, the noise spectrum can be obtained as Fourier transform of C(t).

Analysis of Velocity Fluctuations

A general analysis of the carrier velocity fluctuations in semiconductors can be performed with an EMC technique, by means of the analysis of the autocorrelation function [24]. Let us consider a homogeneous many-valley semiconductor subject to a uniform static electric field $\vec{E}$, in steady-state conditions. Considering for simplicity only the drift velocity direction, we define $v^{(V)}(t)$ as the valley drift velocity, i.e. the mean velocity of electrons in valley V(t) in which the electron under consideration is at time t, and $v^{(\epsilon)}(t)$ as the mean velocity of electrons with energy between ϵ and $\epsilon + d\epsilon$ in valley V(t), where $\epsilon(t)$ is the energy of the considered electron at time t.

The fluctuation of the instantaneous velocity v(t) of each electron over the drift value can then be written as a sum of a number of terms that accounts for the different physical sources of fluctuations:

$$\delta v(t) \equiv v(t) - v_d = \delta v^{(v)}(t) + \delta v^{(\epsilon)}(t) + \delta v^{(k)}(t) \tag{24}$$

where $\delta v^{(V)}(t)=v^{(V)}(t)-v_d$ is the fluctuation associated with the drift velocity of the valley in which an electron lies at time t with respect to the total drift velocity, $\delta v^{(\epsilon)}(t)=v^{(\epsilon)}(t)-v^{(V)}(t)$ is the velocity fluctuation associated with the fluctuation of the electron energy at time t with respect to the valley drift velocity, and $\delta v^{(k)}(t)=v(t)-v^{(\epsilon)}(t)$ is the velocity fluctuation associated with the fluctuation of the electron momentum k(t) with respect to the average velocity of all electrons with the same energy $\epsilon(t)$.

The autocorrelation function of velocity fluctuations becomes, by using Equation (24)

$$C(\tau) = \sum_{i,j=1}^{3} <\delta v_i(t) \cdot \delta v_j(t+\tau)> = \sum_{i,j=1}^{3} C^{(ij)}(\tau) \tag{25}$$

where i,j = V, ϵ , k. The three diagonal terms (i=j) in the above equation are at the origin of intervalley [25] convective [26] and thermal noise, respectively.

Equation (25) shows that off-diagonal terms (i $\neq$ j) may also contribute to the total autocorrelation function and, therefore, to noise and diffusion. The total noise due to velocity fluctuations is given by the sum of the three diagonal contributions only when the relaxation times of the various fluctuating terms have well differentiated values, so that in calculating the off-diagonal terms one of the two fluctuations can be assumed as constant, while the other one averages to zero.

By means of Equation (23) we can associate specific terms in the autocorrelation function with corresponding terms in the diffusion constant and noise, thus making explicit their physical origins.

In the following, EMC results for Si and GaAs as typical examples will be shown. Figure 7 shows the autocorrelation function of velocity fluctuations for electrons in Si at 77K with an electric field E = 10 kV/cm applied along <111> and <100> directions. After an initial sharp decay, a negative part of $C(\tau)$ is present in both cases. When $\vec{E}$ is along a <111> direction, the negative part goes directly to zero, while in the other case a positive slow decaying contribution is present, which also reduces the negative part of $C(\tau)$.

In order to understand the physical origin of these curves, it is useful to analyse the different contributions $C^{(i,j)}(\tau)$. Figure 8a reports the diagonal terms of $C(\tau)$ when $\vec{E}$ is along a <100> direction. The intervalley contribution $C^{(VV)}$ is seen to be responsible for the long tail of the total autocorrelation function; in fact the intervalley transition time is the longest of the characteristic times

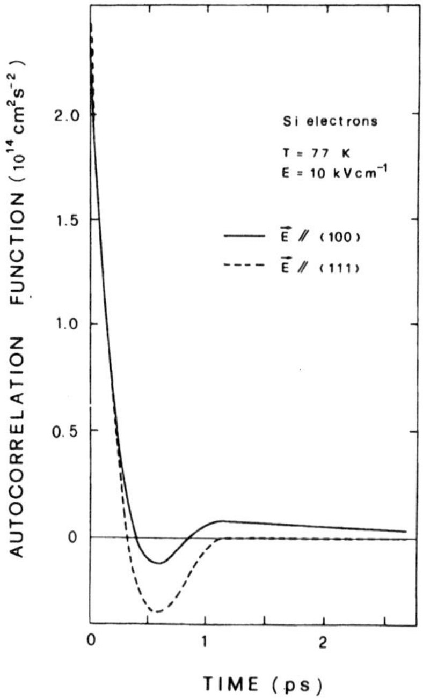

Figure 7. Autocorrelation function of velocity fluctuations for electrons in Si at the indicated temperature and fields.

of the transport process under these conditions of field and temperature.

The convective contribution $C^{(\epsilon\epsilon)}$ has a regular behavior with a small negative part, while the thermal contribution $C^{(kk)}$ has a large negative part to which most of the negative part of the total $C(\tau)$ can be ascribed. Furthermore, the off-diagonal term $C^{(\epsilon k)}$ shown in Figure 8b also contributes appreciably to the negative part of $C(\tau)$, partially compensated by the positive $C^{(k\epsilon)}$ (other off-diagonal terms are negligible in this case).

The particular form of these contributions is related to the energy dependence of the scattering mechanisms. A detailed analysis is given in Reference (24).

In a transverse direction velocity fluctuations are only due to thermal fluctuations ($v_d = v^{(V)} = v^{(\epsilon)} = 0$); their autocorrelation is always positive, since no energy transfer is associated with velocity fluctuations. For this reason $D_t = D_\ell$.

When the electric field is applied along a <111> direction, all the above considerations are still valid with the exception of the intervalley contribution that, in this case, is absent in the longitudinal direction, but present in a transverse direction.

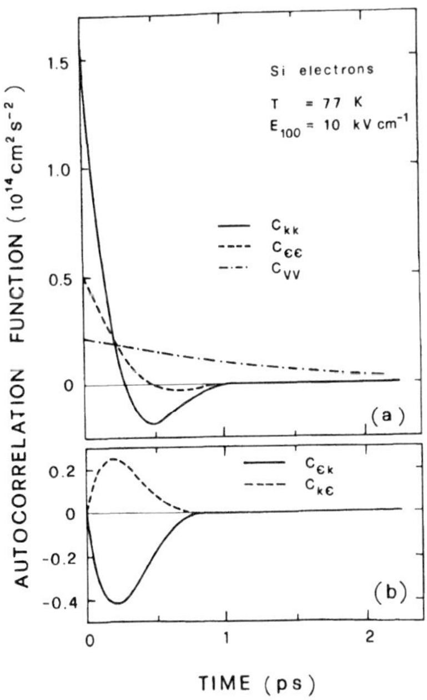

Figure 8. a) Diagonal and b) off-diagonal terms of the autocorrelation function of velocity fluctuations for electrons in Si at the indicated temperature and field.

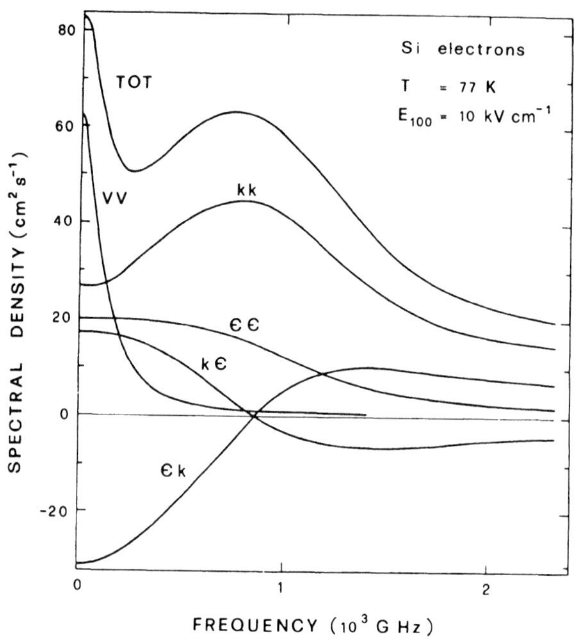

Figure 9. Spectral density of velocity fluctuations and its components for electrons in Si at the indicated temperature and field.

Figure 9 shows the noise spectral density $S_v(\omega)$ for the case of $\vec{E}$ // <100> discussed above. Again, the peculiar shape of the total $S_v(\omega)$ can be understood from the analysis of its partial contributions, also shown in Figure 9.

Figure 10a reports the autocorrelation function of velocity fluctuations for electrons in GaAs at 300 K with E = 10 kV/cm, together with its three diagonal terms. Thermal fluctuations at the chosen field strength are much higher than for the case of Si, owing to the higher electron energy. Figure 10b reports the off-diagonal terms. They are all of the same order of magnitude, but much smaller than the thermal contribution. Again, a detailed interpretation of these results would take us much more space than available in these lectures.

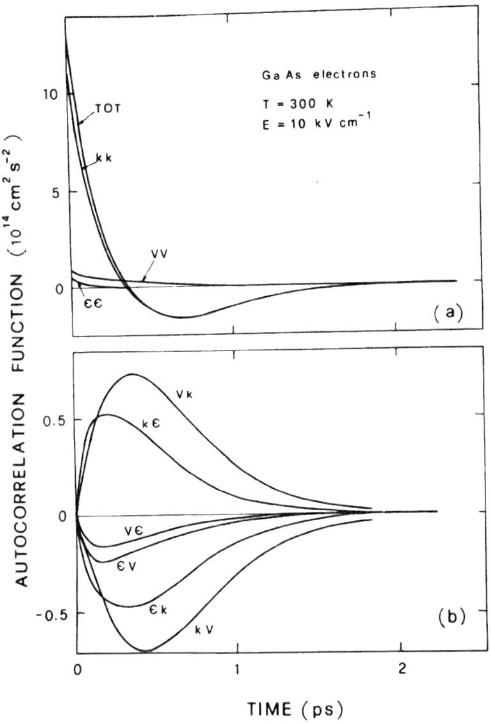

Figure 10. Autocorrelation function of velocity fluctuations for electrons in GaAs at the indicated temperature and field. (a) diagonal and (b) off-diagonal terms.

Transient Diffusion and Autocorrelation Function

The diffusion process of a carrier ensemble comes from the particle space-velocity correlations that arise during the evolution in time of the system. Starting from an initial condition in which the particle positions and velocities are totally uncorrelated, the process which occurs during the time necessary for setting up the correlations may be defined as the correlation transient. Furthermore, we have seen that when a high electric field is applied at a certain time to the electron ensemble, the transport process itself has to pass through a transient regime called transient transport, which is necessary for attaining the stationary distribution $f(\vec{k})$ in $\vec{k}$-space.

In what follows we shall discuss how these two different transients can be analyzed separately, but simultaneously, and how their effect influences the transient of the diffusivity of the electron ensemble.

The definition of the transient diffusion coefficient has been given by a generalization of Equation (16) to arbitrary small times [22, 27]:

$$D(t) = \frac{1}{2}\frac{d}{dt} <(z(t) - <z>_t)^2> \tag{26}$$

From this definition it is easy to obtain [28, 29]

$$D(t) = \int_0^t d\tau\, C_t(\tau) \tag{27}$$

with

$$C_t(\tau) = <\delta v(t)\,\delta v(t-\tau)> , \quad 0 \leq \tau \leq t \tag{28}$$

where we have put $\tau = t - t'$ in Eq. (27). Eq. (27) reduces to Eq. (20) in steady-state conditions ($t \rightarrow \infty$). By comparison of these two expressions, we see that in the transient case:

i) the integration interval does not extend to $+\infty$ but toward the past, back only to the initial conditions; this finite integration brings about the effect of the correlation transient;

ii) the autocorrelation function to be integrated in the transient analysis is not time independent, it is given by the specific ensemble average at a particular time and its shape carries information about the transport transient.

The transient autocorrelation function $C_t(\tau)$ can be calculated with an EMC procedure through the simulation of the dynamics of an ensemble of electrons. At fixed times 0, Δt, $2\Delta t$,........., the direct calculation of the velocity fluctuations is performed with respect to the mean value calculated at the same time over the carrier ensemble. The products $\delta v(i\Delta t)\cdot\delta v(j\Delta t)$, $j=0,1,\ldots i$ are then averaged over the ensemble and give $i+1$ values of the transient autocorrelation function at the time $t = i\Delta t$.

The analysis of the various contributions to the autocorrelation function described in the previous section can also be performed for the transient case in a similar way.

The present analysis holds also if the field is switched on at a time t_s larger than the time $t=0$ of the initial conditions [30]. In this case, the effect of the transport transient is separated in time from the initial correlation transient of zero-field diffusion.

Let us now discuss a few special examples that contain the significant features of most of the interesting cases.

Figure 11 represents results for the second central moment and for the transient longitudinal diffusion coefficient obtained with Eq. (26) as a function of time, for electrons in silicon. The following initial conditions have been taken: electrons are randomly situated in one of the six valleys with equal probability; the velocities are chosen according to an equilibrium distribution, and the electrons are all positioned at $\vec{r} = 0$. The second central moment first increases quadratically with time, as predicted by ballistic behavior. At intermediate (0.2 - 1.0 psec) times an irregular behavior is exhibited by the diffusivity: a tendency to level off followed by an overshoot. Then at sufficiently long times, the second central moment shows the linear dependence on time with the slope corresponding to steady-state diffusion.

This behavior is essentially due to the combined action of the acceleration impressed by the field and to the dissipation of energy

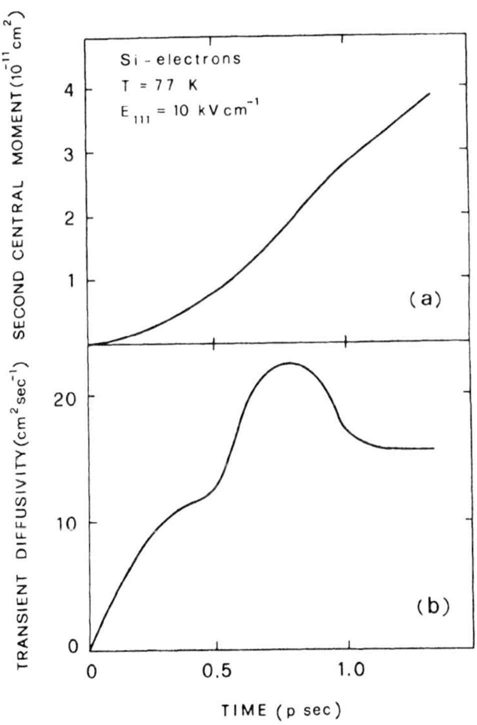

Figure 11. Second central moment (a) and longitudinal diffusion coefficient (b) as functions of time for electrons in Si at 77K and E=10 kV/cm along a <111> direction. The final slope of the curve shown in (a) corresponds to the steady-state value of D of 15 cm^2 sec^{-1} shown in (b) at sufficiently long times.

and momentum associated with intervalley scattering [29].

Figure 12 shows the transient autocorrelation function as a function τ at various times t. The field has been taken along a <100> direction in order to include in the discussion the effect of the intervalley contribution. The curves are interrupted at $\tau = t$, when the correlation with the initial conditions is reached. The area under each curve gives the corresponding value of time t of the transient diffusion coefficient. At very short times this area is very small both because $C_t(\tau)$ starts from low values, and because it is interrupted at short τ. The first effect is a consequence of the transport transient, the electrons being still "cold", while the second effect is due to the fact that at small times the correlations are not yet established (correlation transient). As t increases, the area reaches larger values and t is maximum when the positive part of $C_t(\tau)$ is totally present, and negative correlations are not yet present. This leads to the overshoot of the transient diffusivity. When also negative correlations are established, D(t) decreases toward the steady-state value, which is attained when the shape of $C_t(\tau)$ reproduces the steady-state function, and the integral of the autocorrelation function is extended to all significant values of τ.

It is particularly interesting to reproduce the same analysis for the intervalley contribution alone. Fig. 13 shows the transient autocorrelation function $C_t^{(VV)}(\tau)$ for intervalley velocity fluctuations $\delta v^{(V)}(t)$, defined in the previous section, as a function of τ for various times T. The most striking aspect of this set of curves is the presence of a bump which is shifted toward a greater value of τ as t increases. This phenomenon is due to the velocity overshoot in the two types of valleys which increases the values of $\delta v^{(V)}$ with respect to the steady state. As t increases, this overshoot recedes more and more in the past so that it is seen at larger correlation times until it vanishes, when any memory of the overshoot effect is lost.

Finally, in Fig. 14 an analysis is reported of the transient diffusivity in a case in which the electric field is applied after the onset of the zero-field correlations. Before application of the field, the diffusivity slowly reaches the steady-state value. When the electric field is applied, a transport transient occurs with heating of the carriers on a shorter time scale; during this time new correlations are established and the diffusivity reaches the new lower steady value, passing through a minimum with a region of negative values. This negative region is in great contrast with the overshoot of D seen for the case in Fig. 11. This can be explained by considering that the first electrons that undergo intervalley scattering after the field application are the fastest electrons in the direction of the field; during the first part of the process

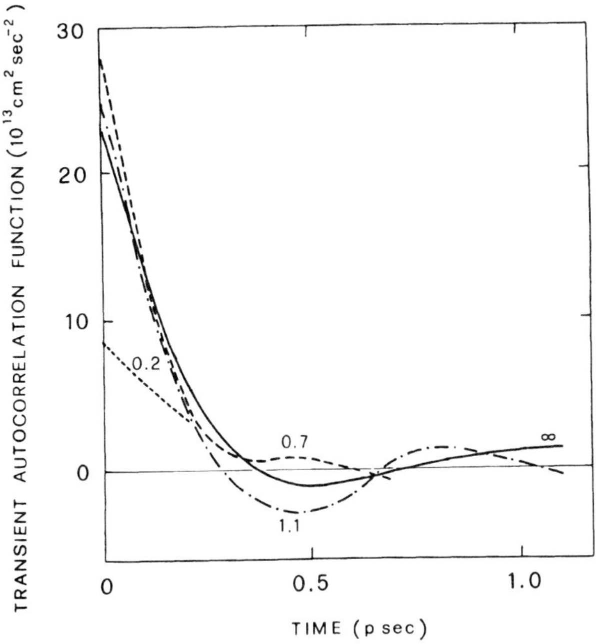

Figure 12. Transient velocity autocorrelation function as a function of time τ at fixed times t, reported in psec over each curve in the figure, for electrons in Si at 77K and E=10 kV/cm along a < 100 > direction. Each curve is interrupted at $\tau = t$ when correlation with the initial condition is reached. The stationary autocorrolation function (continuous line) is shown for comparison.

(zero-field correlation transient) the distributions of carriers have fully developed the space-velocity correlations at the basis of the diffusion process, so that a slowing down of the fastest particles due to intervalley scattering produces a shrinking of the space distribution, corresponding to the negative D. The comparison between the second part of Fig. 14 and Fig. 11 may be useful to appreciate the influence of initial conditions on the transient diffusion.

Electron-electron interaction

Interparticle collisions do not usually affect transport properties to a large extent in semiconductors. In fact, in such an interaction the total momentum and the total energy of the colliding particles is conserved and no dissipation occurs. Momentum and energy are however redistributed among the particles so that the shape of the distribution function $f(\vec{k})$ is influenced by e-e interaction. This fact has been used to state that, for high electron densities, $f(\vec{k})$ assumes a Maxwellian shape far from equilibrium, characterized

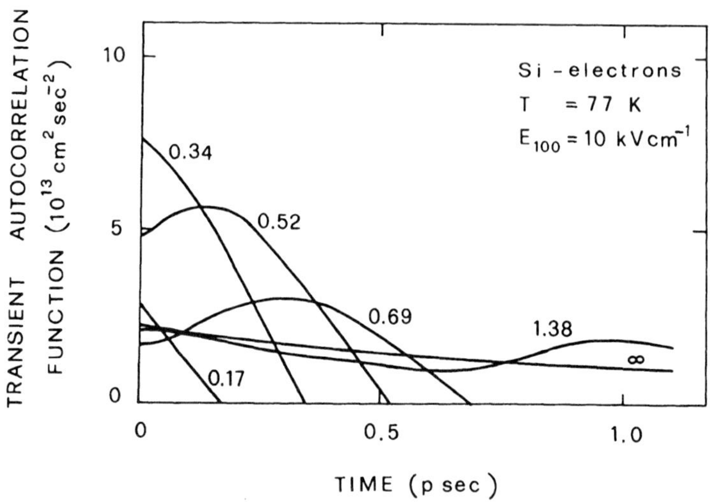

Figure 13. Transient intervalley autocorrelation function as a function of τ at fixed times t (reported in psec over each curve in the figure) for electrons in Si at 77K and E=10 kV/cm along a <100> direction. Each curve is interrupted at τ=t, when the correlation with the initial conditions is reached. The stationary intervalley autocorrelation function ($t \to \infty$) is shown for comparison.

by a mean drift velocity different from zero and an electron temperature T_e higher than the crystal temperature T_o.

From the above considerations it is reasonable to expect that e-e interaction may affect to some extent the microscopic transport quantities which are more sensitive to the particular shape of the distribution function. It has been found both theoretically and in experimental measurements [31-33] that this is the case for energy relaxation time and valley repopulation.

Several attempts have been performed [33-36] to include e-e interaction in MC calculations. A fundamental difficulty must however be faced due to the nonlinear nature of such interaction. In fact, the scattering probability itself depends on the distribution function both through the screening factor of the scattering potential, and through the probability of the sample electron's colliding with another electron of given wave vector $\vec{k}$. Some sort of self-consistent calculation must therefore be performed, in which an expression for $f(\vec{k})$ is used to evaluate scattering probabilities and the same $f(\vec{k})$ results as solution.

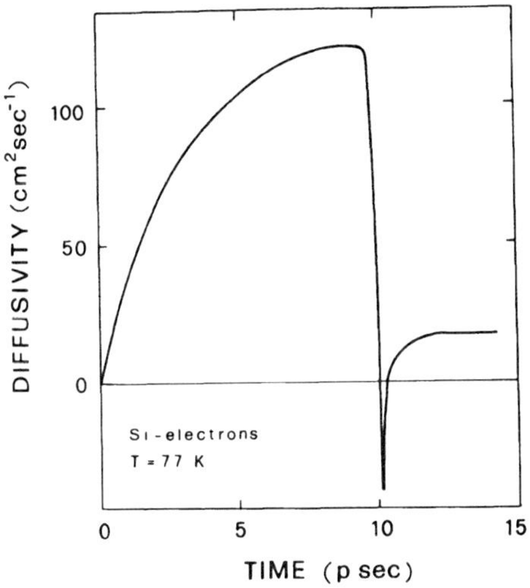

Figure 14. Transient diffusivity as a function of time for electrons in Si at 77K, in a physical situation in which a field E_{100} = 10 kV/cm is switched on at the time t_s = 9.57 psec, after the initial conditions (t=0) of uncorrelated particles.

In the approach proposed by Matulionis et al. [35] two electrons are simulated, each representing the whole system. For each of them memory is kept of the distribution as obtained from the simulation during the most recent past for a fixed time duration. In the simulation of one electron, among the scattering machanisms, an e-e interaction with a screened Coulomb potential is introduced, evaluated by means of the memory of the other electron. At each collision the memory of the simulated particle is renewed by the new state while the memory of the other electron is not changed. When the other electron is considered as MC electron, the memory of the former particle now represents the ambient electron gas. Sufficiently long simulations provide convergence of the results.

In an EMC calculation [36] self-consistency can be searched by selecting at random in the ensemble an electron to use for the collision with the electron considered for the simulation at each moment.

In another attempt a simultaneous simulation of many interacting particles has been tried [33] but difficulties arise from the long-range nature of Coulomb interaction. This approach may take advantage of the experience accumulated in plasma simulations.

Pauli exclusion principle is also a kind of e-e interaction that brings about nonlinear terms in the Boltzman equation. It can be included in a MC calculation with a self-consistent procedure suggested in Ref. 37.

Device Simulation

In recent years EMC has been widely used to study the properties of semiconductor devices [38]. Particular emphasis has been lately attributed to submicron structures, because of their performances in switching and high frequency operations. Once the basic physics involved in the transport of such devices is known, EMC simulation provides a formidable tool to determine their limits and characteristics and can be very helpful in modeling. From this point of view, EMC is preferable to the phenomenological approach, in which the current and continuity equations are solved self-consistently with Poisson's equation for the electrostatic field. In principle, in fact, the former method does not require any a priori knowledge of $\vec{v}_d$-$\vec{E}$ curves, and it allows, as we shall see, a self-consistent solution of the field profile. The larger amount of computations required seems to be a difficulty that can be easily overcome with present-day supercomputers. Together with the determination of the macroscopic properties of a device, EMC also gives a microscopic description of the local electric field, charge density, velocity, distribution, etc.

A particle model, similar to the ones long established in plasma physics and astrophysics, is generally used [11]. For a two-dimensional simulation it combines a MC description of scattering processes in three-dimensional $\vec{k}$ - space, with a two-dimensional description of field and carrier distribution in real space. One-dimensional simulations have also been performed, especially to study ballistic structures of n^+-i-n^+ type [39]. In this case, only one space component is considered.

The basic steps in a device simulation are [40]:

i) Charge assignment. A mesh of cells is set up and the charge of each particle is assigned to a particular mesh point. Since it is not possible to simulate all the electrons present in a real device, each simulated particle represents a cloud of electrons for the purpose of estimating currents, charge and field distributions. For all other purposes, each individual particle carries its elementary charge e. The doping charge is also added to the mesh according to its distribution; charge neutrality is maintained throughout the device.

ii) Potential solution. Poisson's equation is solved to determine the electrostatic potential at the mesh points. This is usually done in a very efficient way by using Fast Fourier Transform techniques. The electrostatic field is then obtained from the potential with a finite-difference algorithm.

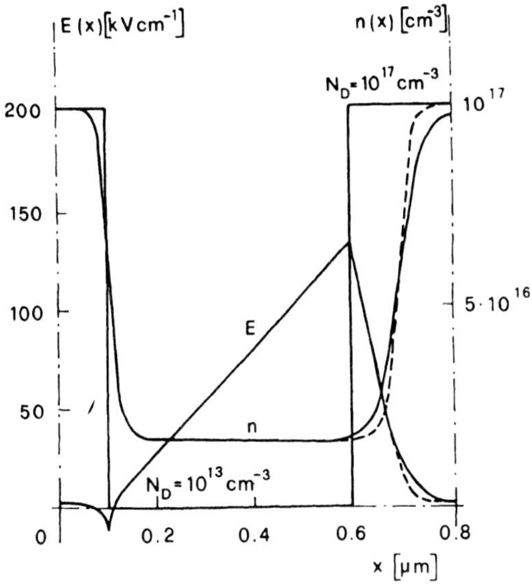

Figure 15. Single carrier method. Free carrier density and electric field profiles. The solid lines represent the Monte Carlo results. The dashed lines represent the electrokinetcs solution [43].

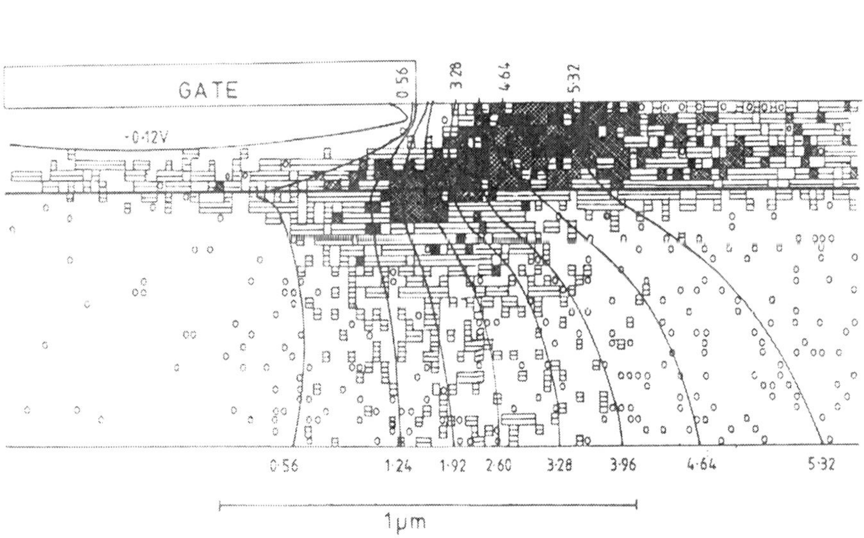

Figure 16. Detail of local heating inside a field-effect transistor ▭ 0.75-2.41 mWμm^{-3} heating; ▨ 2.41-4.08 mWμm^{-3} heating; ■ 4.08-6.55 mWμm^{-3} heating; 0-1.02 mWμm^{-3} cooling [44].

iii) Flights. Each particle, now treated as an individual electron, undergoes the standard MC sequence of scatterings and free flights, subject to the local field previously determined from the solution of Poisson's equation. The MC sequence is stopped at fixed times, when the field is adjusted following the steps described above.

The description of the problem is completed by setting initial and boundary conditions. The initial conditions are not so important, since only the self-consistent steady-state result is usually retained. Boundary conditions are instead crucial, in particular in submicron devices, where contact properties drastically influence the whole behavior of the device [41, 42].

EMC has been used to study a variety of recently developed devices as, for example, the vertical ballistic electron transistor and the superlattice field-effect transistor. To illustrate some of the main features of the method, two examples are now shortly presented. The first one is a one-dimensional simulation of a submicron n^+-n-n^+ structure [43]. Fig. 15 shows the free carrier density and field profile. It should be noticed that the electric field reaches very high values near the n-n^+ junction, mainly as a result of the submicron length of the active region. In this case, the boundary conditions are handled by assuming that the carriers are very close to thermal equilibrium in the high-doping region, far from the active zone.

The second example is taken from a two-dimensional simulation of a GaAs field-effect transistor [44]. Fig. 16 illustrates the kind of microscopic information that the model provides. The heating inside the MESFET is calculated by accounting for the energy lost or gained in the interaction of the carrier with the lattice. The effect is stronger in the region between the gate and the drain, where the local electric field is largest.

REFERENCES

1. T. Kurosawa, Proc. 8th Intern. Conf. Phys. Semic., Kyoto, J. Phys. Soc. Japan, Suppl. 21 424, 1966.
2. H. Budd, Proc. 8th Intern. Conf. Phys. Semic., Kyoto, J. Phys. Soc. Japan, Suppl. 21, 420, 1966.
3. P.J. Price, Proc. 9th Intern. Conf. Phys. Semic., edited by S.M. Ryvkin, Nauka, Leningrad, 1968, p. 753.
4. H.D. Rees, J. Phys. Chem. Solids, 30, 643, 1969.
5. W. Fawcett, D.A. Boardman, and S. Swain, J. Phys. Chem. Solids 31, 1963, 1970.
6. N.P. Buslenko, D.I. Golenko, Yu.A. Shreider and V.G. Sragovich, "The Monte Carlo Method," Edited by Yu.A. Shreider, Pergamon, London 1966.
7. P.J. Price, in Semiconductors and Semimetals, Academic, New York 1979, p. 249.

8. C. Jacoboni and L. Reggiani, Reviews Modern Physics, to be published, 1983.
9. H.D. Rees, Phys. Letters 26A, 416, 1968.
10. V. Borsari and C. Jacoboni, Phys. Stat. Solidi (b) 54,649, 1972.
11. R.W. Hockney and J.W. Eastwood, "Computer Simulation Using Particles" Pergamon 1982.
12. P.A. Lebwohl and P.J. Price, Solid State Comm. 9, 1221, 1971.
13. P.A. Lebwohl and P.J. Price, Appl. Phys. Letters 19, 530, 1971.
14. P.A. Lebwohl, J. Appl. Phys. 44, 1744, 1973.
15. J. Zimmermann, Y. Leroy and E. Constant, J. Appl. Phys. 49, 3378, 1978.
16. L. Reggiani, J.C. Vaissiere, J.P. Nougier and D. Gasquet, Proc. 3rd Intern. Cnf. on Hot Carriers in Semic., Journal de Physique Coll.,Tome 42, C7-357, 1981.
17. T.H. Glissom, C.K. Williams, J.R. Hauser and M.A. Littlejohn, in VLSI Electronics: Microstructure Science, Vol. 4, Academic, New York 1982, p. 99.
18. C. Canali, C. Jacoboni, F. Nava, G. Ottaviani, and A.Alberigi-Quaranta, Phys. Rev. B12, 2265, 1981.
19. R. Brunetti, C. Jacoboni and L. Reggiani, Proc. 3rd Intern. Conf. on Hot Carriers in Semic., J. de Physique Coll., Tome 42, C7-117, 1981.
20. L. Gherardi, A. Pellacani and C. Jacoboni, Lettere Nuovo Cim. 14, 225, 1975.
21. R. Fauquenberque, J. Zimmermann, A. Kaszynsky, E. Constant and G. Microondes, J. Appl. Phys. 51, 1065, 1980.
22. D.K. Ferry and J.R. Barker, J. Appl. Phys. 52, 818, 1981.
23. F. Reif, "Fundamentals of Statistical and Thermal Physics," McGraw Hill, New York 1965.
24. R. Brunetti and C. Jacoboni, Phys. Rev. Letters 50, 1164, 1983.
25. P.J. Price, J. Appl. Phys. 31, 950, 1960.
26. P.J. Price, in "Fluctuation Phenomena in Solids" edited by R.E. Burgess, Academic, New York 1965, p. 355.
27. A. Alberigi-Quaranta, V. Borsari, C. Jacoboni and C. Zanarini, Appl. Phys. Letters 22, 103, 1973.
28. P. Lugli, J. Zimmermann and D.K. Ferry, Proc. 3rd Intern. Conf. on Hot Carriers in Semic., J. de Physique Coll., Tome 42, C7-103, 1981.
29. R. Brunetti and C. Jacoboni, Transient and Stationary properties of Hot Carrier Diffusivity in Semiconductors, in "Semiconductor Processes Probed by Ultrafast Laser Spectroscopy, Academic, New York, to be published.
30. E. Constant and B. Boittiaux, Proc. 3rd Inter. Conf. on Hot Carriers in Semic. J. de Physique Coll., Tome 42, C7-73, 1981.
31. M. Ashe, B.L. Boichenko, V.M. Bondar and O.G. Sarbei, Phys. Stat. Sol. (b) 44, 173, 1971.
32. J.G. Nash and J.W. Holm-Kennedy, Appl. Phys. Lett. 24, 139, 1974.
33. C. Jacoboni, Proc. 13th Intern. Conf. Phys. Semic., edited by G. Fumi, Marves, Roma 1976, p; 1195.

34. L. Bacchelli and C. Jacoboni, Solid State Commun. 10, 71, 1972.
35. A. Matulionis, J. Pozela and A. Reklaitis, Solid State Commun. 16, 1133, 1975.
36. F. Lugli and D.K. Ferry, Proc. 16th Intern. Conf. Phys. Semic., edited by M. Averous, North Holland, Amsterdam 1983, p. 251.
37. S. Bosi and C. Jacoboni, J. Phys. C9, 315, 1976.
38. Proceedings of the Workshop on the Physics of Submicron Devices, Urbana 1982, to be published.
39. Y. Awano, K. Tomizawa, N. Mashizume and M. Kawashima, Electr. Lett. 18, 133, 1982.
40. R.W. Hockney, R.A, Warriner and M. Reiser, Elect. Lett. 10, 484, 1974.
41. C. Hamaguchi, K. Terashima, A. Shibatomi and O. Ryuzan, Proc. 16th Intern. Conf. Phys. Semic., edited by Mr. Averous, North Holland, Amsterdam 1983, p. 238.
42. J.R. Barker, D.K. Ferry, and H.L. Grubin, IEEE ED21, 209, 1980.
43. E. Constant and J. Zimmermann, Solid State Electr. 23, 915, 1980.
44. C. Moglestue, Computer Methods in Applied Mechanics and Engineering 30, 173, 1982.

NOISE AND DIFFUSION IN SUBMICRON STRUCTURES

Lino Reggiani

Gruppo Nazionale di Struttura della Materia e Dipartimento di Fisica Universita'
di Modena, Via Campi 213/A, 41100
Modena, Italy

I. INTRODUCTION AND GENERAL DEFINITIONS

Let us consider a two terminal device of length L and cross sectional area A. We are interested in the noise associated to carrier velocity fluctuations (here also called white-noise), physically analyzed in terms of the noise current-spectral density $S_I(\omega)$ and in its relation with the diffusion coefficient D associated to the spreading in space of an ensemble of carriers.

We recall that through the Wiener-Khintchine theorem $S_I(\omega)$ is related to the autocorrelation function of current fluctuations by

$$S_I(\omega) = 2\int_{-\infty}^{+\infty} \exp(i\omega\tau)\ \overline{\delta I(t)\,\delta I(t+\tau)}\, d\tau \qquad (1)$$

where $\delta I = I(t) - I_0$ is the current fluctuation around the average value I_0 and bar indicates time average (which by ergodicity is equivalent to ensemble average). Steady state conditions, that is invariance with respect to a time translation, ensure that $S_I(\omega)$ is only a function of ω. By neglecting quantum effects (Van Vliet and Van der Ziel, 1977) $S_I(\omega)$ is conveniently represented by an equivalent noise temperature $T_n(\omega)$ defined through the relation (Van der Ziel, 1968):

$$S_I(\omega) = \frac{4K_B T_n(\omega)}{\mathrm{Re}\left[Z(\omega)\right]} \qquad (2)$$

K_B being the Boltzmann constant and $Z(\omega)$ the small signal impedance at the bias point. $Z(\omega)$ relates the voltage and current fluctuations, δV and δI, measured between the terminals of the device through a generalized Ohm's law (Shockley et al, 1966; Nougier, 1973):

$$\delta V = Z(\omega)\delta I \tag{3}$$

$T_n(\omega)$ can be determined experimentally through the measurement of the maximum power dissipated by the noise source on the matched output impedance at frequency $\omega = 2\pi f$ per unit bandwidth Δf (Bareikis et al, 1968; Nougier, 1978):

$$P_n(\omega) = K_B T_n(\omega)\Delta f \tag{4}$$

By inverting equation (1) and setting $\tau = 0$, the power spectrum relationship is found:

$$\int_0^\infty S_I(\omega)\,d\omega = \overline{(\delta I)^2} \tag{5}$$

By simply exchanging current I with voltage V, voltage fluctuations can be defined through equations (1,5). Furthermore, in a theory which accounts only for the first order terms in the fluctuating quantities, the following relation between S_I and S_V can be proven to hold (Nougier, 1975):

$$S_V(\omega) = |Z(\omega)|^2 S_I(\omega) \tag{6}$$

Here we are assuming Re $[Z(\omega)] > 0$, that is the electrical system is stable with respect to fluctations in charge density.

Under equilibrium conditions (i.e. $E = 0$) equation (2) recovers the Nyquist relation, and the use of the Einstein relation:

$$D = \frac{K_B T_0 \mu}{e} \tag{7}$$

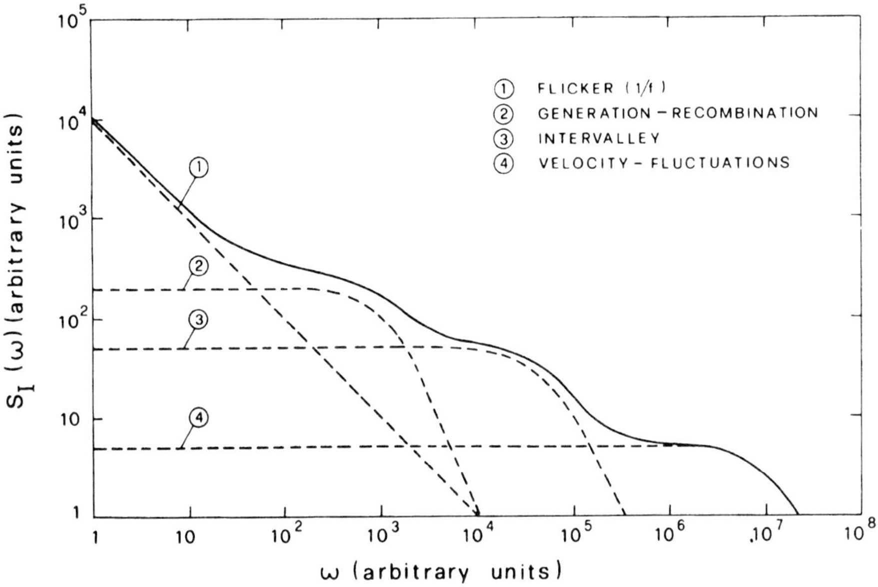

Figure 1. Schematic frequency dependence of the current noise spectral density along the field direction. Steps on the curves correspond to various fluctuation mechanisms characterized by different relaxation times.

μ being the mobility, e the electron charge and T_0 the thermodynamic temperature, establishes an exact relationship between white-noise and diffusion which writes:

$$\boxed{S_I^{eq}(\omega \to 0) = 4e^2 \frac{N}{L^2} D} \tag{8}$$

where N is the number of carriers and $\omega \to 0$ means that we are considering conditions under which $\omega \ll 1/\tau$, τ being a mean scattering time.

Equation (8) is strictly valid only at equilibrium, and refers to a local diffusion coefficient D which is related to Fick's phenomenological law. More generally $D = D(\vec{q}, \omega)$ is a linear response coefficient which exhibits $\vec{q}$ and ω dispersion relation, $\vec{q}$ and ω being the wavevector and angular frequency of the perturbation (Zwanzig, 1964; Jacoboni et al, 1981).

When a small electric field is applied to the device, as shown in **Figure 1**, other sources of noise (typically Flicker and generation recombination) appear as additive contributions in the low frequency range, so that the experimental detection of the contribution due to velocity-fluctuations only would require a frequency analysis of the whole spectrum.

In the case of submicron structures, however, new features may arise as a natural consequence of the small lengths involved (see Table 1), so that equation (8) with D given by equation (7) is no longer valid.

By limiting our interest to a space time domain where a semi-classical approach is justified (see Figure 2), the principal aim of these lectures is to investigate which kind of relationships will replace equation (8) under these new features. To this end three methods of tackling the problem will be presented namely: the Boltzmann equation, the generalized diffusion coefficient, the transfer impedance method. Accordingly in Section 2 the relationship between current fluctuation spectrum and diffusion coefficient is obtained from the kinetic Boltzmann equation under quite general conditions, that is by accounting for particle-particle interaction. In Section 3 a generalized diffusion coefficient which depends upon frequency and wavevector is defined and its dispersion relation is correlated to the frequency dependence of the noise current spectral density. In Section 4 the transfer impedance method is presented and applied to find the relationship between noise and diffusion under space-charge-limited conditions. In Section 5 available results on noise and diffusion are briefly surveyed.

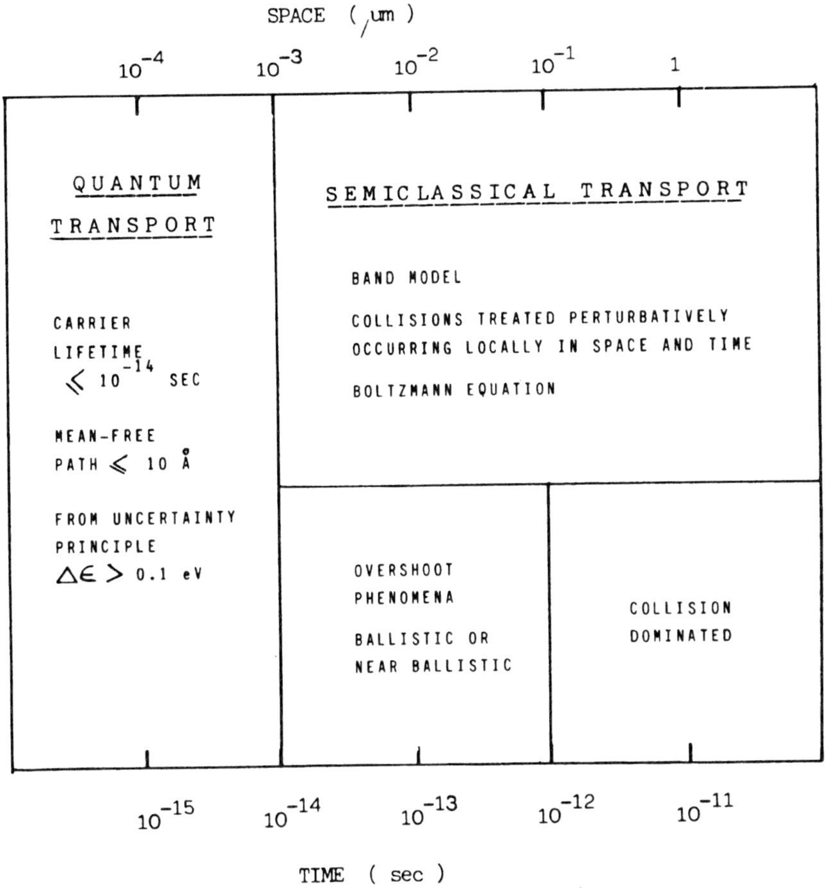

Figure 2. Synoptic table of space-time domain for transport theory in submicron structures. For the length scale a typical value of drift velocity 10^7 cm/sec has been assumed.

2. CURRENT FLUCTUATION SPECTRUM AND DIFFUSION COEFFICIENT IN NON-EQUILIBRIUM STATE

In this Section we shall calculate the relations between current fluctuation spectrum and diffusion coefficient when an electric field of arbitrary strength is applied to the device. Following the original results of Gantsevich et al (1979) the kinetic theory makes use of the Boltzmann equation with inclusion of two-particle collisions.

2.1 THE BOLTZMANN EQUATION WITH THE TWO-PARTICLE COLLISION TERM

The Boltzmann equation for steady-state and spatially homogeneous systems including two-particle collisions writes:

$$\left(e\frac{\vec{E}}{\hbar}\frac{\partial}{\partial\vec{k}}+I_k^{th}\right)\bar{f}(\vec{k})+I_k^{ee}\left[\bar{f},\bar{f}\right]=0 \tag{9}$$

Here the bar denotes ensemble average (in general the interpretation of the single particle distribution function as the ensemble average of a function $f(\vec{k},\vec{r},t)$, defined as the physical density in a six dimensional position ($\vec{r}$) and wavevector ($\vec{k}$) space, enables fluctuations about the average to be treated (Bixon and Zwanzig,1969)). $\vec{v}(\vec{k})$ is the group velocity which is related to the carrier energy ϵ by:

$$\vec{v}(\vec{k})=\frac{1}{\hbar}\frac{\partial}{\partial\vec{k}}\epsilon(\vec{k}) \tag{10}$$

The collision operator I_k^{th} describes the rate of change of $\bar{f}(\vec{k})$ due to scattering with thermal bath via phonon and imperfections. None of the interactions with which we are concerned produces spin-flip and we shall only deal with non-degenerate semiconductors, in which $f\ll 1$. Consequently I_k^{th} is linear in $\bar{f}$ and given by the usual form:

$$I_k^{th}\bar{f}=\frac{V_0}{(2\pi)^3}\int d\vec{k}'\left[W(\vec{k},\vec{k}')\bar{f}(k)-W(\vec{k}',\vec{k})\bar{f}(\vec{k}')\right] \tag{11}$$

where V_0 is the volume of the crystal and $W(\vec{k},\vec{k}')$ the transition rate. The two-particle collision integral is:

$$I_k^{ee}(\bar{f},\bar{f})=\left[\frac{V_0}{(2\pi)^3}\right]^3\int d\vec{k}'\int d\vec{k}_1'\int d\vec{k}_1\left[W(\vec{k},\vec{k}_1;\vec{k}',\vec{k}_1')\bar{f}(\vec{k})\bar{f}(\vec{k}_1)\right.$$
$$\left.-W(\vec{k}',\vec{k}_1';\vec{k},\vec{k}_1)\bar{f}(\vec{k}')\bar{f}(\vec{k}_1')\right] \tag{12}$$

where $W(\vec{k},\vec{k}_1;\vec{k}',\vec{k}_1')$ is the transition rate for particle-particle collision which carries two-particles from the occupied states $\vec{k}$ and $\vec{k}_1$ to the empty states $\vec{k}'$ and $\vec{k}_1'$. Let us note that, owing to lack of distinguishability, it is:

$$W(\vec{k},\vec{k}_1;\vec{k}',\vec{k}_1') = W(\vec{k}_1,\vec{k};\vec{k}',\vec{k}_1') = W(\vec{k},\vec{k}_1;\vec{k}_1',\vec{k}') = W(\vec{k}_1,\vec{k};\vec{k}_1',\vec{k}') \qquad (13)$$

Let us point out that equation (9) is non-linear due to the interparticle collision integral. Furthermore, the two-particle collision integral of equation (12) vanishes after substitution of the displaced Maxwellian distribution function with arbitrary drift momemtum $\hbar\vec{k}_d$ and electron temperature T_e:

$$f(\vec{k}) = \mathcal{A}\exp\left\{-\left[\hbar^2(\vec{k}-\vec{k}_d)^2/(2mK_BT_e)\right]\right\} \qquad (14)$$

where $\mathcal{A}$ is an appropriate normalization constant and m the carrier effective mass.

Indeed the transition rates in equation (12) contain δ-functions ensuring energy and wavevector conservation in a single collision (Umklapp processes are neglected) $\vec{k}+\vec{k}_1 = \vec{k}'+\vec{k}_1'$; $\epsilon(\vec{k})+\epsilon(\vec{k}_1) = \epsilon(\vec{k}')+\epsilon(\vec{k}_1')$ and this leads to the expression:

$$I_k^{ee}(\bar{f},\bar{f}) = \left[\frac{V_0}{(2\pi)^3}\right]\int d\vec{k}_1\,\mathcal{A}^2\exp\left\{-\hbar^2\left[(\vec{k}-\vec{k}_d)^2+(\vec{k}_1-\vec{k}_d)^2\right]/(2mK_BT_e)\right\}$$
$$\times\int d\vec{k}'\int d\vec{k}_1'\left[W(\vec{k},\vec{k}_1;\vec{k}',\vec{k}_1') - W(\vec{k}',\vec{k}_1';\vec{k},\vec{k}_1)\right] \qquad (15)$$

which is equal to zero owing to the Stueckelberg's property (1952) which ensures the zero value of the last integral in the right-hand-side of equation (15).

2.2 FLUCTUATIONS NEAR A NON-EQUILIBRIUM STEADY-STATE

Let us consider a semiconductor in which, along with a strong and constant electric field $\vec{E}$, there also exists a weak variable field $\delta\vec{E}(\vec{r},t)$ of the form:

$$\delta\vec{E}(\vec{r},t) = \delta\vec{E}_{q\omega}\exp(-i\omega t + i\vec{q}\vec{r}) \tag{16}$$

We are interested in evaluating the corresponding correction $\delta f(\vec{k},\vec{r},t)$ to the homogeneuous steady-state distribution function $\bar{f}(\vec{k})$ which satisfies equation (9) and such that:

$$\delta f(\vec{k},\vec{r},t) = \delta f_{q\omega}(\vec{k})\exp(-i\omega t + i\vec{q}\vec{r}) \tag{17}$$

The kinetic equation for $\delta f(\vec{k},\vec{r},t)$ is obtained by linearizing the time and space dependent Boltzmann equation which includes particle-particle collision and which results to be (Gantsevich et al, 1979):

$$\left[\frac{\partial}{\partial t} + \vec{v}\frac{\partial}{\partial \vec{r}} + I_k\right]\delta f(\vec{k},\vec{r},t) + \frac{e}{\hbar}\,\delta\vec{E}(\vec{r},t)\frac{\partial}{\partial \vec{k}}\,\bar{f}(\vec{k}) = 0 \tag{18}$$

where I_k is the linearized collision operator given by (Gantsevich et al, 1970):

$$I_k = \frac{e\vec{E}}{\hbar}\frac{\partial}{\partial \vec{k}} + I_k^{th} + I_k^{ee}(\bar{f}), \tag{19}$$

with the linearized particle-particle (e-e) collision operator $I_k^{ee}(\bar{f})$ given by:

$$I_k^{ee}(\bar{f})\,\phi(\vec{k}) = I_k^{ee}(\bar{f},\phi) + I_k^{ee}(\phi,\bar{f}) =$$

$$\left[\frac{V_0}{(2\pi)^3}\right]^3 \int d\vec{k}_1 \int d\vec{k}' \int d\vec{k}_1' \Big\{ W(\vec{k},\vec{k}_1;\vec{k}',\vec{k}_1')\left[\bar{f}(\vec{k})\phi(\vec{k}_1) + \bar{f}(\vec{k}_1)\phi(\vec{k})\right] \tag{20}$$

$$- W(\vec{k}',\vec{k}_1';\vec{k},\vec{k}_1)\left[\bar{f}(\vec{k}')\phi(\vec{k}_1') + \bar{f}(\vec{k}_1')\phi(\vec{k}')\right]\Big\}$$

Thus the operator I_k^{ee} (and hence I_k) is a functional of the steady-state distribution $\bar{f}(k)$.

In general, by differentiating equation (9) with respect to the total number of electrons N and accounting for the property of the linearized two-particle scattering operator given by equation (20), we obtain:

$$I_k\left[\frac{\partial}{\partial N}\,\bar{f}(\vec{k})\right] = 0 \tag{21}$$

where we use the distribution function normalized to the number of particles N; thus it is:

$$\frac{V_0}{4\pi^3}\int \bar{f}(\vec{k})\,d\vec{k} = N; \qquad \frac{V_0}{4\pi^3}\int \frac{\partial}{\partial N}\,\bar{f}(\vec{k})\,d\vec{k} = 1 \tag{22}$$

Let us point out that the function $\frac{\partial}{\partial N}\,\bar{f}(\vec{k})$ coincides with $\frac{1}{N}\,\bar{f}(\vec{k})$ only in the equilibrium state or when interelectronic collisions can be neglected.

The Poisson equation can be incorporated by representing the perturbing field $\delta\vec{E}(\vec{r},t)$ in the form of a sum of two terms: the external field $\delta\vec{E}^{ext}$ and the self-consistent field $\delta\vec{E}^{scf}$ which arises due to the redistribution of the charges in the semiconductor. It is so:

$$\mathrm{div}\,\delta\vec{E}^{scf}(\vec{r},t) \equiv i\vec{q}\,\delta\vec{E}^{scf}(\vec{r},t) = \frac{e}{\epsilon}\,\delta n(\vec{r},t), \tag{23}$$

where ε is the static dielectric constant of the material and $\delta n(\vec{r},t)$ is the fluctuation of concentration given by:

$$\delta n(\vec{r},t) \equiv \delta n_{q\omega} \exp(-i\omega t + i\vec{q}\vec{r}) = \frac{1}{4\pi^3} \int \delta f(\vec{k},\vec{r},t) d\vec{k} \qquad (24)$$

By substituting equations (16,17,23) in equation (18) we find:

$$(-i\omega + i\vec{q}\vec{v} + I_k)\delta f_{q\omega}(\vec{k}) - i\frac{e^2}{\hbar\epsilon q^2}\delta n_{q\omega}\vec{q}\frac{\partial \bar{f}(\vec{k})}{\partial \vec{k}} + \frac{e}{\hbar}\delta \vec{E}^{ext}_{q\omega}\frac{\partial \bar{f}(\vec{k})}{\partial \vec{k}} = 0 \qquad (25)$$

Equation (25) is the kinetic equation for the space-time Fourier transform of the perturbed distribution function. By setting $\delta\vec{E}^{ext}_{q\omega} = 0$, equation (25) will describe the spontaneous fluctuations of the system around the steady-state.

This case can be treated by using the Langevin method: to this end we shall add on the right-hand side of equation (25) the random term $y_{q\omega}(\vec{k})$ which satisfies the two following properties: (i) it has average value equal to zero:

$$\langle y_{q\omega}(\vec{k})\rangle = 0, \qquad (26)$$

(ii) its spectral density according to Gantsevich et al (1979) is given by:

$$\left[y(\vec{k})\,y(\vec{k}_1)\right]_{q\omega} = (I_k + I_{k_1})\bar{f}(\vec{k})\delta_{k,k_1} - I^{ee}_{k,k_1}(\bar{f},\bar{f}), \qquad (27)$$

where $I^{ee}_{kk_1}$ is the two-particle collision term without including one summation given by:

$$I^{ee}_{kk_1}(\bar{f},\bar{f}) = \left[\frac{V_0}{(2\pi)^3}\right]^2 \int d\vec{k}' \int d\vec{k}_1' \left[W(\vec{k},\vec{k}_1;\vec{k}',\vec{k}_1') \times \bar{f}(\vec{k})\bar{f}(\vec{k}_1) + \right. \qquad (28)$$

$$\left. - W(\vec{k}',\vec{k}_1';\vec{k},\vec{k}_1)\bar{f}(\vec{k}')\bar{f}(\vec{k}_1')\right]$$

Here $(\ )_{q\omega}$ stands for the space-time Fourier transform of the two-particle (with wavevector $\vec{k}$ and $\vec{k}_1$, respectively) correlator $\langle y(\vec{k},\vec{r},t)\ y(\vec{k}_1,\vec{r},t+\tau)\rangle$. When e-e interaction is neglected only the first term in the right-hand-side of equation (27) survives.

The Boltzmann Langevin equation in $\vec{q}$-ω space is thus obtained as:

$$(-i\omega + i\vec{q}\vec{v} + I_k)\delta f_{q\omega}(\vec{k}) - i\frac{e^2}{\hbar\epsilon q^2}\delta n_{q\omega}\vec{q}\frac{\partial f(\vec{k})}{\partial\vec{k}} = y_{q\omega}(\vec{k}) \tag{29}$$

By integrating over $\vec{k}$ and recalling the zero-sum property (see equation (35)) of the random term we obtain the continuity equation in $\vec{q}$-ω space:

$$-i\omega\delta n_{q\omega} + \frac{i}{e}\vec{q}\cdot\delta\vec{j}_{q\omega} = 0\,, \tag{30}$$

where the space-time Fourier transform of the density-current fluctuation $\delta\vec{j}_{q\omega}$ is given by:

$$\delta\vec{j}_{q\omega} = \frac{e}{4\pi^3}\int\vec{v}\,\delta f_{q\omega}(\vec{k})d\vec{k} \tag{31}$$

By defining the external random-current density $\vec{g}_{q\omega}(\vec{k})$ related to the random term $y_{q\omega}(\vec{k})$ as:

$$\vec{g}_{q\omega} = \frac{e}{4\pi^3}\int\vec{v}\,I_k^{-1}y_{q\omega}(\vec{k})\,d\vec{k} \tag{32}$$

Gantsevich et al (1979) demonstrated the following theorem:

$$(\delta j_\alpha\,\delta j_\beta)_{\omega\to 0} = (g_\alpha\,g_\beta)_{q\omega}\,, \tag{33}$$

which reveals the identity of the spectral density of local random current density $\vec{g}$, which provokes the spatially inhomogeneous fluctuations, with the spectral density of the spatially homogeneous density-fluctuations $\delta\vec{j}$ in the limit $\omega \to 0$.

2.3 DEFINITION OF THE KINETIC COEFFICIENTS FROM THE LOW-FREQUENCY AND LONG-RANGE FLUCTUATIONS

In order to obtain $\delta f_{q\omega}(\vec{k})$ and $\delta\vec{j}_{q\omega}$ from equations (29,31) we use the Chapman-Enskog procedure (Chapman and Cowling, 1970) taking advantage from the inequalities $\omega\tau >> 1$, $ql >> 1$ (l being the carrier mean free path). Here we recall that $1/\tau$ gives the order of magnitude of the linearized collision operator I_k.

The idea of Chapman-Eskog method is to reduce the solution of the space and time-dependent Boltzmann equation to the solution of a number of differential equations with the coefficients expressed through the solution of some Boltzmann equations of the type of space-and-time independent response. These last equations have the general form:

$$I_k z(\vec{k}) = x(\vec{k}), \tag{34}$$

I_k being a linear operator.

To manage equation (29) we need now a proper definition of an inverse operator I_k^{-1}. The collisions and the electric field conserve the number of electrons redistributing them in $\vec{k}$-space; thus, while acting on a distribution function $z(\vec{k})$, the operator I_k generates a function with a zero-sum property:

$$\int I_k z(\vec{k}) d\vec{k} = 0 \tag{35}$$

Therefore, from equation (34) we should demand

$$\int x(\vec{k})d\vec{k} = 0 \tag{36}$$

Since the function $\frac{\partial}{\partial N}\bar{f}(\vec{k})$ is the solution of the homogeneous equation (21), the solution of equation (34) is not unique; in fact if $z_0(\vec{k})$ is a particular solution of equation (34), then $z^1(\vec{k}) = z_0(\vec{k}) + \text{const}\frac{\partial}{\partial N}\bar{f}(\vec{k})$ is also a solution. Thus for the present an inverse operator I_k^{-1} is not defined uniquely. To define it uniquely let us require:

$$\int I_k^{-1}x(\vec{k})d\vec{k} = 0 \tag{37}$$

that is, let us name as $I_k^{-1}x(\vec{k})$ the particular solution $z^1(\vec{k})$ obeying the zero-sum property. Such a solution can always be constructed from any particular solution $z_0(\vec{k})$ by choosing const $= -\frac{V_0}{4\pi^3}\int z_0(\vec{k})d\vec{k}$. Thus the operations I_k and I_k^{-1} generate the functions with the zero-sum property, and the operator I_k^{-1} is allowed to act only on the functions with the same property. It is convenient to demand that the operator I_k also would act only on the functions having the zero-sum property. This requirement can always be fulfilled because of the identity:

$$I_k z(\vec{k}) = I_k\left[z(\vec{k}) - \frac{V_0}{4\pi^3}\int z(\vec{k}')d\vec{k}'\frac{\partial}{\partial N}\bar{f}(\vec{k})\right] \tag{38}$$

According to this convention, we should rewrite equation (34) as follows:

$$I_k\left[z(\vec{k}) - \frac{V_0}{4\pi^3}\int z(\vec{k}')d\vec{k}'\frac{\partial}{\partial N}\bar{f}(\vec{k})\right] = x(\vec{k}) \tag{39}$$

Now the general solution of equation (34) can be obtained from equation (39) after multiplying by I_k^{-1} as:

$$z(\vec{k}) = \frac{V_0}{4\pi^3}\int z(\vec{k}')d\vec{k}'\,\frac{\partial}{\partial N}f(\vec{k}) + I_k^{-1}x(\vec{k}) \tag{40}$$

At this stage we are ready to treat the Boltzmann-Langevin equation (29). Multiplying the continuity equation (30) by $V_0 \frac{\partial}{\partial N}\bar{f}(\vec{k})$ and substracting the result from equation (29) we obtain:

$$-i\omega\left[\delta f_{q\omega}(\vec{k}) - V_0\,\delta n_{q\omega}\frac{\partial\bar{f}(\vec{k})}{\partial N}\right] + i\vec{q}\left[\vec{v}\,\delta f_{q\omega}(\vec{k}) - \frac{V_0}{e}\,\delta\vec{j}_{q\omega}\frac{\partial}{\partial N}\bar{f}(\vec{k})\right]$$

$$-i\frac{e^2}{h\epsilon q^2}\,\delta n_{q\omega}\vec{q}\,\frac{\partial\bar{f}(\vec{k})}{\partial\vec{k}} - I_k\delta f_{q\omega}(\vec{k}) = y_{q\omega}(\vec{k}) \tag{41}$$

By rewriting the last term on the left-hand-side of the above equation in accordance with the convention of equation (38) and multiplying by I_k^{-1} we obtain:

$$\delta f_{q\omega}(\vec{k}) = V_0\delta n_{q\omega}\frac{\partial}{\partial N}\bar{f}(\vec{k}) + i\frac{e^2}{\hbar\epsilon q^2}\,\delta n_{q\omega}\,\vec{q}\,I_k^{-1}\frac{\partial}{\partial \vec{k}}\bar{f}(\vec{k})$$

$$-\,i\vec{q}I_k^{-1}\left[\vec{v}\,\delta\bar{f}_{q\omega}(\vec{k}) - \frac{V_0}{e}\,\delta\vec{j}_{q\omega}\frac{\partial}{\partial N}\bar{f}(\vec{k})\right] \tag{42}$$

$$+\,i\omega I_k^{-1}\left[\delta f_{q\omega}(\vec{k}) - V_0\,\delta n_{q\omega}\frac{\partial\bar{f}(\vec{k})}{\partial N}\right] + I_k^{-1}y_{q\omega}(\vec{k})$$

All expressions under the sign I_k^{-1} in equation (42) are functions with the zero-sum property. It allows us to treat them seperately, and this is the reason for our subtracting procedure.

Multiplying equation (42) by $ev_\alpha/(4\pi^3)$, we obtain after integration over $\vec{k}$:

$$\delta j_{q\omega,\alpha} = e v'_{d,\alpha}\,\delta n_{q\omega} + \sigma_{\alpha\beta}\zeta_\beta - i\frac{e}{4\pi^3} q_\beta \int d\vec{k}\, v_\alpha I_k^{-1}\left[v_\beta\,\delta f_{q\omega}(\vec{k}) - \frac{V_0}{e}\,\delta j_{q\omega,\beta}\frac{\partial}{\partial N}\bar{f}(\vec{k})\right] + i\frac{e}{4\pi^3}\omega\int d\vec{k}\, v_\alpha I_k^{-1}\left[\delta f_{q\omega}(\vec{k}) - V_0\,\delta n_{q\omega}\frac{\partial \bar{f}(\vec{k})}{\partial N}\right] + g_{q\omega}, \quad (43)$$

where $v'_{d,\alpha}$ is the differential (with respect to the number of carriers) drift velocity given by:

$$v'_{d,\alpha} = \frac{V_0}{4\pi^3}\int v_\alpha \frac{\partial \bar{f}(\vec{k})}{\partial N}\, d\vec{k}, \quad (44)$$

$\sigma_{\alpha\beta}\zeta_\beta$ is the current density caused by a fluctuational field ζ_β given by:

$$\zeta_\beta = -iq_\beta \frac{e}{\epsilon q^2}\,\delta n_{q\omega}, \quad (45)$$

$\sigma_{\alpha\beta}$ being the differential conductivity given by:

$$\sigma_{\alpha\beta} = \frac{e^2}{4\pi^3\hbar}\int v_\alpha I_k^{-1}\frac{\partial}{\partial k_\beta}\bar{f}(\vec{k})\, d\vec{k} \quad (46)$$

$g_{q\omega,\alpha}(\vec{k})$ is the external random density-current given by equation (32).

Now equations (40,41) are suitable for the Chapman-Enskog like iteration. After the first iteration (i.e. by substitituing $\delta f_{q\omega}(\vec{k}) \cong V_0 \delta n_{q\omega} \frac{\partial}{\partial N} \bar{f}(\vec{k})$ and $\delta \vec{j}_{q\omega} \cong e\vec{v}_d^{\,\prime} \delta n_{q\omega}$) the third term in equation (32) gives the diffusion density current:

$$\frac{ieV_0}{4\pi^3} q_\beta \int d\vec{k} v_\alpha I_k^{-1} (v_\beta - v'_{d,\beta}) \frac{\partial}{\partial N} \bar{f}(\vec{k}) \delta n_{q\omega} = ieq_\beta D_{\alpha\beta} \delta n_{q\omega} , \tag{47}$$

with the diffusion constant $D_{\alpha\beta}$, which accounts for carrier-carrier scattering, defined by:

$$\boxed{D_{\alpha\beta} = \frac{V_0}{4\pi^3} \int d\vec{k} v_\alpha I_k^{-1} (v_\beta - v'_{d,\beta}) \frac{\partial}{\partial N} \bar{f}(\vec{k})} \tag{48}$$

while the fourth term vanishes.

Thus as first approximation in the parameters $\omega\tau$ and ql for the fluctuating density-current we obtain the Fick-Langevin equation:

$$\delta j_{q\omega,\alpha} = ev'_{d,\alpha} \delta n_{q\omega} + \sigma_{\alpha\beta} \zeta_\beta - ieq_\beta D_{\alpha\beta} \delta n_{q\omega} + g_{q\omega,\alpha} \tag{49}$$

According to equations (27,33) the spectral density for the random-density-current $\vec{g}_{q\omega}$ is given by:

$$(g_\alpha g_\beta)_{q\omega} = \left(\frac{e}{4\pi^3}\right)^2 \int d\vec{k} \int d\vec{k}_1 v_\alpha v_{1\beta} I_k^{-1} I_{k1}^{-1} \left[(I_k + I_{k_1}) \bar{f}(\vec{k}) \delta_{k,k_1} - I_{kk_1}^{ee}(\bar{f}, \bar{f}) \right] \tag{50}$$

To make use of the identity $I_k^{-1} I_k = 1$ by accounting for equation (21), we can rewrite equation (50) as follows:

$$(g_\alpha g_\beta)_{q\omega} = \left(\frac{e}{4\pi^3}\right)^2 \int d\vec{k} \int d\vec{k}_1 v_\alpha v_{1\beta} I_k^{-1} I_{k1}^{-1} \Big\{ (I_k + I_{k_1}) N \frac{\partial}{\partial N} \bar{f}(\vec{k}) \left[\delta_{k,k_1} - \frac{\partial}{\partial N} \bar{f}(\vec{k}_1) \right]$$
$$+ \left[(I_k + I_{k_1}) \left(\bar{f}(\vec{k}) - N \frac{\partial}{\partial N} \bar{f}(\vec{k}) \right) \delta_{k,k_1} - I_{kk_1}^{ee}(\bar{f}, \bar{f}) \right] \Big\} \tag{51}$$

Now the first item in curly brackets can be written as:

$$\int d\vec{k} \int d\vec{k}_1 \, v_\alpha \, v_{1\beta} I_k^{-1} I_{k1}^{-1} \left(I_k + I_{k_1}\right) N \frac{\partial}{\partial N} \bar{f}(\vec{k}) \left[\delta_{k,k_1} - \frac{\partial}{\partial N} \bar{f}(\vec{k}_1)\right] =$$

$$\frac{4\pi^3}{V_0} \left\{ N \int d\vec{k} v_\alpha I_k^{-1} (v_\beta - v'_{d,\beta}) \frac{\partial}{\partial N} \bar{f}(\vec{k}) + N \int d\vec{k}_1 v_{1\beta} I_{k_1}^{-1} (v_{1\alpha} - v'_{d,\alpha}) \frac{\partial}{\partial N} \bar{f}(\vec{k}_1) \right\} =$$

$$\left(\frac{4\pi^3}{V_0}\right)^2 N \left(D_{\alpha\beta} + D_{\beta\alpha}\right) \tag{52}$$

and the second term in curly brackets of equations (50) can be written as:

$$\int d\vec{k} \int d\vec{k}_1 \, v_\alpha \, v_{1\beta} I_k^{-1} I_{k1}^{-1} \Big[I_{kk_1}^{ee}(\bar{f}, \bar{f}) - \left(I_k + I_{k_1}\right)\left(\bar{f}(\vec{k}) - N \frac{\partial}{\partial N} \bar{f}(\vec{k}_1)\right) \delta_{k,k_1} \Big] = \left(\frac{4\pi^3}{V_0}\right)^2 N \Theta_{\alpha\beta} \tag{53}$$

Making use of the relation (31), equations (50 - 53) give finally:

$$(g_\alpha \, g_\beta)_{q\omega} \equiv (\delta j_\alpha \, \delta j_\beta)_{\omega \to 0} = \frac{e^2 N}{V_0^2} (D_{\alpha\beta} + D_{\beta\alpha} - \Theta_{\alpha\beta}) \tag{54}$$

For the simple geometry considered, the current fluctuation spectral density measured along a particular direction α is now given by:

$$\boxed{S_{I,\alpha}(E, \omega \to 0) = \frac{4e^2 N}{L^2} \left[D_\alpha(E) - \frac{1}{2} \Theta_\alpha(E) \right]} \tag{55}$$

Thus from equation (55) it is seen that no simple relation exists between diffusion and noise-spectral density when carrier-carrier collisions are effective; the term $\Theta\alpha\beta$ given by equation (53) accounting for these collisions explicitely. Calculations of $\Theta\alpha\beta$ were performed under an electron temperature model, in which case the difference between $\bar{f}(\vec{k})$ and $N \frac{\partial}{\partial N} \bar{f}(\vec{k})$ can be neglected, by

Shulman (1970) and Gantsevich et al (1974). Results show that positive as well as negative values of $\Theta_{\alpha\beta}$ are possible according to different dependence on electron-energy of energy and momentum relaxation times, respectively.

When electron-electron collisions are negligible (i.e. $I^{ee}_{kkl}(\bar{f},\bar{f}) = 0$, $N \frac{\partial \bar{f}}{\partial N} = \bar{f}$), or they are determinant so that the distribution function is a displaced maxwellian one, then $\Theta_{\alpha\beta} = 0$ and we have:

$$\boxed{S_{I\alpha}(E, \omega \to 0) = \frac{4e^2N}{L^2} D_\alpha(E)} \tag{56}$$

which generalizes at high fields equation (8) through a field dependent diffusion coefficient in general given by equation (48).

In thermal equilibrium we always have $\Theta_{\alpha\beta} = 0$ since $f^{eq}(\vec{k}) = N \frac{\partial}{\partial N} \bar{f}(\vec{k})$ and $I^{ee}_{kk}(f^{eq}, f^{eq}) = 0$. Moreover in equilibrium $v_d = 0$ and $\vec{v} f^{eq}(\vec{k}) = - \frac{K_B T_O}{\hbar} \frac{\partial}{\partial \vec{k}} f^{eq}(\vec{k})$, thus recalling equation (46) we get:

$$(\delta j_\alpha \, \delta j_\beta)^{eq}_{\omega \to 0} = \frac{2e^2N}{V_0^2} D^{eq}_{\alpha\beta} = \frac{2K_BT_0}{V_0} \sigma^{eq}_{\alpha\beta} \tag{57}$$

in accordance with the Nyquist and Einstein relations.

3. THE GENERALIZED DIFFUSION COEFFICIENT

In this section we shall confine our attention to the case in which electron-electron interaction is neglected. By briefly surveying various equivalent expressions for the diffusion coefficient, we shall arrive at defining a frequency and wavevector dependent $D(\vec{q},\omega)$. In particular we shall demonstrate the exact relation which exists between $D(\omega)$ and $S_I(\omega)$.

3.1 EQUIVALENT DEFINITIONS OF FICK'S DIFFUSION COEFFICIENT

Because of the prohibitive difficulty in calculating diffusion through equation (48), in the following, three equivalent ways of expressing such a quantity at an hydrodynamic level are reported. For simplicity let us deal with the components of x, v_{do} and v (of position vector x, drift velocity of the space-homogeneous case v_{do} and particle velocity v) in some specific but arbitrary direction.

On defining the m-th moment of n(x,t) as

$$M^{(m)} = \frac{1}{N}\int_{-\infty}^{+\infty} x^m n(x,t)dx \equiv \langle x^m \rangle, \tag{58}$$

N being the total number of particles, by using the diffusion equation:

$$\frac{\partial n(x,t)}{\partial t} = - v_{d_o}(E)\frac{\partial n(x,t)}{\partial x} + D(E)\frac{\partial^2 n(x,t)}{\partial x^2} \tag{59}$$

the following recursion relations are found:

$$M^{(0)} = 1, \tag{60}$$

$$\frac{d}{dt} M^{(1)} = v_{d_o}, \tag{61}$$

$$\frac{d}{dt} M^{(2)} = 2(v_{d_o} M^{(1)} + D), \tag{62}$$

$$\frac{d}{dt} M^{(m\geq 2)} = m v_{d_o} M^{(m-1)} + m(m-1)DM^{(m-2)} \tag{63}$$

For the second central moment $\langle(x - \langle x\rangle)^2\rangle$ we have:

$$D = \lim_{t \to \infty} \frac{d}{dt} \frac{1}{2} \left[\langle x^2(t) \rangle - \langle x(t) \rangle^2 \right] \tag{64}$$

where the long-term limit ensures that Fick's law is satisfied (Van Beijern, 1982). Equation (64) is consistent with the interpretation of diffusion as a spreading in space of some initial particle distribution; diffusion is thus related to a second central moment which increases linearly with time.

By carrying out the time derivative of equation (64) and using the fact that ensemble average commutes with time derivative, an equivalent expression of diffusion is obtained as (Price, 1979; Matulionis et al, 1981):

$$D = \lim_{t \to \infty} \left[\langle x(t) v(t) \rangle - \langle x(t) \rangle \langle v(t) \rangle \right] \tag{65}$$

In this way diffusion is correlated to the achievement of a stationary value for the space velocity covariance of the initial particle distribution.

By writing the trajectory in terms of velocity a third equivalent expression of diffusion is obtained as (Matulionis et al, 1981):

$$D = \lim_{t \to \infty} \left[\int_0^t \langle \delta v(t) \delta v(t-\tau) \rangle d\tau + \langle \delta x(0) \delta v(t) \rangle \right] \tag{66}$$

Equation (66) relates diffusion to the autocorrelation function of velocity fluctuations; it has to be noted that under the long-time limiting condition the integrand becomes a function of τ only, and the last term vanishes.

At very short times, comparable or less than an average scattering time, the time dependence of the quantities inside the limit in equations (64 - 66) can be used to describe a diffusion coefficient which will depend upon time and initial conditions of motion. Figure 3 illustrates the dependence upon time of quantities defined by equations (64 - 66) when there is no applied electric field and each particle is assumed to start at t=0 at the position x=0 with an equilibrium Maxwell-Boltzmann momentum distribution.

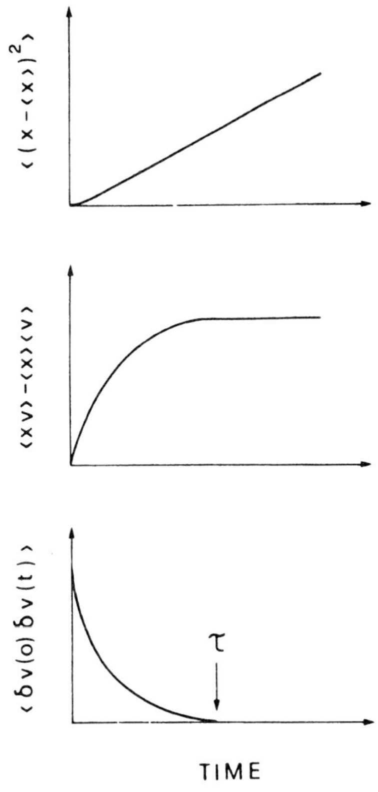

Figure 3. Schematic of the time dependence of spatial spreading, space-velocity covariance and velocity fluctuation autocorrelation function.

In the figure τ indicates the time necessary for the correct space-velocity correlation to be achieved so that Fick's law holds. It is worth noting that since the space-velocity covariance and the velocity fluctuation autocorrelation function achieves a constant value in the long-time limit, they are better suited to check the validity of Fick's law than the second central moment which exhibits a long-time trend.

Figure 4 illustrates the transient behaviour of the spreading in space and of the related diffusion coefficient, which depends upon time, when, for an ensemble of particles with same initial conditions as above, at t=0 an electric field is switched on; curve 1 refers to a small applied electric field and curves 2 and 3 to a high applied electric field so that hot-electron conditions are established. In particular, under transient conditions hot-electron diffusion is shown to exhibit values greater than those under long-time limit conditions (overshoot effect) (Reggiani et al, 1981) as well as negative value (Reggiani et al, 1981; Boittiaux et al, 1982).

3.2 THE GENERALIZED DIFFUSION EQUATION

In real physical systems the particle density n(x,t) is never exactly of Gaussian form (Van Beijern, 1982). If n(x,t) satisfies the diffusion equation it does so only in an asymptotic sense, that is only on a macroscopic time length scale. In the following we present a generalization of the diffusion equation (59) which allows us to define a frequency and wavevector dependent diffusion coefficient $D(q,\omega)$ (Jacoboni et al, 1981).

Let us consider a system of particles subject to an external uniform static electric field E along x. A particle source of sinusoidal type

$$S(x,t) = C\left[1 + \cos(qx - \omega t\right] \tag{67}$$

is then introduced together with a trapping mechanism with a constant rate τ_d^{-1}.

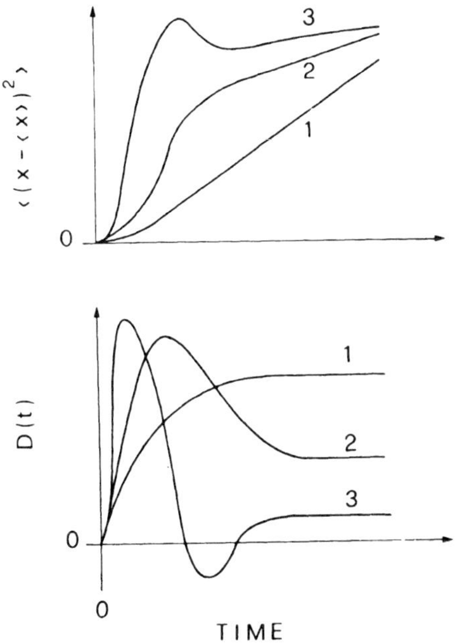

Figure 4. Schematic of the spatial spreading and related diffusion coefficient as a function of time when transient conditions are analyzed. Curve 1 refers to the Ohmic case; curves 2 and 3 refer to possible hot-electron conditions (see text).

The source amplitude C and τ_d^{-1} are connected through the condition of particle conservation over a wavelength λ:

$$\int_\lambda C\left[1+\cos(qx-\omega t)\right]dx \quad - \quad \int_\lambda \tau_d^{-1} n(x,t)\,dx \tag{68}$$

The particle are thus trapped and redistributed in space. By making use of the more convenient complex formalism, the diffusion equation (59) becomes

$$\frac{\partial n}{\partial t} = D\frac{\partial^2 n}{\partial x^2} - v_{d_0}\frac{\partial n}{\partial x} + C\left\{1 + \exp\left[i(qx-\omega t)\right]\right\} - \tau_d^{-1} n \tag{69}$$

We look for a solution of equation (69) of the type

$$n(x,t) = n_0 + n_1 \exp\left[i(qx - \omega t + \phi)\right] \tag{70}$$

where n_0 is a constant average term, n_1 and ϕ are the amplitude and the phase-shift of the harmonic disturbance, respectively. By substituting equation (70) into equation (68) we obtain:

$$C = n_0 \tau_d^{-1} \tag{71}$$

and equation (69) becomes:

$$-i\omega + q^2 D + iqv_{d_0} + \tau_d^{-1} = \frac{n_0}{n_1} \tau_d^{-1} \exp(-i\phi) \tag{72}$$

As long as problems where n is linear are considered, this equation can also be used at frequencies and wavelength so high that Fick's law does not hold; in this case D must be taken as a function of q and ω. Thus from equation (72) we obtain an expression for $D(q,\omega,\tau_d^{-1})$ in terms of the relative amplitude n_1/n_0 and of the phase shift ϕ of the disturbance:

$$D = \frac{1}{q^2}\left[\frac{n_0}{n_1}\tau_d^{-1}\exp(-i\phi) + i(\omega - qv_d) - \tau_d^{-1}\right] \tag{73}$$

By using the path integral method (Chambers, 1952) we can obtain the following integral expression for n(x,t):

$$n(x,t) = n_0 \int_{-\infty}^{t} \frac{dt'}{\tau_d} \int_{-\infty}^{+\infty} dx' \left\{1 + \exp\left[i(qx' - \omega t')\right] P(x', t'; x, t) \exp\left[-\tau_d^{-1}(t-t')\right]\right\}, \tag{74}$$

where $P(x',t';x,t)dx$ is the probability that without trapping a particle in x' at t' will be in dx around x at t.

The integral of the first term in brackets in equation (74) is equal to n_o, since from the time $-\infty$ to t all the particles have been trapped and redistributed by the source. P(x',t';x,t) is a function of x', t', and x, t only through the differences x" = x-x' and t" = t-t' so that equations (70,74) yield:

$$n(x,t) - n_0 = n_0 \exp\left[i(qx - \omega t)\right] \int_0^{\infty} \frac{dt''}{\tau_d} \int_{-\infty}^{+\infty} dx'' \exp\left[-i(qx'' - \omega t'') - t''/\tau_d\right] P(x'', t'') \quad (75)$$

From equations (70,75) we then obtain:

$$R \equiv \frac{n_1}{n_0} \exp(i\phi) = \int_0^{\infty} \frac{dt}{\tau_d} \int_{-\infty}^{+\infty} dx \exp\left[-i(qx - \omega t) - \frac{t}{\tau_d}\right] P(x,t) \quad (76)$$

and equation (72) finally yields:

$$D(q,\omega,\tau_d^{-1}) = \frac{1}{q^2 \tau_d R}\left[1 - R + i\tau_d R(\omega - qv_{do})\right] \quad (77)$$

Equations (76,77) provide an expression for $D(q,\omega,\tau_d^{-1})$, which is in general a complex quantity, valid for any wavevector, frequency, and particle lifetime τ_d. When τ_d is much longer than both ω^{-1} and any microscopic time, $D(q,\omega) = D(q,\omega,o)$ is obtained for steady-state conditions. P(x,t) must be derived from knowledge of the particle dynamics.

Figures 5 and 6 report the results for the real part of $D(q,\omega)$ as a function of q and ω, respectively as obtained from a Monte Carlo simulation for the case of electrons in Si (Jacoboni et al, 1981). As a general trend $D(q,\omega)$ decreases at asymptotically high values of both q and ω because the average over several wavelengths as well as over several periods of carrier concentrations in the past is not affected by contributions from electrons coming from far away (Jacoboni et al, 1981). The peaks of $D(q,\omega)$ at intermediate frequency **(see figure 6)** are related to the presence of high-energy intervalley phonons whose emission at high fields leads to some streaming character of the carrier motion. The peaks in $D(q,\omega)$ of **figure 5** have the same interpretation.

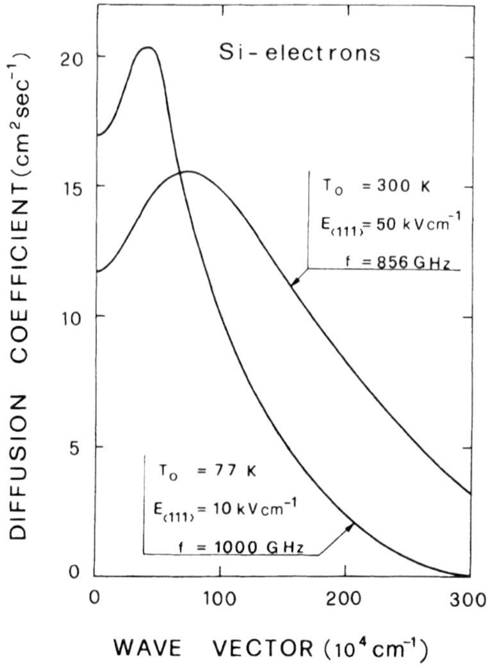

Figure 5. High-field, low frequency diffusion coefficient as as function of wavevector at the indicated temperatures for the case of electrons in Si (Jacoboni et al, 1981).

3.3 COMPARISON WITH PREVIOUS DEFINITIONS FOR D(q) AND D(ω)

The problem of defining a q-dependent diffusion coefficient D(q) =D(q,0) and a ω-dependent diffusion coefficient D(ω) = D(0,ω) has been analyzed in several papers (Price, 1965; Schlup, 1971; Jacoboni, 1974; Jacoboni et al, 1981). As to the limit $\omega \rightarrow 0$ equation (77) gives:

$$D(q,0) = \lim_{\tau_d \to \infty} \frac{1}{q^2 \tau_d R}(1 - R - iq\tau_d v_{d_0} R) \qquad (78)$$

which by using equation (76) and carrying out the limit gives:

$$D(q,0) \equiv D(q) = \frac{1}{q^2}\left(\frac{1}{P_q} - iqv_{d_0}\right) \tag{79}$$

with

$$P_q = \int_0^{\infty} dt \int_{-\infty}^{+\infty} dx \exp(-iqx) P(x,t) \tag{80}$$

which is in agreement with the results of Jacoboni (1974).

As regards the expression of $D(\omega)$, let us consider the limit for q approaching zero of equation (77). We will expand in series of q R given by equation (76) as:

$$R = R_0 + qR_1 + q^2 R_2 + \cdots \tag{81}$$

Expressions for R_0, R_1, and R_2 can be obtained from equation (77) as:

$$R_0 = \int_0^{\infty} \exp\left[-(1 - i\omega\tau_d)\frac{t}{\tau_d}\right] \frac{dt}{\tau_d} = \frac{1}{1 - i\omega\tau_d} \tag{82}$$

$$R_1 = -i\int_0^{\infty} \exp\left[(1 - i\omega\tau_d)\frac{t}{\tau_d}\right] \langle x \rangle_t \frac{dt}{\tau_d} = -\frac{iv_{d_0}\tau_d}{(1 - i\omega\tau_d)^2} \tag{83}$$

$$R_2 = -\frac{1}{2}\int_0^{\infty} \exp\left[-(1 - i\omega\tau_d)\frac{t}{\tau_d}\right] \langle x^2 \rangle_t \frac{dt}{\tau_d} =$$

$$\frac{(v_{d_0}\tau_d)^2}{(1 - i\omega\tau_d)^3} - \frac{\tau_d}{(1 - i\omega\tau_d)^2} \int_0^{\infty} C(t) \exp\left(i\omega t - \frac{t}{\tau_d}\right) dt \tag{84}$$

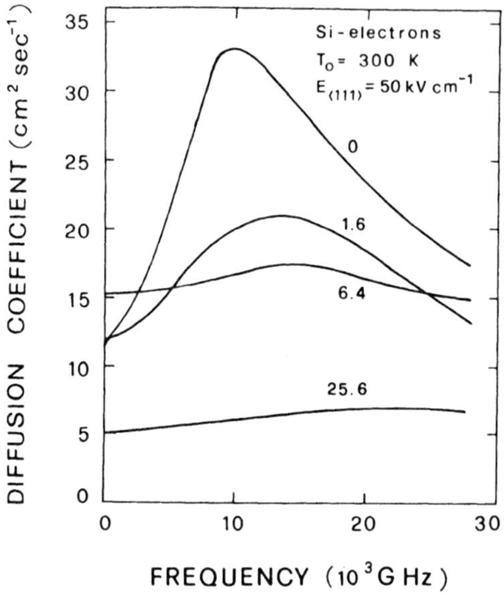

Figure 6. High-field diffusion coefficient as a function of angular frequency ω at the different values of wavevector q indicated by the numbers on the curves in units of 10^5 cm^{-1} for the case of electrons in Si (Jacoboni et al, 1981).

where C(t) is the autocorrelation function of velocity fluctuations:

$$C(t) = \langle \delta v(t) \delta v(0) \rangle \tag{85}$$

By substituting equations (82-84) into equation (77), for vanishing small q one obtains:

$$D(0,\omega) \equiv D(\omega) = \lim_{\tau_d \to \infty} \lim_{q \to 0} D(q, \omega\, \tau_d^{-1}) = \int_0^\infty C(t) \exp(i\omega t) dt \tag{86}$$

Thus it is found that for a density gradient which depends upon time the diffusion coefficient can be obtained from the Fourier transform of the autocorrelation function of velocity fluctuations, in agreement with that proposed by Price (1965) and Schlup (1971).

From the definition of $S_I(\omega)$ given in equation (1), under space- homogeneous conditions and negligible electron-electron interaction it follows:

$$S_I(\omega) = \frac{2e^2N}{L^2}\int_{-\infty}^{+\infty} \exp(i\omega t_d) < \delta v(t)\delta v(0) > dt \tag{87}$$

Thus by comparison with the above equation (86) it is:

$$\boxed{S_{I,\alpha}(E,\omega) = 4\frac{e^2N}{L^2}\,\mathrm{Re}\left[D_\alpha(E,\omega)\right]} \tag{88}$$

which generalizes equation (8) at any frequency and field strength. Therefore, the study of noise spectra near cut-off frequency is equivalent to the study of a frequency dependent diffusion coefficient.

4. THE TRANSFER IMPEDANCE METHOD AND THE SPACE CHARGE LIMITED-CURRENT-CONDITIONS

A general theory for the calculation of the small signal impedance and the noise makes use of the transfer impedance method (Van Vliet et al, 1975; Nougier et al, 1981). This method allows to relate a small perturbation in local density current at frequency $\omega, \delta\vec{j}(\vec{r},t)\exp(i\omega t)$ superimposed to the dc value $\vec{j}(\vec{r})$ with the instantaneous small ac local field $\delta\vec{E}(\vec{r},\omega)\exp(i\omega t)$ superimposed to the dc value $\vec{E}(\vec{r})$. To first order terms in the local equation we have a relation of the form:

$$\hat{L}\delta\vec{E}(\vec{r},\omega) = \vec{\delta j}(\vec{r},\omega), \tag{89}$$

where $\hat{L}$ is a linear operator.

The Green matrix $z(\vec{r},\vec{r}',\omega)$ of the operator $\hat{L}$ is defined by:

$$\hat{L}z(\vec{r},\vec{r}',\omega) = \mathbb{1}\delta(\vec{r}-\vec{r}'), \tag{90}$$

where $\delta(\vec{r}-\vec{r}')$ is the Dirac function and $\mathcal{U}$ the unitary matrix in $\vec{r}$ space. By applying equation (90) to the vector $\delta\vec{j}(\vec{r})$ and integrating over $\vec{r}'$ gives taking into account equation (89).

$$\delta\vec{E}(\vec{r},\omega) = \int_{vol} d^3r' z(\vec{r},\vec{r}',\omega)\,\delta\vec{j}(\vec{r}',\omega), \tag{91}$$

The kernel $z(\vec{r},\vec{r}',\omega)$ is called the transfer impedance matrix. The line integral of the two terms in equation (91) from one electrode to the other gives the induced voltage fluctuation as a function of the current fluctuation, therefore it is found

$$Z(\omega) = \int_0^L d\vec{r}' \int_0^L d\vec{r}\, z(\vec{r},\vec{r}',\omega) \tag{92}$$

$Z(\omega)$ being the ac impedance of the whole device.

When Re $[Z(\omega)]>0$ and carrier-carrier interaction is neglected, exact relations between white-noise and diffusion can be found. In the next section the space-charge-limited current condition is illustrated under the limit $\omega \to 0$.

4.1 THE SPACE-CHARGE-LIMITED CURRENT CASE AND THE GYSOLF-ZIJLSTRA RELATION

Using a one dimensional treatment, justified by the simple geometry here considered, the current equation, with the diffusion term neglected, and the Poisson's equations are, respectively:

$$j = en(x)\mu(x)E(x), \tag{93}$$

$$\frac{d}{dx}E(x) = \frac{e}{\epsilon}n(x) \tag{94}$$

where, owing to steady-state conditions, j is x-independent.

Linearization of equations (93,94) gives:

$$\delta j = \epsilon\left[\frac{dE}{dx}\mu' + \mu E\frac{d}{dx}\right]\delta E \tag{95}$$

It can be easily verified that the Green function giving the transfer impedance is:

$$z(x,x') = \frac{u(x-x')}{A\epsilon\mu E} = \frac{u(x-x')}{I}\frac{dE}{dx}, \tag{96}$$

where $u(x-x')$ is the unit step function. The fact that x must be larger than x for a non-zero result corresponds to field propagation to the right.

Let us evaluate the spectral density of voltage fluctuations:

$$S_V(\omega\to 0) = 4\int_0^L dx_1\int_0^L dx_2\int_0^L dx_1'\int_0^L dx_2'\left[z(x_1,x')z(x_2,x'')\right]\int_0^\infty \overline{\delta I(x',t)\delta I(x'',t+\tau)}\,d\tau \tag{97}$$

Now by considering slices of material of volume $A\Delta x$, providing that Δx remains larger than the carrier mean free path but small enough to justify a local field approximation (Rees, 1973) we argue that carrier velocity fluctuations at x' around $\Delta x'$ and x" around $\Delta x''$ are uncorrelated unless the slices overlap. Under this assumption it is:

$$\int_0^\infty \overline{\delta I(x',t)\delta I(x'',t+\tau)}\,d\tau = e^2 An(x')D(x')\delta(x'-x'') \tag{98}$$

where $n(x')$ is the carrier concentration within a slice of material at position x'.

Substituting equation (98) into equation (97), performing the line integration and using the Poisson equation the Gisolf-Zijlstra formula (1976) is obtained as:

$$S_V(\omega\to 0) = \frac{4e\epsilon A}{I^2}\int_0^{E(L)}\left[E(L)-E\right]^2 D(E)\,dE \tag{99}$$

The small signal impedance being:

$$Z = \frac{LE(L)-V_a}{I} \tag{100}$$

V_a being the voltage applied to the device.

By using equation (6) we obtain for S_I:

$$S_I(\omega \to 0) = \frac{4e\epsilon A}{\left[LE(L)-V_a\right]^2}\int_0^{E(L)}\left[E(L)-E\right]^2 D(E)dE \tag{101}$$

Equations (99,101) relate noise and diffusion under space-charge limited current conditions. Their validity requires that the carrier mean free path remains small with respect to the length of the sample and that diffusion current is negligible.

It is worth pointing out that under Ohmic field conditions (that is $D = K_B T_O \mu/e$) equation (101) recovers the well known result (Van Vliet et al, 1975):

$$S_I(\omega \to 0) = \frac{8K_B T_O}{Re[Z]} \tag{102}$$

which double the classical Nyquist value valid under field homogeneous conditions.

5. RESULTS

A test for confirming the validity of equations (56,88,102), relating white-noise with diffusion, can be pursued by comparing the two independent evaluations of diffusion obtained: (i) from the proper definition of D (see Section 3) and (ii) from its relationship with the noise current spectral-density (see Sections 2 and 4) .

From an experimental point of view the determination of D by means of the time-of-flight and the noise-conductivity techniques (Bareikis et al, 1968; Reggiani et al 1978; Nava et al, 1979; Reggiani, 1980; Brunetti et al, 1981) has offered the main support to theoretical findings. At the same time, simulative calculations performed with the Monte Carlo technique (Hill et al, 1979; Fauquembert et al, 1980) (and to a less extent with iterative methods (Nougier and Rolland, 1973; Vinter, 1973)) have offered a complementary pseudo-experimental technique to test the results so obtained.

An exemplification of different findings, in the limit of small frequencies $\omega \to 0$, is given in **Figures 7-9** where the results of the longitudinal diffusion coefficient and of the equivalent noise temperature are reported for different materials and for the case of electrons and holes.

It is worth pointing out that for the case of GaAs **(see Figure 9)** at fields above the threshold for Gunn effect (i.e. $E \geq 3.5$ kV/cm) a noise temperature cannot be defined since owing to a $\mu' < 0$, the system becomes electrically unstable with respect to fluctuations in charge density (Pamplin, 1970). In all cases the agreement between data obtained with different experimental techniques and between theory and experiments fully support the theoretical results.

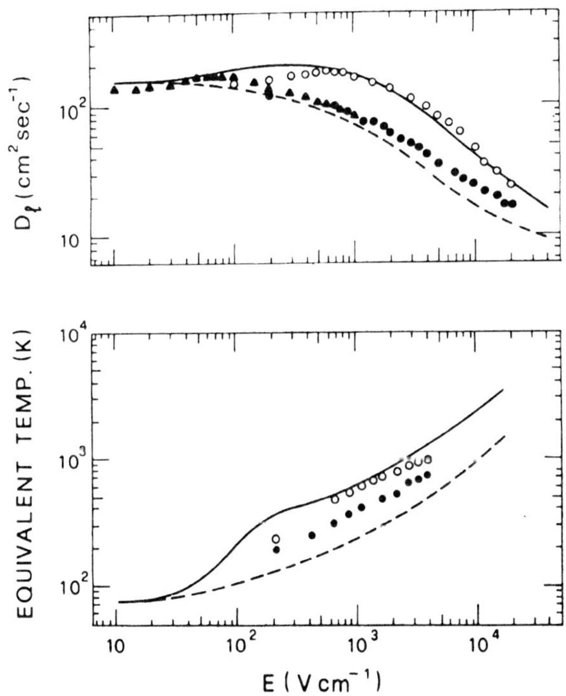

Figure 7. Longitudinal diffusion coefficient and longitudinal noise temperature as a function of electric field strength for the case of electrons in Si at 77K. Circles and triangles refer to experiments and lines to Monte Carlo calculations; full (open) symbols and dashed (continuous) lines refer to a field applied along <111> (<100>) direction, respectively (Reggiani et al, 1981a; Brunetti et al, 1981).

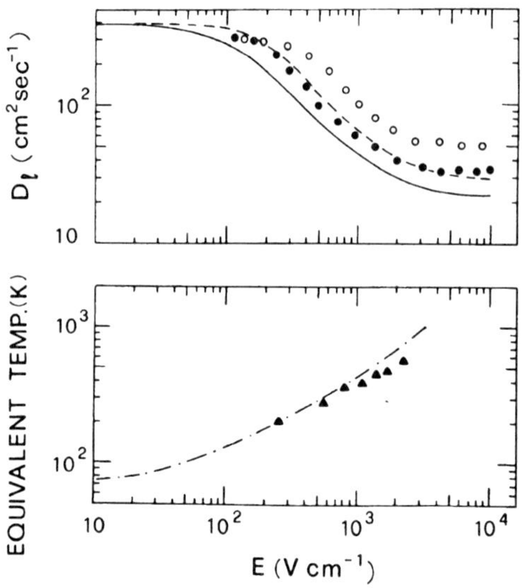

Figure 8. Longitudinal diffusion coefficient and longitudinal noise temperature as a function of electric field strength for the case of holes in Ge at 77K. Circles and triangles refer to experiments and lines to Monte Carlo calculations; full circles and continuous lines refer to a field applied along <111> direction, triangles and dot dashed lines to <110> direction, respectively (Reggiani, 1980).

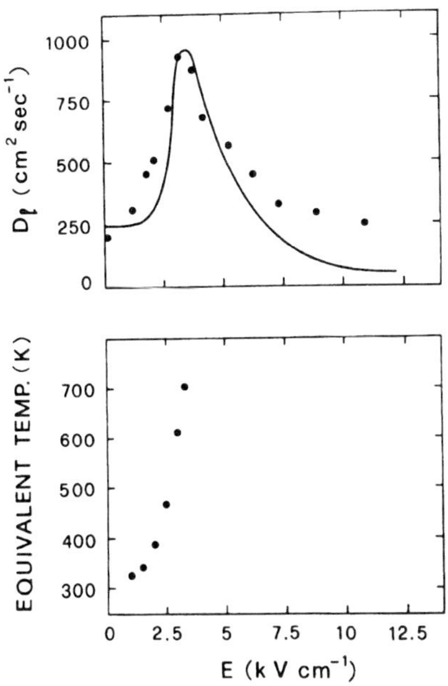

Figure 9. Longitudinal diffusion coefficient and longitudinal noise temperature as a function of electric field strength for the case of electrons in GaAs at 300K. Circles refer to experiments (Ruch and Kino, 1968; Bareikis et al, 1978) and lines to Monte Carlo calculations (Pozhela and Reklaitis, 1978).

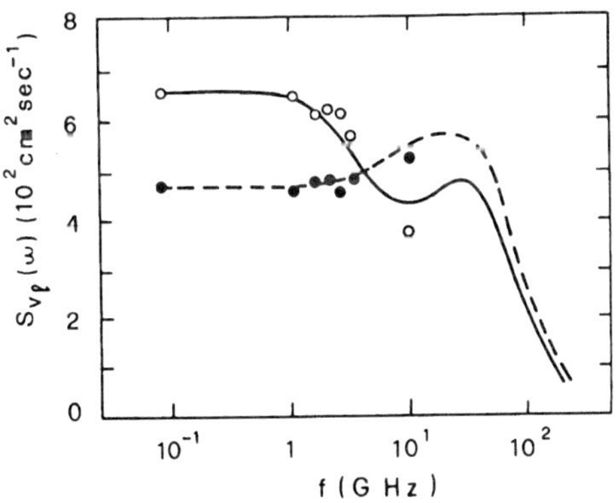

Figue 10. Frequency dependence of the longitudinal velocity-fluctuation spectral-density for n-Si at T = 78K and E= 200 V/cm. Circles refer to experiments and lines to Monte Carlo calculations; full (open) circles and dashed (continuous) lines refer to a field applied along <111> (<100>)directions, respectively (Bareikis et al, 1982).

Figure 10 reports the longitudinal velocity fluctuation spectral density $S_{v\ell}(\omega) = \frac{L^2}{e^2 N} S_{I\ell}(\omega)$ as a function of frequency for the case of electrons in Si; results are well suited to exemplify the dependence with frequency of noise, and thus of diffusion.

Recently Van der Ziel and Bosman (1982) have calculated the current noise spectral density for the case of short n^+-$n^-$$n^+$ or n^+-$p^-$$n^+$ solid state diodes operating under ballistic conditions. By increasing the applied voltage up to values for which the space charge limited current conditions are achieved (but still remaining within the ohmic region) they have predicted a slow decrease of the initial thermal spectral density from its equilibrium value, $S_I^{eq}(0) = 4\ K_B T_O/Z$ till the limiting value $S_I^{ball} = 0.644\ S_I^{eq}$ is still achieved. This value for $S_I^{ball}(\omega \rightarrow 0)$, when compared with the collision limited case discussed in Section 4.1 for which $S_I^{ball}(\omega \rightarrow 0) = 2\ S_I^{eq}(0)$, implies a significant decrease in the noise level under ballistic mode operation.

6. CONCLUSIONS

In these lectures I have focused on the physical aspects of noise/diffusion relationships which may be expected to exist in submicron structures. It is seen that in steady-state and collision dominated processes, general relationships between noise and diffusion valid at any electric field strength can be established provided that electron-electron collisions are negligible or so determinant as to give rise to a drifted maxwellian distribution function. In particular, under homogeneous field conditions and negligible carrier-carrier interactions the frequency dependent diffusion coefficient has been proven to be exactly correlated to the frequency dependent spectrum of noise about the cutt-off region.

Under transient time conditions overshoot effects as well as negative values of the diffusion coefficient can be predicted, but no general relationship with noise is available at present. Under ballistic (i.e. collision free) conditions a decrease in the value of the noise spectral density with respect to the case in which collisions are dominating has been proposed but even in this case no general relationship with diffusion is available at present. Futhermore, the above predicted effects still await experimental evidence.

Helpful suggestions from Drs. R. Brunetti and C. Jacoboni and the Computer Center of Modena University for having provided computer facilities, are greatfully acknowledged.

Table 1 - Physical aspects correlated to submicron structures

High-electric-fields
Short-time and/or high frequency
Strong space-gradient and/or small wavelength
Overshoot phenomena
Ballistic phenomena

REFERENCES

1. Bareikis, V., Galdikas, A., Miliusyte, R., and Victoravichius, V., 1978, in "Abstracts of the 5th Int. Conf. on Noise in Physical Systems", Bad Nauheim, p. 212 unpublished.
2. Bareikis, V., Galdikas, A., and Victoravichius, V., 1982, Fiz. Techn. Populoprovodn. 16, 1968.
3. Bareikis, V., Pozhela, K., and Matulenene, I.B., 1968, in "Proc. 9th Int. Conf. Phys. Semicon." Ed. S.M. Ryvkin (Nauka, Lenigrad) p. 760.
4. Bixon, M., and Zwanzig, R., 1969, Phys. Rev. 187, 267.
5. Boittiaux, B., Constant, E., Reggiani, L., Brunetti, R., and Jacoboni, C., 1982, Appl. Phys. Lett. 40, 407.
6. Brunetti, R., Jacoboni, C., Nava, F., Reggiani, L., Bosman, G., and Zijlstra, R.J.J., 1981, J. Appl. Phys. 52, 6713.
7. Chambers, R.G., 1952, Proc. Phys. Soc. London 65A, 458.
8. Chapman, S., and Cowling, T.G., 1970, in "The Mathematical Theory of Non-Uniform Gases" Cambridge University Press.
9. Fauquembergue, R., Zimmerman, J., Kaszynski, A., and Constant, E., 1980, J. Appl.Phys., 51, 1065.
10. Gantsevich, S.V., Gurevich, V.L., and Katilius, R., 1970, Sov. Phys. JETP, 30, 276.
11. Gantsevich, S.V., Gurevich, V.L., and Katilius, R., 1974, Phys. Cond. Matter 18, 106.
12. Gantsevich, S.V., Gurevich, V.L., and Katilius, R., 1979, Rivista Nuovo Cimento, 2, 1.
13. Gisolf, A., and Zijlstra, R.J.J., 1976, J. Appl. Phys., 47, 2727.
14. Hill, G., Robson, P.N., and Fawcett, W., 1979, J. Appl. Phys. 50, 356.
15. Jacoboni, C., 1974, Phys. Stat. Sol., 65B, 61.
16. Jacoboni, C., Reggiani, L., and Brunetti, R., 1981, in "Proc. 3rd Int. Conf. on Hot Carriers in Semiconductors", J. Phys. Coll. Tome 42, C7-123.
17. Matulionis, A., Pozhela, J., and Starikov, E., 1981, Phys. Stat. Sol, 68A, K149.
18. Nava, F., Canali, C., Reggiani, L., Gasquet, D., Vaissiere, J.C., and Nougier, J.P., 1979, J. Appl, Phys. 50, 922.

19. Nougier, J.P., 1973, Phys. Stat. Sol., 55B, K43.
20. Nougier, J.P., 1975, in "Cours de D.E.A. Signauz et Bruit, Methodes de Calcul du bruit de fond (Montpellier) unpublished.
21. Nougier, J.P., 1978, Appl. Phys. Lett., 32, 671.
22. Nougier, J.P., Vaissiere, J.C., Gasquet, D., and Motadid, A., 1981, J. Appl. Phys. 52, 5683.
23. Pamplin, B.R., 1970, Contemp. Phys., 11, 1.
24. Pozhela, J., and Reklaitis, A., 1978, Sol. State Commun. 27, 1073.
25. Price, P.J., 1965, in "Fluctuation Phenomena in Solids", Ed. R.E. Burgess, Chapter 8 (Academic Press, New York).
26. Price, P.J., 1979, in "Semiconductors and Semimetals" Academic Press, New York, p. 249.
27. Rees, J., 1973, J. Phys. C., 6, 262.
28. Reggiani, L., 1980, J. Phys. Soc. Japan, 49, 317.
29. Reggiani, L., Brunetti, R., and Jacoboni, C., 1981a, in "International Symposium on Noise in Physical Systems", Edts. P.H.E. Meijer, R.D. Mountain, and R.J. Soulen (NBS Special Publications 614, Washington) p. 414.
30. Reggiani, L., Brunetti, R., and Jacoboni, C., 1981b, in "Proc. 3rd Int. Conf. on Hot Carriers in Semiconductors", J. Phys. Coll., Tome 42, C7-111.
31. Reggiani, L., Canali, C., Nava, F., and Alberigi-Quaranta, A., 1978, J. Appl. Phys., 49, 4446.
32. Ruch, G., and Kino, G.S., 1968, Phys. Rev. 174, 921.
33. Schlup, W.A., 1971, Phys. Kondens. Mat. 13, 307.
34. Shockley, W., Copeland, J.A., and James, R.P., 1966, in "Quantum Theory of Atoms, Molecules and Solid State", Academic Press (New York).
35. Shulman, YA., 1970, Sov. Phys. Solid State, 12, 922.
36. Stueckelberg, E.C.G., 1952, Helv. Phys. Acta 25, 577.
37. Van Beijern, H., 1982, Rev. Mod. Phys., 54, 195.
38. Van der Ziel, A., 1968, in "Solid State Physical Electronics" Prentice Hall (New Jersey).
39. Van der Ziel, A., and Bosman, G., 1982, Phys, Stat. Sol. A73, K93.
40. Van Vliet, K.M., Friedman, R.J.J., Zijlstra, R.J.J., Gisolf, A., and Van der Ziel, A., 1975, J. Appl. Phys. 46, 1804; ibidem 46, 1814.
41. Van Vliet, K.M., and Van der Ziel, A., 1977, Solid State Electron., 20, 931.
42. Vinter, B., 1973, Thesis dissertation, Lyngby, unpublished.
43. Zwanzig, R., 1964, Phys. Rev. 133, A50.

SUPERLATTICES

Karl Hess

Department of Electrical Engineering & Coordinated
Science Laboratory
University of Illinois at Urbana-Champaign
Urbana, IL 61801

INTRODUCTION

This is a very one-sided review of superlattices and by no means does justice to this rapidly growing field. My intent is to illustrate possibilities inherent in the variability of boundary conditions in finite superlattices or more generally speaking, multiple heterostructure layers.

In recent years, much progress has been made in the technology and basic understanding of superlattices and lattice-matched III-V compound heterostructure layers in general [1]. Two new epitaxial technologies have emerged --molecular beam epitaxy (MBE) [2] and metalorganic chemical vapor deposition (MOCVD) [3] -- which have opened a variety of new possibilities, including superlattice transport as discussed by Esaki, Tsu and Chang [4-6], the fabrication of materials exhibiting extremely high mobilities [7], quantum well heterostructure lasers [8,9], planar doped barrier structures [10] and real space transfer switching [11]. More recent research has also concentrated on strained (lattice mismatch) superlattices [12] and on type II superlattices [6].

Section 2 gives a short review of the band structure of superlattices (layered heterostructures). In Sections 3 to 5 some of the most important effects of heterojunction geometries on electronic transport are discussed and Section 5 illustrates these effects in the context of possible device applications.

BAND STRUCTURE OF SEMICONDUCTOR HETEROLAYERS

Among all possible combinations of semiconductors, the lattice matched (or at least closely matched) combinations have attracted most interest. The reason is, of course, the possibility of ideal interfaces if the lattice constant of the constituents is the same.

A simple model of the electronic structure of the heterojunctions can be obtained by employing the effective mass theorem. This approach views the semiconductor as homogeneous with an additional potential superimposed on the crystal potential as shown in Figure 1. The effective mass theorem tells us then that for electrons at band minima (maxima) we can replace the Hamiltonian

$$P^2/2m_o + V(\vec{r}) + \phi(z) \tag{1}$$

by

$$P^2/2m^* + \phi(z) \tag{2}$$

and therefore transform away the periodic crystal potential $V(\vec{r})$ (provided that $\phi(z)$ and its Fourier components fulfill certain conditions). In other words, we have reduced the heterojunction problem to a well-defined potential problem for a quasi-free electron and we can use all the well-known solutions for single (square well solution) and multiple (Kronig-Penney model) quantum wells. The E(k) relation for a periodic square well potential is easily derived and given in k_z direction by

$$E_z = \tfrac{1}{2}E_o(1 - \cos k_z L_z) \tag{3}$$

Here, z is the direction perpendicular to the heterolayers, L_z is the layer thickness, and E_o is the width of the "miniband" generated by the super imposition of a new period on top of the crystal potential.

Typically, E_o is of the order of 100 meV. Therefore, it is plausible that collision broadening effects will not be negligible. A rough estimate of collision broadening can be obtained from the rate of polar optical phonon emission which is 10^{13} s^{-1} giving a collision broadening of $\sim$6.58 meV. Indeed, it was shown [12] that the inclusion of collision broadening is necessary for a selfconsistent calculation of the electron phonon interaction in superstructures. It is also important to notice that the minibands are not separated by an energy gap as they would be in a truly three-dimensional superstructure. Perpendicular to the layers we have a quasi-continuum of states in which the electrons can be accelerated much in the same way as they can be without the superstructure in z-direction.

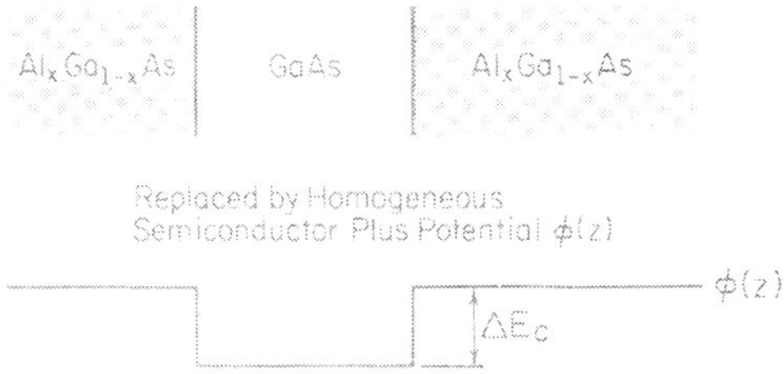

Figure 1. Semiconductor Heterostructure and model replacement.

The effective mass approximation in the form described above makes it necessary to deal with the envelope functions which are supposed to match smoothly across the interface. However, it is the wave functions and derivatives which are continuous, not necessarily the envelope functions. This problem area has been discussed most recently by Zhu and Kroemer [13] and the reader is referred to this reference for details.

In some instances, the form of the band structure away from the minima play a role, or the layer width is very small and of the order of the lattice constant itself. Then more powerful methods have to be used to calculate the electronic band structure. Several tight binding approaches showed excellent agreement with experimental data [14-16].

During his tight binding calculations, Mon [16, 17] noticed an interesting phenomenon for superlattices composed of two indirect materials such as Si-GaP: For certain layer widths of the order of a few lattice constants the material becomes direct, which means that the X-minima along the superlattice directions are folded into the zone center, i.e., the superlattice forms a direct semiconductor. Recent experimental investigations in this direction are discussed in [18].

REAL SPACE EFFECTS SPECIFIC TO SUPERLATTICES

Distribution of Electrons and Electronic Transport

Consider a GaAs quantum well heterojunction as shown in Figure 1. Assume that the structure is modulation doped [7], i.e., the GaAs is not intentionally doped and the dopant resides entirely in the $Al_xGa_{1-x}As$. The electrons will, of course, leave their parent donors and reside in the GaAs, which is the place of lowest potential energy. As a consequence, a depletion width forms in the $Al_xGa_{1-x}As$.

If the $Al_xGa_{1-x}As$ is homogeneously doped the depletion width is, as usual, given by

$$W = \sqrt{\frac{2\varepsilon\varepsilon_o V_{ba}}{eN_D}} \tag{4}$$

where $\varepsilon\varepsilon_o$ is the dielectric constant, N_D the donor concentration, and V_{ba} an effective barrier height which is approximately equal to the band edge discontinuity for low electron densities (non-degenerate) in the GaAs. Notice, however, that in many practical cases the structures are close to a surface and the pinning of the Fermi level at the surface can contribute significantly in creating depletion layers.

The separation of electrons and donors leads to the unique possibility of controlling electron and impurity density separately in a system with almost ideal interfaces. As a consequence of this control, ultra-high mobilities can be achieved at high carrier concentrations. This can easily be understood from a "model matrix element" which is obtained for a two-dimensional sheet of electrons (x-y direction) and an impurity distribution with density N(z) where z is the coordinate perpendicular to the electron sheet which is located at z = 0. The absolute square of the matrix element for the transition of an electron from $\vec{k}_{//}$ to $\vec{k}'_{//}$ is then given by [19]

$$|M(\vec{k},\vec{k}')|^2 = \frac{e^4}{4A^2L_z(\bar{\varepsilon}\varepsilon_o)^2(q+S_1)^2} \times$$

$$\int_a^b \exp(-2q|z|)N(z)dz \tag{5}$$

Here A is the area of the electron sheet, $\bar{\varepsilon}$ the average dielectric constant of the two neighboring media, $q = |\vec{k}_{//} - \vec{k}'_{//}|$, and S_1 is the two-dimensional screening constant which for high electron densities is equal to $2/a_B^*$ where a_B^* is the effective Bohr radius. a and b are the boundaries of the sample in the z-direction.

Since the donor impurities are separated by the distance z_o, the scattering rate will be decreased by the exponent in Equation 5. The background impurities still give rise to scattering but are effectively screened by the high density of electrons in the GaAs which give rise to screening constant S_1. A detailed description is given in [19, 20]. Although much work has been done on modulation doped structures, there is still much room for innovations. Some possibilities are discussed in the next section.

Let us now turn to the dynamics of electron collection and emission in and out of wider regions of GaAs in a modulation doped structure. The classical case of emission and collection is well described by the thermionic emission theory of Bethe [21]. For the current from the GaAs(1) to the $Al_xGa_{1-x}As(2)$ we have (assuming no band bending as shown in Figure 1) [22]:

$$J_{1-2} = A^* T_c^2 \exp((E_{F1} - \Delta E_c)kT_c) \quad (6)$$

Here, A^* is the effective Richardson constant, ΔE_c is the band edge discontinuity, T_c is the temperature, and E_{F1} is the quasi-Fermi energy in the GaAs. An analogous equation can be derived for the current J_{2-1} from the $Al_xGa_{1-x}As$ back to the GaAs. In 1979 I realized that by separate control of the electron temperatures T_c in the GaAs and the $Al_xGa_{1-x}As$, the electrons can be switched from one layer to the other. For example, if T_c in GaAs goes to infinity, almost all electrons will transfer in steady state to the $Al_xGa_{1-x}As$, as can be seen from the balance $J_{1-2} = J_{2-1}$. Estimates showed that this switching is very fast and effective [22], device applications have been proposed and the effect has been termed real space transfer effect.

This real space transfer effect can be generalized in the following way: As electrons or holes are accelerated (e.g., by DC or AC electric fields) in the three or two-dimensional continuum of a semiconductor (or insulator), their average energy increases. It is therefore possible that the electrons (holes) reach states at such high energies that propagation to a neighboring semiconductor can occur even if a substantial band edge discontinuity has to be overcome. In the case of a superlattice, the electrons can be energized from the lowest confined states to higher minibands in which they may then propagate in any direction. In a single quantum well electrons can be accelerated in the two-dimensional continuum and then at high energies propagate out of the well region after a scattering event which supplies the necessary momentum. Of course, the effect can also occur in quasi-one-dimensional systems where the electrons can be accelerated in a one-dimensional continuum and then leave the one-dimensional "wire" if they reach high enough energies. The consequence of this effect is negative differential resistance, switching and storage.

Experimental verifications of this effect have been given, some of which will be discussed below. A more detailed discussion of transport in modulation doped structures including electron-phonon interactions can be found in [19, 20].

Combination of Heterolayers and Doping Superlattices

Doping superlattices [23] are structures consisting of one type of semiconductor which have built-in ultrathin layers of doping (e.g., n-type, intrinsic, p-type, intrinsic (nipi) layers). Since the original work on doping superlattices by Doehler [23], a number of investigations have revealed many interesting effects specific to these structures. The claim has been made [23] that nipi superlattices represent a whole new class of semiconductors because of their tunable electronic properties. The term "tunable" means that the basic electronic properties such as the band gap, the electron hole concentration, and any other property which depends on these parameters are no longer fixed but may be varied over a wide range by external perturbation. This unique feature is a consequence of an indirect band gap in real space which causes a spatial separation between electrons and holes in these structures.

Modulation doped structures can in some respect, be regarded as a combination of a doping superlattice and a heterolayer structure and it is clear that one can think of many other such combinations.

Holonyak [24] has proposed that it may be advantageous to substitute doped AlGaAs with a binary GaAs-AlAs superlattice and dope the GaAs only in order to avoid the complications of doping an alloy, which are especially pronounced above the direct indirect transition. The energy level of the donor (acceptor) in the GaAs layers can be varied almost at will by adjusting the thickness of the layers. By simply reducing the GaAs quantum well size over a few or gradually over many layers, the effect of a graded bandgap can be achieved in addition to the n (or p) doping [24].

One still can construct many other combinations of heterolayers and doping superlattices to achieve various doping and transport properties and research in this area should be very stimulating.

$\vec{k}$-space Effects Specific to Superlattices

Research in superlattice transport was originally stimulated by the possibility of achieving high-frequency Bloch oscillations [4]. Using Equation 3 and the equation for k_z (no scattering)

$$\hbar \dot{k}_z = e F_z \qquad (7)$$

where F_z is the electric field in z-direction, and using the velocity $v_z = \frac{1}{\hbar} \frac{\partial E_z}{\partial z}$, one can see immediately that a DC electric field generates an AC current. The effect is still under discussion. It is clear that real structures are considerably more complicated because of the presence of scattering and collision broadening, which causes a distribution of electrons all over the minibands.

Recently other $\vec{k}$-space effects have been discussed. These are connected with zone folding effects. A prominent example is the above-mentioned transition to an optically direct material from indirect constituents by zone folding, i.e., the superlattice formed by two indirect semiconductors can be direct. The possibility of "engineering" electronic states has also been discussed in connection with impact ionization-resonance phenomena. It is known that spin-orbit resonance effects can be achieved in GaAlSb when the spin orbit splitting is equal to the band edge energy. Similar effects can also be created by superlattice zone-folding but in a much wider range of parameters [24].

New Device Concepts Involving Heterolayer Structures

In this section I will discuss how the effects described before can be used advantageously in semiconductor devices. Three examples will be given.

Modulation Doping and the High Electron Mobility Transistor (HEMT)

It is obvious that the extremely high mobilities which can be achieved by using modulation doping result in speed advantages when combined with field effect transistor (FET) technology. Soon after the experimental realization of modulation doping [7] a transistor structure with extremely high speed capabilities has been reported by Mimura et al. [26]. Figure 2 shows the basic geometry of the HEMT, a typical profile of the electric field in this structure, and a sketch of the current voltage characteristic in bulk GaAs and modulation doped structures. At low and intermediate electric fields as they occur on the source side the electrons still stay in the GaAs layer(s) and maintain a high mobility although much lower than for very low electric fields. The dependence of mobility on the electric field for moderate fields can be deduced from Figure 2b. Although the mobility decreases substantially (for ultrahigh mobilities already at very modest fields), the drift velocity (mobility) is still much higher than in bulk GaAs, especially at low temperatures ($T \lesssim 77K$). This results in a lower source access resistance superior transistor performance. In the region under the gate the electric field can be extremely high and real space transfer can occur, although it may be weakened by the repulsing gate field. In this region the

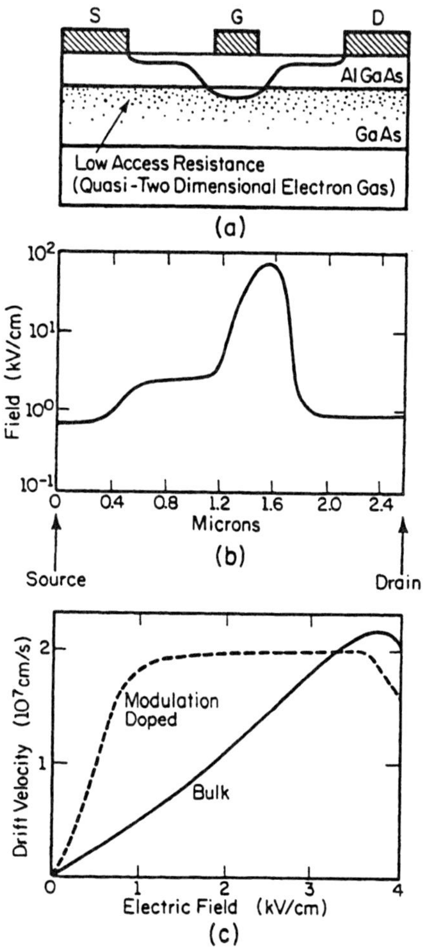

Figure 2. High mobility transistor (HEMT)
(a) Structure
(b) Schematic field distribution (for high drain voltages, ~3 V)
(c) Schematic velocity-field characteristic for bulk GaAs and modulation doped material. The peak velocities in bulk GaAs are lowered if impurity scattering becomes very important.

modulation doping offers no advantage as the electrons can reunite with their parent donors and the mobility deteriorates substantially. Also, electrons may be caught in deep traps at the interface or in the AlGaAs, leading to charge accumulation and problematic threshold shifts.

The hot electron real space transfer effect offers therefore only disadvantages for the HEMT and should be avoided by proper design. However, real space transfer of hot electrons can also result in novel effects with high potential for application.

Real Space Transfer -- Fast Switching and Negative Differential Resistance

I have discussed in Section 3 the possibility of negative differetial resistance by the transfer of hot electrons in real space. An oscillator on this basis has been produced and described before, extensively [11].

It has also been suggested that the electrons can be selectively switched between layers and stored in neighboring layers as schematically shown in Figure 3. The effect has meanwhile been verified experimentally [27]. The time constants for the switching have been estimated to be of the order of 1 - 5 ps. The switching time needed is determined by the time constant of the heating of the electron gas and the time needed to traverse the layers. The latter time constant is of the order of the layer width divided by thermal velocity, which is extremely short (less than 1 ps). Notice that these effects make use of transport perpendicular to the layers over distances which can be controlled and engineered down to dimensions of 10 Å using MBE or MOCVD crystal growth.

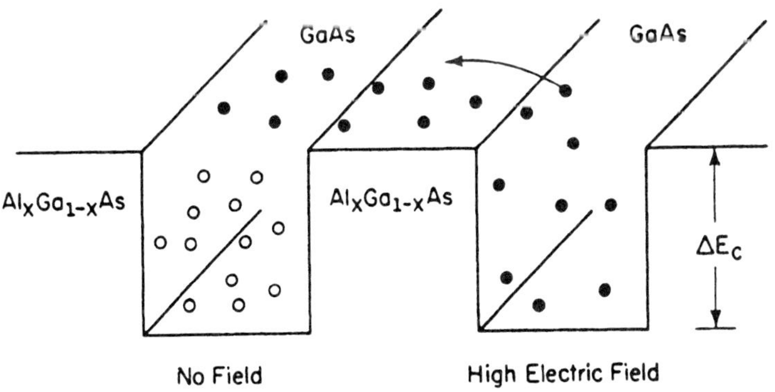

Figure 3. Switching of electrons between heterolayers by selective application of high electric fields. The electrons can be stored in layers when no field is applied.

IMPACT IONIZATION IN HETEROLAYERS AND THE SOLID STATE PHOTOMULTIPLIER

Impact ionization in the presence of band edge discontinuities can be substantially different from ionization in bulk semiconductors [28]. The basic physical idea is the following:

The impact ionization rate depends exponentially on the impact ionization threshold and on the energy from which the electron (hole) starts to be accelerated. Therefore, a step-like band structure can enhance or decrease the ionization rate. An electron propagating from a AlAs layer to a GaAs layer finds itself suddenly far above the GaAs band edge and needs only little additional acceleration to reach the ionization threshold. Of course, when propagating out of the GaAs toward the AlAs layer, the inverse happens. The process, however, is strongly nonlinear, resulting in a new increase of the electron ionization rate. The hole ionization rate β is not substantially increased because the valence band discontinuity is much smaller. Thus, the asymmetry of the superlattice band structure with respect to electrons and holes introduces an asymmetry in the impact ionization rates. It has been proposed by Chin et al. [28] that this asymmetry can be used advantageously in avalanche photodiodes which have superior noise properties if one ionization rate dominates. Enhancements of the α/β ratio of 10 and more have been found for a model superlattice avalanche photodiode. Capasso et al. verified this result experimentally [29] and also suggested the development of similar structures with the goal of developing a "solid state photomultiplier".

Acknowledgements

I would like to thank N. Holonyak for many valuable discussions. The work was supported by the Office of Naval Research and the Army Research Office.

REFERENCES

1. See, e.g., the reviews by R. Dingle, Confined Carrier Quantum States in Ultrathin Semiconductor Heterostructures, in Festköperprobleme XV, pp. 21-47 (1975) or by K. Hess and N. Holonyak, Jr., Höt-Electron and Phonon Effects in Layered Semiconductor Structures and Heterostructures, Comments Solid State Phys. 10, pp. 67-84 (1981).

2. See the papers by J.R. Arthur, J. Appl. Phys. 39, 4032 (1968) and A.Y. Cho, J. Vac. Sci. Technol. 8, 531 (1971); A.Y. Cho, Aook. Phys. Lett. 19, 467 (1971).

3. H.M. Manasevit, J. Electrochem. Soc. 118, 647 (1971); see also R.D. Dupuis and P.D. Dapkus, Proceedings of the Seventh International Symposium on GaAs and Related Compounds, St. Louis, September 1978, edited by C.M. Wolfe, Conference Series 45 (Institute of Physics, London, 1979), pp. 1-9.

4. L. Esaki and R. Tsu, IBM Journal of Res. 14, 61 (1970).

5. L. Esaki and L.L. Chang, Phys. Rev. Lett. 33, 495 (1974).

6. More recently, interest has been focused on Type II systems such as InAs-GaSb; see, e.g., G.A. Sai-Halasz, L.L. Chang, J.M. Weter, C.-A. Chang and L. Esaki, Solid State Commun. 27, 935 (1978).

7. R. Dingle, H.L. Störmer, A.C. Gossard and W. Wiegmann, Appl. Phys. Lett. 3, 665 (1978).

8. N. Holonyak, Jr., R.M. Kolbas, R.D. Dupuis and P.D. Dapkus, Appl. Phys. Lett. 33, 73 (1978).

9. W.T. Tsang, Appl. Phys. Lett. 39, 786 (1981).

10. R.J. Malik, J.R. AuCoin, R.L. Ross, K. Board, C.E. Wood and L.F. Eastman, Electron. Lett. 16, 836 (1980).

11. K. Hess, Journal de Physique C7, C7-3 (1981).

12. R.K. Reich, R.O. Grondin and D.K. Ferry, Phys. Rev. B 27, 3483 (1983).

13. Qi-Gao Zhu and Herbert Kroemer, Phys. Rev. B 27, 3519 (1983).

14. J.N. Schulman and T.C. McGill, Phys. Rev. B 19, 6341 (1979).

15. J.N. Schulman and Y.C. Chang, Phys. Rev. B 24, 4445 (1981).

16. K.K. Mon, Solid State Communications 41, 699 (1982).

17. K.K. Mon, Private communication, October, 1981.

18. N. Holonyak, Jr. and K. Hess, Quantum Well Heterostructure Lasers (Academic Press, New York, 1983) and references therein.

19. K. Hess, Appl. Phys. Lett. 35, 489 (1979).

20. K. Hess, Aspects of High-Field Transport in Semiconductor Heterolayers and Semiconductor Devices, in Advances in Electronics and Electron Physics, vol. 59, ed. by P.W. Hawkes (Academic Press, New York, 1982).

21. See, e.g., S. Sze, Physics of Semiconductor Devices (Wiley-Interscience, New York, 1981).

22. H. Shichijo, K. Hess and B.G. Streetman, Solid State Electronics 23, 817 (1980).

23. G.H. Döhler, Phys. Stat. Sol. (b) 52, 79 and 533 (1972).

24. N. Holonyak, Jr., Proceedings of the Third Workshop on the Physics of Submicron Structures, Urbana, Illinois (1982), (Plenum Press, to be published).

25. K.K. Mon and K. Hess, Solid State Electronics 25, 491 (1982).

26. T. Mimura, S. Hiyamizu, T. Fujii and K. Nanbu, Jap. J. of Appl. Phys. 19, L225 (1980).

27. See, e.g., K. Hess, Proceedings of the XVI International Conference on the Physics of Semiconductors, Montpellier (1982).

28. R. Chin, N. Holonyak, Jr., G.E. Stillman, J.Y. Tank and K. Hess, Electronics Lett. 16, 467 (1980).

29. F. Capasso, W.T. Tsang and A. Hutchinson, Proc. Int. Electron Devices Meeting (1981), p. 284.

SUBMICRON LITHOGRAPHY

C.D.W. Wilkinson and S.P. Beaumont

Department of Electronics and Electrical Engineering
The University of Glasgow
Glasgow, G12 8QQ

INTRODUCTION

Lithography plays a central role in the fabrication of electronic devices. The minimum feature size required in the device is important in the choice of lithographic method. Generally for linewidths above 1μm linewidth optical lithography is used whilst for smaller dimensions it is necessary to use electron, ion beam or X-ray lithography. The linewidths in current production VLSI circuits range between 1 and 3μm; there are programs in the U.S. and the U.K. to reduce the linewidth to 0.5μm over the next few years[(1)]. Single conventional semiconducting electronic devices have been made with gates as small as 0.1μm and some experimental superconducting devices employ features with sizes in the 10-30nm range.

The performance of field effect transistors depends on the width of the gate and marked deviation from conventional behaviour can be expected in silicon for gate widths of the order of 0.1 to 0.3μm and in GaAs for gate widths less than 0.1μm.

Very small structures with dimensions of the order of tens of atoms exhibit effects not seen in larger scale structures.

Overall, the challenge of making very fast electronic devices and of exploring the behavious of material at sizes where quantum mechanical behaviour is directly observable, has led to wide international interest in very high resolution lithography.

In the following pages, an overview of submicron lithography will be given and the limits of resolution discussed.

BASIC LITHOGRAPHIC PROCESS

Lithography is used to delineate a pattern in metal or in a dielectric or in doping concentration on a semiconducting substrate. A radiation sensitive layer (often polymeric) is deposited upon the substrate and exposed. If in the subsequent development the areas exposed to the radiation are removed the resist is said to have a positive tone (see fig. 1(a)), whereas if the exposed areas are hardened by exposure and the unexposed region removed in the developer the resist is then used in the subsequent procession in one of two ways (etching).

In subtractive processing (etching) (see fig. 1 (d)) the resist is used to protect the material underneath it, while any unprotected material is removed by chemical action (acids or alkalis) or by etching in a reactive gas or by inert gas ion bombardment.

In additive processing (lift-off or stencil processing) (see fig. 1)c)) the material to be patterned is deposited onto the patterned resist and underlying substrate and the resist is dissolved away, taking with it the material deposited on the resist but leaving behind the metal deposited on the substrate. For lift-off to be successful, the resist solvent must be able to penetrate into the resist and thus, ideally, the metal coating on the resist must not be continuous with the metal coating on the substrate (see fig. 2). Lift-off is most readily performed when the resist profile is slightly undercut and when the material is deposited in a collimated beam. Plating through a resist pattern on an underlying plating base deposit is an interesting variation of additive processing which does not require an undercut resist profile.

Lift-off processing allows the definition of a high resolution pattern in materials for which no high resolution etching technique is available. Lift-off processing in electron beam lithography on thin substrates[2] has been demonstrated to have very high resolution (16nm lines). For lift-off processing to be successful the undercut profile established in the resist during exposure and development must be preserved during the deposition of metal. Using polymeric resists this implies that the resist temperature during metal deposition must be maintained well below the glass transition temperature of the polymer.

EXPOSURE METHODS

Optical Exposure

Photolithography is the workhorse of the silicon microelectronics industry. However, the resolution of optical lithography is limited

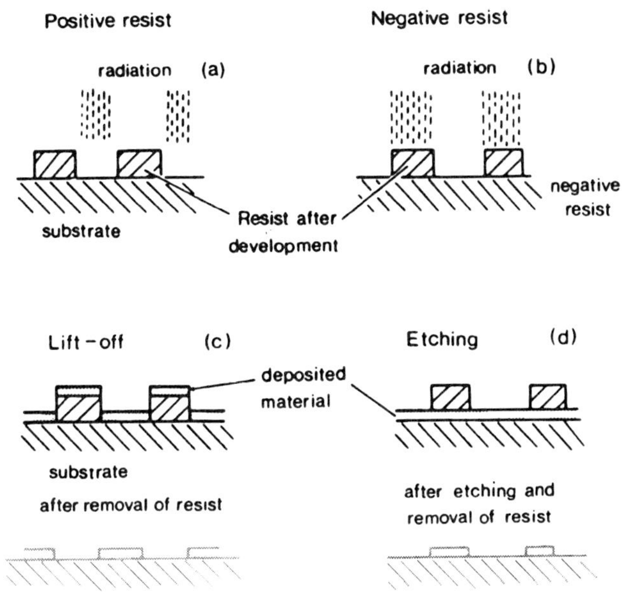

Figure 1a. In positive tone resists exposed material is removed by developer.

1b. In negative tone resist exposed material is hardened by exposire and unexposed material is removed by developer.

1c. Liftoff or additive processing.

1d. Subtractive or etching processing.

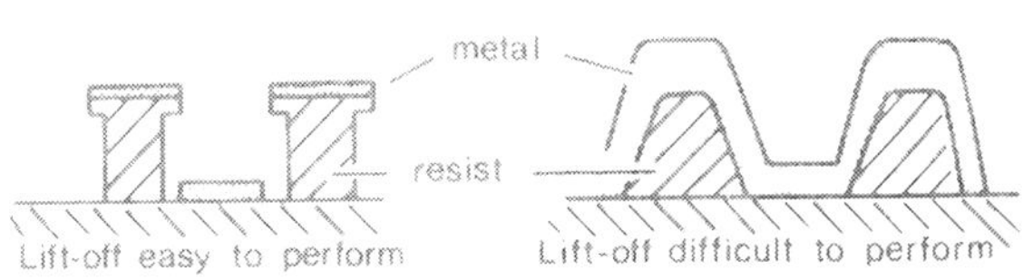

Figure 2. Liftoff ideally requires an undercut resist profile.

by the wavelength of light and the practical limit at the moment appears to be around 0.5μm lines with 0.5μm gaps printed in positive photoresist.

As this review is concerned with submicron lithography, optical lithography will not be discussed further.

Electron Beam Exposure

An electron beam can be focussed to a spot of diameter less than a nanometre and so provides a potentially very high resolution means of exposing an electron resist. An electron beam lithography machine consists of a source of electrons, a series of condensor lenses to collect the electrons and an objective lens to focus the electron beam in a narrow probe onto the resist(3). Deflection coils are also provided so that the electron probe can be scanned over the specimen and some means of turning the electron beam on and off (blanking) is essential so that a pattern can be exposed in the resist. The deflection and blanking of the beam are normally under the control of a computer. The area which can be exposed by deflection of the electron beam is limited; to expose larger areas accurate movements (often under interferometric control using a laser) of the specimen on a stage are necessary. The machine is similar in overall design to a scanning electron microscope (SEM), and as in an SEM, secondary and backscattered electrons from the specimen can be used for forming an image on a CRT.

Many electron beam lithographic machines with different design goals have been built in the past ten years. Until recently the achievement of the ultimate in resolution has not been a major aim, rather the target has been to make a machine which while it has resolution at or a little beyond the optical limit, can write a pattern over a whole wafer in as short a time as possible while retaining the flexibility in pattern design given by direct computer control. Two approaches in improving the writing speed of electron beam lithographic machines have been used 1) the use of very high speed deflection of a round electron beam (up to 300 Mb/s(4)) and 2) the use of a variably shaped beam (square or rectangular beam at will) so that a larger area can be written at each exposure of the beam(5,6). The beam size employed in such machines is commensurate with the final linewidth desired and so is rarely less than 50nm.

For very high resolution (linewidths below 0.1μm), modified SEMs and Scanning Transmission Electron Microscopes (STEM S) have been employed(7,8,9). Speed of writing is not of the essence but a spot size smaller than the narrowest line to be written is clearly vital. In a SEM the specimen is placed at the focus of the electron beam which is some tens of millimetres from the objective lens pole piece and in many SEM's large objects (such as 2" diameter wafers) can be accomodated on the stage. With such an objective lens the minimum probe size which can be obtained is of the order of 2-10nm (depending particularly on the choice of electron source-tungsten filament, LaB_6, etc.).

The STEM is a modification of a conventional transmission electron microscope (TEM) and as with a TEM it is customary to use specimens which are transparent to an electron beam. After passing through the specimen electrons impinge onto a detector. In order to form the smallest possible probe, an immersion objective lens in which the specimen is placed in the middle of two coils is commonly employed. In such an objective lens a probe of diameter 0.5-2nm can be formed; however, space limitations often restrict the specimen size to 10mm or less.

Ion Beam Exposure

Beams of ions can be focused and used to expose resist. In recent years the molten metal tipped source[10] has been used to produce focused ion beams of high brightness. As the mass of a metallic ion is much greater than that of an electron, electrostatic rather than magnetic lenses must be employed. Focused ion beams can be used to expose resist or to implant dopants directly into material without the need for any masking material.

The minimum focused beam size that has been obtained to date is of the order of 50nm. Improvement on this figure will require limitation of the spread of energies in ion beam (which is higher than in an electron beam) and better electrostatic lens designs.

X-Ray Exposure

While focused electron and ion beam exposure provide a means to define a high resolution pattern, X-ray exposure (and flood electron or ion beam exposure) gives a route to the replication of high resolution masks. The basic concept is very similar to optical printing of masks onto silicon. An X-ray mask consists of a relief pattern in a X-ray absorbing material formed on top of an X-ray transparent substrate. This mask is placed a distance d above a resist coated wafer and exposed to X-rays of wavelength between 0.2 and 5nm[11,12,13]. Two geometric causes of pattern distortion are present[12] as shown in figure 3. The distortion can be eliminated by decreasing the spacing d to zero.

The X-ray source may be a fixed or rotating metal target bombarded by electrons, a hot plasma or synchrotron radiation from a storage ring[14]. Most high resolution work has been done with electron bombarded targets which usually comprise an electron beam evaporation system run at low power so the target is not melted. The X-rays strike the resist and if absorbed give rise to photo-electrons of maximum energy approximately equal to the input photon energy. The choice

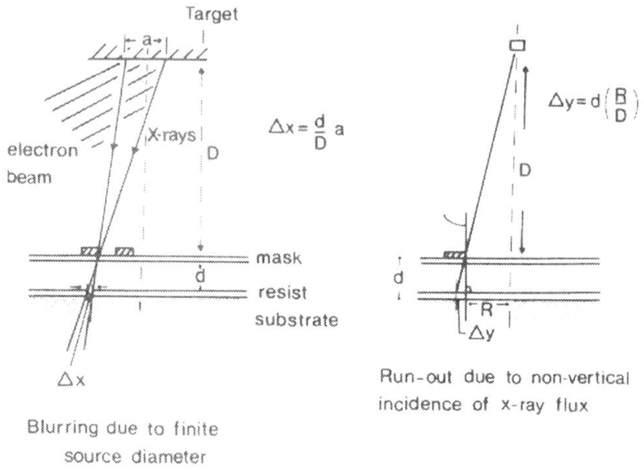

Figure 3. Run-out caused by finite spacing between mask and substrate. Penumbra errors caused by finite width of source.

of X-ray wavelength depends on the balance between two factors (a) the range of the photoelectrons and (b) the effect of diffraction. The range of photoelectrons in the resist increases with increasing energy and so the range varies as the inverse of the X-ray wavelength X, while dittrachon effects vary in proportion to the wavelength; thus a compromise has to be struck.

The mask has to be transparent to X-rays. The absorption of X-rays in solids varies as a function of wavelength, showing jumps at wavelengths corresponding to characteristic emission wavelengths of the absorbing material. In general, the absorption increases with atomic number Z; so masks must be thin membranes of materials of low atomic number. Membranes of Boron-doped silicon, silicon nitride[15], Boron nitride, polyimide[16] (a tough polymeric material used as an encapsulant in microelectronics) and carbon[14] have been used.

RESISTS

Electron Beam Resists (a) Positive

Electron irradiation of positive polymeric resists causes degradation of the resist and the formation of fragments of lower molecular weight. The development step consists of the removal of the degraded material by dissolution in a solvent/nonsolvent solution which does not dissolve unexposed resist. Under a fixed set of conditions of beam energy, resist thickness, substrate material and development positive resists are characterised by a plot of the remaining resist thickness versus the log of the dosage. The contrast γ of a resist is a concept borrowed from photographic practice, a low γ photographic film reproducing a wide range of greys while a high γ film is used to give sharp white-black transitions. The contrast γ is often defined[17] as the slope of the linear portion of the thickness versus log (dosage) curve where Q_o and Q_i are found as shown in fig. 4. However, γ should only be regarded as a rough guide to the behavior of the material - the complete thickness-dosage curve contains much additional information. Clearly for lithographic purposes a high γ is desirable, so that a sharply defined resist pattern can be obtained from a less sharply defined electron dosage distribution.

A widely used polymeric electron resist is poly(methylmethacrylate) - PMMA[18]. It is a glassy solid below 120° C (the glass transition temperature - T_g).

The number of monomer molecules polymerised to form one macromolecule depends on the method of polymerisation. If all macromolecules consist of the same number of monomer molecules, the polymer sample is said to be monodisperse. The degree of dispersivity is defined as M_w/M_n where M_w is the weight average molecular weight and M_n is the number average molecular weight. Commercial PMMA is not

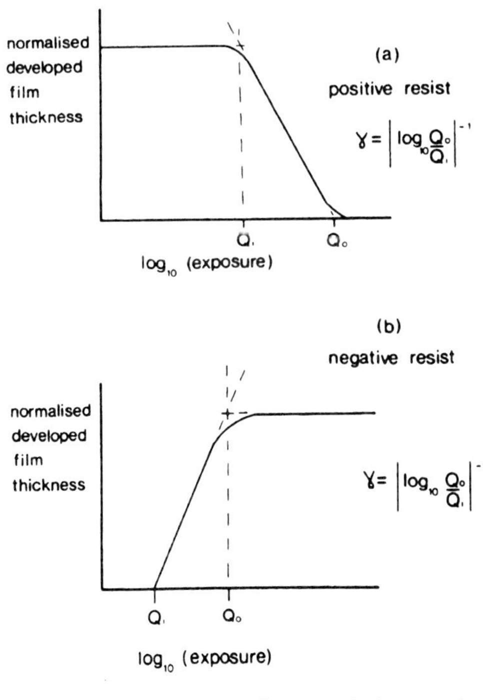

Figure 4. Definition of contrast for positive and negative resist.

monodisperse, typically having a M_w/M_n greater than 2.

PMMA is normally spin coated from solution onto the substrate. The uniformity of the coating depends on the rheology of the polymer-solvent system and on the parameters of the spinner (speed, acceleration rate, etc.)[19]. After spinning the PMMA is baked above the glass transition temperature for a number of hours to promote adhesion to the substrate.

Upon exposure to an electron beam the polymer is broken up into fragments of lower molecular weight, the secondary electrons generated by the incident electron beam breaking bonds within the polymer. The molecular weight distribution of PMMA (Elvacite 2041) as determined by gel permeation chromatography is shown in fig. 5 [21], before and after exposure to a 20kV electron dosage of $50X10^{-6}C/cm^2$. The effect of the electron irradiation is to reduce the average molecular weight by a factor of five, permitting development of a relief pattern by dissolution of the degraded polymer in a suitable solvent/nonsolvent solution. A typical developer[22] is 1:3 Methyl isobutyl ketone (solvent):Isopropyl alcohol (nonsolvent). The contrast of PMMA varies with the strength of the developer being lower with stronger developers (1:1 MIBK:IPA). However, with weaker developers the development rate is not linear with time for low exposures - indeed residual material (probably a higher molecular weight fraction) remains after prolonged development[39].

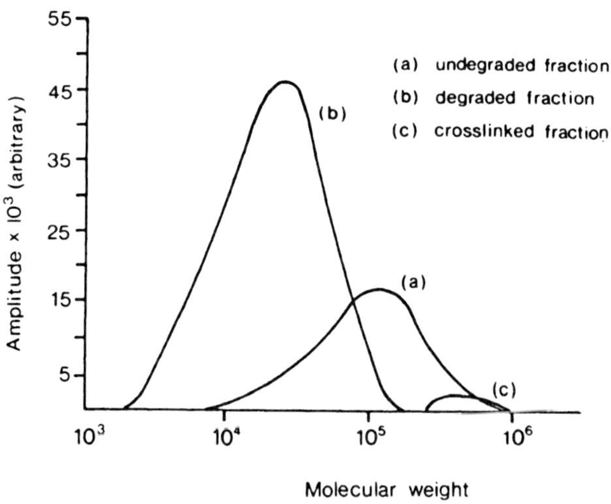

Figure 5. Molecular weight of PMMA after exposure to $50\mu C/cm^2$ of 20kV electrons. Note that while the major effect of exposure is to reduce the average molecular weight. A small fraction (2%) of the resist is crosslinked.

After development PMMA is used directly as a resist against chemical attack or, for liftoff processing is dissolved in a strong solvent, after metallisation.

The requirements for a resist are:
1) high sensitivity (to reduce exposure time)
2) good resolution (for desired linewidths)
3) high gamma
4) good dry etch resistance and substrate adhesion (in CF_4, CCl_4 etc.)

PMMA is widely used as a resist for sub 0.2μm applications as it satisfies requirements 2 and 3. More sensitive positive resists are available with sensitivities as low as 1μC/cm^2; however, with such resists the minimum line width achievable is greater than 0.1μm compared to less than 20nm for PMMA. Again, such very fast resists use very dilute developers and the development rate is very sensitive to temperature (which is not surprising - high sensitivity implies low activation energy). High dry etch resistance can be obtained using aromatic compounds but it has been observed that higher sensitivity implies lower dry etch resistance in positive resists[23].

Very high resolution positive resist behaviour has been observed in inorganic materials under high current electron beam irradiation. 1.5nm wide lines have been etched into a thin (<50nm) film of NaCl using a 0.5nm diameter electron probe[9], the salt being vapourised by the beam. The dosages used were in the range 5 - 50 C/cm^2. Holes 2nm in diameter and lines 2nm wide have been drilled into metal beta - B-alumino-silicates[24]. The 2nm hole can penetrate a alumino silicakes 200nm thick film albeit at a dosage of 5x10^{6}C/cm^2.

(b) Negative resists

Electron beam irradiation of negative resists causes toughening of the exposed area so that during development the unhardened material is removed. In polymeric negative resists, crosslinking of a low molecular weight polymer occurs. As a chemical groups with high radiation sensitivity such as vinyls and epoxies can be introduced into the resist, negative resists are usually more sensitive than positive. However, in general they exhibit lower contrast (γ) and also lower resolution. Many negative resists are susceptible to swelling during development; this can lead to pattern distortion. High resolution can be obtained with negative resists if very low molecular weight is used but at the expense of sensitivity[25]. An interesting limiting case of high resolution, low sensitivity and low molecular weight is the use of hydrocarbon contamination in a scanning transmission electron microscope as a negative resist. 8nm wide lines[8] were etched in W-Re using this resist.

Nonpolymeric negative resist systems have been investigated and found capable of high resolution. The chalcogenide glasses As-S and Ge-Se can be deposited by evaporation or sputtering, sensitized by a silver layer (of Ag_2S or Ag_2Se) formed by immersion in silver nitrate or silver cyanide and upon exposure to an electron beam, regions of silver rich glass are formed which are insoluble in a mild alkali developer. These systems can also be dry developed in CF_4 on SF_6 by reactive ion etching. The contrast is very high and 30nm wide lines[26] can be obtained in the As_2S_3/Ag_2S system. The sensitivity is low ($5x10^{-3}C/cm^2$) but acceptable for very high resolution applications.

Multiple Level Resists

It is clear that it is very difficult to combine high sensitivity, good resolution, high gamma and high dry etch resistance in a single resist material. If a single layer of resist is to be used compromises must be made. Moreover, it is often necessary to coat with resist a substrate which is not flat, for example in the final metallisation step in NMOS fabrication. Under these circumstances in order to obtain proper coverage the resist film must be thick and thick films often have reduced resolution on account of diffraction or electron scattering effects and because the contrast is never infinite.

Multilayer resist systems have been introduced to overcome this problem[27,28]. A single level resist has two roles to play; as an image forming layer and as a layer resistant to or topographically suitable for further processing. In multilayer resist systems, the layers are chosen so that their properties are optimum for only one of these roles. Thus, a thin, high resolution high sensitivity but low etch resistant layer will be used for image formation and this image will be transferred to a thicker underlying layer which is chosen specifically for its etch resistant properties. If it is not possible to deposit the two layers directly on top of each other, an intervening separating layer is often employed (see fig. 6).

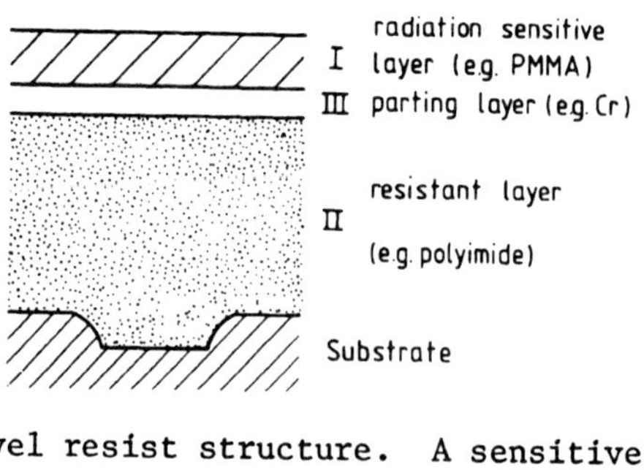

Figure 6. Tri-level resist structure. A sensitive thin upper layer of resist is deposited on separating layer under which is a thick layer of polymer material.

Figure 7. Metallisation and resist profiles prior to liftoff in formation of a narrow aluminum gate on silicon. Note the build-up of metal at resist neck. Markers are 1.0 microns.

Exposure by high energy electrons ions or X-rays is a two-step process. Initially, the beam interacts with the resist and then secondary electrons are formed which expose the resist. Howard and Prober point out(38) that another statistical limit relevant to the exposure process is the average number of secondary electrons created in a minimum resolution volume. Making reasonable assumptions and using experimental dose figures for PMMA exposure they calculated that approximately 400 secondary electrons are created in a $(10nm)^3$ volume by a 50keV electron or ion beam and 220 by an X-ray exposure. Given a 50nm thickness of resist, the number of secondary electrons created would appear to be adequate. As in this case, further statistical fluctuations will not be added to those caused by the number of primary particles.

ELECTRON BEAM WRITING

Interaction of Electron Beams with Resist

When a tightly focused beam of electrons hits a solid material,

Howard et al.[31] used a PMMA top layer, 25nm Ge parting layer on a 280nm layer of methylmethacrylate and methacrylic acid copolymer. After exposure and development of the PMMA, the germanium layer was etched in CF_4 by reactive ion etching and the germanium used as a mask for reactive ion etching in oxygen of the copolymer. Polymeric materials etch in oxygen RIE in a markedly anisotropic manner, giving vertical walls. The resulting copolymer structure was used as a liftoff mask to linewidths down to 25nm.

Using inorganic resist as the top layer, it is possible to avoid the use of a parting layer, as silver doped As_2S_3 glass is itself very resistant to oxygen RIE. 30nm wide lines in a 120nm high polyimide underlying layer have been made using a 80nm thick layer of As_2S_3 glass as the imaging layer[41].

In very high resolution electron beam lithography the need to obtain an undercut profile with good liftoff of thick metal patterns has led to the use of double layer resists. A simple example is the use of two layers of PMMA of different molecular weight, to obtain an under-cut profile suitable for liftoff[29]. In order to obtain a 0.1μm aluminium gate for a silicon FET[30], a faster resist (BDH M_w=85K) was coated to a thickness of 200nm, baked and a 100nm layer of Elvacite 2041 (M_w=361K) deposited on top. After exposure both resists were developed simultaneously in 1:3 MIBK:IPA. Figure 7 shows a micrograph of the resist and metallisation prior to liftoff. The build-up of metal at the resist opening leads to a triangular profile in the metal gate. Figure 8 shows a 0.12μm wide aluminium line, 0.16μm high.

The need to obtain high aspect ratio patterns of submicron sizes on substrates which already have topographic relief patterns engraved in them will inevitably lead to the greater employment of multiple layer resist systems despite the greater process complication.

Statistical Considerations in Image Formation

Suppose that in order to expose a given area of resist, N primary electrons, X-ray photons or ions are required, then the statistical nature of the emission of the particles implies that the fluctuation in the number of particles will be of the order of $\sqrt{N}$. Clearly, for adequate definition of the final image N must be large, certainly not less than a few hundred. For the 16nm wide lines, made by e-beam exposure at 50kV, shown in figure 12, an exposure dose of $3 \times 10^{-4} C/cm^2$ was used implying that 2000 electrons were used to exposure a 10 by 10nm area. Clearly, this resist would not be suitable for the definition of a 1 by 1nm pattern even if all other constraints on resolution were removed.

In fig. 16 the number of particles per picture element is plotted against the linewidth for resists of various sensitivity. It can be

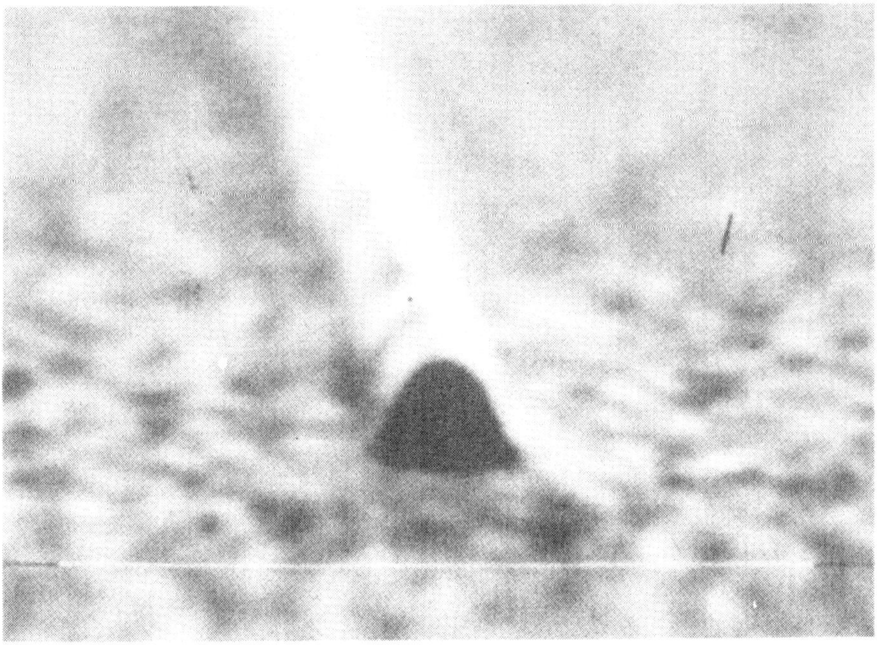

Figure 8. Aluminum line profile. Aluminum thickness 0.18 microns. Markers are 1.0 microns.

seen that statistical considerations play a much greater role in ion beam lithography at sub 50nm linewidths.

it is scattered both elastically and inelastically. The elastic collisions change only the direction of the electrons while in the inelastic collisions, energy is exchanged with the material and secondary electrons are formed. The secondary electrons are vital as they are responsible for exposing the resist. The incident electrons will spread out as they penetrate the solid until all their energy is lost or they leave the surface after back scattering interactions. A Monte Carlo simulation of the path of 50 electrons initially incident as a point beam of 50kV is shown in a 0.3um thick film on a silicon substrate (see figure 9). Some electrons penetrate 10 microns into the material while others are scattered back, re-emerging in the resist up to 10 microns away from the entrance point of the beam. For comparison the trajectories of a 10kV electron beam is shown in Figure 10. Here, more of the original energy is lost in the resist film and the initial beam spreads more widely in the resist.

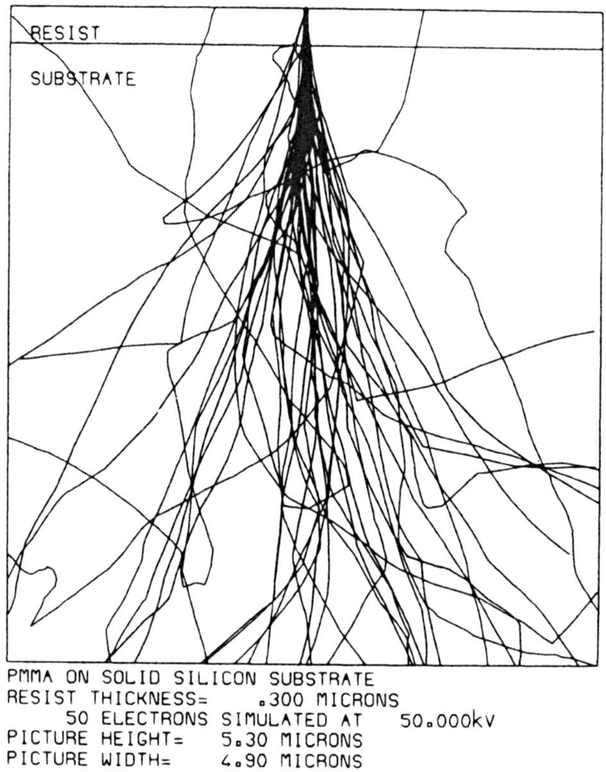

Figure 9. Monte Carlo simulation of trajectories of 100 electrons initially incident as a point beam of 50kV. Resist thickness is 0.3 micron.

Figure 10. Trajectories of 10kV electrons.

The absorbed energy density in the resist can be usefully approximated in some cases by[32]

$$\varepsilon(r)=\kappa[\exp(-r^2/\beta_f^2) + \eta_e \cdot \frac{\beta_f^2}{\beta_e^2} \exp(-r^2/\beta_e^2)]$$

where $\varepsilon(r)$ is the energy density at distance r from the entrance point of a very narrow beam and β_f and β_e are the characteristic half widths of the forward and backscattered distributions and η_e is the ratio of the contribution to the total energy dissipation of the back scattered to the forward-scattered electrons. This model is by no means exact, but does enable useful conclusions to be drawn.

At 20kV for a 0.5 m film of PMMA on Si, $\beta_f \sim 0.08\mu m$ and $\beta_e \sim 2\mu m$ while at 50kV $\beta_f \sim 0.04\mu m$ $\beta_e \sim 9\mu m$

For thinner resist β_f is smaller while β_e is essentially unaffected.

The effect of the back-scattered electrons is to reduce the overall contrast in the latent image formed by electron exposure in the resist. Broers[34] has used this model to predict the contrast in the latent image for electron exposure of equal lines and spaces in PMMA on a silicon substrate. For a 1 μm thick resist layer on silicon exposed at 20kV, the effects of the back-scattered electrons wipe out the contrast in the image for equal spaced lines and gaps than 0.15μm. Using thin resist on a thick substrate, the back scattered electrons are widely spread and a contrast of 0.5 can be obtained down to a linewidth of 50nm. If the effects of the back-scattered electrons are removed entirely (and this can be done by employing a membrane as a substrate), excellent contrast should be obtained for linewidths of 40nm or more.

The double gaussian model also predicts that very narrow lines should be more easily obtained on a solid substrate at higher voltages so both the width of the forward scattering is reduced and the width of the back-scattered contribution is increased.

Very High Resolution Lithography on Membrane Substrates

It is clear that elimination of the back-scattered electrons by using a very thin substrate opens a route to the fabrication of 20nm linewidth structures. Beaumont et al.[2] exposed PMMA on a thin membrane (50nm) of carbon using a 50kV probe 8nm in diameter. After development, the sample was overcoated with 10nm of platimum/palladium and lift-off processing was used to produce metal lines on a carbon substrate which were examined in a transmission electron microscope, (see fig.11). (One advantage of using thin substrates is that the samples are immediately ready for examination by TEM at a few Å resolution). The figure shows a group of three lines all having the same width of 16nm but on different centre to centre spacings. It will be seen that the narrowest centre to centre spacing that can be achieved is 55nm.

Broers[33] using PMMA on a Si_3N_4 membrane, fabricated well defined 12nm wide isolated lines using a 3nm diameter probe.

For both these experiments, performed on thin substrates, electron back scattering has been eliminated. However, in both cases, the linewidth obtained was some 8-9nm wider than the diameter of the elec-

Figure 11. TEM micrograph of 10nm thick Pt/Pd lines exposed by a 50kV 8nm electron probe on a single layer of 60nm of PMMA on a 50nm carbon substrate. Lines are 16 1nm wide, minimum pitch for successful liftoff is 53nm.

tron probe. A major cause of this broadening is the effects of the low energy secondary electrons produced by the 50kV beam. The range of low energy electrons has recently been measured directly[35]. A low-voltage electron beam was used to expose PMMA through a wire mesh as shown in fig. 12. The depth after development of the exposed regions can be taken as a measure of the range of these low-energy electrons (see fig. 13). The exposure range above 400eV follows closely calculated values for the range of electrons in polyethylene. At lower voltages, the range falls to approximately 4 ± 2nm at an energy of 50eV. In order to calculate the exposure dosage across the width of the line, it is necessary not only to know the range of a low energy secondary electron, but also the distribution of secondary electron energy. It is possible to measure the distribution of energy of the primary beam after passing through a membrane of PMMA - a technique known as energy loss spectrometry. Work is proceeding to evaluate the effective dosage distribution from such measurements.

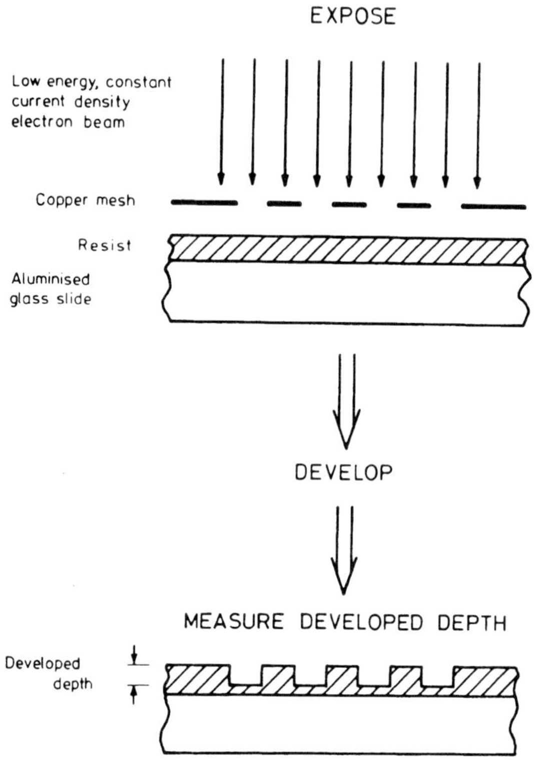

Figure 12. Exposure of PMMA by low energy electron beam.

While the range of the secondary electrons is a major factor determining the ultimate resolution of electron beam lithography, it is probable that resist dependent factors also play a role. The minimum centre to centre spacing of lines which can be achieved on thin substrates is 45-55nm[2] which is somewhat wider than twice the minimum linewidth, the expected value if the secondary range were the only factor. It is believed that when PMMA is spin-coated onto a substrate from solution, the polymer chains are not left randomly intertwined but they tend to coil so in any small volume (1nm cube) only a limited number of polymer chains will be found. When the exposure of a small volume has reduced the molecular weight of this group of chains sufficiently, all the affected polymers will develop. The present evidence is that such a molecular size limitation does exist in positive polymeric resists - but much further study is required to elucidate the details.

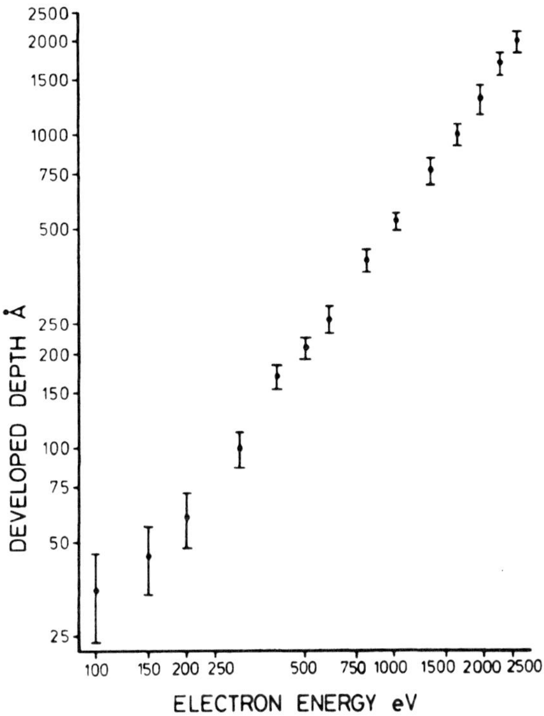

Figure 13. Depth of pit in resist as measured by Talystep after exposure to low energy electrons.
(Exposure 5 x 10^{-5} C cm^{-2})

Very High Resolution Electron Beam Lithography on Solid Substrates

On solid substrates coated with a thin layer of resist, back scattered electrons will reduce the contrast in the latent image (see above sections). It has proved possible to compare directly the linewidths of lines written on a solid and a thin substrate[36]. A 60nm thick membrane of Si_3N_4 is prepared by selective etching of silicon as shown in figure 15 and after coating with PMMA, the exposing beam crosses from solid silicon to thin substrate. After metallisation and lift-off, the resulting pattern is protected by resist while a second etch of the silicon, leaves the whole pattern on a thin membrane of Si_3N_4 and thus ready for examination in a TEM.

In figure 14, 22nm wide lines are shown. It will be observed for these widely spaced (1μm apart) lines that there is no change in linewidth on going from a thick to a thin substrate. On the other hand if lines are written on 70nm centre to centre spacing, the linewidths are different on thin and thick substrates. Figure 15 shows the line-

Figure 14. TEM micrograph of 22nm wide lines exposed on a thin Si_3N_4 substrate and on a thick silicon substrate. The transition from thick to thin substrate is marked by the dark square area on the micrograph; below and to the right-hand bottom edge of the micrograph exposure was performed on a solid surface; above and to the left-hand top edge on a thin substrate.

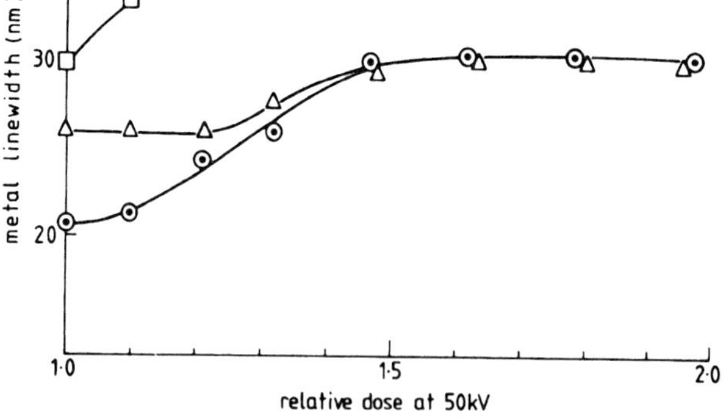

Figure 15. Linewidth as a function of exposure dose for lines on 1 micron and 70 nm spacings on thin and thick substrates.

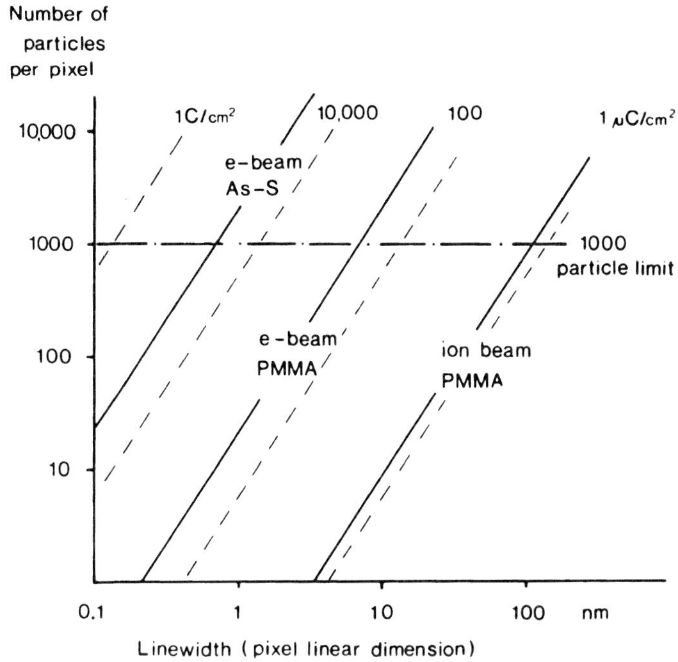

Figure 16. Number of particles per picture element (pixel) against linewidth for resists of different sensitivity. One pixel is taken as the area (linewidth)^2.

widths as a function of exposure dose (normalised to the minimum dose required to form a pattern) for thick and thin substrates. It should be noted that on a thick substrate and for slosely spaced lines, the linewidth is a strong function of exposure dosage and lift-off fails completely at 1.3 times the minimum dose while for widely spaced lines the linewidth is the same over the exposure range explored for both thin and thick substrates.

Howard et al. [(7)] have produced 20nm wide lines on 70nm centre to centre spacing at 20kV and 15nm wide lines on 50nm centres at 120 kV in gold-palladium using lift-off processing of PMMA on a solid GaAs solid substrate. A STEM with a 2nm electron probe was used. It should be noted that improved resolution was obtained at the higher electron beam voltages.

The advantages and disadvantages of thin and thick substrates are summarised in Table I.

Semiconducting materials can be formed into thin membranes which allow active devices to be made using very high resolution

TABLE I

	Thin Substrates	Thick Substrates
Advantages	Wide exposure latitude Good linewidth control Examination in TEM immediately possible Other examination techniques (1nm spot microprobe X-ray analysis, energy loss spectroscopy, etc.) also available. Very high resolution alignment demonstrated.	Samples directly compatible with process line. No need to thin specimen.
Disadvantages	Not all materials can be formed in membrane form.	Examination in TEM requires much sample preparation. Exposure latitude small. Line-width control difficult. Focusing difficult.

electron beam lithography and for immediate TEM examination of these devices. Membranes of active GaAs 150μm in diameter and 50nm thick have been made by selective etching of multi-layered wafers of GaAs/GaAlAs on a GaAs substrate[37].

In the production of semiconductor devices, some means of aligning the different patterns required in the many process steps must be provided. While re-alignment to 0.1μm at 3σ confidence limit has been demonstrated in electron beam writing machines designed for 0.1μm linewidths, alignment to a quarter or better of the minimum linewidth of 20nm presents difficulties on solid substrates. However, re-alignment to better than 4nm has been achieved on thin substrates[38] using a 8nm probe size. The technique used allows resolution to better than half the probe size and so the excellent contrast obtainable on a STEM should permit alignment if required to 0.5nm.

Ion Beam Writing

The behaviour of 50kV ions and electrons on hitting a solid material are very different. As the Monte Carlo picture shows, an electron penetrates many micrometres, is often deflected through large angles and sometimes is returned to the surface. On the other hand an 50kV ion makes a ponderous but majestic advance, creating many secondary electrons before finally coming to rest a few hunderd nm below the surface. As an ion creates more secondary electrons than an electron, the exposure dosage necessary is much lower. Again, as the ions have a very limited lateral straggle, it is possible to write patterns on a solid substrate in thick resist without allowing for sideways scatter of the ions so proximity correction is not important.

Using a 55keV gallium beam PMMA has been exposed and 40nm wide features subsequently ion etched into a gold film[10]. The focused beam was less than 100nm wide. As pointed out earlier ion-ion interactions and aberations due to velocity spread of the beam make it very difficult to achieve sub 50nm ion beam probe sizes.

Direct ion beam lithography using resists would appear to show promise in the 50nm to 300nm linewidth regime offering reasonable speed (but see ref. 38 for a comparison with electron beam machine writing speeds) and excellent spatial fidelity. For very high resolution, the difficulties of forming very small probes and statistical considerations seem to preclude direct ion beam lithography.

However, maskless ion implantation seems likely to prove important in the future for the fabrication of very small devices (although not for volume production).

X-Ray Printing

X-Ray lithography is a printing process using radiation of wavelength 0.4 to 4.5nm. It is potentially an important tool for high resolution lithography offering high speed replication of electron beam written master masks. Machines offering 0.5μm linewidth resolution have been made[11,13], using Rh (0.46nm) and Si(0.83nm) targets. However, for very high resolution applications the range of the secondary electrons requires the use of longer wavelength radiation. Spiller and Feder[14] give the secondary electron mean free path in PMMA as 5nm for 4.4nm Carbon Kα radiation, 35nm for Aluminium Kα radiation and 65nm for Rhodium L radiation.

However, a longer wavelength implies that X-Ray diffraction effects become more important. For diffraction to be limited in extent, the gap between mask and resist plus the resist thickness must be less than the Airy distance; that is to faithfully reproduce a line of width d, the total spacing must be less than $d^2/$. For Carbon K radiation (λ = 4.5nm) and a minimum linewidth of 20nm, the total spacing must be brought into intimate contact which can be done using electrostatic hold-down techniques[11].

Using a mask fabricated by shadowing techniques, Flanders[39] has demonstrated that 17.5nm lines and spaces can be reproduced using X-Ray lithography. The mask consisted of sets of two tungsten lines, 17.5nm wide and approximately 100nm high, spaced by 20nm of carbon. The mask was made by successive shadow evaporations of tungsten and carbon onto a square wave relief pattern of period 180nm formed in a thin polyimide membrane. The resulting mask showed a high transmission (contract) ratio for the Carbon Kα radiation of 90% (Carbon) to 18% (Tungsten). This high contrast ratio arises from the high X-Ray attenuation of tungsten and the extreme aspect ratio of metal lines which are available using shadowing techniques.

In replication of arbitrary patterns written by electron beam lithography, metal lines in the original mask down to 68nm in width were replicated into gaps between the metal on the reproduced metal pattern 90 to 50nm wide. The metal mask lines were gold palladium 55nm thick which had a transmission of 0.6 at Carbon Kα wavelength, giving a relatively low contrast image.

Achievement of narrow lines of an arbitrary pattern replicated by X-Rays on a solid substrate requires very high aspect ratio lines on the mask. It seems unlikely that X-Ray replication of e-beam written masks will be an easily-used technique for linewidths less than 30nm in the near future.

CONCLUSION

Very exciting progress has been made in the past five years in the fabrication of devices with dimensions less than 100nm. Ultimately, the atomic nature of matter itself will present a final limitation; a pattern can only be defined to within one atomic spacing. Further progress towards this goal of the ultimate in resolution will doubtless be made and in all probability some important applications of this emerging technology have not yet even been envisaged.

ACKNOWLEDGEMENTS

Our thanks go to our colleagues at Glasgow University: in the Department of Electronics and Electrical Engineering, Professor John Lamb, Dr. Bhanwar Singh, Dr. Toshiaki Tamamura (now with NTT Electrical Communication Laboratory, Ibaraki, Japan), Mr. Colin Binnie, Mr. Stuart Mackie, Mr. Steve Rishton, Mr. Will Patrick and Mr. Calum MacGregor; and in the Department of Natural Philosophy, Professor Bob Ferrier and Dr. John Chapman. The technical assistance of Mr. John Crichton and Mr. Frank Regan was important to the work. The authors would like to thank Professor Hank Smith of MIT, Dr. Rich Howard of BTL, Holmdel, Professor Fabian Pease of Stanford University, Dr. Dick Pethrick of Strathclyde University and Professor Dan Prober of Yale University for useful conversations. The work at Glasgow University is supported by the Science and Engineering Research Council (U.K.).

REFERENCES

1. J.H. Hains, The Government Role in VLSI Chap. 7 of VLSI Electronics Microstructure Science, Vol. 1, ed. N.G. Einspruch, Academic Press, N.Y., 1981.
2. S.P. Beaumont, P.G. Bower, T. Tamamura and C.D.W. Wilkinson, Sub 20nm Wide Metal Lines by Electron Beam Exposure of Thin PMMA Films and Liftoff, Appl. Phys. Lett. 38, 436, 1981.
3. G.R. Brewer, Electron Beam Technology in Microelectronic Fabrication, Academic Press, New York, 1980.
4. J. Kelly, T. Groves and H.P. Kuo, A High-Current High-Speed Electron Beam Lithography Column, J.Vac.Sci. Technol., 19, 936-940 (1981)
5. R.D. Moore, G.A. Caccoma, H.C. Pfeiffer, E.V. Weber and O.C. Woodward, EL-3 A High Throughput High Resolution Lithography Tool, J. Vac. Sci. Technol., 19, 950-952 (1981).
6. M. Fujinami, T. Matsuda, K. Takamoto, H. Yoda, T. Ishiga, N. Saitu and T. Komoda, Variably Shaped Electron Beam Lithography System EB-55, J.Vac. Sci. Technol., 19, 941-945,(1981).
7. H.G. Craighead, R.E. Howard, L.D. Jackel and P.M. Mankiewich, 10 nm Linewidth Electron Beam Lithography on GaAs, Appl. Phys. Letts., 42, 38-40 (1983).

8. A.N. Broers, W.W. Molzen, J.J. Cuomo and N.D. Wittels, Electron Beam Fabrication of 80 Metal Structures, Appl. Phys. Lett., 29, 596-598 (1976).
9. M. Isaacson and A. Murray, In-situ Vapourization of Very Low Molecular Weight Resists Using 1/2 nm Diameter Electron Beams, J. Vac. Sci. Technol., 19, 1117-20, (1981).
10. R.L. Seliger, J.W. Ward, V. Wang and R.S. Kubena, A High Intensity Scanning Ion Probe with Submicrometer Spot Size, Apply. Phys. Lett., 34, 310-312, (1979).
11. H.I. Smith, P.L. Spears and S.E. Bernacki, X-Ray Lithography: A Complementary Technique to Electron Beam Lithography, J. Vac. Sci. Tech., 10, 913-917, (1973).
12. S.E. Bernacki and H.I. Smith, Fabrication of Silicon MOS Devices Using X-Ray Lithography, IEEE Trans. El. Devices, ED-22, 421-428 (1975).
13. D. Maydan, G.A. Coquin, J.R. Maldonado, S. Somekh, D.Y. Lou and G.N. Taylor, High Speed Replication of Submicron Features on Large Areas by X-Ray Lithography, IEEE Trans, El. Devices, ED-22 429-433, (1975).
14. W.D. Grobman, Synchrotron Radiation X-Ray Lithography, in Handbook on Synchrotron Radiation Vol. I, ed. E.E. Koch, D.E. Eastman and Y. Farge, North Holand (1980).
15. M. Sekimoto, H. Yoshira, T. Ohtaubo and Y. Saiton, Silicon Nitride Single-Layer X-Ray Maskm Jap J. Appl. Phys. 20, L667-672, (1981).
16. D.C. Flanders and H.I. Smith, Polyimide Membrane X-Ray Lithography Masks - Fabrication and Distortion Measurements, J. Vac. Sci. Technol., 15, 995-997, (1978).
17. M.J. Bowden, Electron Irradiation of Polymers and its Application to Resists for Electron Beam Lithography, CRC ritical Reviews, Solid State Sciences 8 223-264, (1979).
18. M. Hatzakis, Electron Resists for Microcircuit and Mask Production, J. Electrochem Soc. 116, 1033-37, (1969). Also: R.A. Harris, Polymethyl Methacrylate as an Electron Sensitive Resist, J. Electrochem. Soc. 120, 270-274, (1974).
19. W.J. Daughton and F.L. Givens, An Investigation of the Thickness Variation of Spun-on Thin Films, J. Electrochem. Soc. 129, 173-179, (1982).
20. J.M. Shaw and M. Hatzakis, Developer Temperature Effects on e-beam and Optically Exposed Positive Photoresist, J. Electrochem, Soc. 1266, 2026-2031, (1979).
21. V.K. Sharma, R.A. Pethrick and S. Affrossman, Polymethyl methacrylate): Influence of Tacticity on its Use as an Electron Resist, Polymer 23, 1732-36, (1982).
22. J.S. Greeneich, Developer Characteristics of Poly(methyl Methacrylate) Electron Resist, J. Electrochem. Soc. 122, 970-976, (1975).
23. K. Harada, O. Kogure and K. Murase, Poly(Phenylmethacrylate-co-Methacrylic Acid) as a Dry-Etching Durable Positive Electron Resist, IEEE Trans. on Elect. Dev., ED-29, 518-524 (1982).
24. M.E. Mochel, C.J. Humphreys, J.A. Eades, J.M. Mochel and A.M. Petford, Electron Beam Writing on a 20 Scale in Metal - Aluminas, Appl. Phys. Lett. 42, 392-394, (1983).

25. T. Tamamura, K. Sukegawa and S. Sugaward, Resolution Limit of Negative Electron Resist Exposed on a Thin Film Substrate, J. Electrochem. Soc. 129, 1831-35, (1982).
26. B. Singh, S.P. Beaumont, P.G. Bower and C.D.W. Wilkinson, New Inorganic Electron Resist System for High Resolution Lithography, Appl. Phys. Lett. 41, 889-891, (1982).
27. J.M. Moran, High Resolution Resist Patterning Using Reactive Ion Etching Techniques, Solid State Technol. 24(4), 195-198, (1981).
28. M. Hatzakis, J. Paraszczak and J. Shaw, Double Layer Resist Systems for High Resolution Lithography, Proceedings of Microcircuit Engineering 81, Laussane, Switzerland, p.386-396.
29. S.P. Beaumont, T. Tamamura and C.D.W. Wilkinson, A Two-Layer Resist System for Efficient Lift-Off in Very High Resolution Electron Beam Lithography, Proc. Microcircuit Engineering, 381-388, Amsterdam, Delft University Press, (1981).
30. C.E. Binnie, S.P. Beaumont, C.D.W. Wilkinson and J.C. White, The Fabrication of Very Short Gate Length n-Channel Mosfet's by Direction Electron Beam Exposure, Proc. Microcircuit Engineering 82, Grenoble, France, (1982).
31. L.D. Jackel, R.E. Howard, E.L. Hu, D.M. Tennant and P. Grabbe, 50nm Silicon Structures Fabricated with Tri-level Electron Beam Resist and Reactive-Ion Etching, Appl. Phys. Lett. 39, 268-270, (1981).
32. T.H.P. Chang, J.Vac.Sci. Technol. 12, 1271, (1975).
33. A. Broers, Resolution Limits of PMMA resist for Exposure with 50kV Electrons, J. Electrochem, Soc. 128, 166-1970, (1981).
34. S.A. Rishton, S.P. Beaumont and C.D.W. Wilkinson, Measurement of the Effect of Secondary Electrons on the Resolution Limit of PMMA, Proc. Microcircuit Engineering 82, Grenoble, (1982).
35. S.P. Beaumont, B. Singh and C.D.W. Wilkinson, Very high Resolution Electron Beam Lithography - thin films on solid substrates? Proc. 10th Electron and Ion Beam Sci. Technol. Conf., Montreal, Canada, (1982).
36. W.S. Mackie, S.P. Beaumont, C.D.W. Wilkinson and J.S. Roberts, High Resolution Lithography on Thin Active Semiconductor Membranes, Proc.10th Electron and Ion Beam Sci. Technol. Conf. Montreal, Canada (1982).
37. M.P. Lepselter and W.T. Lynch, Resolution Limitations for Submicron Technology in VSLI Electronics, ed. N.G. Einspruch, Academic Press, N.Y., 1981.
38. D.C. Flander, Replication of 175 Å Lines and Spaces in Polymethylmethcrylate Using X-Ray Lithography, Appl. Phys. Lett. 36, 93-96, (1980).
39. S.P. Beaumont, P.G. Bower, T. Tamamura and C.C.W. Wilkinson, Replication of Very High Resolution e-beam Written Masks by Carbon k X-ray Contact Printing, Proc. Microcircuit Eng. 81, Lausanne, (1981).

40. R. Feder, E. Spiller and J. Topalian, Replication of 0.1 Micron Geometries with X-ray Lithography, J. Vac. Sci. Tech. 12, 1332-35 (1975).

41. B. Singh, S.P. Beaumont, A.Webb, P.A. Bower and C.D.W. Wilkinson High Resolution Patterning with Ag_2S/As_2S_3 inorganic Electron Beam Resist and Reactive ion etching Proc. 11th Electron, Ion Photon Beam Sci. and Technology Conference, Los Angeles (1983).

QUANTUM EFFECTS IN DEVICE STRUCTURES DUE TO SUBMICRON CONFINEMENT IN ONE DIMENSION

B.D. McCombe

Department of Physics & Astronomy

SUNY at Buffalo

I. Introduction

The advent of improved materials growth technologies (e.g., Molecular Beam Epitaxy (MBE) and Metallorganic Chemical Vapor Deposition (MOCVD)) combined with improved device processing and sophisticated fabrication techniques has made possible a number of physical investigations that could not have been carried out just a few years ago. In concert with this, the thrust toward submicron and ultrasubmicron device feature sizes has also brought into focus several problem areas that were not previously considered, and has unearthed interesting new phenomena, both of which result from spatial confinement on a submicron scale in one dimension i.e., the direction perpendicular to the surface of the structure. Thus, the submicron dimension is not associated with lateral feature size. The structures to be discussed behave in many ways as two dimensional systems from the point of view of their electronic properties. However, the results have more general significance. These investigations can be viewed as studies of presently available submicron systems whose goal is an improved understanding of all aspects of the electronic properties of structures that are confined in one or more dimensions (either physically or electrically) on a submicron scale. In fact, Si MOS devices have recently been fabricated in which one of the two lateral dimensions is also confined on a submicron scale (1000 A° to 8000 A°) by definition of the device gate, and in which the device quality is sufficiently high to observe apparent quantum size effects at low temperatures[1]. These devices behave as nearly one dimensional structures at very low temperature.

The present lectures will focus on two major subjects: 1) studies of the quantized subbands in Si MOS devices due to the carrier confinement by applied gate voltage and the resulting band bending; and 2) investigations of the effect of submicron confinement in one dimension on the binding energy of "impurities" in Si MOS structures and in GaAs-$Ga_{1-x}Al_xAs$ quantum wells fabricated by MBE. In particular, the emphasis will be on the use of far infrared (FIR) optical techniques for the investigation of these structures.

Silicon MOS devices comprise a particularly advantageous system for the study of a number of interesting phenomena including many-body effects, interface scattering, effects of oxide charge, 2D carrier localization, etc. A major advantage of this system is the ease with which several of the most important parameters, e.g., charge carrier density, electric field, positive oxide charge (impurities) near the interface, and extent of the carrier wave function perpendicular to the interface can be varied. These parameter changes can generally be accomplished simply by turning a knob which controls an appropriate voltage. Another related system that provides extremely high quality electronic properties is the "selectively doped": GaAs-$Ga_{1-x}Al_xAs$ heterostructure system. Several other semiconductors (InSb, InAs, InP, $Hg_{1-x}Cd_xTe$, PbTe, Ge, for example) have been fabricated in metal-insulator-semiconductor (MIS) configurations with interface quality adequate to carry out studies of the quantized levels. An excellent review of the overall field through mid-1981 can be found in reference 2.

II. General Background

Two types of structures will be considered in these lectures. The first is the Si-MOS Field Effect Transistor. The devices investigated were all fabricated on p-type substrates with a conducting n-channel (inversion layer). In this arrangement a positive voltage, V_G, is applied to the metal gate electrode with respect to the ohmic substrate contact (usually shorted to either the source or the drain contact). In equilibrium, the electric field developed across the thin insulating layer (SiO_2) is terminated by negative charge in the Si, and the bands are bent downward. Initially, the charge is fixed in the depletion region and is formed by ionized acceptors. When the bands in the Si are bent far enough, a conducting channel of mobile electrons (minority carriers), the inversion layer, is formed in the Si at the Si-SiO_2 interface. This situation is shown schematically in **figure 1.**

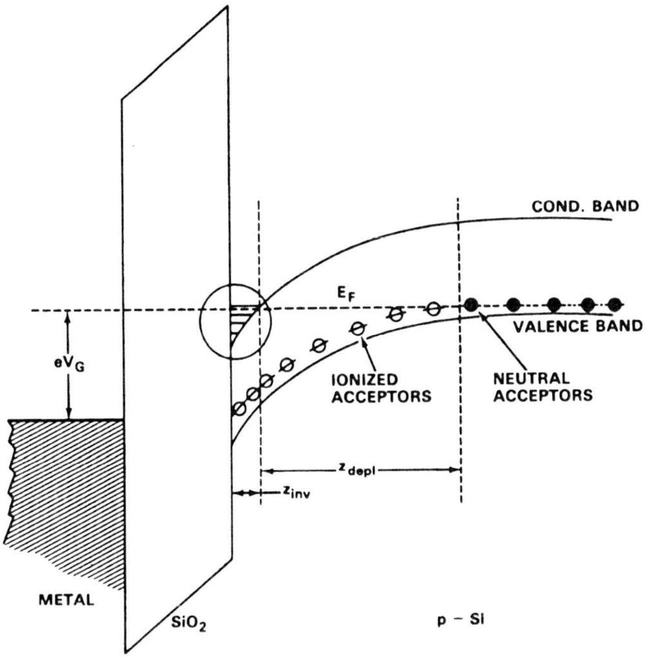

Figure 1. Schematic band-bending diagram for an n-channel Si MOSFET fabricated on a p-type substrate. The situation shown is for an applied positive gate voltage, V_G.

An analagous system can be formed by selectively doping a GaAs -$Ga_{1-x}Al_xAs$ heterostructure. In this case, $Ga_{1-x}Al_xAs$ is heavily doped with an appropriate donor (usually Si) during growth and electrons from the donors transfer to the GaAs layer in equilibrium leaving a positively charged depletion layer in the $Ga_{1-x}Al_xAs$ and negatively charged depletion layer and mobile negative charge in an inversion layer or accumulation layer (depending on the background doping in the GaAs) in the GaAs. The resulting band bending is shown schematically in **figure 2.**

A. Simple Model of the MOS Inversion Layer

The following discussion is confined for simplicity to the MOS case; the situation for the heterojunctions is obtained by a relatively simple extension.

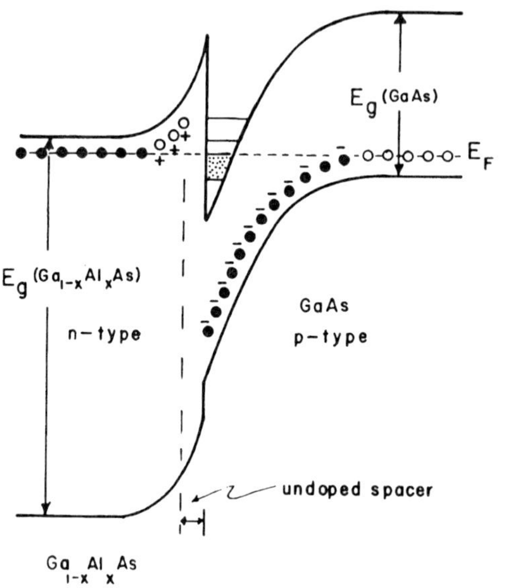

Figure 2. Schematic band-bending diagram for a selectively doped GaAs-$Ga_{1-x}Al_xAs$ heterostructure.

In order to determine the band bending, and hence the shape of the potential at the interface, the 1D Poisson equation,

$$\frac{\partial^2 \Phi}{\partial z^2} = \frac{-4\pi\rho}{\varepsilon_s} \quad , \tag{1}$$

must be solved with appropriate boundary conditions. Here, Φ is the electrostatic potential, ρ is the charge density per unit volume, ε_s is the static dielectric constant of the semiconductor, and the z-direction is taken perpendicular to the interface.

For the situation where the density of donor impurities, N_D, is much less than the density of acceptors, N_A, and at very low temperatures such that there are no free electrons or holes in the bulk, the charge density for bands bent just prior to formation of inversion layer is given simply by $\rho = -qN_A$, where q is the magnitude of the electronic charge.

In this case,

$$\frac{\partial^2 \Phi}{\partial z^2} = \frac{4\pi q N_A}{\varepsilon_s} \quad . \tag{2}$$

Integrating twice from the edge of the depletion region in the bulk, (where $z = z_{depl}$ and $\phi = 0$), toward the interface we have for the potential energy,

$$V = -q\Phi = \frac{2\pi q^2 N_A}{\varepsilon_s} [z - z_{depl}]^2 \quad z < z_{depl} \qquad (3)$$

Typically z_{depl} is large compared to the extent of the inversion layer, and a linear approximation to equation (3) can be made near z=0 (the circled region in figure 1)

$$V \approx qF_s \qquad z \; z > 0$$

$$= \infty \quad , \qquad \underline{z} < \qquad (4)$$

where $F_s \equiv$ surface electric field $= \frac{4\pi q N_A z_{depl}}{\varepsilon_s}$, $N_{depl} = N_A z_{depl}$, and the zero of energy has been arbitrarily shifted by an amount $+2\pi q^2 N_A z^2_{depl}/\varepsilon_s$.

A qualitative feeling for the magnitude of the quantum effects can be obtained by solving the Shrodinger equation for a single electron moving in the potential of equation (4), i.e.,

$$\frac{\hbar^2}{2m_z} \frac{\partial^2 \Psi_z}{\partial z^2} - [qF_s z - E] \Psi_z = 0 \quad , \qquad (5)$$

where m_z is the effective mass for motion perpendicular to the interface, and Ψ_z is the wave-function for z-motion. Motion in the plane of the interface is separated out and is free-electron-like, i.e., $\Psi_x \propto e^{ik_x x}$, $\Psi_y \propto e^{ik_y y}$. Equation (5) can be rewritten

$$\frac{\partial^2 \Psi_\xi}{\partial \xi^2} - [\xi - \xi_0] \Psi_\xi = 0 \quad , \qquad (6)$$

where the dimensionless coordinate $\xi = z/\alpha$, with $\alpha = [\hbar^2/2m_z F_S]^{1/2}$ and $\xi_0 = E/[\hbar^2 q^2 F_s{}^2/2m_z]^{1/2}$. The general solutions of equation (6) are Airy functions. An approximate solution to equation (6) can also be obtained by using the WKB approximation with a single classical turning point (the other turning point is eliminated by the infinite barrier). This yields the following expressions for the energy eigenvalues, which are very close to the exact ones;

$$E_i = \left(\frac{\hbar^2}{2m_z}\right)^{1/3} \left(\frac{3\pi}{2} q\ F_s\right)^{2/3} (i + 3/4)^{2/3}, \quad i=0,\ 1,\ 2\ \ldots \qquad (7)$$

and for the total energy,

$$E_{tot} = E_i + \hbar^2 \frac{(k_x{}^2 + k_y{}^2)}{2m_z} \qquad (8)$$

where m_z is the effective mass for motion parallel to the interface (for simplicity assumed to be isotropic in the x-y plane).

The quantum states form a series of subbands, with i the so-called subband index. Motion perpendicular to the interface is quantized, while motion parallel to the interface is free-electron like. At sufficiently low temperatures, with regard to motion in the plane, electrons behave as a 2D gas.

An improved triangular well approximation can be obtained including the effects of the inversion charge (see reference 2) by taking $V(z)=eFz$ for $z> 0$, where

$$F = \frac{4\pi q(N_{depl} + fn_s)}{\varepsilon_s} \quad . \qquad (9)$$

Here, F is the "effective" electric field, n_s is the inversion layer charge density, and f is a numerical coefficient, $0\leq f\leq 1$. Taking f=1 gives the field at the interface, f=0 gives the depletion contribution alone, and f=1/2 gives the average field in the inversion layer.

Although equation (7) results from a number of simplifying approximations, it gives a good qualitative description of the important dependences of the eigen-energies (on effective mass, surface electric field, and subband index), particularly at low electron densities. Typical values of "effective" electric field lie in the range 3×10^4V/cm to 3×10^5V/cm (10^2 to 10^3esu); and for {100} Si where $m_z \approx m_0$ (m_0 is the free electron mass), this corresponds to a range of energies, E_1-E_0, between 12meV and 56meV, close to the energies determined experimentally. Similarly, for $GaAs-Ga_{1-x}Al_xAs$ selectively doped heterostructures, energy splittings calculated from equations (7) and (9) are typically in the range 17-35meV.

This simple model is sufficient to give the qualitative features of the quantum effects. Discussions of more realistic models which include self consistency, image forces, and many body effects are deferred to the section concerned with subband spectroscopy. The two dimensional (2D) nature of the electronic structure of these systems modifies a number of their properties compared with the usual 3D results. In particular, the electronic density of states is changed greatly. For a 3D system it is simple to show that in zero magnetic field the density of states per unit energy per unit volume is given by

$$\rho_{3D}(E) = \frac{\sqrt{2}(m^*)^{3/2}}{\pi^2 h^3} E^{1/2} \quad , \tag{10}$$

where a factor of 2 has been included for spin. For a 2D system, the result is

$$\rho_{2D}(E) = \frac{m^*}{\pi\hbar^2} \quad , \tag{11}$$

<u>independent</u> of energy (provided m* is independent of energy). Thus, the density of states for the energy spectrum of equation (8) is a series of step functions with onsets at the bottom of each of the subbands.

In the presence of a magnetic field along the z-direction the motion perpendicular to z is quantized into a series of Landau levels, and the motion along the field is unaffected; thus, the energy eigen-values for a 3D system are given by (neglecting spin)

$$E_\ell(k_z) = (\ell+1/2)\, h\omega_c + \frac{\hbar^2 k_z^2}{2m^*} \quad , \tag{12}$$

where $\omega_c = eB/m_c c$, $\omega_c \equiv$ cyclotron frequency, and ℓ is the Landau quantum number. The density of states is

$$\rho_{3D}(E,B) = \frac{2m^*}{4\pi^2\hbar} \left(\frac{eB}{c\hbar}\right) \left\{\sum_\ell [E - (\ell+1/2)\, \hbar\omega_c]\right\}^{-1/2} \tag{13}$$

Again a factor of 2 has been included for spin.

For a 2D system such as that described by equation (8), the motion is completely quantized, and the energy eigenvalues are given by

$$E_{\ell,i} = E_i + \hbar\omega_c(\ell+1/2) \quad . \tag{14}$$

In this case, there are no allowed states between Landau levels and the density of states is even more singular than in 3D,

$$\rho_{2D}(E,B) = \frac{m^*\omega_c}{\pi\hbar}\,\delta[E - (\ell+1/2)\,\hbar\omega_c] \quad , \tag{15}$$

i.e., a series of delta functions at each of the Landau level positions. This singular density of states has important implications for oscillatory magneto-transport and cyclotron resonance at high densities. A comparison of the densities of states for 3D and 2D is shown schematically in **figure 3.**

In zero magnetic field, in the present simple model with isotropic constant energy surfaces, optical transitions are allowed from occupied states in any subband to unoccupied states in any other subband (parity is not a good quantum number for this asymmetric potential) for the optical electric field polarized perpendicular to the interface. The selection rule is $\Delta i \neq 0$.

The above estimates of the quantized subband energies were carried out for typical substrate doping densities that have been investigated (N_A in the range from mid $10^{14}cm^{-3}$ to low $10^{15}cm^{-3}$). It is worth noting in the context of this advanced study institute that since scaling of lateral device dimensions into the submicron regime requires much higher substrate doping than has been utilized in these estimates, the quantized subband energies will be concomitantly higher (equations (7) and (9)). Thus, for submicron (Si) devices, the quantum effects are expected to be important at room temperature. Estimates of the depletion charge and subband splittings are ilustrated in **figures 4 and 5.** The splittings for a substrate doping in the range N_A=$2\times10^{16}cm^{-3}$ -$10^{17}cm^{-3}$ are substantially greater than the thermal energy at room temperature. It is also worth mentioning that even for the relatively low-doped Si substrates that have been studied, clearly resolved subband transitions have been observed at room temperature[3].

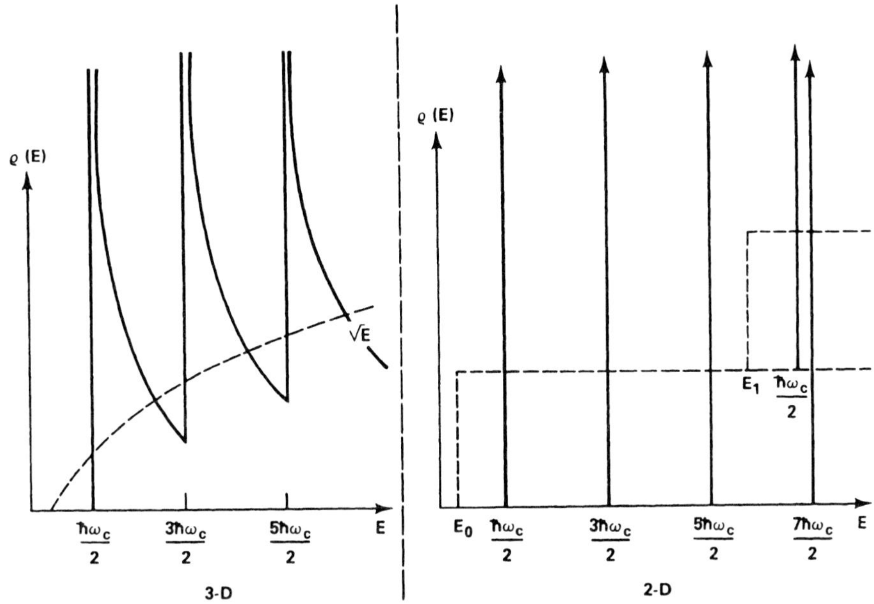

Figure 3. Density of states for 2D and 3D systems (as indicated) in zero magnetic field (dashed lines) and in a large magnetic field (solid lines).

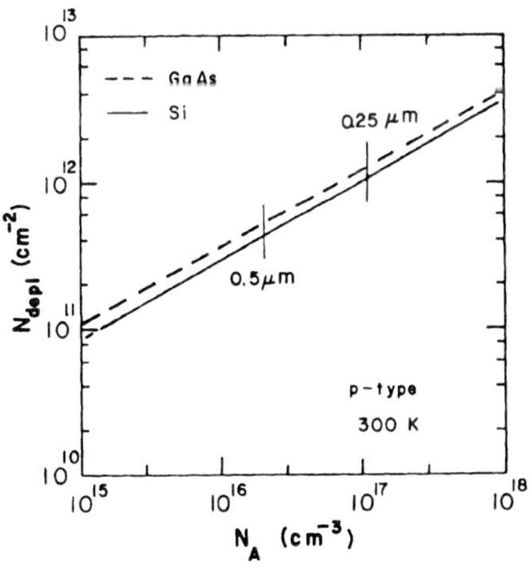

Figure 4. Depletion charge, N_{depl}, vs. acceptor doping density, N_A, for Si and GaAs. Typical values fo N_A for MOS devices scaled to 0.5μm and 0.25μm gate length are indicated.

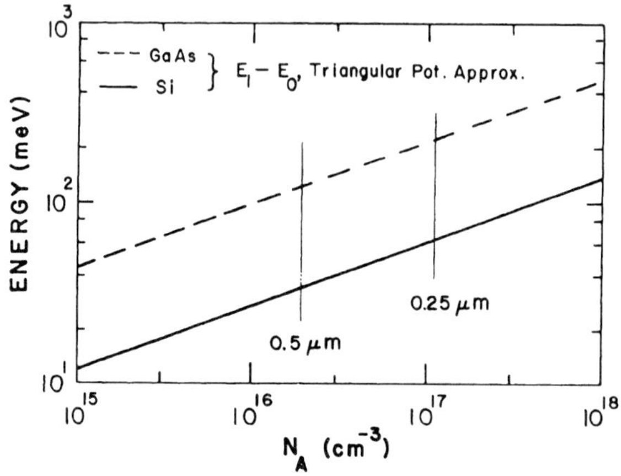

Figure 5. Plot of the energy separation between the ground and first excited subband in the triangular potential approximation as a function of substrate acceptor doping density, N_A. The two vertical lines indicate typical values of N_A for scaling MOS device dimensions to 0.5μm and 0.25μm.

B. Experimental Considerations

As discussed in the previous section, typical energy separations for the quantized subbands occur in a range that corresponds to the far infrared (FIR) region of the electromagnetic spectrum. This is a particularly difficult region in which to carry out spectroscopic studies, and it is worth a few comments on the various approaches. In addition, there are some special requirements placed on devices for these spectroscopic investigations, and the device configurations are also described briefly in this section, as well as methods for coupling IR radiation into the space charge layer with the proper polarization.

Two approaches are currently being used for spectroscopic studies, FIR laser spectrometers and Fourier Transform Spectrometers. The present work has made use of a Fourier transform spectrometer; a block diagram of this system is shown in **figure 6.**

To obtain adequate signal to noise ratio for the absorption signals, some of which are as little as 0.01%, a differential gate modulation technique has been employed in conjunction with the Fourier Transform Spectrometer. The device gate is square wave modulated from a value slightly below threshold (no inversion layer carriers) to some desired gate voltage (which corresponds to a particular n_s), at a frequency between 100 and 1000H_z, and the resulting detector signal phase detected and amplified. The dc output of the lock-in amplifier is thus proportional to the difference between the detector signal with the light transmitted through the sample and coupling arrangement for no inversion layer carriers, and that transmitted with some desired carrier density. The lock-in-derived analog signal is then signal averaged and fast Fourier transformed with an on-line mini-computer to obtain the frequency spectrum. This spectrum is ratioed with a background spectrum obtained with chopped light. External light pipe optics are used to transmit the light for the source to the sample/detector arrangement.

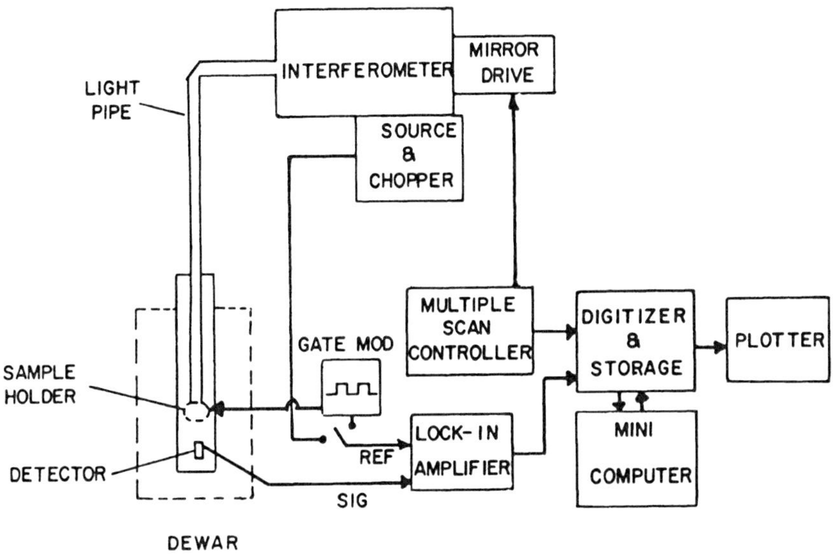

Figure 6. Block diagram of the Fourier Transform Spectrometer system used in the MOS studies described in the text.

In order to observe various optical transitions experimentally, it is necessary to couple the FIR EM radiation into the inversion layer with appropriate polarizations of the IR electric field. Probing optical transitions associated with the dynamical response of the system in the plane of the interface is rather straight forward. The required polarization is parallel to the interface, and a simple transmission geometry with semi-transparent metal gate electrode is adequate.

It is somewhat more difficult to couple radiation into the inversion layer with E_{IR} polarized perpendicular to the interface, the geometry required to observe intersubband transitions in simple systems. In the present work, a prism coupling scheme has been used. This is illustrated in **figure 7.**

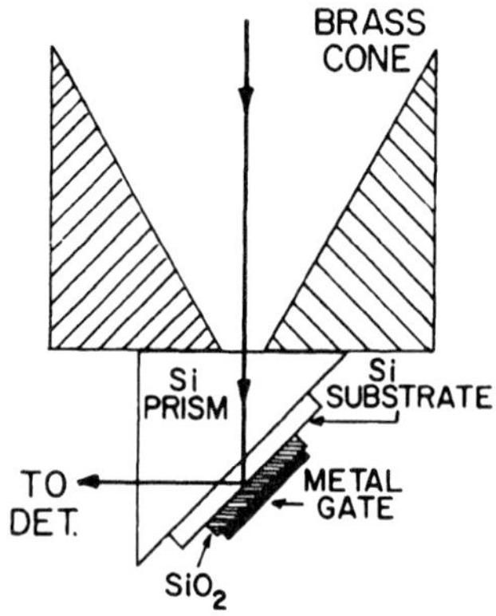

Figure 7. Schematic diagram of the prism coupling scheme used to obtain light polarized perpendicular to the MOS interface for the intersubband optical absorption studies.

In this approach, the Si substrate of the device to be studied is glued backside down to the large surface of the Si prism. Light is incident normal to one of the small faces of the prism, is totally internally reflected at the space charge layer ($Si-SiO_2$ interface) where it is coupled to the space charge layer, and then passes out through the other small face of the prism whence it is directed to an appropriate IR detector. The other approach makes use of a transmission line arrangement[4].

In both cases, it is important that a thick metal gate electrode be used, and that the oxide layer be relatively thin. Otherwise, the polarization properties of the structures, largely determined by the shorting out of the parallel component of electric field by the thick metal gate in close proximity to the space charge layer, will not be very good. The shorting out of the parallel electric field results from the boundary condition on the tangential component of electric field. These effects have been verified by detailed model calculations[5].

In addition to absorption measurements, photoresponse[6] and emission[7] have been used to study quantum levels in Si MOS structures. To date, the only investigations of the quantized subbands in $GaAs-Ga_{1-x}Al_xAs$ heterostructures have made use of resonant Raman scattering to excite and detect the intersubband transitions[8].

For the FIR investigations large area samples are required; the lateral dimensions of the active region of typical devices are 2.5mm x 2.5mm, far from submicron feature sizes. The MOSFETs investigated were fabricated by standard processing techniques. Various gate and contact metallizations were used, depending upon the type of experiment. For transmission measurements, light must pass through the gate metallization, and 50-100 A° of Cr was used. As discussed above, in order to have light polarized perpendicular to the interface, a thick gate metallization is required; either 2000A° of Al or 50-100A° of Cr with a 2000A° Au overlayer was used.

The MOS devices are usually characterized by capacitance-voltage (C-V) measurements at room temperature or below and liquid nitrogen temperature measurements of the threshold, V_T, from conductance or transconductance.

From these measurements one can obtain the inversion layer density-gate voltage relationship

$$n_s = \frac{C_{ox}}{qA}(V_G - V_T) = \frac{\varepsilon_{ox}}{qd_{ox}}(V_G - V_T) \quad , \tag{16}$$

where C_{ox} is the oxide capacitance, A is the gate area and d_{ox} is the oxide thickness.

A particularly useful technique for these 2D systems is the oscillatory magnetoconductance at low temperatures. The number of electrons per unit area per Landau level (obtained from equation (15) including possible valley, g_v and spin, g_s, degeneracy is

$$\eta_\ell = g_s g_v \left(\frac{eB}{ch}\right) \quad . \tag{17}$$

From this, it is clear that one can write

$$n_s = (\ell'+1)\eta_\ell + \Delta n_s \quad , \tag{18}$$

where ℓ' is the Landau quantum number of the highest completely filled Landau level, and Δn_s is the density of additional electrons in the next partially filled Landau level. By combining equations (18) and (16), and noting that the diagonal magnetoconductivity (or resistivity) will be a minimum whenever an integral number of Landau levels are exactly filled ($\Delta n_s = 0$ one obtains the following expression (for the conductivity or resistivity minima)

$$(V_G - V_T) = \left(\frac{qd_{ox}}{\varepsilon_{ox}}\right) g_s g_v \left(\frac{eB}{c\hbar}\right) (\ell'+1) \quad , \qquad (19)$$

from which the degeneracy, $g_s g_v$, and V_t can be determined, provided d_{ox} is known. Alternatively for structures such as $GaAs-Ga_{1-x}Al_xAs$ where there is frequently no gate and no simple way of determining n_s, the magnetic field can be swept and n_s determined from equation (17) from the period of the oscillations in 1/B.

The equilibrium depletion charge, an important parameter indetermining subband separations, can be found from C-V measurements and/or substrate bias measurements at liquid nitrogen temperature. In addition, the net positive oxide charge, which has a strong influence on the low temperature mobility and, in some cases, localization of inversion layer carriers, is determined by C-V or V_T measurements.

III. Subband Spectroscopy

The most direct means of determining the quantized subband energies in these confined structures is through optical absorption measurements. However, the model developed in the previous section is far too simplified to be useful in any quantitative studies. In fact, several very interesting features have been ignored. In particular, the triangular potential approximation is expected to fail when the inversion layer density of mobile electrons is comparable to or greater than the fixed depletion charge density, and a self-consistent solution is required in this case. In addition, band structure effects, e.g., the multivalley nature and the anisotropy of the conduction bands in Si, image force effects due to the dissimilar dielectric constants of Si and SiO_2, and many body effects (exchange and correlation) all have to be taken into account. These effects are outlined below.

The conduction band of bulk Si as noted above is multivalley in nature and consists of 6 equivalent minima located at 0.85 of the distance to the Brillouin zone boundary along the <100> directions. The energy surfaces are ellipsoids of revolution with long axes along the <100> directions, and longitudinal and transverse masses given by $m_l=0.916m_o$, and $m_t=0.190m_o$, respectively. The effective masses for the 3 principal surfaces of Si are given in Table I[9]. Here m_z is the mass perpendicular to the surface,

TABLE I. Effective masses and valley degeneracies for the three principle surface orientations of Si. After reference 2.

Surface Orientation	m_x	m_y	m_z	Valley Degeneracy g_v	Numerical Vaules m_z
{100}	m_t	m_t	m_ℓ	2	$0.916m_o$
{110}	m_t	$(m_t + m_\ell)/2$	$(2m_t m/(m_t + m_\ell)$	4	$0.315m_o$
	m_t	m_ℓ	m_t	2	$0.910m_o$
{111}	m_t	$(m_t + 2m_\ell)/3$	$(3m_t m_\ell/(m_t + 2m_\ell)$	6	$0.258m_o$

$$m_t = 0.910m_o$$

$$m_N = 0.916m_o$$

and m_x and m_y are the principal masses of the constant energy ellipse in the surface. Note that according to equation (7), where there are inequivalent valleys those with the heaviest mass will have subbands that lie lowest in energy, e.g., for the {100} surface the ground subband of the 2-fold degenerate valleys with $m_z = m_\ell$ lie lower than the corresponding ground subband for the 4-fold degenerate valleys with $m_z = m_t$. For simplicity, we initially specialize to the {100} surface. The bulk Si energy surfaces and their projection on to a {100} plane are shown schematically in **figure 8(a).**

For this case, we write the Schrodinger equation for motion perpendicular to the surface in the effective mass approximation as

$$\frac{d^2\Psi_z^i}{dz^2} + \frac{2m_z}{\hbar^2} [E_i + V_{eff}]\Psi_z^i = 0 \quad , \qquad (20)$$

where the effective one electron potential energy,

$$V_{eff} = V_{depl} + V_{image} + V_{Hartree} \quad , \qquad (21)$$

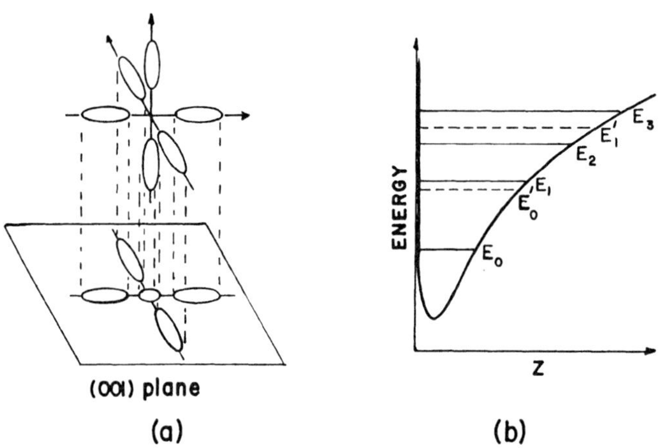

Figure 8(a). Surfaces of constant energy for electrons in Si and a projection onto the (001) plane; **(b)** Schematic diagram of the effective one-electron potential for the (001) plane and the bottoms of several subbands showing the two sets of subbands (unprimed and primed) corresponding to m_z=0.916m_0 and 0.19m_0, respectively.

with

$$V_{depl} \approx \frac{4\pi q^2 N_A z_{depl}}{\varepsilon_s} \quad z \tag{22}$$

$$V_{image} = \frac{(\varepsilon_s - \varepsilon_{ox})\, q^2}{4\varepsilon_s(\varepsilon_s + \varepsilon_{ox})} \quad \frac{}{z} \quad , \tag{23}$$

and

$$V_{Hartee} = -\frac{4\pi q^2}{\varepsilon_s} \int_0^z dz' \int_0^{z'} dz'' \; n(z) \quad , \tag{24}$$

from integrating the Poisson equation. The total wave function is given by $\Psi = \Psi_x \Psi_y \Psi_z u_{k_o}(\underset{\sim}{r})$, where $\Psi_x \;\alpha\; e^{ik_x x}$, $\Psi_y \propto \alpha\; e^{ik_y y}$, and $u_{k_o}(\underset{\sim}{r})$ is the band edge function appropriate to the particular conduction band valley.

A self-consistent solution of equations (20) and (24) must be carried out since n(z) is determined by

$$n(z) = 2g_v \sum_{i,\ \text{occupied}\ \underset{\sim}{k}} \left| \Psi_z^i \right|^2 \quad , \tag{25}$$

where g_v is the valley degeneracy, and the factor of 2 is for spin.

The major effect of the inversion electrons is to sharpen and deepen the potential well in a narrow region near the interface. This has a dominant effect on the ground state and serves to increase the energy separation between the ground and excited states as would be expected from simple considerations, i.e., the surface electric field is proportional to $(n_{inv}+N_{depl})$.

Thus far, the discussion has been confined to single-particle states, and electron-electron interactions via their mutual Coulomb repulsion have been neglected. The parameter that measures the importance of Coulomb effects is r_s, the average interparticle separation in units of the effective Bohr radius. For typical electron densities, n_s, the 2D r_s varies from 8.1 to 0.81. Thus, many body effects are expected to be important in the Si MOS system (at low temperatures). They are expected to be less important for GaAs (r_s=0.54-0.24) for typical values of n_s (10^{11}-$5\times10^{11}cm^{-2}$).

The effects of exchange and correlation on the subband energies can be taken into account approximately by adding an effective one-electron exchange-correlation potential, V_{xc}, to equations (20) and (21), with a density functional formalism used to determine V_{xc}[10]. The major effect of V_{xc} is a lowering of the ground state subband energy with respect to the excited states; this is largely due to the exchange interaction. The effective potential and the energies of the bottoms of several subbands for {100} Si are shown schematically in **figure 8(b).**

The theoretical treatment becomes considerably more complicated when other surfaces of Si are considered. This is discussed in more detail after data for Si {100} surfaces are presented and compared with theory.

For GaAs, the discussion above also applies, but with some simplifications. Since the dielectric constants of GaAs and $Ga_{1-x}Al_xAs$ are nearly the same, the image term can be neglected. In addition, as mentioned above, V_{xc} can also be omitted to a good approximation since exchange and correlation are expected to be much less important.

A. Experimental Results for Si (100)

Experimental intersubband spectra can be used to test the theoretical models described above. The measurements allow the determination of the energy vs. density (n_s) relationship for the subband transitions over a wide range densities at any desired discrete values of n_s. Relative intensities and line widths of the various transitions can also be measured.

A great deal of work over the past several years, both experimental and theoretical, has gone into understanding the basic physics of the subbands in semiconductor space charge layers, particularly in the case of the Si {100} surface.

Koch and co-workers in Munich have provided the major body of experimental work in this area (largely via laser absorption studies), and Ando has developed much of the theory. A review of this work is given in reference 2. Detailed understanding of the subbands has advanced considerably through a series of theoretical and experimental iterations, and it is now well-accepted that many body effects play a substantial role in determining the subband energies. In addition, two additional effects must be considered if one is to make a comparison of the experimental optical resonance positions with theory. These are: 1) the so-called depolarization effect; and 2) the final state interaction or excitation-like effect (vertex corrections in diagrammatic language). The first of these, the depolarization effect, can be understood classically. It is a shift in the resonance position to energies higher than the subband separation due to the fact that the space charge layer is a thin polarizable medium, and the electrons respond dynamically to the internal IR electric field, not the applied field. For a two level system (ground state (i=0)) and one excited state(i=1)), the resonance occurs at a frequency[11]

$$\omega_{res} = \omega_{01}(1 + \alpha)^{1/2} \tag{26}$$

where α is proportional to the product of n_s and the oscillator strenth of the transition, and ω_{01} is the frequency separation of the ground and excited state.

The final state interaction is strictly a quantum mechanical, many body effect resulting from the Coulomb interaction of the optically excited electron with the hole it left behind in the Fermi sea. This is a negative correction that lowers the transition energy. Again, for a two level system, it has been shown[12], that the resonance frequency including both effects can be written,

$$\omega_{res} = \omega_{01}(1 + \alpha - \beta)^{1/2} \tag{27}$$

where β is a parameter that characterizes the final state interaction. Ando[12] has also shown that for $\{100\}$ Si α and β are nearly equal over a range of electron densities, and $\omega_{res} \approx \omega_{01}$. For a multilevel system, in the absence of such near cancellation, there is a coupling among the various transitions and a concomitant shift in frequency and intensity.

An example of intersubband spectra for a high quality Si MOSFET (peak effective mobility at 4.2°K 11,000 cm^2/v-sec) is shown in **figure 9.** The normal to the sample surface is tilted by approximately 1.5° from the $\{100\}$ direction toward the $\{111\}$ direction; this small tilt angle is not important for the present purposes and will not be discussed further. This data was taken with the spectroscopic arrangement of figure 6 and the prism coupling scheme of **figure 7.** The light transmitted through this prism arrangement (reflected at the Si-SiO_2 interface) is filtered by sapphire and black polyethylene and detected by a Ge:Ga photoconductive detector. The Si prism prevents band gap light from reaching the device while allowing through sufficient room temperature background radiation to allow a nearly equilibrium depletion layer to be formed under the device gate. The ratioed differential absorption spectrum is proportional to the power absorbed in the inversion layer at a given electron density. A very strong and sharp (FWHM≈1 meV) 0-1 transition is evident as well as a much weaker and broader 0-2 transition at higher energies. The peak positions can be determined with high precision.

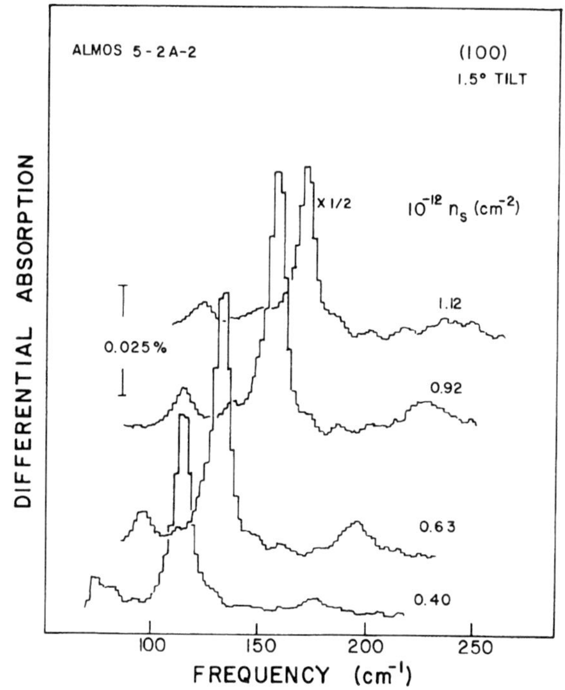

Figure 9. Differential absorption spectra for an inversion layer sample tilted 1.5° away from the <100> toward a <111> axis. The 0-1 transition is the dominant absorption line. Data were taken at 4.2°.

Some results from a different sample ((100 surface) are shown in **figure 10.** Both 0-1 and 0-2 transitions are plotted. The experimental points are compared with Ando's theoretical calculations[13] which take into account the various effects mentioned above. The theoretical results have been fit to one point for the 0-1 transitions by adjusting the depletion charge (and the depletion contribution to the transition energy) within the range of experimental uncertainty. The overall fit is excellent considering the approximate nature of the theory. The discrepancy for the 0-2 transition could be largely removed while maintaining a good fit to the 0-1 data by lowering the depletion charge slightly. The 0-2 and higher transitions are much more sensitive to the depletion potential than the 0-1 transition.

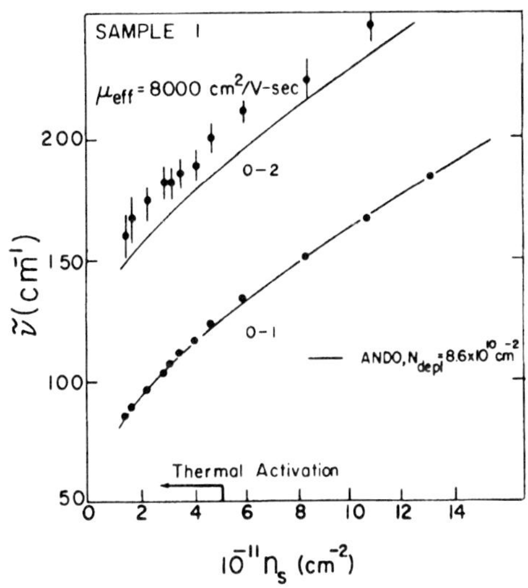

Figure 10. Plot of peak absorption frequency vs. electron density for a (100) inversion layer with N_{depl} 9×10^{10} cm^{-2}. Experimental data;--theory.

This excellent fit combined with the observation that the 0-1 transition is always much more intense than the higher transitions (indicating little coupling among transitions) is evidence that the theory is adequate for the Si (100) surface.

B. Experimental Results for Si {110} and {111}

Electron space charge layers on {111} and {110} Si surfaces are neither as completely studied nor as well understood as those for {100} Si. Until recently, it appeared that for both surfaces, the observed valley degeneracies and effective masses could be different from those predicted by effective mass theory. A study of the intersubband transitions is expected to provide different information than is contained in transport and magnetotransport measurements and thus help to increase the understanding of the electronic properties of these surfaces. In this section, we present some results on inversion layers on {100} and {110} surfaces of Si and discuss them in terms of current theoretical models, some of which were proposed to explain other experimental observations. The most detailed and extensive theoretical calculations concerned with the quantized subbands and intersubband transitions on {111} and {110} surfaces of Si is the recent work of Yi and Quinn[14]; and we will compare the intersubband measurements with their results.

As discussed in the previous section, Si has six conduction band minima located along the <100> directions which project onto a {111} surface to give six equivalent valleys, 12 with spin included. In the independent particle effective mass approximation, the surface electric field does not remove this degeneracy, and six-fold valley degeneracy is expected since m_z is identical for each of the ellipsoids (see Table I). However, a number of magnetotransport measurements[15] have found a valley degeneracy of two for all principal surfaces of Si. Later experiments by Tsui and Kaminsky[16] demonstrated that the valley degeneracy, obtained in the above manner, depends on the processing steps used to prepare the samples, and valley degeneracies of both two and six have now been obtained by a number of experimenters.

For the {110} surface of silicon, the six conduction band minima in the presence of the surface electric field give rise to four degenerate valleys (eight with spin), which are lowest in energy and two degenerate valleys, (four with spin) somewhat higher in energy (Table I). Again, early transport experiments[15] also gave a valley degeneracy of two for this surface. However, more recent experiments have shown that in most samples, including those investigated here, the valley degeneracy is <u>four</u> and the spin degeneracy is lifted at the fields at which the experiments are done[17]. Thus it now appears, that the perplexing problems concerned with the anomalous valley degeneracy are confined to the {111} surface.

Several models have been proposed to explain the different valley degeneracies observed in these surfaces. The most straightforward of these makes recourse to large, local, inhomogeneous stresses at the oxide-semiconductor interface as the cause of the lower degeneracy observed in some samples. In this model, the large stress splits the six-fold degeneracy, lowering two of the valleys substantially below the other four. On the other hand, other workers[18] have concluded from measurements of mobility anisotropy and Shubnikov-de Haas oscillations that such large stresses are unlikely. They argued that such a large splitting required much larger stresses than were applied in their experiments, while the mobility anisotropy measurements indicated that the local inhomogeneous stresses and the applied stress were of the same order.

Two possible means of resolving this dilemma have been proposed. The first of these is based solely on many-body effects. Simply stated, the argument is that the ground occupied state (lowest subband) is lowered due to exchange and correlation, and the amount of lowering is proportional to the Fermi energy at constant electron density ($E_F \propto n_s/g_s g_v$, a transition to a lower degeneracy ground state can become favorable at densities low enough that the energy lowering due to many-body effects is greater than the corresponding gain in kinetic energy. However, this effect is much too small[19] (~1-2 meV) to account for large valley splittings. Bloss et al.,[19] have suggested that this effect may be enhanced by small amounts of strain on the {111} surface to the extent that the transition takes place at relatively high carrier densities. They estimate that only about one third as much stress is required with exchange effects included to explain the apparent splitting between valleys. Another model which accounts for the observed "valley" degeneracies: without the need for large amounts of strain has been proposed by Vinter and Overhauser[18]. They suggest that the valley degeneracy deduced from the period of the Shubnikov-de Haas oscillations can be two even when the valleys are only split slightly in energy provided that the strain splits the valleys into three pairs, and the splitting is more than the width of the levels. This implies that all samples are "nearly" six-fold degenerate. Their argument is based on the assertion that the Fermi energy for the 2D electron gas is not constant, but rather the electron density in the layer is constant for a given gate voltage. Hence, for sharp Landau levels, the Fermi energy is locked to the highest filled Landau level and jumps to the next available Landau level when it is filled. The minima in the Shubnikov-de Haas oscillations occur whenever a Landau level is exactly filled; thus, the period in n_s (directly related to the gate voltages change through the device capacitance) is characteristic of a valley degeneracy of two since each 2D Landau level holds $2g_v qB/ch$ electrons and $g_v=2$ in this model. When there is substantial overlap of the broadened Landau levels from the three pirs of valleys, the measured g_v from the period will be six. Since the Landau levels are relatively broad for the {111} and {110} samples (maximum mobilities of 3000 cm^2/V-sec), it is expected that unless the splittings are actually quite large, there should be some component of six-fold degenerate period in the measurements on "two-fold" samples. Careful studies by two groups[19] have failed

TABLE II. Polarization selection rules and transition frequencies for the 0-1 transition for the Si (110) surface. After reference 14.

Valley Degeneracy / Polarization	4-fold	2-fold
{110}	$\omega_{Res} = \omega_{10}(1+\alpha_{11}-\beta_{11})^{1/2}$	$\omega_{Res} = \omega_{10}(1+\alpha_{11}-\beta_{11})^{1/2}$
{1$\bar{1}$0}	$\omega_{Res} = \omega_{10}(1-\beta_{11})^{1/2}$	$\omega_{Res} = \omega_{10}(1+\alpha_{11}-\beta_{11})^{1/2}$
{001}	No Resonance	No Resonance

TABLE III. Polarization selection rules and transition frequencies for the 0-1 transition for the Si (111) surface. After reference 14.

Valley Degeneracy / Polarization	6-fold	12-fold	
		[112] oriented valleys	Pairs of valleys tilted with respect to
(111)	$\omega_{Res}=\omega_{10}(1+\alpha_{11}-\beta_{11})^{1/2}$	$\omega_{Res}=\omega_{10}(1+\alpha_{11}-\beta_{11})^{1/2}$	$\omega_{Res}=\omega_{10}(1+\alpha_{11}-\beta_{11})^{1/2}$
$(1\bar{1}0)$	$\omega_{Res}=\omega_{10}(1-\beta_{11})^{1/2}$	No Resonance	$\omega_{Res}=\omega_{10}(1+\alpha_{11}-\beta_{11})^{1/2}$
$(11\bar{2})$	$\omega_{Res}=\omega_{10}(1-\beta'_{11})^{1/2}$	$\omega_{Res}=\omega_{10}(1+\alpha_{11}-\beta_{11})^{1/2}$	$\omega_{Res}=\omega_{10}(1+\alpha_{11}-\beta_{11})^{1/2}$

to reveal this feature and substantiate the fact that there are actually two-fold and six-fold degenerate samples for the {111} surface. In the two-fold samples, the energy splitting must be greater than the Fermi energy at zero magnetic field.

An additional mechanism proposed to account for the lifting of the valley degeneracy is the occurrence of a charge density wave ground state[20]. For example, phonon mediated intervalley exchange interaction between pairs of valleys can cause two valleys to be lowered relative to the other four in {111} inversion layers (thus a two-fold degenerate ground state). However, it now appears that the magnitude of the intervalley exchange coupling is much too small to lead to such a ground state.

Subband transitions on the {111} and {110} surfaces are complicated by the fact that the principle axes of the constant energy ellipsoids are, in general, tilted with respect to these planes. Thus, the effective mass tensors have off-diagonal components, and there can be currents induced by incident IR electric fields in directions perpendicular to the direction of electric field polarization. In fact, for both these surfaces IR electric fields polarized in the plane of the surface can induce intersubband transitions (unlike the {100} case). Theoretical treatment including the anisotropy of the ellipsoids, as well as the various factors discussed in the previous section, is quite complicated, and the lack of such a detailed theory hampered interpretation of earlier published experimental work[21].

The recent work of Yi and Quinn[14] has rectified this situation. Their results for the polarization selection rules and resonant frequencies for the 0-1 transition are summarized in Tables II and III. These results include both depolarization and final state interactions but assume no coupling among the subband oscillators.

Some experimental data for the (110) surface is shown in **figures 11 and 12** for the infrared electric field polarized parallel and perpendicular to the surface, respectively. It has been verified in separate measurements on chips from these same wafers that the valley degeneracy, $g_V = 4$ for these samples[17]. In both sets of data approximately the full equilibrium depletion charge was established; this can be seen from the line shapes, where the 0-2 transition is well separated and clearly resolved from the 0-1 transition. These data are summarized in **figure 13,** and the calculations of Ando are shown for comparison. Ando's calculations are for two-fold degenerate samples; there will be a small shift (1-2meV) to lower energies for the four-fold case due to the smaller Fermi energy for a given n_s and the resulting smaller exchange lowering of the ground subband relative to the excited subbands. There is a rather substantial shift to lower energies for the E∥ surface polarization compared with the E⊥ surface polarization. As can be seen from Table II, this is in agreement with theory for four-fold degenerate samples, because the E resonance contains both depolarization shift and final state interaction effects, whereas the E∥ resonance is shifted from the subband energy only by the final state interaction. In principle, the magnitude of the depolarization shift can be obtained from these data; however, due to uncertainties in the value of depletion charge (equilibrium is not completely established), and the approximate nature of the sample results shown in Table II (coupling among subband oscillators is neglected), numerical values cannot be obtained with any confidence. Recent results of Koch and co-workers in accumulation samples exhibit large shifts to high energies for E⊥ polarization, and polarization studies within the plane (i.e., E∥ [110] and E∥ [001] are in agreement with the results shown in Table II.

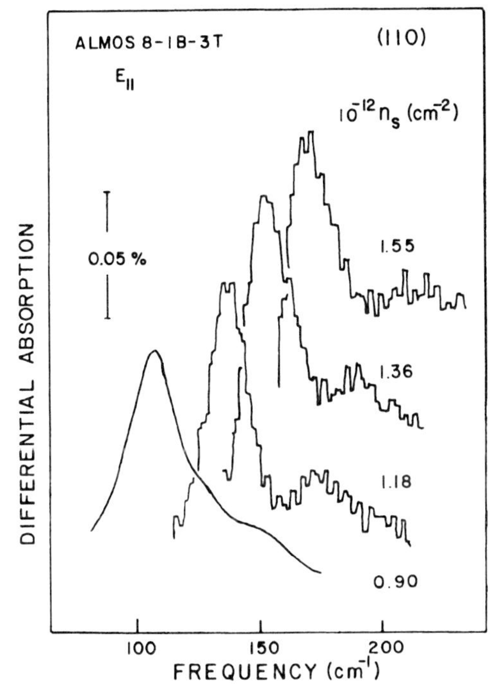

Figure 11. Differential absorption spectra for a (110) sample with the IR electric field polarized parallel to the surface. The "noise" on the three highest density spectra are a result of multiple internal reflection interference. This has been smoothed on the lowest density curve. This sample had nearly equilibrium depletion charge established.

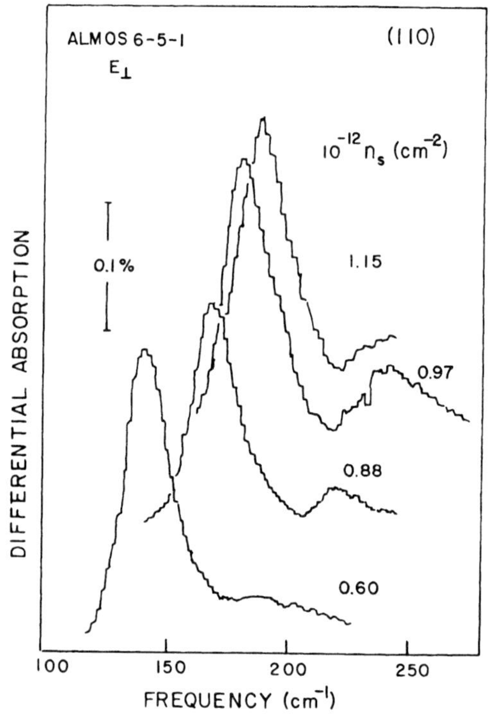

Figure 12. Differential absorption spectra for a (110) sample with the IR electric field polarized perpendicular to the surface. ($\|$ to [110]). Nearly equilibrium depletion charge was established in this sample.

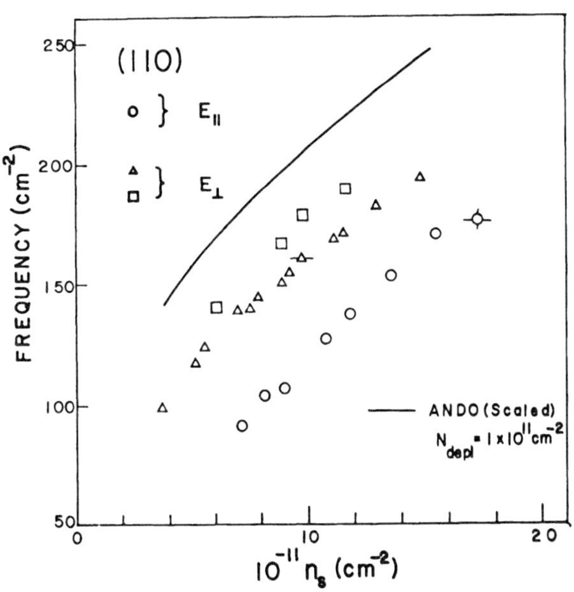

Figure 13. Summary of data for (110) Si for the 0-1 transition for E∥ and E⊥ to the surface. Absorption lines are shifted by approximately 30-40% to higher frequencies for E⊥ compared with E∥. Data denoted by the squares were for a sample with slightly higher depletion charge.

Some experimental data for a Si (111) sample having two-fold valley degeneracy is shown in **figure 14.** The valley degeneracy was determined from magnetoconductance oscillations on Hall bars taken from adjacent chips. A polarization study at two different electron densities is shown. A small shift to higher energies for the E∥ data is exhibited. Data for two samples taken in several runs over a period of several months is summarized in **figure 15.** For comparison, the results of Yi and Quinn are also shown. As can be seen, there is some scatter in the data (of the order of 1.5meV) due to small charge in depletion charge. With the scatter there is no shift of the resonance between E ∥ (E ∥ [110] and E⊥

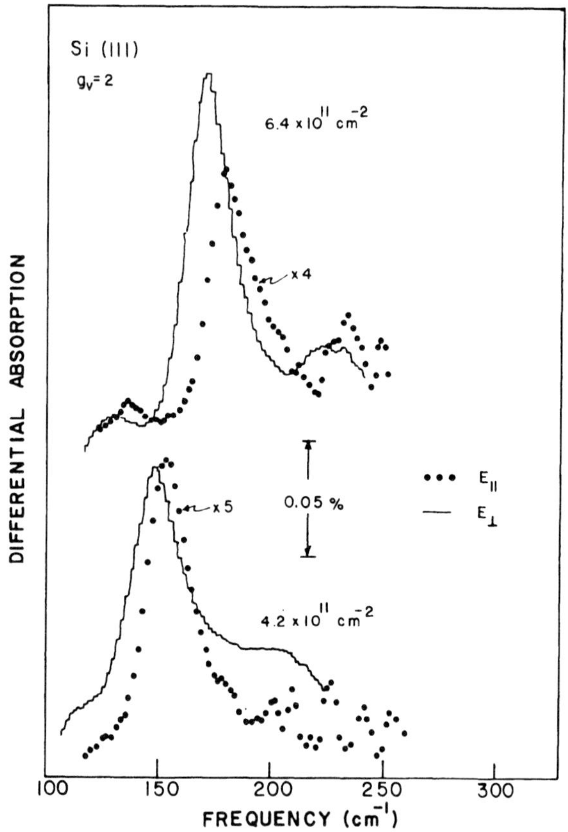

Figure 14. Differential absorptions spectra for a (111) Si MOSFET with valley degeneracy, g_V=2 for the IR electric field polarized parallel (——) and perpendicular (•••) to the surface. There is a very slight shift to higher energies for the E∥ data, but it appears that these data are also for a slightly higher depletion charge.

for these samples. This is only consistent with the result of Table III for two-fold samples having a pair of valleys lowest in energy whose principle axes are tilted with respect to the [110] or [112] axes of the crystal.

Several samples that underwent special Ar annealing treatments and exhibited six-fold valley degeneracy were also investigated. These samples all showed 0-1 transition energies that were 8-10meV lower in energy than the data of **figure 15**, far below the calculated transition energy for six-fold samples. The samples were also consistently of poorer quality than the two-fold samples, as determined from V_T and the peak effective mobility at 4.2K ($\mu_{peak} \lesssim 1000\ cm^2/V\text{-sec}$). In addition, little or no shift between $E_\parallel$ and $E_\perp$ data was observed, contrary to the theoretical results summarized in Table III. These data are not well-understood at this time, but it appears that interface degradation due to the Ar annealing step may be playing a role.

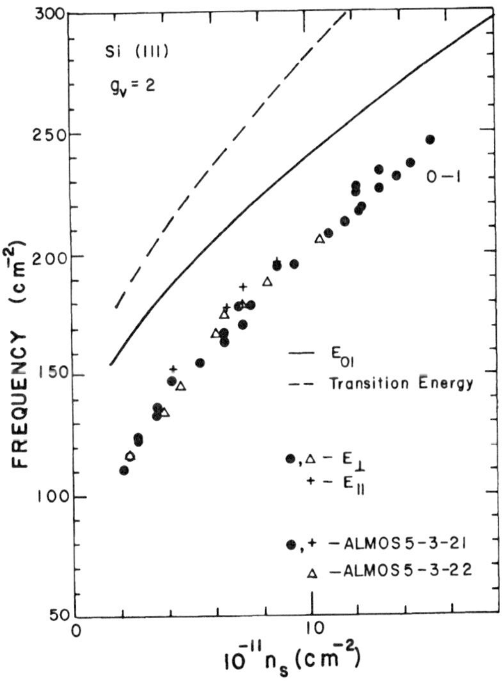

Figure 15. Comparison of experimental data for the 0-1 transition taken from 2 Si (111) samples (symbols) with theoretical calculations for the subband separation (solid line) and the transition energy (dashed line). The theoretical results are from Ref.14.

In summary, the optical measurements of the subband transitions provide an alternative probe that is sensitive to valley degeneracy for the more complicated surfaces of Si; and the results are generally consistent with theory and other measurements. However, the perplexing problem of the origin of the valley degeneracy on Si {111} surfaces and its modification by processing treatments is still not solved.

IV. CONFINED "IMPURITIES"

A. Background

Bound states associated with impurities and other defects in semiconductors and semiconductor device structures are important in determining many of the electrical and optical properties of these materials and devices. As a result, impurities in semiconductors have received considerable attention, both theoretically and experimentally, for a number of years. Many shallow impurities in the bulk can be rather well-described by a simple hydrogen-like model in which a single charge is bound to an oppositely charged fixed impurity center through their mutual Coulomb attraction. The semiconductor host crystal is taken into account through the use of an effective mass for the orbiting charged particle (effective mass approximation) and a background dielectric constant ε_s. For a simple parabolic and isotropic conduction band, the hydrogen atom result can simply be taken over by replacing the free electron mass by the effective mass, and $-q^2/r$ by $-q^2/\varepsilon r$. This yields the following energy spectrum

$$E_n = -\frac{m^* q^4}{2\varepsilon_s^2 \hbar^2 n^2} \quad , \quad n = 1, 2 \ldots \tag{28}$$

measured with respect to the conduction band edge. The effective Bohr radius in this case is given by

$$a_o^* = \frac{\hbar^2 \varepsilon_s}{q^2 m^*} \quad . \tag{29}$$

For donors in GaAs the binding energy is 5.8 meV and the effective Bohr radius is ~100A°. For donors in bulk Si (using $m^*=m_t$ for simplicity) the binding energy is 18.6 meV and the effective Bohr Radius is 33A°.

In general, we would expect that if the impurities were confined on a scale of the order of or less than the effective Bohr radius, there should be modifications to these simple results for the binding energies. In the limit where the width of the confinement goes to zero, the system becomes two dimensional (2D hydrogen atom), and the energy levels of the system are given by[23]

$$E_n = -\frac{m^* q^4}{2\varepsilon_s^2 \hbar^2 (n-1/2)^2} \qquad (30)$$

with the effective radius of the ground state wave function equal to $a_0^*/2$. From equations (28) and (30), the binding energy in the 2D limit is four times the binding enrgy in the 3D limit.

Prior to discussing the effects of more general confining potentials on the binding of impurities, for the case of an impurity located near the boundary of two materials with dissimilar dielectric constants (such as encountered in the Si-SiO_2 system), it is necessary to take into account this discontinuity. If an impurity is at the boundary, and there is a large potential barrier that excludes the envelope wave function of the electron (or hole) and image effects and possible band bending are ignored, the energy levels are given by

$$E_n = -\frac{m^* q^4}{2\bar{\varepsilon}_s^2 \hbar^2 (n^2)}, \quad n = 2,3, \cdots\cdots \qquad (31)$$

where $\bar{\varepsilon}_s$ is the average dielectric constant of the two materials. (For the GaAs-$Ga_{1x}Al_xAs$ case discussed below, the dielectric constant are nearly identical, and this effect can be ignored to a good approximation). Note that this expression excludes s-states due to the boundary condition on the envelope wave-function; so the principle quantum number, n, begins with 2. Thus, the binding energy in this case is 1/4 the usual 3D binding energy.

B. Impurities in Si MOS and Semiconductor Quantum Well Structures: Theory

The simplest approach to treating the case of impurities in some sort of external potential (that produced by an applied voltage in the MOS case, and that due to the heterjunction discontinuity in the case of quantum wells formed by layers of different semiconductors) is to make use of the one band effective mass approximation for a single isolated impurity. The validity of this approximation must be considered on a case by case basis. Lee and Lin have treated Wannier excitons in a quantum well making use of a perturbation approach[25].

The effective mass approach has been used for Na+ "impurities" in {100} Si MOS structures[26] and for donor impurities in GaAs quantum wells [27,28] **(see figure 16).** A general effective mass Hamiltonian for a single electron can be written

$$H = -\frac{\hbar^2}{2m}\left(\frac{\partial^2}{\partial x^2} + \frac{\partial^2}{\partial y^2}\right) - \frac{\hbar^2}{2m_z}\frac{\partial^2}{\partial z^2} + V(\underset{\sim}{r}) + V_I(z) + V_{ext}(z) , \qquad (32)$$

where m_z is an appropriate effective mass for motion parallel to the interface, V(r) is the impurity potential (which, in principle, should take into account the interaction of the electron with the impurity ion, with, the image of the ion as discussed in conjunction with equation (31) and also screening of the impurity ion potential by any mobile electrons.); $V_I(z)$ is the image term for the electron (see equation (1)), and $V_{ext}(z)$ is the appropriate external potential.

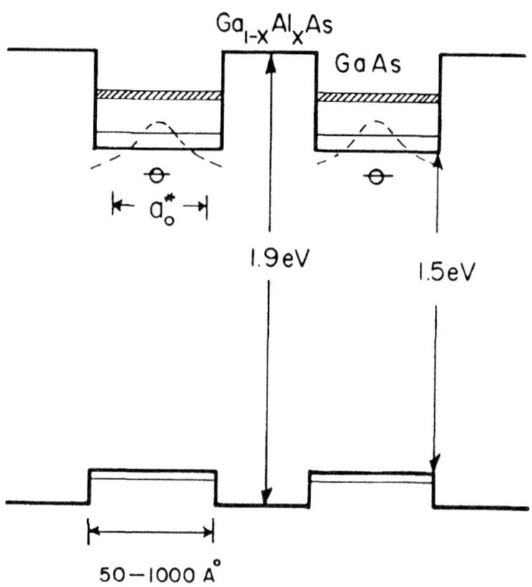

Figure 16. Schematic diagram of a GaAs-$Ga_{1-x}Al_xAs$ quantum well structure with donor impuities in the centers of the GaAs wells. The dashed line indicates the extent of the impurity ground state wave-function.

For the case of Na^+ "impurities" in the oxide at the interface in Si MOS structures (neglecting screening)

$$V(\mathbf{r}) = \frac{-2q^2}{\bar{\varepsilon}(x^2+y^2+z^2)^{1/2}} \quad , \tag{33}$$

$$V_I(z) = \left(\frac{\varepsilon_s-\varepsilon_{ox}}{\varepsilon_s+\varepsilon_{ox}}\right).q^2/4\varepsilon_s z \quad , \tag{34}$$

and

$$\begin{aligned} V_{ext}(z) &= qFz \quad , \ z>0 \\ &= \infty \quad , \ z \leq 0 \quad , \end{aligned} \tag{35}$$

in the triangular potential approximation. Here, z is measured from the SiO_2-Si interface.

For the case of GaAs-$Ga_{1-x}Al_xAs$ heterojunction quantum wells, the image term can be neglected to a good approximation over the range of alloy compositions of interest and

$$H = -\frac{\hbar^2}{2m_1}\left(\frac{\partial^2}{\partial x^2} + \frac{\partial^2}{\partial y^2} - \frac{\partial^2}{\partial z^2}\right) - \frac{q^2}{\varepsilon_{s1}(x^2+y^2+(z-z_i)^2)^{1/2}} \ , \ |z|<a \tag{36}$$

and

$$H = -\frac{\hbar^2}{2m_2}\left(\frac{\partial^2}{\partial x^2} + \frac{\partial^2}{\partial y^2} + \frac{\partial^2}{\partial z^2}\right) - \frac{q^2}{\varepsilon_{s2}(x^2+y^2+)z-z_i)^2)^{1/2}} + V_o \ , \ |z|>a \tag{37}$$

where z_i is the position of the impurity ion, z is measured from the center of the GaAs well of width 2a, and m_1 is the electron effective mass in GaAs, and m_2 is the electron effective mass in $Ga_{1-x}Al_xAs$. A simpler Hamiltonian is obtained for an infinite barrier, $V_o = \infty$ for Z >a. These Hamiltonia are not separable in general, and are not susceptible to analytic solution. Existing theoretical work has made use of variational approaches with various trial wave functions [26,27,28]. Some results of Martin and Wallis[26] and Lipari[29] for the case of Na^+ in Si MOS are shown in **figure 17**, taken from reference 30. These calculations differ in complexity of the trial wave functions. Screening is ignored in both calculations. Vinter[31] has taken screening into account in the metallic regime and finds very small binding energies in this regime. The experimental points are inferred from transport measurements[30]. The binding energy of a donor impurity located in the center of a GaAs quantum well is shown in **figure 18** as a function of well thickness for several different alloy compositions (barrier heights). Rather large shifts in binding energy are obtained. There are also substantial shifts in binding energy as a function of position of the impurity ion in the well (compare equations (28) and (31)), and the optical properties of a uniformly doped layer are expected to exhibit a double peaked structure[27].

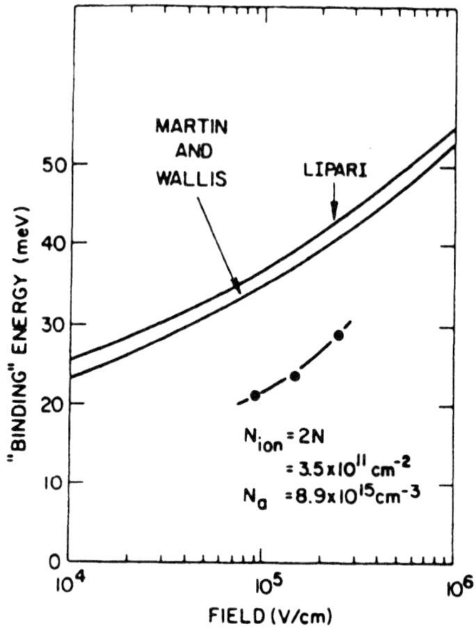

Figure 17. Binding energies of Na^+ "impurities" in Si MOS structures from two calculations, Martin and Walis, and Lipari. The experimental points are from low temperature thermally activated transport measurements. (After A.B. Fowler and A. Hartstein, Phil. Mag., B 42, 949 (1980)).

C. Impurities in Si MOS and Semiconductor Quantum Well Structures: Experiment

Effects of mobile sodium (Na^+) ions in the oxide of Si MOS devices have been investigated extensively for a number of years due to their importance in limiting device performance. More recently, Na^+ ions have been used as a tool to probe the fundamental physics of confinement of impurities. Both thermally activated transport[30] in the low inversion layer density regime and Shubnikov-deHaas oscillations in the region of metallic densities[32] have been studied on {100} Si as a function of positive oxide charge density (N_{ox}) varied by drifting Na^+ ions to the interfacial region. The low temperature activated conductivity results have yielded evidence of a rather narrow band of bound "impurity" states about 20meV below the bottom of the lowest electric field subband edge. These results are in reasonable agreement with the calculations described above[26,29].

These Na^+ "impurities" should have an effect on the optical spectra of the inversion layer, e.g., resonances corresponding to ground to excited state, or ground state to continuum of inversion layer electrons bound to these positive interfacial charges could be observable.[26,33]. Such studies can provide additional, independent information about the physics of this system.

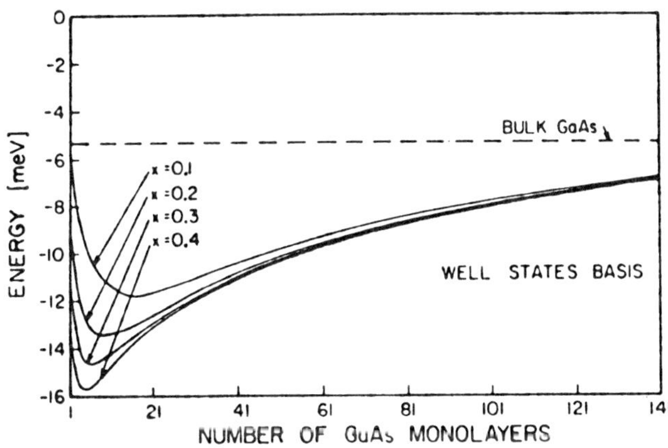

Figure 18. Binding energy of a donor impurity located at the center of the well vs. number of monolayers (a monolayer is 2.83 A°) for a GaAs quantum well in $Ga_{1-x}Al_xAs$. The binding energy is shown for several different alloy compostions (well heights). After Mailhiot et al., Surface Science 113, 161 (1982).

The initial optical studies of this system were carried out by McCombe and Schafer[34]. Some far IR optical absorption spectra on a Si MOS devide intentionally contaminated with Na^+ ions with the IR electric field polarized perpendicular to the surface are shown in **figure 19.** The data were taken with the spectrometer of figure 6. These data (at an intermediate value of N_{ox}, a measure of the density of positive ions near the interface) exhibit 0 - 1 and

0 – 2 "continuum" transitions as well as an additional intermediate energy transition indicated by the arrows. The identification of the "continuum" lines is based on the measured depletion charge and comparisons with other uncontaminated samples as well as comparisons with theory.

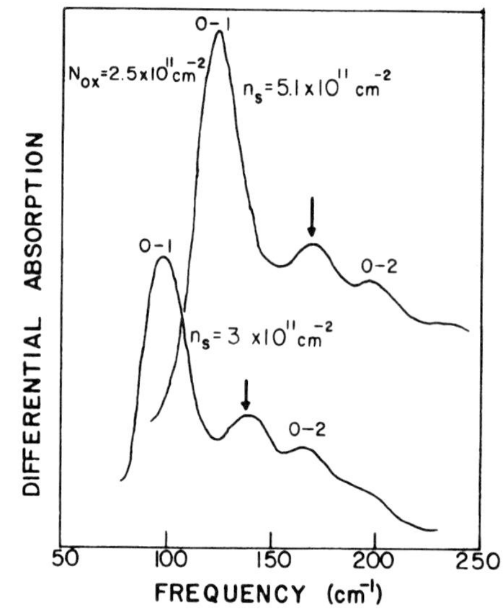

Figure 19. Differential absorption spectra for a Si (100) MOSFET sample intentionally contaminated with Na^+ ions in the oxide. The arrows indicate an additional transition that appears between 0-1 and 0-2 upon increasing the oxide charge.

By drifting mobile Na^+ ions up to the interface under bias stress net positive oxide charge densities between $9\times10^{10}cm^{-2}$ and $1.6\times10^{12}cm^{-2}$ were obtained on single devices. These values correspond to peak effective mobilities between 5050 cm^2/V-sec and 870 cm^2/V-sec at 4.2K. Values of N_{ox} were determined from device capacitance (room temperature) and device threshold (near liquid nitrogen temperature) measurements.

The strong line in figure 19, which is identified as the $0\rightarrow1$ continuum subband transition at the lowest oxide charge densities, shifts to higher frequencies and broadens substantially as the oxide charge is increased. At higher values of n_s and relatively low oxide charge, the intermediate energy line disappears and only the 0-2 transition is observed at higher energies.

A compilation of these and other data are shown in **figure 20** along with the theoretical predictions of Ando for the designated continuum subband transition. As can be seen, for the two lowest values of N_{ox} the lowest energy (strong) line is in reasonable agreement with Ando's calculations for 0-1 (considering the difference in N_{depl}). The intermediate energy weak transition discussed above occurs between 0-1 and 0-2 for N_{ox}=1.6-3x10^{11}cm^{-2} and n_s <8x10^{11}cm^{-2} this line disappears, and only a single line (0-2) is observed above n_s = 1x10^{12}cm^{-2}. For N_{ox}=7.4x10^{11}cm^{-2}, the lowest energy line is a broad (FWHM = 5.5meV) line occurring at energies near the intermediate line. As n_s increases, this line moves up in frequency and approaches the continuum 0-1 transition (observed at low valued of N_{ox}) at large values of n_s.

The additional line observed for N_{ox} between 1.6 and 3x10^{11-2} and the broad lines at higher values of N_{ox} occurring at energies above 0→1 are attributed to transitions from a band of bound states associated with the i=o subband to bound states associated with i=1.

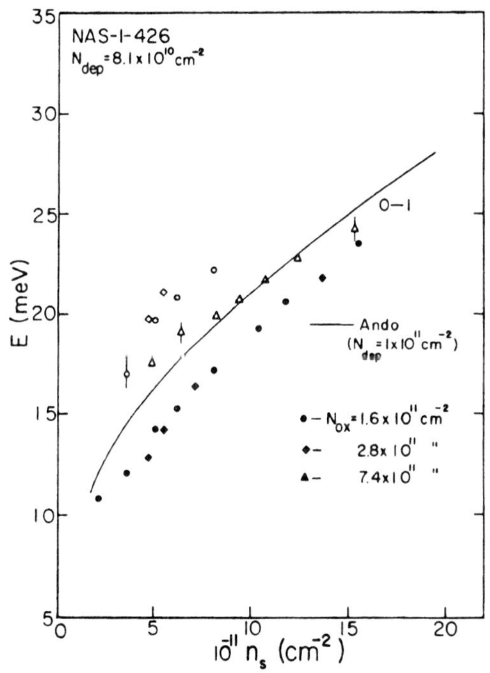

Figure 20. Summary of data for the sample of figure 19 for three values of oxide charge. At a value of N_{ox}=7.4x10^{11}cm^{-3} all of the oscillator strength has shifted to an "impurity" transition.

At sufficiently low N_{ox} the impurity bound states are relatively few and are screened out at moderate values of n_s ($>8\times10^{11}cm^{-2}$). For $n_s>>N_{ox}$ the dominant transitions are the continuum subband transitions broadened by the screened Coulomb scattering centers due to N_{ox}. At larger values of N_{ox} large numbers of inversion layer electrons can be accommodated in localized states, and screening is less important over the range of n_s studied. The observed approach of this transition to the 0→1 continuum transition with increasing N_s (for $N_{ox}=7.4\times10^{11}cm^{-2}$) is in good qualitative agreement with the results of Kramer and Wallis[26]. A comparison of the experimental results with these calculations is shown in **figure 21.** Plotted here are the ratios of the 0-1 "impurity" bound state transition energies to the 0-1 continuum transition energies as a function of electron density. The experimental points are determined directly from the observed resonances for $N_{ox}=7.4\times10^{11}cm^{-2}$ (impurity-shifted) and for N_{ox} $<1\times10^{11}cm^{-2}$ (continuum). The theoretical curve is obtained from - the results of Kramer and Wallis by calculating the electric field corresponding to a given n_s from equation (9) and determining the impurity shifted and continuum subband transition energies at these electric fields form reference 26. The agreement in magnitude and trend is qualitatively satisfactory. The experimental magnitudes are somewhat smaller as would be expected if screening were playing a role, and the smaller measured transition energies are also consistent with the discrepancy between the theoretical

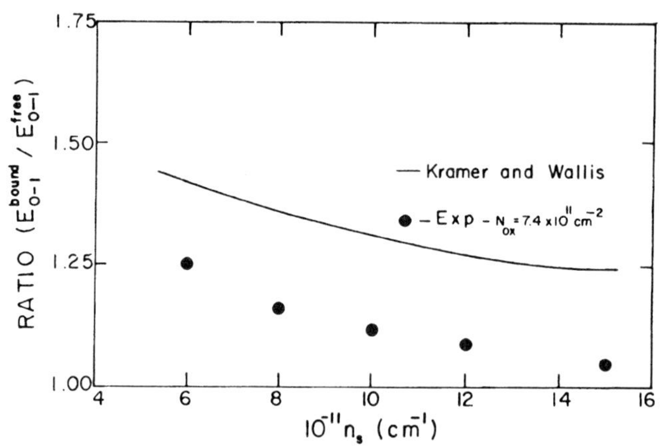

Figure 21. Comparison of the 0 - 1 impurity shifted intersubband transition energy (as a ratio to the free carrier intersubband transition) with the theoretical results of Kramer and Wallis.

binding energies and the transport results shown in figure 17. In addition, these results are also consistent with theoretical calculations of the optical absorption due to transitions between bound states associated with the i=0 and i=1 subbands[33].

Experimental results for the case of impurities in quantum wells are sketchy indeed. The only published studies at present are the work of Miller et al.,[37] and Lambert et al.,[38]. Both of these investigations have utilized photoluminescence as the experimental probe of the impurities, and both are concerned with GaAs - $Ga_{1-x}Al_xAs$ quantum wells. Miller et al have identified bound acceptor recombination lines, and find reasonable agreement with the simple model of reference 27. The experimental lines fall between the theoretical curves for impurities localized at the edge of a well and at the center of a well as a function of well width. Lambert et al., claim to have also identified luminescent recombination through bound donor states in quantum wells. However, the published evidence is somewhat less convincing than the acceptor data (of both reference 37 and reference 38).

The fractional effects on the binding energy should be much larger for donors than acceptors for easily achievable quantum well widths, and a concerted effort to investigate the donors in quantum wells is underway at the Naval Research Laboratory an SUNY at Buffalo. Structures are being grown at NRL and at the National Research and Resource Facility for Submicron Structures at Cornell University. Photoluminescence, far IR optical absorption and magnetoabsorption, and transport measurements are being carried out on these structures. Preliminary results are promising[39].

V. SUMMARY

These lecture notes have provided a brief review of some aspects of quantum effects in structures confined in one dimension on a submicron scale. Two major topics were considered. The current status of our understanding of the quantized subbands in MOS devices fabricated on the three principal surfaces of Si was described, concentrating on optical measurements of the subband energies. In spite of substantial activity, there remain gaps in our understanding, particularly concerning the Si {100} surface. An outline of the theory of confinement effects on the boundstates of impurities in Si MOS and $GaAs-Ga_{1x}Al_xAs$ quantum well structures was presented, and some experimental results were compared with the simple theoretical models. Much of this work is in the early stages, and there should be considerable effort in this area over the next few years and corresponding improvements in our understanding of these important effects.

Note Added June, 1986

Since these lectures were prepared, considerable progress has been made in experimental studies of shallow donor impurities in quantum well structures; See e.g. N.C. Jarosik, et al., Phys. Rev. Letters 54, 1283 (1985); and B.D. McCombe et al., in Proceedings of the IVth Int'l. Winterschool on New Developments in Solid State Physics, Two Dimensional Systems: Physics and Devices, Ed. by G. Bauer, H. Heinrich and F. Kuchar (Springer Verlag, New York, 1986).

REFERENCES

1. R.G. Wheeler, K.K. Chou, A. Goel, R. Wisnieff and D.E. prober, Phys. Rev. Letters 49, 1674 (1982).
2. W.J. Skocpol, L.D. Jackel, and R.E. Howard, E.L. Hu, and L.A. Fetter, Physica 117B and 118B, 667 (1983).
3. Tsuneya Anso, Alan B. Fowler, and Frank Stern, Rev. Mod. Physics 54, 437 (1982).
4. F. Schaffler and F. Koch, Solid State Commun. 37, 365 (1981).
5. See e.g., J.F. Koch, Surface Science 58, 104 (1976).
6. B.D. McCombe, R. Holm and D.E. Schafer, Solid State Commun. 32, 603 (1979).
7. See e.g., C.C. Hu, J. pearse, K.M. Cham, and R.G. Wheeler, Surface Science 73, 207 (1978); and references therein.
8. E. Gornik and D.C. Tsui, Surface Science 73, 217 (1978).
9. See e.g., A. Pinczuk and J.M. Worlock, Surface Science 113, 69 (1982).
10. F. Stern and W.E. Howard, Phys. Rev. 163, 816 (1967).
11. T. Ando, Phys. Rev. B13, 3468 (1976).
12. See e.g., S.J. Allen, D.C. Tsui, and B. Vinter, Solid State Commun. 20, 425 (1976).
13. T. Ando, Solid State Commun. 21, 133 (1977).
14. T. Ando, Z.Phys. B 26, 263 (1977).
15. K.S. Yi and J.J. Quinn, Phys. Rev. B 27, 2396 (1983). (See references to earlier work in this paper.)
16. A.A. Lakhani and P.J. Stiles, Phys. Lett. 51A, 117 (1975); T. Neugebauer, K. Vonklitzing, G. Landwehr, and G. Dorda, Solid State Commun. 17, 295 (1975).
17. D.C. Tsui and G. Kaminsky, Phys. Rev. Letters 42, 595 (1979).
18. K.C. Woo and P.J. Stiles, Submitted to Phys. Rev. B.
19. B. Vinter and A. Overhauser, Phys. Rev. Letters 44, 47 (1980).
20. K.C. Woo and P.J. Stiles, Surface Science 113, 278 (1982); T. Englert, Proceedings of the International Conference on the Application of High Magnetic Fields in Semiconductor Physics, Japan, 1980.
21. M.Kelly and L. Falicov, Phys. Rev. B 15, 1974 (1977) ibid p. 1983.

22. B.D. McCombe and T. Cole, Surface Sci. 98, 469 (1980); T. Cole and B.D. McCombe, Proc. of the 15th International Conference on the Physics of Semiconductors, Kyoto, J. Phys. Soc. Japan 49, Suppl. A., 959 (1980); T. Cole, Surface Science 113, 41 (1982).
23. S.M. Mie, U. Claessen, and F. Koch, submitted to Phys Rev B.
24. S. Fluggeand H. Marschall, Rechnemethoden der Quantentheorie, 2nd ed. (Springer, Berlin) 1952.
25. This general problem was first considered by J.D. Levine, Phys. Rev. 140, A568 (1965).
26. Y.C. Lee and D.L. Lin, Phys. Rev. B 19, 1982 (1979).
27. See e.g., B.G. Martin and R.F. Wallis, Phys. Rev. B 18, 5644 (1978); G.M. Kramer and R.F. Wallis, Surface Science 113, 148 (1982); ans references therein.
28. See e.g., G. Bastard, Phys. Rev. B 24, 4715 (1981); G. Bastard, E.E. Mendez, L.L. Chang, an L. Esaki, Solid State Commun. 45, 367 (1983).
29. C. Mailhiot, Y.C. Chang, and T.C. McGill, Phys. Rev. B 26, 4449 (1982).
30. N.O. Lipari, J. Vac. Sci. Technol. 15, 1412 (1978).
31. A.B. Fowler and A. Hartstein, Phil. Mag. B 42, 949 (1980).
32. B. Vinter, Solid State Commun. 28, 861 (1978); Surface Science 98, 197 (1980).
33. See e.g., F.F. Fang, A.B. Fowler, and A. Hartstein, Surface Science 73, 269 (1978).
34. B.Vinter, Surface Science 113, 140 (1982).
35. B.D. McCombe and D.E. Schafer, Inst. of Phys. Conf. Ser. 43, 1227 (1978).
36. H.R. Chang and F. Koch, Surface Science 113, 144 (1982).
37. R.C. Miller, A.C. Gossard, W.T. Tsang, and O. Munteaunu. Phys. Rev. B 25, 3871 (1982); R.C. Miller, A.C. Gossard, and W.T. Tsang, Physica 117B and 118B, 717 (1983).
38. B. Lambert, B. Devaud, A. Regreny, and G. Talaleff, Physica 117 B and 118 B, 714 (1983).
39. B.V. Shanabrook and J. Comas, Submitted to EP2DS V.

PHYSICS OF HETEROSTRUCTURES

AND HETEROSTRUCTURE DEVICES

Peter J. Price

IBM T.J. Watson Research Center
Yorktown Heights
New York, NY 10598, USA

1. INTRODUCTION

This topic belongs to the broader one of scale-dependent properties of inhomogeneous semiconductors. Most semiconductor devices depend on characteristic lengths, such as screening constant, carrier recombination length, inverse of pair creation rate, and their functioning is based on spatial inhomogeneities. The bipolar junction transistor is made by inhomogeneous doping of a semiconductor; and the field-effect transistor has a conducting channel controlled by the "gate" field at the interface of the semiconductor with an insulator, commonly a silicon/(silicon oxide) interface. In these -- as in the semiconductor laser -- a controlled inhomogeneity supports semiconductor phenomena that depend on this spatial structure.

Those devices are no longer fresh innovations. The recent development which provides our subject is a proliferating family of "custom tailored", increasingly complex, inhomogeneous semiconductor structures. The instigating technique for these is *Molecular Beam Epitaxy (MBE)*, though it is not the only process in use. In the late fifties, vapor-phase epitaxy was used to grow germanium on a gallium arsenide substrate. The Ge and GaAs crystals have almost the same lattice constant. Consequently, it was possible to grow single-crystal germanium on a GaAs crystal, having a lattice which continued the substrate crystal lattice, at an

*Note: This article was written in Spring 1983. I have made no changes in proof, except for substantially replacing the original paragraph containing (17) and (18) by a new version.

almost abrupt interface with little disorder -- that is, *epitaxial* growth. For the conduction electrons, the two crystal halves acted as a single semiconductor, with an internal boundary at which the principal discontinuity was the "step" in band-edge energy. A new kind of electronic junction resulted (Anderson 1962), with the principal inhomogeneity in the semiconductor itself rather than just in the superposed doping -- a *heterojunction*.

The possibility of progressing to *heterostructures* in general was provided by MBE. In suitable conditions, molecular beams impinging on a solid surface will grow a single-crystal solid from the beam constituents. The intensity of a beam can be varied by control of the source (thermal effusion cell). Separate beams of different components, including dopants, can be directed at the same surface. Their intensities can be separately varied (in particular, with shutters) to change the composition quickly during growth. With growth rates on the order of one Ångstrom per second, the composition of successive layers can be varied on scales down to the atomic scale. The chemistry of the growth process is intricate, and at present MBE is a difficult empirical art. A number of review articles are available (for example, Cho and Arthur 1975). The first heterostructures made by MBE had alternating layers (thickness usually in the range 5-50 lattice constants) of GaAs and of the $Ga_{1-X}Al_XAs$ alloy formed by substituting Al atoms for a certain fraction, X, of Ga atoms. It has been found that semiconductor alloys have conduction and valence bands with substantially the properties of a true periodic crystal lattice in the idealized Brillouin-Bloch theory; only some scattering due to the atomic disorder is superposed on an effective band structure which is a function of composition. In the present case, the conduction-band-edge energy is almost a linear function,

$$E_C(X) = E_C(0) + E_A X \tag{1}$$

where $E_A \simeq 1$ eV, for $X<0.4$. (Above about 40% Al, the Γ point energy minimum lies above the X and L minima, and no longer is the E_C value.)

For the conduction electrons, the energy difference $\Delta E_C = E_C(X) - E_C(0)$ acts similarly to a potential step for free electrons. Hence a suitably narrow layer of GaAs embedded in (Ga,Al)As alloy functions as a one-dimensional "potential well", and quantizes the motion of the electron in the direction perpendicular to the layer planes (the z direction). The (x,y) motion normally remains itinerant, so that the total energy for an electron trapped in the "well" is accordingly

$$E = E_C(0) + E_n + (\hbar^2/2m^*)(k_x^2 + k_y^2) \tag{2}$$

where m* is the GaAs effective mass. The quantum energy levels E_n can be obtained from a Schrödinger equation applied to the envelope function, the GaAs layer being an "allowed" region and the adjoining alloy "forbidden"; we discuss this later. The electrons occupying the states represented by (2) are itinerant in the (x,y) plane but are bound in the normal direction by the quantization. They form a "two-dimensional gas" in the GaAs layer. (There is a similar effect for holes, with a shallower "well" in the GaAs.) Motion in the z direction, in contrast, has the character of polarization, as in a giant atom. The appropriate order of magnitude of the well widths is indicated by the result that (for the electron m* in GaAs) a wavevector π/W gives an energy $(\hbar^2/2m^*)(\pi/W)^2$ of 0.1eV when $W \simeq 7.5$ nm (i.e. 75Å).

Figure 1a shows the typical band/potential diagram. The curvature of the band-edge lines arises from the fields due to the space charge of electrons and ionized impurities. In Figure 1b, showing the case of a single interface, this curvature makes possible an accumulation layer which similarly acts as a quantizing potential well, and also traps a two-dimensionally itinerant system of heterolayer electrons. Thus this case resembles an inversion layer at a silicon/(silicon oxide) interface. The GaAs is lightly doped p type; and the alloy is doped n type. Insofar as the variation of the field ξ due to space charge may be disregarded, there is a characteristic relationship in this case between ξ, the quantization energy E_n, and the moment $< z > = \int z\rho dz$ where $\rho(z)$ is the linear electron density (square of wavefunction) and the interface is at $z = 0$. By combining the virial theorem and perturbation theory, one finds that E_n is equal to (3/2) $e\xi < z >$ and is proportional to $\xi^{2/3}$ and to $< z >^{-2}$. We may infer that E_n is equal to a dimensionless constant times $\hbar^2/m^* < z >^2$, and hence

$$E_n = c_n(\hbar^2/2m^*)^{1/3}(e\xi)^{2/3} \tag{3}$$

where the c_n are numerical constants. If the wavefunction vanishes at the interface, then we have $c_1 = 2.34$ for the ground level. For $E_1 = 0.1$ eV, in GaAs the uniform field would be 1.2×10^5V/cm, which corresponds to a sheet charge density of 8.1×10^{11} electrons/cm^2. These are representative numbers, although the effect of electron space charge on E_1 can be appreciable (Ando 1982a).

These heterostructures are obtained by control of the semiconductor composition during MBE growth. The doping, which is also provided by one or more molecular beams during growth, can be varied in the same way. The resulting inhomogeneity is called *Modulation Doping* (MD). An important use of MD has been to

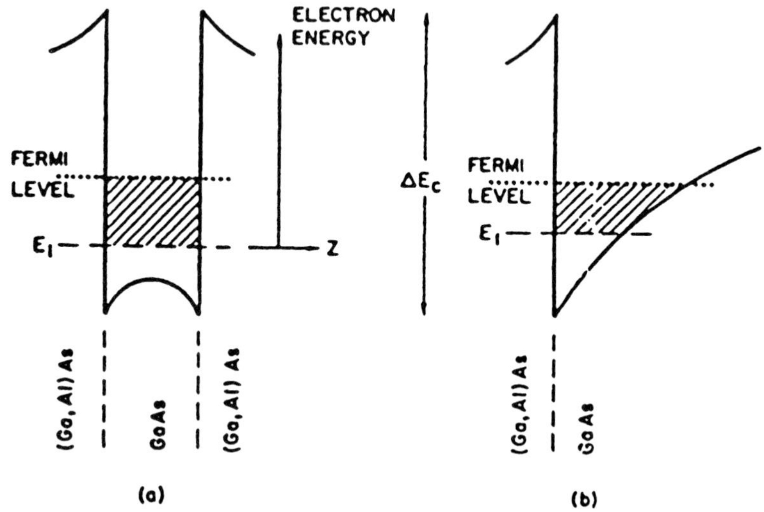

Figure 1a. and 1b. Heterolayer potential diagrams.

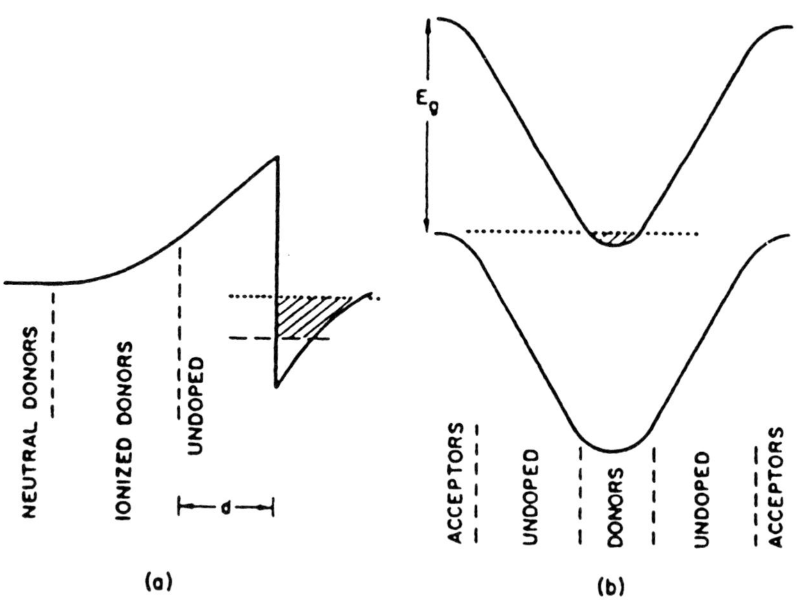

Figure 2. Modulation Doping applied to (a) a heterolayer; (b) a "nipi" structure.

separate the heterolayer electrons from the donor ions, by leaving a layer of undoped (Ga,Al)As adjoining the GaAs and then continuing MBE growth with alloy which is doped *n* type, so that the latter is separated from the GaAs by an almost pure "spacer". The energy diagram for this structure is shown in Figure 2a. Very high mobilities may be obtained at low temperatures (Drummond et al. 1982, Hyamizu and Mimura 1982). Lattice-scattering electron mobilities in the heterolayer plane, for GaAs, are not greatly different from those of the homogeneous semiconductor, but without MD the actual mobility would fall off sharply with decreasing temperature because of ionic scattering. With the MD isolation of the donors, the heterolayer mobility instead levels off at low temperatures. (It can reach a few times $10^6 cm^2/V.sec$ -- still not up to the low temperature lattice-scattering values, but not necessarily limited by *ionic* elastic scattering alone.)

Evidently, *p-n* junctions can be made by MD. In the "nipi" structures (Künzel et al. 1982), *n* and *p* layers alternate, but are separated by high purity (near-intrinsic) layers. With a homogeneous semiconductor, a potential diagram as in Figure 2b can be achieved at suitably high doping densities and doping-layer widths, giving alternating electron and hole heterolayers. This structure is remarkable for the control of the carrier concentrations and energy differences that can be obtained. Ohmic contacts, placed on the ends of the grown crystal, connect electrically only to the *n* layers or only to the *p* layers, and hence a potential difference can be applied to change the relative energy difference of the heterolayers, by having ohmic contacts of differing polarities at the two ends. Recently it has been demonstrated (Künzel et al. 1983) that the "nipi" scheme can be applied to a superlattice of alternating GaAs and (Ga,Al)As layers; the doping is in the alloy layers, and the electron and hole two-dimensional gases are quantized in alternate GaAs layers. For the electrons, fairly high low-temperature mobilities are obtained by the modulation-doping effect.

Variation of semiconductor composition can be used, as one would expect, to intentionally grow graded rather than discontinuous inhomogeneities (Allyn et al. 1980). A gradient of band edge energy implies a force $-dE_C/dz$ on the electrons, equivalent to an electric field of strength $(E_A/e)\ dX/dz$, according to (1); evidently this pseudofield strength can exceed 10^5 V/cm. One may thus get effects of a field without needing space charge to sustain it. More generally, one has variation of the whole energy function $E(\mathbf{k})$ with composition, and expects a quasi-classical force $-(\partial E(\mathbf{k},X)/\partial X)\nabla X$ on the carriers. (The boundary condition for the envelope function at a composition discontinuity similarly involves $\Delta E(\mathbf{k})$.) For $Ga_{1-X}Al_XAs$ alloy,

$$m^*_C(X)/m^*_C(0) \simeq 1 + 1.2\ X \tag{4}$$

The dependence of m^* on position, through $m^*(X)$, can lead to nonlinear response to high frequency fields (Grinberg and Kastal'skii 1972). It is interesting also to consider possible effects of graded doping as an inhomogeneity of scattering rates. Besides providing the forbidden range surrounding a "well", MBE-grown alloy layers can function as quantum tunneling barriers between regions of lower band edge energy (Chang et al. 1974).

In the foregoing aspects, as also with superlattice structures (Esaki and Chang 1976), one is concerned with carrier motion in the z direction normal to the planes of constant composition (the layer planes). Then, as remarked above, the length scale is a critical consideration. When the scale is short enough (in the range below 100 lattice constants, say) one has quantum levels and tunneling transitions. One may then have a "quantum transistor", a unipolar device in which electrons can make a direct transition from a source region to a drain region, or be stopped in an intermediate base region, according to the potentials applied to the intervening barriers. Again, the barriers may be realized by layers of semiconductor with higher band edge energy (Heiblum 1981) or generated by regions of high doping (Shannon and Gill 1981). At the other extreme of scale, one has macroscopic concentration and transport properties subject to gradients of semiconductor parameters. An intermediate scale would be large compared to de Broglie wavelengths, so that quasi-classical particle motion in the z direction occurs, but not large compared to the scattering mean free paths. One may then realize "ballistic" phenomena (Adachi et al. 1982), in which carriers traverse a distinct semiconductor region with a distribution that does not come to equilibrium with the local semiconductor substrate. (Possible values -- for an electron with energy 2 or 3 times 0.1 eV above the band edge, in GaAs -- would be a de Broglie wavelength of 10^{-6}cm and a mean free path of 10^{-5}cm.)

In the foregoing, we have always referred specifically to Ga-Al-As. That is the most developed semiconductor system for heterostructures, but by no means the only one. Other III-V systems, ternary and quaternary, are being investigated. This is in part to obtain a desired range of parameter values; for example (Ga,In)As, which can be grown epitaxially with (Al,In)As and with InP, has a smaller energy gap than GaAs. With InAs/GaSb heterostructures, however, there is a new situation: The conduction band edge of InAs lies lower in energy than the valence band edge of GaSb, when they are brought together at an epitaxial interface. Consequently, as shown in Figure 3, one can have a heterostructure that is a (two-dimensional) semimetal, with the Fermi level lying between the two band edges (Chang and Esaki 1982). Systems such as Ge-Si and Pb-Te-Sn are also being investigated. For the rest of this article, numerical values which pertain to a particular semiconductor, not specified, refer to GaAs.

More elaborate planar-symmetry heterostructures, and heterostructures of lower spatial symmetry, are at present being investigated, while the electronic and optical properties of the simpler systems continue to be studied. The solid-state physicist is concerned with these in three aspects:

1. Properties such as the mobility in a heterolayer, which are modified and differ in detail from the case of bulk, homogenous semiconductor but where the essential phenomenon is common to both.

2. Phenomena that are peculiar to the heterostructures. For example superlattice band structure, and effects of transfer of momentum, energy, or the carriers themselves between different heterolayers.

3. Questions of the applicability of band particle and envelope-wavefunction concepts to the electrons and holes in the heterostructures, which incorporate atomic disorder and which are characterized by spatial scales down to atomic scale.

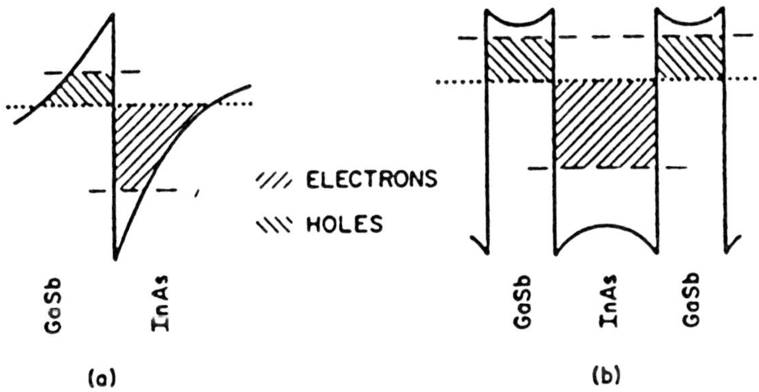

Figure 3. Heterostructures of the semimetal type.

Some topics in each of the forgoing categories will be discussed. In a little more than a decade this has become an extensive subject. Its rapid growth can be observed in every issue of Applied Physics Letters, and elsewhere. The present discussion does not cover all aspects. Nothing is said about self-energy effects (Ando 1982a), for example, or the collective excitations of multilayer systems (Das Sarma and Quinn 1982), and almost nothing about spectroscopy and lasers. An extensive and detailed account of the physics of inversion and accumulation layers, including a bibliography of the subject through 1981 will be found in the review article by Ando, Fowler and Stern (1982).

2. HETEROLAYER CONDUCTION

Electrical conduction parallel to the heterostructure plane, by itinerant carriers occupying a single heterolayer, is of interest for both pure and applied physics. One can have drift velocities of 10^7cm/sec or more and sheet carrier densities $\lesssim 10^{12}$cm^{-2}, and hence currents ~1amp/cm. In the lattice-scattering range, the two-dimensional ohmic mobility $\mu(T)$ normally for GaAs does not differ much from the bulk values. The phenomenon is distinct, however, in several respects:

1. There can be several energy levels in the normal-direction quantization, each the bottom of a two-dimensional sub-band. Conduction can then be a many-band process, with both intraband and interband scattering modes. The density of states is a constant -- in the parabolic energy range -- for each sub-band, so the total density of states becomes a step function of the carrier energy. (At an energy belonging to many sub-bands, it would approach the three-dimensional density of states for the same energy.)

2. The scattering rates are a function of the layer geometry. They depend on bQ where $\mathbf{Q} = \mathbf{k}_2 - \mathbf{k}_1$ is the two-dimensional wavevector change in scattering and b is a measure of the layer width. (And b/w, where w is the distance between classical turning points, depends on the actual shape of the density curve $\rho(z)$, as in eq. (17).)

3. In the neighborhood of the interface the lattice modes will be modified, and there can be interaction with surface phonons ("surfons") split off from the bulk modes (Ezawa et al. 1975). This could be important in cases like silicon/(silicon oxide) where the two solids are lattice-mechanically, as well as electronically, disparate.

4. Screening of the scattering forces, by the carrier system, is quite different. In three dimensions, the scattering matrix elements are reduced by a factor $Q^2/(Q^2 + P_{III})$ where P_{III} is a constant proportional to d(carrier density)/d(Fermi energy) for small Q, but for larger Q values can vary in the manner of the Lindhard function and its generalizations. In the two-dimensional case the screening factor is of form (for a single occupied sub-band)

$$S = Q/(Q + P_{II}\, H(Q)) \tag{5}$$

where P_{II} (the subscript is dropped, below) is the two-dimensional equivalent of P_{III} above, with dimensions [L^{-1}] instead of [L^{-2}], and the additional factor $H(Q)$ represents the electrostatic action of the screening charge on itself in the remaining direction (z). For small Q we can set H equal to $H(0) = 1$, but $H(Q)$ falls off for $1/Q \lesssim$ layer width.

We have, in practice, several constants which are ~1/(100Å) in GaAs. Because the areal density of states is a constant ($m^*/\pi\hbar^2$, equal to $2.8\times10^{13}\text{cm}^{-2}$ per eV) we normally have for P a fixed value, $P_0 = (e^2/\epsilon)(2m^*/\hbar^2)$, equal to $2.0\times10^6\text{cm}^{-1}$ (with Q less than the Fermi wavevector (Stern 1967), frequencies less than scattering rates, temperatures less than degeneracy temperature). The Fermi wavevector is $k_F = (2\pi n)^{1/2}$ where n is the "sheet density" of carriers; so $k_F = 2\times10^6\text{cm}^{-1}$ for $n = 6.4\times10^{11}\text{cm}^{-2}$, a representative value for actual heterolayers in GaAs. With $P \sim k_F$, the screening effect in the scattering will range, with scattering angle, between strong and weak. The Q values at which $H(Q)$ falls off are typically of the same order of magnitude as P_0 and k_F. Finally, the Q values in optical-mode scattering depend on $k_0 = (2m^*\omega_0/\hbar)^{1/2}$, equal to $2.5\times10^6\text{cm}^{-1}$. (For given initial k_1 the geometric mean of maximum and minimum Q values, $k_1 \pm k_2$, is k_0.)

The mobility theory (Mori and Ando 1980, Price 1982a, Ando 1982b) is complicated in detail; and the experimental data now covers many cases and a range of values. Of special interest is the low temperature conduction, ohmic and hot-electron. The degeneracy temperature T_D is proportional to n; for $n = 6\times10^{11}\text{cm}^{-2}$ it is 250 K. For ohmic conduction, therefore, the electron system is degenerate. At very low temperatures the inelastic scattering is so weak, near thermal equilibrium, that hot electron effects appear at quite small fields (Price 1982b). Subsequently as the electron temperature rises (it is reasonable for these high effective electron densities, except at the lowest temperatures, to assume a displaced-Fermi distribution, with a definite temperature T_e), the dissipation is principally by optical-mode phonons until electrons begin to escape from the heterolayer. Thus the momentum and energy dissipation rates, λ_m and λ_{op}, are given by

$$\left.\begin{aligned} m^*u\,\lambda_m &= e\xi \\ \hbar\omega_0\,\lambda_{op} &= e\xi u \end{aligned}\right\} \qquad (6)$$

where u is drift velocity, ξ is field strength, $\hbar\omega_0$ is optical-mode quantum. The numerical values from some representative data at 4.2 K for $n = 10^{12}\text{cm}^{-2}$ (Inoue et al. 1983; similar data have been published by the Illinois and Bell Laboratories groups) are as follows:

Table 1

$\xi(10^2 V/cm)$	1.	2.	3.	4.	5.	6.	7.	8.	9.
$u(10^7 cm/sec)$	1.27	1.80	2.05	2.23	2.38	2.49	2.60	2.73	2.88
$\lambda_m(10^{11} sec^{-1})$	2.08	2.93	3.86	4.73	5.54	6.36	7.10	7.73	8.24
$\lambda_{op}(10^{11} sec^{-1})$	0.35	0.99	1.70	2.46	3.29	4.13	5.03	6.03	7.16

Thus the optical-mode scattering rate increases from a small fraction of the total to become almost equal to λ_m. When $\lambda_{op} = \lambda_m$, then u reaches the velocity $u_0 = (\hbar\omega_0/m^*)^{1/2}$, equal to 3.1×10^7 cm/sec. For isotropic scattering this would be a true saturation velocity. It is not, in the present case, because the *polar* optical-mode scattering is not in general isotropic. (A leveling off of $u(\xi)$ at u values near u_0 would suggest that most of the scattering was from initial energies not much above $\hbar\omega_0$. A further rise would then be expected at higher fields, when the prevailing k_2/k_1 values were no longer small.) The detailed theory for hot electrons is not discussed here, but it is appropriate to point out that

(a) It is quite wrong to apply the *three-dimensional* theory, as though the carriers could move quasi-classically in the normal-direction, either in calculating the hot distribution function or (Price 1983) calculating normal-direction "escape" and transfer effects.

(b) The assumption of a displaced-Maxwell (or displaced-Fermi) distribution, often uncritically used in analyses of the three-dimensional hot electron phenomenon, is liable to be adequate here.

There is naturally an interest in FETs based on these structures, with a gate electrode, on the (Ga,Al)As layer, covering the source-drain path between the heterolayer contact electrodes. If we assume the drift velocity u to be independent of gate potential V_G, then the transconductance g_m becomes $(u/L)C_G$ where C_G is the gate-channel capacity and L is the gate length. (It is usual to specify g_m/W, where W is the width of the gate.) A natural measure of the "switching time" is C_G/g_m, which with this formula for g_m becomes L/u -- the carrier *transit time* (Solomon 1982). How is this result modified by the field effect on mobility? In the L/u formula, u should be replaced by $u' = u + Q(du/dQ)$ where Q (not to be confused with wavevector change in scattering) is the carrier charge under the gate. With a single interface, the two terms of u' are of opposite sign because increased Q entails increased normal-direction field at the interface, ξ_n, and hence increased elastic scattering from the interface. It was suggested by Sakaki (1982) that with a heterolayer of the type of Figure 1(a), with two interfaces, a higher mobility could be associated with one side of the channel than the other, so that the field effect could be enhanced. He pointed out that if the space-charge "bowing" of the potential in the heterolayer is just sufficient to split the ground level slightly, then a small change in the normal field can appreciably shift the electron density from one side to the other. In fact it is known that in a structure of this type the first-grown interface, which here will be the one further from the gate, is associated with much lower mobilities. Therefore we could have a sharp increase of drift velocity as the channel wavefunction tips to the gate side, with $du/dQ \gg u/Q$ and hence $u' \gg u$.

The mobility field effect is an instance of a natural characteristic of the heterolayers -- anisotropic properties. A normal-direction field produces normal-direction space charge polarization; a driving field parallel to the layer plane produces current in that field direction; and there are the two cross effects, one being the mobility field effect. With a single-interface heterolayer having two (or more) quantum levels and sub-bands, or the "tipped" channel discussed above, hot-electron excitation of carriers from the ground sub-band to the upper sub-band will result in normal-direction polarization, since the moment $< z > = \int z\rho(z)dz$ will differ for the two levels. This is the second cross effect. The lowering of space symmetry can result also in a linear magneto-resistance for the Hall geometry (Ohno and Sakaki 1982). As another example, it has been suggested (Suris and Fedirko 1978) that a high frequency field in a planar direction could be detected by a resulting enhanced conductivity in the *normal* direction, due to the "spill over" of hot electrons at energies above the alloy band edge. (This would be subject to the frequency dependence of the planar conduction, as discussed below.) The energy levels of the normal-direction quantization, and the transition oscillator strengths, and hence the optical properties of heterolayers, are of course subject to a field effect. As is indicated by eq. (3), the field effects will be much larger than atomic Stark effects, by the scaling with space dimension. One would expect applications in opto-electronics. For example the frequency and/or intensity of lasers and similar light sources might be modulated (Yamanishi and Suemune 1983, Kroemer and Griffiths 1983).

As indicated in Table 1, with modulation doping one can have momentum and energy relaxation times as large as several picoseconds (and bigger values are possible). Also, in the saturation range, the carrier transit time between "source" and "drain" contacts spaced $\lesssim 1\mu m$ apart can be as small as a few picoseconds. In the ohmic range, we have for the mean free path

$$\ell = (\hbar/e)\ \mu k_F \qquad (7)$$

With $n = 6\text{x}10^{11}\text{cm}^{-2}$ and $\mu = 10^5\text{cm}^2/\text{V.sec}$, (7) gives $\ell = 1.28\mu\text{m}$. The Fermi-surface velocity at this sheet density is $v_F = 3.37\text{x}10^7$ cm/sec. The diffusion constant, $D = \ell v_F/2$, is then equal to $2.15\text{x}10^3$ cm^2/sec; the time to diffuse a distance $L = 1\mu\text{m}$ is therefore $L^2/D = 4.6$ picoseconds. Some interesting possibilities are suggested by these numbers.

For electromagnetic radiation incident in the normal direction, the fraction of its energy absorbed by the layer is $(4\pi/\nu c)\sigma_{II}$ where ν here is refractive index and $\sigma_{II}(\omega) = en\mu'_\omega$ is the two-dimensional ("per square") conductivity where μ'_ω is the real (in phase) part of the high-frequency mobility. Numerical estimates for this free-carrier-absorption effect show that the fraction of energy absorbed should be small (as is not necessarily so with the radiation propagating along the layer, in the "laser" geometry). The effect of the incident energy which is absorbed by the layer electrons may be appreciable, however. We can expect a rise in electron temperature, to $T_e > T$, given by

$$\lambda_e\ K(T_e - T) = e\ \mu'_\omega\ (\xi^2)_\omega \qquad (8)$$

where K is Boltzmann's constant, λ_e is the applicable energy relaxation rate and $(\xi^2)_\omega$ means the time average of the square of the high-frequency field(s) with frequency $\simeq \omega/2\pi$. If we assume the Drude formula $\mu'_\omega = (e/m^*)\lambda_m/(\lambda_m{}^2 + \omega^2)$ then, for a given frequency, λ_m might be chosen (such as by control of doping) to be $\simeq \omega$, bringing μ'_ω near to the maximum value $e/2m^*\omega$. For $\omega = 1\text{x}10^{13}$ sec^{-1}, this is $1.32\text{x}10^3\text{cm}^2/\text{V.sec}$; and we accordingly take $\mu'_\omega = 1000$ $\text{cm}^2/\text{V.sec}$. [Since $\hbar\omega/E_F = 0.3$, with the assumed numbers, the frequency condition for the quasi-classical treatment is adequately (even though not well) met. With more than one sub-band, however, there should also be (Spector 1983) a second-order process combining intraband optical and interband scattering interactions, which can be enhanced by a "resonant denominator" when $\hbar\omega$ approaches the interband energy $E_2 - E_1$. For moderately low temperatures (say 10-100K) with the assumed $n = 6\text{x}10^{11}\text{cm}^{-2}$, energy relaxation is by absorption and emission of acoustic phonons in "equipartition" modes (occupation numbers $\simeq KT/\hbar\omega$). From the theory for this phenomenon (Price 1982b) one finds that a representative value for λ_e would be $2\text{x}10^8\text{sec}^{-1}$. For $(\xi^2)_\omega \simeq 10^3$ $(\text{V/cm})^2$ (incident energy flux of about 10 watt/cm²), Eq. (8) with $\mu'_\omega = 10^3\text{cm}^2/\text{V.sec}$ then gives $K(T_e - T) \simeq 5\text{x}10^{-3}\text{eV}$; i.e., $T_e - T \simeq 60$ kelvin -- sufficient for a substantial change in the static mobility.

As indicated above, the high mobilities at low temperatures obtained by modulation doping imply mean free paths, $\ell = \tau v_F$, greater than a micron. Experimental FET devices made for these structures can have source-drain current paths of length (L) less than ℓ. Then there will be *longitudinal* size effects in the conduction. We may estimate the resulting change in the static conducivity by the following model, for $L << \ell$. With zero applied voltage V_{SD} between source and drain, there is a carrier flux $n(v_F/\pi)\ \delta E/E_F$ in each direction, for the carriers in a small energy range δE around E_F. For $\ell >> L$ they mostly go all the way across without scattering. If the effect of the applied potential is simply to displace the Fermi energies for the S and D carrier sources relative to each other by eV_{SD} (as in the tunnel diode, but not obviously correct for the present case) then the net flux is $n(v_F/\pi)\ eV_{SD}/E_F$ (independent of L, so long as $L >> \ell$). Therefore the conductivity is given by an effective mobility equal to $(2/\pi)(L/\ell)$ times the heterolayer mobility for large L -- i.e., as though the mean path were $(2/\pi)L$.

We may also obtain this "effusion" result, and generalize it to high frequency conduction, by assuming the scattering to be much greater outside the channel range $0 \le x \le L$ of the current path (or that the high-frequency field does not penetrate the source and drain regions), and calculating the conductivity at frequency $\omega/2\pi$ averaged over the position coordinate x in $(0,L)$. The averaged conductivity is found to be proportional to the angle average (for $v_x = v_F \cos\theta$) of $v_x^2[\tau/(1 + i\omega\tau)]F(\tau|v_x|/L(1 + i\omega\tau))$ where $F(u) = 1 - u(1 - \exp(1/u))$. For L both small compared to ℓ and small compared to v_F/ω, this angle average becomes $v_F L/\pi$, which gives the same result as above for the static conductivity with $L << \ell$. On the other hand for $\omega >> v_F/L$ (ie $\omega\tau >> \ell/L$) the real-part conductivity becomes the ordinary Drude conductivity times a factor $1 + (\ell/L)\Omega$ where $\Omega = (8/3\pi)(\omega^2\tau^2 - 1)/(\omega^2\tau^2 + 1)$. The above estimate of the high-frequency warm electron effect would need to be modified accordingly for these conditions. Another size effect should occur when the Larmor orbits in a normal-direction magnetic field become smaller than the current path length L. (For $n = 6\text{x}10^{11}\text{cm}^{-2}$, the Larmor radius of electrons at the Fermi energy is equal to 1μm at about 0.1 T.) We do not consider here the size effects for hot electrons, other than the above high-frequency "warm electron" case.

3. MULTILAYER STRUCTURES

As we have seen, with MBE techniques especially, multilayer structures, and graded layer configurations, can be grown on scales down to atomic. Although still structured in only one space dimension, these can have distinctive new properties. With separate ohmic contacts to different layers (as has been achieved for electron and hole heterolayers, in "nipi" structures) there can be more complex electronic functions. In particular we can have flows of charge in the normal direction perpendicular to the layer planes; and interlayer transfers of energy and momentum without charge transfer, both by carrier-carrier coulomb interaction (see below) and by interaction with electromagnetic and lattice-mode fields.

A primitive interlayer effect is the escape of hot carriers from a heterolayer into the adjacent barrier material. This is not a textbook thermal effusion process, because the carriers in the heterolayer are not itinerant in the normal-direction. The normal-direction component of the wavefunction has to change from the localized function in the heterolayer to an itinerant-state function which extends over the barrier region and is only somewhat modified within the heterolayer, while the planar component of the wavevector also changes. To escape, a carrier must (a) have enough planar itinerant energy (the last term of eq. (2)) to get over the barrier band edge; (b) make this transition induced by a symmetry-breaking force, a kind of scattering. By tedious but straight-forward calculation of this process, one finds (Price 1983) in particular for GaAs/(Ga,Al)As that the phonon scattering at normal temperatures is enough to give substantial escape rates. The lattice-scattering escape rate is roughly proportional to $(KT_e/\Delta E^2)\exp(-\Delta E/KT_e)$ where ΔE is the barrier height. Then a representative figure, for $KT_e = 0.1$ eV and $\Delta E = 0.3$ eV, is $2 \times 10^{10}\,\mathrm{sec}^{-1}$ at 300 K. (A timely theoretical problem is the corresponding calculation with carrier-carrier Coulomb scattering as the escape process.)

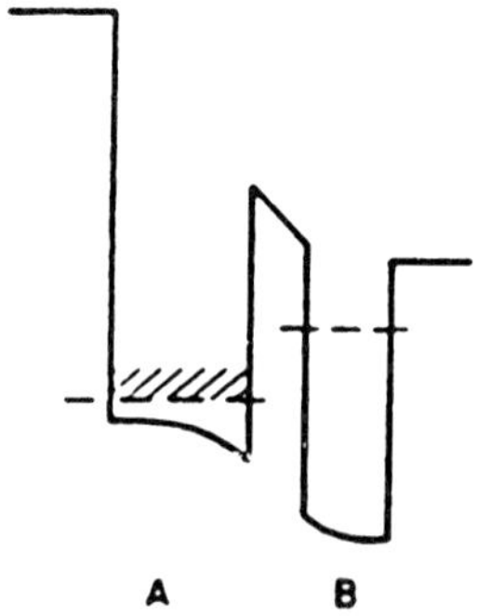

Figure 4. A heterostructure for transferred-electron effects.

These numbers confirm that one could readily get hot carriers flowing into the alloy layer. Since the mobility there may be assumed to be lower, one thinks of the analogy of the classical transferred-electron effect, especially in GaAs, and hence the possibility of a field range of negative differential mobility (Hess et al. 1979), which may have been observed (Coleman et al. 1982). An essential question, however, is what happens to the carriers once they are in the alloy, and how quickly (and reversibly) they can get back into the heterolayer. An alternative concept is to have the roles of the lower and upper band minima in the traditional transferred electron effect performed by two heterolayers of differing widths (Price 1981). Figure 4 shows one version of this. (One might, alternatively, have one or both of the heterolayers of the single-interface type.) With two heterolayers appropriately close, carriers with energy above the threshold can make quantum transitions between them, by a kind of tunneling process, without having to reside in the imbedding medium between escape and capture events. Mobilities in the heterolayers may be varied by modulation doping. Thus we have, with this structure, a transferred electron effect in which the band-edge energy difference, mobilities and transition rate might be "tailored" to desired values. (A hybrid arrangement might have a stepped structure in which a GaAs layer and an alloy layer adjoin, and both are imbedded in an alloy of higher concentration X; the lower sub-band would be provided by a single level of the GaAs "inner well", the upper by a second level extending over both layers.) As discussed below, when two close heterolayers are both occupied by carriers, the mutual equilibriating effects of carrier-carrier Coulomb scattering can be quite strong.

The direct interlayer transition is a scattering (as well as tunneling) process, in which the planar wavevector changes. We may estimate its rate, in a simple way without distinguishing between scattering mechanisms, as follows: Let the wavefunctions associated with each layer in isolation be $\phi_A(z)$ and $\phi_B(z)$. Because of penetration of the barrier, these "orbitals" will be hybridized to some extent in the combined stationary states in absence of scattering. That is,

$$\left.\begin{aligned} \psi_1 &= c\phi_A + \bar{c}\phi_B \\ \psi_2 &= \bar{c}\phi_A - c\phi_B \end{aligned}\right\} \qquad (9)$$

where $c^2 + \bar{c}^2 = 1$. Taking the scattering forces as operating on either ϕ_A or ϕ_B, the resulting channel mobilities will be somewhat changed, to μ_1 and μ_2 where

$$\left.\begin{aligned} 1/\mu_1 &= c^2/\mu_A + \bar{c}^2/\mu_B \\ 1/\mu_2 &= \bar{c}^2/\mu_A + c^2/\mu_B \end{aligned}\right\} \tag{10}$$

The orbitals' scattering rates (corresponding to μ_A and μ_B) being λ_A and λ_B, the transition rate between the hybridized sub-bands (9) should be given by

$$\lambda_{12} = (c\ \bar{c})^2(\lambda_A + \lambda_B) \tag{11}$$

If the barrier height and width are made such that $\bar{c}/c$ is small enough for μ_2/μ_1 to be small when μ_B/μ_A is small, then the first factor on the right of (11), equal to the square of $(\bar{c}/c)/(1 + (\bar{c}/c)^2))$, must be small. (For example, if $(\bar{c}/c)^2 = 0.1$ and $\mu_A/\mu_B = 5$ then $\mu_1/\mu_2 = 3.4$ and $(c\bar{c})^2 = 0.0826$; so if $\lambda_B/\lambda_A = \mu_A/\mu_B$ we have $\lambda_{12} = 0.5\ \lambda_A$.) In terms of a transfer Hamiltonian matrix element, M_{AB}, the hybridization is given by $c\bar{c} = |M_{AB}|/(E_B - E_A)$. We would expect to be able to modulate the parameters of the transferred electron effect, especially by changing the energy difference $E_B - E_A$ of the layer states, which might be accomplished by a potential difference applied through separate contacts to the two heterolayers, or by an added "gate" electrode.

As indicated above, because the heterolayers can be located so close together, their carriers can be coupled by "action at a distance" effects. (The electrons and holes of the structures shown in Figure 3 are, of course, especially close.) Phonons or photons generated by an excited (hot electron) layer state may induce electron transitions in, and hence excitation of, a neighboring layer. Similarly, the charge fluctuations in the layers are a means of interaction tending to equalize drift velocities and electron temperatures in adjoining layers -- this is the "coulomb mutual scattering" (CMS) effect. In the CMS process, electrons in two heterolayers "1" and "2" scatter each other through their coulomb interaction: $\mathbf{k}_1 \rightarrow \mathbf{k}_1' = \mathbf{k}_1 + \mathbf{Q}$ and $\mathbf{k}_2 \rightarrow \mathbf{k}_2' = \mathbf{k}_2 - \mathbf{Q}$. The scattering matrix element is $(2\pi/A)\ (e^2/\epsilon)Q^{-1}R_{12}(Q)$ where

$$R_{12}(Q) = \int\int dz_1 dz_2\ \rho_1(z_1)\rho_2(z_2)\ \exp(-\ Q|z_1 - z_2|) \tag{12}$$

A is normalization area, ϵ is dielectric constant and $\rho_1(z)$ and $\rho_2(z)$ are the squares of the normal-direction wavefunctions $(\int\rho(z)dz = 1))$. This R_{12} is $\simeq 1$ when the layers are close on the scale of $1/Q$, and is $\simeq \exp(-Qd)$ when they are at a distance $d \gg 1/Q$. With energy conservation and equal masses m^*, the locus of the vector $\mathbf{Q}$, for given $\mathbf{k}_1$ and $\mathbf{k}_2$, is a circle of diameter $\mathbf{k}_1-\mathbf{k}_2$, with the vector origin at the $\mathbf{k}_1$ end of this diameter. The overhead bar in the formulas below indicates averaging over Q uniformly on this circle. The density of states for the transitions is $(m^*/4\pi\hbar^2)A$ (half the density of states for a single layer scattering). For Boltzmann statistics, the total scattering rate is

$$\frac{2\pi}{\hbar}\frac{m^*\pi}{\hbar^2}\frac{N_1N_2}{A}\left(\frac{e^2}{\epsilon}\right)^2 \int d^2\mathbf{k}_1 f_1(\mathbf{k}_1) \int d^2\mathbf{k}_2 f_2(\mathbf{k}_2)\ \overline{\left(\frac{S(Q)R_{12}(Q)}{Q}\right)^2} \qquad (13)$$

where N_1, N_2 are the numbers of carriers, f_1 and f_2 are the distribution functions ($\int f(\mathbf{k})d^2\mathbf{k} = 1$) and $S(Q)$ is the appropriate screening function. (For Fermi statistics, there are additional factors $(1 - f_1(\mathbf{k}_1 + \mathbf{Q}))$ and $(1 - f_2(\mathbf{k}_2 - \mathbf{Q}))$ in the integrand.)

The rate of transfer of energy is obtained by including an additional factor $E(\mathbf{k}_1) - E(\mathbf{k}_1+\mathbf{Q}) = (\hbar^2/2m^*)Q^2((\mathbf{k}_1+\mathbf{k}_2)\cdot\mathbf{Q}/(\mathbf{k}_1-\mathbf{k}_2)\cdot\mathbf{Q})$ in the integrand. It is found (Price 1983) that this interlayer energy flow is quite large; between a hot layer and cold one ($T_{e1} \gg T_{e2}$) the rate per electron would be typically in the range 0.1 - 1 erg/sec. Consequently neighboring layers will tend to equalize electron temperatures when their mobilities, driving fields ξ, etc., differ. The momentum exchange rate also turns out to be quite strong in practice. The additional factor required in the integrand of (13), to give the transfer rate of the momentum $\hbar\mathbf{k}$, is $\hbar\mathbf{Q}$. If we assume the $f(\mathbf{k})$ to be displaced Maxwellians, const. $\exp(-(E - \hbar\mathbf{u}\bullet\mathbf{k})1/KT_e)$, and expand these to first order in $\mathbf{u}$, then the total rate of transfer of momentum is

$$\frac{\pi^2 m^*}{2\hbar}\left(\frac{e^2}{\epsilon}\right)^2 \frac{N_1N_2}{A}\left(\frac{\mathbf{u}_2}{KT_{e2}} - \frac{\mathbf{u}_1}{KT_{e1}}\right) \ll \overline{(R_{12}S)^2} \gg \qquad (14)$$

where $\ll \ldots \gg$ means average over the two (undisplaced) distributions with electron temperatures T_{e1}, T_{e2}. Since the latter will in any case tend to equalize, we look at the outcome when they are equal. The momentum balance equations are then

$$\left.\begin{aligned} p_1\xi_1 - \mathbf{u}_1/\mu_1 &= en_2\ (\mathbf{u}_1 - \mathbf{u}_2)\ Z \\ p_2\xi_2 - \mathbf{u}_2/\mu_2 &= en_1\ (\mathbf{u}_2 - \mathbf{u}_1)\ Z \end{aligned}\right\} \qquad (15)$$

where

$$Z = (e/\epsilon)^2\ (\pi^2 m^*/2\hbar KT_e)\ \langle\langle\overline{(R_{12}S)^2}\rangle\rangle = Z_0\langle\langle\ldots\rangle\rangle \qquad (16)$$

and the p factors are -1 for electrons and +1 for holes. Since we are allowing for electrons and holes in the two heterolayers, however, we should also let the effective masses differ: say m_1 and m_2. The generalization of (15) and (16) is given by substituting the density-of-states mass $m_d = 2m_1m_2/(m_1 + m_2)$ for m^* in (16) and replacing $\mathbf{u}_2 - \mathbf{u}_1$ on the right of the eqs. (15) by $(m_2\mathbf{u}_2 - m_1\mathbf{u}_1)/m_d$. The diameter of the $\mathbf{Q}$ circle, to be used in the averages ($\overline{\quad}$), is $|(m_d/m_2)\mathbf{k}_2 - (m_d/m_1)\mathbf{k}_1|$ instead of just $|\mathbf{k}_2-\mathbf{k}_1|$.

At T_e = 300 kelvin, Z_0 equals 8.6 kilohms (with m^* the GaAs electron mass). The final factor $\langle\langle\ldots\rangle\rangle$ need not be small for close heterolayers. As a representative figure we may take Z = 1 kilohm. The effect of the right side of (15), in displacing $\mathbf{u}_1$ and $\mathbf{u}_2$ towards each other, is large when $e(n_1\mu_2 + n_2\mu_1)\ Z$ is large. (This condition is marginally attainable, since $\mu = 2\times10^4 \mathrm{cm}^2/\mathrm{V.sec}$ and $n = 3\times10^{11} \mathrm{cm}^{-2}$ gives $e\mu n \simeq 1$ kilohm^{-1}.) If in addition $n_1\mu_2 \gg n_2\mu_1$, both u_1 and u_2 tend to $\mu_1 p_1 \xi_1$ (and similarly interchanging 1 and 2). For electrons and holes ($p_1 = \pm 1$, $p_2 = \mp 1$) with $n_1 = n_2$, the large-Z limit has zero net current; this could apply to the "semimetal" situation of Figure 3 and the "nipi" situation of Figure 2(b). For either polarity case, the combined current for a given $\xi_1 = \xi_2$ is reduced by the CMS term on the right of (15). Thus the transferred electron effect should be modified in the direction of being accentuated by the CMS momentum flow, though one would expect it to be diminished by the effect of the energy flow. The CMS energy transfer also provides a possibility of inter-layer modulation, since when the carriers in one layer are in a hot-electron state they are no longer a dissipative drain on the carrier energies in a neighboring layer, and tend to raise the latter's electron temperature.

For completely degenerate distributions there will be no CMS effect, since all possible $f(\mathbf{k})(1 - f(\mathbf{k} + \mathbf{Q}))$ factors will be zero for one of the layers. For non-zero T_e values small compared to the degeneracy temperatures, there will be a smaller effect than in the non-degenerate range; this situation was investigated long ago for the two partly occupied bands in a transition metal (Baber 1937), and it is known that the mutual scattering rate is proportional to T_e^2. The low-temperature hot-electron effect discussed in Section 2 should provide an instance of this, with heterolayers close together. The CMS effects for a heterolayer close to a "gate" (bulk or film) is a theory problem waiting for consideration.

The Joule heat of a conducting heterolayer will normally be dissipated by a flux of phonons. How much enhancement of the phonon densities should result, and with what consequences? This can be investigated by a somewhat heuristic theory (Price 1985). The power per electron, $w = eu\xi$, is distributed among the phonon modes participating in the electron scattering, so that the effective power per mode is w/η where η is the ratio (number of participating modes)/(number of electrons). For near-degenerate electrons, in homogeneous material the phonon wave-vectors are comparable with the Fermi wavevector and hence $\eta \sim 1$. In heterolayers the situation is more complicated, but it can be made plausible that again for near-degenerate electrons $\eta \sim 1$. As w increases, the average phonon occupation number, N, for those phonon modes participating in the scattering of the electrons, increases from N_L towards N_e, where N_L is the Planck function of $\hbar\omega_0/KT$ (i.e., the thermal-equilibrium value of N) and N_e is the Planck function of $\hbar\omega_0/KT_e$, where T_e is the heterolayer electron temperature. This increase of the N values reduces the dissipation rate w for given T_e. We find

$$N = \frac{w_e N_e + w_p N_L}{w_e + w_p} \tag{17}$$

and

$$w = \frac{w_e w_p}{w_e + w_p} [N_e - N_L] \tag{18}$$

where the constant w_e is the coefficient in the dissipation formula $w = w_e(N_e - N)$, approximately equal to $\hbar\omega_0$ times the spontaneous emission LO scattering rate at electron energies a little above $\hbar\omega_0$. The constant w_p is equal to $\eta\hbar\omega_0/\tau_0$, where τ_0 is the lifetime for decay of the LO phonons (found to be about 7 picoseconds, at 77K or below, in GaAs). Typically $w_p \ll w_e$, so that the first factor on the right of Eq. (18) is approximately equal to w_p, instead of the w_e which would apply if the dissipation were not buffered by the "hot phonon" effect indicated by Eq. (17). The result (18) appears to be consistent with recent luminescence data for GaAs quantum wells at low temperature.

The forgoing does suggest that one should further investigate heterolayer coupling to the lattice fields. In addition to the phenomena for a single heterolayer, exchange of energy and momentum between neighboring heterolayers is obviously of interest. Similarly, sharing and exchange of electromagnetic energy between layers may be important, especially with heterostructure lasers, for which it may provide a mode of device action (Levine et al. 1983). Also, a laser may be modulated (Tsang et al. 1981) by integrated optical coupling to a layer with a modulated absorption.

Another category of device structure uses a controlled potential profile (from space charge, band-edge energy $E_C(X(z))$, or a combination) as a barrier to carrier motion. For the transistor structures mentioned in Section 1 (Heiblum 1981), the barrier appropriately is narrow. In another type of application, the barrier potential $U(z)$ is more slowly varying, so that the carrier motion in it can be quasi-classical. In the *Camel FET* (Thorne et al. 1982) a "tailored" gate replaces the Schottky barrier. A graded band gap can be used to accelerate carriers, by the force $-dU/dz$, across a device region (Levine et al. 1982), for example, in a photodetector to speed the photo-carriers to the electrodes (Capasso et al. 1982). An imaginative suggestion (Cooper et al. 1982) to obtain a large carrier drift velocity is a periodic structure of alternating large and small fields dU/dz of the same sign.* Obviously, these structures characterized by a predetermined and/or externally controlled $U(z)$ could have many special properties.

4. LOWER DIMENSIONALITY

MBE structures are naturally layered. Their composition varies in the growth direction, normal to the surface. On the other hand, semiconductor devices have structure in the surface plane, made by lithographic and beam techniques. The practical limit of resolution for the latter now lies between 10^{-4} and 10^{-5}cm; and efforts to push this downward in scale are widespread. Fabricated metal lines of width ~ 10^{-6}cm have been demonstrated (Laibowitz et al. 1979). So far as one knows, solid structures with scale ~ 100 Å or less in two or in all three dimensions are possible -- that is, once in existence, they would be metallurgically stable -- and we may freely invent such objects on paper and theorize about their properties.

---

*An interesting question of physics is posed by such a periodic "saw tooth" potential, in the case where the average of dU/dz over z is zero, while $U(z)$ is not symmetric -- i.e. the functions $U(z)$ and $U(-z)$ differ: Viewing this as a quantum superlattice, the Kramers degeneracy apparently implies that it will have a reversible current-voltage characteristic (when the voltage changes sign but not magnitude, the current does the same). But when conduction can be treated as a localized phenomenon (carrier mean free path small compared to "superlattice" period) one expects the characteristic to be not reversible -- to exhibit rectification. How are these aspects to be reconciled? More generally, the properties of such a macroscopically polar semiconductor should be of interest, and invite investigation.

A proposed device of this kind is the two-dimensional superlattice (Reich et al. 1982): a mosaic of two different solids, with "lattice constant" $\sim 10^{-6}$cm, still uniform in the third direction. A hoped-for property of conventional (one-dimensionally periodic) superlattices is a negative differential mobility, at fields above some minimum value. This is a demonstrable consequence of the existence of an upper bound of the carrier energy $E(k_z)$ in a *single* superlattice band. The weakness of the notion is that there actually will be multiple bands of electron energy $E_1(k_z)$, $E_2(k_z)$.... Then the energy associated with the other two directions (like the $(\hbar^2/2m^*)(k_x^2 + k_y^2)$ of eq. (2)) could allow the electrons to scatter between bands. Only for periodic potentials in one dimension are the allowed energy ranges $E_n(k)$ of Bloch bands necessarily separated by forbidden ranges, rather than overlapping. For a two-dimensional superlattice squeezed in the third direction by a gate-like field (Iafrate et al. 1982), bands completely separated in energy are possible. A companion idea is that a superlattice structure may be obtained by superposing a grid, at the surface, on an inversion or accumulation layer (Kotthaus and Heitman 1982).

A notion persistently associated with superlattices is the *Bloch resonance*. In a uniform electric field ξ, for sufficiently weak scattering, the quasi-classical motion of the carriers will have wavevector k increasing at a uniform rate $e\xi/\hbar$. Since $E(k)$, and hence velocity $dE/d(\hbar k)$, are periodic with period $2\pi/d$ where d is the space period of the lattice, the carrier state including its velocity oscillates with frequency $\nu_B = e\xi d/h$. (The scattering rate has to be smaller than ν_B. For $\nu_B = 10^{12}$ hertz, $\xi d = 4.1$ milli-V.) In quantum terms, one finds that the band electron states form a "Stark ladder" of levels with energy spacing $e\xi d$, which is just $h\nu_B$; it appears then that the level spacing and hence the periodicity will be unchanged in the more realistic multi-band case. To extract power at ν_B -- to have a "unipolar laser" type of device based on an extended, periodic, structure -- one must, of course, have a population inversion or some dynamical equivalent.

A further step in the Less is More direction might be a one-dimensional conductor, on the same nanometer scale. Interesting fabrication attempts have been made recently. A narrow "wire" was laid down on an inversion-layer structure, and biassed so as to accumulate carriers in a one-dimensional channel (Fowler et al. 1982). Anisotropic etching, undercutting a deposited mask stripe, was used to narrow a heterolayer laterally to this same scale (Petroff et al. 1982). A similar proposal (Sakaki 1980) was to etch a V-groove through a heterolayer, lay down a metal gate in the groove, and electrostatically accumulate carriers on each side of it. These thin-cylinder structures should have electron states which are quantized in two directions and itinerant in the third (axis) direction. As with modulation doped heterolayers, it is hoped to keep mobilities in the axis direction high by confining the ions to a distance from this carrier channel. With a degenerate distribution, the Fermi wavevector k_F for a *linear* carrier density n is $(\pi/2)n$. A value $1\text{x}10^6\text{cm}^{-1}$ for n may be realistic; it gives $k_F = 1.57\text{x}10^6\text{cm}^{-1}$ and hence (for GaAs) $E_F = 16.2$ milli eV. With $n = 10^6\text{cm}^{-1}$ and $\mu = 10^5\text{cm}^2/\text{V.sec}$, the resistance per unit length would be 6.2 ohms per nanometer.

One could argue that mobilities should be small because every scattering counts double: the $(1 - \cos\theta)$ factor equals 2. A possibly more compelling consideration is that the carriers have no way around obstacles, so that the resistance tends to be dominated by the strongest scatterers, or highest barriers, along the channel. (For a random distribution of scatterers, this aspect has been quantified by Landauer (1970).) On the other hand, calculation of scattering rates, in particular for distanced ions, by conventional Born approximation can be seen to give an over-estimate: The wavefunction for a slowly varying potential is almost that of the WKB formula, which has no scattering in the allowed energy range. Therefore the scattering comes only from corrections to WKB. Let the wavefunctions be $\phi(x)\eta(y,z)$, and the scattering potential after averaging over η^2 be $U(x)$, so that the applicable wave equation is $-(\hbar^2/2m^*)d^2\phi/dx^2 + U\phi = E\phi$. The WKB solution for ϕ is

$$u(x)\exp\ (\pm i\int^x k(x')dx'),$$

where $(\hbar^2/2m^*)k^2 = E-U(x)$ and $u = (k_0/\text{k})^{1/2}$ where k_0 is the value of k for $x \to \pm\infty$. By substituting this WKB approximation in the wave equation and obtaining the correction term, we find that the rate of scattering (from $\exp + i\int$ to $\exp - i\int$) is that given by the Born approximation for the potential

$$U_{eff}(x) = (\hbar^2/2m^*)\ u\ d^2u/dx^2$$

With an ion at a distance R from the channel, it appears that this would be smaller than the bare potential U roughly for $2kR > 1$. In the case of the two-dimensional heterolayers it was found, however, that in practice scattering by remote ions was critically dependent on screening of the ion fields by the heterolayer space charge. We need to investigate the effects of screening in the present case.

At a two-dimensional distance $\mathbf{R} = (y,z)$ from an ion, the Coulomb potential is $\sim 1/r$ where $r^2 = R^2 + x^2$. This is

Fourier analyzed by

$$\int_{-\infty}^{\infty} \frac{1}{r} \exp\,(iQx)\; dx = 2\; K_0(QR) \tag{19}$$

where $K_0(\ )$ is a cylinder function (a solution of $(\nabla^2 - Q^2)\; G(\mathbf{R}) = 0$, in two dimensions). The screening function, corresponding to (5), is found to be

$$S(Q) = 1/(1 + P_I H(Q)) \tag{20}$$

where $P_I = (2e^2/\epsilon)\; d(\text{linear carrier density})/d(\text{Fermi energy})$ and

$$H(Q) = \int\int d^2\mathbf{R}_1 d^2\mathbf{R}_2\;\; \rho(1)\rho(2)K_0(R_{12}Q) \tag{21}$$

where $\rho(\mathbf{R})$ is here the normalized square of the two-dimensional channel wavefunction $\eta(y,z)$ and R_{12} is $|\mathbf{R}_1 - \mathbf{R}_2|$. The screening parameter P_I, of zero dimensionality in this case, is equal to n_0/n for a degenerate distribution, where $n_0 = (2m^*/\epsilon)(2e/\pi\hbar)^2 = 8.0\text{x}10^6$ cm^{-1}; so in practice P_I will be >1, and screening is important unless H is small. The dependence of H on Q is singular; by assuming a Gaussian function for the electron density, $\rho(\mathbf{R}) = (\alpha^2/\pi)\exp(-\alpha^2\mathbf{R}^2)$, we find

$$H(Q) = \frac{1}{2}\int_0^{\infty} \frac{\exp\,(-(Q^2/2\alpha^2)t)}{1 + t}\; dt \tag{22}$$

which varies as $\ln\,(1/Q)$ when $Q \to 0$. With degeneracy, however, only a single finite value, $Q = 2k_F$, applies in the scattering (though the plasma-mode $\omega(Q)$ involves the $H(Q)$ function). Then for channel width ~10^{-6}cm and hence ~n^{-1}, it appears reasonable to take $H \sim 1$. The calculation will not be carried any further here. A laboratory realization of these one dimensional conductors, especially of some reproducible uniformity, would certainly be of interest.

5. LATTICE SCALE EFFECTS

The structures discussed above are really on an intermediate scale, small enough for quantum phenomena based on their linear dimensions, but still large compared to lattice constants. One may call this the *meso-scale*. The underlying truly microscopic scale, of atomic diameters and bond lengths and lattice constants, may have effects: (a) As the structure's dimensions are made smaller, modification of the electron states by lattice-scale corrections can become appreciable. (b) The interface between distinct semiconductors itself belongs to the lattice scale, and needs a special prescription for the way it connects the meso-scale electron states on the two sides. We should also note (but will not discuss here) that an alloy like (Ga,Al)As is not truly an ordered crystal. Underlying its "averaged" effective band structure is the disorder of the possible configurations of Ga and Al atoms on the cation sites. In small structures with electron states extending over a modest number of lattice cells, this might have significant effects.

The "wavefunctions" imputed to electrons and holes in the preceding sections, as in almost all the literature, are actually *envelope functions*; and their wave equation is not the true Schrödinger equation for an electron in a real potential. It differs especially by the substitution of the effective mass m^* for the free electron mass in the kinetic energy operator $-(\hbar^2/2m^*)\nabla^2$. The usual term "effective mass approximation" is misleading, however, in that the equation can be derived from a precise formulation, the *Envelope Representation*, and it has a more general form than that of the parabolic $E(\mathbf{k}) \sim k^2$. Let the Bloch states for a homogeneous crystal lattice have wavefunctions $\phi_{\mathbf{k}}(\mathbf{r})$, satisfying $H_0\phi_{\mathbf{k}} = E_B(\mathbf{k})\phi_{\mathbf{k}}$, and let the wavefunction ψ of some electron state be given by the expansion

$$\psi(\mathbf{r}) = \Sigma_{\mathbf{k}}A_{\mathbf{k}}\phi_{\mathbf{k}} \tag{23}$$

when the $\phi_{\mathbf{k}}$ are normalized in unit volume. Then (Kittel and Mitchell 1954) the envelope function is

$$\Psi(\mathbf{r}) = \Sigma_{\mathbf{k}}A_{\mathbf{k}}\exp(i\mathbf{k}\cdot\mathbf{r}) \tag{24}$$

For simplicity we assume below that the summation in (23) and (24) is over one whole Brillouin zone. The envelope function is related to the Wannier representation, since (23) and (24) give

$$\psi(\mathbf{r}) = \upsilon^{1/2}\Sigma_{\mathbf{R}}\Omega(\mathbf{r} - \mathbf{R})\Psi(\mathbf{R}) \tag{25}$$

where the summation is over all lattice cells, υ is the volume of a cell, and $\Omega(\ )$ is the Wannier function for this band. Another method (Koster and Slater 1954) is direct solution of the difference equations which are obtained with the Wannier Representation.

To obtain the envelope equivalent of the Schrödinger equation, we require the transform $\tilde{H}$ of the Hamiltonian operator H, such that $(\Psi_1|\tilde{H}|\Psi_2) = (\psi_1|H|\psi_2)$. For the "unperturbed" homogeneous crystal, it is easily seen from (23), (24) that $\tilde{H}_0 = E_B(-i\nabla)$; the usual

$-(\hbar^2/2m^*)\nabla^2$ follows directly from this, for the parabolic range. To get the potential term for $H = H_0 + U(\mathbf{r})$, we make use of

$$\psi(\mathbf{r}) = \int \Gamma(\mathbf{r},\mathbf{r}')\Psi(\mathbf{r}')d^3\mathbf{r}' \tag{26}$$

where

$$\begin{aligned}\Gamma(\mathbf{r}_1,\mathbf{r}_2) &= V^{-1}\Sigma_{\mathbf{k}}\phi_{\mathbf{k}}(\mathbf{r}_1)\exp(-\,i\mathbf{k}\cdot\mathbf{r}_2)\\ &= \Sigma_{\mathbf{R}}\Omega(\mathbf{r}_1 - \mathbf{R})\Omega_0(\mathbf{r}_2 - \mathbf{R})\end{aligned} \tag{27}$$

in which $\Omega_0(\ )$ is the Wannier function for the "empty lattice" (i.e., equal to a sum of plane waves rather than Bloch functions over the same Brillouin zone). Then the envelope wave equation becomes

$$\begin{aligned}E\Psi &= (\tilde{H}_0 + \tilde{U})\Psi\\ &= E_B(-i\nabla)\Psi + \int\int \Gamma(\mathbf{r},\mathbf{r}_1)\ U(\mathbf{r}_1)\Gamma(\mathbf{r}_1,\mathbf{r}_2)\Psi(\mathbf{r}_2)d^3\mathbf{r}_1 d^3\mathbf{r}_2\end{aligned} \tag{28}$$

for eigenvalue E. The double kernal $\mathbf{K}(\mathbf{r};\ \mathbf{r}_1,\ \mathbf{r}_2) = \Gamma(\mathbf{r},\ \mathbf{r}_1)\Gamma(\mathbf{r}_1,\mathbf{r}_2)$ in (28) localizes $U(1)$ and $\Psi(2)$ in $\gtrsim$ a lattice-constant distance around $\mathbf{r}$ (and the integral of $K(\ ;1,2)$ over $\mathbf{r}_1$- space and $\mathbf{r}_2$-space equals 1). So for Ψ varying slowly on this scale the right-hand side of (28) reduces to $E\Psi = E_B(-i\nabla)\Psi + U_{eff}(\mathbf{r})\Psi$ where U_{eff} is a smoothed potential: $U_{eff}(\mathbf{r}) = \int J(\mathbf{r},\ \mathbf{r}_1)U(\mathbf{r}_1)d^3\mathbf{r}_1$, in which the kernel is $J(\mathbf{r},\ \mathbf{r}_1) = \int K(\mathbf{r};\ \mathbf{r}_1,\ \mathbf{r}_2)d^3\mathbf{r}_2$, also with a range $\gtrsim$ a lattice constant for $|\mathbf{r}_1 - \mathbf{r}|$. Obviously, correction terms based on Taylor expansion of $U(\mathbf{r}_1)$ and $\Psi(\mathbf{r}_2)$ about $\mathbf{r}$ may be obtained from (28).

This procedure works just as well if the summations (23),(24) are over part of a Brillouin zone (a valley sector) or over several bands. The general form is a set of summations, each associated with an envelope function, and resulting coupled wave equations for these component functions Ψ_n. In fact, for each such Ψ_n the transformation $\psi \to \Psi$ amounts to operating on the quantum state vector with a projection operator, say P. Then quantum operators transform by $\tilde{A} = PAP^\dagger$. (The Boltzmann equation, in fact, derives from this transformation applied to the density matrix.) The coupling between the equations for these projected-out components completes the scheme. For the degenerate valence bands of the "diamond type" semiconductors, the coupled envelope equations describe the shallow acceptor states, as was shown by Kittel and Mitchell (1954).

For a graded semiconductor we have to deal with a space-dependent

H_0. A convenient model uses a set of homogeneous (not space-dependent) Wannier functions, with matrix elements

$$E_W(\mathbf{R}_1,\mathbf{R}_2) = \int \Omega(\mathbf{r} - \mathbf{R}_1)H_0\Omega(\mathbf{r} - \mathbf{R}_2)d^3\mathbf{r} \tag{29}$$

which do not depend only on $\mathbf{R}_1 - \mathbf{R}_2$. (That is, we neglect inter-band coupling effects of the grading.) Then we may apply a kind of Wigner-Moyal transformation to the Bloch representation of H_0,

$$E_B(\mathbf{k}_1,\mathbf{k}_2) = N^{-1}\Sigma_{\mathbf{R}_1}\Sigma_{\mathbf{R}_2}E_W(\mathbf{R}_1,\mathbf{R}_2)\ \exp[i(\mathbf{R}_1\cdot\mathbf{k}_1 - \mathbf{R}_2\cdot\mathbf{k}_2)] \tag{30}$$

(where $N = V/v$ is the number of terms in each lattice sum), to obtain the Fourier transform on the variable $(\mathbf{k}_1 - \mathbf{k}_2)$ at fixed $\mathbf{k} \equiv (\mathbf{k}_1 + \mathbf{k}_2)/2$:

$$\hat{E}(\mathbf{r},\mathbf{k}) = v^{1/2}\Sigma_{\mathbf{R}_1}\Sigma_{\mathbf{R}_2}E_W(\mathbf{R}_1,\mathbf{R}_2)\Omega_0(\mathbf{r}-(\mathbf{R}_1 + \mathbf{R}_2)/2)\ \exp\,[i(\mathbf{R}_1 - \mathbf{R}_2)\cdot\mathbf{k}] \tag{31}$$

The envelope-function operator $\tilde{H}_0$ is then $\hat{E}(\mathbf{r}, -i\nabla)$. (In the parabolic range this will be proportional to a symmetrized product of $1/m^*(\mathbf{r})$ and $\nabla\cdot\nabla = \nabla^2$.)

Calculation explicitly in terms of the Koster-Slater equations, for the electron states in the Wannier representation, may be useful on the scale of a small number of lattice constants. The Koster-Slater equation may be solved directly in numerical or analytical terms. Its "finite difference" form has an affinity with the Schrödinger equation; for example, it has an analog of the WKB solution of the latter. It is misleading to refer to this scheme as the "tight binding approximation", as it is not limited to the idealized case in which the only non-zero off-diagonal elements (29) are the nearest-neighbor ones (corresponding to a sinusoidal form of $E_B(\mathbf{k})$), for which the Wannier functions do not extend far beyond a single lattice cell. The nearest-neighbor idealization is useful in modeling particular problems, such as the application below.

At an abrupt heterostructure boundary, the actual electron wavefunction ψ will change continuously, from one side to the other, but there is *a priori* no reason to expect Ψ to be continuous. The boundary conditions can usefully be modeled in the Wannier representation. It is sufficient to consider only one dimension, normal to the interface. By (25) the $\Psi(z = na) \equiv C_n$, where a is the lattice constant and n an integer, are coefficients in the Wannier expansion of ψ. In the minimal model, only the diagonal and nearest-neighbor elements, $E_W(n, n) = F$ and $E_W(n, n \pm 1) = G$, are non-zero; so the Koster-Slater equation becomes

$$(E - F)C_n = G(C_{n-1} + C_{n+1})$$

for energy eigenvalue E. We take the lattice constant as the same on both sides of the interface, but F and G equal to F_- and G_- for $z < 0$, and to F_+ and G_+ for $z > 0$. Between the last cell on one side and the first on the other, a distinct value of G represents the interface: $G = G_I$. We consider the nearest cell on each side as at a distance $a/2$ from the interface, so that, to a sufficient approximation,

$$C_1 = \Psi_+(0) + (1/2)\Phi_+(0)$$

and

$$C_2 = \Psi_+(0) + (3/2)\Phi_+(0)$$

where $\Phi(z) \equiv a\, d\Psi(z)/dz$, and similarly for C_{-1} and C_{-2} in terms of Ψ_- and Φ_-. On neglecting $E\text{-}F_-$ compared to G_- and $E\text{-}F_+$ compared to G_+, we then find that Ψ and its derivative Φ/a are connected at the interface by the linear relations

$$\left.\begin{aligned} \Psi_+ &= \alpha\Psi_- + (\beta/2)\Phi_- \\ \Phi_+ &= (2\beta)\Psi_- + \alpha\Phi_- \end{aligned}\right\} \tag{32}$$

where

$$\alpha = \frac{1}{2}\left(p_- + \frac{1}{p_+}\right), \qquad \beta = \frac{1}{2}\left(p_- - \frac{1}{p_+}\right), \qquad p_\pm \equiv \frac{G_\pm}{G_I} \tag{33}$$

The determinant of the coefficients in (32), $\alpha^2 - \beta^2 = p_-/p_+$, is equal to m^*_+/m^*_-, as current conservation requires. The condition for the "off diagonal" coefficient β to vanish, $G_+G_- = G_I{}^2$, is also the condition in this model (Price 1962, Kroemer and Zhu 1982) for the reflection probability of a Bloch wave, incident on the interface from either side, to be zero. For the conventional continuity relations ($\Psi_+ = \Psi_-$, $\Phi_+ = \Phi_-$) to apply, however, we require $G_+ = G_- = G_I$.

The forgoing single-band model is too crude to represent semiconductors of interest in detail. A more sophisticated multi-band (multi-orbital) version has been developed by Ando and Mori (1982) among others. For GaAs/(Ga,Al)As, a better approximation than (32), (33) may be $\Psi_+ = \Psi_-$ and $(1/m^*_+)\Phi_+ = (1/m^*_-)\Phi_-$. A more general point of view is as follows: For the "flat band" condition, or not too strong interface fields, an *ordered* interface can be characterized by coefficients $A(\mathbf{k})$ and $B(\mathbf{k})$ in the linear combination $A\phi_{\mathbf{k}'} + B\phi_{\mathbf{k}''}$ of incident ($\mathbf{k}'$) and reflected ($\mathbf{k}''$) Bloch waves on the left that "match" (belong to the same state as) a transmitted wave $\phi_{\mathbf{k}}$ on the right. If only one band is involved, these coefficients give a complete description of the wavefunction matching; and since they apply equally to the linear relations between plane waves as Fourier components of the envelope functions $\Psi(\mathbf{r})$, they give boundary-matching conditions for the latter. (At an interface with an effectively semi-infinite semiconductor in which E is in the forbidden range, we must have $B/A = \exp i\phi(E)$ and hence $d\Psi/dz = (k_z \tan(\phi/2))\Psi$, where the value of ϕ is such that the amplitude of the growing wave, beyond the interface, is zero.) For the multi-component situation, the A and B coefficients become matrix arrays interconnecting the components. (These in general can include contributions from bands for which E is outside the band energy range, with imaginary or complex $\mathbf{k}$ components, which might in some circumstances be important.) One may suppose that there are possibilities here for a combination of theoretical and experimental physics to determine the properties of the interfaces from observations on heterostructure electron states; but this prospect presumably depends on the extent to which ordered and reproducible interfaces can be grown.

REFERENCES

1. Adachi, S., Kawashima, M., Kumabe, K., Yokoyama, K. and Tomizawa, K., 1982, IEEE Electron Device Letters EDL-3: 409.

2. Allyn, C.L., Gossard, A.C. and Wiegmann, W., 1980, Appl. Phys. Lett. 36:3.

3. Anderson, R.L., 1962, Solid-State Electronics 5:341.

4. Ando, T., 1982a, J. Phys. Soc. Japan 51:3893.

5. Ando, T., 1982b, J. Phys. Soc. Japan 51:3900.

6. Ando, T., Fowler, A.B. and Stern, F., 1982, Revs. Mod. Phys. 54:437.

7. Baber, W.G., 1937, Proc. Roy. Soc. A158:383.

8. Capasso, F., Tsang, W.T., Bethea, C.G., Hutchinson, A.L. and Levine, B.F., Appl. Phys. Lett. 42:93.

9. Chang, L.L. and Esaki, L., 1980, Surface Science 98:70.

10. Chang, L.L., Esaki, L. and Tsu, R., 1974, Appl. Phys. Lett. 24:593.

11. Cho, A.Y. and Arthur, J.R., 1975, Prog. Solid-State Chem. 10:157.

12. Coleman, P.D., Freeman, J., Morkoc, H., Hess, K., Streetman, B. and Keever, M., 1982, Appl. Phys. Lett. 40:493.

13. Cooper, J.A., Capasso, F. and Thornber, K.K., 1982, IEEE Electron Device Letters EDL-3:407.

14. Das Sarma, S., and Quinn, J.J., 1982, Phys. Rev. B25:7603.

15. Drummond, T.J., Kopp, W., Keever, M., Morkoc, H. and Cho, A.Y., 1982, J. Appl. Phys. 53:1023.

16. Esaki, L. and Chang, L.L., 1976, Thin Solid Films 36:285.

17. Ezawa, H., Kawaji, S. and Nakamura, K., 1974, Japan. J .Appl. Phys. 13:126; ibid. 14:921.

18. Fowler, A.B., Hartstein, A., and Webb, R.A., 1982, Phys. Rev. Lett. 48:196.

19. Grinberg, A.A. and Kastal'skii, A.A., 1972, Sov. Phys. -Semicond. 5:1768.

20. Heiblum, M., 1981, Solid-State Electronics 24:343.

21. Hess, K., Morkoc, H., Shichijo, H. and Streetman, B., 1979, Appl. Phys. Lett. 35:469.

22. Hiyamizu, S. and Mimura, T., 1982, J. Crystal Crowth 56:455.

23. Iafrate, G.J., Ferry, D.K. and Reich, R.K., 1982, Surface Science 113:485.

24. Inoue, M., Hiyamizu, S., Inayama, M. and Inuishi, Y., 1983, Proc. 1982 Tokyo Conf. Solid State Devices: Japan J. Appl. Phys. (Supplement).

25. Kittel, C. and Mitchell, A.H., 1954, Phys.Rev. 96:1488.

26. Koster, G.F. and Slater, J.C., 1954, Phys.Rev. 95:1167.

27. Kotthaus, J.P. and Heitmann, D., 1982, Surface Science 113:481.

28. Kroemer, H. and Griffiths, G., 1983, IEEE Electron Device Letters EDL-4:2.

29. Kroemer, H., and Zhu, Q.-G., 1982, J. Vac. Sci. Technol. 21:551.

30. Kunzel, H., Dohler, G.H. and Ploog,K., 1982, Appl.Phys. A27:1.

31. Kunzel, H., Fischer, A., Knecht, J. and Ploog, K., 1983, Appl. Phys. A 30:73.

32. Laibowitz, R.B., Broers, A.N., Yeh, J.T.C., and Viggiano, J.M., 1979, Appl. Phys. Lett. 35:891.

33. Landauer, R., 1970, Phil. Mag. 21:863.

34. Levine, B.F., Logan, R.A., Tsang, W.T., Bethea, C.G. and Merritt, F.R., 1983, Appl. Phys. Lett. 42:339.

35. Mori, S. and Ando, T., 1980, J. Phys. Soc. Japan 48:865.

36. Ohno, H. and Sakaki, H., 1982, Appl. Phys. Lett. 40:893.

37. Petroff, P.M., Gossard, A.C., Logan, R.A. and Wiegmann, W., 1982, Appl. Phys. Lett. 41:635.

38. Price, P.J., 1962, Proc. 6th Int. Conf. Physics of Semicond., Exeter (Institute of Physics, London): 99.

39. Price, P.J., 1981, IEEE Trans. Electron Devices ED-28:911.

40. Price, P.J., 1982a, Surface Science 113:199.

41. Price, P.J., 1982b, J. Appl. Phys. 53:6863.

42. Price, P.J., 1983, Proc. 16th Int. Conf. Physics of Semicond., Montpellier 1982 -- Physica 117B/118B:750.

43. Price, P.J., 1985 Physica 134B:164.

44. Reich, R.K., Grondin, R.O., Ferry, D.K. and Iafrate, G.J., 1982, IEEE Electron Device Letters, EDL-3:381.

45. Sakaki, H., 1980, Japan J. Appl. Phys. 19:L735.

46. Shannon, J.M., and Gill, A., 1981, Electronics Lett. 17:620.

47. Solomon, P.M., 1982, Proc. IEEE 79:489.

48. Spector, H., 1983, Phys. Rev., B28:971.

49. Stern,F., 1967, Phys. Rev. Lett. 18:546.

50. Suris, R.A. and Fedirko, V.A., 1978, Sov. Phys.-Semicond. 12:629.

51. Thorne, R.E., Su, S.L., Kopp, W., Fischer, R., Drummond, T.J. and Morkoc, H., 1982, J. Appl. Phys. 53:5951.

52. Tsang, D.Z., Walpole, J.N., Groves, S.H., Hsieh, J.J. and Donnelly, J.P., 1981, Appl. Phys. Lett. 38:120.

53. Yamanishi, M. and Suemune, I., 1983, Japan J. Appl. Phys. 22:L22.

CORRELATION EFFECTS IN SHORT TIME,

NONSTATIONARY TRANSPORT

J. J. Niez

Leti - Ceng
Grenoble, France

I. Introduction

The hot electron problem can be approached, like every problem of statistical physics, through either the Liouville equation for the density matrix or the Heisenberg equations for the mechanical variables of interest. For thermodynamical purposes, these two approaches are equivalent since the mean value of every observable 0 at time t is given by either

$$\langle 0\rangle^t = \mathrm{Tr}\{\rho(t)\ 0\}$$

or

$$\langle 0\rangle^t = \mathrm{Tr}\{\rho(t=0)\ 0(t)\}\ .$$

For technical reasons, the Liouville equation is very often easier to handle, but can be less rich than the Heisenberg approach. As a matter of fact, by starting from the Heisenberg equations we can try to derive a Langevin equation for the variables of interest which will separate the dissipationless motion, the dissipative motion, and the fluctuating forces precisely responsible for this dissipation.

The hot electron problem has been treated for a long time within the framework of the Liouville equation by first utilizing an approximation of the latter[1-3] (the Boltzmann equation) and more recently through a more complex treatment of the nonequilibrium problem at hand[4-6]. There is a need for a Langevin equation, which would give a complete description of the build-up of the fluctuations during the evolution of the carrier system toward the nonequilibrium steady state and would give an understanding of the features of the random forces in the steady state. In fact, we can say that this need is two-fold. First, we would like to know how to build a bridge between the correlation functions, which we may determine through experimentation, and the fluctuating forces. Second, we would like to be able, for practical reasons, to predict the features of the noise we may have in submicron devices, working in either the hot electron regime or the so-called "ballistic regime."

Therefore, the objective of these lectures is to show how to build a Langevin equation for the variables of interest. Even though our presentation is consistent with a pioneer work by Grabert[7] on nonequilibrium thermodynamics, our approach differs in the point of view.

We first define the variables of interest and a quasi-equilibrium statistical distribution (Section II). Then we discuss the dissipationless motion and introduce the relevant projection operator (Sections III and IV). From this, we derive the Langevin equation of our thermodymanical varibles and build a bridge with previous work[5,6] based on the Liouville equation (Sections V and VI). Finally, in Section VII, we focus on the fluctuating forces.

II. The Quasi-Equilibrium Statistical Distribution

In the approach to be used here, we treat an electronic system in a semiconductor lattice. One cannot decouple this system from the lattice if a complete and accurate representation is to be achieved. We, therefore, take the Hamiltonian of the system to be

$$H = H_e + H_{e-f} + H_{e-L} + H_L \quad , \tag{1}$$

where H_e is the complete electronic Hamiltonian (including the electron-electron interaction), H_{e-L} is the electron-lattice interaction, H_{e-f} is the time independent electric field term, and H_L is the Hamiltonian of the lattice variables (usually limited to those terms representing the phonon field).

Following earlier work in this approach[4-10], we introduce a set of thermodynamically relevant variables (the macrovariables) of the problem. There is no absolute rule governing such an approach. Rather, one must be led in this by physical intuition, and the aim of this approach is to be able to describe in a deterministic manner all the important thermodynamic properties of the system. We note also that the choice of the pertinent macrovariables used is critical as this choice and the definition of the entropy operator for the far-from-equilibrium system are two aspects of the same problem.

In the problem at hand, we choose to use the electronic energy H_e, the total momentum of the electronic system $\vec{P}_e$, the energy of the lattice system H_L, and the momentum of the lattice system (which is identically zero). In order to maintain a grand canonical framework, we add to the previous set the number operator N_e for the electrons.

Many people approach the treatment of this system by the introduction of a "drifted Maxwellian" distribution function, in which the density matrix of the system is assumed to be not too far from[11]

$$\rho_{L,0} = \exp\left\{-\phi_0 - \sum_m F_m(t)\, P_m\right\}, \tag{2}$$

where the individual P_m's are the previously defined macrovariables and the F_m's are the thermodynamically conjugate forces associated with the macrovariables. In addition, the quantity $\phi_0 = \ln(\mathrm{Tr}\,\rho_{L,0})$ assures the proper normalization of the system by introducing a quantity similar to a partition function.

However, the above choice of $\rho_{L,0}$ is not convenient, primarily because it does not give the proper equilibrium density matrix when $\vec{F}=0$ ($\langle\vec{P}\rangle=0$ as well) and when $\beta_e=\beta_L$ (β_e and β_L are the inverse temperatures of the electronic and lattice systems, respectively). In other words, if we write the Liouville operator equation with the total Hamiltonian of (1) but without the electric field, we would like any irreversibility to arise only from the fact that the trial density matrix does not obey the Liouville equation when the system is not in equilibrium. If we write the density matrix in the linear response form $\rho=\rho_L+\Delta\rho$, the Liouville equation for the deviation term is

$$i\,\frac{\partial \Delta\rho}{\partial t} - [H,\Delta\rho] = [H,\rho_L] - i\,\frac{\partial \rho_L}{\partial t}\;. \tag{3}$$

Thus, in order to fit the above requirements, we must introduce terms in H_{e-L} into $\rho_{L,0}$.

To accomplish the above goals, we introduce the term H_{e-L} as a thermodynamic variable as well. To be consistent, however, we assign it the conjugate force β_L^{eq}. We can then define the quasi-equilibrium statistical distribution (QESD) as

$$\rho_L = \exp\left\{-\Phi - \sum_m F_m(t)\, P_m\right\}, \tag{4}$$

where

$$\{P_m\} = \{\vec{P}_e, H_e, N_e, H_L, H_{e-L}\}$$

and

$$\{F_m\} = \{-\beta_e\vec{V}_D, \beta_e, -\beta_e (\mu-1/2mV_D^2), \beta_L, \beta_L^{eq}\}$$

The values of the F_m's are such as to make $\langle P_m\rangle = Tr\{\rho_L P_m\}$ for the first four of the macrovariables. In the following, we shall use the reduced chemical potential μ' for the quantity $\mu - mV_D^2/2)$. We further define $\sigma(t)$ as

$$\sigma(t) = \phi + \sum_m P_m F_m (t) \quad .$$

We will find it useful in the following first to derive two important properties of the mean values that are calculated with the help of ρ_L the quasi-equilibrium statistical distribution (QESD). These are ($\hbar$ = 1)

$$\langle \dot{P}_{m,L}\rangle_{\rho_L} = i\ Tr\{\rho\ [H_{e-L}, P_m]\} = 0, \qquad (5)$$

$$\sum_m \langle \dot{P}_m\rangle_{\rho_L} F_m = i \sum_m F_m\ Tr\{\rho_L\ [H, P_m]\} = 0, \qquad (6)$$

where the time to which ρ_L and F_m refer has been dropped. Here, the notation $\dot{P}_{m,L}$ means the time variation introduced by $i[H_{e-L}, P_m]$. We first note that the property (5) is trivial if $P_m = H_{e-L}$. To demonstrate (5) for the other four macro-variables, we note that

$$[\rho_L, H_{e-L}] = \frac{1}{\beta_L{}^{eq}} [\rho_L, -\beta_e H_e +$$

$$\beta_e \vec{V}_D \vec{P}_e + \beta_e \mu' N_e - \beta_L H_L] \quad .$$

Now, by writing $\langle \dot{P}_{m,L} \rangle = i \, Tr \{P_m [\rho_L, H_{e-L}]\}$ and using the above expression for $[\rho_L, H_{e-L}]$, it becomes easy to demonstrate (5) by recognizing that the first four macrovariables mutually commute. The property (5) implies that we can prepare our system in a state defined by the QESD (4) so that the physical irreversibility which arises from H_{e-L} has no effect on the motion, at least during the time in which the system resides in a stationary state. We will see below that (5) is essential for the derivation of a Langevin equation describing the motion of the first four macrovariables.

We can also easily establish (6). We first write the average of the time variation of the quantity

$$\sum_m \langle \dot{P}_m \rangle_{\rho_L} F_m = i \sum_m F_m \, Tr \{H[P_m, \rho_L]\} = i \, Tr \{H[\sum_m F_m P_m, \rho_L]\}$$

The result (6) arises from the evident communtation of ρ_L and $\sum_m F_m P_m$.

We pointed out earlier that the macrovariables are defined through the averaging process $\langle P_m \rangle = Tr\{\rho P_m\} = Tr\{\rho_L P_m\}$. Now, by differentiating ρ_L with respect to the F_n's, we get

$$\delta \langle P_m \rangle = -\sum_n (P_m;P_n)^t \, \delta F_n \quad , \tag{7a}$$

and

$$\langle \dot{P}_m \rangle = -\sum_n (P_m;P_n)^t \, \dot{F}_n \tag{7b}$$

where the correlation function $(P_m;P_n)^t$ is defined by

$$(P_m;P_n)^t = (P_n;P_m)^t = \int_0^1 d\tau \, Tr \{P_m e^{-\tau\sigma(t)} (P_n - \langle P_n \rangle^t) e^{(\tau-1)\sigma(t)}\} \, . \tag{8}$$

Recall that the formulae (7) and (8) are easily derived if we remember the expansion of $e^{-(A+\delta A)}$ in powers of δA (limited to the first order)

$$e^{-(A+\delta A)} = \{1 - \int_0^1 d\tau\, e^{-A\tau}\delta A\, e^{A\tau}\}e^{-A} \;. \tag{9}$$

In the following, we will use this correlation function extensively, and will employ the same formal definition even if P_m and P_n are time dependent quantities. Equations (7a) and (7b) are here restricted to the first four P_m's.

Finally, to round out this paragraph, we note one more interesting property. If P_n and P_m are two of the macrovariables in the hot electron problem, we obtain

$$(\dot{P}_{n,L};\, P_m)^t = 0 \;. \tag{10}$$

This relation is readily obtained from (5), which we can apply both for ρ_L (associated with the near-equilibrium force F_m) and for $\rho_L + \delta\rho_L$ (associated with the modified force F_m and δF_m). Using this combination, we then obtain

$$\langle \dot{P}_{m,L} \rangle_{\rho_L+\delta\rho_L} = \langle \dot{P}_{m,L} \rangle_{\rho_L} + \sum_n (\dot{P}_{m,L};\, P_n)\, \delta F_n \;,$$

From which (10) is an obvious result.

III. The Dissipationless Motion

The Langevin equation describes the relaxation of mechanical variables of a given system near equilibrium. It was first formally derived with the help of a projection operator technique by Mori[12]. In the following, we give a derivation of the Langevin equation, of Mori-type, to emphasize the physical priciples we will use in the derivation of the Langevin equation for the P_m's in a far-from-equilibrium situation.

Let us call A a set of coupled mechanical variables of the system and L its Liouville operator, such as

$$\dot{A} = iLA$$

In the framework of the derivation of the Langevin equation, the system is assumed to be in a non-dissipative state at the initial time t=0. Any initial value of A at this time is assumed to arise from a static field applied in the far distant past. Moreover, this field is turned off at t=0. At time t = 0, the "speed" of the variables A is given by

$$\langle \dot{A} \rangle^0 = \text{Tr} \{\rho(t = 0) \dot{A}\} \quad , \tag{11}$$

where

$$\rho(t = 0) = \exp \{-\phi - \beta H - \sum_j \mu_j A_j\} \quad , \tag{12}$$

and ϕ assumes the proper normalization conditions of ρ. The μ_j describe the fields which hold the system in the non-dissipative state for t=0. Being in the framework of linear response theory, the μ_j are supposed to be small.

The linear expansion of $\rho(t = 0)$ gives (see (9))

$$\rho(t = 0) = \rho_o - \sum_j \mu_j \int_o^1 d\tau\, e^{-\tau\beta H} A_j\, e^{\tau\beta H} \rho_o \quad , \tag{13}$$

where ρ_o corresponds to $\rho(t=0)$ with $\mu_j = 0$. In the above formula, we have supposed that Tr $\{\rho_o \dot{A}_j\} = 0$. The expression for $\langle \dot{A}_i \rangle^0$ is now straightforward, and we get

$$\langle \dot{A}_i \rangle^0 = -\sum_j \mu_j \, (\dot{A}_i, A_j) \quad , \tag{14}$$

in which

$$(\dot{A}_i, A_j) = \int_o^1 d\tau \,\text{Tr} \{\rho_o \dot{A}_i\, e^{-\tau\beta H} A_j\, e^{\tau\beta H}\} \quad , \tag{15}$$

a similar definition as the one used in (8). The expression for $\langle \dot{A}_i \rangle^0$ is not exactly what we want because of the presence of the μ_j's. However, we can again use the expansion of $\rho(t = 0)$ to evaluate $\langle A \rangle^0$ through

$$\langle A_i \rangle^0 = \text{Tr} \{\rho(t = 0)\, A_i\} \quad , \tag{16}$$

and we get

$$\langle A_i \rangle^0 = - \sum_j \mu_j (A_i ; A_j) \quad . \tag{17}$$

This allows us to obtain the μ_j's, and finally

$$\langle \dot{A} \rangle^0 = \Omega \, \langle A \rangle^0 \quad , \tag{18}$$

where the matrix Ω is defined by

$$\Omega = (\dot{A};A)\,(A;A)^{-1} \quad . \tag{19}$$

From the above, we can now define a projection operator D through

$$DB = (B;A)\,(A;A)^{-1}\,A \quad , \tag{20}$$

and write, for t = 0,

$$\langle\dot{A}\rangle^{0} = \langle i\ D\ LA\rangle^{0} \quad . \tag{21}$$

There is no dissipation associated with this particular motion. This simple, but physical, fact is clear if we recall that at t = 0, when the applied field is switched off, the system still has time reversal symmetry and the fluctuations cannot, by themselves, induce any dissipation. Let us now write the Liouville equation in the following way:

$$\dot{A} = e^{iLt} DiLA + e^{iLt}(1-D)iLA \quad . \tag{22}$$

The first term of (22) obviously characterizes the dissipationless motion of the variable A at time t. The second term can be evaluated with the help of the following identity:[13]

$$e^{iLt} = e^{(1-D)iLt} + \int_0^t ds\ e^{iL(t-s)} DiLe^{(1-D)iLs} \tag{23}$$

From (22) and (23), we obtain the Langevin equation

$$\dot{A} = e^{iLt} DiLA + e^{(1-D)iLt}(1-D)iLA$$
$$+ \int_0^t ds\ e^{iL(t-s)} DiLe^{i(1-D)Ls}(1-D)iLA \quad , \tag{24}$$

in which the first term is the dissipationless motion by construction.

We now look at the second term. By performing the expansion of the exponential, we can show that

$$f^{(2)} = e^{(1-D)iLt}(1-D)iLA = (1-D)\, f^{(2)} \quad , \qquad (25)$$

for D is a projector. To understand what this force is, we have to look at its mean value which is

$$\langle f^{(2)}\rangle = \mathrm{Tr}\ \{\rho(t = 0)\ f^{(2)}\} \quad .$$

Writing this last equality with the help of D, we obtain

$$\langle f^{(2)}\rangle = -\sum_j \mu_j\ (A_j;(1-D)\ f^{(2)}) \quad , \qquad (26)$$

which is zero, for D is a projector associated with the scalar product (8), so that DA = A. We have just shown that $f^{(2)}$ is a fluctuating form.

Writing the third term $f^{(3)}$ of (24) explicitly is a matter of technical algebra, and we obtain

$$f^{(3)} = \int_o^t ds \sum_k \phi_{jk}(s)\ A_k(t-s) \quad , \qquad (27)$$

with

$$\phi_{jk}(s) = (A_k;\ i\ Lf_k^{(2)}) \quad . \qquad (28)$$

Thus, $f^{(3)}$ characterizes the relaxation of the process.

We see from this derivation of the Mori-type Langevin equation that the crucial point is to be able to define a projection operator. This is achieved here by a careful analysis of the dissipationless motion. We are now going to use the above argument to construct the Langevin equation for the relevant mechanical variables P_m's of our hot-electron problem. However, we introduce some slight differences due to the existence of the far- from-equilibrium stationary state in the present problem. We begin by preparing our system at t=0 in a thermodynamic state in which the density matrix is given by $\rho_L(t=0)$ i.e., $\rho(t=0)=\rho_L(t=0)$. We know from (5) that the source of the irreversibility in the system does not induce any dissipation into the motion of $\langle P_m\rangle$ at this initial time t=0. Further, a given P_m can be written ($\langle P_m\rangle$ is assumed to be $\langle P_m\rangle^0$)

$$P_m = \langle P_m\rangle + (P_m - \langle P_m\rangle) \quad . \qquad (29)$$

If we seek the particular evolution of P_m, for which there is no dissipation at t = 0, we suppose that the evolution of the fluctuation $(P_m - \langle P_m \rangle)$ is the same as that for $\langle \delta P_m \rangle$ induced by some generalized force δF_m (t = 0) (this is a form of an Onsager hypothesis and is consistent with the fact that δP_m should be given by δ_{ρ_L} at t = 0 to insure dissipationless motion). We therefore introduce a dissipationless motion at t = 0 as

$$\left.\frac{dP_m}{dt}\right|_{D\cdot L} = \mathrm{Tr}\,\{\rho_L\, \dot{P}_m\} + \sum_n \Omega_{mn}\,(P_n - \langle P_n \rangle) \quad , \tag{30}$$

where the matrix Ω is given by

$$\Omega = (\dot{P};P)\,(P;P)^{-1} \quad . \tag{31}$$

As before, the correlation functions are given by (8). The thermodynamical parameters which enter these correlation functions are given by our initial condition ρ_L (t = 0). From (30), it is easy to see that the difference between

$$\frac{dP_m}{dt} \text{ and } \left.\frac{dP_m}{dt}\right|_{D\cdot L}$$

is just a fluctuation term.

We are now in a position to define the relevant projection operator, at least at t = 0. This becomes

$$\Pi(0)\, X = \mathrm{Tr}\,\{\rho_L\,(t = 0)\, X\} + \sum_{j,k} (X;P_j)\,(P;P)^{-1}_{jk}\,(P_k - \langle P_k \rangle), \tag{32}$$

and at t=0

$$\left.\frac{dP}{dt}\right|_{D\cdot L} = \Pi(0)\; i\; L\; P \quad . \tag{33}$$

Here, and in the following, we shall take P to be the column matrix whose elements are the individual P_m's. Thus, all equations involving P (without an index) have to be interpreted as having matrix products and sums.

Let us now turn to the $t \neq 0$ case. With the previous considerations in mind, we will assume that the dissipationless motion of the P_m is given by an expression analogous to (30), as

$$\left.\frac{dP_m}{dt}\right|_{D\cdot L} = \mathrm{Tr}\,\{\rho_L\,(t)\, P_m\} + \sum_n \Omega_{mn}\,(P_n\,(t) - \langle P_n \rangle^t) \quad , \tag{34}$$

where the matrix Ω has the definition (31), in which the correlation functions are calculated with the appropriate thermodynamical parameters at time t. By defining $\Pi(t)$ as

$$\Pi(t)X=\mathrm{Tr}\ \{\rho_L(t)X\}+\sum_{j,k}(X;P_j{}^t)\ \{(P;P)^t\}^{-1}_{jk}(P_k-\langle P_k\rangle^t), \qquad (35)$$

we can write

$$\left.\frac{dP}{dt}\right|_{D\cdot L} = e^{iLt}\ i\ \Pi(t)\ L\ P\ . \qquad (36)$$

It is obvious from (35) that $\Pi(t)$ is a projection operator. This operator corresponds to one introduced in a somewhat different manner by Grabert[7]. We are now going to examine its properties.

IV. Properties of $\Pi(t)$

To clarify our task, let us list the properties we will use in the following discussion:

$$\Pi(t)X = \mathrm{Tr}\ \{\rho_L(t)X\} + \sum_k \mathrm{Tr}\ \{\frac{\partial\rho_L(t)}{\partial\langle P_k\rangle}X\}\ [P_k - \langle P_k\rangle^t] \qquad (37)$$

$$\dot{\Pi}(t)X = \sum_{j,k}\mathrm{Tr}\ \{\frac{\partial^2\rho_L(t)}{\partial\langle P_k\rangle\partial\langle P_j\rangle}X\}\ \langle\dot{P}_j\rangle\ [P_k - \langle P_k\rangle^t\ ] \qquad (38)$$

a $\quad \Pi(t)\ \Pi(t') = \Pi(t')$ (39)

b $\quad [1-\Pi(t)]\ [1-\Pi(t')] = [1-\Pi(t)]$

a $\quad \mathrm{Tr}\ \{\rho(t)\ \Pi(t)X\} = \mathrm{Tr}\ \{\rho_L(t)X\}$ (40)

b $\quad \mathrm{Tr}\ \{\rho_L(t)\ \Pi(t)X\} = \mathrm{Tr}\ \{\rho_L(t)X\}$

c $\quad \mathrm{Tr}\ \{\rho_L(t)\ [1-\Pi(t)]X\} = 0$

$$(X;\ \Pi(t)Y)^t = (\Pi(t)X;\ Y)^t \qquad (41)$$

Property (37) comes from the definition (35), if we recognize that

$$\frac{\partial \rho_L}{\partial \langle P_K \rangle} = \sum_j \frac{\partial \rho_L}{\partial F_j} \frac{\partial F_j}{\partial P_K} ,$$

and if we recall (7) and (9). Property (38) is easily stated from (37) by differentiation.

Property (40) comes straightforwardly from the definition of $\Pi(t)$ and the identity $\langle P_k \rangle^t = \mathrm{Tr}\,\{\rho(t)P_K\}$. Property (41) is a consequence of the fact that $\Pi(t)$ is a projection operator associated with the scalar product defined by (8).

The only property that is hard to prove is property (39). However, even though it is quite messy, this is just a matter of technical algebra. We turn now to the derivation of the Langevin equation for the P_m's.

V. The Langevin Equation

We are mainly going to follow the scheme of section III. We first write

$$\frac{dP}{dt} = e^{iLt}\, i\, \Pi(t)\, L\, P + e^{iLt}\, i\, [1 - \Pi(t)]\, L\, P\, . \tag{42}$$

Introducing the operator $G(s,t)$ as a two-time Green's function given by [7,14]

$$G(s,t) = T_-\, \exp\left\{i \int_s^t du\, L\, (1 - \Pi(u))\right\} ,$$

and satisfying the differential equations

$$\frac{dG(s,t)}{dt} = i\, G(s,t)\, L\, [1 - \Pi(t)] , \tag{43a}$$

$$\frac{dG(s,t)}{ds} = -i\, L\, [1 - \Pi(s)]\, G(s,t) , \tag{43b}$$

it is easy to establish the following identity similar to (23)

$$G(s,t) = e^{iL(t-s)} - \int_s^t ds'\, e^{-iL(s-s')}\, i\, L\, \Pi(s')G(s',t)\, . \tag{44}$$

The property given by (44) can now be used to express the

operator exp(iLt) in the second term of (42) within the framework usually utilized for the derivation of a Langevin equation (see (24)), and

$$e^{iLt}[1 - \Pi(t)]\, L\, P = G\,(0,t)[1 - \Pi(t)]\, L\, P \qquad (45)$$

$$+ \int_0^t ds' e^{iLs'}\, i\, L\, \Pi(s')G(s',t)[1 - \Pi(t)]\, L\, P \quad .$$

In order to examine the characteristic features of (45), we have to define precisely our initial condition. For this, we take $\rho(t = 0) = \rho_0$, where ρ_0 is the density matrix corresponding to the equilibrium situation where the field is zero. We therefore assume that the field is applied at t = 0 (when $\beta_e = \beta_L$, $\langle\vec{V}_e\rangle = 0$). This initial condition corresponds to a particular ρ_L. Now, it appears that the first term of (45) is not a fluctuating force for this particular initial condition. Nevertheless, we can always write it as

$$G(0,t)[1 - \Pi(t)]\, L\, P = [1 - \Pi(0)]\, G(0,t)[1 - \Pi(t)]\, L\, P \qquad (46)$$

$$+ \Pi(0)\, G(0,t)[1 - \Pi(t)]\, L\, P\ ,$$

where the first term on the right is a fluctuating force for our intial condition (see (40)). The second term on the right of (45) is now integrated by parts, and the first term of (45) is replaced using (46), so that we arrive at

$$\dot{P}(t) = e^{iLt}\Pi(t)\dot{P} + \int_0^t ds'\ e^{iLs'}\Pi(s')i\ L[1-\Pi(s')]G(s',t)\dot{P} \qquad (47)$$

$$+ [1 - \Pi(0)]G(0,t)\ \dot{P} + \int_0^t ds'\ e^{iLs'}\dot{\Pi}(s')G(s',t)[1 - \Pi(t)]\ \dot{P}\ ,$$

where the first term gives the dissipationless motion (which may be compared to the t = 0 situation). The second term on the right of (47) can now be divided into two parts, one of which characterizes the dissipation and the other of which leads to a fluctuating force caused by the irreversibility induced by the H_{e-L} coupling, and the nonequilibrium nature of the system (we return to this latter term in section VII below). The third and fourth terms are two additional fluctuating forces, the former of which is Mori-like[12, 14] in nature and the latter characterizes the fluctuations induced by the "speed" of macrovariables during the transient regime.

Equation (47) is now a Langevin equation for the macrovariables, and its derivation does not include any approximations. In the next section, we focus our attention on the first two terms on the RHS of (47) and establish the evolution equation of the $\langle P_m \rangle$'s. This will allow us to obtain a closed set of equations for the F_n's which is necessary in order to evaluate ρ_L and the correlation functions involved in the theory.

VI. The Thermodynamical Motion

From the form of (45), it is easy to determine the average motion of the macrovariables. This can be written as

$$\langle \dot{P}(t) \rangle = \langle e^{iLt} \Pi(t) \dot{P} \rangle + \langle \int_0^t ds' e^{iLs'} \Pi(s') i\, L[1 - \Pi(s')] G(s',t)\, \dot{P} \rangle , \tag{48}$$

where

$$\langle X \rangle = \mathrm{Tr}\, \{\rho_o X\} .$$

We are now going to show that up to second order in electron-lattice interaction, (48) takes the form [5, 6]

$$\langle \dot{P}_m(t) \rangle = \sum_n \alpha_{mn} \langle P_n(t) \rangle + \sum_n \int_0^t dt' (\dot{P}_{m,L}; \dot{P}_{n,L}(-t'))^t \, F_n(t-t') , \tag{49}$$

where $\dot{P}_{m,L}$ was defined in section II and corresponds to the effect of H_{e-L} on the time derivative of P_m. In (49), the α_{mn} arise from the fact that the individual macrovariables are coupled in the hot electron case. Here, the Liouville equation

$$\dot{P}_m = i\, L_o\, P_m ,$$

where L_o is the Liouville operator in the absence of the interaction between the electrons and the phonons, can be written

$$\dot{P}_m = \sum_n \alpha_{mn} P_n . \tag{50}$$

More precisely, if $H_{e-L} = 0$, the definition of α_{mn} is given by

$$\dot{\vec{P}}_e = e N_e \vec{F} \quad , \quad \dot{N}_e = 0 \; ,$$
$$\dot{H}_e = \frac{e\vec{F}}{m} \vec{P}_e \quad , \quad \dot{H}_L = 0 \; . \tag{51}$$

Here, $\vec{F}$ is the applied electric field.

To prove the equivalence of (48) and (49), we first look at the two first terms. The equality of these terms is easily established through (5). Let us now turn to the dissipation terms, which we shall examine only up to second order in the interaction H_{e-L}. As a starting point, we take the term from (48) as

$$\begin{aligned} T &= \langle \int_o^t ds' e^{iLs} \Pi(s') \, i \, L \, [1 - \Pi(s')] \, G \, (s',t) \, \dot{P} \rangle \\ &= \int_o^t ds' \, \mathrm{Tr} \, \rho_L(s') \, i \, L \, [1 - \Pi(s')] \, G \, (s',t) \, \dot{P} \end{aligned} \tag{52}$$

where the last equality comes from section IV. To establish the connection with the equivalent form in (49), we proceed in three steps. We first show that

$$T = \sum_n \int_o^t ds' \, F_n(s')([1 - \Pi(s')] \, G \, (s',t) \, \dot{P}; \, \dot{P}_n)^{s'} \quad . \tag{53}$$

This can be established by recognizing that the properties of the operator L under a trace operation allows us to write (52) as

$$T = \sum_n \int_o^t ds' \mathrm{Tr}\{\{-iL\rho_L(s')\} \, [1 - \Pi(s')] \, G \, (s',t) \, \dot{P} \, \} \; . \tag{54}$$

Using the Kubo formula for the commutator, T takes the form

$$\begin{aligned} T = \sum_n \int_o^t ds' \, F_n(s') \int_o^1 d\alpha \, \mathrm{Tr} \, \{ e^{-\alpha \sum_j F_j(s') P_j} \, \dot{P}_n e^{\alpha \sum_k F_k(s') P_k} \times \ldots \\ \times \ldots \rho_L(s')[1 - \Pi(s')] \, G \, (s',t) \, \dot{P}\} \quad , \end{aligned} \tag{55}$$

which can be written as (53) by using (6) and the definition (8) for the correlation functions.

For the second part of the proof, we want to show that (53) can be written in the form

$$T=\sum_n \int_0^t ds' \; F_n(s') \; (G(s',t)[1 - \Pi(t)] \; \dot{P}; \; [1 - \Pi(s')] \; \dot{P}_n)^{s'} \; , \qquad (56)$$

This can be established from (53) by recognizing that the operator G(s',t) has the general property (which can be established by its expansion and the properties of Section IV)

$$[1 - \Pi(s')] \; G(s',t) \; \dot{P} = [1 - \Pi(s')] \; G(s',t)[1 - \Pi(t)]\dot{P} \; . \qquad (57)$$

We then note that since $\Pi(s')$ is a projection operator associated with the scalar product $(A;B)^{s'}$, defined by (8), the result (56) follows immediately.

We now show that

$$T = \sum_n \int_0^t dt' \; (\dot{P}_{,L};\dot{P}_{n,L}(-t'))^t F_n(t-t') \; , \qquad (58)$$

where $\dot{P}_{,L}$ is a general notation for the $\dot{P}_{n,L}$'s. The value of T given by (58) is the expected result that insures the equivalence of (48) and (49). This final result can be established by making use of an expansion in H_{e-L} up to second order, an approximation not made up to now. We start by noticing that for the first four P_m

$$[1 - \Pi(s')]\dot{P}_n = \dot{P}_{n,L} \; , \qquad (59)$$

which is easily stated with the help of (10). If $P_n = H_{e-L}$, we define $\dot{P}_{n,L}$ as $\dot{P}_{n,L} = -\dot{H}_{e,L} - \dot{H}_{L,L}$ and we get

$$T = \sum_n \int_0^t ds' \; F_n(s') \; (G(s',t) \; \dot{P}_{,L}; \; \dot{P}_{n,L})^{s'} \qquad (60)$$

In order to determine T up to second order in H_{e-L}, it is adequate to obtain the expansion of $G(s',t)\dot{P}_{,L}$ only up to zero order in H_{e-L}. This follows from the examination of the form of (60). In the appendix we establish that this may be achieved by writing

$$G(s',t)\,\dot{P}_{,L} = \dot{P}_{,L}(t-s') \;.$$

It, therefore, turns out that T can now be written,

$$T = \sum_{n} \int_{o}^{t} ds' \,(\dot{P}_{,L}(t-s');\, \dot{P}_{n,L})^{s'} F_{n}(s') \;,$$

which is not exactly the form (58). This is because the correlation functions are not evaluated at the same time. However, we can write $t - s' = u$, so that the correlation function we have to evaluate is

$$(\dot{P}_{,L}(u);\dot{P}_{n,L})^{t-u} = \int_{o}^{1} d\alpha\, \mathrm{Tr}\{\dot{P}_{,L}(u)e^{-\alpha\sigma(t-u)}\dot{P}_{n,L}e^{\alpha\sigma(t-u)}\rho_{L}(t-u)\} \quad (61)$$

where the product $\langle\dot{P}_{,L}\rangle\langle\dot{P}_{n,L}\rangle$ has been omitted since it is of fourth order in the interaction. The evaluation of (61) to second order in the interaction requires that our knowledge of ρ_L and $\exp(\alpha\sigma)$ only has to be up to zero order. Up to this order, we can write (here u is always limited to a characteristic time)

$$\rho_{L}(t) = e^{-iHu}\;\rho_{L}(t-u)\;e^{iHu} \;, \quad (62a)$$

$$e^{\alpha\sigma(t)} = e^{-iHu}\;e^{\alpha\sigma(t-u)}\;e^{iHu} \quad (62b)$$

We can now write (61) as

$$(\dot{P}_{,L}(u);\;\dot{P}_{n,L})^{t-u} = (\dot{P}_{,L};\;\dot{P}_{n,1}(-u))^{t} \;,$$

which gives (58) for T.

Having shown that (48) leads to (49) in the lowest order in H_{e-L}, we can derive a closed set of equations describing the evolution of the F_n at the same level of approximation. Using (49) and (7b), we get[5]

$$\dot{F}_n = - \sum_{ml} \{(P;P)^t\}^{-1}_{nm} \alpha_{ml} \langle P_l \rangle^t - \tag{63a}$$

$$- \sum_{ml} \{(P;\ P)^t\}^{-1}_{nm} \int_o^t dt' \ (\dot{P}_{m,L};\ \dot{P}_{l,L}(-t'))^t \ F_l(t-t')$$

These equations form a highly nonlinear set, since the F_n's define the values of $\langle P_l \rangle^t$, $(P;P)^t{}_{nm}$, and $(\dot{P}_{m,L};\ \dot{P}_{l,L}(-t'))^t$. In principle, (63a) allows us to obtain any quantity we may want to evaluate, through the knowledge of the thermodynamic parameter F_n.

It is possible to write (63a) in a different way by using the $\langle P_m \rangle$'s as variables. By integrating (7b), we get, after some algebra,

$$\langle \dot{P}_m \rangle(t) = \sum_n \alpha_{mn} \langle P_n \rangle(t) + \sum_n [F_n(0) +$$

$$+ \sum_l \{(P;\ P)^o\}^{-1}_{nl} \langle P_l \rangle(0)] \int_o^t ds' \ (\dot{P}_{m,L};\dot{P}_{n;L}(s'-t))^t +$$

$$+\sum_{nl} \int_o^t ds' \ \langle P_l \rangle(s') \ \frac{d}{ds'} \ [\{(P;\ P)^{s'}\}^{-1}_{nl} \ \int_{s'}^t ds'' \ (\dot{P}_{m,L};\dot{P}_{n,L}(s''-t))^t]-$$

$$-\beta_L^{eq} \int_o^t ds' \ (\dot{P}_{m,L};\dot{H}_{e,L}(s'-t) + \dot{H}_{L,L}(s'-t)) \tag{63b}$$

In (63b), m and n are such as $m, n \leq 4$. If we add the fluctuating forces to the RHS of (63b), we obtain an equation for $\dot{P}_m(t)$, which bears some similarity to the usual Langevin equation near equilibrium.

VII. Fluctuations and Correlations

Having considered the thermodynamical motion of the macrovariables, we now come back to (47) and analyze its purely fluctuating component. It is easy to realize that there are four random forces buried in this equation. Let's call them f_1 (corresponding to the second term in (47)), f_2 (corresponding to the fourth term in

(47)), f_3 (corresponding to the first term in (47)) and f_4 (the Mori-like fluctuating force).

Here, f_1, f_2, f_3, f_4 are understood in a vectorial notiation, and we obtain

$$f_1 = \sum_n \int_o^t ds' \, K_n^{(1)} (s', t) \, (P_n(s') - \langle P_n \rangle^{s'}) \, , \qquad (64a)$$

$$f_2 = \sum_n \int_o^t ds' \, K_n^{(2)} (s', t) \, (P_n(s') - \langle P_n \rangle^{s'}) \, , \qquad (64b)$$

$$f_{3,m} = \sum_n \alpha_{mn} \, (P_n(t) - \langle P_n \rangle^t) \, , \qquad (64c)$$

$$f_4 = [1 - \Pi(0)] \, G(0,t) \, \dot{P} \, , \qquad (64d)$$

where the kernels $K_n^{(1)}$ and $K_n^{(2)}$ take the form

$$K_n^{(1)}(s',t) = \sum_j (i \, L \, (1-\Pi(s')) \, G(s',t) \, \dot{P};P_j)^{s'} \{(P;P)^{s'}\}^{-1}_{jn} \, , \qquad (65a)$$

$$K_n^{(2)}(s',t) = \sum_j \mathrm{Tr} \left\{ \frac{\partial^2 \rho_L(s')}{\partial \langle P_j \rangle \, \partial \langle P_n \rangle} \, G(s',t) \, [1 - \Pi(t)] \, \dot{P} \right\} \langle \dot{P}_j \rangle \, . \qquad (65b)$$

Notice that the lowest order in H_{e-L} is second for f_1 and f_2, first for f_4 and zero for f_3. Now, the kernels $K^{(1)}$ and $K^{(2)}$ characterize the evolution of the fluctuating forces f_1 and f_2 when the system goes from its initial state (defined by $\rho_L(0)$) towards the steady state. When this steady state is reached, $K_n^{(2)}$ goes to zero and $K_n^{(1)}$ becomes a function of t-s' only. These kernels, which bear the memory of the fluctuating forces f_1 and f_2, are extremely important if we want to analyze the noise of submicron devices during the so called "ballistic regime."

They can be determined through the evolution equations of a well-chosen set of correlation functions - the $H_{lm}(t)$ defined by

$$H_{lm}(t) = \theta(t)\ (P_l;P_m(t))^o\ . \tag{66}$$

The initial state at t=0 is defined by $\rho_L(t=0)$. If $\rho_L(t=0)$ corresponds to a thermodynamical equilibrium, $H_{lm}(0) = \delta_{lm}(P_l,\ P_l)^0$; this situation is the usual one for the "ballistic regime." Now, the $H_{lm}(t)$ obey the following set of equations.

$$\frac{d}{dt} H_{lm}(t) = \delta(t)\ H_{lm}(0) + \sum_n \alpha_{mn}\ H_{ln}(t) + \sum_n \int_o^t ds'\ [K^{(1)}_{mn}(s',t) + K^{(2)}_{mn}(s',t)]\ H_{ln}(s')\ , \tag{67}$$

where the $K^{(1)}_{mn}$ and $K^{(2)}_{mn}$ are the components of the $K^{(1)}_n$ and $K^{(2)}_n$. These equations are general and do not presume any approximation. They can be used to obtain $K^{(1)}$ and $K^{(2)}$ through the knowledge of the $H_{lm}(t)$ during the nonequilibrium, nonstationary regime of a packet of carriers in a device. Notice that the choice of the H_{lm}'s is such that the Mori-force has no influence on their equation of motion.

If we are interested in $K^{(1)}$ during the stationary regime, we notice that the $H_{lm}(t)$ are no longer the relevant functions, instead we must use the $G_{lm}(t)$ defined by

$$G_{lm}(t) = (P_l(t_o);\ P_m(t_o + t))^o\ , \tag{68}$$

where t=0 is again the initial time, but where t_o is a time sufficiently large to allow the system to be in the steady state. These functions satisfy a set of differential equations which can be solved by Fourier analysis. However, they contain an extra source term which arises from the Mori-like fluctuating force that does not leave.

We would like now to be more specific and to study the Langevin equation of the momentum P_e of the carriers[8]. To do so, we will neglect the cross-correlation functions in the dissipation terms of (47) (which has been shown earlier to be an acceptable procedure if mV_D^2 is small compared to $\langle H_e\rangle$.[5] We will also neglect the cross-correlations in the fluctuating force in order to be consistent. By using the notation of (65) and the definition of the parameter that is thermodynamically conjugate to P_e, i.e., $\beta_e V_D$, we can use (47) and (49) for the component P_e along the electric field

$$\frac{dP_e}{dt} = e F N_e - \int dt' \beta_e(t-t') V_D(t-t')(\dot{P}_{e,L}; \dot{P}_{e,L}(-t'))^t$$

$$+ \int_o^t dt' [K_{ee}^{(1)}(t',t) + K_{ee}^{(2)}(t',t)] (P_e(t') - mN_e V_D(t'))$$

$$+ [1 - \Pi(0)] G(0,t) \dot{P}_e \quad . \tag{69}$$

The kernel $K_{ee}^{(1)}$ can now be examined by expanding the terms up to second order in interaction. We can write, form (65a),

$$K_{ee}^{(1)}(t',t) =$$

$$\frac{1}{(P_e;P_e)^{t'}} \int_o^t d\alpha Tr\{\rho_L(t')[H,\{1-\Pi(t')\}G(t',t)\dot{P}_e]e^{-\alpha\sigma(t')}P_e e^{\alpha\sigma(t')}. \tag{70}$$

If H comutes with $\rho_L(t')$ (and hence with $e^{\pm\alpha\sigma(t')}$), then $K_{ee}^{(1)}(t', t)$ reduces to

$$K_{ee}^{(1)}(t',t) = \frac{-1}{(P_e;P_e)^{t'}} ([1-\Pi(t')] G(t',t) \dot{P}e; \dot{P}e)^{t'} \tag{71}$$

$$= \frac{-1}{(P_e;P_e)^{t'}} (\dot{P}_{e,L}; \dot{P}_{e,L}(t'-t))^t,$$

where the last equality is valid up to second order in the interaction.

However, because dissipation is involved in the problem at hand, ρ_L does not commute with the Hamiltonian H, and we must write

$$K_{ee}^{(1)}(t',t) = \frac{-1}{(P_e;P_e)t'} [-(\dot{P}_{e,L}; \dot{P}_{e,L}(t'-t))^t + I_R(t',t)] \quad (72)$$

where I_R can be evaluated in the classical limit (in which quantum fluctuations are ignored) up to second order as

$$I_R(t',t) = i \text{ Tr } \{[\rho_L(t'), H][1 - \Pi(t')] \dot{P}_e(t-t') P_e\} . \quad (73)$$

The correlation function $(P_e; P_e)^{t'}$ in (72) can be evaluated with (7a) and we get

$$(P_e; P_e)^{t'} = N m \beta_e^{-1}(t') . \quad (74)$$

Our Langevin equation for P_e is now

$$\frac{dP_e}{dt} = eFN_e - \frac{1}{mN_e} \int_o^t dt' \beta_e(t-t') P_e(t-t')(\dot{P}_{e,L}(\dot{P}_{e,L}(-t'))^t$$
$$+ \int_o^t dt' [P_e(t-t') - N_e mV_D(t-t')][I_R(t-t',t) + K_{ee}^{(2)}(t-t',t)] \quad (75)$$
$$+ [1 - \Pi(0)] G(0,t) \dot{P}_e .$$

The first terms in (75) are the expected ones in view of Section 5. The third term arises from the dissipation and the nonstationary part of the dynamics. It does not disappear in the steady state (because of I_R) and makes (75) different from the near equilibrium Langevin equations.[12,14] However, this third term in (75) can be shown to be of first order in the field, for the near equilibrium situations while the fluctuating forces coming from the second and fourth term are of zero order. Therefore, in the near equilibrium situation, the third term can be dropped and we come back to the usual Langevin equation. We conclude that this third term is relevant only in the ballistic and hot-electron regime.

VIII. Conclusions

In these lectures, we have shown how to derive a Langevin equation for the relevant variables of an electronic system that is coupled to a phonon bath in a semiconductor for the case of high electric field. The presence of this high field leads to a far-from-equilibrium situation. We have analyzed the thermodynamical motion of the system and have carefully looked at the random forces. Two of them have memory of the past due to irreversibility and nonstationary behavior. The kernels of these forces were analyzed and linked with a complete set of correlation functions.

Finally, in the framework of reasonable assumptions we have derived a Langevin equation for the electronic momentum itself. This equation has been proven to be different from the usual one near equilibrium, because of extra fluctuating forces which arise from the nonequilibrium part of the dynamics.

APPENDIX

We want to show that the equality

$$G(s,t)\ \dot{P}_{n,L} = \dot{P}_{n,L}\ (t - s) \tag{A.1}$$

is valid to first order in H_{e-L}, and the higher orders are insignificant in the present context. From the definition (43), we can write

$$G(s,t)\dot{P}_{n,L} = \left[1 + \sum_{m=1}^{\infty} \int \dots \int L\ [1 - \Pi(t_1)] \dots L\ [1 - \Pi(t_m)]\ dt_1 \dots \dots dt_m\right] \dot{P}_{n,L}\ , \tag{A.2}$$

where $t_1 \dots t_m$ are time ordered so that $t_1 < t_2 \dots < t_m$.

If we want to evaluate $G(s,t)\ P_{n,L}$ to first order in H_{e-L}, we need to replace L by L_o in (A.2). We notice that

$$[1 - \Pi(t_m)]\ \dot{P}_{n,L} = \dot{P}_{n,L}\ ,$$

thanks to (59). We have now to study the terms

$$[1-\Pi(t_{m-1})]L_o\dot{P}_{n,L}$$

in order to evaluate $G(s,t)\ P_{n,L}$ from (A.2). It appears that $\Pi(t)L_o\ \dot{P}_{n,L}$ is of second order in H_{e-L} because $[\rho_L, H_o]$ is of first order in H_{e-L} and also $(\dot{P}_{n,L}; L_o\ P_m)$ can be shown to be zero, by (51) and (10). Therefore

$$[1 - \Pi(t_{m-1})]\ L_o\ \dot{P}_{n,L} = L_o\ \dot{P}_{n,L} \tag{A.3}$$

This proof can be extended to all orders in L_o so that

$$[1 - \Pi(t_j)]L_o^k\ \dot{P}_{n,L} = L_o^k\ \dot{P}_{n,L} \tag{A.4}$$

to first order in H_{e-L}. Consequently, up to this same order, it is trivial to write

$$G(s,t)\ \dot{P}_{n,L} = e^{iL_o(t-s)}\ \dot{P}_{n,L}\ , \tag{A.5}$$

from which (A.1) is a natural extension if only the first order in H_{e-L} is considered.

BIBLIOGRAPHY

1. Price, P.: in Fluctuation Phenomena in Solids, R.E. Burgess, Ed., Academic Press, NY, 355-380 (1965).
2. Conwell, E.: High Field Transport in Semiconductors, Academic Press, NY (1967).
3. Ferry, D.K.: in Handbook of Semiconductors, W. Paul, Ed., North-Holland (1982).
4. Zubarev, D.: Nonequilibrium Statistical Thermodynamics, Plenum Press, NY., (1974).
5. Kalashnikov, V.: Physica, 48, 93. (1970).
6. Ferry, D.K.: J. Physique, 42 (Suppl. 10), C7-253 (1981).
7. Grabert, H.: Journ. Stat. Phys., 19, 479, (1978).
8. Niez, J.J. and Ferry, D.K.: Phys. Rev. B, (1983).
9. MacLennan, J.: Phys. Fluids, 4, 1319, (1961).
10. Procaccia, I., Ronis, D., Collins, M.A., Ross, J.: Phys. Rev., A 19, 1290 (1979).
11. Landau, L. and Lifshitz, E.: Statistical Physics, Pergamon Press, NY, 11-14 (1980).
12. Mori, H.: Prog. Theor. Phys., 33, 423, (1965).
13. Grabert, H.: Z. Physik B 26, 79, (1977).
14. Lovesey, S.: Condensed Matter Physics, Benjamin Press, Reading, MA, 56-99, (1980).

DEVICE-DEVICE INTERACTIONS

D.K. Ferry

Department of Electrical Engineering
Colorado State University
Fort Collins, CO 80523

INTRODUCTION

Over the last two decades, the growth of very large scale integration (VLSI) has been phenomenal and the application of special integrated circuits and microprocessors has blossomed. In fact, the complexity, as defined by the number of devices on an individual integrated circuit chip has approximately doubled each year over this time span, an effect known as Moore's law (1975). Even so, technological momentum is pushing device dimensions to even smaller scales, with current research devices having effective gate lengths less than 0.1 μm in both Si MOSFETSs and Ga MESFETs. Moreover, the technology is becoming available to fabricate really small devices. Present day research results in electron-beam lithography and PMMA photoresist technology is capable of addressing linewidths (and therefore feature sizes) of the order of 250Å (Broers et al., 1978), although finer structures have been prepared with alternative resist techniques. The central feature of transport in such very small devices is that the device dynamics cannot be treated in isolation and must be considered in conjunction with the surrounding environment.

Such very small semiconductor devices (L_g<0.1μm) are controlled by the short spatial scales, very fast temporal response, very high electric fields (>100 kV/cm), high carrier densities, and strong size related effects. These latter arise from the coupling to the environment of contacts, interfaces/boundaries/surfaces, interconnects and other devices. The ultra fast transit time within any single device precludes any significant dissipation of energy within the single device. Rather, coupling of the device to its environment leads to dissipation that occurs over an extended region surrounding the device.

Many people have examined the limit to which semiconductor devices can be scaled downward, and while small devices in the range 0.03–0.3μm gate length have been made, problems such as interconnections, electromigration and thermo-migration of metallization, and power density within the device strongly affect the packing density and device size that can be achieved (Keyes, 1979 and references contained therein). Yet, no one has evaluated the operation and performance of very small semiconductor devices, i.e. - those that can be conceived in the sub-0.1 μm size range, and the interactions within arrays of such devices cannot be treated in isolation. Because of the size of such a very-small device, it is coupled strongly to the environment in which it is located, as was pointed out above. The basic transport equations cannot be separated from their causal boundary conditions and both of these factors must be modified to account for the influence of the device environment.

In conventional descriptions of LSI/VLSI circuits, each device is assumed to behave in the same manner within the total system as it does when it is isolated. The full function of the system (or IC) is determined solely by the interconnection metallization specified to join the individual devices together. A different function can only be assigned to the system by redesigning the interconnection metallizations -- a practical impossibility for most systems, but in practice accomplished in some PLA modules. The conventional clear separation of device design from system design thus depends ultimately on being able to isolate each individual device from the environment of the other devices except for planned effects occurring through the interconnection matrix. This simplification is likely to be seriously in error for submicron-configured VLSI systems, where the isolation of one device from another (any by generalization, from the surrounding environment of metallizations) will be difficult to achieve.

The possible device-device coupling mechanisms are numerous and include such effects as capacitive coupling, of which line-to-line parasitic capacitance is one example, and wave-function penetration (tunneling and charge spillover) from one device to another. The former is significant, as many people restrict device-device interactions to only those regions where characteristic lengths are of the scale of the carrier wavelength. This would be a serious error. Indeed, for device sizes below 0.3 μm, the line-to-line parasitic capacitance begins to dominate the direct line capacitance in determining the total capacitance of a logic gate (Ferry, 1982). This parasitic capacitance leads to a *direct device-device interaction outside the normal circuit or architectural design*. The constraints imposed by device-device interactions will have to be included in future architectural design of compact VLSI systems, and this will most easily be accomplished if these constraints are reflected in the system theory description of the architecture itself.

In this chapter, the current (albeit poorly understood) theory of the device-device interaction is treated along with its architectural limitations. We first show via a BBGKY heirarchy that it is the pair-wise nearest neighbor interaction that dominates the device-device effects. Then we discuss the lowest two orders of this interaction.

Nearest Neighbor Dominance

At the total circuit (or system) level, control over the array of devices that compose the VLSI structure is exercised by means of control fields (or voltages) and input or feedback currents (or voltages). These make up a set of generalized forces $F_{i,ext}$, i=1,...,N, where N is the total number of devices and may exceed 10^6 in modern VLSI. The applied generalized forces are screened by a variety of interactions, but lead to a set of local applied forces F_i, which must be found by a total self-consistent method, just as is done in present two- (and three-) dimensional device modeling. In the coupled N-device system, we can write the total Hamiltonian as

$$H = \sum_i H(x_i, F_i) + \sum_{i<j} H_E(x_i, x_j, F_k) \tag{1}$$

where

$$H_E = H_e + H_{ed} \tag{2}$$

is the environment plus device-environment terms and thus represents the coupling between devices i and j (which is assumed to be an instantaneous pair-wise interaction for simplicity). The variables x_i, x_j refer to complete sets of dynamical variables for the ith and the jth devices, respectively. In the absence of interdevice coupling, the second term on the RHS of (1) is zero and we recover the Liouville equation result for each individual device treated separately. The terms in $H(x_i,F_i)$ are assumed moreover to be time-dependent through the coupling to the generalized, time-dependent forces F_i. We note finally that the set of generalized parameters x_i must also include any local, spatially delineated structure such as contacts and local interconnects.

In the following, we adopt an approach similar to that of a BBGKY heirarchy, which illustrates the various correlations between devices as they become important. We assume that

$$H(x_i, F_i) \equiv p_i^2/2m + \phi_F(x_i) \tag{3}$$

where all external fields are reflected into the generalized local potential ϕ_F. Moreover, we assert that H_E is a pair-wise potential interaction. We have separated the generalized parameters into generalized coordinates x_i and conjugate momenta p_i.

In the appendix, it is shown that the reduced density matrix $\rho_1(x_1)$ for a single device, in the presence of the other devices, can be written as the solution to the Liouville equation

$$i\hbar \frac{\partial \rho_1}{\partial t} = \hat{H}_1 \rho_1(x_1) + \frac{(N-1)}{v} Tr_2 \left\{ \hat{H}_E(|x_1 - x_2|) \rho_1(x_2) \right\} \rho_1(x_1) - \frac{(N-1)}{v} Tr_2 \left\{ \hat{H}_E(|x_1 - x_2|)\, g_2(x_1, x_2) \right\} \quad (4)$$

where

$$g_2(x, x_2) = \rho_1(x_1)\rho_1(x_2) - \rho_2(x_1, x_2) \quad (5)$$

is the two-device correlation function (Barker and Ferry, 1979). In (4), the last term represents high (higher than zero) order effects that arise due to correlated effects of one device upon another. On the other hand, the RHS second term represents an effect due to the *average* presence of all other devices. Thus, this term is analogous to the Hartree potential for electrons. If the correlation term g_2 is set to zero, the first two terms on the RHS lead to a Hamiltonian for an equivalent Hartree equation for the one-device performance, in which each individual device can be treated in isolation if an average mean-field background potential is evaluated. Clearly, the deviation from this simpler approach arises when correlation between devices begins to become important, a case that already arises when line-to-line capacitance begins to significantly affect circuit behavior. *There are in fact two effects here. The first is a global modification that occurs when the Hartree potential differs from zero, for example when a substrate current begins to produce global shifts in substrate bias levels. The second effect is the correlated device behavior.*

The global modification, a zero-order device-device interaction, can arise from substrate leakage. The leakage current of one transistor, flowing through the substrate, to the substrate bias supply leads to a resistive voltage drop. This latter drop changes the substrate bias seen by other devices, and this in turn causes a threshold voltage shift. Such effects can critically alter the performance of VLSI systems, especially when variations in threshold voltage are already a limiting factor (Keyes, 1975). On the other hand, device to device leakage, such as can occur in high-density dynamic MOS memory (Chatterjee *et al.*, 1979) is an example of the first-order interaction. This is shown in Figure 1.

A Circuit-Theoretic Approach

From the above, we begin to perceive that the onset of device-device interactions can occur even though spatial scales are large compared to an electron wavelength. Indeed, global shifts of bias levels (induced by other device variations) and capacitance coupling are the forerunner effects that can be expected. We want to follow this approach by first using a simple circuit approach to treat the first-order effects. In the following, the basic principles of a system restructuring that can occur are illustrated by a simple circuit theoretic analogy utilizing component-connection type approaches. First, a special case of isolated devices is developed. Then a connection function (such as the network of desired metallizations for interconnecting logic gates) is introduced to describe the system in terms of the devices and to show how the properties of the connection can alter the system's dynamics. As the individual device dynamics and connections will be nonlinear, it may be expected that although the equations used here are linear, the general result will admit of synergetic responses for the system. As an example, we apply this formalism to the ring oscillator, a standard digital test circuit.

First, the state equations are examined for an isolated integrator with input-output conditioning. The applicability of this example is examined later. The state equations for the ith device are

$$\dot{u}_i = a_i u_i + b_i y_i \ , \qquad z_i = c_i u_i \tag{6}$$

where u_i is the state variable and y_i and z_i are the input and output variables for the single device, respectively. For an ensemble of N devices, these become

$$\dot{\underline{U}} = \underline{AU} + \underline{BY} \ , \qquad \underline{Z} = \underline{CU} \tag{7}$$

where $\underline{A}$, $\underline{B}$, and $\underline{C}$ are square diagonal matrices and $\underline{U}$, $\underline{Y}$, and $\underline{Z}$ are column matrices. Solving for the transfer function of the single devices gives (in the Laplace transform domain with relaxed initial state)

$$\underline{Z} = \underline{C}(s\underline{I} - \underline{A})^{-1}\underline{BY} \tag{8}$$

So far, it has been considered that each device was isolated from the others. If we describe the connection matrix $\underline{F}$ through the ansatz (Ransom and Saeks, 1975)

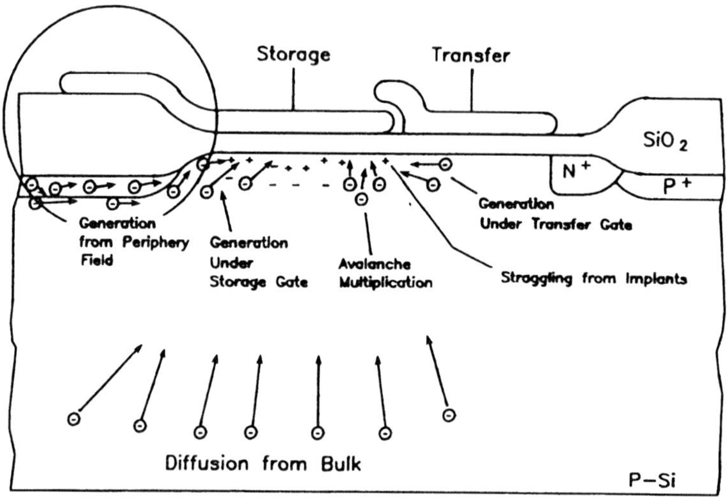

Figure 1. Sources of leakage in dynamic Ram cells.

$$\underline{Y} = \underline{FZ} + \underline{LG}, \quad \underline{H} = \underline{MZ} \tag{9}$$

where $\underline{G}$ and $\underline{H}$ are the total system input and output matrices and $\underline{L}$ and $\underline{M}$ are conditioning matrices. The connection matrix $\underline{F}$ generally describes how the input of a particular device is related to the outputs of other devices, and is thus the system representation of the metallization interconnect chosen for a particular VLSI chip. Equations (8) and (9) can then be combined to yield the system transfer function as

$$\underline{H} = \underline{MCB}(s\underline{I} - \underline{A} - \underline{FCB})^{-1}\underline{LG} \tag{10}$$

where we have used the fact that $\underline{A}$ and hence $(s\underline{I} - A)^{-1}$ is diagonal.

Equation (10) for the system transfer function is a special case of the connection function theory of systems applicable to the integrator. Although we have used this special case, the approach is far more general and is applicable to arbitrary circuits. Further, even though we have assumed an analog signal approach by employing the Laplace transformation for the time variation, the technique is currently extendable to a class of digital circuits, linear sequential circuits, through the description of the systems dynamics in the abstract mathematical concept of an extension field. However, this simple case is adequate to illustrate the major points discussed above.

The quantity $(s\underline{I} - \underline{A} - \underline{FCB})^{-1}$ plays the conceptual role of a resolvent for the system and the zeroes of $\det(\underline{S})=\det[(s\underline{I}-A-\underline{FCB})]$

define the various modes of the operation expected. Since $\underline{B}$ and $\underline{C}$ are diagonal, any deviations of the system response from that defined by $\underline{A}$ must arise through the structure of $\underline{F}$. For example, if we consider that the system is logically connected, i.e. y_i is connected only to z_j if $j<i$, then $\underline{F}$ has elements only in the lower triangle below the main diagonal. Since $\underline{A}$ is diagonal, $\underline{F}$ does not modify the modes determined by $\underline{A}$, i.e. $\det(\underline{S})=\det(s\underline{I}-\underline{A})$. Only when $\underline{F}$ has entries across the main diagonal does this change. For example, if $y_i=z_{i-1}$, then $\underline{F}$ has entries along the diagonal just below the main diagonal. If now the last stage is fed back to the first stage, an entry appears in the upper right corner of $\underline{F}$ and one new mode is generated, the collective ring-oscillator mode. This last follows as the last entry altered the topology of the directed graph of $\underline{S}$, allowing a new, strongly connected graph to arise. If $\underline{C} = \underline{B} = \underline{I}$ is assumed (for simplicity), we can write

$$s\underline{I} - \underline{A} - \underline{FCB} = \begin{bmatrix} (s-a_{11}) & 0 & \cdots & 1 \\ f_{21} & (s-a_{22}) & \cdots & 0 \\ \cdot & \cdot & & \cdot \\ \cdot & \cdot & & \cdot \\ \cdot & \cdot & & \cdot \\ 0 & 0 & & (s-a_{nn}) \end{bmatrix} \tag{11}$$

and the determinant is

$$\det(S) = (-1)^{n-1} \prod_{i=2}^{n-1} (s-a_{ii}) + (s-a_{11}) \prod_{i=2}^{n} (s-a_{ii}) \tag{12}$$

$$= \left\{ \sum_{i=2}^{n-1} (s-a_{ii}) \right\} \left\{ (s-a_{11})(s-a_{nn}) + (-1)^{n-1} \right\}$$

We see that the ring oscillator has changed only two modes, the a_{11} and a_{nn}. The new eigenvalues are given by

$$s^2 + (a_{11} + a_{nn})s + a_{11}a_{nn} + (-1)^{n-1} = 0 \tag{13}$$

and

$$s = \frac{1}{2}(a_{11} + a_{nn}) \pm \left\{ \frac{1}{4}(a_{11} - a_{nn})^2 - (-1)^{n-1} \right\}^{1/2} \tag{14}$$

In order for the system to show oscillations, we must have

$$(-1)^{n-1} > \frac{1}{4}(a_{11} - a_{nn})^2 \tag{15}$$

which requires n to be odd and $(a_{11} - a_{nn})$ to be sufficiently small.

If the system is tightly coupled, i.e. each device is tied to its nearest neighbor, so that $\underline{S}$ is tridiagonal, all modes are changed and the new eigenvalues lie in a band of allowed values. This case corresponds to what we have called the functional superlattice (Ferry, 1982).

In general, the connection function $\underline{F}$ can be divided into two parts, $\underline{F}_1$ and $\underline{F}_2$, where $\underline{F}_1$ is the portion of $\underline{F}$ that represents the desired metallizations, i.e. the designed architectural circuit yielding

$$\underline{S}_1 = s\underline{I} - \underline{A} - \underline{F}_1\underline{CB} \tag{16}$$

Then $\underline{F}_2$ represents the parasitic interactions, such as the parasitic device-device couplings that arise from e.g. the line-to-line coupling capacitance. Thus, a new resolvent $\underline{S}_2$,

$$\underline{S}_2 = \underline{S}_1 - \underline{F}_2\underline{CB} \tag{17}$$

arises with a new set of eigenmodes given by $\det(\underline{S}_2)$. Thus, the structure of the system is altered in the presence of $\underline{F}_2$. As $\underline{F}_2$ depends upon the states of $\underline{U}$ (voltages, for example), as well as the inputs $\underline{G}$, it is entirely conceivable that the system is now strongly nonlinear. In large scale systems, where sizes are more than a micron in scale, $\underline{F}_2$ may reasonably be assumed to be negligible. In future VLSI and ULSI systems of submicron dimensions, this is no longer the case, and the presence of $\underline{F}_2$ will have to be accounted for in the design.

The terms in $\underline{F}_2$ play the role of the extra terms of (4). Contributions to $\underline{F}_2$ that do not depend on particular devices enter the effective Hartree potential, while device dependent terms enter the correlation g_2.

First Order Interaction

When device sizes begin to shrink toward the 0.1μm or less region, the line-to-line capacitance in dense device arrays begins to dominate the total node capacitance (Ferry, 1982). This parasitic capacitance leads to a direct device-device interaction outside of the normal circuit or architectural design. In conventional descriptions of

LSI circuits, each device is assumed to behave in the same manner within the total system as it does when it is isolated. In the dense arrays discussed here, this will no longer be the case.

The possible device-device coupling mechanisms are numerous and include such effects as the capacitive coupling mentioned above, but also include such effects as wave function penetration or tunneling and charge spill-over. Formally, however, one may describe these effects on system and device behavior by assuming the simplest form of coupling. Arrays of devices, interacting in this manner, form a lateral surface superlattice (Reich and Ferry, 1983). Lateral superlattices, in which the superstructure lies in a surface or heterostructure layer, offer considerable advantages for obtaining superlattice effects in planar technology. While a surface MOS structure is formally similar to an array of CCD devices, superlattices can also be fabricated through the use of electron and ion beam lithography and selective area epitaxial growth (Iafrate et al., 1982). If the coupling is capacitive, then the limitation to a spacing less than the de Broglie wavelength is removed. We have examined transport in such lateral surface superlattices (LSSL) through a Wigner function approach. Before proceeding however, it is worth noting that the circuit theory view of LSSLs presented above is generically that of cellular automata (Burks, 1970). Many of the image processing applications proposed for LSSLs arise from the 'games' aspect of cellular automata.

Potential barriers within the device can play a significant role in the quantum ballistic transport of carriers through it. Such barriers are found in very small devices, for example, to confine carriers to the active region, and are an intimate part of devices such as the planar-doped-barrier transistor (Malik et al., 1981), real-space transfer devices (Keever et al., 1981), or superlattice avalanche photodiodes (Capasso et al., 1982). When a barrier is present at the contact region, care must be taken to adequately handle turning-point reflection of the electrons from this barrier. Even when the electron has sufficient energy to pass over the barrier, there is a well-known quantum mechanical reflection at the barrier interface. If the potential barrier is smooth, i.e. -- introduces a transition over many wavelengths, the reflection and wave function matching can be handled by well-known approximation techniques such as the WKB approximation in which the potential barriers represent turning points for a near-classical path. In the very small device, however, the barriers are expected to be sharp on the scale of the electron wavelength and care must be exercised in matching wavefunctions and determining reflections. The reflection problem is further complicated in real-space transfer devices due to the different band structure on either side of the barrier. Here, additional terms arise due to the spatial variation of the effective mass.

A number of interesting factors arise. The potential is generic, in the sense that it is similar in form not only to the devices mentioned above, but also to the sinusoidal force term in the Josephson junction devices. In the latter case, interaction between devices can lead to a parametric pumping of the potential which yields period-doubling bifurcations and chaotic behavior. We can illustrate this behavior with a simple example. In Figure 2 a single memory cell is shown embedded in an array of read/write transfer lines. Charge passing along one of these lines has a Coulomb interaction with the charge in the well. The passage of the charge "rocks" the potential well of the memory cell, and if this interaction is sufficiently strong the memory charge will be delocalized. We first consider the memory cell by itself.

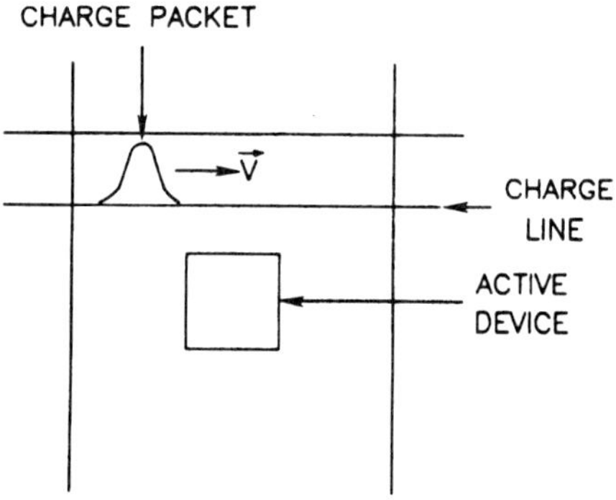

Figure 2. Concept of an isolated RAM cell localized within an array of read/write control lines.

If the potential barriers are slowly varying on the scale of the wave packet, the trajectories are largely those of the classical motion. Even if this is not the case, as we expect for the very small device, nearly semiclassical trajectories can be expected if the variation of the action is limited to a few low-order derivatives (Schulman, 1981). In semiclassical systems, the phase space of the classical motion forms a natural framework in which to examine problems such as these. While classical one-dimensional transport appears to be basically simple, there exists recent work that suggests that even this simple problem contains a number of unexpected subtleties (Jorna, 1978). Here, we can investigate the role of barriers in submicron devices through the use of such finite, area-preserving phase-space mappings of classical dynamics. The use of such mappings reveals a variety of complicated structure. The results that we obtain indicate that if these mappings are applicable to the VLSI scale, then present concepts of submicron transport may require substantial generalization.

The types of subtleties to which we are referring are best illustrated by Figure 3. There, we are using an area-preserving mapping of

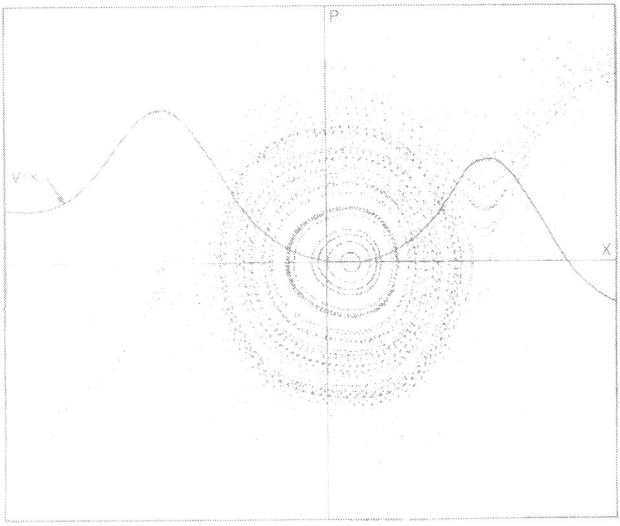

Figure 3. Phase trajectories of electrons within a single potential, where V(x) is the well potential in the presence of an applied electric field.

particle position (horizontal) and momentum (vertical) within a device active region bounded by two Gaussian potentials V(x) (overlaid in the figure), which represents the bounding potentials of the memory cell. In addition, an electric field has been applied. The canonical mappings are area-preserving since we are considering a collisionless (conservative) system and looking at the ballistic transport. The equations of motion are

$$\dot{p} = -\frac{\partial V(x)}{\partial x}, \qquad \dot{x} = -p/m \tag{18}$$

which lead to the discreet mapping

$$p_{n+1} = p_n - T\left.\frac{\partial V}{\partial x}\right|_{x_{n+1}}, \qquad x_{n+1} = x_n + Tp_n/m \tag{19}$$

In Figure 3, the potential has been scaled so that the two potentials have a weak overlap. For small values of the total energy, resonant orbits occupy the central part of the figure. These orbits are in the region of the classically integrable motion. For larger values of the energy, however, the orbits are such that the particles are swept out of the well by the field. An energy dissipating collision can drop a high energy particle into the resonant region, thus trapping it within the structure, and this is the basis of charge storage. We expect these particles to contribute a diffusive component of current. Large angle scattering, however, can move a particle to a back-flowing orbit, which effectively causes a reflection of particles from the device input. The system of equations (18) is not truly generic, as the coordinates (p,x) are not in proper action angle form.

We can now examine the role of the additional charge of the read/write lines. This produces an additional force, of the form

$$\frac{eC_L V_L(t)}{2\pi\epsilon d} \tag{20}$$

where C_L is the capacitance per unit length of the line, $V_L(t)$ is the line potential, and d is the distance to the line. This is a simple approximation, but will give an indication of the magnitude of the effect. Since $V_L(t)$ is time varying, we can affect its potential pumping, with dissipation, by replacing the first equation of (19) with

$$p_{n+1} = p_n - T\frac{\partial V_0(x_{n+1})}{\partial x} - T\gamma p_n + TF_0 \sin(\Omega t) \tag{21}$$

where V_0 is the set of Gaussian potentials, $F_0 = eC_L V_{L0}/2\pi\epsilon d$ is the amplitude of the force in (20), and Ω is the clock frequency of the system (read/write cycle time). The factor γ is an effective damping factor. In Figure 4, we plot the normalized (F_0', Ω') plane results. The curve is a separatrix below which a stable device results. Above the curve, the device is unstable. No period-doubling bifurcations within a single device are found.

The problem we have in Figures 3, 4 is that the axes in each case are fully normlized. We can, however, invertt the normalization. The relationship between F_0' and (20) leads to

$$F_0' = 0.0586\,\frac{C_L}{4\epsilon}\left(\frac{V_L}{V_0}\right) \tag{22}$$

where V_0 is the well depth and the numerical factor incorporates the constants and numerical factors from the simulation. The frequency Ω' scales to very high ranges and $\Omega'=1$ corresponds to $\Omega=3.3$ THz in a 1.0 μm Si well with $V_0=1V$. For all practical pusposes, only the low frequency is of interest, where $F_0'=1$ is the limit. Thus, for SiO_2 insulation, $V_L=V_0=1V$, we find that

$$C_L < 68.4\epsilon = 23\,\text{pF/cm} \tag{23}$$

or 2.3fF/μm. Integrated circuits today have values of line capacitances well below this value by a considerable amount.

The limit (23) tells us that if we faithfully scale our circuits (see Dennard et al., 1974), the line capacitance will not change. While this will have a detrimental effect upon RC delays, the effects observed here will not occur. However, there are trends to move away from such faithful scaling, either to maintain higher potentials or to reduce RC delays and losses. In this case, the usual move is to reduce the line width while maintaining the field oxide thickness (Sinha et al., 1982, Ferry, 1982). If this occurs, line-to-line capacitance begins to dominate C_L, and it is this effect which leads to the charge delocalization discussed here. Indeed, if we examine the data of Sinha et al., (1982) for a field oxide thickness of 0.35μm, we find that (23) is isolated at a 0.1μm design rule, a size that can be expected before the turn of the century.

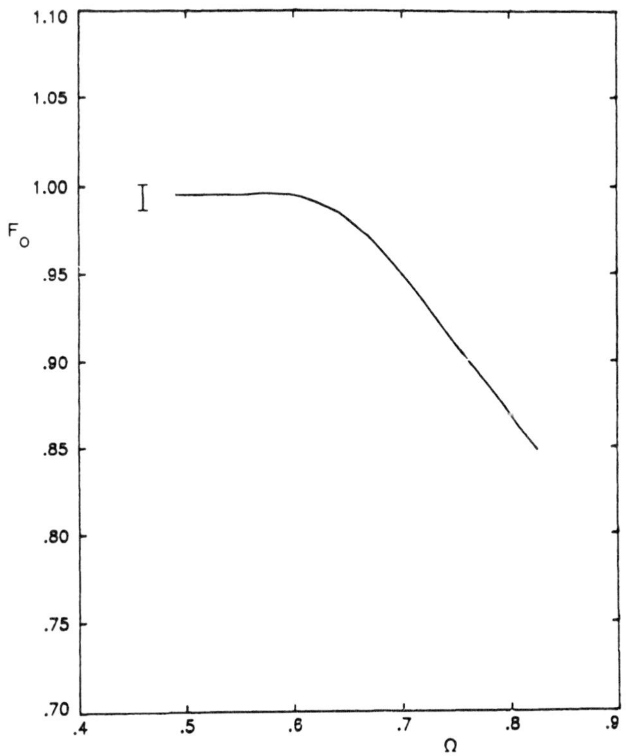

Figure 4. Parameter space of stability for the single potential well when subjected to a periodic pumping potential. The axes are normalized quantities (see text).

References

Barker, J.R. and Ferry, D.K., 1979, in Proc. 1979 Intern. Conf. Cybernetics and Society, IEEE Press 79CH3424-15MC, pp. 762-9.

Broers, A.N., Harper, J.M.E., and Molzen, W.W., 1978, Appl. Phys. Letters 33:392.

Burks, A.W., 1970, Essays on Cellular Automata, Univ. Illinois Press, Urbana.

Capasso, F., Tsang, W.T., Hutchinson, A.L., and Williams, G.F., 1981, Appl. Phys. Letters 40:38.

Chatterjee, P.K., Taylor, G.W., Tasch, A.F., and Fu, H.-S., 1979, IEEE Trans. Electron Dev. ED-26:564.

Dennard, R.H., Gaensslen, F.H., Yu, H.N., Rideout, V.L., Bassous, E., and LeBlanc, A.R., 1974, IEEE J. Sol. -State Circuits SC-9:256.

Dirac, P.A.M., 1958, The Principles of Quantum Mechanics, Oxford U.P., 4th Edition.

Ferry, D.K., 1982, in Adv. Electron and Electron Physics, C. Marton, Ed., Academic Press, New York, 58:311.

Iafrate, G.J., Ferry, D.K., and Reich, R.K., 1982, Surf. Sci. 113:485.

Jorna, S., 1978, Topics in Nonlinear Dynamics, Amer. Inst. Phys., New York, No. 46.

Keever, M., Schichijo, H., Hess, K., Banerjee, S., Witkowski, L., Morkoc, H., and Streetman, G.B., 1981, Appl. Phys. Letters 38:36.

Keyes, R.W., 1975, Proc. IEEE, 63:740.

Keyes, R.W., 1979, IEEE Trans. Electron Devices, ED-26:271

Malik, R.J., Hollis, M.A., Eastman, L.F., Woodard, D.W., Wood, C.E.C., and AuCoin, T.R., 1981, in Proc. Eighth Bienniel Cornell E.E. Conf., Ithaca.

Moore, G.E., 1975, in Proc. Intern. Electron Dev. Mtg., IEEE Press, New York, p. 11.

Ransom, M.N., and Saeks, R., 1975, Circ. Theory Appl. 3:5.

Reich, R., Grondin, R.O., and Ferry, D.K., 1983, Phys, Rev. B, 27:3483.

Schulman, L.S., 1981, Techniques and Applications of Path Integration, Wiley, New York.

Sinha, A.K., Cooper, J.A., Jr., and Levinstein, H.J., 1982, IEEE Electron Dev. Letters EDL-3:90.

Appendix

The total system density matrix satisfies

$$i\hbar \frac{\partial \rho}{\partial t} = H\rho \tag{A.1}$$

where $\rho = \rho(x_1, \ldots, x_N; t)$ is an interacting N-device density matrix and $\hat{H}$ is the super-operator corresponding to the Hamiltonian of (1). Let us now introduce the reduced density matrices $\rho_s(x_1, \ldots, x_s; t)$, which are symmetrical with regard to the set $x_1, \ldots, x_s$, time dependent, and normalized in such a way that the expression $\rho_s dx_1 \ldots dx_s / V^s$ gives the probability that the dynamical states of the group of s devices are located, respectively, in the infinitesimal volume elements $dx_1, \ldots, dx_s$ around the points $x_1, \ldots, x_s$ at the time t. Thus, we can write

$$\rho_s(x_1, \ldots, x_s; t) = V^s \operatorname{Tr}_{s+1, \ldots, N} \left\{ \rho(x_1, \ldots, x_N; t) \right\} \tag{A.2}$$

where $1<s<N$. Thus, we are producing the reduced density matrices by a trace over the residual set $(s+1, \ldots, N)$, just as described in any mixed state discussion (Dirac, 1958). The equation of usual interest is the equation for the distribution function of a single device $\rho_1(x_1, t)$. It is also sometimes desirable to know $\rho_2(x_1, x_2; t)$, the correlation function for the distribution of the dynamic states of pairs of devices. However, our main attention is given to ρ_1.

We shall then proceed from (A.1), and using (1) represent the Liouville equation in the form

$$i\hbar \frac{\partial \rho}{\partial t} = \sum_{i=1}^{N} \hat{H}(x_i)\rho + \sum_{i<j=1}^{N} \hat{H}_E(r_{ij})\rho \tag{A.3}$$

where $r_{ij} = |x_i - x_j|$ is the vector magnitude of the separation in phase space. In general, we will seek a development for ρ_s, the s-device density matrix. Multiplying both sides of (A.3) by V^s and taking the partial trace over the x+1,...,N states, we obtain

$$i\hbar \frac{\partial \rho_s}{\partial t} = \sum_{i=1}^{N} V^s \operatorname{Tr}_{s+1, \ldots, N} \left\{ \hat{H}(x_i)\rho \right\} + \sum_{i<j=1}^{N} V^s \operatorname{Tr}_{s+1, \ldots, N} \left\{ \hat{H}_E(r_{ij})\rho \right\} \tag{A.4}$$

Now, we can make use of the identities

$$\mathrm{Tr}\left\{\left[H(\dot{x}_i),\rho\right]\right\} = \mathrm{Tr}\left\{H(x_i)\rho\right\} - \mathrm{Tr}\left\{\rho H(x_i)\right\} = 0$$

$$\mathrm{Tr}_{i,j}\left\{\left[H_E(r_{ij}),\rho\right]\right\} = \mathrm{Tr}_{i,j}\left\{H_E(r_{ij})\rho\right\} - \mathrm{Tr}_{i,j}\left\{\rho H_E(r_{ij})\right\} = 0 \tag{A.5}$$

which follow from the cyclic properties of the trace operation. We can further use the relations ($i,j\leq s$)

$$V^s \mathrm{Tr}_{s+1,\ldots,N}\left\{\hat{H}(x_i)\rho\right\} = \hat{H}(x_i)V^s \mathrm{Tr}_{s+1,\ldots,N}\left\{\rho\right\}$$

$$V^s \mathrm{Tr}_{s+1,\ldots,N}\left\{\hat{H}_E(r_{ij})\rho\right\} = \hat{H}_E(r_{ij})V^s \mathrm{Tr}_{s+1,\ldots,N}\left\{\rho\right\} \tag{A.6}$$

By considering the symmetry of the density matrix ρ with regard to the variables $x_1,\ldots,x_N$, we can also write

$$\sum_{i=1}^{s}\sum_{j=s+1}^{N} V^s \mathrm{Tr}_{s+1,\ldots,N}\left\{\hat{H}_E(r_{ij})\rho\right\} = (N-s)\sum_{i=1}^{s} V^s \mathrm{Tr}_{s+1,\ldots,N}\left\{\hat{H}_E(|x_i - x_{s+1}|)\rho\right\}$$

$$= \frac{N-s}{V}\sum_{i=1}^{s} \mathrm{Tr}_{s+1}\left\{\hat{H}_E(|x_i - x_j|)\rho_{s+1}\right\} \tag{A.7}$$

We can now use (A.5)-(A.7) in (A.4) so that we obtain

$$i\hbar\frac{\partial\rho_s}{\partial t} = H_s\rho_s + \frac{N-s}{V}\sum_{i=1}^{s} \mathrm{Tr}_{s+1}\left\{\hat{H}_E(|x_i - x_j|)\rho_{s+1}\right\} \tag{A.8}$$

where

$$H_s = \sum_{i=1}^{s}\left\{p_i^2\, 2/m + \phi_F(x_i)\right\} + \sum_{i<j=1}^{s} H_E(r_{ij}) \tag{A.9}$$

is the Hamiltonian of an s-device system with the same dynamics as (1). These equations, however, are still rather complex, since in order to calculate the single-device density matrix ρ_1, we must know ρ_2, the two-device density matrix, which is in turn coupled to all higher correlation functions through the heirarchy (A.8). However, in many physical situations it is possible to approximate the second or some higher correlation function, or to ignore them altogether as is usually done in the isolated device case. This truncates the heirarchy and allows a closed result to be obtained. For example, a common approximation used earlier for electrons in the Boltzmann equation is to set

$$\rho_2(x_1,x_2) = \rho_1(x_1)\rho_1(x_2)$$

In particular, we wish now to look at the one-device density matrix. To ease this, we introduce the two-device correlation function $g_2(x_1,x_2)$ through

$$g_2(x_1,x_2) = \rho_1(x_1)\rho_1(x_2) - \rho_2(x_1,x_2) \tag{A.10}$$

Then

$$\begin{aligned} i\hbar\frac{\partial \rho_1}{\partial t} = &\left[\hat{H}_1 + \frac{N-1}{V}\mathrm{Tr}_2\left\{\hat{H}_E(|x_1-x_2|)\rho_1(x_2)\right\}\right]\rho_1(x_1) \\ &- \frac{N-1}{V}\mathrm{Tr}_2\left\{\hat{H}_E(|x_1-x_2|)\,g_2(x_1,x_2)\right\} \end{aligned} \tag{A.11}$$

The second term in the square brackets is analogous to the Hartree energy for electrons. The potential seen by each device is determined from the average distribution over all other devices ($\mathrm{Tr}_2[\ldots\rho_1(x_2)]$).

Therefore, if we set $g_2=0$, the terms in square brackets lead to the Hamiltonian for an equivalent Hartree equation for the one-device performance, in which each individual device can be treated in isolation if an average mean-field background potential is evaluated.

QUANTUM TRANSPORT AND THE WIGNER FUNCTION

G.J. Iafrate

U.S. Army Electronics Technology and Devices Laboratory

Fort Monmouth, New Jersey 07703

I. TRANSPORT BASED ON THE BOLTZMANN EQUATION

Classical transport is based on the Boltzmann transport equation:

$$\frac{\partial f}{\partial t} + \mathbf{v}\cdot\vec{\nabla}_{\mathbf{r}} f + \mathbf{F}\cdot\vec{\nabla}_{\mathbf{p}} f = \left(\frac{\partial f}{\partial t}\right)_{\text{collision}} \tag{1}$$

where

$$\mathbf{v} = \vec{\nabla}_{\mathbf{p}} E(\mathbf{p}); \quad \mathbf{p} = \hbar\mathbf{k}; \quad \mathbf{F} = qE \tag{2}$$

and $f=f(\mathbf{r},\mathbf{p},t)$ is the probability of finding a typical particle within a unit volume around the point **r**, in momentum state p, at time t, and

$$\left(\frac{\partial f}{\partial t}\right)_{\text{collision}} = \int d\mathbf{p}'[f(\mathbf{p}')W(\mathbf{p}',\mathbf{p}) - f(\mathbf{p})W(\mathbf{p},\mathbf{p}')] \tag{3}$$

Here $W(\mathbf{p}',\mathbf{p})$ is the transition rate from **p** to p′ and depends on the source of scasttering.

The critical assumptions are: (1) Carriers are point particles with well-defined position and momentum. (2) Quantum effects are included implicitly in the effective mass and transition rates. (3) Carriers are considered to be uncorrelated and single particle distribution functions are valid. (4) Collisions are instantaneous in space and time.

As semiconductor technology continues to pursue the scaling down of IC device dimensions into the submicron / ultrasubmicron regime, the relevant space and time scales, depending upon the materials under consideration (silicon, gallium arsenide, etc.), will be comparable to the de Broglie wavelengths and the collision duration; thus semiconductor transport in the ultrasubmicron regime approaches the category of quantum transport (see Fig. 1). Note in Fig. 1 that, within the effective mass approximation, the thermal de Broglie wavelength for electrons in semiconductors is of the order of ultrasubmicron dimensions. Thus we will discuss basic transport methodology.

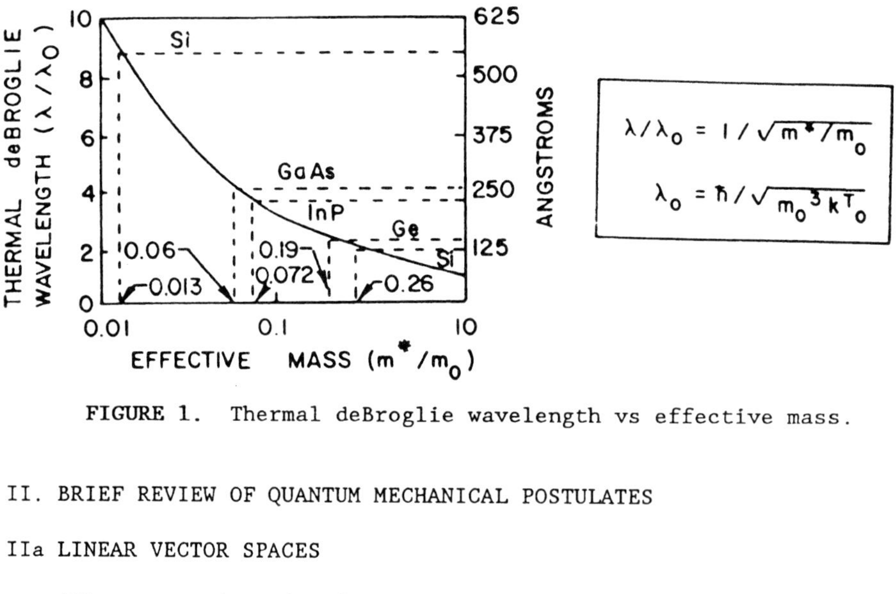

FIGURE 1. Thermal deBroglie wavelength vs effective mass.

II. BRIEF REVIEW OF QUANTUM MECHANICAL POSTULATES

IIa LINEAR VECTOR SPACES

(1) Assumption of infinite dimensions.

(2) The set of eigenvectors of every Hermetian operator forms a basis.

(3) If A has a unique inverse, then $AA^{-1} = A^{-1}A = 1$

(4) For U to be unitary, $UU^{+} = 1 = U^{+}U$

(5) Assume that $|\langle U|U\rangle|^2 \geq 0$, for quantum mechanics

IIb POSTULATES OF QUANTUM MECHANICS

(1) Every physical system is associated with either a finite or an infinite linear vector space.

(2) Each possible state of the physical system is characterized by a vector in the vector space.

(3) Any vector $e^{i\alpha}|a\rangle$ where α is real, describes the same physical state as $|a\rangle$.

(4) To every observable quantity, there is associated with it a Hermetian operator, such that $\langle b|A'|a\rangle = \langle a|A|b\rangle^*$.

(5) The eigenvectors of an observable A constitute a complete set (basis).

(6) The only possible results of measuring an observable are its eigenvalues, which are real.

(7) The spectrum of eigenvalues of A consist of a discrete portion $\langle A'|A''\rangle = \delta_{A'A''}$ and a continuous portion $\langle A'|A''\rangle = \delta(A'-A'')$ implying that $|A'\rangle$ is a continuous function of A'.

III. BRIEF REVIEW OF QUANTUM TRANSPORT METHODOLOGY

IIIa.RULES FOR COMMUTATORS

$$[A,A] = 0 \tag{4}$$

$$[A,c] = 0 \tag{5}$$

where c is a constant or "c" number.

$$[A+B,C] = [A,C] + [B,C] \tag{6}$$

$$[A,BC] = [A,B]C + B[A,C] \tag{7}$$

$$[A,[B,C]] = [B,[C,A]] + [C,[A,B]] = 0 \tag{8}$$

$$[x_i, x_j] = [p_i, p_j] = 0 \text{ for all } i,j \tag{9}$$

$$[x_i, p_j] = i\hbar\delta_{ij} \tag{10}$$

$$[x, p^{\ell}] = i\hbar(\ell p^{\ell-1}) = i\hbar\frac{\partial}{\partial p}(p)^{\ell} \tag{11}$$

$$[p, x^{\ell}] = -i\hbar(\ell x^{\ell-1}) = -i\hbar\frac{\partial}{\partial x}(x)^{\ell} \tag{12}$$

$$[x, F(p)] = i\hbar\frac{\partial}{\partial p}F(p) \tag{13}$$

$$[p, G(x)] = -i\hbar\frac{\partial}{\partial x}G(x) \tag{14}$$

IIb. REPRESENTATIONS: DIRAC FORMALISM FOR REPRESENTING AN OBSERVABLE

Coordinate Representation:

$$\hat{x}|x'\rangle = x'|x'\rangle \quad -\infty<x'<\infty; \tag{15}$$

$$\int dx'|x'\rangle\langle x'| = 1 \tag{16}$$

$$\langle x''|x'\rangle = \delta(x''-x') \tag{17}$$

If

$$[\hat{x}, \hat{p}_x] = i\hbar \tag{18}$$

then

$$\langle x''|\hat{p}|x'\rangle = -i\hbar\frac{\partial}{\partial x'}\delta(x''-x') \tag{19}$$

so

$$\langle\hat{p}\rangle_x \longrightarrow -i\hbar\frac{\partial}{\partial x'} \tag{20}$$

Also,

$$\Psi(x') = \langle x'|\Psi\rangle \tag{21}$$

Momentum Representation

$$\hat{p}|p'\rangle = p'|p'\rangle, \quad -\infty \leq p' \leq \infty \tag{22}$$

$$\langle p''|p'\rangle = \delta(p'-p') \tag{23}$$

if

$$\int dp'\,|p'\rangle,p'| = 1 \tag{24}$$

and

$$[\hat{x},\hat{p}_x] = i\hbar \tag{25}$$

then

$$\langle p''|\hat{x}|p'\rangle = -i\hbar\frac{\partial}{\partial p'}\delta(p''-p') \tag{26}$$

$$\langle\hat{x}\rangle p \longrightarrow i\hbar\frac{\partial}{\partial p'} \tag{27}$$

Energy Representation

$$H|E'\rangle = E'|E'\rangle, \quad \hat{H} = \hat{H}(\hat{x},\hat{p}) \tag{28}$$

To go from the energy representation to the coordinate representation:

$$|E'\rangle = \int dx'\,|x'\rangle\langle x'|E'\rangle \tag{29}$$

where $\langle x'|E'\rangle$ is the wavefunction $\Psi(x')$. To go from $|E'\rangle$ to $|x'\rangle$ we need $\langle x'|E'\rangle$. How do we get it? We insert the complete set of states (Eq. 29) into Eq. 28 and then take the scalar product with $\langle x''|$ to get

$$\int dx' \langle x''|H|x'\rangle\langle x'|E'\rangle = E'\langle x''|E'\rangle \tag{30}$$

Now $H = H(\hat{x},\hat{p})$ so

$$\langle x''|H(\hat{x},\hat{p})|x'\rangle = H\left(x', -i\hbar\frac{\partial}{\partial x'}\right)\delta(x''-x') \tag{31}$$

and

$$\int dx' \langle x''|H|x'\rangle\langle x'|E'\rangle = H\left(x'', -i\hbar\frac{\partial}{\partial x''}\right)\langle x''|E'\rangle \tag{32}$$

thus we get,

$$\boxed{H\left(x'', -i\hbar\frac{\partial}{\partial x''}\right)\Psi_{E'}(x'') = E'\Psi_{E'}(x'')} \tag{33}$$

where

$$\Psi_{E'}(x'') = \langle x''|E'\rangle \tag{34}$$

Equation 33 is the time independent Schrodinger equation in the coordinate representation.

To go from the momentum to coordinate representation the following transformation is used

$$|p'\rangle = \int dx' |x'\rangle\langle x'|p'\rangle = \int dx' |x'\rangle\Psi_{p'}(x'); \tag{35}$$

with $|p'\rangle$ given by Eq. 35, the scalar product of Eq. 22 with $|x''\rangle$ yields

$$\int dx' \langle x''|p|x'\rangle\langle x'|p'\rangle = p'\langle x''|p'\rangle \tag{36}$$

recalling Eq. 19 we get

$$-i\hbar\frac{\partial}{\partial x''}\Psi_{p'}(x'') = p'\Psi_{p'}(x'') \tag{37}$$

with solution

$$\Psi_{p'}(x'') = \frac{1}{\sqrt{2\pi\hbar}} e^{\frac{i}{\hbar}p'x''} \tag{38}$$

which is the matrix element connecting the momentum and coordinate representations

IIIc. TIME EVOLUTION (time dependence of the state of the system)

Time evolution means

$$|a,t_2> = T(t_2,t_1)|a,t_1> \tag{39}$$

where $T(t_2,t_1)$ is the time evolution operator. $T((t_2,t_1)$ is unitary, i.e. preserves normalization:

$$T^+T = TT^+ = 1 \tag{40}$$

Now for

$$|a,t> = T(t,0)|a,0> \tag{41}$$

It can be shown that $T(t) \equiv T(t,0)$ satisfies the equation

$$H(t)T(t) = i\hbar\frac{dT}{dt} \tag{42}$$

so that,

$$i\hbar\frac{\partial}{\partial t}|a,t> = H(t)|a,t> \tag{43}$$

Now if $H \neq H(t)$, explicitly, then

$$\frac{\partial H}{\partial t} = 0 \tag{44}$$

and

$$T(t) = e^{-\frac{i}{\hbar}\hat{H}t} \tag{45}$$

III d. QUANTUM MECHANICAL "PICTURES"

IIId1: Schrodinger Picture	IIId2: Heisenberg Picture
1. Operators and basis eigenvectors are time independent	1. Operators and basis vectors are time dependent
2. State vectors are time dependent	2. State vectors are time independent
$\hat{A}\|A'\rangle = A'\|A'\rangle$ (S1)	$A_H(t) = T^{+}A_S T;$ (H1)
$\frac{\partial \hat{A}_S}{\partial t} = 0; \quad \frac{\partial \|A'\rangle_S}{\partial t} = 0$ (S2)	$\|a\rangle_H = T^{+}\|a,t\rangle_S$ (H2)
$i\hbar\frac{\partial}{\partial t}\|a,t\rangle_S = H\|a,t\rangle_S$ (S3)	$\frac{d}{dt}\|a_H\rangle = 0$ (H3)
$\frac{d}{dt}\langle\hat{A}\rangle = \frac{1}{i\hbar}\langle[A,H]\rangle + \frac{\partial\langle\hat{A}\rangle}{\partial t}$ (S4)	$\frac{d\hat{A}_H}{dt} = \frac{1}{i\hbar}[\hat{A}_H,\hat{H}] + \frac{\partial \hat{A}_H}{\partial t}$ (H4)

Connection between Schrodinger and Heisenberg pictures

$$|a,t\rangle_S = T|a,0\rangle_S \tag{SH1}$$

thus,

$$|a\rangle_H = |a,0\rangle_S \tag{SH2}$$

Note that in terms of the complete set $|\mathbf{k}'\rangle$

$$|a,t\rangle = \sum_{\mathbf{k}'}|\mathbf{k}'\rangle\langle\mathbf{k}'|a,t\rangle \tag{46}$$

and,

$$|\langle\mathbf{k}'|a,t\rangle|^2 \tag{47}$$

is the probability of finding $\mathbf{k}'$ for $\hat{\mathbf{k}}$ in state "a" at time "t".

IIId3. Interaction Picture (mixed picture)

This is a convenient picture when H can be broken into two parts,

$$H \equiv H_0 + H_1 \tag{48}$$

where

$$\frac{\partial H_o}{\partial t} = 0, \tag{49}$$

H_1 may or may not be time dependent.

In the Heisenberg representation; from Eq. H2

$$(|a\rangle H = T^{+}(t)|a,t\rangle S) \tag{50}$$

In the interaction picture

$$|a,t\rangle_I = T_o^{+}|a,t\rangle_S; \quad T_o = e^{-\frac{i}{\hbar}H_o t} \tag{51}$$

where T_o is governed by H_o, and $|a,t\rangle S$ is governed by the total Hamiltonian as given by Eq. (48). At t = 0

$$|a,0\rangle_I = |a,0\rangle_S \tag{52}$$

for t > 0

$$A_I(t) = T_O^{+} A_S T_O \tag{53}$$

Note,

$$H_{OI} = T_O^{+} H_{OS} T_O = H_{OS} \tag{54}$$

and

$$\frac{dA_I}{dt} = \frac{1}{i\hbar}[A_I, H_O] + \frac{\partial A_I}{\partial t} \tag{55}$$

Implying that the time dependence of the operator is governed by H_o. But,

$$i\hbar\frac{\partial}{\partial t}|a,t\rangle_I = (H_1)_I|a,t\rangle_I \tag{56}$$

or $|a,t\rangle_I$, is governed by $(H_1)_I$ alone.

IV. DENSITY MATRIX

The expectation value of a quantum mechanical operator is

$$\langle A\rangle = \int\Psi^*(\mathbf{x})\hat{A}(\mathbf{x})\Psi(\mathbf{x})d^3\mathbf{x} = \langle\Psi|\hat{A}|\Psi\rangle \tag{57}$$

where $\Psi(\mathbf{x})$ is the wavefunction for the isolated system. An isolated system possesses a wavefunction of the coordinates only of the system; this wavefunction determines the state of the system.

A system on the other hand, coupled to an external environment is described by a wavefunction, which is a function of the coordinates of the system, say "x", and coordinates of the environment, say "q". Under such circumstances, we may say that the system having coordinates "x" does not possess a wavefunction depending on "x" alone. A system that does not have such a wavefunction is said to be in a mixed state. An isolated system that does have a wavefunction is said to be in a pure state.

If $\Psi(x,q)$ is the wavefunction for an isolated system plus environment, then the expectation waves of the operators associated with the isolated system must be averaged over the coordinates of the environment or "ensemble" averaged. Then:

$$\langle\bar{A}\rangle \equiv \langle A\rangle_{\substack{\text{ensemble}\\ \text{averaged}}} = \int d^3\mathbf{x} \int dq\Psi^*(\mathbf{x},q)A(\mathbf{x})\Psi(\mathbf{x},q) \tag{58}$$

By defining

$$\boxed{\rho(\mathbf{x},\mathbf{x}') = \int dq\Psi^*(\mathbf{x}',q)\Psi(\mathbf{x},q)} \tag{59}$$

as the density matrix in the coordinate representation, we can rewrite $\langle A\rangle$ as

$$\langle\bar{A}\rangle = \int d^3\mathbf{x}[\hat{A}(x)\rho(\mathbf{x},\mathbf{x}')]_{\mathbf{x}=\mathbf{x}'} \tag{60a}$$

$$= \int d^3\mathbf{x}' \int d^3\mathbf{x}[\tilde{A}(\mathbf{x})\rho(\mathbf{x},\mathbf{x}')]\delta(\mathbf{x}'-\mathbf{x}) \tag{60b}$$

We can establish the density operator matrix by expanding $\Psi(\mathbf{x},t,q)$ in a convenient basis set, say: $[U_n(\mathbf{x})]$ where $(U_n,U_m) = \delta_{nm}$. Then

$$\Psi(\mathbf{x},t,q) = \sum_n C_n(t,q)U_n(\mathbf{x}) \tag{61}$$

so that

$$\rho(\mathbf{x},\mathbf{x}') = \int dq \Psi^*(\mathbf{x}',q)\Psi(\mathbf{x},q) = \overline{\Psi(\mathbf{x}',q)\Psi(\mathbf{x},q)} \tag{62a}$$

$$= \sum_{m,n} \overline{C_m^*(q,t)C_n(q,t)}\ U_m^*(\mathbf{x}')U_n(\mathbf{x}) \tag{62b}$$

$$= \sum_{m,n} \overline{C_m^* C_n} <\mathbf{x}'|m>^* <\mathbf{x}|n> \tag{62c}$$

$$= \sum_{m,n} C_m^* C_n <\mathbf{x}|n>\ <m|\mathbf{x}'> \tag{62d}$$

$$\equiv <\mathbf{x}|\hat{\rho}|\mathbf{x}'> \tag{62e}$$

where

$$\hat{\rho} = \sum_{m,n} C_m^* C_n |n>\ <m| \tag{63}$$

Letting

$$\rho_{nm} = \overline{C_m^* C_n} \tag{64}$$

where the products imply probabilistic quantities, we obtain the density operator:

$$\boxed{\hat{\rho} = \sum_{m,n} |n>\ \rho_{nm}\ <m|} \tag{65}$$

V. GENERAL PROPERTIES OF THE DENSITY MATRIX

1.

$$<\bar{A}> = \text{trace}\ \hat{\rho}\hat{A} = \text{tr}\ \rho A \tag{66}$$

Proof:

$$\overline{\langle A\rangle} = \overline{(\Psi, A\Psi)} = \sum_{m,n} \overline{C_m^* C_n}\ (U_m, AU_n) \tag{67a}$$

$$= \sum_{m,n} \rho_{nm} A_{mn} = \sum_n \left(\sum_m \rho_{nm} A_{mn} \right) \tag{67b}$$

$$= \mathrm{Tr}\rho A \tag{67c}$$

1a.

$$\mathrm{Tr}\rho = 1 \tag{68}$$

Proof: Let

$$\hat{A} = \hat{I} \tag{69}$$

and Eq. 68 is satisfied.

2.

$$\overline{\langle A\rangle} = \mathrm{Tr}\rho A \tag{70}$$

is invariant under unitary transformations so that $\langle A\rangle$ is independent of basis.

Proof

$$\overline{\langle A\rangle} = \mathrm{Tr}\rho A \tag{71a}$$

$$= \mathrm{Tr}U^{-1}U\rho U^{-1}UA \tag{71b}$$

But $\mathrm{Tr}\hat{B}\hat{F} = \mathrm{Tr}\hat{F}\hat{B}$, implying

$$\overline{\langle A\rangle} = \mathrm{Tr}U\rho U^{-1}UAU^{-1} \tag{71c}$$

$$= \mathrm{Tr}\rho' A' \tag{71d}$$

$$= \overline{\langle A\rangle}' \tag{71e}$$

Thus, while ρ and A are basis dependent, the trace is invariant under unitary transformations.

In the above we have used the properties of the unitary operator, U:

$$U^{-1}U = I = UU^{-1} \tag{72a}$$

$$U^{-1} = U^{+} \tag{72b}$$

$$U^{+}U = UU^{+} = I \tag{72c}$$

3. Equation of motion for $\hat{\rho}(t)$. The time evolution for $|m,t\rangle$ is

$$|m,t\rangle = T(t)|m,0\rangle \tag{73}$$

Recalling

$$i\hbar\frac{dT}{dt} = HT$$

and noting

$$-i\hbar\frac{dT^{+}}{dt} = T^{+}H \tag{74}$$

Then,

$$\hat{\rho}(t) = \sum_{m,n} |n,t\rangle\, \rho_{nm}\, \langle m,t| \tag{75a}$$

$$= \sum_{m,n} T(t)|n,0\rangle\, \rho_{mn}\, \langle m,0|T^{+}(t) \tag{75b}$$

$$= T(t)\hat{\rho}(0)T^{+}(t) \tag{75c}$$

So

$$\frac{\partial\hat{\rho}}{\partial t} = \frac{\partial T}{\partial t}\,\rho(0)T^{+} + T\rho(0)\frac{\partial T^{+}}{\partial t} \tag{76a}$$

$$= \frac{1}{i\hbar}\,[H\underbrace{T\hat{\rho}(0)T^{+}}_{\hat{\rho}(t)} - \underbrace{T\hat{\rho}(0)T^{+}}_{\hat{\rho}(t)}H] \tag{75b}$$

Thus

$$\boxed{i\hbar\frac{\partial\hat{\rho}}{\partial t} = [H,\hat{\rho}(t)]} \tag{76c}$$

which is the analog of the Liouville equation.

4. $\hat{\rho}$ is Hermetian:

Proof, Note:

$$\rho_{nm} = \overline{C_m^* C_n} \longrightarrow \rho_{mn} = \overline{C_n^* C_m} \tag{77a}$$

Thus

$$\rho_{mn}{}^* = \overline{C_m^* C_n} = \rho_{nm} \tag{77b}$$

which represents Hermetian elements. Then

$$\rho = \sum_{m,n} |n\rangle\, \rho_{nm} \langle m| \tag{78}$$

and

$$\rho^+ = \sum_{m,n} |m\rangle\, \rho_{nm}{}^* \langle n| = \sum_{m,n} |m\rangle\, \rho_{mn} \langle n| = \rho \tag{79}$$

Thus,

$$\boxed{\rho^+ = \rho} \tag{80}$$

4a. Magnitude of the density matrix

Since

$$\sum_n \rho_{nn} = 1 \tag{81}$$

and

$$\rho_{mn}{}^* = \rho_{nm} \longrightarrow \rho_{nn}{}^* = \rho_{nn} \tag{82}$$

then the diagonal elements of the density matrix are real, and:

$$\rho_{nn} = |C_n|^2 > 0 \tag{83}$$

and

$$0 \le \rho_{nn} \le 1 \tag{84}$$

5. Positive definite properties of $\hat{\rho}$.

Consider

$$\hat{\rho}^2 = \hat{\rho}\hat{\rho} \tag{85a}$$

$$= \sum_{m,n} \sum_{k,\ell} |n\rangle \rho_{nm} \underbrace{\langle m|k\rangle}_{\delta mk} \rho_{k\ell} \langle \ell| \tag{85b}$$

$$= \sum_{n} \sum_{k,\ell} |n\rangle \rho_{nk} \rho_{k\ell} \langle \ell| \tag{85c}$$

$$= \sum_{n,\ell} |n\rangle \left(\sum_{k} \rho_{nk} \rho_{k\ell}\right) \langle \ell| \tag{85d}$$

Thus,

$$\boxed{\rho^2 = \sum_{m,n} |n\rangle \left(\sum_{k} \rho_{nk} \rho_{km}\right) \langle m|} \tag{85e}$$

We note,

$$\rho^2 \ge 0 \tag{86}$$

so that

$$S_{nm} = \underbrace{\left(\sum_{k} \rho_{nk} \rho_{km}\right)}_{(\rho^2)_{nm}} \ge 0 \tag{87}$$

5a. For pure states,

$$\rho^2 = \rho \quad \text{or} \quad (\rho)_{nm} = (\rho^2)_{nm} \tag{88}$$

Proof: For pure states

$$\rho_{nm} = C_m^* C_n \tag{89}$$

with no ensemble average. Then

$$S_{nm} = \sum_k (C_k^* C_n)(C_m^* C_k) \tag{90a}$$

$$= C_m C_n \sum_k C_k^* C_k = C_m^* C_n \sum_k |C_k|^2 \tag{90b}$$

$$= \rho_{nm} \underbrace{(\text{trace } \rho)}_{=1} = \rho_{nm} = (\rho^2)_{nm} \tag{90c}$$

Thus

$$\rho^2 = \sum_{m,n} |n\rangle \, \rho_{nm} \, \langle m| = \rho \tag{91}$$

ρ is idempotent and has eigenvalues "0" or "1".

5b. Expectation value of the density matrix operator.

Starting from

$$\overline{\langle A \rangle} = \mathrm{tr}\hat{\rho}\hat{A}$$

with

$$\hat{A} = \hat{\rho} \tag{92}$$

we obtain

$$\langle\bar{\rho}\rangle = \sum_{m,n} \rho_{nm} \, \rho_{mn} = \sum_n \left(\sum_m \rho_{nm} \, \rho_{mn}\right) = \sum_n (\rho)^2_{nm} \tag{93a}$$

$$= \sum_n (\rho)_{nn} \le \left(\sum \rho_{nn}\right)^2 = 1 \tag{93b}$$

Thus

$$<\bar{\rho}> = \sum_n (\rho^2)_{nn} \leq 1 \tag{93c}$$

Equation (93c) places severe restrictions on the density matrix. Note: since ρ is invariant under unitary transformations, this result is true for any representation. For pure states, the equality holds since

$$\sum_{m,n} \rho_{nm}\, \rho_{mn} = \sum_n \rho_{nn} = 1 \tag{94}$$

VII. EXAMPLE OF THE USE OF THE DENSITY MATRIX

Consider a simple example of the use of the density matrix in calculating quantum mechanical transport parameters. We treat electrons moving in a solid within the framework of the one band, free electron approximation, subject to uniform electric fields and scattering defects. Let

$$H = H_0 + H' + H_E \tag{95}$$

where

$$H_0 = \frac{p^2}{2m^*} \tag{96}$$

for free electrons,

$$H' = \sum_{i=1}^{N} \phi(\mathbf{r}-\mathbf{r}_i) = \lambda V \tag{97}$$

for impurities, and

$$H_E = -eE_\alpha X_\alpha = -e\mathbf{E}\cdot\mathbf{X} \tag{98}$$

for the external electric field.

The eigenvalue equation for free electrons is,

$$H_0 \Psi_{\mathbf{k}} = \epsilon^0_{\mathbf{k}} \Psi_{\mathbf{k}} \tag{99}$$

with eigenfunctions

$$\Psi_{\mathbf{k}} = \frac{1}{\sqrt{\Omega'}} e^{i\mathbf{k}\cdot\mathbf{r}} \tag{100}$$

The eigenvalue equation can also be written as,

$$H_0|\mathbf{k}\rangle = \epsilon^0_{\mathbf{k}}|\mathbf{k}\rangle; \tag{101}$$

where

$$\epsilon^0_{\mathbf{k}} = \frac{\hbar^2}{2m^*} \tag{102}$$

and

$$\langle\mathbf{k}'|\mathbf{k}\rangle = \delta_{\mathbf{k}'\mathbf{k}} \tag{103}$$

Assume, as illustrated in Fig. 2, that the field is turned on adiabatically.

$$E_\alpha = E^0_\alpha e^{st} \tag{104}$$

so that as t approaches $-\infty$, E_α approaches zero.

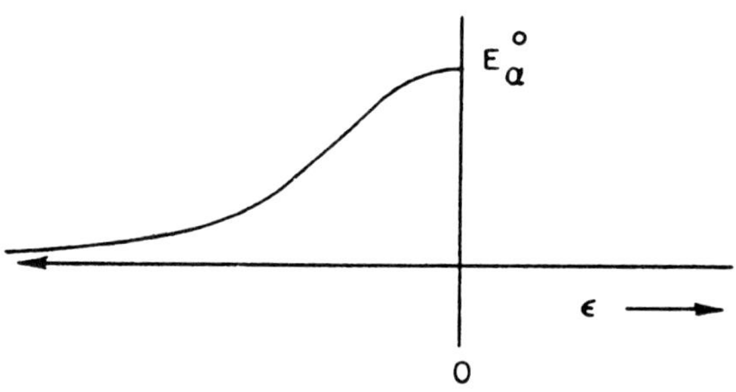

FIGURE 2. Time evolution of field.

To connect quantum theory with statistical mechanics we use the canonical density matrix (in correspondence) with the "canonical" distribution of classical theory.

$$\hat{\rho} = Ke^{-\beta H} \qquad \beta = \frac{1}{kT} \tag{105}$$

for systems in equilibrium with a heat bath at temperature T. In the above, the normalization can be chosen so that $\mathrm{Tr}\hat{\rho} = 1$. More generally, for systems with a variable number of particles, n_i, we use the " grand canonical " density matrix,

$$\hat{\rho} = \exp(-q+\beta\sum_i n_i U_i - \beta H) \tag{106}$$

where q and U are determined by the constraints:

$$\mathrm{Tr}\hat{\rho} = 1$$

$$\sum n_i = N \tag{107}$$

Let's consider the canonical ensemble. To verify the corrrespondence with classical theory, let's choose to evaluate $\hat{\rho}(H)$ in a representation in which H is diagonal (the energy basis). Let

$$H|m\rangle = \epsilon_m|m\rangle \qquad \langle m|n\rangle = \delta_{mn} \tag{108}$$

then

$$\rho_{nm} = \langle n|\hat{\rho}|m\rangle = K\langle n|e^{-\beta H}|m\rangle \tag{109a}$$

$$= K\langle n|m\rangle e^{-\beta\epsilon_m} \tag{109b}$$

and

$$\rho_{nm} = K\delta_{nm}e^{-\beta\epsilon_m} \tag{109c}$$

and the density matrix is diagonal. With this result,

$$\hat{\rho} = \sum_{m,n} |n\rangle\ \rho_{nm}\ \langle m| \tag{110a}$$

$$= K\sum_{m,n} |n\rangle\delta_{nm}\exp(-\beta\epsilon_m)\ \langle m| \tag{110b}$$

so,

$$\boxed{\hat{\rho} = K\sum_m |m\rangle|\exp(-\beta\epsilon_m)\ \langle m|} \tag{110c}$$

To obtain the normalization constant K of Eq. (110), we use the constraint of $Tr\hat{\rho} = 1$ yielding

$$K\sum_{m}\exp(-\beta\epsilon_m) = 1 \tag{111}$$

or

$$\boxed{K^{-1} = \sum_{m}\exp(-\beta\epsilon_m)} \tag{112}$$

Thus for operators, in the canonical ensemble

$$\hat{\rho}(H) = Ke^{-\beta H} \tag{113}$$

where,

$$K^{-1} = Tr\hat{\rho} \tag{114}$$

For the simple example of free electrons subject to uniform fields and defect scattering, we will assume a canonical ensemble for the initial distribution. That is

$$\hat{\rho}(-\infty) = Ke^{-\beta H}; \tag{115}$$

where, $H = H_0 + H' + H_E$, we note that for $t = -\infty$, $H \longrightarrow H_0 + H'$ and $\rho(-\infty)$ is the equilibrium density matrix for electrons plus defects. Also, observables are calculated using the density matrix through use of $\langle\bar{A}\rangle = Tr(\hat{\rho}\hat{A})$.

To find observables such as current, velocity, acceleration, etc., we need $\hat{\rho}$. We saw that $\hat{\rho}$ satisfies,

$$i\hbar\frac{\partial\hat{\rho}}{\partial t} = [\hat{H},\hat{\rho}]$$

We look for $\hat{\rho}$ in the representation in which H_0 is diagonal. For simplicity, we look for ρ to first order in field:

$$\rho = \rho_0 + \Delta\rho \tag{116}$$

Then

$$[H,\rho] = [H_0 + H' + H_E\ ,\quad \rho_0 + \Delta\rho] \tag{117a}$$

$$= [H_O + H' + H_E \ , \ \rho_O] + [H_O + H' + H_E \ , \ \Delta\rho] \tag{117b}$$

$$= [H_O \ , \ \rho_O] + [H' + H_E \ , \ \rho_O] + [H_O + H' \ , \ \Delta\rho] + \cdots \tag{117c}$$

where ••• denotes higher order contributions. Thus

$$i\hbar\left(\frac{\partial \rho_O}{\partial t} + \frac{\partial \Delta\rho}{\partial t}\right) = [H_O,\rho_O] + [H',\rho_O] + [H_E,\rho_O] + [(H + H'),\Delta\rho] \tag{118}$$

To zeroth order:

$$i\hbar\frac{\partial \rho_O}{\partial t} = [(H_O + H'),\rho_O] \tag{119}$$

To first order:

$$i\hbar\frac{\partial \Delta\rho}{\partial t} = [(H_O + H'),\Delta\rho] + [H_E,\Delta\rho] \tag{120}$$

where the time evolution of the inhomogenerous part of the above equation is determined only to zeroth order. Now

$$\rho = \rho_O + \Delta\rho \xrightarrow[\substack{E\to 0 \\ t\to -\infty}]{} \rho_O = K\exp(-\beta(H_O + H') \tag{121}$$

If

$$\rho(-\infty) = \rho_O \tag{122}$$

for all time t, then

$$[H_O + H' \ , \ \rho_O] = 0 \tag{123}$$

and the time derivative of ρ_o is zero. Thus we look for solutions,

$$\boxed{i\hbar\frac{\partial \Delta\rho}{\partial t} = [H_O + H',\Delta\rho] + [H_E,\rho_O} \tag{124}$$

with

$$\Delta\rho(t=-\infty) = 0 \tag{125}$$

We now solve Eq. (124) with

$$H_E = \left(-eE_\alpha^0 X_\alpha\right) e^{st} \tag{126}$$

and

$$\rho_0 = \rho_0(H_0 + H') \tag{127}$$

We look for the solution of the form

$$\Delta\rho = fe^{st} \tag{128}$$

to get

$$i\hbar sf = [H_0 + H', f] + [H_1, \rho_0] \tag{129a}$$

$$= [H_0, f] + [H', f] + [H_1, \rho_0] \tag{129b}$$

Taking the matrix elements of f in the $|\mathbf{k}\rangle$ representation, we get

$$i\hbar sf_{\mathbf{kk}'} = [H_0, f]_{\mathbf{kk}'} + [H', f]_{\mathbf{kk}'} + [H_1, \rho_0]_{\mathbf{kk}'} \tag{130}$$

Now,

$$[H_0, f]_{\mathbf{kk}'} = (\epsilon_{\mathbf{k}}^0 - \epsilon_{\mathbf{k}'}^0) \equiv \hbar\omega_{\mathbf{kk}'}^0 f_{\mathbf{kk}'} \tag{131}$$

$$[\rho_0, H_1]_{\mathbf{kk}'} = C_{\mathbf{kk}'} \tag{132}$$

to be discussed shortly,

$$[H', f]_{\mathbf{kk}'} = \langle\mathbf{k}|(H'f - fH')|\mathbf{k}'\rangle$$

$$= \sum_{\mathbf{k}''} (\langle\mathbf{k}|H'|\mathbf{k}''\rangle\langle\mathbf{k}''|f|\mathbf{k}'\rangle - \langle\mathbf{k}|f|\mathbf{k}''\rangle\langle\mathbf{k}''|H'|\mathbf{k}' \tag{133a}$$

$$= \sum_{\mathbf{k}''} (H'_{\mathbf{kk}''} f_{\mathbf{k}''\mathbf{k}'} - f_{\mathbf{kk}''} H'_{\mathbf{k}''\mathbf{k}'}) \tag{133b}$$

to get

$$\boxed{\hbar(\omega^0_{\mathbf{kk}'} - is)f_{\mathbf{kk}'} = C_{\mathbf{kk}'} + \sum_{\mathbf{k}''}(f_{\mathbf{kk}'}H'_{\mathbf{k}''\mathbf{k}'} - H'_{\mathbf{kk}''}f_{\mathbf{k}''\mathbf{k}'})} \tag{134}$$

Now,

$$H' = \sum_{i=1}^{''} \phi(\mathbf{r}-\mathbf{r}) = \lambda V \tag{135}$$

$H'_{\mathbf{kk}'}$ in the coordinate representation is,

$$H'_{\mathbf{kk}'} = (\Psi_{\mathbf{k}}, H'\Psi_{\mathbf{k}'}) \tag{136a}$$

$$= \frac{1}{\Omega}\int_{\Omega} e^{(\mathbf{k}-\mathbf{k}')\cdot\mathbf{r}} \sum \phi(\mathbf{r}-\mathbf{r}_i)d\mathbf{r} \tag{136b}$$

$$= \frac{1}{\Omega}\sum_{i=1}^{N}\int_{\Omega} e^{(\mathbf{k}-\mathbf{k}')\cdot\mathbf{r}} \phi(\mathbf{r}-\mathbf{r}_i)d\mathbf{r} \tag{136c}$$

setting $\mathbf{r} = \mathbf{r} - \mathbf{r}_i$, we get

$$H'_{\mathbf{kk}'} = \frac{1}{\Omega}\sum_{i=1}^{N} e^{-i(\mathbf{k}-\mathbf{k}')\cdot\mathbf{r}_i} \int_{\Omega} e^{-i(\mathbf{k}-\mathbf{k}')\cdot\mathbf{r}} \phi(\mathbf{r})d\mathbf{r} \tag{137}$$

Thus

$$H'_{\mathbf{kk}'} = \frac{\phi_{\mathbf{kk}'}}{\Omega}\sum_{i=1}^{N} e^{-i(\mathbf{k}-\mathbf{k}')\cdot\mathbf{r}_i}; \qquad \phi_{\mathbf{kk}'} \equiv \int_{\Omega} e^{-i(\mathbf{k}-\mathbf{k}')\cdot\mathbf{r}} \phi(\mathbf{r})d\mathbf{r} \tag{138}$$

In particular, the diagonal elements are

$$H'_{\mathbf{kk}} = \frac{N}{\Omega}\phi_{\mathbf{kk}} \tag{139}$$

In addition

$$C_{\mathbf{kk}'} = [\rho_0(H_0 + H'), H_1]_{\mathbf{kk}'} \tag{140}$$

where $H' = \lambda V$ can be expanded in a power series in λ so that,

$$C_{\mathbf{kk}'} = C_{\mathbf{kk}'}^{(0)} + C_{\mathbf{kk}'}^{(1)}\lambda + C_{\mathbf{kk}'}^{(2)}\lambda^2 + \cdots \tag{141}$$

To evaluate $C_{\mathbf{kk}'}$ in powers of λ we digress to show a worthwhile technique for expanding $\exp(\hat{A}+\hat{B})$ in powers of $\hat{B}$ We introduce

$$F(\lambda) = e^{\lambda(A + B)} \tag{142}$$

where

$$\frac{dF}{d\lambda} = (A + B)F(\lambda), \quad F(0) = 1 \tag{143}$$

Now transform $F(\lambda)$ so that

$$F(\lambda) = e^{\lambda A}\, G(\lambda), \qquad G(0) = 1 \tag{144}$$

Then,

$$\frac{dF}{d\lambda} = Ae^{\lambda A}\, G(\lambda) + e^{\lambda A}\,\frac{dG}{d\lambda} = (A + B)e^{\lambda A}G \tag{145}$$

Solving for $dG/d\lambda$, we get

$$\frac{dG}{d\lambda} = \left(e^{-\lambda A}\, Be^{\lambda A}\right)G(\lambda) \tag{146}$$

which upon integration gives,

$$G(\lambda) = 1 + \int_0^\lambda e^{-\lambda' A}\, Be^{\lambda' A}\, G(\lambda')d\lambda' \tag{147}$$

This expression can be iterated so that,

$$G(\lambda) = 1 + \int_0^\lambda e^{-\lambda' A}\, Be^{\lambda' A}\, d\lambda^+ + \cdots \tag{148}$$

Therefore,

$$e^{(A+B)} = F(1) = e^{A}G(1) = e^{A} + \int_0^1 e^{(1-\lambda)A}\, Be^{\lambda A}\, d\lambda + \cdots \tag{149}$$

The matrix elements of exp(A+B) in the representation in which $\hat{A}$ is diagonal, i.e.

$$<\hat{A}|a> = a|a>, <a|a'> = \delta aa' \tag{150}$$

are

$$\left(e^{A+B}\right)_{a'a''} = e^{a'}\delta a'a'' + \int_0^1 <a''|\left(e^{(1-\lambda)A}\, Be^{\lambda A}|a'>d\lambda\right) + \cdots \tag{151}$$

Now,

$$<a''|\left(e^{(1-\lambda)A}\, Be^{\lambda A}\right)|a'> = e^{(1-\lambda)a''}\, <a''|B|a'>e^{\lambda a'} = e^{(1-\lambda)a''}\, B_{a'a''}e^{\lambda a'} \tag{152}$$

so

$$\int_0^1 <a''|\left(e^{(1-\lambda)A}\, Be^{\lambda A}\right)|a'>d\lambda = B_{a'a''}e^{a}\int_0^1 e^{(a'-a'')\lambda}\, d\lambda \tag{153a}$$

$$= B_{a'a''}e^{a''}\,\frac{\left(e(a'-a'')_{-1}\right)}{(a'-a'')} \tag{153b}$$

$$= B_{a'a''}\,\frac{e^{a'}-e^{a''}}{a'-a''} \tag{153c}$$

so

$$\left(e^{A+B}\right)_{a'a''} = e^{a'}\delta_{a',a''} + \frac{e^{a'}-e^{a''}}{(a'-a'')}B_{a'a''} + \cdots \tag{154}$$

also,

$$\mathrm{Tr}\left(e^{A+B}\right) = \sum_{a'}\left(e^{A+B}\right)_{a'a'} = \sum_{a'} e^{a'} + \cdots \tag{155}$$

Thus,

$$\left(e^{-\beta(H_0+H')}\right)_{\mathbf{k}'\mathbf{k}''} = e^{-\beta\epsilon^0_{\mathbf{k}'}} \delta_{\mathbf{k}'\mathbf{k}''} + \frac{e^{-\beta\epsilon^0_{\mathbf{k}'}} - e^{-\beta\epsilon^0_{\mathbf{k}''}}}{(e^0_{\mathbf{k}'}-e^0_{\mathbf{k}''})} H'_{\mathbf{k}'\mathbf{k}''} + O(\lambda^2) \tag{156}$$

and

$$\mathrm{Tr}\left(e^{-\beta(H_0+H')}\right) = \sum e^{-\beta\epsilon^0_{\mathbf{k}'}} + O(\lambda^2) \tag{157}$$

Remember that,

$$C_{\mathbf{k}'\mathbf{k}''} = C^{(0)}_{\mathbf{kk}'} + C^{(1)}_{\mathbf{kk}'}\lambda + \cdots \tag{158}$$

To carry the results to zeroth order in $C_{\mathbf{kk}'}$, we simply take

$$\rho_0(H_0 + H') \sim K_0 e^{-\beta H_0}; \quad K_0^{-1} = \mathrm{Tr} e^{-\beta H_0} \tag{159}$$

To carry the results to first order in $C_{\mathbf{kk}'}$ we let

$$\rho_0(H_0 + H') \sim K_0\left(e^{-\beta H_0} + \int_0^1 e^{-(1-\lambda)\beta H_0} (-\beta H_1 e^{-\lambda\beta H_0} d\lambda\right) \tag{160}$$

So $C^{(0)}_{\mathbf{kk}'}$ is diagonal and $C^{(1)}_{\mathbf{kk}'}$ is off diagonal

The reason for looking at $H_{\mathbf{k}'\mathbf{k}}$ in detail and $C_{\mathbf{kk}'}$ in powers of λ is that we are going to look for solutions to matrix element equations for $f_{\mathbf{k}'\mathbf{k}}$ in low order in λ. Recall $H_1 = -eE^0_\alpha X_\alpha$

$$C^{(0)}_{\mathbf{kk}'} = [\rho_0(H_0), H_1]_{\mathbf{kk}'} = -eE^0_\alpha[\rho_0(H_0), X_\alpha]_{\mathbf{kk}'} \tag{161}$$

To evaluate equation (161), we have

$$[\rho_0(H_0), X_\alpha) = \frac{1}{K_0}[e^{-\beta p^2/2m}, X_\alpha] \tag{162}$$

$$= -i\hbar\frac{\partial}{\partial p_\alpha}\left(\frac{1}{K_0}e^{-\beta p^2/2m}\right) = -i\hbar\frac{\partial}{\partial p_\alpha}(\rho_0(H_0) = -i\hbar\frac{\beta}{m}\rho_0(H_0) \tag{163}$$

In the above we have used

$$[X, F(p)] = i\hbar \frac{\partial F(p)}{\partial p} \tag{164}$$

so,

$$[\rho_0(H_0), X_\alpha]_{\mathbf{kk}'} = -i\left(\frac{\partial}{\partial \mathbf{k}_\alpha}\rho_{\mathbf{k}}\right)\delta_{\mathbf{kk}'} = i\frac{\partial \rho_{\mathbf{k}}}{\partial \mathbf{k}_\alpha}\delta_{\mathbf{kk}'} \tag{165}$$

Note:

$$\rho_{\mathbf{k}} = Ke^{-\beta\epsilon^{o}_{\mathbf{k}}}; \qquad K_0^{-1} = \sum e\text{-}\epsilon^{-\beta\epsilon^{o}_{\mathbf{k}}} \tag{166}$$

Thus

$$\boxed{C^{(0)}_{\mathbf{kk}'} = ieE^0_\alpha \frac{\partial \rho_{\mathbf{k}}}{\partial \mathbf{k}_\alpha}\delta_{\mathbf{kk}'}} \tag{167}$$

We now look for solutions to

$$\hbar(\omega^0_{\mathbf{kk}'} - is)f_{\mathbf{kk}'} = C_{\mathbf{kk}'} + \sum_{\mathbf{k}''}(f_{\mathbf{kk}''}H'_{\mathbf{k}''\mathbf{k}'} - H'_{\mathbf{kk}''}f_{\mathbf{k}''\mathbf{k}'}) \tag{168}$$

We look for solutions by splitting the equations into diagonal and off-diagonal components:

For the condition

$$\mathbf{k} = \mathbf{k}' \longrightarrow \omega^0_{\mathbf{kk}'} = 0 \tag{169}$$

we obtain

$$\sum_{\mathbf{k}''}(f_{\mathbf{kk}''}H'_{\mathbf{k}''\mathbf{k}'} - H'_{\mathbf{kk}''}f_{\mathbf{k}''\mathbf{k}'})$$

$$= (f_{\mathbf{kk}}H'_{\mathbf{kk}'} - H'_{\mathbf{k}'\mathbf{k}}f_{\mathbf{kk}}) + \sum_{\mathbf{k}''}(f_{\mathbf{kk}''}H'_{\mathbf{k}''\mathbf{k}'} - H'_{\mathbf{kk}''}f_{\mathbf{k}''\mathbf{k}'}) \tag{170}$$

yielding

$$-i\hbar s f_{\mathbf{k}} = C_{\mathbf{k}} + \sum_{\substack{\mathbf{k}' \\ (\mathbf{k}'\neq\mathbf{k})}} (f_{\mathbf{k}\mathbf{k}'} H'_{\mathbf{k}'\mathbf{k}} - H'_{\mathbf{k}\mathbf{k}'} f_{\mathbf{k}'\mathbf{k}}) \tag{171}$$

which relates the diagonal part of f to H_1, H' and $f_{\mathbf{k}\mathbf{k}'}$. Note here:

$$f_{\mathbf{k}} \equiv f_{\mathbf{k}\mathbf{k}} , \quad C_{\mathbf{k}} \equiv C_{\mathbf{k}\mathbf{k}} \tag{172}$$

For

$$\mathbf{k} \neq \mathbf{k}' \tag{173}$$

we get,

$$\begin{aligned} \hbar(\omega^0_{\mathbf{k}\mathbf{k}'} - is) f_{\mathbf{k}\mathbf{k}'} &= C_{\mathbf{k}\mathbf{k}'} + (f_{\mathbf{k}\mathbf{k}} H'_{\mathbf{k}\mathbf{k}'} - H'_{\mathbf{k}\mathbf{k}'} f_{\mathbf{k}'\mathbf{k}'}) \\ &+ (f_{\mathbf{k}\mathbf{k}'} H'_{\mathbf{k}'\mathbf{k}'} - H'_{\mathbf{k}\mathbf{k}} f_{\mathbf{k}\mathbf{k}'}) + \sum_{\substack{\mathbf{k}'' \\ (\mathbf{k}''\neq\mathbf{k},\mathbf{k}')}} (f_{\mathbf{k}\mathbf{k}''} H'_{\mathbf{k}''\mathbf{k}'} - H'_{\mathbf{k}\mathbf{k}''} f_{\mathbf{k}''\mathbf{k}'}) \end{aligned} \tag{174}$$

Now diagonal terms in H' can be incorprated into the left side by letting

$$\epsilon_{\mathbf{k}} = \epsilon^0_{\mathbf{k}} + H'_{\mathbf{k}\mathbf{k}} ; \quad \epsilon^0_{\mathbf{k}'} + H'_{\mathbf{k}'\mathbf{k}'} , \quad \hbar\omega_{\mathbf{k}\mathbf{k}'} = \epsilon_{\mathbf{k}} - \epsilon_{\mathbf{k}'} \tag{175}$$

so that we get

$$\hbar(\omega_{\mathbf{k}\mathbf{k}'} - is) f_{\mathbf{k}\mathbf{k}'} = C_{\mathbf{k}\mathbf{k}'} + (f_{\mathbf{k}} - f_{\mathbf{k}'}) H'_{\mathbf{k}\mathbf{k}'} + \sum_{\substack{\mathbf{k}'' \\ (\mathbf{k}''\neq\mathbf{k},\mathbf{k}')}} (f_{\mathbf{k}\mathbf{k}'} H'_{\mathbf{k}''\mathbf{k}'} - H'_{\mathbf{k}\mathbf{k}''} f_{\mathbf{k}''\mathbf{k}'}) \tag{176}$$

In solving Eqs. (171) and (176) we look for solutions in lowest order of the defect perturbation λ. Note: $C_{\mathbf{k}} \equiv C_{\mathbf{k}\mathbf{k}}$ is defined in zeroth order and $C_{\mathbf{kk}'}(\mathbf{k}\neq\mathbf{k}')$ is defined in first order. Also

$$f_{\mathbf{k}} \sim \lambda^{-2} \qquad f_{\mathbf{kk}'} \sim \lambda^{-1} \tag{177}$$

So to lowest order in λ, the solution to Eq. (176) is

$$\hbar(\omega_{\mathbf{kk}'}-is)f_{\mathbf{kk}'} \sim (f_{\mathbf{k}}-f_{\mathbf{k}'})H'_{\mathbf{kk}'} \tag{178}$$

or

$$f_{\mathbf{k},\mathbf{k}'} = \frac{(f_{\mathbf{k}}-f_{\mathbf{k}'})H'_{\mathbf{kk}'}}{\hbar(\omega_{\mathbf{kk}'}-is)} \tag{179}$$

Then inserting this result for $f_{\mathbf{kk}'}$ into Eq. (171), we obtain to lowest order

$$\left[-isf_{\mathbf{k}} = ieE^0_\alpha \frac{\partial\rho_{\mathbf{k}}}{\partial k_\alpha} + \sum_{\substack{\mathbf{k}'\\(\mathbf{k}'\neq\mathbf{k})}} \left(\frac{(f_{\mathbf{k}}-f_{\mathbf{k}'})|(H')_{\mathbf{kk}'}|^2}{\hbar(\omega_{\mathbf{kk}'}-is)} - \frac{(f_{\mathbf{k}'}-f_{\mathbf{k}})|(H')_{\mathbf{kk}'}|^2}{\hbar(\alpha_{\mathbf{kk}'}-is)}\right)\right] \tag{180}$$

or

$$-isf_{\mathbf{k}} = ieE^0_\alpha \frac{\partial\rho_{\mathbf{k}}}{\partial k_\alpha} + \sum_{\substack{\mathbf{k}'\\(\mathbf{k}'\neq\mathbf{k})}} |(H')_{\mathbf{kk}'}|^2 \left[(f_{\mathbf{k}}-f_{\mathbf{k}'})\left(\frac{1}{\hbar(\omega_{\mathbf{kk}'}-is)} - \frac{1}{\hbar(\omega_{\mathbf{kk}}-is)}\right)\right] \tag{181}$$

as the equation for $f_{\mathbf{k}}$ alone to low order in λ.

Now

$$H'_{\mathbf{kk}'} = \frac{\phi_{\mathbf{k},\mathbf{k}'}}{\Omega}\sum_{i=1}^{N} e^{-(\mathbf{k}-\mathbf{k})\cdot\mathbf{r}_i} \tag{182}$$

Then

$$|(H'_{\mathbf{kk}'})|^2 = \frac{|\phi_{\mathbf{k},\mathbf{k}'}|^2}{\Omega^2}\sum_{i,j}^{N} e^{-i(\mathbf{r}_i-\mathbf{r}_j)\cdot\mathbf{q}}\ ; \qquad \mathbf{q} = \mathbf{k}-\mathbf{k}' \neq 0 \tag{183}$$

An ensemble average of the impurity distribution is

$$<|(H'_{\mathbf{kk}'})|^2> = \frac{|\phi_{\mathbf{kk}'}|^2}{\Omega^2} \underbrace{<|\sum_{ij} e^{-i(\mathbf{r}_i-\mathbf{r}_j)\cdot\mathbf{q}}|>}_{\sim N} \tag{184}$$

$$= \frac{|\phi_{\mathbf{kk}'}|^2}{\Omega^2} N \tag{185}$$

(See Kohn & Luttinger)

Then we get

$$\boxed{-isf_{\mathbf{k}} = ieE_\alpha^0 \frac{\partial \rho_{\mathbf{k}}}{\partial k_\alpha} + \sum \frac{N|\phi_{\mathbf{kk}'}|^2}{\Omega^2} \left[(f_{\mathbf{k}}-f_{\mathbf{k}'})\left(\frac{1}{\hbar(\omega_{\mathbf{kk}'}-is)} - \frac{1}{\hbar(\omega_{\mathbf{kk}'}-is)}\right)\right]} \tag{186}$$

What about "s"? There is a large range of "s" values for which the solution to Eq. (186) is practically independent of s. This range is given by

$$s \ll \frac{1}{\tau_r}, \frac{1}{\tau_a} \tag{187}$$

where τ_r is the order of a relaxation time. τ_a is an atomic time $\sim \hbar/\epsilon$, where $\hat{\epsilon}$ is an average energy of the carriers. When "s" in this range $sf_{\mathbf{k}}$ in Eq. (186) may be omitted. Also if $s > \Delta\epsilon/\hbar$, with $\Delta\epsilon$ of the order of the electronic energy bands then

$$\sum_{\mathbf{k}'} \longrightarrow \frac{\Omega}{(2\pi)^3} \int d\mathbf{k}' \tag{188}$$

Once we assume that $s \ll 1/\tau a$ then we can use the well-known relation:

$$\lim_{s\to 0^+} \left(\frac{1}{x-is}\right) = P\left(\frac{1}{x}\right) + i\pi\delta(x) \tag{189a}$$

or

$$\lim_{s\to 0} \left(\frac{1}{x-is} - \frac{1}{x-is}\right) = \lim_{s\to 0} 2i\left(\frac{S}{x^2+s^2}\right) = 2\pi i\delta(x)$$

$$\text{note: } \pi\delta(x) = \lim_{s\to 0} \frac{s}{x^2+s^2} \tag{189b}$$

and Eq. (186) becomes

$$0 = eE_\alpha^0 \frac{\partial \rho_{\mathbf{k}}}{\partial k_\alpha} + 2\pi \sum_{\substack{\mathbf{k}' \\ (\mathbf{k}=\mathbf{k}')}} \frac{N|\phi_{\mathbf{kk}'}|^2}{\Omega^2} \delta(\hbar\omega_{\mathbf{kk}'})(f_{\mathbf{k}}-f_{\mathbf{k}'}) \tag{190}$$

Note: The transition probability $w_{\mathbf{k},\mathbf{k}'}^{(0)}$ per unit time for transitions from the state $\mathbf{k}'$ to $\mathbf{k}$ due to single impurity scatterers is, in the first Born approximation

$$w_{\mathbf{kk}'}^{(0)} = 2\pi \frac{|\phi_{\mathbf{kk}'}|^2}{\Omega^2} \delta(\hbar\omega_{\mathbf{kk}'}) \tag{191}$$

Then we get

$$0 = eE_\alpha^0 \frac{\partial \rho_{\mathbf{k}}}{\partial k_\alpha} + \sum_{\substack{\mathbf{k}' \\ (\mathbf{k}\neq\mathbf{k}')}} (W_{\mathbf{k}'\mathbf{k}}^{(0)} f_{\mathbf{k}} - W_{\mathbf{kk}'}^{(0)} f_{\mathbf{k}'}) \tag{192}$$

where $W_{\mathbf{k}'\mathbf{k}}^{(0)}$ is the total transistion probability as the sum of those occuring over each of the impurities $W_{\mathbf{k}'\mathbf{k}}^{(0)} = Nw_{\mathbf{k}'\mathbf{k}}^{(0)}$. This is the customary transport equation used in the theory of impurity conductivity.

Now let's calculate some key expectation values using this density matrix approach for this problem. We note $\langle\overline{A}\rangle \equiv \overline{A} = \mathrm{tr}(\hat{\rho}_T\hat{A})$, where for this problem

In this problem

$$\rho_T = \rho + f \tag{193}$$

so that the expectation value of A is

$$\overline{A} = \mathrm{Tr}(\rho A) + \mathrm{Tr}(fA) \tag{194}$$

or

$$\overline{A} = \overline{A}_0 + \overline{A}_E \tag{195}$$

In the above the first term is the equilibrium value of A and the second term is the correction due to the electric field. Let's ignore $\overline{A}_0$ and concentrate on $\overline{A}_E$.

In the $|\mathbf{k}\rangle$ representation

$$\bar{A}_E = \sum_{\mathbf{k}}(fA)_{\mathbf{kk}} = \sum_{\mathbf{k}}\langle\mathbf{k}|(fA)|\mathbf{k}\rangle = \sum_{\mathbf{k}}\sum_{\mathbf{k}'}\langle\mathbf{k}|f|\mathbf{k}'\rangle\langle\mathbf{k}'|A|\mathbf{k}\rangle \tag{196}$$

or

$$\bar{A}_E = \sum_{\mathbf{k},\mathbf{k}'} f_{\mathbf{kk}'}A_{\mathbf{k}'\mathbf{k}} \tag{197}$$

Breaking $\bar{A}_E$ into diagonal and off-diagonal parts

$$\bar{A}_E = \sum_{\mathbf{k}} f_{\mathbf{k}}A_{\mathbf{kk}} + \sum_{\substack{\mathbf{k},\mathbf{k}' \\ (\mathbf{k}\neq\mathbf{k}')}} f_{\mathbf{kk}'}A_{\mathbf{k}'\mathbf{k}} \tag{198}$$

For calculating the currents, we need the velocity v_β. Then $A=v_\beta$, and

$$v_\beta = \frac{i}{\hbar}[H_T, x_\beta] = p_\beta/m \tag{199}$$

with $H_T = p^2/2m + H' + H_E$. Then

$$(v_\beta)_{\mathbf{kk}'} = \frac{\hbar k_\beta}{m}\delta_{\mathbf{kk}'} \tag{200}$$

which is diagonal. Thus $f_{\mathbf{kk}'}$ plays no role in A_E. Note

$$\bar{v}_E = \sum_{\mathbf{k}} \frac{\hbar}{m}k_\beta f_{\mathbf{k}'} \tag{201}$$

And in lowest order, only the diagonal elements of f play roles in the velocity for this problem.

For acceleration

$$a_\beta = \frac{i}{\hbar}[H_T, v_\beta] = \frac{i}{\hbar}\left[H_T, \frac{p_\beta}{m}\right] \tag{202a}$$

$$= \frac{i}{\hbar}\left(\left[H_E, \frac{p_\beta}{m}\right] + \left[H', \frac{p_\beta}{m}\right]\right) \tag{202b}$$

$$= \frac{i}{\hbar}\left(-eE_\alpha^0 \underbrace{\left(x_\alpha, \frac{p_\beta}{m}\right)}_{\frac{i\hbar}{m}\delta_{\alpha\beta}} + \left(H', \frac{p_\beta}{m}\right)\right) \tag{202c}$$

$$= \left(eE_\beta^0 + \frac{i}{\hbar}[H', p_\beta]\right)/m \tag{202d}$$

Now

$$(a_\beta)_{\mathbf{kk}'} = \left(-eE_\beta^0 \delta_{\mathbf{k},\mathbf{k}'} + \frac{i}{\hbar}[H', \hbar k_\beta]_{\mathbf{kk}'}\right)/m \tag{203a}$$

$$= \left(-eE_\beta^0 \delta_{\mathbf{k},\mathbf{k}'} + i(k_\beta'-k_\beta)H'_{\mathbf{kk}'}\right)/m \tag{203b}$$

Thus for the expectation value of acceleration

$$\bar{a}_\beta = \mathrm{Tr}(a_\beta f) \tag{204a}$$

$$= \sum_{\mathbf{k}} f_{\mathbf{k}}(a_\beta)_{\mathbf{kk}} + \sum_{\substack{\mathbf{k},\mathbf{k}' \\ (\mathbf{k}\neq\mathbf{k}')}} f_{\mathbf{kk}'}(a_\beta)_{\mathbf{k}'\mathbf{k}} \tag{204b}$$

And

$$\boxed{\bar{a}_\beta = \left(\frac{1}{m}eE_\beta^{(0)} + \frac{i}{m}\sum_{\substack{\mathbf{kk}' \\ (\mathbf{k}\neq\mathbf{k}')}} (k_\beta-k_\beta')f_{\mathbf{kk}'}H'_{\mathbf{k}'\mathbf{k}}\right)} \tag{205}$$

Or, to lowest order, $\bar{a}_\beta$ depends on the off-diagonal elements of f. To show the role of the off-diagonal parts, remember that to lowest order in λ

$$f_{\mathbf{kk}'} = \frac{(f_{\mathbf{k}}-f_{\mathbf{k}'})H'_{\mathbf{kk}'}}{\hbar(\omega_{\mathbf{kk}'}-is)} \tag{206}$$

so that

$$i \sum_{\substack{\mathbf{k},\mathbf{k}' \\ (\mathbf{k}\neq\mathbf{k}')}} (k_\beta - k_\beta) f_{\mathbf{kk}'} H_{\mathbf{kk}'} = i \sum_{\substack{\mathbf{k},\mathbf{k}' \\ (\mathbf{k}\neq\mathbf{k}')}} (k_\beta - k'_\beta) \frac{(f_\mathbf{k} - f_{\mathbf{k}'})}{\hbar(\omega_{\mathbf{kk}'} - is)} |H'_{\mathbf{kk}'}|^2 \tag{207a}$$

$$= i \sum_{\substack{\mathbf{k},\mathbf{k}' \\ (\mathbf{k}\neq\mathbf{k}')}} k_\beta (f_\mathbf{k} - f_{\mathbf{k}'}) |H'_{\mathbf{kk}'}|^2 \left(\frac{1}{(\omega_{\mathbf{kk}'} - is)\hbar} - \frac{1}{\hbar(\omega_{\mathbf{kk}'} + is)} \right) \tag{207b}$$

In the following limit, the right-hand side of Eq. (206) becomes

$$\xrightarrow[s \to 0^+]{} i \sum_{\substack{\mathbf{k},\mathbf{k}' \\ (\mathbf{k}\neq\mathbf{k}')}} k_\beta (f_\mathbf{k} - f_{\mathbf{k}'}) |H'_{\mathbf{kk}'}|^2 (2\pi i) \delta(\hbar\omega_{\mathbf{kk}'}) \tag{207c}$$

$$= eE_\alpha^{(0)} \sum_\mathbf{k} k_\beta \frac{\partial \rho_\mathbf{k}}{\partial k_\alpha} \tag{207d}$$

$$= eE_\alpha^{(0)} \sum \delta_{\alpha\beta} = -eE_\beta^{(0)} \tag{207e}$$

Thus $\bar{a}_\beta = 0$ to lowest order.

We also note that the above comments hold using the Bloch representation: $H_0 = (p^2/2m) + U(\mathbf{r})$, $U(\mathbf{r}) = U(\mathbf{r}+\mathbf{a})$.

So far the quantum effects have not surfaced. The problem under consideration can be pushed to higher order. If the process is continued for two more iterations, new effects begin to appear that don't usually surface in conventional transport:

(1) Effects or terms appear which account for interference between the change in the distribution function due to the electric field and the change due to collisions with the impurities.

(2) As we go to the limit of very few impurities (average distance between impurities is much greater that the mean wavelength of the electrons) then single impurity scattering dominates.

(3) In transitions between energy states, energy is not "exactly" conserved. It is conserved up to an energy of the order of the average reciprocal lifetime of the states in question.

TIME DEPENDENT TRANSPORT

So far we have been dealing with the steady state equilibrium situation (by virtue of the way we chose $s, s \ll 1/\tau_a$). We now outline a method which has the advantages of permitting the study of transport in the transient (approach to steady state) as well as steady state regimes.

Let's consider the same problem (in low order perturbation theory) except now we imagine the electrons plus impurities to be in equilibrium until a time t=o when a uniform electric field is turned on. Then at $t = 0$, $\rho = \rho_o(H^o+H')$

To first order in the field, we had the time-dependent Eq. (124) for the density matrix operator. To study the time dependence, we look for the Laplace transform solution:

$$\Delta\hat{\rho}(t) = \int_0^t e^{-st} f(s) ds \tag{208}$$

to get

$$i\hbar s\hat{f} = [\hat{H}_o, \hat{f}] + [\hat{H}', \hat{f}] + \hat{H}_E, \hat{\rho}_o] \tag{209}$$

which is exactly the equation we generated earlier (Eq. (129)). We then obtain the first order Eq. (186).

Laplace transforming Eq. (186) back to the time domain with the use of

$$sf_{\mathbf{k}}(s) \longrightarrow \frac{\partial \Delta\rho(t)}{\partial t} \tag{210}$$

and using the convolution theorem

$$\int_0^t g_1(\tau) g_2(t-\tau)\, d\tau \longrightarrow F_1(s) F_2(s) \tag{211}$$

for

$$\left[(f_{\mathbf{k}}-f_{\mathbf{k}'})\left(\frac{1}{\hbar(\omega_{\mathbf{kk}'}-is)} - \frac{1}{\hbar(\omega_{\mathbf{kk}'}+is)}\right)\right] = 2i(f_{\mathbf{k}} - f_{\mathbf{k}'})\frac{s}{(\hbar\omega_{\mathbf{kk}'})^2 + s^2} \tag{212}$$

which, when transformed back, yields

$$\longrightarrow 2i\int_0^t \left[\Delta\rho_{\mathbf{k}}(t-\tau) - \Delta\rho_{\mathbf{k}'}(t-\tau)\right] \cos(\hbar\omega_{\mathbf{kk}'}\tau)d\tau \tag{213}$$

and we get

$$\frac{\partial\Delta\rho_{\mathbf{k}}}{\partial t} = - eE_\alpha^0 \frac{\partial\rho_{\mathbf{k}}}{\partial k_\alpha} - \frac{2N}{\Omega^2}\sum_{\mathbf{k}'}|\phi_{\mathbf{kk}'}|^2 \int_0^t \left[\Delta\rho_{\mathbf{k}}(t-\tau) - \Delta\rho_{\mathbf{k}'}(t-\tau)\right] \cos(\hbar\omega_{\mathbf{kk}'}\tau)d\tau \tag{214}$$

In conventional Boltzmann equation theory, the rate of change of the distribution function at any instant depends on the distribution function at that instant; here, the distribution function depends on the value of the distribution function at all previous times up to the time in question.

Also, before, Eq. (186) in place of the memory term on the right hand side of Eq. (214) we had

$$\frac{2\pi N}{\Omega^2}\sum_{\substack{\mathbf{k}' \\ (\mathbf{k}\neq\mathbf{k}')}} |\phi_{\mathbf{kk}'}|^2 \delta(\hbar\omega_{\mathbf{kk}'})(f_{\mathbf{k}}-f_{\mathbf{k}'}) \tag{215}$$

whereby using

$$W_{\mathbf{kk}'}^{(0)} = \frac{2\pi N}{\Omega^2}|\phi_{\mathbf{kk}'}|^2 \delta(\hbar\omega_{\mathbf{kk}'}) \tag{216}$$

we obtained the familiar collision term. Let's see how the time dependent term in Eq. (214) reduces to this result. Assume "t" is much longer than $\tau_a \sim \hbar/\omega_{\mathbf{kk}'}$ and assume we can formally expand

$\Delta\rho_{\mathbf{k}}/\Delta\rho_{\mathbf{k}'}$ in powers of τ. Then

$$\int_0^t \Delta\rho_{\mathbf{k}}(t-\tau)\cos(\hbar\omega_{\mathbf{k}\mathbf{k}'}\tau)d\tau \cong \Delta\rho_{\mathbf{k}}(t)\int_0^t \cos(\hbar\omega_{\mathbf{k}\mathbf{k}'}\tau)d\tau + \frac{\partial\Delta\rho_{\mathbf{k}\mathbf{k}'}}{\partial t}\int_0^t \tau\cos(\hbar\omega_{\mathbf{k}\mathbf{k}'}\tau)d\tau + O(\tau^2) \tag{217}$$

Now

$$\int_0^t \cos(\hbar\omega_{\mathbf{k}\mathbf{k}'}\tau)d\tau = \frac{\sin \hbar\omega_{\mathbf{k}\mathbf{k}'}t}{\hbar\omega_{\mathbf{k}\mathbf{k}'}} \xrightarrow[t\gg(\omega_{\mathbf{k}\mathbf{k}'})]{} \pi\delta(\hbar\omega_{\mathbf{k}\mathbf{k}'}) \tag{218}$$

Thus we retrieve the familiar result as times get much longer than the atomic time.

WIGNER DISTRIBUTION FUNCTION

Thus far we have been discussing the density matrix approach to quantum transport and we saw the similarity between equations for the density matrix elements and the classical probability distribution function.

There is yet another approach to the formulation of quantum transport based on the construction of a Wigner distribution function. We will see that this distribution function has no simple interpretation in the sense of probability theory, but in lieu of its special propoerties (which we will show), can be used to calculate observables in a manner quite analogous to that of classical theory, i.e., by integrating the product of the observable and the Wigner distribution function over all phase space.

Let's introduce the subject by defining the Wigner function as was originally done by Wigner (Phys Rev. 40, 749 (1932)). Then we will fill-in and expand on the subject accordingly.

The Wigner function, P_W, is generally defined in terms of all the generalized coordinates and momenta of the system as

$$P_W(x_1\ldots x_n,\ p_1\ldots p_n)$$

$$= \frac{1}{(2\pi\hbar)^n}\int_{-\infty}^{\infty} dy_1\ldots dy_n\ \Psi^*\left(x_1+\frac{y_1}{2},\ldots x_n+\frac{y_n}{2}\right)\Psi\left(x_1-\frac{y_1}{2},\ldots x_n-\frac{y_n}{2}\right) \cdot e^{i(p_1y_1,\ldots p_ny_n)/\hbar} \tag{219}$$

For purposes of simplicity in this discussion we consider the Wigner function of a single coordinate and momenta

$$P_W(x,p) = \frac{1}{2\pi\hbar}\int_{-\infty}^{\infty} dy\Psi^*\left(x+\frac{y}{2}\right)\Psi\left(x-\frac{y}{2}\right)e^{ipy/\hbar} \tag{220}$$

where $\Psi(x)$ is the time dependent state of the system in the coordinate representation.

The above $P_W(x,p)$ is defined for pure states; a $P_W(x,p)$ can be defined for mixed states. As such

$$P_W(x,p) = \sum_n p_n\left(\frac{1}{2\pi\hbar}\int_{-\infty}^{\infty} dy\Psi_n^*\left(x+\frac{y}{2}\right)\Psi_n\left(x-\frac{y}{2}\right)e^{ipy/\hbar}\right) \tag{221}$$

where p_n is the probabality to be in the state "n". (For example, in the canonical ensemble, $p_n = e^{-\epsilon_n/kT}$). Let's consider P_W for pure states only:

Interesting sum rules

$$\int_{-\infty}^{\infty} P_W(x,p)dp = \Psi^*(x)\Psi(x) \tag{222}$$

Proof:

$$\int_{-\infty}^{\infty} P_W(x,p)dp = \frac{1}{2\pi\hbar}\int_{-\infty}^{\infty} dy\Psi^*\left(x+\frac{y}{2}\right)\Psi\left(x-\frac{y}{2}\right)\underbrace{\int_{-\infty}^{\infty} e^{ipy/\hbar}dp}_{2\pi\hbar\delta(y)} \tag{223a}$$

$$= \Psi^*(x)\Psi(x) \tag{223b}$$

Also

$$\int_{-\infty}^{\infty} P_W(x,p)dx = \phi^*(p)\phi(p) \tag{224}$$

where

$$\phi(p) = \frac{1}{\sqrt{2\pi\hbar}}\int_{-\infty}^{\infty} e^{-ipx/\hbar}\,\Psi(x)d(x) \tag{225}$$

proof:

$$\int_{-\infty}^{\infty} P_W(x,p)\,dx = \frac{1}{2\pi\hbar}\int_{-\infty}^{\infty} dy e^{ipy/\hbar} \int_{-\infty}^{\infty} dx \Psi^*\left(x+\frac{y}{2}\right)\Psi\left(x-\frac{y}{2}\right) \tag{226a}$$

$$= \frac{1}{2\pi\hbar}\int_{-\infty}^{\infty}\int_{-\infty}^{\infty} dy dx e^{ipy/\hbar}\, \Psi^*\left(x+\frac{y}{2}\right)\Psi\left(x-\frac{y}{2}\right) \tag{226b}$$

Let

$$\xi_1 = x + \frac{y}{2}; \qquad \xi_2 = x - \frac{y}{2} \tag{227a}$$

where

$$x = \frac{1}{2}(\xi_1+\xi_2), \qquad y = \xi_1 - \xi_2 \tag{227b}$$

so

$$dxdy = \begin{vmatrix} \frac{\partial y}{\partial \xi_1} & \frac{\partial y}{\partial \xi_2} \\ \frac{\partial x}{\partial \xi_1} & \frac{\partial x}{\partial \xi_2} \end{vmatrix} d\xi_1 d\xi_2 = \begin{vmatrix} 1 & -1 \\ \frac{1}{2} & \frac{1}{2} \end{vmatrix} d\xi_1 d\xi_2 = d\xi_1 d\xi_2 \tag{227c}$$

So

$$\int_{-\infty}^{\infty} P_W(x,p)\,dx = \frac{1}{2\pi\hbar}\int_{-\infty}^{\infty} d\xi_1 \int_{\infty}^{\infty} d\xi_2\; e^{ip(\xi_1-\xi_2)/\hbar}\Psi^*(\xi_1)\Psi(\xi_2) \tag{228}$$

with $\phi(p)$ defined as Eq. (225), Eq. (224) is obtained. Therefore, with the above two sum rules, it follows that for an observable $\overline{A}(x,p)$ which is a function of $\hat{x}$ alone, $\hat{p}$ alone, or any additional sum of the two (certain Hamiltonians), the expectation value of $\hat{A}$ is

$$\langle\hat{A}\rangle = \iint \hat{A} P_W(x,p)\,dx\,dp \tag{229}$$

which is analogous to the classical expression for the average value.

Herein lies the interesting aspect of the Wigner distribution function; The $\langle A\rangle$ above suggests that it is possible to transfer many of the results of classical transport into quantum transport by simply replacing the classical distribution function by the Wigner function.

TIME EVOLUTION OF THE WIGNER DISTRIBUTION

Further resemblance of the Wigner distribution function to the classical distribution function is apparent by examining the time evolution of $P_W(x,p)$.

Assuming $H = (p^2/2m) + V(x)$, $i\hbar\ (\partial\Psi/\partial t) = H\Psi$, it can readily be shown that

$$\frac{\partial P_W}{\partial t} + \left(\frac{p}{m}\right)\frac{\partial P_W}{\partial x} + \Theta\cdot P_W = 0 \tag{230}$$

where

$$\Theta\cdot P_W = -\frac{2}{\hbar}\sum_{n=0}^{\infty}(-1)^n\frac{(\hbar/2)^{2n+1}}{(2n+1)!}\frac{\partial^{2n+1}V(x)}{\partial x^{2n+1}}\frac{\partial^{2n+1}P_W(x,p)}{\partial p^{2n+1}}$$

$$\xrightarrow[\hbar\to 0]{} -\left(\frac{\partial V}{\partial x}\right)\left(\frac{\partial P_W}{\partial p}\right) \tag{231}$$

so that as $\hbar \to 0$, Eq.(231) reduces to the classical collsionless Boltzmann equation.

Proof (due to Wigner):

$$\frac{\partial P_W}{\partial t} = \frac{1}{2\pi\hbar}\int_{-\infty}^{\infty}\frac{\partial}{\partial t}\left(\Psi^*\left(x+\frac{y}{2}\right)\Psi\left(x-\frac{y}{2}\right)\right)e^{ipy/\hbar}dy \tag{232}$$

To look for time evolution we study

$$\frac{\partial}{\partial t}\left(\Psi^*\left(x+\frac{y}{2}\right)\Psi\left(x-\frac{y}{2}\right)\right) \equiv \frac{\partial}{\partial t}\left((T\Psi)^*(T^+\Psi)\right) \tag{233}$$

where we have used

$$T = e^{-i\frac{\hat{p}}{\hbar}\frac{y}{2}} \equiv e^{\frac{\partial}{\partial x}\frac{y}{2}} \tag{234a}$$

with

$$T^{+} = T^{-1} \tag{234b}$$

and

$$Tf(x) = f\left(x+ \frac{y}{2}\right) \tag{234c}$$

$$T^{+}f(x) = f\left(x- \frac{y}{2}\right) \tag{234d}$$

Thus

$$\Psi\left(x- \frac{y}{2}\right) = T^{+}\Psi(x) \tag{235a}$$

$$\Psi\left(x+ \frac{y}{2}\right) = T\Psi(x) \tag{235b}$$

and

$$i\hbar\frac{\partial}{\partial t}[T\Psi(x)] = \frac{p^2}{2m}[T\Psi(x)] + V\left(x+ \frac{y}{2}\right)[T\Psi(x)] \tag{236a}$$

$$-i\hbar\frac{\partial}{\partial t}[T^{+}\Psi(x)]^{*} = \frac{p^2}{2m}[T^{+}\Psi]^{*} + V\left(x- \frac{y}{2}\right)[T^{+}\Psi(x)]^{*} \tag{236b}$$

so that

$$\frac{\partial}{\partial t}\left(\Psi^{*}\left(x+ \frac{y}{2}\right)\Psi\left(x- \frac{y}{2}\right)\right) = \frac{i\hbar}{2m}\left(\Psi^{*}\left(x+ \frac{y}{2}\right)\frac{\partial^2\Psi\left(x- \frac{y}{2}\right)}{\partial x^2} - \frac{\partial^2\Psi^{*}\left(x+ \frac{y}{2}\right)}{\partial x^2}\Psi\left(x- \frac{y}{2}\right)\right)$$
$$+ \frac{1}{i\hbar}\left\{\Psi^{*}\left(x+ \frac{y}{2}\right)\Psi\left(x- \frac{y}{2}\right)\left(V\left(x- \frac{y}{2}\right) - V\left(x+ \frac{y}{2}\right)\right)\right\} \tag{237}$$

which gives

$$\frac{\partial p_W}{\partial t} = C + S \tag{238}$$

where

$$C = \frac{1}{2\pi\hbar}\int dy\; e^{ipy/\hbar}\left\{\frac{i\hbar}{2m}\left(\Psi^*\left(x+\frac{y}{2}\right)\frac{\partial^2\Psi\left(x-\frac{y}{2}\right)}{\partial x^2} - \frac{\partial^2\Psi^*\left(x+\frac{y}{2}\right)}{\partial x^2}\Psi\left(x-\frac{y}{2}\right)\right)\right\} \tag{239}$$

and

$$S = \frac{1}{2\pi\hbar}\int dy\; e^{ipy/\hbar}\left\{\frac{1}{i\hbar}\left(\Psi^*\left(x+\frac{y}{2}\right)\Psi\left(x-\frac{y}{2}\right)\left[V\left(x-\frac{y}{2}\right) - V\left(x+\frac{y}{2}\right)\right]\right)\right\} \tag{240}$$

First let's discuss S

$$V\left(x-\frac{y}{2}\right) - V\left(x+\frac{y}{2}\right) = T^+V(x) - TV(x) = (T^+-T)V(x) \tag{241}$$

Now

$$T = e^{i\frac{p}{\hbar}\frac{y}{2}} = e^{\frac{y}{2}\frac{d}{dx}} = \sum_{n=o}^{\infty}\frac{1}{n!}\left(\frac{y}{2}\right)^n\left(\frac{d}{dx}\right)^n \equiv 1 + \alpha + \frac{\alpha^2}{2!} + \frac{\alpha^3}{3!} + \cdots \tag{242}$$

$$T^+ = e^{-i\frac{p}{\hbar}\frac{y}{2}} = e^{-\frac{y}{2}\frac{d}{dx}} = \sum_{n=o}^{\infty}\frac{(-1)^n}{n!}\left(\frac{y}{2}\right)^n\left(\frac{d}{dx}\right)^n \equiv 1 - \alpha + \frac{\alpha^2}{2!} - \frac{\alpha^3}{3!} + \cdots \tag{243}$$

so

$$T^+ - T = \left(1 - \alpha + \frac{\alpha^2}{2!} - \frac{\alpha^3}{3!} + \cdots\right) - \left(1 + \alpha + \frac{\alpha^2}{2!} + \frac{\alpha^3}{3!} + \cdots\right) \tag{244a}$$

$$= -2\left(\alpha + \frac{\alpha^3}{3!} + \frac{\alpha^5}{5!} + \cdots\right) = -2\sum_{n=o}^{\infty}\frac{1}{(2n+1)!}\left(\frac{y}{2}\right)^{2n+1}\left(\frac{d}{dx}\right)^{2n+1} \tag{244b}$$

Thus

$$S = -\frac{2}{i\hbar}\sum_{n=o}^{\infty}\frac{1}{(2n+1)!}\left(\frac{1}{\alpha}\right)^{2n+1}\frac{\partial^{2n+1}V(x)}{\partial x^{2n+1}} \tag{245a}$$

$$\frac{1}{2\pi\hbar}\int_{-\infty}^{\infty}dy\Psi^*\left(x+\frac{y}{2}\right)\Psi\left(x-\frac{y}{2}\right)\frac{y^{2n+1}\; e^{ipy/\hbar}}{\left(\frac{\hbar}{i}\right)^{2n+1}\frac{\partial^{2n+1}}{\partial p^{2n+1}}(e^{ipy/\hbar})}$$

$$= -\frac{2}{i\hbar}\sum_{n=o}^{\infty}\frac{\left(\frac{\hbar}{2i}\right)^{2n+1}}{(2n+1)!}\frac{\partial^{2n+1}V(x)}{\partial x^{2n+1}}\frac{\partial^{2n+1}P_w(x,p,t)}{\partial p^{2n+1}} \xrightarrow[\hbar\to 0]{} +\frac{\partial V}{\partial x}\frac{\partial P_w}{\partial p} \tag{245b}$$

Now since,

$$\frac{-1}{i}\left(\frac{1}{i}\right)^{2n+1} = \left(\frac{1}{i}\right)^{2n} = \left(\frac{1}{i^2}\right)^{n} = (-1)^n$$

and

$$\sin x = \sum_{n=o}^{\infty}\frac{(-1)}{(2n+1)!}x^{2n+1} ,$$

S can formally be written as

$$S = \frac{2}{\hbar}\left(\sin\frac{\hbar}{2}\left(\frac{\partial}{\partial x}\ \frac{\partial}{\partial p}\right)\right)V(x)P_w(x,p) \tag{246}$$

where it is understood that $\partial/\partial x$ operates on V(x) alone.

Now let's discuss, C Eq. (239), which can be written as

$$C = \frac{\partial}{\partial x}\left\{\frac{1}{2\pi\hbar}\int_{-\infty}^{\infty}dy\ e^{ipy/\hbar}\left(\frac{i\hbar}{2m}\right)\left(\Psi^*\left(x+\frac{y}{2}\right)\frac{\partial}{\partial x}\Psi\left(x-\frac{y}{2}\right) - \frac{\partial}{\partial x}\Psi^*\left(x+\frac{y}{2}\right)\Psi\left(x-\frac{y}{2}\right)\right)\right\} \tag{247}$$

Using the fact that

$$\frac{\partial f}{\partial x}\left(x-\frac{a}{2}\right) = -2\frac{\partial f}{\partial a}\left(x-\frac{a}{2}\right);\qquad \frac{\partial f}{\partial x}\left(x-\frac{a}{2}\right) = -2\frac{\partial f}{\partial a}\left(x-\frac{a}{2}\right) \tag{248}$$

"C" becomes

$$C = \frac{i\hbar}{2m}\frac{\partial}{\partial x}\left\{\frac{1}{2\pi\hbar}\int_{-\infty}^{\infty}dy(-2)\left(\Psi^*\left(x+\frac{y}{2}\right)\frac{\partial}{\partial y}\Psi\left(x-\frac{y}{2}\right) + \frac{\partial}{\partial y}\Psi^*\left(x+\frac{y}{2}\right)\Psi\left(x-\frac{y}{2}\right)\right)e^{ipy/\hbar}\right\} \tag{249a}$$

$$= \frac{i\hbar}{2m}(-2)\frac{\partial}{\partial x}\underbrace{\left\{\frac{1}{2\pi\hbar}\int_{-\infty}^{\infty}dy\frac{\partial}{\partial y}\left(\Psi^*\left(x+\frac{y}{2}\right)\Psi\left(x-\frac{y}{2}\right)\right)e^{ipy/\hbar}\right\}}_{I} \tag{249b}$$

To evaluate I let

$$U = e^{ipy/\hbar} \tag{250a}$$

$$dU = \frac{ip}{\hbar} e^{ipy/\hbar} dy \tag{250b}$$

and let

$$dV = \frac{\partial}{\partial y}\left(\Psi^*\left(x+\frac{y}{2}\right)\Psi\left(x-\frac{y}{2}\right)dy \tag{251a}$$

$$V = \Psi^*\left(x+\frac{y}{2}\right)\Psi\left(x-\frac{y}{2}\right)\Bigg) \tag{251b}$$

so that

$$I = \frac{1}{2\pi\hbar}\Psi^*\left(x+\frac{y}{2}\right)\Psi\left(x-\frac{y}{2}\right)e^{ipy/\hbar}\Bigg|_{-\infty}^{\infty} - \frac{ip}{\hbar}\frac{1}{2\pi\hbar}\int_{-\infty}^{\infty}\Psi^*\left(x+\frac{y}{2}\right)\Psi\left(x-\frac{y}{2}\right)e^{ipy/\hbar}dy \tag{252}$$

Therefore

$$I = -\frac{ip}{\hbar} P_w(x,p,t) \tag{253}$$

so that

$$C = \frac{i\hbar}{2m}(-2)\frac{\partial}{\partial x}I \tag{254a}$$

or

$$\boxed{C = -\frac{p}{m}\frac{\partial}{\partial x}P_w(x,p,t)} \tag{254b}$$

Thus

$$\frac{\partial P_w}{\partial t} = -\frac{p}{m}\frac{\partial}{\partial x}P_w(x,p,t) + \frac{2}{\hbar}\sum_{n=0}^{\infty}(-1)^n\frac{(\hbar/2)^{2n+1}}{(2n+1)!}\frac{\partial^{2n+1}}{\partial x^{2n+1}}V(x)\frac{\partial^{2n+1}}{\partial p^{2n+1}}P_w \tag{255}$$

or

$$\boxed{\frac{\partial P_W}{\partial t} + \frac{p}{m}\frac{\partial P_W}{\partial x} + \theta\cdot P_W = 0} \tag{256}$$

where

$$\theta\cdot P_W = -\frac{2}{\hbar}\sum_{n=0}^{\infty}(-1)^n \frac{(\hbar/2)^{2n+1}}{(2n+1)!}\frac{\partial^{2n+1}}{\partial x^{2n+1}}V(x)\frac{\partial^{2n+1}}{\partial p^{2n+1}}P_W \tag{257a}$$

$$\equiv -\frac{2}{\hbar}\left(\sin\frac{\hbar}{2}\left(\frac{\partial}{\partial x}\frac{\partial}{\partial p}\right)\right)V(x)P_W(x,p,t) \tag{257b}$$

PROBLEMS WITH THE WIGNER FUNCTION

Unlike the density matrix, the Wigner distribution function itself cannot be viewed as the quantum analog of the classical distribution function since: (1) it is not positivie definite–no probability interpretation; and (2) it is not unique: $P_W(x,t)$ is not the only bilinear expression in $\Psi(x,t)$ that satisfies the above sum rules.

The last part of this lecture series is concerned with certain properties and notions concerning the use of the Wigner function in quantum transport.

GENISIS OF WIGNER FUNCTIONS

The Wigner distribution function can be obtained as the Fourier inversion of the expectation value of the operator

$$e^{i(\tau\hat{p}+\theta\hat{x})} \tag{258}$$

with respect to $\Psi(x)$.

Proof:

As such

$$P_W(x,p) = \frac{1}{4\pi^2}\int_{-\infty}^{\infty}\int_{-\infty}^{\infty} C_\omega(\tau,\theta)e^{-i(\tau p+\theta x)}d\tau d\theta \tag{259}$$

where

$$C_W(\tau,\theta) = \int_{-\infty}^{\infty} \Psi^*(x) e^{i(\tau\hat{p}+\theta\hat{x})} \Psi(x)\,dx \tag{260}$$

Note: the Baker-Hausdorff theorem

$$e^A e^B = e^{A+B+\frac{1}{2}[A,B]} \tag{261}$$

provided A,B commute with [A,B]. So

$$e^{A+B} = e^A e^B e^{-\frac{1}{2}[A,B]} = e^B e^A e^{\frac{1}{2}[A,B]} = e^{B+A} \tag{262}$$

Let

$$\hat{A} = i\tau\hat{p} \;; \qquad B = i\theta\hat{x} \tag{263}$$

Then

$$[A,B] = i^2\tau\theta[p,x] = \tau\theta\hbar \tag{264}$$

Also

$$e^{i(\tau\hat{p}+\theta\hat{x})} = e^{i\theta\hat{x}}\, e^{i\tau\hat{p}}\, e^{\tau\theta\hbar/2}$$

$$= e^{i\theta\hat{x}}\, e^{i\tau\hat{p}/2}\, e^{i\tau\hat{p}/2}\, e^{\tau\theta\hbar/2} \tag{265}$$

Now

$$e^{i\theta\hat{x}}\, e^{i\tau\hat{p}/2} = e^{i\tau\hat{p}/2}\, e^{i\theta\hat{x}}\, e^{-i(\hbar/2)\theta\tau} \tag{266}$$

Thus

$$e^{i(\tau\hat{p}+\theta\hat{x})} = e^{i(\tau/2)\hat{p}}\, e^{i\theta\hat{x}}\, e^{i(\tau/2)\hat{p}} \tag{267}$$

and

$$C_W(\tau,\theta) = \int_{-\infty}^{\infty} \Psi^*(x) e^{i(\tau/2)\hat{p}}\, e^{i\theta\hat{x}}\, e^{i(\tau/2)\hat{p}}\, \Psi(x)\,dx \tag{268a}$$

$$= \int_{-\infty}^{\infty} \left[e^{-i(\tau/2)\hat{p}}\ \Psi(x)\right]^* e^{i\theta\hat{x}} \left[e^{i(\tau/2)\hat{p}}\ \Psi(x)\right] dx \tag{268b}$$

$$= \int_{-\infty}^{\infty} \Psi^*\left(x - \frac{\tau}{2}\hbar\right)\ e^{i\theta\hat{x}}\ \Psi\left(x + \frac{\tau}{2}\hbar\right) dx \tag{268c}$$

Putting $C(\tau,\theta)$ into $P_W(x,p)$, we get

$$P_W(x,p) = \frac{1}{4\pi^2} \int_{-\infty}^{\infty}\int_{-\infty}^{\infty} d\tau\ d\theta \left[\int_{-\infty}^{\infty} dx'\Psi^*\left(x' - \frac{\tau\hbar}{2}\right)\ e^{i\theta x'}\ \Psi\left(x' + \frac{\tau\hbar}{2}\right)\right] e^{-i(\tau p)} \tag{269a}$$

$$= \frac{1}{4\pi^2} \int_{-\infty}^{\infty}\int_{-\infty}^{\infty}\int_{-\infty}^{\infty} d\tau\ d\theta\ dx'\Psi^*\left(x' - \frac{\tau\hbar}{2}\right)\ \Psi\left(x' + \frac{\tau\hbar}{2}\right) e^{i\theta(x'-x)}\ e^{-i(\tau p)} \tag{269b}$$

Now

$$\int_{-\infty}^{\infty} d\theta e^{i\theta(x'-x)} = 2\pi\delta(x'-x) \tag{270}$$

So

$$P_W(x,p) = \frac{1}{2\pi} \int_{-\infty}^{\infty} d\tau\Psi^*\left(x' - \frac{\tau\hbar}{2}\right)\Psi\left(x' + \frac{\tau\hbar}{2}\right) e^{-i(\tau p)} \tag{271}$$

Setting $\tau \to -\tau$, $\tau\hbar \to y$ we get

$$P_W = \frac{1}{2\pi\hbar} \int_{-\infty}^{\infty} dy\Psi^*\left(x + \frac{y}{2}\right)\Psi\left(x - \frac{y}{2}\right) e^{i\tau p/\hbar} \tag{272}$$

which is the desired result.

Additionally, the Wigner function is the Fourier transform of the characteristic function for the operator $(\tau\hat{p} + \theta\hat{x})$ in the state $\Psi(x)$. To develop this, we digress to talk about characteristic functions. We note that the method outlined above asserts that the Wigner function is based on the notion of a characteristic function.

The characteristic function of an observable $\hat{A}$ with respect to the state $|\Psi\rangle$ (here Dirac notation is used for purposes of generality) is defined as

$$C_A(\xi) = \langle\Psi|e^{i\xi\hat{A}}|\Psi\rangle \tag{273}$$

where ξ is real. Assuming $A|A'\rangle = A'|A'\rangle$, C_A can be evaluated in the $|A'\rangle$ representation as

$$C_A(\xi) = \int dA' \int dA'' \langle\Psi|A'\rangle \underbrace{\langle A'|e^{i\xi\hat{A}}|A''\rangle}_{e^{i\xi A''}\delta(A'-A'')} \langle A''|\Psi\rangle \tag{274a}$$

$$= \int dA' \; e^{i\xi A} \; |\Psi_{A'}|^2 \tag{274b}$$

where

$$|\Psi_{A'}|^2 = |\langle A'|\Psi\rangle|^2 = P(A') \tag{275}$$

which is the probability to measure A' while in $|\Psi\rangle$. Thus

$$P(A') = \frac{1}{2\pi} \int C_A(\xi) e^{-i\xi A'} d\xi \tag{276}$$

The Fourier transform of the characteristic function is the probability distribution function P(A'). So it is not unreasonable to take double Fourier transforms to find

$$P(A',B) = \frac{1}{(2\pi)^2} \int \int C_{AB}(\xi_1,\xi_2)(\epsilon)e^{-i(\xi_1 A'+\xi_2 B')} d\xi_1 d\xi_2 \tag{277}$$

as Wigner does, with

$$C_{AB}(\xi_1,\xi_2) = \langle\Psi|e^{i(\xi_1\hat{A}+\xi_2\hat{B})}|\Psi\rangle \tag{278}$$

In view of the connection between the probability distribution function and the characteristic function for a given observable, this approach seems to be a natural way of obtaining the distribution function for momenum and position. Unfortunatly, the non-commutative nature of the two observables destroys the convenient probability interpretation of the characteristic function implicit in the above discussion. In order to demonstrate this point, assume $\hat{A},\hat{B}$ to be non-commuting

$$[\hat{A},\hat{B}] = ic \tag{279}$$

where c = "c" number. From above, Eq. (278) with

$$A|A'\rangle = A'|A'\rangle, \quad B|B'\rangle = B'|B'\rangle$$

and

$$[A,(A,B)] = [B,(B,C)] = 0 \tag{280}$$

we may use

$$e^{i(\xi_1\hat{A}+\xi_2\hat{B})} = e^{i\xi_1\hat{A}} e^{i\xi_2\hat{B}} e^{-\xi_1\xi_2[\hat{A},\hat{B}]/2} \tag{281}$$

to find

$$C_{AB}(\xi_1,\xi_2) = e^{-\xi_1\xi_2[\hat{A},\hat{B}]/2} \int dA' \int dB' \; e^{i(\xi_1A'+\xi_2B')} \langle\Psi|A'\rangle \langle A'|B'\rangle \langle B'|\Psi\rangle \tag{282}$$

Define

$$F(A',B') \equiv \langle\Psi|A'\rangle \langle A'|B'\rangle \langle B'|\Psi\rangle \tag{283}$$

so that

$$F(A',B') = \frac{1}{(2\pi)^2} \int d\xi_1 \int d\xi_2 \; e^{\xi_1\xi_2[\hat{A},\hat{B}]/2} \; C^{-(\xi_1\xi_2)} \; e^{-i(\xi_1A'+\xi_2B')} \tag{284}$$

$F(A',B')$ is a function bilinear in Ψ, satisfies the "Wigner" sum-rules, yet is not P_W. The phase factor $\exp[\xi_1\xi_2[A,B]/2]$ precludes F from being P_W; if $[\hat{A},B] = 0$, $F = P_W$ classical distribution function.

Properties of $F(A',B')$

$$\int F(A',B')dA' \equiv |\langle B'|\Psi\rangle|^2 = \frac{1}{(2\pi)^2} \int d\xi_2 \; C_{AB}(0,\xi_2) \; e^{-i\xi_2B'} \tag{285}$$

$$\int F(A',B')dB' = |\langle A'|\Psi\rangle|^2 = \frac{1}{(2\pi)^2} \int d\xi^1 \; C_{AB}(\xi_1,0) \; e^{-i\xi_1 A'} \qquad (286)$$

If $\hat{A} = \hat{x}$, $\hat{B} = \hat{p}$, then

$$F(x,p) = \langle\Psi|x\rangle \; \langle x|p\rangle \; \langle p|\Psi\rangle \qquad (287a)$$

and

$$= \Psi^*(x) \left(\frac{1}{\sqrt{2\pi\hbar}} \; e^{ip_x\hbar} \right) \phi(p) \qquad (287b)$$

There is no simple probability interpretation of F(A'B') because of the necessary overlap between the state of the non-commuting observables. However, if $\hat{A}$ and $\hat{B}$ are made to commute so that $|A'\rangle$, $|B'\rangle$ are a common set of eigenstates, the function F(A',B') reduces to the probability distribution function for $\hat{A}$ and $\hat{B}$.

This exercise points to the nonuniqueness of the Wigner function. The nonuniqueness here arises from the sensitivity of the Fourier transform to operator ordering in the characteristic function. Note when

$$\hat{C} = e^{i(\tau\hat{p}+\theta\hat{x})} \longrightarrow P_W(x,p) = \frac{1}{2\pi\hbar} \int_{-\infty}^{\infty} dy \Psi^*\left(x - \frac{y}{2}\right) \Psi\left(x + \frac{y}{2}\right) e^{ipy/\hbar} \qquad (288)$$

when

$$\hat{C} = e^{i\theta\hat{x}} \; e^{i\tau p} = e^{-i\hbar\tau\theta/2} \; C_W \longrightarrow P_1 = \frac{1}{2\pi\hbar} \int_{-\infty}^{\infty} dy \Psi^*(x) \Psi(x-y) e^{ipy/\hbar} \qquad (289)$$

when

$$\hat{C} = e^{i\tau p} \; e^{i\theta x} = e^{i\hbar\tau\theta/2} \; C_W \longrightarrow P_2 = \frac{1}{2\pi\hbar} \int_{-\infty}^{\infty} dy \Psi^*(x+y) \Psi(x) e^{ipy/\hbar} \qquad (290)$$

in fact if

$$\hat{C} = e^{in\tau\theta\hbar/2} \; \hat{C}_W \longrightarrow P_n = \frac{1}{2\pi\hbar} \int_{-\infty}^{\infty} dy \Psi^*\left(x-(n-1) \; \frac{y}{2}\right) \Psi\left(x-(n+1) \; \frac{y}{2}\right) e^{ipy/\hbar} \qquad (291)$$

Note

$$P_W(A',B') = \sum_{A''} \langle\Psi|A''\rangle \langle A''|B'\rangle \langle B'|2A'-A''\rangle \langle 2A'-A''|\Psi\rangle \qquad (292)$$

is the result for

$$C = C_W = e^{i(\xi_1 A+\xi_2 B)} \qquad (293)$$

Although the Wigner function is non-probabilistic and nonunique, integating it over position and momentum gives a physical average. So we look at the momentum moments of the Wigner equation to look for quantum corrections.

MOMENTS OF THE WIGNER FUNCTION

Consider the moments of the "Wigner-Boltzmann" type transport equation

$$\frac{\partial P_W}{\partial t} + \frac{p}{m}\frac{\partial P_W}{\partial x} + \theta\cdot P_W = \left(\frac{\partial P_W}{\partial t}\right)_{coll} \qquad (294)$$

where $\theta\cdot P_W$ is given by Eq. (257).

This equation is constructed to include an ad hoc collision term which may not necessarily express the same phenomenology as that of the classical Boltzmann equation, since P_W is not a true probability distribution function. These problems are conceptually reduced when dealing with moments in a relaxation approximation.

To obtain the moments, we multiply Eq. (294) by an arbitrary function of P, $\phi(p)$ and get

$$\frac{\partial\langle\phi\rangle}{\partial t} + \frac{1}{m}\frac{\partial}{\partial x}\langle\phi p\rangle$$

$$-\sum_{n=o}^{\infty}\left(\frac{\hbar}{2i}\right)^{2n}\frac{1}{(2n+1)!}\;\frac{\partial^{2n+1}V(x)}{\partial x^{2n+1}}\;\frac{\partial^{2n+1}\phi}{\partial p^{2n+1}} = \langle\phi\left(\frac{\partial p}{\partial t}\right)_{coll}\rangle \qquad (295)$$

where < > refers to an integration over momentum. In making the assumptions that $\phi(p)$ be an analytic funtion of momentum and that

$P_W(x,p)$ vanish at the momentum limits, it follows that

$$\int_{-\infty}^{\infty} \phi(p)\frac{\partial^{2n+1} P_W}{\partial p^{2n+1}}dp = (-1)^{2n+1} \int \left(\frac{\partial^{2n+1}\phi}{\partial p^{2n+1}}\right) P_W dp \tag{296}$$

to yield

$$\frac{\partial\langle\phi\rangle}{\partial t} + \frac{1}{m}\frac{\partial}{\partial x}\langle\phi P\rangle$$

$$-\sum_{n=o}^{\infty}\left(\frac{\hbar}{2i}\right)^{2n}\frac{1}{(2n+1)}\frac{\partial^{2n+1}V(x)}{\partial x^{2n+1}}(-1)^{2n+1}\left\langle\frac{\partial^{2n+1}\phi}{\partial P^{2n+1}}\right\rangle = \left\langle\phi\left(\frac{\partial P}{\partial t}\right)_{coll}\right\rangle \tag{297}$$

Let's consider moments for $\phi = p^o$, $p, p^2/2m$. For $\phi = 1$

$$\frac{\partial\phi}{\partial p} = 0 \ , \qquad \frac{\partial\rho}{\partial t} + \frac{1}{m}\frac{\partial}{\partial x}\langle p\rangle = 0 \tag{298}$$

where $\langle p^o\rangle = \psi^*(x)\Psi(x)$. For $\phi = p$

$$\frac{\partial\phi}{\partial p} = 1 \qquad \frac{\partial^2\phi}{\partial p^2} = 0 \ , \tag{299}$$

$$\frac{\partial\langle P\rangle}{\partial t} + \frac{1}{m}\frac{\partial}{\partial x}\langle p^2\rangle - \frac{\partial V}{\partial x}\rho = \left\langle p\left(\frac{\partial P_W}{\partial t}\right)\right\rangle_{coll} \equiv -\frac{\langle P\rangle}{\tau_m} \tag{300}$$

For $\phi = \dfrac{p^2}{2m}$

$$\frac{1}{2m}\frac{\partial}{\partial x}\langle p^2\rangle + \frac{1}{2m^2}\frac{\partial}{\partial x}\langle p^3\rangle + \frac{\langle P\rangle}{m}\frac{\partial V}{\partial x} = \left\langle\frac{p^2}{2m}\left(\frac{\partial P_W}{\partial t}\right)\right\rangle_{coll} \equiv -\frac{(\epsilon-\epsilon_2)}{\tau_\epsilon} \tag{301}$$

Now

$$\langle p^n\rangle = \frac{1}{2\pi\hbar}\int_{-\infty}^{\infty}\Psi^*\left(x+\frac{y}{2}\right)\Psi\left(x-\frac{y}{2}\right)\int_{-\infty}^{\infty}p^n\, e^{ipy/\hbar}dydp \tag{302a}$$

$$= \frac{1}{2\pi\hbar}\int_{-\infty}^{\infty} dy\Psi^*\left(x+\frac{y}{2}\right)\Psi\left(x-\frac{y}{2}\right)\left(\frac{\hbar}{i}\right)^n\frac{\partial^n}{\partial y^n}\underbrace{\left\{\int_{-\infty}^{\infty} e^{ipy/\hbar}dp\right\}}_{2\pi\hbar\delta(y)} \tag{302b}$$

$$= \int_{-\infty}^{\infty}\Psi^*\left(x+\frac{y}{2}\right)\Psi\left(x-\frac{y}{2}\right)\left(\frac{\hbar}{i}\right)^n\left(\frac{\partial^n}{\partial y^n}\,\delta(y)\right)dp \tag{302c}$$

But

$$\int_{-\infty}^{\infty} f(y)\frac{\partial^n}{\partial y^n}\delta(y)dy = (-1)^n\left.\frac{\partial^n f}{\partial y^n}\right|_{y=o} \tag{303}$$

So

$$<p^n> = (-1)^n\left(\frac{\hbar}{i}\right)^n\frac{\partial^n}{\partial y^n}\left.\left\{\Psi^*\left(x+\frac{y}{2}\right)\Psi\left(x-\frac{y}{2}\right)\right\}\right|_{y=o} \tag{304}$$

Explicitly

$$<p^o> = \Psi^*\Psi = \rho \tag{305}$$

$$<p> = \frac{\hbar}{2i}\left(\Psi^*\frac{\partial\Psi}{\partial x} - \Psi\frac{\partial\Psi^*}{\partial x}\right) \tag{306}$$

but

$$J = \frac{\hbar}{2mi}\left(\Psi^*\frac{\partial\Psi}{\partial x} - \Psi\frac{\partial\Psi^*}{\partial x}\right) \tag{307}$$

Therefore $<p> = mJ$. Also

$$<p^2> = \left(\frac{\hbar}{2i}\right)^2\left(\Psi^*\frac{\partial^2\Psi}{\partial x^2} - 2\frac{\partial\Psi^*}{\partial x}\frac{\partial\Psi}{\partial x} + \frac{\partial^2\Psi^*}{\partial x^2}\Psi\right) \tag{308}$$

$$<p^3> = \left(\frac{\hbar}{2i}\right)^3\left(\Psi^*\frac{\partial^3\Psi}{\partial x^3} - 3\frac{\partial\Psi^*}{\partial x}\frac{\partial^2\Psi}{\partial x^2} + 3\frac{\partial^2\Psi^*}{\partial x^2}\frac{\partial\Psi^*}{\partial x} - \frac{\partial^3\Psi^*}{\partial x^3}\Psi\right) \tag{309}$$

The above moments are expressed in terms of wavefunctions. In order to see the transition between the quantum and classical regimes, we invoke a wavefunction of the form

$$\Psi(x,t) = A(x,t)e^{iS(x,t)/\hbar} \tag{310}$$

where

$$A^2(x,t) = \Psi^*\Psi = \rho(x,t); \qquad \frac{1}{m}\frac{\partial S(x,t)}{\partial x} = v(x,t) \tag{311}$$

Then the moments become

$$\langle p^o \rangle = \rho(x,t) \tag{312a}$$

$$\langle p^1 \rangle = mv\rho \tag{312b}$$

$$\langle p^2 \rangle = (mv)^2\rho - \frac{\hbar^2}{4m}\,\rho\frac{\partial^2}{\partial x^2}\ln\rho \tag{312c}$$

$$\langle p^3 \rangle = (mv)^3\rho - \frac{\hbar^2}{4m}\,\rho\left(3mv\frac{\partial^2}{\partial x^2}\ln\rho + \frac{\partial^2}{\partial x^2}(mv)\right) \tag{312d}$$

It is evident that the terms possessing an explicit dependence on $\hbar$ are quantum corrections to the classical momentum density moments. As $\hbar \to 0$ these moments reduce to the classical expressions.

The logaritmic derivative appearing in $\langle p^2 \rangle$ and $\langle p^3 \rangle$ is not uncommon; such a term appears in the real part of the Schrodinger equation when Eq. 310 is used. In the limit where this term is negligible the Schrodinger equation goes over to the Hamilton-Jacobi equation.

Proof: With

$$i\hbar\frac{\partial\Psi}{\partial t} = -\frac{\hbar^2}{2m}\Psi'' + V(x)\Psi, \qquad \Psi = Ae^{iS/\hbar}$$

$$i\hbar\frac{\partial\Psi}{\partial t} = \left(i\hbar\frac{\partial A}{\partial t} - \frac{\partial S}{\partial t}A\right)e^{iS/\hbar} \tag{313}$$

$$-\frac{\hbar^2}{2m}\Psi'' = \left\{-\frac{\hbar^2}{2m}\left(A'' - \left(\frac{S'}{\hbar}\right)^2 A\right) - \frac{\hbar^2}{2m}\,i\left(\frac{2A'S'}{\hbar} + \frac{AS''}{\hbar}\right)\right\}e^{iS/\hbar} \tag{314}$$

The imaginary part satisfies

$$\hbar\frac{\partial A}{\partial t} = -\frac{\hbar^2}{2m}\left(\frac{2A'S'}{\hbar} + \frac{AS''}{\hbar}\right) \rightarrow \frac{\partial A}{\partial t} + \frac{1}{2m}\left(2A'S' + AS''\right) = 0 \tag{315}$$

Multiplying by 2A we get

$$2A\frac{\partial A}{\partial t} + \frac{1}{m}\left(AA'S' + A\frac{\partial}{\partial x}(AS')\right) = 0 \qquad (316)$$

or

$$\frac{\partial}{\partial t}(A^2) + \frac{1}{m}\left(\frac{\partial}{\partial x}(A^2S')\right) = 0 \qquad (317)$$

If

$$A^2 = \rho \ , \quad \frac{S'}{m} = v \qquad J = \rho v \qquad (318)$$

we get the continuity equation

$$\frac{\partial \rho}{\partial t} + \frac{\partial}{\partial x}J = 0 \qquad (319)$$

The real part satisfies

$$-\dot{S}A = -\frac{\hbar^2}{2m}\left(A'' - \left(\frac{S'}{\hbar}\right)^2 A\right) + VA \qquad (320)$$

or

$$-\dot{S} = -\frac{\hbar^2}{2m}\left(\frac{A''}{A} - \left(\frac{S'}{\hbar}\right)^2 A\right) - V \qquad (321)$$

or

$$-\dot{S} + \frac{(S')^2}{2m} + V(x) = \frac{\hbar^2}{2m}\frac{A''}{A} \ , \qquad A = \sqrt{\rho} \qquad (322)$$

$$\xrightarrow[\hbar\to 0]{} = 0 \qquad (323)$$

REFERENCES

1. W. Kohn and J. Luttinger, Phys. Rev. 108 (590), 1957.

2. E. Wigner, Phys Rev. 40 (749), 1932.

3. J. Moyal, Prec. Cambridge Phil. Soc. 45 (1173) 1951.

4. J. Kruger and A Potlyn, Physica 85A (84) 1976.

5. K. van Vliet, et al., J. Math Phys. 23 (318) 1982 [and ref. therein].

6. E. Wigner, "Perspectives in Quantum Theory" (edited by W. Yourgran and A. van derMerwe) Dover Press.

7. J. Barker, "Physics of Nonlinear Transport in Semiconductors," (edited by D. Ferry, J. Barker, and C. Jacoboni) P.127 - Quantum Transport.

8. J. Zubarev, "Nonequilibrium Statistical Thermodynamics," Consultants Bureau, New York.

9. L. Kadanoff and G. Baym, "Quantum Statistical Mechanics", Benjamin.

10. J. Ziman, "Elements of Advanced Quantum Theory," Cambridge Press.

11. D. Blokhintsev, "Principles of Quantum Mechanics", Allyn and Bacon.

12. A Davydov, "Quantum Mechanics", Pergamon Press.

FAR INFRARED MEASUREMENTS OF VELOCITY OVERSHOOT AND HOT ELECTRON DYNAMICS IN SEMICONDUCTOR DEVICES

S.J. Allen, Jr.

Bell Communications Research, Inc.
Murray Hill
New Jersey 07974

ABSTRACT

Broad band measurements of the dynamical conductivity of semiconductors or devices in the presence of large electrical fields can provide direct measure of small signal velocity overshoot. Energy and momentum scattering rates determine the relevant time scales which dictate that spectroscopic measurements be made in the far infrared. The efficacy of this approach is revealed by studying the dispersion of the saturated current in micron size GaAs n^+-n-n^+ diodes. These experiments directly expose the effective energy and momentum relaxation rates in the hot electron regime. Velocity overshoot appears in the frequency domain as a recovery of the small signal device conductance at high frequency.

I. Hot electron dynamics play a fundamental role in the behavior of submicron devices and establish intrinsic high frequency limits of microwave devices.[1] Because modest voltages in small devices lead to large fields, the electron transport occurs in the presence of strong carrier heating. At the same time the device transit time is sufficiently short that the carriers may not relax either their energy or momentum. Under these conditions the carriers will be transported in large electric fields without appreciable heating or elastic scattering. Although this phenomena is particularly interesting in small submicron devices, the basic phenomena is controlled by the time scale for energy or momentum relaxation, τ_ε and τ_p. As a result the underlying physics can be probed in samples or devices that do not necessarily satisfy the condition that the transit time be short if we probe at sufficiently high frequency or short time.

At room temperature energy relaxation rates are typically of the order of $\lesssim 10^{-12}$ secs while momentum scattering may be several times shorter. This means that intrinsic processes must be probed on time scales of the order of $\lesssim$ picoseconds or frequency scales of the order of $\gtrsim$ 100 GHz. If we restrict our discussion to small signal linear response, time domain and broad band frequency domain experiments contain the same information. Of course, small signal or linear response does not provide information about hot electron dynamics unless we simultaneously apply a large heating electric field that biases the semiconductor or device into a non-ohmic regime. However, under these conditions the frequency dependent conductivity when measured over a sufficiently wide frequency range can directly expose the relevant relaxation processes.

In Section II we briefly describe the philosophy of the technique, while in Section III we give details about its implementation. In Section IV we describe results on GaAs n^+-n-n^+ devices. Conclusions are drawn in Section V.

II. Hot electron dynamics have been extensively studied in the past by measuring differential conductance at microwave frequencies in the presence of large electric fields.[2] In Fig. 1 we schematically show a nonlinear current-voltage characteristic. If the device is biased to a voltage, V_0, in the nonlinear part of the curve and a small a.c. voltage, V_1, superimposed upon it, we can measure the dynamical conductance which will be given by the induced a.c. current. At very low frequencies the dynamical conductance will be given by the slope of the curve at the point V_0. However, at high frequencies the response will increase as the frequency exceeds the effective energy relaxation rate. At frequencies which exceed the momentum scattering time, inertial effects will begin to dominate the response and the conductance will fall. In Fig. 2a we schematically indicate this frequency dependent response.

Previous measurements of this sort were restricted by microwave techniques to frequencies $\lesssim$ 100 GHz. What sets the current work apart is the breadth of the frequency range that can be covered by broad band far infrared techniques. In the following we report measurements from 100 GHz to 1200 GHz. This enables us to effectively probe both energy and momentum relaxation in the hot electron regime.

Before examining how we implement this measurement we point out in Fig. 2b the current response to a voltage step that corresponds to the dynamical conductance in Fig. 2a. There is a one-to-one correspondence between velocity or current overshoot in the time domain and the peak response seen in the dynamical conductance.[3-5] The time scale for momentum scattering or ballistic transport is given by the early time behavior in Fig. 2b, and this can be related directly to the roll off the conductance at high frequency, shown in Fig. 2a.

III. The system used to measure the dynamical response of electrically excited planar devices is shown in Fig. 3.[6,7] The sheet conduc-

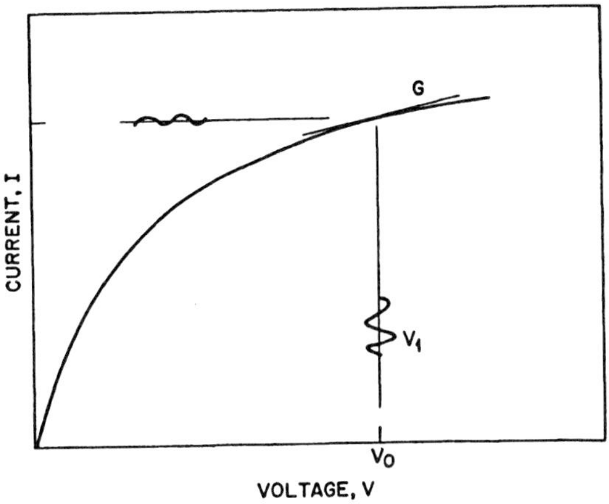

Figure 1. Nonlinear current-voltage relation. We presume that the mobility drops at large voltage due to carrier heating. A small a.c. voltage, V_1, superimposed on the bias voltage, V_0, measures the dynamical conductance G.

tance of the device is determined by measuring the absorption of far infrared radiation transmitted through it. Broad band radiation is produced by a black body source and its spectral content, after passing through the system and sample and falling on the detector, is measured with an interferometer.

During the experiment a pulse of voltage is applied which electrically excites the sample. By using a fast but sensitive InSb hot electric bolometer we can measure the change in radiation transmitted through the device as the device conductance is altered by the applied field. (Pulsed conditions are required to avoid catastrophic heating of the sample.) If we refer to Fig. 1, the pulsed field is alternately changing the bias from zero to V_0. The infrared or submillimeter wave radiation is providing the a.c. signal V_1 and the change in absorption is related to the change in conductance at high frequency as we switch from zero to the bias point V_0. The specific relation between change in transmitted power and change in conductance is determined by device geometry, and the wave admittance of the semiconductor substrate and metallization.

This technique has been used on two different systems. In the first instance we measured velocity overshoot in Si inversion layers in a Si metal-oxide-semiconductor field-effect-transistor.[7] The second involved a measurement of the dispersion of the saturated current in GaAs n^+-n-n^+ diodes at room temperature.[8] We focus on the latter. Not only is it technically more interesting, but it exposes

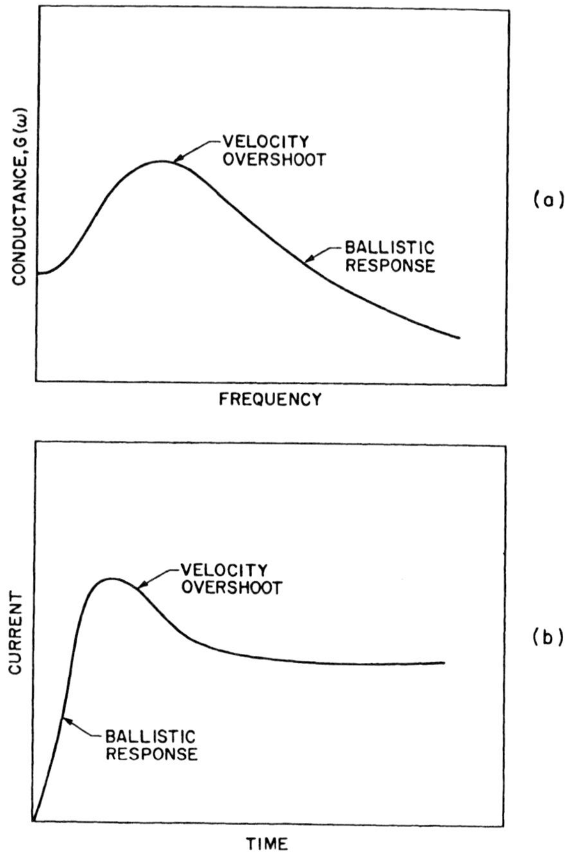

Figure 2. (a) The dynamical conductance, $G(\omega)$, versus frequency. A maximum occurs at frequencies which are higher than the energy relaxation rate but less than the momentum scattering rate.

(b) The current response to a small voltage step superimposed on the bias voltage, V_0. Current or velocity overshoot appears on a time scale long compared to the momentum relaxation time but short compared with the energy relaxation time.

a technique for coupling broad band radiation into devices that are many times smaller than the wavelengths of the radiation we are using as a probe.

IV. We have measured the dynamical conductance of small n^+-n-n^+ planar diodes shown in Fig. 4. The device was fabricated using techniques that are employed for microwave Schottky barrier metal-semiconductor-field-effect-transistors,[9] with the obvious exception that there is no gate. The active layer is 3000 Å thick with a nominal doping of $n \sim 10^{17}/cm^3$ and was produced by VPE growth on a Cr doped substrate. It is difficult to couple broad band radiation into a single diode as shown in Fig. 4. To be sure at a particular frequency, suitable microwave circuits and antennae could be devised that will enable us to couple effectively into this device. But our goal is broad band measurements of the device response.

An approach that alleviates this problem is to construct a two-dimensional, 2D, array of these devices as shown in Fig. 5.[10] If the periodicity of the array is small enough, then the sheet admittance presented to an incident plane wave will be just the dynamical conductance of the diode in the gaps in the metallization. How small is small enough is determined by the inductance of the metal strip leading up to the diode and the capacitance of the gap. Figure 6a shows the equivalent 2D array of resistors, inductors and capacitors.

The inductance L and capacitance C can be estimated as follows[11]

$$L \approx \frac{\mu_0 a}{2\pi} \ln\left[\csc\left(\frac{\pi w}{2a}\right)\right]$$
$$C \approx \frac{(\varepsilon_0+\varepsilon_s)}{\pi} w \ln\left[\csc\left(\frac{\pi g}{2a}\right)\right] \tag{1}$$

where a is the period of the square array, w the width of the metallization, and g the length of the gap. μ_0 is the permeability of free space and ε_0 and ε_s the dielectric permittivity of free space and GaAs respectively. "Small enough" requires that the grid resonance, $\omega_g = \frac{1}{\sqrt{LC}}$, be larger than the frequencies of interest and ωL be substantially smaller than the conductance of the device at frequencies of interest.

The actual L and C can be measured on a grid with no electrically activated layer. There we find $L \sim 7.6\times10^{-12} h/\square$ and $C \sim 3.7\times10^{-16} f/\square$. This gives a resonance in the grid response at 3000 GHz. Furthermore, the inductance is sufficiently low that the radiation is short circuited for the electric vector oriented parallel to the continuous wires. For the opposite polarization the effective sheet conductance is just the device conductance for frequencies well below the grid resonance.

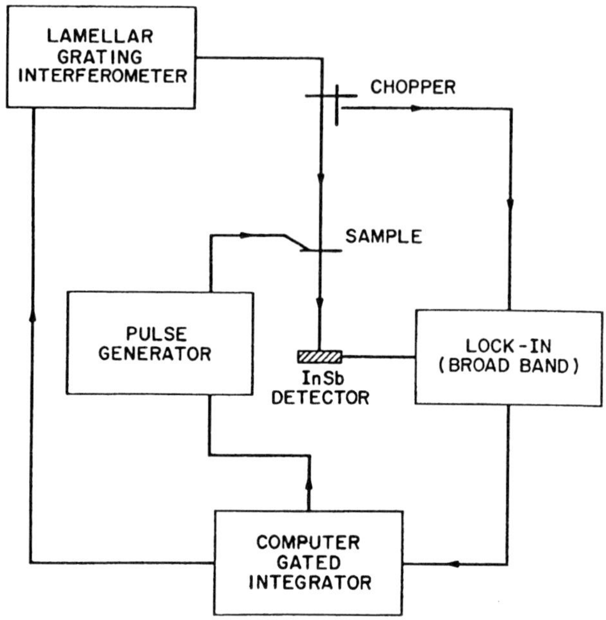

Figure 3. Submillimeter wave system used to measure the change in far infrared radiation transmitted through the device as it is electrically excited by voltage pulses.

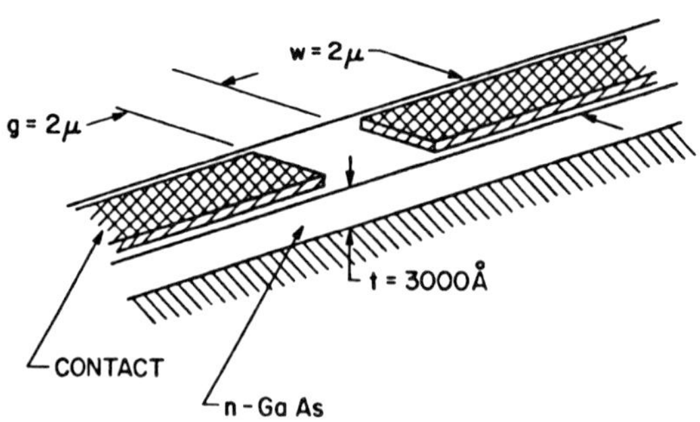

Figure 4. The n^+-n-n^+ GaAs diode.

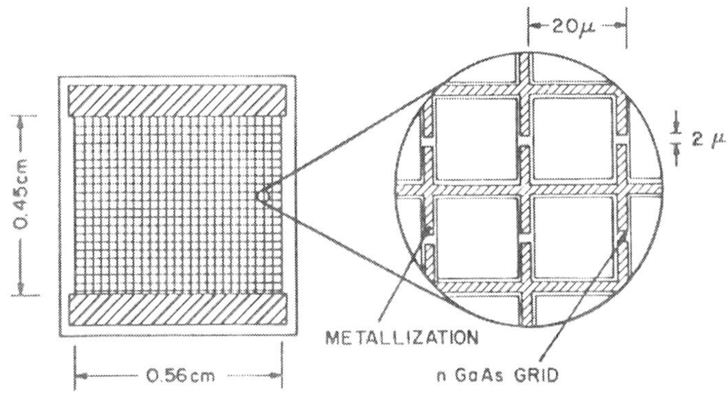

Figure 5. Two dimensional array of n^+-n-n^+ diodes. Radiation is transmitted normal to the page.

(a)

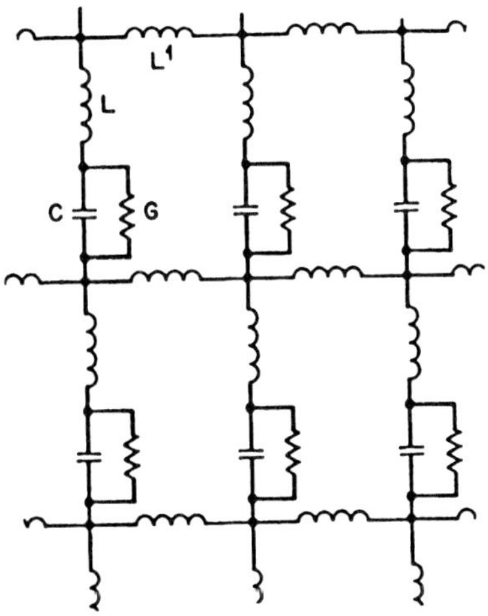

(b)

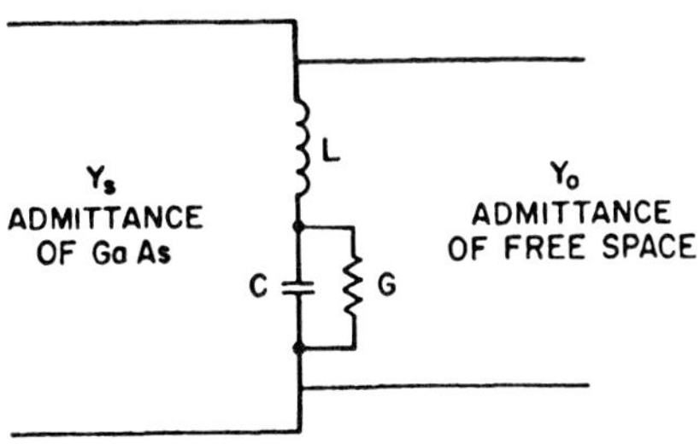

Figure 6. (a) Equivalent two dimensional array of induction capacitors and resistors. L' is approximately equal to L.
(b) Transmission line equivalent for the two dimensional array at the GaAs/free space interface.

The change in device conductance at frequency ω, induced by the applied field, can be related to the fractional change in transmission ΔT/T, by reference to Fig. 6b.

$$\frac{\Delta T}{T} \approx -2 \frac{\Delta\left(R_e(G(\omega))\right)}{Y_0+Y_s} \qquad (2)$$

where Δ Re(G(ω)) is the change in the real part of the device conductance at frequency ω. Y_0 and Y_s are the wave admittance of free space and the GaAs substrate respectively. This approximate expression is valid for small ΔT/T.

The d.c. current-voltage relation for the device shown in Fig. 5 is shown in Fig. 7. Since the measurement is performed on a square array, Fig. 7 may be taken as the average response of a single element. At low voltages the device is ohmic with nearly half of the voltage dropped in the contact region. At high voltages the current saturates at a value that corresponds to carrier velocities of the order of 10^7cm/sec.

Far infrared transmission through the grid is modulated by pulsing the grid or device from zero volts into current saturation. The fractional change in real device conductance obtained in this manner is shown in Fig. 8a. At the largest applied voltages 2.66 volts/element the conductance change at low freauencies approaches the change in slope of the I-V curve, $\sim .6\times10^{-3}$ ℧/□. At high frequencies less change in device conductance is observed.

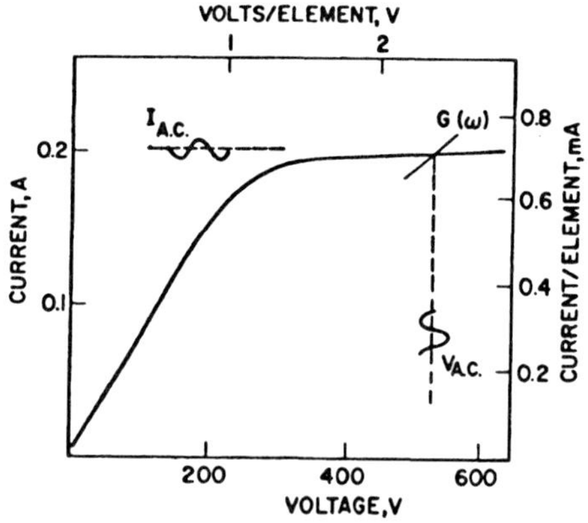

Figure 7. Current voltage characteristic for the device shown in Figure 5. The far-infrared radiation superimposes V_{AC} and the attendant loss measures G(ω), the dynamical conductance.

The parameter of interest is not the change in conductance but the conductance at the applied voltage. To obtain this we simply add the data shown in Fig. 8a to the response at V = 0 and obtain the central result in Figure 8b. This represents the frequency dependent device conductance with applied voltage as a parameter.

At low frequency the device conductance approaches zero for large voltage. Indeed the slope shown in Fig. 7 for large voltage is zero. At high frequencies, however, we see the device conductance rise which signals small signal velocity or current overshoot in the frequency domain as we had anticipated in the earlier discussion. The current is driven above and below its saturated value at frequencies which are so high that the electron distributions in real and k-space cannot follow. Figure 8b indicates that at frequencies above 100 GHz any quasistatic description of the electron transport in GaAs is not valid. This essentially agrees with theoretical estimates of effective energy relaxation rates in GaAs by Rees.[12]

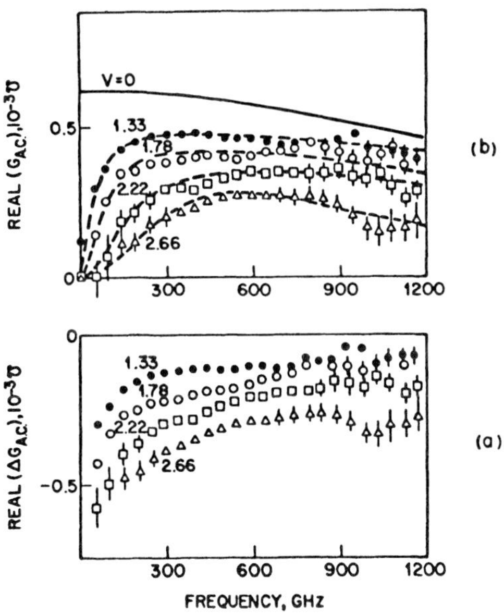

Figure 8. (a) Fractional change in real device conductance vs frequency with voltage per element as a parameter.
(b) Solid curve is the Drude relaxation of the low field conductance. (a) added to this solid line gives the relaxation of the real conductance at the indicated bias. The dashed lines are fits with a series R-L-C.

To model the dispersion shown in Fig. 8b we first note that heavily doped GaAs in the presence of large electric fields suffers an absolute instability.[13-17] A high field anode domain develops whose width and average electric field increases with the voltage in current saturation. As a result, our device breaks up into three parts: the contact resistance, the electrically cold ohmic part which has the same conductance as the device at V = 0, and the high field anode domain which contains electrons moving in electric fields of the order of 10^4 V/cm at electron temperatures of the order of 10^{3} °K. The series connection of these three components can be described by a series R, L, C circuit shown in Fig. 9.

We may estimate the width of the anode domain as follows:

$$x_a \sim \left[2(V-V_0)/k_0(E_T-E_0)\right]^{\frac{1}{2}} \tag{3}$$

where V is the applied voltage and V_0 is the position of the knee of the curve in Figure 5. E_T is the threshold field for GaAs which we take as 4000 V/cm and E_0 is the electric field in the low field region. k_0 is defined by

$$k_0 = (\sigma/D\varepsilon_s)^{\frac{1}{2}} \tag{4}$$

where σ is the low field conductance and D the diffusion coefficient.

We can now remove the contribution of the contact and low field region from the conductance in Fig. 8b to reveal the dynamical conductance of the high field anode domain shown in Fig. 10. On the right hand side of the figure we indicate the estimated values of domain width and average electric field for the four different curves generated in Fig. 8b.

From published calculations[18] of the average mass, energy and velocity versus electric field, we can calculate the dynamical conductance of the anode domain following Shur.[19] We explicitly ignore the motion of the domain wall which may be a fatal defect in this calculation. (At d.c. it is the growth of the high field domain that keeps the current constant.) Further, we assume the entire high field domain to be at the average field. The result for the conductance of the high field domain is

$$g_a = n\frac{e^2}{m}\frac{wt}{x_a} \times \frac{i\omega + \omega_\varepsilon - v_s^2(\partial m/\partial\varepsilon)\ (i\omega + \omega_p)}{(i\omega+\omega_p^0)\ (i\omega+\omega_\varepsilon) + (q\langle E_a\rangle/m)v_s(\partial m/\partial\varepsilon)\ (i\omega+\omega_\rho)} \tag{5}$$

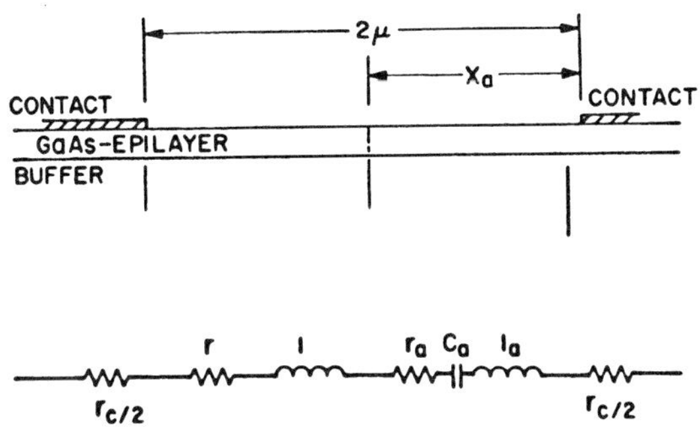

Figure 9. Model of a.c. impedance used to fit data in Fig. 8b. It is decomposed into contact resistance r_c without dispersion, low field region (r,ℓ) and anode domain (r_a, ℓ_a, C_a). X_a is the length of the anode domain.

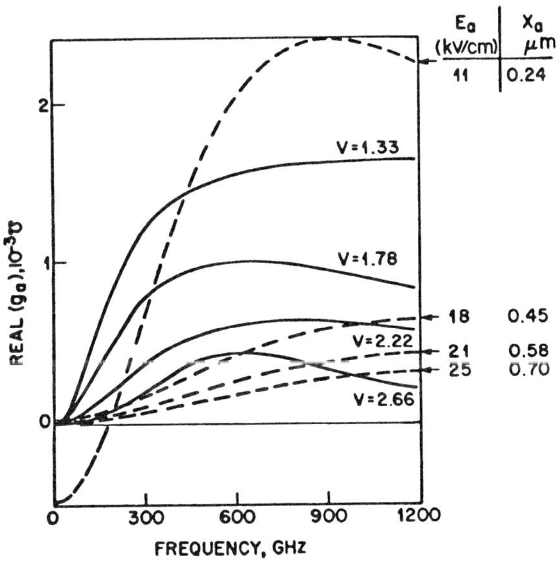

Figure 10. Real conductance of the anode domain deduced from the fit in Fig. 8b and model in Fig. 9. Dashed curves are predictions based on Refs. 27 and 18 assuming that the anode domain can be characterized by the average field $\langle E_a \rangle$ and width X_a.

Here n, e and m are the electron density, charge and mass in the anode domain ε_0 and V_s are the average energy and drift velocity. ω_ε, ω_p and ω_p are the effective energy and momentum relaxation rate and momentum relaxation rates respectively. w and t are the width of the device and the electrical thickness of the active layer.

The dashed curves in Figure 9 indicate qualitative agreement with experiment. However, it should be clear that the model calculation that we have performed is rather crude. It does, however, indicate that the data can be used as a critical test of more sophisticated models of hot electron transport on very short time scales.[18-27]

V. We have demonstrated a technique for probing hot electron dynamics on a time scale which covers energy and momentum relaxation times. Short time phenomena which play a role in transport in submicron devices can be directly exposed. In particular, small signal velocity overshoot can be seen as a recovery of device conductance as a maximum in differential conductance at high frequency. The dispersion of nonlinear current flow such as the data generated here should prove a critical testing ground for sophisticated theories of transport on short time scales.

Two dimensional arrays of small devices allow us to make these broad band measurements on devices as small as we can make them. Presumably our efforts should be directed to carrying out these measurements on high mobility small devices where we can investigate not only the effects due to elastic and inelastic scattering but where transit time effects play an important role.

REFERENCES

1. H. Kroemer, Solid State Electron. 21, 61 (1978).
2. E.M. Conwell, Solid State Physics: Advances in Research and Applications, Supp. 9, Academic Press (1967).
3. P. Das and D.K. Ferry, Solid-State Electronics 19, 851 (1976).
4. D.K. Ferry and P. Das, Solid-State Electronics 20, 355 (1977).
5. P. Das, D.K. Ferry and H. Grubin, Solid State Commun. 38, 537 (1981).
6. D.C. Tsui, S.J. Allen, Jr., R.A. Logan, A. Kamgar and S.N. Coppersmith, Sur. Sci. 73, 419 (1978).
7. S.J. Allen, Jr., D.C. Tsui, F. DeRosa, K.K. Thornber and B.A. Wilson, J. Phys. Coll. C7, Supp. No. 10 42, 369 (1981).
8. S.J. Allen, Jr., C.L. Allyn, H.M. Cox, F. DeRosa and G.E. Mahoney, Appl. Phys. Lett. 42, 96 (1983).
9. "GaAs FET Principles and Technology", ed. J.V. Dilorenzo and D.D. Khandelwal, Artech House, Inc., Dedham, Mass., 1982.
10. R. Ulrich, Appl. Opt. 10, 1987 (1968).

11. N. Marcuvitz, "Waveguide Hardbook", McGraw-Hill, pp. 280-289, (1951).
12. H.D. Rees, IBM J. Res. Develop. 13, 537 (1969).
13. P. Gueret, Phys. Rev. Lett. 27, 256 (1971).
14. P. Gueret and M. Reiser, Appl. Phys. Lett. 20, 60 (1972).
15. E. Conwell, IEEE Trans. Electron Devices Ed-17, 262 (1970).
16. P. Jeppesen and B.I. Jeppsson, IEEE Trans. Electron Devices ED-20, 371 (1973).
17. G. Dohler, IEEE Trans. Electron Devices ED-18, 1190 (1971).
18. B. Carnez, A. Cappy, A. Kaszynski, E. Constant and C. Salmer, J. Appl. Phys. 51, 784 (1980).
19. M. Shur, Electron. Letter. 12, 615 (1976).
20. T.J. Maloney and J. Frey, J. apply. Phys. 48,781 (1977).
21. H.L. Grubin, D.K. Ferry and K.R. Gleason, Solid State Electron. 23, 157 (1980).
22. N. Ino and M. Ohmori, IEEE Trans. Microwave Theory Tech. MTT-28, 456 (1980).
23. M. Deblock, R. Fauquemberque, E. Constant and B. Boittiaux, Appl. Phys. Lett. 36, 756 (1980).
24. H.L. Grubin, IEEE Trans. Microwave Theory Tech MTT-28, 422 (1980).
25. R. Bosch and N.W. Thim, IEEE Trans. Electron Devices ED-21, 16 (1974).
26. P.A. Lebwohl, J. Appl. Phys. 44, 1744 (1973).
27. J. Zimmerman, Y. Leroy and E. Constant, J. Appl. Phys. 49, 3378 (1978).

THE INFLUENCE OF CONTACTS ON THE BEHAVIOR OF NEAR AND SUB-MICRON InP DEVICES

Peter A. Blakey and Jack East

Department of Electrical and Computer Engineering
University of Michigan
Ann Arbor, Michigan

and

Melvin P. Shaw

Department of Electrical and Computer Engineering
Wayne State University
Detroit, Michigan

ABSTRACT

We discuss the status of the preliminary results of: 1) numerical calculations of submicron ($\leq$ 0.5 μm) two-terminal InP N^+-N-N^+ transferred electron devices; 2) experiments on near micron InP ($\geq$ 1.0 μm) devices. Our calculations used both Monte Carlo techniques and a static velocity-field parameter model. Contact and interface effects were emphasized. Besides obtaining known, well-documented results such as the effect of Debye tail diffusion, we also found that the anode contact region can become important in short device structures. Experimentally, we have observed both conventional and unconventional behavior, such as the existence of anomalous modes of amplification.

INTRODUCTION

The different ways in which contacts can affect semiconductor device performance, and the reasons why "contact effects" are likely to be particularly important in submicron semiconductor devices, will first be explained in terms of a simple model device, an N-type semiconductor resistor. More useful devices (transistors, transferred electron devices, etc.) will embody the same basic mechanisms; consideration of a simple resistor permits us to focus first on the basic mechanisms without obscuring the underlying picture with additional complications.

The designer of a semiconductor resistor will probably seek to obtain a desired bulk resistance by adjusting the length, ℓ, cross sectional area, A, and doping of a piece of semiconductor (Figs. 1(a) and 1(b)). In order to use the resistor in a discrete circuit it must be provided with contacts. This will normally be done by adding highly doped regions with metal deposited on, and alloyed to them (Fig. 1(c)) in such a way as to form low resistance linear contacts, i.e. contacts which allow unrestricted current flow in both directions with only small potential variations across the semiconductor-metal junctions.[1,2,3] Here the contact resistance, R_c, is much less than the bulk resistance, R_b. The device is <u>bulk-dominated</u> if its terminal properties are essentially those of a resistor with a resistance close to $R = \rho\ell/A$ (at high frequencies the resistivity ρ can become frequency dependent, and the parallel geometric capacitance $C = \varepsilon A/\ell$, where ε is the permittivity, may become significant). If adding contacts alters the terminal characteristics significantly the device properties become contact <u>influenced</u> and, in extreme cases, can become contact <u>dominated</u>. Contact influence and domination can arise from one or more of several possible mechanisms which will now be described.

Contact Influence Mechanisms

Consider the equilibrium band diagram of the idealized contacted device shown in Fig. 2. This assumes linear contacts operating by field emission or thermionic-field emission between the degenerate semiconductor material and the metal,[1] and an abrupt, well-behaved transition between N and N^+ material. Five general mechanisms can be identified by which the contacts may influence the terminal properties. These will be described in terms of electron motion from left (cathode) to right (anode), following application of a positive potential on the right hand side metal contact. Four of the mechanisms are associated with electron transport across particular planes, and a fifth with modification of the properties of the "bulk" region, BC. The mechanisms are as follows:

1. Carrier motion from the metal to the semiconductor across AA', by field and/or thermionic-field emission.

2. Carrier injection from the N+ semiconductor to the N semiconductor, over the potential barrier centered on BB'

3. Carrier collection by the N+ material following acceleration down the potential barrier centered on CC'.

4. Collection by the metal across DD', by some combination of field emission, thermionic-field emission, diffusion, etc.

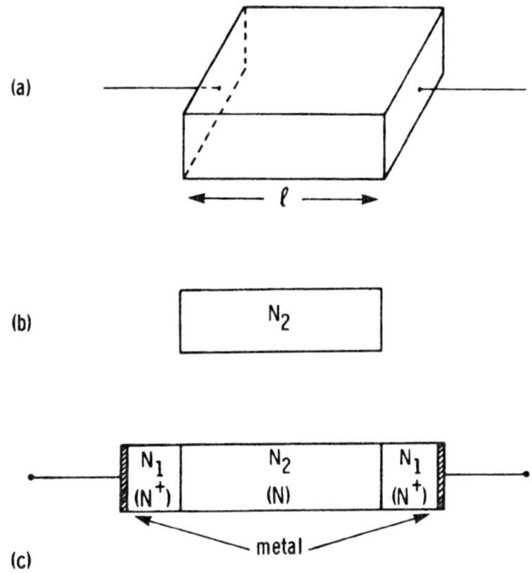

Figure 1. a) Semiconductor resistor of length l. b) bulk cross-section of resistor with N-type doping level N_2. c) cross-section including N^+ (N_1) and metallized contacts to the resistor.

5. Elimination of the space-charge-neutral bulk region due to merging of the Debye tails and potential barriers centered on BB' and CC'.

Contact Influence Conditions

It is useful to establish the conditions (mainly spatial dimensions) for which mechanisms 1 through 5 will or will not significantly influence the device properties. This will now be done for each mechanism in turn.

1. In order to isolate the bulk region from the metal contact at the cathode plane AA' we estimate that the length AB of N^+ material should be made longer than the depletion width associated with the barrier Φ_b, plus about 10 extrinsic Debye lengths L_D.[1,4] ($L_D = \{\frac{\varepsilon kT}{q^2 N}\}^{1/2}$, where q is the electronic charge and kT the thermal energy). About five Debye lengths are associated with the transition from depleted to neutral material, and five with the N^+ side of the transition from N^+ to N material. Thus, the condition that the metal contact at AA' not significantly influence device properties can be written

$$AB \gtrsim \left(\frac{2\varepsilon\phi_b}{N_1 q}\right)^{\frac{1}{2}} + \left(\frac{100\varepsilon kT}{q^2 N_1}\right)^{\frac{1}{2}} \qquad (1)$$

Providing Eq. (1) is well met, carrier injection into the bulk is effectively controlled by injection from the degenerate electron sea between A and B over the potential barrier centered on B.

2. The potential barrier at B is low (using Boltzmann statistics its height at equilibrium is $\frac{kT}{q} \ln (N_1/N_2)$ - this value will be slightly modified when degeneracy is allowed for). Small reductions in barrier height allow for copious injection of electrons. Provided that the current density requirements can be met (i.e. $J \ll Nqv_{th}$ where v_{th} is the thermal electron velocity[5]), AB and BC are both long enough to permit quasi-space charge neutrality away from BB' and the mobility is well behaved at this plane, this barrier should have a minimal affect on the terminal properties if space charge limited currents do not develop.

3. The potential barrier at CC' serves to accelerate carriers into the N^+ material between C and D. This will normally not be important, except when hot electron effects in the bulk are significant. In this case, hot electron effects (e.g. intervalley transfer[6]) will be accentuated by the accelerating potential.

4. If there are no significant hot electron effects in the bulk material the condition that the bulk be isolated from the collecting anode contact at DD' is the same as for mechanism 1, i.e. CD should be greater than the depletion width associated with the barrier, plus about 10 Debye lengths. If there are hot electron effects a much more restrictive condition may apply; the contact properties may be influenced by the fact that some of the incident electrons are "hot." In addition there will be thermal stresses at this contact as the hot electrons give up their "excess" energy. Avoidance of these effects will require that the length CD is long with respect to the thermalization distance of the hot electrons. In GaAs and InP, where significant intervalley transfer occurs, the thermalization distance can be a significant fraction of a micron.

5. The Debye tails associated with the N^+N junctions at B and C can merge if the distance BC is less than about 10 L_D, where L_D is the extrinsic Debye length in the N material. In this case there is no space-charge-neutral bulk material; the electron concentration between B and C is everywhere higher than the background doping.

NUMERICAL CALCULATIONS

The conditions that contacts have negligible influence on device properties are summarized in Table I, along with examples of the

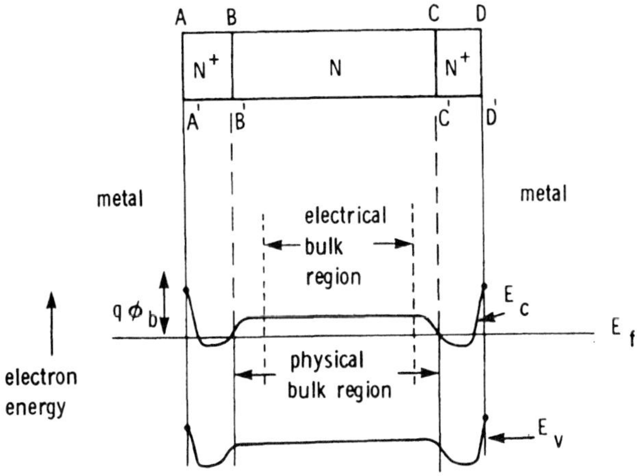

Figure 2. Equilibrium band diagram for the semiconductor resistor with contacts. $q\phi_b$ is the metal-semiconductor barrier height and E_c, E_f and E_v denote the conduction band minimum, Fermi level and valence band maximum, respectively.

characteristic dimensions involved. If any of the conditions are violated the device properties may be strongly influenced by contact effects. Most of the dimensions involved are small fractions of a micron, distances which are negligible when minimum feature sizes are greater than about a micron. However, they may all become significant in submicron dimensioned devices. In particular, mechanisms 4 and 5 (collecting contacts in the presence of hot electrons, and Debye tail merging in the active region) usually insignificant in longer structures, now appear to be particularly important. In contrast, in longer devices or those having interfacial mobility anomalies,[7] mechanisms 1 and 2 (the cathode contact) are usually considered the most probable source of contact influence.[1,6]

Mechanism IV (the collecting contact) is expected to be important in submicron structures when hot electron effects, such as intervalley transfer, are present. It is these effects which are exploited in high frequency modes of operation of transferred-electron devices. Collecting contact effects can thus be expected to be particularly important in transferred electron devices with submicron length N^+ anodes, and may provide an explanation for some anomalous features displayed by these devices. For example, we have performed Monte Carlo calculations for a variety of device component lengths AB, BC and CD to predict the fundamental features of both GaAs and InP structures. Figures 3 and 4 show representative results for an InP structure whose total length is 0.6 μm and whose N active region is 0.2 μm long. Note in

TABLE I

Conditions for Avoiding Contact Influence

The typical distances quoted for 1 and 5 satisfy the equality given for

$\varepsilon = 10^{-12}$ F/cm, T = 300 K, $\phi_b = 0.2$V, $N_1 = 10^{19}\,cm^{-3}$, $N_2 = 10^{16}\,cm^{-3}$.

Mechanism	Condition for Avoiding Influence	Typical Distance (μm)
1	$AB \gtrsim \{\frac{2\varepsilon\phi_b}{N_1 q}\}^{\frac{1}{2}} + \{\frac{100\varepsilon kT}{q^2 N_1}\}^{\frac{1}{2}}$	0.02
2	1, 5, $J < N_1\, q\, v_{th}$	-
3	No hot electron effects in bulk ($V_o < \frac{\Delta}{q}$)	-
4	1 and 3 otherwise CD $\gtrsim$ thermalization distance	∿ 1
5	$BD \gtrsim \left[\frac{100\varepsilon kT}{q^2 N_1}\right]^{\frac{1}{2}}$	0.40

Fig. 4 that most of the intervalley transfer takes place in the N^+ anode region CD because the N region BC is too narrow to have the electrons accelerated to sufficiently high energies. Note also that the Debye tails determine the equilibrium carrier density in region BC and that the electric field is nonuniform and peaks near the CC' plane.

In order to compare our Monte Carlo results with other techniques for calculating the features displayed in figs. 3 and 4, we also modelled the structure in another way and explored two different lengths BC and mobility models. These structures have 0.2 micron N^+ contacts doped at $10^{17}/cm^3$ The "short" structure has a 0.2 micron N center region (BC) doped at $10^{16}/cm^3$. The "long" structure has a 0.5 micron long center region doped at $10^{16}/cm^3$. Two transport models were used. The first model uses a field-independent mobility of 3200 cm^2/V-sec (the low field mobility given by the Monte Carlo program).

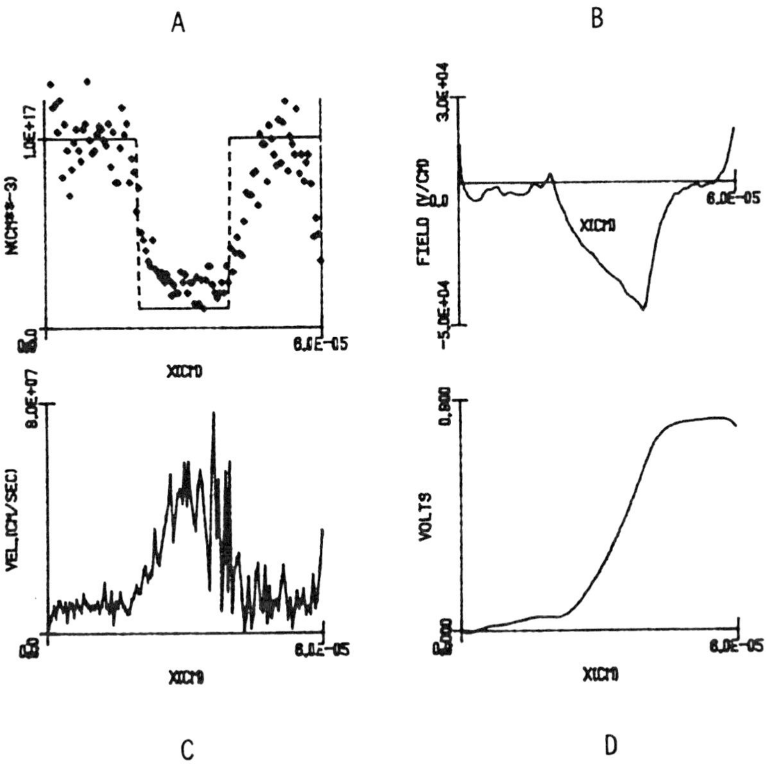

Figure 3. Monte Carlo calculation of the spatial distribution 2 psec after 0.7V is applied to an InP structure. 1 = 0.2 μm. a) carrier density (doping distribution is dashed). b) field. c) velocity. d) potential. The current density in 14.8 kA/cm^2.

The diffusion coefficient is given by the Einstein relation. The second transport model uses the drift-diffusion approximation with the static InP velocity-field and diffusion-field curves. The results are plotted vs. normalized distance. The electric field, voltage and electron concentration vs. distance are shown in Fig. 5 for an InP velocity-field, v(E), curve.

The dashed line in the field plots is the zero field line. (In Fig. 5 we are plotting the negative of the field). Note that Figs. 3 and 5 are quite similar. By way of comparison, Fig. 6 shows the 0.5 micron structure with a constant mobility model. This structure approximates a simple resistor. The electron concentration in the N region is equal to the background doping value. Here, even under high applied bias conditions the electron concentration is

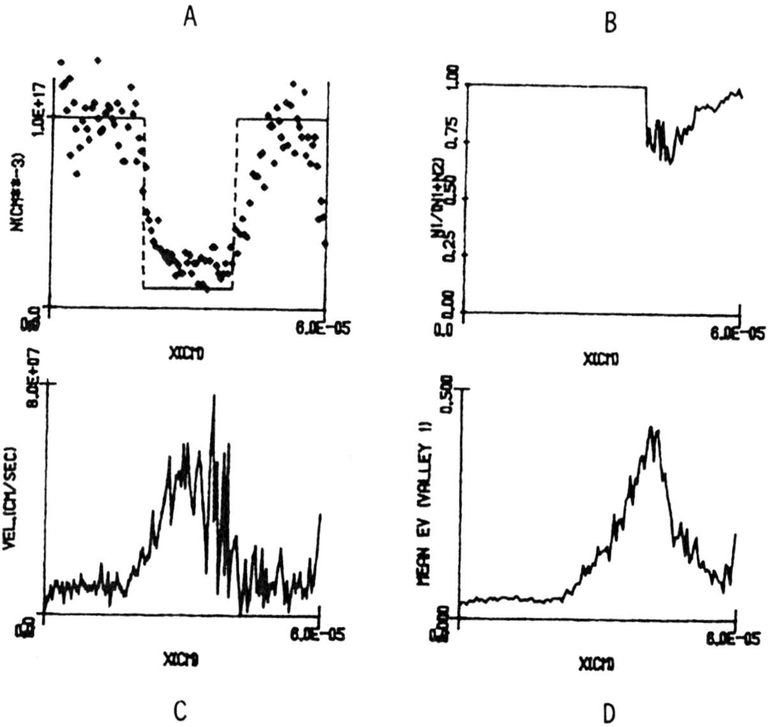

Figure 4. b) ratio of upper to lower valley electrons. d) mean energy of electrons in the upper valley (1)

approximately constant and the current density voltage, J(V), curve is approximately linear. Figure 7 shows the 0.5 micron structure using the InP static v(E) parameters. In these curves the high fields in the N region reduce the mobility and lower the current density as compared to the constant mobility case. At a bias of 0.4 volts the constant mobility current is approximately 25% higher than the static parameter current; sublinearities in J(V) develop.

In comparing the long and short structures, we see that in the short structure the Debye tails meet near the center of region BC. The extra electrons in the channel increase the electric field in the channel. In the long structure, the channel field is relatively uniform and depends on the applied bias. In the short structure the field is larger, with a larger component due to the excess electrons in the channel. The electron concentration in the channel also depends strongly on the applied bias. Applying a bias increases the minimum electron concentration from 1.3 times the background doping to almost 1.5 times the background doping. The constant mobility J(V)

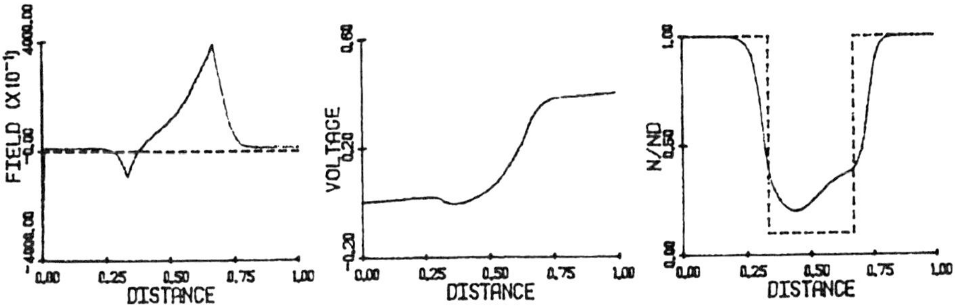

A B C

Figure 5. InP static v(E) curve parameter model spatial distribution results l = 0.2 μm. a) negative of the field. b) voltage. c) carrier density (doping distribution is dashed). The total length of the sample is 0.9 μm. V = 0.4 volts.

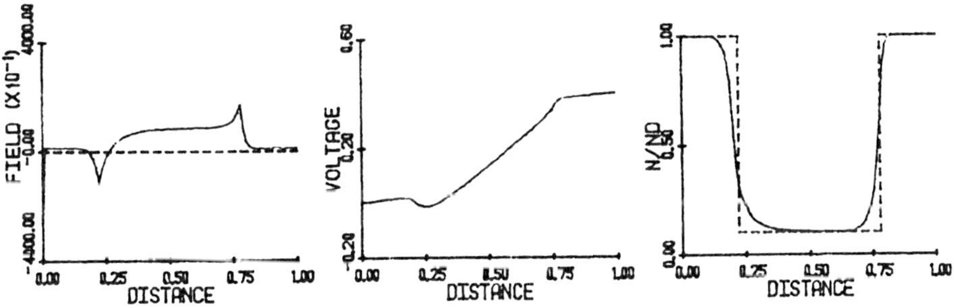

A B C

Figure 6. Spatial distribution results for a constant mobility model. l = 0.5 μm. a) negative of the field. b) voltage. c) carrier density (doping distribution is dashed). The total length of the sample is 0.9 μm. V = 0.6 volts.

characteristic is approximately linear. However, the InP static parameter J(V) curve begins to saturate at low applied bias levels for both structures, but the sublinearity is more pronounced in the shorter structure. Even at low applied bias levels the high field in the channel reduces the electron mobility and current density. At 0.4 volts applied bias the constant mobility current is twice the static parameter current.

In comparing these results with the Monte Carlo results, we expect that the constant mobility curve will match the Monte Carlo results for the short structure at low fields. The reason for the drop in mobility in InP at reasonable fields is intervalley transfer. For static v(E) curves this occurs at about 10 kV/cm. However, transfer requires that the central valley electrons gain approximately 0.6 volts of energy. With no scattering in the central valley a 0.6 volt bias (discounting a small thermal energy) is needed to produce transfer. For lower bias voltages, regardless of the field distribution, the electrons will be in the central valley. Central valley electrons can be described reasonably well with a constant mobility.

EXPERIMENTAL ASPECTS

At the present time we have not been able to investigate samples whose lengths ℓ (region BC) were below 1 μm long, so a profitable comparison between experiment and numerical calculations cannot yet be made. (Detailed calculations for larger samples are now being initiated.) However, a preliminary study of the characteristics of epitaxial InP samples between 1 and 3 μm thick has already been made.[7] In order to provide a qualitative understanding of the influence of contacts on the manifestation of current instabilities in such samples, investigations have been made on: 1) the form of the pre-threshold current-voltage, I(V), characteristic; the nature of the post-threshold oscillation in different microwave circuits; the pre and post-threshold reflection-amplification response. In this section we will outline the results, discuss how they fit into the framework of our present understanding of microwave devices, and present some new results with speculations as to their origins. With regard to the latter, the preliminary results indicate that: 1) the v(E) curve may not be readily extractable from near micron or submicron device data; 2) anomalous low-voltage amplification modes exist in specific device structures; 3) mobility "notches" at smoothly graded N-N^+ interfaces can exert a strong influence on device behavior.

Low duty-cycle pulsed-mode I(V) characteristics were investigated when the devices were packaged and held in a $J(k_u$; 12-18 GHz) band circuit. Every sample (over 40) showed some form of asymmetry in its I(V) curve. In some cases one of the metal contacts was a relatively high resistance Schottky barrier, and, for the polarity when the Schottky barrier was in reverse bias, often contributed a high, nonlinear resistance to the characteristics. However, the nonlinearity could often

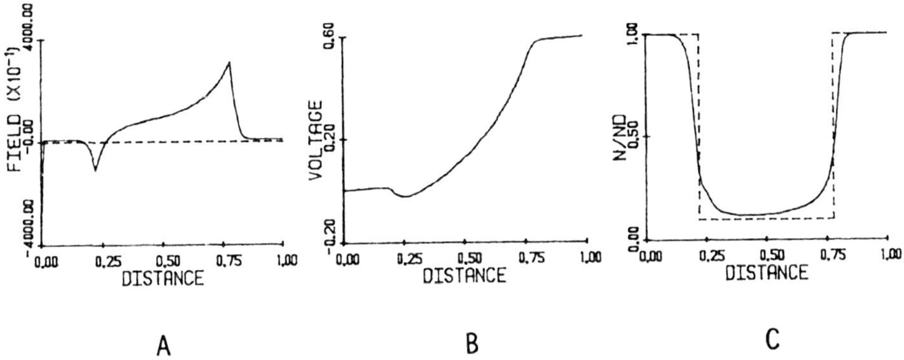

Figure 7. Spatial distribution results for an InP static v(E) curve parameter model. l = 0.5 μm. a) negative of the field. b) voltage. c) carrier density (doping distribution is dashed). The total length of the sample is 0.9 μm. V = 0.6 volts.

be "healed" with time - after the first few cycles of operation, the reverse-bias resistance of the structure would decrease to a steady state "running" value. This behavior suggests that an MIS structure may have been involved, with punch-thru breakdown of an insulator indicated.

The most symmetric characteristics were obtained with samples which provided the data shown in Fig. 8. Note the nonlinearity in the curves. (We expect a contribution to the nonlinearity from both contact and size effects.) These characteristics are typical of most that were observed, except that the asymmetry in the (+) and (-) polarities is more pronounced in general in other samples. On the I(V) curve we have noted the oscillation threshold, V_o, the amplification threshold, V_A, and the threshold for anomalous amplification, V'_A. These a.c. features will be discussed later. Here we consider the pre-threshold characteristics and their meaning.

For uniform fields across the N region, which we assume to be uniformly doped, and for cases where the resistance of the N region is much larger than any other series resistance in the system, the threshold field will be given by

$$E_T = \frac{V_o}{\ell} .$$

For Fig. 8 this yields $E_T \cong 3.16 \times 10^4$V/cm (for $\ell = 1.5$ μm), a

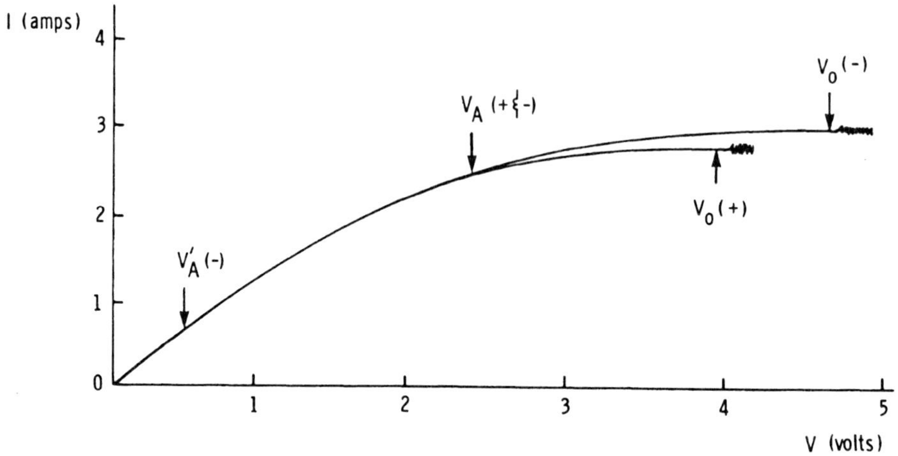

Figure 8. Experimental I(V) characteristics for an InP sample having $l \approx 1.5$ μm. Both polarities, + and −, are shown. V_o is the voltage required for oscillation. V_a is the voltage required for amplification. V_A' is the voltage required for the observation of anomalous amplification.

value greater than twice that expected for the field at peak velocity, E_p, in InP. Without exception, all samples showed these high values of E_T. We therefore can conclude immediately that the field distribution within each sample was decidedly nonuniform and that $E_T \neq E_p$. This conclusion is reinforced by the asymmetric characteristics. Since the instability is a critical current phenomenon,[6] and this current is usually determined primarily by the condition at the cathode contact, asymmetries will manifest themselves in asymmetric I(V) characteristics. Indeed, such asymmetries are present.

A more significant representation of a lower-limit for E_p can be obtained by determining the voltage at which reflection amplification initiates in such structures. If a subcritical part of the sample reaches E_p first, and this region develops a negative resistance that reduces the net positive resistance of the remainder of the structure at the frequency of interest; reflection amplification will be observed. This occurs for the sample of Fig. 8 at about the same voltage in both polarities, and produces a lower-limit value of E_p near 1.6×10^4 V/cm, more in line with what might be expected for a near-micron size sample of InP.

Once the point E_p is determined, it is next vital to estimate $v(E_p)$, since this parameter often controls the ultimate device efficiency that can be expected from any structure. If we assume a nonuniform structure, then the field profile when biased below V_A probably looks like that shown in Fig. 9. Current continuity requires

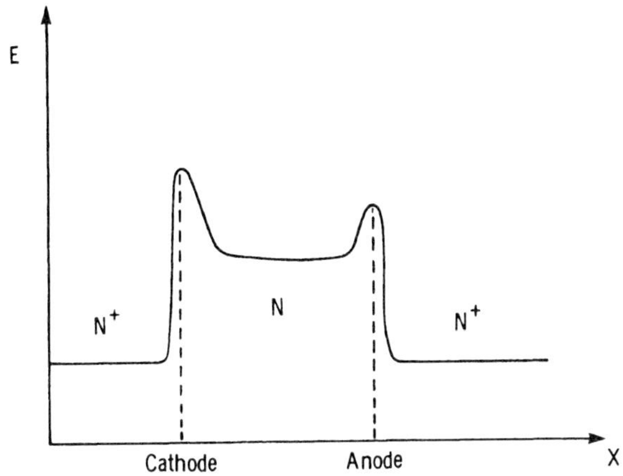

Figure 9. Suggested field profile (negative) for the sample that produced the characteristics shown in Fig. 8.

that J be given by Nqv, where N and v denote the mobile carrier density and the velocity respectively, in the region of uniform field. As the bias is increased we first reach a condition where a sufficiently wide part of the cathode region exceeds E_p such that reflection amplification occurs. This is the point V_A in Fig. 8. At this point $J(V_A)=Nqv_A(<v(E_p))$. With further increases in bias J tends strongly towards saturation. The high field region near the cathode widens into the bulk to accommodate the excess voltage, but the $n\ell'$ product of the region remains subcritical. (ℓ' is the length of its region of high field adjacent to the cathode). Finally, when $n\ell'$ becomes critical, an instability occurs (at V_o). Note that even when this critical value of current is reached, v_A will be below $v(E_p)$. Hence a lower-limit on $v(E_p)$ can be obtained by determining the velocity at V_o, which is

$$v(E_p) \cong \frac{I(V_o)}{AN\,q} \quad . \tag{2}$$

For the sample of Fig. 8, this results in $v(E_p)\cong 4 \times 10^7$cm/sec. Since $v(E_p)$ for long samples of InP is closer to 3×10^7 cm/sec, this value appears rather high. However, values in the same manner obtained for other longer samples ($\ell \cong 2.5$ μm) were considerably lower than this, closer to 2.6×10^7 cm/sec. Furthermore, some samples ($\ell \cong 2.0$ μm) yielded values of $v(E_p)$ near 3.3×10^7 cm/sec. Thus, a size effect is probably being observed - $J(E_p)$ increases for near micron samples, perhaps because of the effect of the Debye tails, which suggests that the doping profiles are not uniform since our calculations indicate that the Debye tail effect should not be present in these samples.

Impedance changes suggestive of, and what we term, amplification, have also been observed, but a direct measure of greater than unity prethreshold gain has yet to be made. However, for amplification in the pre-instability regime, the parasitic resistance effect of the uniform-field region of the sample will tend to reduce the net overall gain.

Qualitative amplification characteristics of these structures have been studied to date primarily at J band. Aside from the anomalous mode, the normal amplification characteristics are similar to those observed in typical nonuniform n-GaAs structures. The common occurrence is weak prethreshold amplification followed by oscillation and amplification. With further increases in bias the oscillation can sometimes be made to cease and maximum amplification observed. In the InP samples studied, however, the samples were not biased much above V_o, in order to protect against burnout. Thus, the stable-unstable-stable amplification mode was not studied.[6] It is expected that a search for it in the future will be fruitful.

Prethreshold impedance changes and amplification are associated with field nonuniformities either at the cathode or anode. Anode nonuniformities, however, generally do not result in instabilities, hence we expect that the observed effects are associated primarily with cathode originated nonuniformities. (For any given polarity the upstream nonuniformity tends to dominate when both are of similar magnitude, which is what we expect based on the observed I(V) curves). Cathode originated effects have been analyzed in detail for GaAs and the results are predictive of what we observe for InP: amplification occurs at the lower end of the band (about 12-13 GHZ); amplification occurs in narrow subbands within a band; the lowest frequency mode exhibits the strongest amplification.

In order to optimize device performance it is of fundamental importance not only to understand the amplification modes, but even more significantly, the oscillation modes. At present the oscillating characteristics in a J-band guide and in a guide covering the frequency range 90-120 GHz have been studied.[7] Well defined oscillations are observed in some samples in this latter frequency band, and in order to determine the mode of oscillation (transit or circuit dominated, fundamental or harmonic frequencies) we plan to set up a program that studies the sample over many bands beginning at J-band and extending up to 120 GHz. At the present time we are unable to study a given device in the various circuits because the packages for several bands often differ. However, design modifications will be made so that the same device can be studied over the frequency range of interest.

When most of the samples are packaged for J-band circuits, they almost invariably oscillate at frequencies substantially higher. Often two modes are observed, one near 25 GHz, the other near 35 GHz. These modes are both weakly voltage tunable, and do not respond to the tuning of the J-band circuit. It is possible that they are associated with narrow-domain transit-time effects, but more likely are EVO-type oscillations.[6] Here, when an instability nucleates, the field rearranges such that the highest field is near the anode and the oscillating mode is a combined pulsing of the anode layer and transit of an accumulation layer from the cathode. In essence the accumulation layer leaves the cathode before the prior layer has drained at the anode. This mode is, as most are, a hybrid, but often locks at a frequency that cannot be controlled by the external circuit above an upper limit. At this point it is clear that much more data must be obtained on the devices in various circuits before reliable mode identification can be made.

CONCLUSIONS

Most of the experimental work we have discussed here is in its preliminary stages. Comparatively little quantitative data have been garnered, particularly on the amplifying and oscillating characteristics of the devices in microwave circuits tuned to different frequencies. Further, only a small population of devices has been studied, much more data is required. However, the initial results, both experimental and theoretical, are satisfying. First, the early evidence suggests that the $J(E_p)$ point increases for sufficiently small samples; to our knowledge this is the first observation of such an effect. This result is probably determined by the boundary conditions, or doping nonuniformities, again, since recent theoretical work,[8] including our own, has shown that our present understanding of transport in short structures has not been completely resolved. In particular, it will be very difficult to separate velocity data from suspected carrier distributions because the mobile carrier density in the active region of short-channel structures will be substantially enhanced by Debye tail diffusion from the adjacent heavily doped regions. Second, an anomalous mode of amplification at low voltage has been observed which may perhaps be involved with a novel class of microwave-Schottky diode amplifying devices. Next, oscillations have been observed at over 100 GHz, and perhaps very weak amplification at these frequencies. The study of amplification at these frequencies is crucial to the determination of an upper frequency limit for the existence of negative differential mobility. Finally, the experimental support for the existence of a mobility notch can provide us with a starting point for the optimization of device efficiency in the oscillator mode. It is known that for specific device modes a precise dynamic cathode boundary field (or control characteristic) is often desired for optimum performance. Optimistically, once the mobility notch is understood and controlled, a "best" notch can be built into the system for each specific circuit. Further, since mobility notches often dominate device performance, their presence should be considered in the next round of numerical calculations; and will attack this problem in the near future.

We are grateful to Norman Apsley of RSRE for both providing the InP samples and his fundamental contributions during the early stages of this work.

REFERENCES

1. M.P. Shaw, Handbook on Semiconductors, 4 1 (North-Holland, Amsterdam, 1981).
2. N. Braslau, J. Vac. Sci. Tech. 19, 803 (1981).
3. B.L. Sharma, Semicond. and Semimetals, 15, 1 (Academic Press, NY, Ed. Willardson and Beer, 1981).
4. H.L. Grubin, D.K. Ferry, G.J. Iafrate and J.R. Barker, VLSI Electronics: Microstructure Science 3, 198 (Academic Press, NY, 1982).
5. G. Persky, Sol. St. Elec. 15, 1345 (1972).
6. M.P. Shaw, H.L. Grubin and P.R. Solomon, The Gunn-Hilsum Effect (Academic Press, NY, 1979).
7. N. Apsley and M.P. Shaw, RSRE Internal Report 9-23-79.
8. J.J. Rosenberg, E.J. Yoffa and M.I. Nathan, IEEE Trans. ED-28, 941 (1981). P.E. Schmidt, M. Octavio and P.D. Esquida, IEEE Trans. Elec. Dev. Lett. EDL-2, 205 (1981). P. Hesto, J.F. Pone and R. Castagne, Proc. 3rd Int. Conf. Hot Carriers in Semicond., Montpellier, France, July 1981.

MONTE CARLO SIMULATION OF TRANSPORT IN SUBMICRON STRUCTURES

Jacques Zimmermann, Anne Ghis and Bernard Boittiaux

Centre Hyperfréquences et Semiconductors, L.A. au CNRS n° 287 and Gréco Microndes, Université de Lille 1
59655 - Villeneuve D'Ascq Cedex France

I. INTRODUCTION

The use of Monte Carlo method designed for the study of semiconductor materials and devices has known a very wide development in the last ten years. Initially the MC method was thought of as a way of solving the semiclassical Boltzmann Transport Equation in the frame of well accepted carrier dynamic models for interactions with phonons, ionized impurities, etc. [1,2]. Since then, the wide capability of MC method for solving complicated systems has appeared. The initial impetus in this way was given by Ruch [3], as concerns the possible existence of drift velocity overshoot in short devices, and by Lebwohl and Price [4], as concerns the use of MC method for simulating a complete and real device, including the effects of the contacts. At the same time, the rapid development of high-speed computers allowed more and more sophisticated models of carrier dynamics to be checked [5] and more refined and realistic, in particular submicron sized, devices to be modeled [6]. Indeed with the MC method very small sized device simulation is easy, since in these submicron devices the total number of carriers to be simulated is relatively small, and this renders the simulation possible with reasonable computation times. In this paper we evoke several possible kinds of application of the MC method: i) the study of hot electron diffusion noise in Gunn-like $N^+\nu N^+$, Si and GaAs structures. As diffusion noise represents probably the most fundamental limitation to the working of these devices, as the physical source of this noise is the velocity fluctuations inside the structure and these fluctuations are a direct outcome of the MC simulation, the study of diffusion noise can be interesting and relatively easy. Care must be taken with contact problems and possible displacement currents in order to compute correctly the noise power emission spectrum of the device. The body of the paper will center on this problem. ii) in the same spirit, the dynamic impedance of the $N^+\nu N^+$ structures with very short ν channels can be computed. This gives an idea of the equivalent electric circuit of the device, depending on the bias voltage conditions. iii) the influence of a magnetic field on the transient and static properties of semiconductor materials can be studied by adding the Laplace force to the effect of the internal electric field. iv) in very small devices, size effects can become important, in some cases more important than bulk effects. This is typically the case in inversion layers of MOSFET's or in MESFET channels at pinch-off. Then the carriers move in close contact with surfaces and there space-charge effects may become essential to the determination of the mobility of carriers. Basically heterojunctions present the same features, and very anisotropic impurity concentrations in devices are always associated with them. In the following we illustrate all these topics with precise examples.

II. HOT CARRIER DIFFUSION NOISE IN UNIPOLAR $N^+\nu N^+$ DEVICES

The generic structure of the devices studied here is shown in **figure 1.** This represents the simple structure of a Gunn-device in which the width of the "active" region (i.e., the low N_D region) can be made short enough that velocity overshoot may occur at rather low bias voltages. Far enough from the homojunctions, the $N_D{}^+$ regions are provided with (supposed perfect) ohmic contacts. If we are to see only diffusion noise, we have to get rid of any other sources of noise that could exist due to the contacts. With ohmic contacts, no additional noise other than thermal noise comes into play. From the point of view of the simulation, perfect ohmic contacts can be simulated in the following way: any carrier crossing one of the contacts (at x=o or x=W) is immediately reinjected at the other contact with the same $\vec{k}$-state and energy (carrier position is simply displaced from 0 to W or from W to 0). This is especially true when these conditions are applied far enough from the homojunctions, in the heavily doped regions where steady state conditions are very likely to be reached. With these conditions, conservation of the total charge is preserved, and the number of carriers to be simulated is given by,

$$n_e = A\int_0^W N_D(x)\ dx, \qquad (1)$$

where A is the cross-section area of the device: that is the number of carriers equals the total number of ionized centers that we suppose fully ionized. Moveover, we assume that the symmetry of the system allows a one dimensional treatment along the direction of the current.

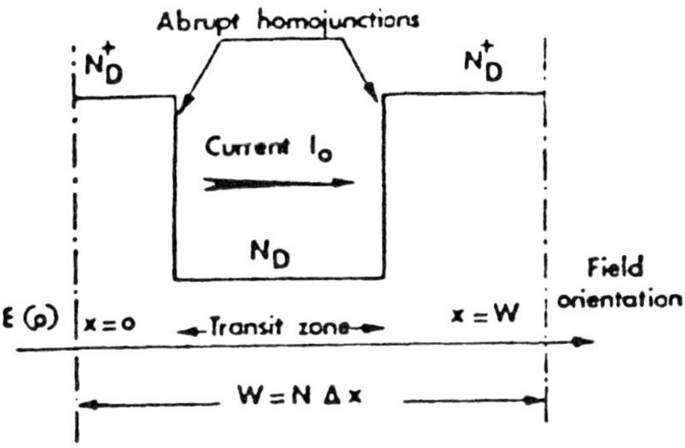

Figure 1. Sketch of the devices studied in the paper. The field orientation is <111> for n-Silicon and <100> for n-GaAs. Abrupt homojunctions are assumed for the sake of simplicity.

The Monte Carlo procedure used here has already been described in another paper, and the reader is referred to [7] for the details. The device simulation is based on a count of the number dn of carriers located between any x and x+dx in the device, at every time t. Thus, in practice, we devide the component into N slices of equal width Δx, and the time scale is divided into M meshes of equal duration δt. At each $t_i = i\,\delta t$, and at each $x_j = j\,\Delta x$, the state of the n_e particles is recorded. If the count of particles gives $N_{j,i}$ between (j-1) x and j x at some time t_i, then the actual carrier density in the j-th slice is

$$n(x_j, t_i) = \frac{N_{j,i}}{A\Delta x} \tag{2}$$

$n(x_j,\ t_i)$ is the fundamental outcome of the simulation. Next, the internal electric field is readily computed using Poisson's equation with suitable boundary condition at x=0 for instance. Of course, because of relation (1), one has

$$E(W) = E(0) \tag{3}$$

Integrated again, $E(x)$ gives the voltage drop in the device. Any average value as a function of space position can be computed at the same time: let α be the estimator given by the state of a particle. At a time $i\delta t$, between $(j-1)\Delta x$ and $j\Delta x$ are located $N_{j,i}$ particles, each with a numerical value α_k and α, and the average value is,

$$A.V.(\alpha;\ j,i) = \frac{1}{N_{j,i}} \sum_{k=1}^{N_{j,i}} \alpha_k = \langle \alpha(x_j, t_i) \rangle \qquad (4)$$

Insofar as we are involved with average characteristics of the device, the basic method just described is enough. If we are to calculate noise in the device and diffusion noise in particular, we have to deal with fluctuations around average values, and those fluctuations are all functions of time. In other words we have to account for internal field (and voltage) time fluctuations, that is the possible existence of displacement current that should be added to the particle current:

$$J_{total}(t) = qn(x,t)\langle v(x,t)\rangle + \varepsilon_o \frac{\partial E(x,t)}{\partial t} \qquad (5)$$

Due to current continuity, $J_{total}(t)$ should not depend on position. Poisson's equation is implicitly present in (5). Moreover, the term $qn\langle v\rangle$ indeed represents the particle current, because it contains the conduction current <u>and</u> the diffusion current due to the fluctuations of $\langle v(x,t)\rangle$ around its average value. We can now describe the two conventional modes of observation of noise: voltage noise operation and current noise operation.

A - VOLTAGE NOISE OPERATION

In this case, the total current density $J_{total}\ (t)$ is constant in time, and we observe the voltage fluctuations at the ends of the device. With these conditions it can easily be shown that we have

$$V(t) = V(o)\ \frac{I_o W t}{\varepsilon_o A} - \frac{q}{\varepsilon_o A} \sum_{i=1}^{n_e} \int_o^t v_i(t')dt', \qquad (6)$$

where I_o is the bias current and $v_i(t)$ is the component of the velocity of a carrier parallel to the current direction. As we know the v_i's at each time, $V(t)$ is readily carried out and up to 30,000 values are recorded in order to compute the noise power spectrum. To do this we use a rather conventional method through the autocorrelation function of $V(t)$ and the Wiener-Khinchin theorem relation function of V(t) and the Wiener-Khinchin theorem

$$\Phi_V(t) = \overline{(V(u) - \bar{V})(V(t + u) - \bar{V})}, \tag{7}$$

$$S_V(f) = 4 \int_0^\infty \Phi_V(t) \cos (2\pi f t) dt, \tag{8}$$

where the big bar indicates an average over the time labelled u. This is not the unique way of doing: $S_V(f)$ can be directly carried out from the V(t)- recording [8] with $\Phi_V(t)$ as an extra outcome.

A very illustrative, although rather hypothetical example of application of the above model can be shown: the simulation of equilibrium thermal noise in a semiconductor resistor. A piece of silicon with homogeneous doping density N_D provided with ohmic contacts at both ends : it is very easy [9] to calculate analytically the $S_V(f)$ when the component is not biased (i.e., I_o=0),

$$S_V(f) = \frac{4K_B T_L R_o}{\left(1 - \frac{f^2}{f_p^2}\right)^2 + \left(\frac{f}{f_o}\right)^2}, \tag{9}$$

where R_o is the D.C. ohmic resistance, T_L is the lattice temperature, k_B is Boltzmann's constant; f_o is the dielectric frequency given by

$$f_o = \frac{qN_D \mu_o}{2\pi\varepsilon_o},$$

and f_p is the plasma frequency of the material given by

$$f_p = \frac{q}{2\pi} \left(\frac{N_D}{m\varepsilon_o}\right)^{1/2}.$$

The only basic assumption to be made in order to derive (9) is that the A.C. mobility can be written as

$$\mu(f) = \frac{\mu_o}{1 + j2\pi f \tau_m} \quad . \tag{10}$$

Figure 2 (left) compares the reduced voltage autocorrelation function $\Phi_V(t)/\Phi_V(0)$ given by the MC simulation with the inverse cosine-Fourier transform of (9). The negative part of $\Phi_V(t)$ appears when $N_D > 5.3\times10^{15}cm^{-3}$, which is a result of the competition between mean-momentum relaxation frequency $f_m=1/2\pi\tau_m$ and dielectric frequency of the material f_o. In the range of N_D studied here - at room temperature f_m is not much affected by N_D, but f_o varies as N_D, and this may change strongly the shape of $S_V(f)$. In particular at $N_D>10^{16}cm^{-3}$ the $S_V(f)$ peaks at some frequency, although below $10^{16}cm^{-3}$ the $S_V(f)$ is flat.

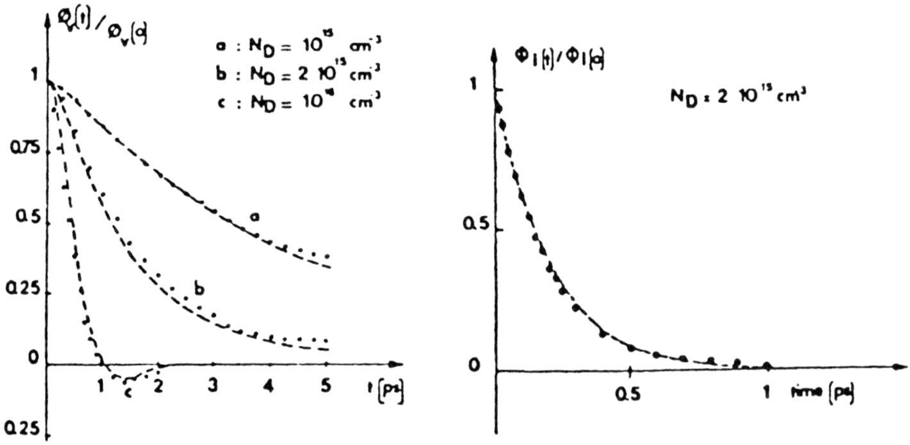

Figure 2. Reduced correlation functions of thermal noise at equilibrium. On the left, the influence of doping density N_D is illustrated by the reduced voltage autocorrelation function $\Phi_V(t)$. Note the change of behavior when N_D approaches $10^{16}cm^{-3}$. On the right, the reduced current autocorrelation function $\Phi_I(t)$ fits the function exp $(-\frac{t}{\tau_m})$ where τ_m is the mean momentum relaxation time. Dots: MC simulation results; dashes: analytical solution.

B - CURRENT NOISE OPERATION

The total current in the device is now allowed to fluctuate freely but the voltage at the terminals remains unchanged in time. From (5), written in terms of finite differences, and with the extra condition

$$\int_0^W E(x,t)dx = V(t) = V_o \quad , \qquad (11)$$

at any time t, one can show that

$$I(t) = \frac{q}{W} u(t) \quad , \qquad (12)$$

where u (t) is the sum of velocities of the n_e carriers in the current direction. Of course, (12) is a statement of the celebrated Ramo-Shockley theorem. The current autocorrelation function $\Phi_I(t)$ and the corresponding $S_I(f)$ are computed according to (7) and (8).

In case of equilibrium thermal noise in the Si - resistor that is when $V_o=0$, one finds that [9],

$$S_I(f) = \frac{4K_BT_L}{R_o\left(1 + \left(\frac{f}{fm}\right)^2\right)} \quad , \qquad (13)$$

Figure 2 (right) compares the reduced current autocorrelation function $\Phi_I(t)/\Phi_I(0)$ given by the MC simulation, with the theoretical result which from (13) is $\exp(-t/\tau_m)$.

C - HOT ELECTRON DIFFUSION NOISE IN GaAs N^+N^+ STRUCTURES

One can find a number of results concerning GaAs-$N^+\nu N^+$ Monte Carlo simulations in the literature [10,11]. All these concern static characteristics of devices, and in particular the way in which velocity overshoot may occur in such devices. Many results can also be found in [6]. Here we are specifically interested with noise. The device we are looking at now has a low-doped "active region" of thickness $L=0.48\mu m$ with $N_D=10^{14}cm^{-3}$, sandwiched between two N_D^+ layers with $N_D=5\times10^{16}cm^{-3}$, the total length $W=1\mu m$. The number of particles simulated here is $n_e=1040$ and constant bias voltage condition is considered with V_o=0 and V_o=0.5V. With the latter V_o one can observe velocity overshoot in the device at 77 and 300K as well [11]. An example is shown in **figure 3** at 77K. It displays the electric field in the device described above, and the velocity response as a function of abcissa. Velocity overshoot is particularly marked near x=0.7 µm where it peaks to about $5x10^7$ cm/s which is almost five times the static drift velocity corresponding to the local electric field. Similar results have been obtained with silicon, but the velocity overshoot effects are much less impressive [9,12].

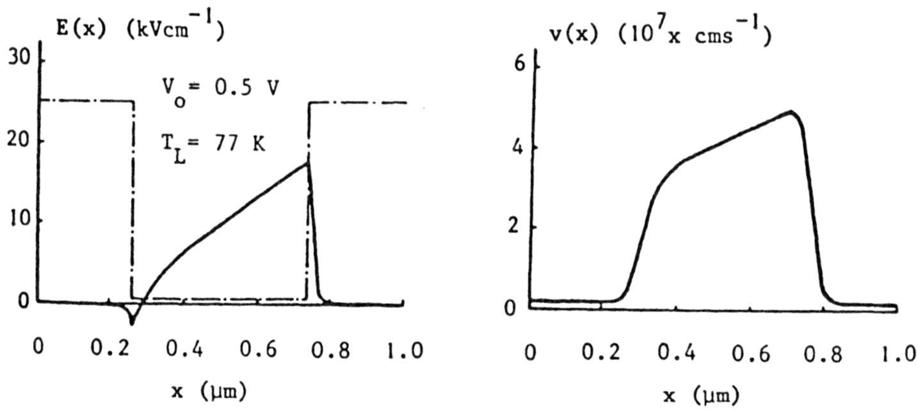

Figure 3. Plots of electric field and drift velocity as function of abcissa in the GaAs - $N^+\nu N^+$ structure described in Section II-C. The dash-dotted line is the doping contour.

All the results presented in the following are obtained when the bias voltage is constant and the transient regime has vanished. In **figure 4** we show a typical result of a current autocorrelation function obtained at V_o=0.5V, for various values of time mesh duration δt. The same is shown in **figure 5** for various values of space mesh Δx. These clearly show the parasitic effect of the finiteness of the δt and Δx. We get rid of this effect by choosing values small enough that correlation functions remain unchanged. In the present case, this is obtained when δt<0.01 ps and δx<50Å. We then observe a correlation function which presents a very significant negative part until it reaches zero at longer times. In fact what we are observing here is just the correlation of velocity fluctuations in non steady-state regime and inhomogeneous field conditions. The negative downswing of the correlation functions reflects a quite general feature of systems out of equilibrium as already observed by several authors [13-16], and this oscillatory behavior of fluctuation correlation function is always associated with velocity overshoot. However, at present, the analysis is complicted by the fact that we are observing noise in a device, and it can be expected that other phenomena are present to explain the effect observed here. First the cut-off frequency of the carrier mobility itself, which depends on momentum and energy relaxation times. These relaxation times are fundamentally field dependent through the energy of the carriers, and this phenomenon is most easily seen in the steady-state case in presence of high field [7,14,17], but here we are in presence of an out-of-equilibrium system. Second, the dielectric relaxation time of the material used for making the device plays a role, as mentioned above. The doping level in the device determines in fact the relative importance of the capacitor between the metallic contacts with respect to the losses of the device: this effect is implicitly present in the simulation since we use (12) for the current, and the field at each point is calculated in such a way that (11) is met. Third, the carrier transist-time τ_t through the active region plays also a role. In the limit of low losses in the active region, a carrier behaves as an unattenuated plane wave and the effect of transit- time is fundamental to the determination of the impedance of the device and on its dependence with the frequency [18]. Something happens when $2\pi f \tau_t \sim 1$, and with correlation functions as shown in **figures 4 and 5,** the whole frequency spectrum is involved. All of the three effects described here are reflected in the correlation functions but they are mixed together and it is extremely difficult at present to ascribe some particular feature to each of them. Indeed, from (5), one can find that

$$E(x,t+\delta t) = E(x,t) + \frac{\delta t}{\varepsilon_0}\left[\frac{I(t)}{A} - qn(x,t)\langle v(x,t)\rangle\right], \qquad (14)$$

in terms of finite differences, and with the help of (11) we can write

$$I(t) = \frac{qA}{W}\int_0^W n(x,t)\langle v(x,t)\rangle dx \quad . \qquad (15)$$

This expression shows the source of the three effects mentioned above. First the contribution of <v(x,t)>, second the contribution of n(x,t) which strongly depends on the doping levels in the device, and third the effect of transit-time coming from the integral.

The current noise emission spectrum corresponding to **figures 4 and 5** is shown in figure 6. Because of the oscillatory shape of $\phi_I(t)$, the noise spectrum is expected to peak at some frequency. This structure is clearly seen in **figure 6** where the $S_I(f)$ is shown in a log-log plot. The most interesting part of $S_I(f)$ is near f=0 where the spectrum is almost "white". There, at low frequency, the noise power can be measured experimentally and compared to the generalized Nyquist formula.

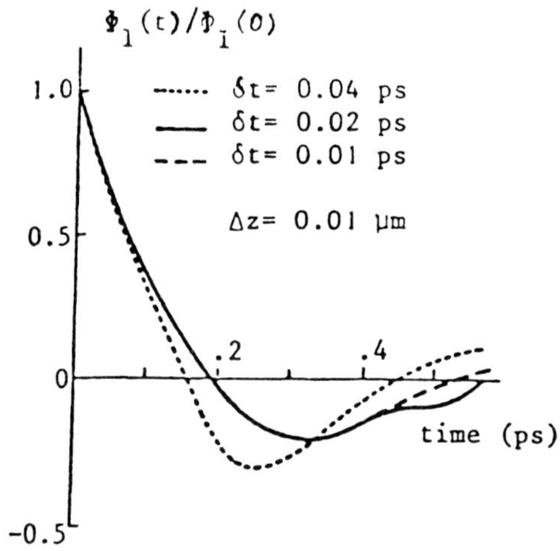

Figure 4. Reduced current correlation functions: effect of the δt in the simulation, Δz is fixed.

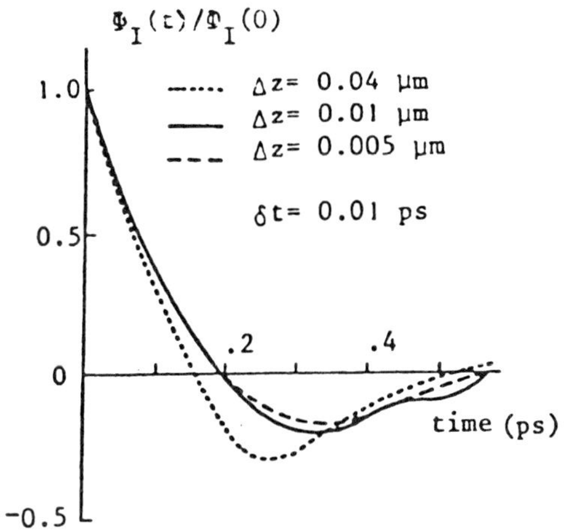

Figure 5. Reduced current correlation functions: effect of the Δz in the simulation, δt is fixed.

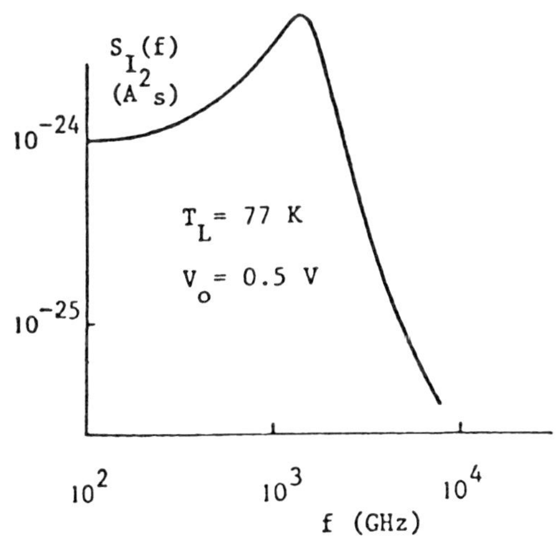

Figure 6. Current noise emission spectrum deduced from the correlation functions of figures 4 and 5.

$$S_I(0) = 4\ k_B\ T_N\ G(V_o;f=0) \quad , \qquad (16)$$

where $G(V_o,f=0)$ is the D.C. small signal conductance of the device at low frequency, and T_N is the equivalent noise temperature. From the current equivalent of (8) we have

$$S_I(0) = 4 \int_0^{\infty} \Phi_I(t)dt \quad , \qquad (17)$$

and a good estimate of $S_I(0)$ requires at first a good estimate of $\Phi_I(t)$ over a quite large length of time. Generally, formula (17) leads to over-estimated values of $S_I(0)$.

III. DYNAMIC IMPEDANCE OF N^+ N^+ GaAs STRUCTURES

We suppose now that the voltage applied to the device has the form

$$V(t) = V_o + V_1 \sin(2\pi f_o t) \quad . \qquad (18)$$

Then the current flowing through the structure is calculated according to (5), but the requirement for the electric field is now

$$\int_0^W E(x,t)dx = V_o + V_1 \sin(2\pi f_o t) \quad . \qquad (19)$$

This again is a consequence of the total current continuity. In response to the above V(t), a recording of I(t) is performed and the amplitude $I_1{*}$ at the fundamental frequency f_o is taken and the dynamic impedance is calculated as

$$Z(f_o) = \frac{V_1}{I_1^*} \quad . \qquad (20)$$

It will be found in general that $Z(f_o)$ has a complex value, and if the simulation is long enough (a big number of periods of the incident signal V(t)), I_1^* can be obtained with a reasonably good accuracy. Indeed the fundamental limit of f_o is imposed by the

δt used in the simulation (a maximum value of δt is 0.01 ps, as seen above) and the total number of periods of V(t) that we can run on the computer in a reasonable time. That is why we show some results obtained at f_o = 100 GHz (the response I(t) covers 40 periods of the input V(t)) and at f_o=625 GHz (230 periods). Several authors have suggested impedance equivalent circuits to represent nearly ballistic effects on the dynamic impedance of the device [19,20]. A very simple model [19] shows that the impedance can be cast in the form

$$\frac{1}{Z(\omega)} = \frac{1}{\dfrac{W'}{qn\mu A} + j\omega \dfrac{W'm}{q^2 nA}} + j\omega \frac{\varepsilon_o A}{W'} \quad , \qquad (21)$$

where n is the carrier density in the active region (width W'), the regions where N_D is high, near the contacts, playing the role of series resistances. The form (21) is probably true in a linear regime where $V_1 \ll V_o$ and if the plasma frequency $f_p = (\frac{n}{\varepsilon_o m})^{1/2} \frac{q}{2\pi}$ is small enough that μ is not frequency dependent between 0 and f_p. In the case at hand, with V_o=0.2V, one finds that $n \simeq 1.5 \times 10^{15} cm^{-3}$ and $f_p \simeq 380$ GHz. The Monte Carlo result is shown in Table 1 for the case where $V_o = V_1 = 0.2V$, at f_o=625 GHz. We compare this result, after subtraction of the effect of the capacitance $C' = \frac{\varepsilon_o A}{W'}$, with the theoretical values given by (21). At 100 GHz the MC values and the formula (21) compare favorably, but at 625 GHz we are over the plasma frequency and formula (21) probably fails. However in any case we find that nearly ballistic motion in the active region is reflected in the impedance through an inductive effect.

TABLE 1

FREQUENCY	R_{MC} Ω	$\frac{W'}{qn\mu A}$ Ω	$L\omega_{MC}$ Ω	$\left.\omega \frac{W'm}{q^2 nA}\right\|$ Ω
100 GHz	77,000	63,000	8,100	12,000
625 GHz	92,000	63,000	32,000	120,000

IV. INFLUENCE OF A MAGNETIC FIELD ON VELOCITY OVERSHOOT EFFECT

We wish now to study the influence of a magnetic field upon the transport properties in $N^+\nu N^+$ devices. A magnetic field is easily taken into account in a Monte Carlo simulation by adding simply the Laplace force to the external electric field, that is

$$h \frac{d\vec{k}}{dt} = q\ (\vec{E} + \vec{v} \times \vec{B}) \quad \text{[MKS units]} \quad , \qquad (22)$$

where $\vec{k}$, $\vec{v}$ are the instantaneous wave vector and velocity of the carrier respectively. As $\vec{v}(t)$ is known at each time, the Laplace force is easily computed. We have studied the case of crossed $\vec{E}$ and $\vec{B}$, with B ranging from 0 to 3 Tesla. With B=3T, one finds a sensitive effect on the velocity , but in conditions where modifications in the band structure of the material are very weak, especially above 77K. In other words, we have $B<m\, k_BT/(hq)$. Moreover the device is supposed to be large enough that the transverse Hall voltage remains negligible. The effect of $\vec{B}$ is to bend the electron trajectories transverse to the current direction, thus reducing the average electron velocity induced by $\vec{E}$.

At first we have studied the influence of $\vec{B}$ on the v-E characteristic in the bulk material. A result is presented in **figure 7** for intrinsic GaAs at 77K. The main feature is that the negative differential mobility disappears for B>1T. However at very high electric fields, the Hall angle tends to zero (that is $\vec{E} \gg \vec{v}\times\vec{B}$), and the saturation velocity remains almost unsensitive to the magnetic field.

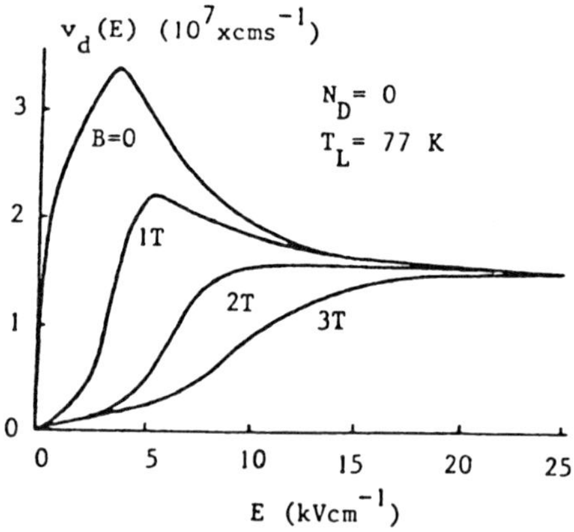

Figure 7. Effect of a magnetic field on the v-E characteristic of bulk intrinsic-GaAs in the steady-state at 77K.

Similar results can be obtained when observing the system in velocity overshoot or TDR regime. These are shown in **figure 8.** Initially the system of carriers is at equilibrium with a Maxwell-Boltzmann distribution. At t=0, the system is subjected to a high electric field (E-10kV/cm), and we observe the ensemble drift velocity as it evolves in time, in response to the field. This is performed at various magnetic field strengths, assuming that it was applied at t=-∞. We find that with B=3T, the velocity overshoot is strongly depressed and, moreover, exhibits large damped oscillations. Basically, the oscillations come from the cyclotron effect due to the magnetic field. If there were no collision between the crystal and the carriers, the oscillations would be unattenuated and the ensemble velocity could be periodically negative. The fact that carriers undergo collisions introduces the damping of these oscillations. At B=3T, the cyclotron resonance period would be τ_c=2m/qB=0.8ps. This figure agrees very well with the pseudo-period of the oscillations as seen in **figure 8** at B=3T.

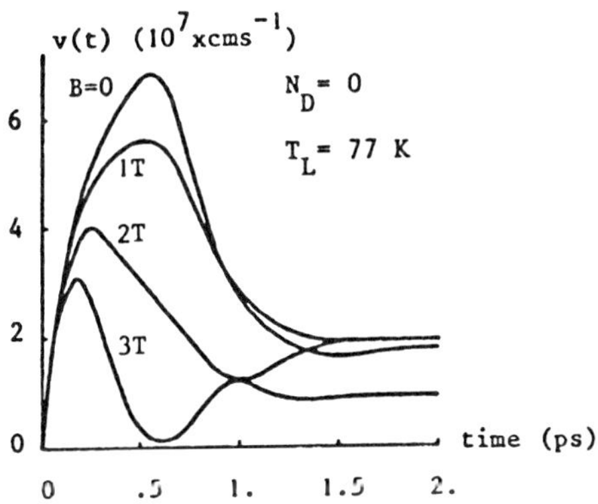

Figure 8. Effect of a magnetic field on velocity overshoot in bulk GaAs at 77K. The amplitude of the electric field is 10kV/cm.

V. STUDY OF SOME SIZE EFFECTS WITH THE MONTE CARLO METHOD

One of the most important cases where size effects come into play is in electron transport in field-effect-transistors. In these devices, electrons move in close contact with interfaces between the semiconductor material and an insulator, as in a MOSFET, or in a narrow conductive channel as in MESFET at pinch-off. In this section, we are looking at a situation where electrons are in an inversion regime in a MOSFET. The problem of MESFET is considered in an other paper [6].

The usual and widely accepted picture describing carrier motion at the surface in an inversion regime considers that the carriers are confined in a potential well due to the band bending near the surface because of the external gate bias [21,22]. If the bands are strongly bent, as in strong inversion, this potential well formed by the Si-SiO_2 interface and the electrostatic potential in the semiconductor can be narrow enough that quantum mechanical effects become important, and the system of carriers behaves as a quasi two-dimensional gas. When source-drain bias is applied and the field is high enough, a hot two-dimensional gas of electrons may exist. Effects related to non-ohmic conduction have had extensive theoretical investigations [23], especially in view of calculating the hot electron drift velocity. Interesting Monte Carlo calculations of two-dimensional gas performed by Basu [24] at high field include complete details such as surface-oxide charge and inter-subband transitions. At high lattice temperatures, however, it can be expected that many subbands are already populated even at weak field, due to the extension of the energy density over about $2k_BT_L$ which is roughly four times, on average, the separation between the subbands. A high fields, the extension grows to $2k_BT_e$ and T_e, the electron temperature, is the higher when the driving field approaches that of velocity saturation. In this case, many subbands are filled, even in strong inversion, and this completely screens the effect of quantization. Then carrier dynamics becomes three-dimensional again, the carriers are much more bulk-like than at low temperature, only the interference with the interface may affect the dynamics.

A. A THREE-LAYER MODEL OF THE Si-SiO_2 INTERFACE

Due to the considerable use of Si-MOS devices in practical applications, the constitution and stoichiometry of the interface between Si and SiO_2 has received considerable attention. According to Helms [25], to accurately describe a real Si-SiO_2 system it may be necessary to consider up to five separate regions: bulk Si, a disordered Si layer near the metallurgical interface, a nonstoichiometric SiO_x at the interface, strained SiO_2 near the interface, and a bulk SiO_2. The estimated width of the transition region, although still controverted, is ~3 to 10Å between pure Si substrate and the pure SiO_2 matrix. Moreover, fixed charges are located in this transition region with a surface density Q_{ss} uniformly scattered in the SiO_x section. In our model, and according to Hubner [26], the transition of band-gap from Si (1.1 eV) to SiO_2 (8.4 eV) is supposed to vary linearly across the SiO_x layer. This means that as a carrier penetrates the interface, it looses gradually a part of its energy corresponding to the movement perpendicular to the interface. The electron encountering the potential barrier has a two-dimensional motion along the interface since it has lost its energy perpendicular to the interface until an interaction with the crystal lattice drives its trajectory away from the barrier, making the motion three-dimensional again. Space-charge field effects due to accumulation of electrons at the interface are taken into account by solving Poisson's equation in the direction perpendicular to the interface. The boundary condition is imposed at the SiO_2-SiO_x interface by the continuity of the field in the oxide layer due to the gate voltage. However the driving electric field along the interface is taken uniform.

B. RESULTS

Now, we give some results calculated for the interface in the <100> direction showing the evolution of inversion electron drift velocity as a function of the driving electric field E along the <011> axis. Various surface fields are considered. The three parameters needed in the simulation are Q_{SS}, d, the width of the SiO_x layer an E_S, the surface field at the SiO_2-SiO_x interface. These three parameters are generally given by the experiment [27,28].

In **figure 9,** we compare the computed results with the experiment of Cohen [27] and Nelson [28] at 300K, when d equals 5Å. The theory compares favourably with the experiment by Nelson, and shows in particular that at high driving fields, the saturated velocity of inversion electrons is very close to that of bulk electrons $v \simeq 10^7 cms^{-1}$, contrary to what was found by Cohen. **Figure 10** shows the evolution of the surface field transverse to the interface. This field is responsible for the band bending at the interface. However, its evolution strongly depends on the strength of the driving field. When the latter is high (corresponding to velocity saturation regime), channeling tends to disappear and the inversion electrons are much more bulk-like carriers. This could be an argument, first, to explain why the saturated velocity is almost equal to the value in the bulk material; second, to support the assumption of the semiclassical model used here.

At low temperature however, a quantized model is necessary [21,23,24]. **Figure 11** presents some results at 115K obtained [29] with a model similar to [24]. These show the influence of surface roughness scattering on the drift velocity of the inversion electrons by varying the roughness parameters Δ and L, respectively the depth and correlation length of roughness potential.

Electron transport in heterojunctions (GaAlAs/GaAs and GaInAs/InP) can be approached using the same ideas as developed here.

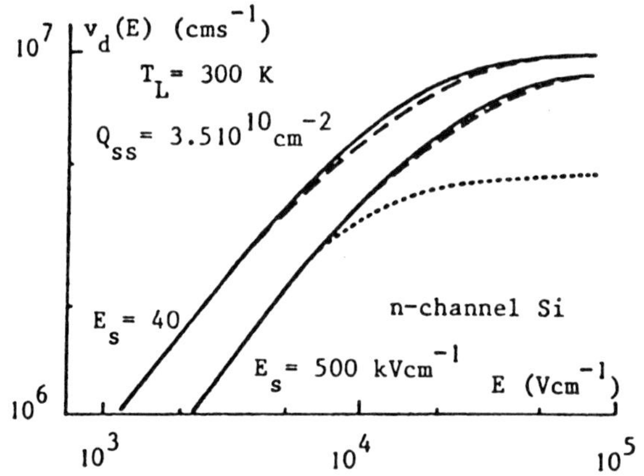

Figure 9. Three layer model. Drift velocity of inversion electrons at a Si-SiO_2 interface showing the effect of surface field E_s. Full lines: Monte Carlo simulation. Dashed lines: experimental results by Nelson [28]. Dots: experimental results by Cohen [27].

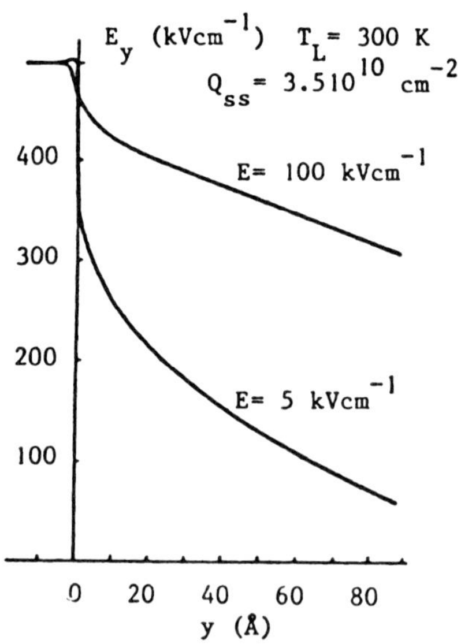

Figure 10. Three layer model. Space evolution of the transverse field E_y. The driving field E is along the <011> direction.

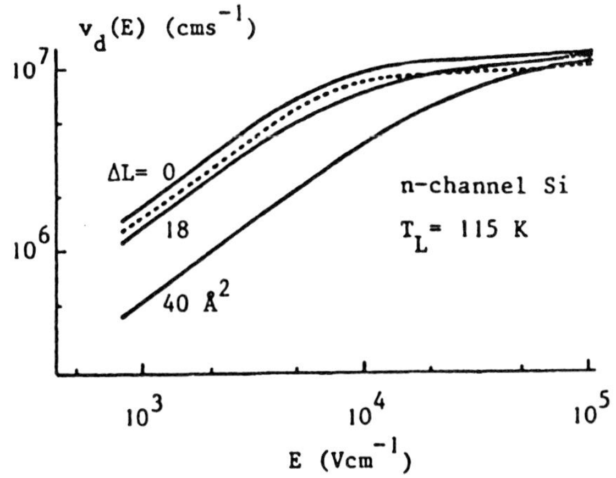

Figure 11. Inversion electron quantized model. Drift velocity as a function of the driving field at 115K. The effect of surface roughness is shown. Dots: experimental results by Cohen [27].

VI. CONCLUSION

In this paper we have tried to show the wide capability of Monte Carlo method for solving a variety of carrier transport problems in semiconductors, and in particular when dealing with complicated systems like submicron devices. The main advantage of the method is that it can take into account accumulation of carriers, space-charge reaction, strong field grandients, very easily. Moreover fluctuations in the systems are computed in a realistic way and uneasy questions like diffusion noise can be treated at no extra cost.

REFERENCES

1. W. Fawcett, A.D. Boardman, and S.Swain, J. Phys. Chem. Solids, 31, 1963 (1970).

2. J.G. Ruch, and W. Fawcett, J. Appl. Phys., 41, 3843 (1970).

3 J.G. Ruch, IEEE Trans. Electron Dev., ED-19, 652 (1972).

4. P.A. Lebwohl, and P.J. Price, Appl. Phys. Lett., 19, 530 (1971).

5. Many aspects of this chapter are closely related to other parts of these proceedings: see, for instance, the papers by C. Jacoboni, by L. Reggiani, and also by H.L. Grubin (These proceedings).

6. See for instance, the paper by E. Constant (these proceedings.)

7. J. Zimmermann, Y. Leroy, and E. Constant, J. Appl. Phys., 49, 3378 (1978).

8. R.O. Grondin, P.A. Blakey, J.R. East, and E.D. Rothman, IEEE Trans. Electron Dev., ED-28, 914 (1981).

9. J. Zimmermann, and E. Constant, Solid-State Electron., 23, 915 (1980).

10. K. Tomizawa, Y. Awano, N. Hashizume, and M. Kawashima, IEE Proc., 129-1, 131 (1982).

11. A. Ghis, E. Constant, and B. Bottiaus, J. Appl. Phys., 54, 214 (1983).

12. J. Zimmermann, A. Kaszynski, Y. Leroy, and B. Carnez, Springer Series in Electrophysics, Volume 2, 101 (Springer-Verlag, Berlin 1978).

13. P.J. Price, in Fluctuation Phenomena in Solids, R.E. Burgess, Ed., (Academic Press, New York, 1965).

14. D.K. Ferry, and J.R. Barker, J. Appl. Phys., 52, 818 (1981).

15. J. Zimmerman, P. Lugli, and D.K. Ferry, J. Physique, 42, (Suppl. 10), c7-95 (1981).

16. J. Zimmerman, P. Lugli, and D.K. Ferry, Solid-State Electron. 26, 233 (1983).

17. P.A. Lebwohl, J. Appl. Phys., 44, 1744 (1973).

18. D. Dascalu, in Transit-time Effects in Unipolar Solid-State Devices, (Abacus Press, Bucarest, 1972).

19. W.R. Frensley, IEEE Electron Device letters, EDL-1,137 (1980).

20. J.R. Barker, D.K. Ferry, and H.L. Grubin, IEEE Electron Device letters, EDL-1, 209 (1980).

21. F. Stern, and W.E. Howard, Phys. Rev., 163, 816 (1967).

22 F. Stern, Phys. Rev., 135, 4391 (1972).

23. D.K. Ferry, Solid-State Electron., 21, 115 (1978).

24. P.K. Basu, Sol. State Comm., 27, 657 (1978).

25. C.R. Helmes, J. Vac. Sci. Technol., 16, 608 (1979).

26. K. Hubner, Phys. Stat. Sol.b, 94, K35 (1979).

27. R.W. Cohen, and R.J. Muller, Solid-State Electron., 23, 35 (1980).

28. D.F. Nelson, and J.A. Cooper, Proceedings of the Second Workshop on the Physics of Submicron Structures, Urbana, IL., June 1982 (Plenum Press, New York, 1983) in the press.

29. M. Charef, Thesis, University of Lille (1983), available on request.

TWO DIMENSIONAL ELECTRON GAS FET

Nuyen T. Linh

Thomson CSF Central Research Laboratory
91401 Orsay
France

INTRODUCTION

The progress which has been made by GaAs MESFETs during the past decade is impressive in both microwave and high-speed logic applications. An example of this progress is shown in Table I. It is quite impressive but it is not doubtful that, there will be a limit in the performance of this device, as there is a limit in Si one. These limitations are due to two main factors:

- physical properties of the basic materials, such as electron velocity, relaxation times, ionization coefficient, etc.

- technology: fine line lithography and processing, contact resistance parasitic capacitances, etc.

Since III - V compounds offer a large variety of possibilities, many researchers are thinking about other binary (for example InP) or ternary alloys (for example GaInAs) as a novel material which exhibits more interesting physical properties, in particular higher

Table I: Evolution of GaAs MESFET Technology

1973		1983	
Microwave	Logic	Microwave	Logic
3GHz	No	60GHz	1K RAM

electron velocity. But it seems that this approach which needs new technologies, would present some difficulties in its industrial development. The other approach is to use heterojunctions with GaAs as a basic material. GaAs/AlGaAs heterojunction is a well known couple in opto-electronics (LEDs, lasers, etc...).

The more recent so-called modulation doped heterostructures which exhibit two-dimensional electron gas properties, is a new way to use heterojunctions.

In this paper, the author will present the state-of-the-art of two-dimensional electron gas FET (TEGFET) and will try to show what is understood and what is not.

TRANSPORT PROPERTIES IN THE HETEROJUNCTION TWO-DIMENSIONAL ELECTRON GAS

In 1978, Dingle et al. realized the modulation-doped superlattice, Figure 1, i.e., alternate GaAs/Al_xGa_{1-x} As layers where only the Al_xGa_{1-x} As is n-doped (1). They observed that electrons all accumulated in the unintentionally doped GaAs layers. Moreover, they showed that modulation-doped superlattices present high electron mobility. This high electron mobility was interpreted as due to the separation of electrons from their parent donor impurities which reduces the Coulombian scattering process.

Since this pioneer work on modulation-doped structures, many studies have been devoted to transport properties in modulation-doped heterostructures. The most important results are:

1) The spatial separation of electrons and impurities is not limited to superlattices. It is also valid for single heterojunctions. Figure 2.

2) The electron accumulation layer has the properties of a two-dimensional electron gas (2 DEG): first, because of the weak interaction between the electrons and the lattice and second, because the movement perpendicular to the heterojunction is hindered due to the potential well in which electrons are located. The size of the potential well leads to the quantum mechanical effects which, while interesting for fundamental semiconductor physics, have only a slight influence in practical applications, particularly at room temperature.

Mobility in 2DEG

Mobilities as high as 9000, 140 000 and 2 x 10^6 $cm^2v^{-1}s^{-1}$ have been obtained at 300, 77 and 4 K respectively [2]. These high mobilities confirm the theoretical considerations on Coulombian scattering and clearly show the extremely high quality of the epitaxial layers.

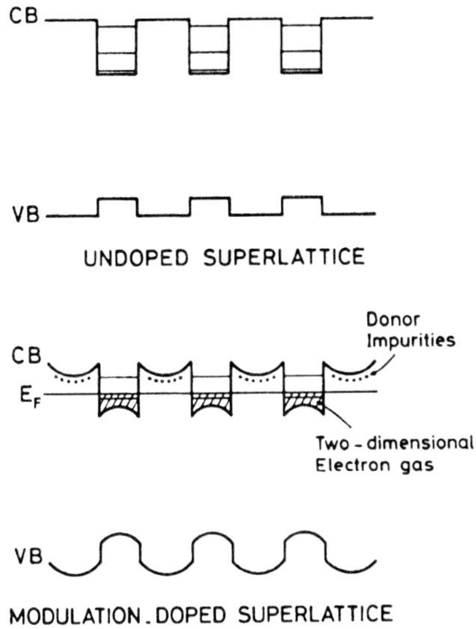

Fig. 1. Undoped and Modulation-doped superlattice

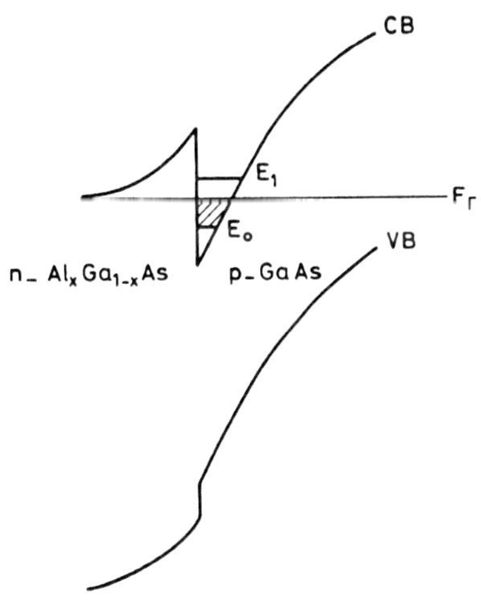

Fig. 2. AlGaAs/GaAs heterostructure band diagram

In addition to the reduced Coulombian scattering, screening effect also plays an important role in the high 2DEG mobility, since the density of electrons is much higher than that of ionized impurities (3). Figure 3 shows that due to screening effect, modulation doped heterostructures which have a background impurity concentration of about $10^{14}cm^{-3}$, present higher low temperature mobility than a single GaAs epilayer whose background impurity concentration is only $5x10^{13}$ cm^{-3}.

Even at room temperature, screening effect can be observed, by noting the dependence of electron mobility versus sheet carrier concentration n_s in the 2DEG. It has been found by Delagebeaudeuf et al. (4) that this dependence can be expressed as:

$$\mu = \mu_o(n_s)^k \qquad k\sim 1.5 \qquad (1)$$

Electron velocity in 2DEG

All the results reported above concern low electric field mobility. The influence of electric field on mobility has been studied by Drummond et al. (5). Figure 4 shows that mobilities decrease as the field increases, particularly at low temperature. This mobility drop is attributed to polar optical phonon scattering. Due to this scattering mode, the enhancement in electron velocity is not as high as in mobility.

Drummond et al. (6) have tried to determine the electron velocity in the modulation doped system by Monte Carlo calculations and have found that the enhancement of electron velocity, compared n-doped bulk GaAs, is approximately 60% at 300K and 200% at 77K. In fact, these calculations have been applied to a high purity GaAs material but not a 2DEG system. So, question is still opened for the exact determination of the electron velocity in this latter case.

Drummond et al. (6) have also tried to experimentally determine the electron velocity in the 2DEG. It seems that at 77K, velocity reaches the value of $2.3x10^7$ cm/s in good agreement with Monte Carlo calculations. But, velocity at room temperature could not be measured at sufficiently high field, because of some problem of thermal dissipation. Since there is some agreement between experimental data and calculated results at 77K, one can suppose that this agreement is still valid at room temperature and then assumes that 2DEG electron velocity at 300K, is effectively 60% higher than in n-doped GaAs.

However, it cannot be concluded that a TEGFET would be just 1.6 times faster than a GaAs FET, because in short gate length FETs, the nonstationary (or dynamic) regime has a strong effect on the electron velocity (7,8). The overshoot effect which is a long story and will not be treated in this paper. Let us just mention that this aspect

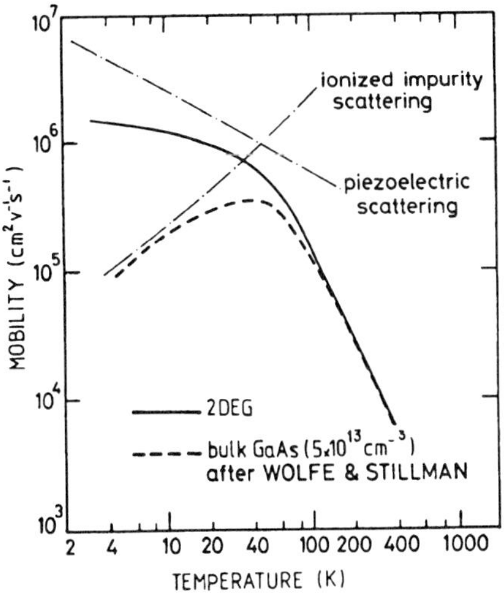

Fig. 3. Mobility of a 2DEG

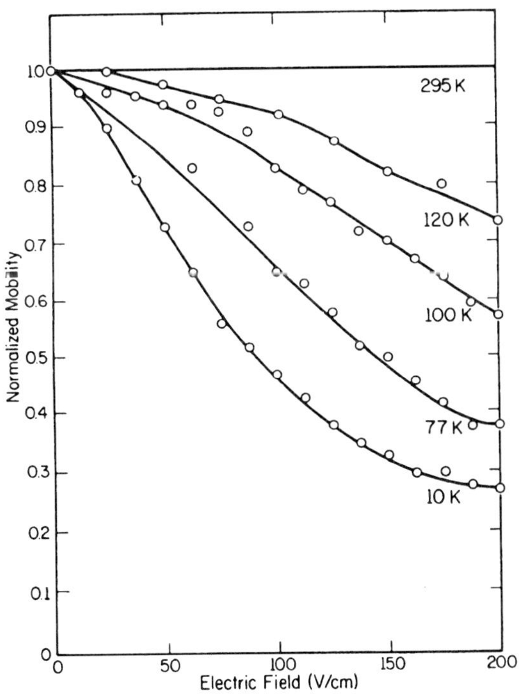

Fig. 4. Mobility versus electric field

of the problem has been treated by CAPPY et al. (9) and MUDARES et al. (10). CAPPY et al. have calculated that overshoot effect in TEGFET is more important than in GaAs FET. Who knows exactly?

DEVICE CONSIDERATIONS AND THEORETICAL APPROACH

TEGFET Structures

The TEGFET, which is essentially a MESFET, can be made with two different structures, Figure 5: one presents $Al_xGa_{1-x}As$ as its uppermost layer; the other, which is inverted, has the low-doped GaAs layer on top. It has been shown that, due to epitaxial growth problems, the inverted structure exhibits low mobility unless particular multilayer structures are grown (11). Moreover, the conduction in the embedded $Al_xGa_{1-x}As$ n-layer leads to undesirable parasitic conductance (12).

For these reasons, only TEGFETs with $Al_xGa_{1-x}As$ as their uppermost layer will be examined. This structure has an additional advantage, related to the low potential of $Al_xGa_{1-x}As$(13).

Before entering into the calculations of the TEGFET model, let us consider the band diagrams of the device at equilibrium and under gate bias Figure 6. The diagrams show that the charge control of the 2DEG is performed through the variation of the potential well position with respect to the Fermi level. This level of charge control is similar to that of an NMOS and is quite different from a conventional MESFET, for which the Schottky gate bias induces a variation of the depletion depth. Therefore, it can be deduced that the gate capacitance of a TEGFET is roughly constant whereas in a GaAs MESFET it varies with gate bias.

Analytical Model

The analytical models of the two types of TEGFETs (normal and inverted structures) have been described in recent papers (4,14-16) only the main results are reviewed in this paper.

Figure 7 represents the schematic band diagram of the heterojunction and gives the notation used. We start from the assumption of a quasi-constant electric field ζ_{10} in the potential well, i.e., we assume a triangular potential well. The solution for the longitudinal quantized energy is then well approximated by the formula:

$$E_n = \left(\frac{\hbar}{2m_\ell^*}\right)^{1/3} \left(\frac{3}{2}\pi q \zeta_{10}\right)^{2/3} \left(n + \frac{3}{4}\right)^{2/3} \quad (2)$$

n = 0, 1, 2...

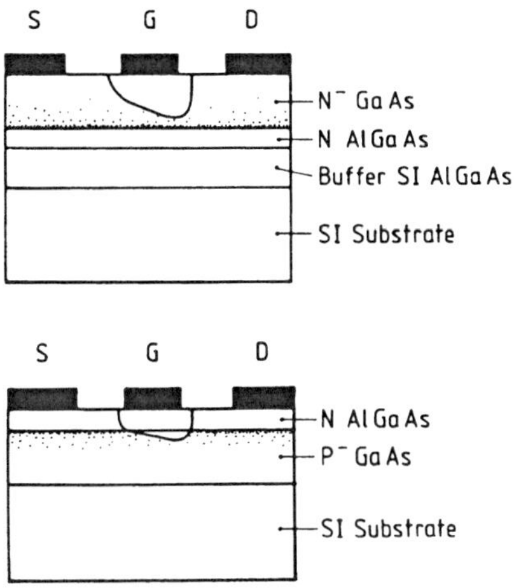

Fig. 5. Sketch of two different TEGFETs

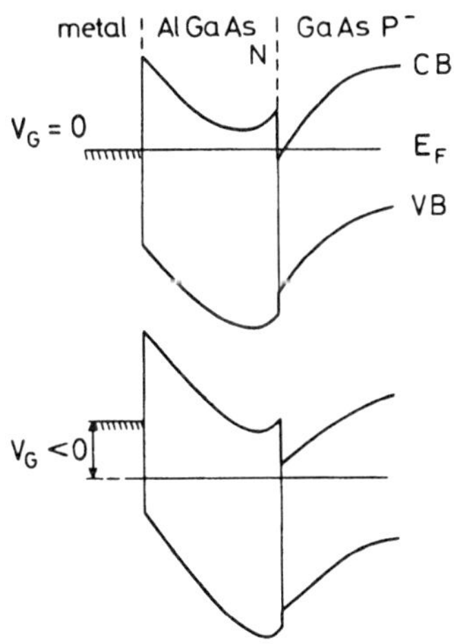

Fig. 6. Band diagram of the M/N-AlGaAs/P-GaAs under equilibrium nonequilibrium conditions

where m_ℓ^* is the longitudinal effective mass $\hbar$ the Planck constant and q the electron charge.

The charge control law of the 2DEG by the Schottky gate has been analytically calculated by

- solving the Poisson equation

$$\frac{d\xi_1}{dx} = -\frac{q}{\varepsilon_1}\,[n(x) + N] \tag{3}$$

where ε_1 is the dielectric constant of GaAs, n(x) the free carrier concentration and N the ionized acceptor concentration (the low-doped GaAs is p^- type) with $n(x) \gg N$. The appropriate boundary condition is $\xi_1 = \xi_{10}$ at the heterojunction interface and $\xi_1 = 0$ far from the interface

- and considering the interpenetration of two space charge regions: the Schottky depletion and the heterojunction.

The final results are expressed as follows:

$$I_{Ds} \sim g_m\,(V_G - V_T)$$

with
$$g_m = \frac{g_{mo}}{1 + R_s g_{mo}} \tag{4}$$

$$g_{mo} = \frac{\varepsilon_2 v_s Z}{d'_2}$$

where I_{Ds} is the drain current, V_G the gate voltage, V_T the pinch off voltage, g_m the transconductance, g_{mo} the intrinsic transconductance, R_s the source resistance, ε_2 the AlGaAs dielectric constant, Z the gate width and v_s the peak electron velocity in the 2DEG, d'_2 the effective thickness of the AlGaAs layer.

The above equations are valid in the gate voltage regions which are not too close to the pinch-off voltage. By using the above equations, we have determined the electron velocity and we found that in most cases, this velocity is approximately 1.1×10^7 cm/s; but in some cases the values of 1.5 to 2×10^7 cm/s have been determined. Is it overshoot effect?

The author does not think so, since these values seem not to depend on gate length. Maybe it is a problem of heterojunction interface.

EXPERIMENTAL DEVICE PERFORMANCES

Low noise discrete devices

Low noise TEGFETs were first fabricated by Laviron et al. (17, 18). Both E-mode and D-mode transistors were obtained and studied. With a gate length of 0.8µm and a FET geometry without recessed gate, microwave performances were promising. At 10GHz, the noise figure was 2.3dB and the associated gains were 10.3 and 7.5dB for D-Mode and E-mode transistors respectively.

Niori et al. (19) working with recessed gate transistors have achieved better performances with a 0.5µm gate length and 2µm source-to-drain spacing. Their transistors present: at 8, 11.3 and 20GHz, noise figures of 1.3, 1.7 and 3.1 dB respectively and associated gains of 13, 11.2 and 7.5 dB.

A four stage amplifier using the HEMT has given, at 20GHz, a noise figure of 3.9 dB with a power gain of 30 dB.

More recently Fujitsu's group reported a noise figure of 1.4 dB at 12 GHz for a 0.5µm gate length device (20). This represents a good result compared to conventional GaAs FETs.

Our latest results on 0.55µm gate length TEGFETs are figure 8.

10 GHz (300 K):
- NF = 1.26 dB
- G_{ass} = 12 dB
- G_{max} = 14.5 dB

10 GHz (120 K):
- NF = 0.25 dB
- G_{ass} = 15 dB
- G_{ass} = 24 dB

17,5 GHz:
- NF = 2.3 dB
- G_{ass} = 7.1 dB
- G_{max} = 9 dB

These performances, particularly at 17.5GHz are better than that of conventional GaAs MESFETs. Moreover, it has to be noticed that the TEGFET is not optimized yet. One of the problems which remains to be solved is the source resistance. For unknown reasons, the source resistance of a TEGFET is higher than expected (and higher than in conventional GaAs FET). To estimate the optimum performance of the TEGFET we can use the semiempirical Fukui equation which gives minimum noise figure dependence on parasitic elements:

$$F_{min} = 1 + 2\pi C_{gs}\ f k_f \sqrt{\frac{R_s + R_g}{g_{mo}}}$$

where C_{gs} is the gate to source capacitance, f the frequency and k_f is a fitting factor which is related to the material quality. In conventional GaAs FET, $K_f \geqslant 2.5$. Our determinations of K_f on TEGFET, and those performed in Fujitsu's Group, have shown that K_f (TEGFET)$\lesssim 2$ and can be as low as 1.5. This finding is difficult to explain.

A priori the TEGFET would present lower noise that GaAs FET because the electron velocity in the 2DEG is higher, therefore g_{mo} is higher. But the Fukui equation shows that the main improvement results from the K_f factor. Since this factor is deduced from a semi-empirical formula, it is then difficult to give an explanation based on fundamental physics.

To have a more fundamental approach of the noise in a TEGFET, one can use the noise model developed by CARNEZ et al. (21) for GaAs MESFET. Through this model we can make the following remark:

The noise in a FET results mainly from the fluctuation of the current in the channel (Baechtold et al. (22), Brewitt Taylar, et al. (23) and Delagebeaudeuf (24) have found the same conclusion by using an analytical model of noise in FETS). This fluctuation can be interpreted as due to a fluctuation of the electron velocity. Then the question is: how is the electron velocity fluctuation in a 2DEG?

The noise spectral density of the current is also related to the longitudinal diffusion coefficient. Then the question is: how is the diffusion coefficient in a 2DEG?

Digital Integrated Circuits

High electron mobility and velocity of the 2DEG lead to think that the TEGFET is well adapted to high speed and low power integrated circuits. (Figure 9).

Table II summarizes and compares the propagation delay time and

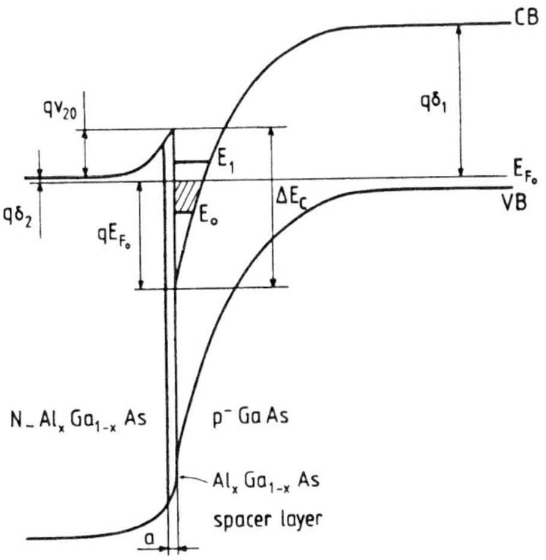

Fig. 7 Schematic band diagram of a heterostructure

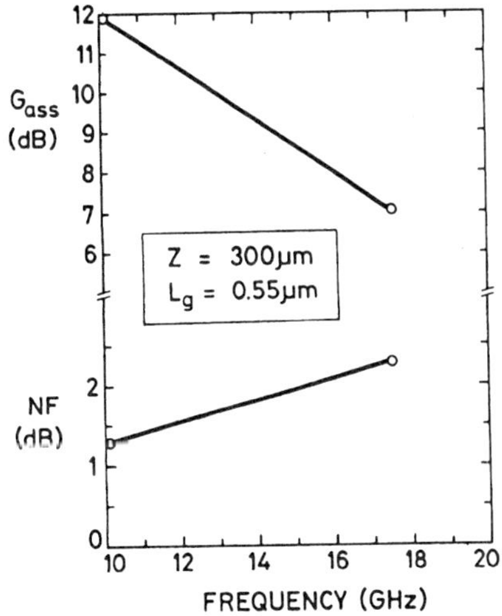

Fig. 8 Gain and Noise - Figure versus frequency

power-delay product at 300K of the best GaAs MESFETs and TEGFETs, ring oscillator being the test vehicle. We can note that GaAs MESFET delay time is approaching that of TEGFET but the figure of merit power-delay product is much better in the latter case (25-27). This figure of merit is a very important parameter in VLSI since the power dissipation can be a limiting factor.

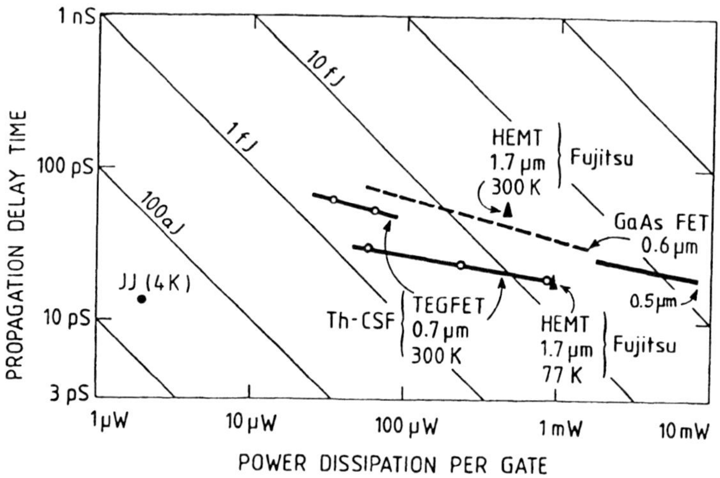

Fig. 9 Delay time versus procedures, per gate for TEGFETs

Table II State-of-the-art of GaAs MESFET AND TEGFET Ring Oscillators at 300K

	SOURCE	APPROACH	DELAY TIME (ps)	POWER DELAY Product (fj)
GaAs MESFET	HUGHES	BFL	34	1394
	ROCKWELL	SDFL	62	65
	THOMSON-CSF	LPFL	95	235
		DCFL(JFET)	150	60
	SONY	DCFL(JFET)	43	83
	NTT	DCFL	30	57
		DCFL	19.6	168.5
	NEC	DCFL	37.2	9.7
	SUMITOMO	DCFL	18.1	105
	BELL	DCFL	20.9	75
	CORNELL	DCFL	15	84
TEGFET	FUJITSU	DCFL	56	26
		-id-	16.7	334
	THOMSON-CSF	DCFL	18.4	16.6
		-id-	34	1.4
	ROCKWELL	DCFL	12.2	13.6

At low temperature (77K) the TEGFET performance increases by a factor 2 to 3, while that of GaAs MESFET is low.

More complicated and realistic circuits have recently been realized with TEGFETs: frequency dividers. Once more, TEGFETs are faster than GaAs MESFETs. With a gate length of $\sim$1.5μm TEGFET frequency divider already operates up at 5.5GHz (28) while the best conventional GaAs device with 0.6μm gate length works at the same frequency (29).

Who is sure that this improvement is only due to the enhancement of electron velocity in the 2DEG? Why the TEGFET consumes less power than GaAs FET?

CONCLUSION

There are many experimental evidences showing that the TEGFET is an excellent microwave transistor (low noise, high out-off frequency) and a fantastic high speed device for large scale integration.

The high electron mobility and velocity of the 2DEG do not explain all the obtained results. What else? Two dimensionality?

Avanti, Physicists!

ACKNOWLEDGMENTS

This work is supported by the DRET. The author would like to thank all of his co-workers for their contribution, CHS's Group at Lille University for stimulating discussions and MUDARES for communication of results.

REFERENCES

1. R. Dingle, H.L. Störmer, H.L. Gossard and W. Wiegmann; Appl. Phys. Lett. 33, pp. 665-667 Oct. 1978.
2. S. Hiyamizu 2nd Internat. Symp. MBE and CST Tokyo, Aug. 1982.
3. Nuyen T. Linh, German Physical Society Annual Meeting, Freudenstadt March 1983 to be published in Freudenstadt, March 1983. To be published in Festkörperprobleme XXIII.
4. D. Delagebeaudeuf, P. Delescluse, M. Laviron, P.N. Tung, J. Chaplart and N.T. Linh; 10th Internat. Symp. GaAs and Related Compounds, Albuquerque (USA) Sept. 1982. Inst. Phys. Conf. Ser n° 65 chap. 5 pp. 393-398.
5. T.J. Drummond, M. Keever, W. Kopp, H. Morkoc, K. Hess and B.G.Streetman; Electron. Lett. 17, 545 (1981).
6. T.J. Drummond, W.Kopp, H. Morkoc and M. Keever; Appl. Phys. Lett. 41, 277 (1982).

7. E. Constant "Modelling of Submicron Devices" Inst. Phys. Conf. Serv n° 57 pp. 141-168 (1981).
8. T.J. Maloney and J. Frey; J. Appl. Phys. 48, 781 (1977).
9. C. Cappy, C. Vernaeyen, A. Vanoverschelde, G. Salmer, D. Delagebeaudeuf, N.T. Linh and M. Laviron; GaAs IC Symp. New Orleans, Nov. 1982.
10. M.A.R. Al-Mudares and K.W.H. Foulds 8th European Specialist Workshop on "Active Microwave Semiconductor Devices" Maidenhead; May 1983.
11. T.J. Drummond, J. Klem, D. Arnold, R. Fischer, R.E. Thorne, W.G. Lyons and H. Morkog; Appl. Phys. Lett. 42, 615 (1983).
12. D. Delagebeaudeuf, P. Delescluse, P. Etienne, M. Laviron, J. Chaplart and N.T. Linh; Electron Lett. 16, pp. 667-668 Aug. 1980.
13, D. Delagebeaudeuf, M. Laviron, P. Delescure, P.N. Tung, J. Chaplart and N.T. Linh; Electron. Lett. 18, pp. 103-105, Jan. 1982.
14. D. Delagebeaudeuf and N.T. Linh; IEEE Trans. Electron Dev. Vol. ED-28, pp. 790-795, July 1981.
15. D. Delagebequdeuf and N.T. Linh; IEEE Trans. Electron Dev. Vol. ED-29, pp. 955-960, June 1982.
16. K. Lee, M. Shur, T.J. Drummond, H. Morkog; J. Appl. Phys. 54, 2093 (1983).
17. M. Laviron, D. Delagebequdeuf, P. Delescluse, J. Chaplart and N.T. Linh; Electron. Lett. 17, pp. 536-537, July 1981.
18. M. Laviron, D. Delagebequdeuf, P. Delescluse, P. Etienne, J. Chaplart and N.T. Linh; Appl. Phys. Lett. 40, pp. 530-532, March 1982.
19. M. Niori, T. Saito, S. Joshin and T. Mimura, presented at the ISSCC; Feb. 1983.
20. K. Joshin, T. Mimura, M. Niori, Y. Yamashita, K. Kosemura and J. Saito; MTT Boston 1983.
21. B. Carnez, A. Capyy, R. Fauquemberque, E. Constant and G. Salmer; IEEE Trans. Electron. Dev. ED-28, 784 (1981).
22. W. Baechtold; IEEE Trans. Ed-18, 97 (1971).
23. C.R. Brewitt-Taylor, P.N. Robson, J.E. Sitch; IEEE Proc. 127, 1 (1980).
24. D. Delagebeaudeuf and Nuyen T. Linh; unpublished.
25. L.H. Camitz, W.L. Jones, J.Lee, L.F. Eastman; 8th European Specialist Workshop on "Active Microwave Semiconductor Devices" Maidenhead, May 1983.
26. C.P. Lee, D.L. Miller, D. Hou and R.J. Anderson; 41st Annual Device Research Conf. Burlington, June 1983.
27. P.N. Tung, P. Delescluse and Nuyen T. Linh; unpublished.
28. K. Nishiuchi, T. Mimura, S. Kuroda, S. Hiyamizu, H. Nishi and M. Abe; 41st Annual Device Research Conf. Burlington, June 1983.
29. M. Cathelin, M. Gavant and M. Rocchi; IEE Proc. I. Solid-State & Electron Devices 127 270 (1980).

HOT ELECTRON TRANSFER AMPLIFIERS

Mordehai Heiblum and Marshall I. Nathan

I.B.M. Thomas J. Watson Research Center
Yorktown Heights, New York 10598

Abstract

The principles of operation of a hot electron transistor is presented. Device proposals and experimental results are reviewed.

Introduction

In 1960 Mead proposed a tunneling transistor which is potentially much faster than the existing bipolar transistor (MOMOM)[1]. Electrons tunnel between two metals (emitter and base) separated by an oxide, traverse the second metal (base) which is thin enough to avoid thermalization, and surmount a barrier created by a second oxide, to be collected by a third metal (collector). The first barrier supports the applied voltage needed to heat up the injected electrons while the second barrier prevents the equilibrium electrons of the base from flowing to the collector. Such a device is potentially very fast because the base is very thin so that transit time is short, and the metal layers have a high conductivity so that RC time constants can be small. The energy diagram shown in **figure 1** describes a general tunneling transistor. The poor quality of interfaces and materials led to a very small collection efficiency in the devices made by Mead[2]. In the sixties other hot electron devices were also studied, for example:

Metal-Oxide-Metal-Semiconductor (MOMS). Hot electrons generated by a tunneling MOM are collected by a reversely biased Schottky barrier[3]. Because of a lower collector barrier than in the MOMOM, the hot electrons in the metal base need not be very hot and suffer fewer collisions, resulting in an expected higher collection efficiency.

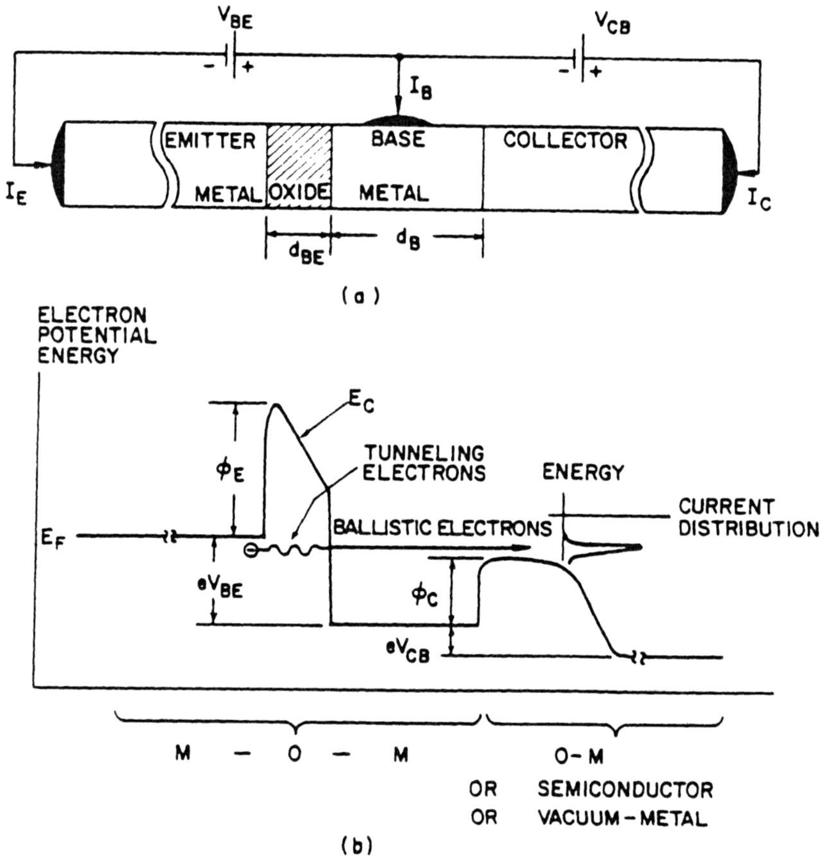

Figure 1. A general structure and band diagram of a hot electron transistor. While the emitter is a tunneling MOM the collector can be a Schottky barrier or another MOM.

Semiconductor-Metal-Semiconductor (SMS). This double Schottky transistor is commonly called the "Metal Base Transistor"[4]. Hot electrons are generated by thermal emission from the forward biased Schottky (emitter) and collected by the reversely biased Schottky (collector). The emitter semiconductor - metal interfaces could not be avoided. This structure had the highest collection efficiency among all hot electron devices at its time (α 0.5)

Since the early work of Mead and others materials, and device technologies have progressed dramatically. Techniques such as molecular beam epitaxy (MBE), organo-metallic chemical vapor depositor (OMCVD), ion implantation, and selective reactive ion etching have reached a mature enough stage to allow the fabrication of fast devices. In particular hot electron devices have become a reality. In this paper we shall discuss these devices and some proposals for others:

(a) The Heterojunction Hot Electron Transistor

(b) The CAMEL Transistor

(c) The Planar Barrier Transistor

(a) The Heterojunction Hot Electron Transistor

The modern work on hot electrons began with the realization that the device could be made totally in semiconductor material. The metallic-like regions can be heavily doped semiconductors. The barriers can be made either by doping with the opposite type impurity or by a heterojunction with a larger band gap material.

Several III-V compound semiconductors, in particular GaAs offer considerable advantage over Si for hot electron devices. First, the conduction band structure is more favorable. The central minimum (000) which is lowest, has a small effective mass, therefore, the electron mobility is high, and the mean free path is long. This will result in a larger collection efficiency for a given base width. This advantage of low mass and long mean free path occurs only over a limited energy range until the next higher minima (at (111)) become populated. This restricts the operating voltage range of the device and the usable energy barrier heights. The mean free path for the (111) minima is actually smaller than in Si because of the strength of polar optical mode scattering . The separation between the central minimum and that higher minima is about 0.3 eV. A second advantage of GaAs is the availability of MBE as a crystal growth technique. While MBE can hand has been done in Si, the situation is more favorable in GaAs because of the higher vapor pressures of the constituents, and the technique is far more developed. The advantage of MBE is that multilayer structures with precise impurity profiles can be obtained. Similar structures can be obtained. Similar structures can also be made by chemical vapor deposition, but to date much less work has been done.

This particular structure has been proposed but not yet demonstrated[5]. Here a barrier is formed by exploiting the band discontinuity between GaAs and AlGaAs. 85% of the discontinuity occurs in the conduction band, amounting to a barrier for electrons (ϕ),ϕ = 1.06x for x< 0.45, where x is the mole fraction of Al in $Al_xGa_{1-x}As$. For x > 0.45 the ALGaAs becomes an indirect material, and ϕ changes much more slowly reaching ≈0.6eV for x=1. Band discontinuity for an n-n heterojunction is shown in **figure 2(a),** for x=0.4. When an undoped AlGaAs is sufficiently thin and sandwiched between two lightly doped GaAs layers, a nearly square barrier prevents the flow of electrons between the two GaAs layers, as shown in **figure 2(b)** . When the GaAs layers are degenerate, the effective barrier for electrons depends on the location of E_F. Upon biasing with V_{BE} electrons can traverse the barrier via tunneling the barrier with excess kinetic energy (ΔE) relative to the thermal electrons in the second n+ GaAs layer (approximately eV_{BE} or ϕ_E) in tunneling or thermal emission cases, respectively). A second barrier (ϕ_C), with $\phi_C<\phi_E$ will prevent the collection of the thermal electrons of the common layer (base) but still allow the collection of the hot electrons, providing $(\Delta E)_x>\phi_C$, where $(\Delta E)_x$ is the normal to the barrier component of the excess kinetic energy of the hot electrons **figure 3(a).**

The traversing hot electron beam should be as monoenergetic as possible and suffer a minimum number of collisions in the base, by impurities, phonons, other electrons and plasmons. Electron-electron collisions are inelastic and can be minimized for eV_{BE}, while

energy beam spreading is minimal for $eV_{BE} \simeq \phi_E$, as shown in **figure 4.** Those curves are taken from reference 5, and represent tunneling in MOM; We expect qualitatively similar results. Phonon and impurity scattering cause the hot electrons to deviate only slightly from their original direction after each collision. Plasmon scattering is more controversial and some more proof of its existence has to be supplied yet[6]. It appears that an emitter barrier of (0.3 - 0.4)eV, and a barrier thickness of (100 - 200) Å is optimum. To minimize quantum mechanical reflections from the collector barrier, ϕ_C has

to be substantially lower than $eV_{BE} \simeq \phi_E$ and its onset from the base has to be slightly graded (a step - function will generate relatively large reflections). The number of states and traps in the AlGaAs - GaAs interface is presumed to be small from the high mobilities which are measured in selectively doped structures, but traps

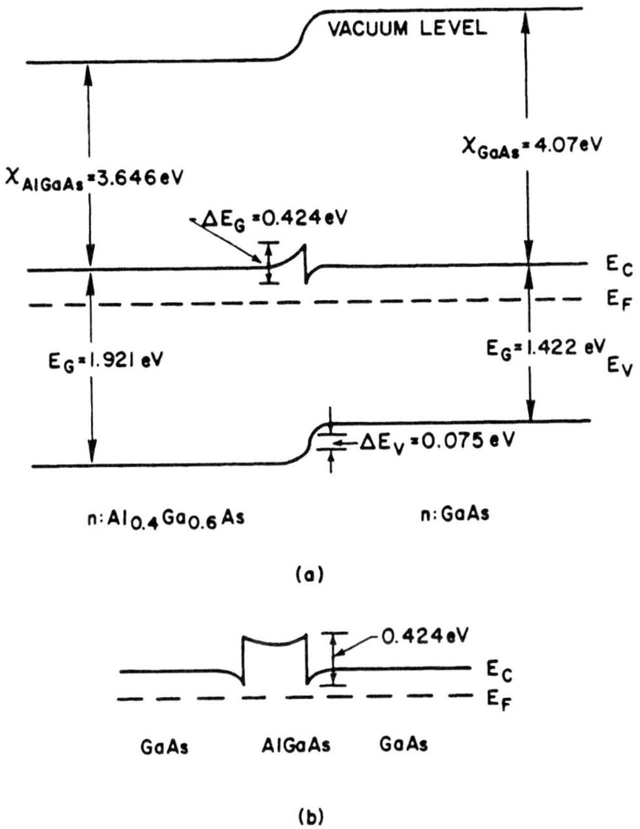

Figure 2. Heterojunctions of n:GaAs-nAlGaAs.

(a) A junction between GaAs and $Al_{0.4}Ga_{0.6}As$, exhibiting a barrier for electrons of 0.42cV.

(b) A thin AlGaAs barrier (few hundred Angstroms) sandwiched between two GaAs layers creating a square barrier for electrons.

which are charged negatively are presumed to exist in the AlGaAs with the quality available at present[7]. If ϕ_C is chosen~0.25eV, and the AlGaAs barrier is free of trapped charges, its thickness could be chosen large enough as to climinate thermal electrons in the base from reaching the collector, but still allow all the hot electrons, which surmount ϕ_C at the base-collector interface, to be collected under a slight V_{CB} bias, as shown in figure 3(a).

But if immobile, negative charges exist in the AlGaAs barrier, the barrier maximum is displaced toward the center of the barrier and for thick enough barriers, many of the hot electrons will thermalize and turn back toward the base **figure 3(b).** In this case the barrier should be thinner, lower x should be chosen and a larger V_{CB} will be necessary.

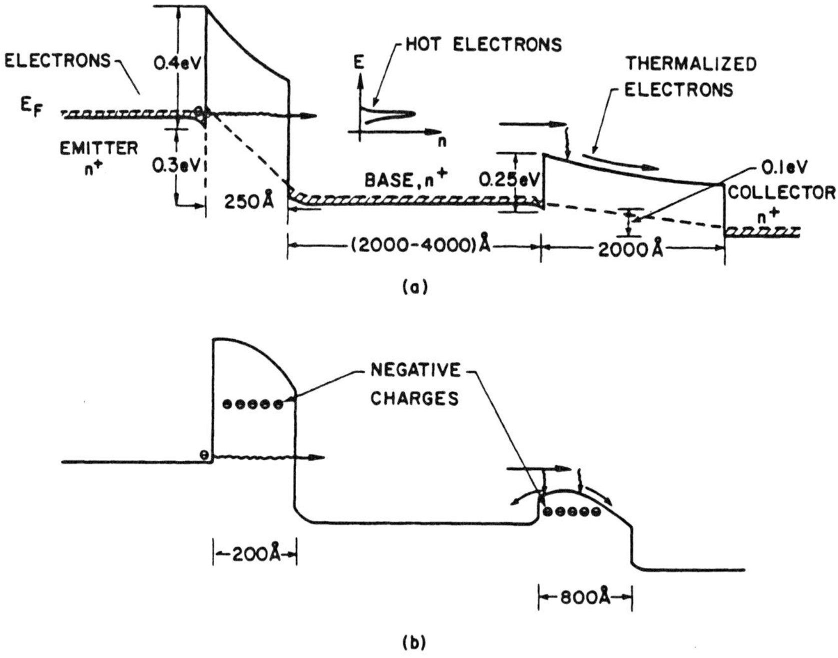

Figure 3. A double heterojunction transistor.

(a) Emitter barrier is formed by x≃0.4, the collectors by x≃0.25. Bias $V_{BE}>\Phi C$ is applied, and the dimensions present a reasonable guess for optimum values.

(b) A similar structure with trapped negative changes in the AlGaAs.

Even though the transit time through a thin base (≤1000Å) is very short(~1ps), the base spreading resistance will limit the high frequency performance of the device, and of course will be more critical for small transfer ratios (α). This factor and processing limitations will put a lower limit on the base thickness. We note for comparison purposes, that in the MOMOM with a 100Å base thickness, the best sheet resistance might be 5Ω/□ ,

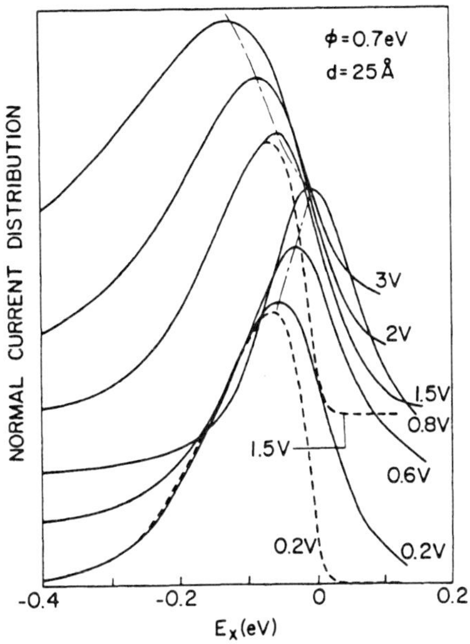

Figure 4. A normal energy distribution of tunneling electrons between two metals sparated by an oxide with 0.7eV barrier height and 25Å thick. The narrowest energy distribution is achieved for a biasing voltage close to Φ, which is the onset of the Fowler-Nordheim region. At this condition the energy peak is also the least displaced from the Fermi-energy.

while in a heavily doped GaAs, best base sheet resistance might be 50Ω/□ for a 1000Å base, which is still much better than any available Si or GaAs bipolar devices. The ionized donors in the base could be eliminated by doping the AlGaAs barriers, and the spill-over electrons would "modulation dope" the base. Under these conditions the base resistance will be very low at low base currents (due to the very high mobility), and degrade at higher base currents.

It is difficult to predict at this point the α of the device due to many loss mechanisms (many of them are material quality related): (a) initial energy spread of the tunneling electrons (can be ≈50meV at low temperatures with a degenerate emitter, due to the small $E_F - E_C$), (b) scattering in the base, (c) QM reflections from base-collector interface, (d) trapping at this interface, and (e) scattering in the barrier collector. From the maximum α of the metal base transistor reported earlier[8], (α≅0.5) we may hope to obtain α's of at least 0.8.

It is not clear at the moment which emitter is superior, tunneling or thermal. The tunneling emitter is faster but has stricter tolerance, only the reduction to practice will favor one of them.

We'll turn now to describe briefly the two "modern" hot electron devices which had been fabricated as of today: the camel and the planar doped barrier transistors, in both the barriers were realized by doping rather than by heterojunctions.

(b) The Camel Transistor

In 1979 Shannon[9] reported realization of a majority carrier camel diode. It consists of an n-layer followed by a thin p+ followed by an n+ layer. The potential diagram is shown in **figure 5.** The doping thickness product in the p+ region is not sufficiently large to give rise to free holes, but it does present a barrier to electrons. The current across the barrier is due mainly to thermionic emission of electrons. It has an asymmetric rectification characteristic because of the unequal doping on the two sides of the barrier as indicated the figure.

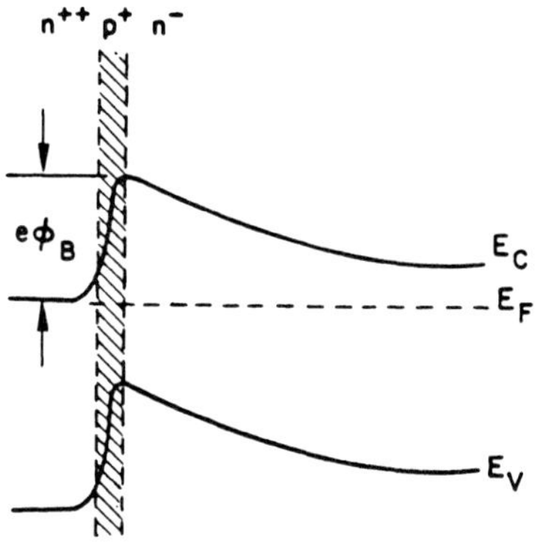

Figure 5. Energy band diagram of a camel diode.

Shannon fabricated camel diodes by implanting boron or indium and arsenic, or antimony into silicon. He found ideality factors betweel.1 and 1.5.

It is apparent that by making two barriers in series in close proximity, one can make a double barrier structure similar to figure 1. Shannon, in his initial efforts to make a hot electron camel transistor, fabricated only a single camel barrier and used a Schottky contact to the n+ region for the emitter[10], as shown in **figure 6.** He used arsenic and boron to form the camel diode, and estimated that the depth of the collector (by employing shallow implantation)., was 50Å below the surface. The fabricated device which is shown in **figure 7** had a best value of β (ratio of collector to base current) of 3.

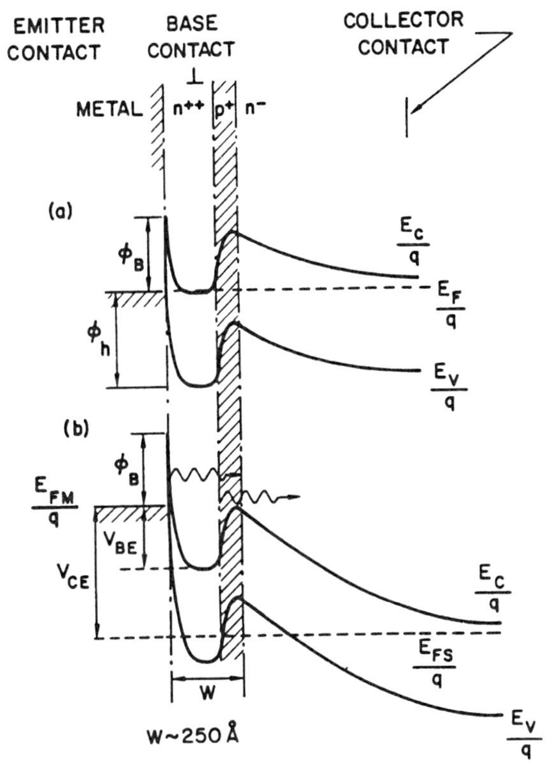

Figure 6. Energy band diagram of a camel transistor with a Schottky emitter barrier.

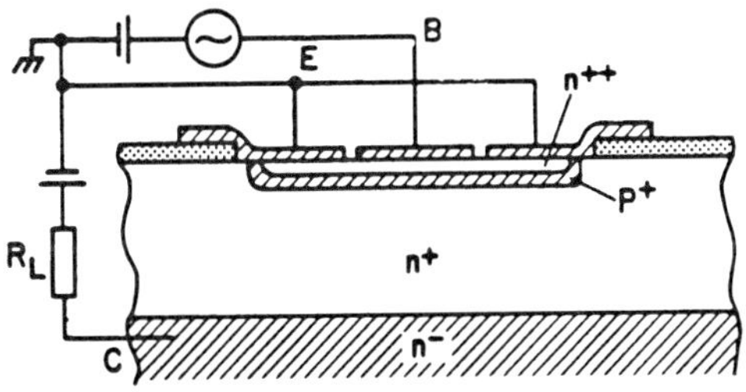

Figure 7. Layout of a camel transistor.

Shannon and Gill realized that they could improve device performance by using a second camel barrier for the emitter, which would improve emitter efficiency by reducing leakage, and in addition would reduce emitter capacitance, and thus improve the high frequency performance[11]. The proposed structures are shown in **figure 8.** The structure in **figure 8a** has half a camel diode for the emitter, where a p+ region enhances the barrier of the Schottky. This gives the benefit of a reduced leakage current but not a reduced capacitance. Because of the difficulties and inflexibility of ion implantation, Shannon was able to fabricate only the structure of figure 8a, by using two implantations of boron. Despite the fact that the profile of impurity density is not ideal, he obtained a β of 19 over a limited emitter voltage range.

Reference 11 points out that parasitic resistances could give rise to a spuriously high values to β, which the authors attempted to climinate, however, it is not clear yet that they have no effect. Thus the real β's may be somewhat lower than Shannon and Gill reported.

(c) Planar Doped Barrier Transistor

Malik et al[12] fabricated and studied a hot electron barrier, which they called a planar doped barier, by MBE of GaAs. Subsequently Malik and coworkers[13] fabricated a planar barrier transistor, which is structured like that in **figure 8b.** The planar doped barrier actually uses a somewhat more elaborate doping profile than the camel barrier as shown in **figure 9.** The p+ region is sandwiched between two undoped regions of different lengths to give the asymmetry to the I-V charcteristics. While MBE gives doping flexibility, it does not permit any masking steps to be performed flexibility, it does not permit any masking steps to be performed between the deposition of the layers. Hence fabrication of the device is considerably more difficult, requiring the precise etching of their layers so that contact to the base can be performed. Moreover, the only proven ohmic contact to an n-typed GaAs is made by alloying Au Ge eutectic mixture with GaAs. Care must be taken to ensure that the alloying is not done too deeply thus shorting the base and collector. Initially Malik and coworkers reported a β of 30 at 77°K for a device with a base thickness of 1000Å.

Recently, more results on the planar barrier doped transistor have been obtained by Hollis, Palmateer, Maki, and Eastman[6]. Hollis reported that the values of β obtained by Malik et al,[13] were spurious and due to the parasitic resistance effects which were alluded to earlier Palmateer, and coworkers reported that

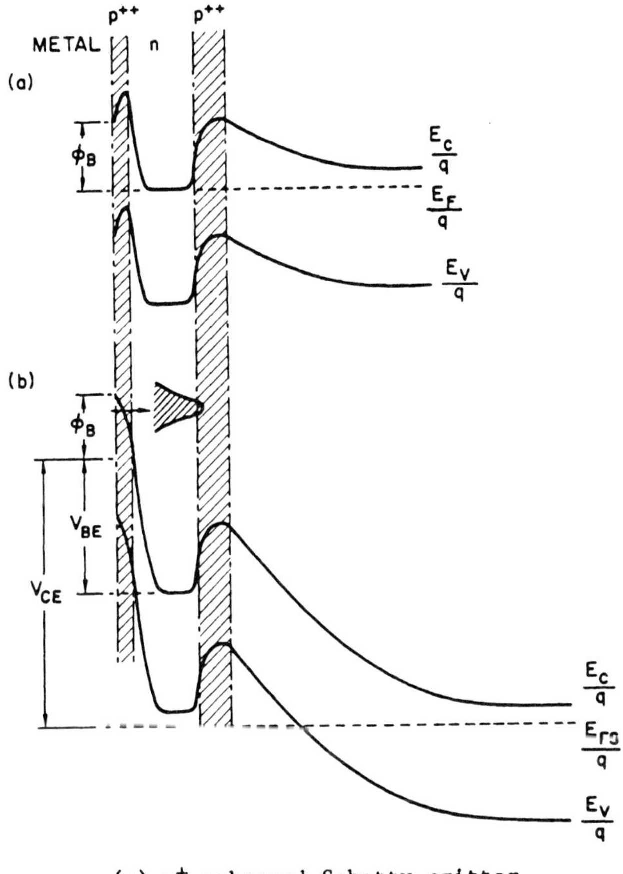

(a) p^+ enhanced Schotty emitter.

Figure 8(a,b). Energy band diagrams of camel transistors.

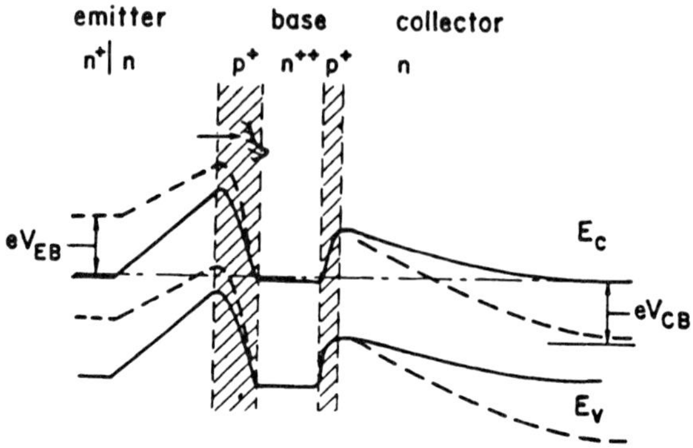

8(b) Double camel barrier transistor.

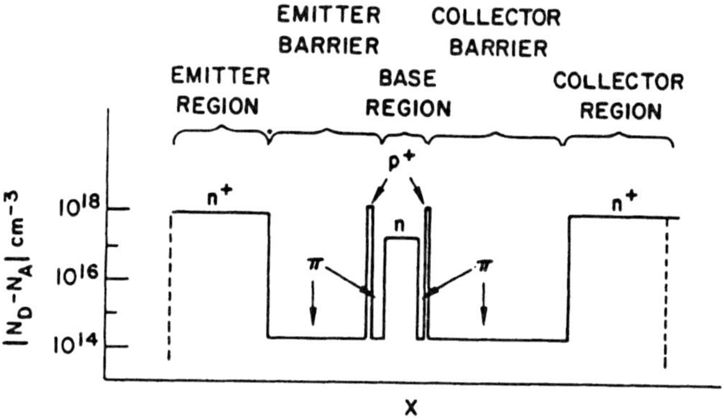

Figure 9. Doping vs. distance of a planar transistor.

unreproducible barriers were obtained but these effects can be eliminated by heat treating the substrates[14]. The highest value of β the Cornell group was able to achieve was 4.5. Hollis suggested that an electron scattering mechanism previously not considered, plasmon scattering, is responsible for the low values of β, however, it seems that more work is needed to verify this point.

REFERENCES

1. C.A. Mead, Proc IRE 48, 359 (1960), J. Appl. Phys. 32, 646 (1961).

2. C.A. Mead, J. Appl. Phys. 32, 646 (1961).

3. J.P. Spratt, R.F. Schwarz, and W.M. Kane, Phys. Rev. Lett., 6, 341 (1961).

4. D.V. Geppert, Proc. IRE 50, 1527 (1962), D. Kahng, Proc. IRE 50. 1534 (1962).

5. M. Heiblum, Solid St. Electron. 24, 343 (1980).

6. M. Hollis, Ph.D. thesis, Cornell University , E.E. Department (1983).

7. P.M. Solomon, T.W. Hickmott, H. Morkoc, and R. Fischer, Appl. Phys. Lett. 42, 821 (1983).

8. S.M. Sze, C.R. Crowel, G.P. Carey, and E.E.>LaBate, J. Appl. Phys. 37, 2690 (1966).

9. J.M. Shannon, Appl. Phys. Lett. 35, 63 (1979).

10. J.M. Shannon, Solid St. Elect. Device, 3, 142 (1979).

11. J.M. Shannon and A. Gill, Elect. Lett. 17, 621 (1981).

12. R.J. Malik, K. Board, L.F. Eastman, C.E.C. Wood, T.R. AuCoin and R. J. Ross, Gallium Arsenide and Related Compounds, 1980, Inst. Phys. Conf. Ser. No. 56: Ch. 9, p. 697.

13. R.J. Malik, M.A. Hollis, L.F. Eastman, D.w. Woodward, C.E.C. Wood and T.R. AuCoin, Proc. COnf. on Active Microwave Devices, Cornell Univ. (1981).

14. S.C. Palmateer, P.A. Maki, M.A. Hollis, L.F. Eastman, C. Hitzman and I. Ward, to be published. Referred to in Ref. 6.

NEW GRADED BAND GAP AND SUPERLATTICE STRUCTURES AND THEIR APPLICATIONS TO PHOTODETECTORS, BIPOLAR TRANSISTORS AND HIGH-SPEED DEVICES

F. Capasso

Bell Laboratories
Murray Hill
New Jersey 07974

INTRODUCTION

This paper reviews recent device applications of compositionally graded materials. Band gap grading represents a powerful tool to engineer the energy band diagram of a device and thus to modify the transport properties (band-gap engineering)[1]. Ultrahigh speed phototransistors and high gain transistors with graded gap base have been reported. Heterojunction bipolar transistors with graded emitter-base interface have also been studied in the AlGaAs/GaAs and AlInGaAs material systems; parabolic grading eliminates the collector-emitter offset voltage and maximizes the injection efficiency. Grading of the high field region and superlattices in an avalanche photodiode have been used to enhance the ionization rates ratio. Multilayer graded gap structures can be used to implement the solid state analog of a photomultiplier and repeated velocity overshoot devices. Finally a new interdigitated pn junction device (channeling diode) is discussed. This structure has many possible applications such as a low capacitance large sensitive area photodiode, low noise avalanche detector and a new type of varactor.

HIGH-SPEED GRADED BASE PHOTOTRANSISTORS

The idea of introducing a quasi-electric field in the base of a transistor to reduce the transit time of minority carriers was introduced by Kroemer in 1957[2]. This effect was demonstrated by Capasso et al.[3] in a wide gap emitter phototransistor with a graded base having response time $t_r \lesssim 20$ps and by Levine et al[4]. via a measurement of the velocity of electrons in a graded gap p^+ AlGaAs layer. These authors reported a velocity as high as 1.8×10^7 cm/sec in a quasi-field of $\simeq 8.8\times10^3$V/cm for a 0.4μm thick layer. They find that the velocity can be approximated by the relationship

$$v = \mu \varepsilon \quad (1)$$

where μ is the ambipolar mobility in comparably doped GaAs and ε the quasi-electric field[4]. Equation 1 is only valid for base transit times greater than the momentum relaxion time.

The phototransistor structure was grown by molecular beam epitaxy on a Si-doped ($\simeq 4\times10^{18}cm^{-3}$) n^+-GaAs substrate. A buffer layer of n^+-GaAs was subsequently grown followed by a Sn-doped, n-type ($\simeq 10^{15}/cm^3$) 1.5μm-thick-GaAs collector layer. The 0.45μm thick base layer was compositionally graded from GaAs (on the collector layer side) to $Al_{0.20}Ga_{0.80}As$ (E_g=1.8eV) and heavily doped with Be($p^+ \cong 5\times10^{18}/cm^3$). The abrupt wide gap emitter consists of an $Al_{0.45}Ga_{0.55}As$ (E_g=2.0eV) 1.5μm-thick window layer n doped with Sn in the range $\cong 2\times10^{15}$-$5\times10^{15}/cm^3$.

The total capacitance at zero bias measured at 1 MHz is ≅0.15-0.25pF; the device area is in the $(5\times10^{-5}$-$1\times10^{-4})cm^2$ range.

Figure 1 shows the energy band diagram of the phototransistor under bias. To study the effect of grading in the base on the speed of the device, 4 ps wide laser pulses were used. The wavelength (λ=6400Å) was chosen so that the light is absorbed only in the base layer. The incident power was kept relatively high (100 mW) to minimize the effective emitter charging time. Under these conditions the speed limiting factors are the RC time constant and the base transit time. Figure 2(a) and (b) show the pulse response of the device mounted in a microwave stripline package as monitored by a fast sampling scope. In Fig. 2(b) the response was signal averaged. Note the symmetric rise and fall time and the absence of long tails; this is normally very difficult to achieve in picosecond photodetectors. From the observed 10-90% response time of 30 ps a sum of the squares approximation was used to estimate an intrinsic detector response time of ≅20ps. In absence of quasi-electric field in the heavily doped p^+ base, a broadening of the response due to slow diffusion processes is expected. The diffusion time t_D is given by W^2/D, where W is the base thickness and D is the diffusion coefficient. For a phototransistor with a GaAs $p+=10^{18}/cm^3$ base $D \cong 16cm^2/s$. In our structure D is likely to be significantly smaller because of the AlGaAs graded base, which has a lower mobility than GaAs and the higher doping. Thus, for our structure t_D > 100 ps. In conclusion, the fact that no pulse broadening is observed indicates that the quasi-electric field in the base sweeps out the electrons in a time small compared with the diffusion time. This is consistent with the prediction by Kroemer.[2]

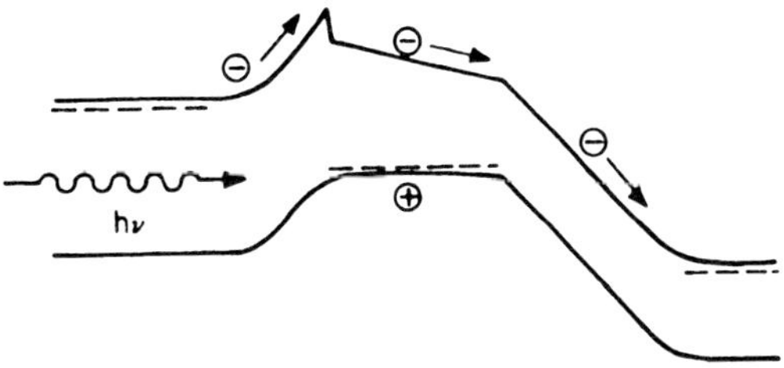

Figure 1. Energy band diagram of graded base phototransistor under operating conditions.

Finally the combination of the graded gap base with the abrupt wide gap emitter (Fig. 1) suggests a new high-speed ballistic transistor. In fact, the conduction band discontinuity can be used to ballistically launch electrons into the base with an initial velocity $=10^8$cm/s; the quasifield in the base will maintain an average velocity substantially higher than 10^7 cm/s[3].

BIPOLAR TRANSISTORS WITH GRADED GAP BASE

Recently, a bipolar transistor with compositionally graded base has also been reported[5]. The device grown by MBE incorporated a wide gap $Al_{0.35}Ga_{0.65}As$ emitter ($n=2x10^{16}/cm^3$) and a 0.4 m thick p^+ ($=2x10^{18}/cm^3$) base graded from $Al_{0.20}Ga_{0.80}As$ to GaAs. DC current gain of 35 with flat, nearly ideal, collector characteristics are observed. Incorporation of a graded gap base gives much faster base transit times due to the induced quasi-electric field for electrons, thus allowing a precious tradeoff against the base resistance.

Figure 3 shows the energy band diagram of the transistor at zero bias. The grading of the emitter-base interface removes a large part of the conduction band spike allowing most of the band-gap difference to fall across the valence band blocking the unwanted injection of holes from the base[6].

It can easily be shown that the base transit time in a graded base transistor, compared to a transistor with nongraded base and diffusion limited base transit time is shortened by the factor.

$$\frac{t_b}{t'_b} = \frac{E_{g1}-E_{g2}}{2KT} \qquad (2)$$

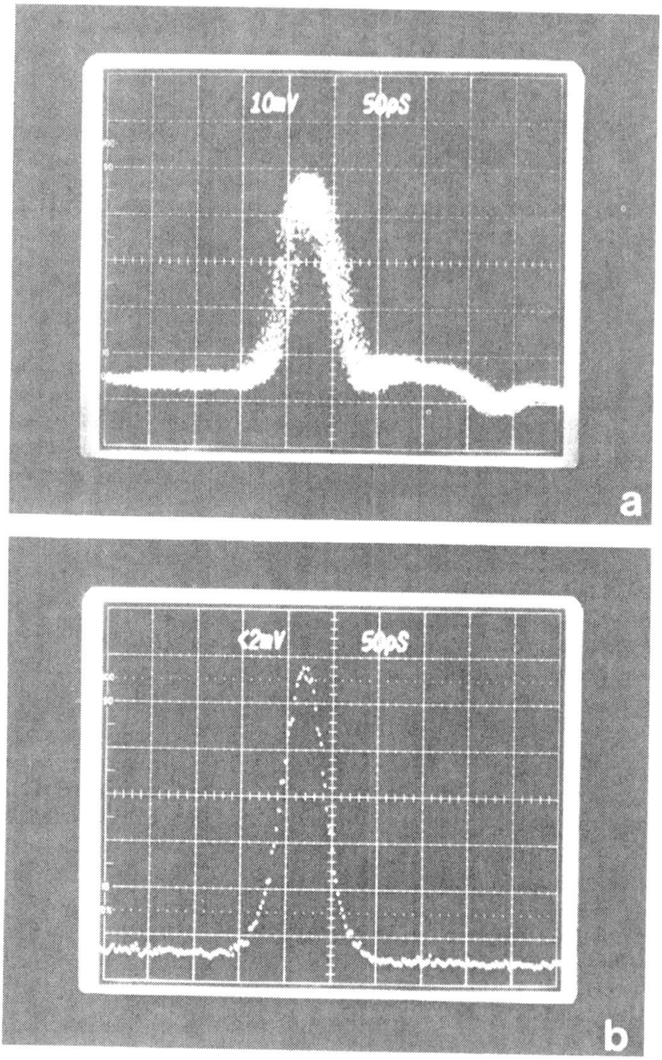

Figure 2. (a) Pulse response of a typical graded gap phototransistor operated at zero bias, to a 5-ps laser (λ =6200Å).
(b) Pulse response of another unit at lower incident power. The pulse from the S4 sampling head was signal averaged.

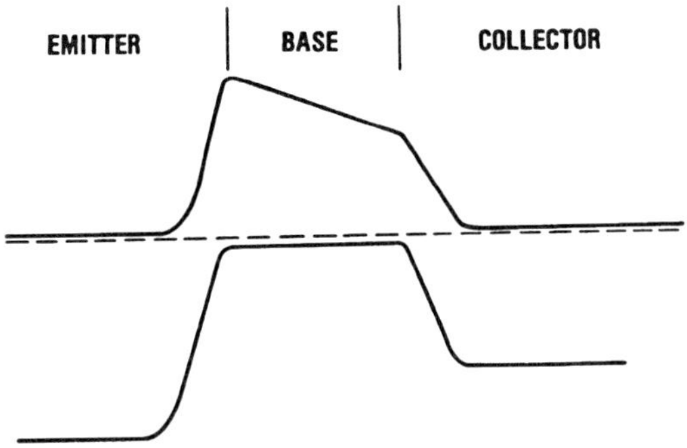

Figure 3. Energy band diagram of a bipolar transistor with compositionally graded base and wide gap emitter at zero bias.

where t_b and t_b' are the base transit times for an ungraded and graded transistor respectively. Although Eq. 2 is rigorous only in the limit $E_{g1}-E_{g2}>>kT$, it can be employed as a useful rule of thumb in cases where $E_{g1}-E_{g2}$ is several times kT. Thus the band-gap difference must be made as large as possible without exceeding the intervalley energy separation ($\Delta E_{\Gamma L}$ of the material with gap E_{g2} which would result in a strong reduction of the electron velocity and in the nonvalidity of Eq.1). Indeed the high velocity reported in Ref. 5 was measured in a sample where $E_{g1}-E_{g2} \lesssim \Delta E_{\Gamma L}$ and was interpreted as due to central valley dominated transport. This is characterized by a lower effective mass than the satellite valleys and hence a higher velocity. Using $E_{g1}-E_{g2}=0.2eV$ as in the transistor, the transit time is reduced by a factor of ≅4 at 300 K over a bipolar with an ungraded base of the same thickness. This allows a precious tradeoff against the base resistance (R_b) making possible an increase of the base thickness and a consequent reduction of R_b, while still keeping a reasonable base transit time. For the present transistor, one can keep the same transit time of an ungraded transistor but with twice the base thickness, while reducing the base resistance by a factor of two. Reduction of the base resistance in turn will reduce the switching time and increase the maximum oscillation frequency of the transistor.

The transistor of Ref. 5 was a test structure with a relatively large emitter area (≅$10^{-4}cm^2$). For operation at microwave frequencies the emitter area must be reduced. Recently, we have been fabricating graded base bipolars with smaller emitter area ($2x10^{-6}cm^2$) designed to operate at high frequency (f_T≅20GHz)[7]. A schematic diagram of the structure is shown in Fig. 4. The base was graded from $Al_{0.20}Ga_{0.80}As$ to GaAs over 2700Å corresponding to a quasi-electric field of ≅6kV/cm and a base transit time of ≅2ps. The dc common emitter characteristics(Fig. 5) indicate a high current gain (β≳150). The apparent negative resistance is a consequence of heating due to the high current[5].

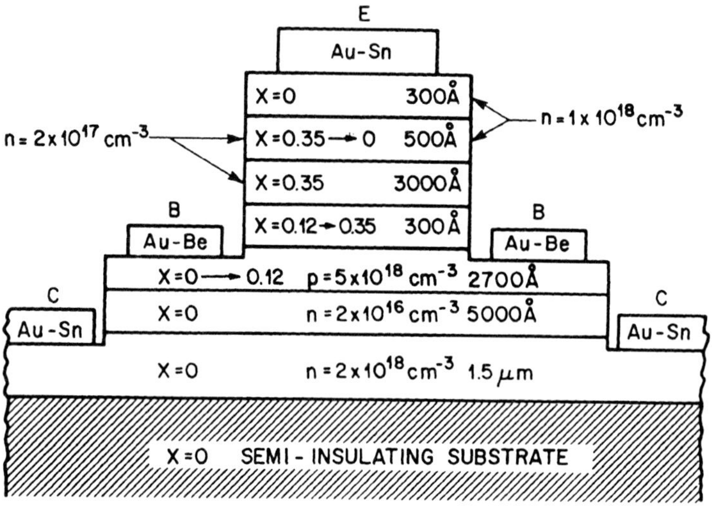

Figure 4. Schematic diagram of the layers comprising the graded base transistor.

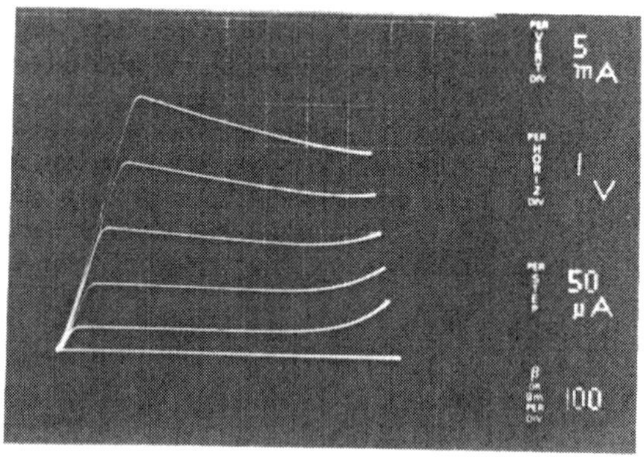

Figure 5. Common emitter characteristics of a typical graded base heterojunction bipolar transistor.

Recently, Miller et al.[8] have reported a bipolar transistor with an 800Å thick base compositionally graded from $Al_{0.025}Ga_{0.925}As$ to GaAs having a cutoff frequency of 16 GHz.

EMITTER GRADING IN HETEROJUNCTION BIPOLAR TRANSISTORS

The essential feature of the heterojunction bipolar transistor relies upon a wide band gap emitter wherein part of the energy band gap difference between the emitter and base is used to suppress hole injection. This allows the base to be more heavily doped than the emitter leading to a low base resistance and emitter-base capacitance both of which are necessary for high frequency operation, while still maintaining a high emitter injection efficiency[6].

Most of the work on MBE-grown heterojunction bipolar transistors has concentrated on the AlGaAs/GaAs system. Recently the first verticle Npn $Al_{0.48}In_{0.52}As$ / $Ga_{0.47}In_{0.53}As$ heterojunction bipolar transistors grown by MBE with high current gain have been reported[9].

The (Al,In)As/(Ga,In)As layers were grown by molecular beam epitaxy (MBE) lattice-matched to a Fe-doped semi-insulating InP substrate. Two HJBT's structures were grown; the first with an abrupt emitter of $Al_{0.48}In_{0.52}As$ on a $Ga_{0.47}In_{0.53}As$ base, and a second with a graded emitter comprised of a quaternary layer of ACluGaAs of width 600Å linearly graded between the two ternary layers. Grading from $Ga_{0.47}In_{0.52}As$ to $Al_{0.48}In_{0.52}As$ was achieved by simultaneously lowering the Ga and raising the Al oven temperatures in such a manner as to keep the total Group III flux constant during the transition.

It should be noted that this is the first use of a graded quaternary alloy in a device structure. The energy band diagram for the abrupt and graded transistors are shown in Figure 6a and 6b respectively. It is seen that the effect of the grading is to eliminate the conduction band notch in the emitter junction. This in turn leads to a larger emitter base valence band difference under forward bias injection. The following material parameters were used in both types of transistors. The $Al_{0.48}In_{0.52}As$ emitter and $Ga_{0.47}In_{0.53}As$ collector were doped n-type with Sn at levels of $5x10^{17}cm^{-3}$ and $5x10^{16}cm^{-3}$, respectively. The $Ga_{0.47}In_{0.53}As$ base was doped p-type with Be to a level of $5x10^{18}cm^{-3}$. A 1μm thick buffer layer of $Ga_{0.47}$ $I_{n\ 0.53}$ As doped to $2x10^{18}/cm^3$ was used to contact the collector and 1000Å of the same materials was used to cap the emitter. The base and collector widths were 2500Å and 5000Å, respectively.

The common emitter characteristics for the abrupt and graded emitter transistors are shown in Figs. 7a and 7b, respectively.

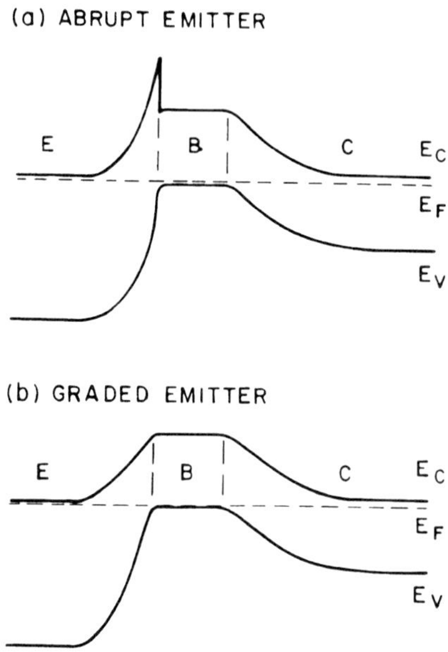

Figure 6. Energy band diagrams under equilibrium for the $Al_{0.48}In_{0.52}As/Ga_{0.47}In_{0.53}As$ heterojunction bipolar transistor with (a) abrupt emitter and (b) graded emitter. Note the elimination of the conduction band notch through the use of a graded emitter.

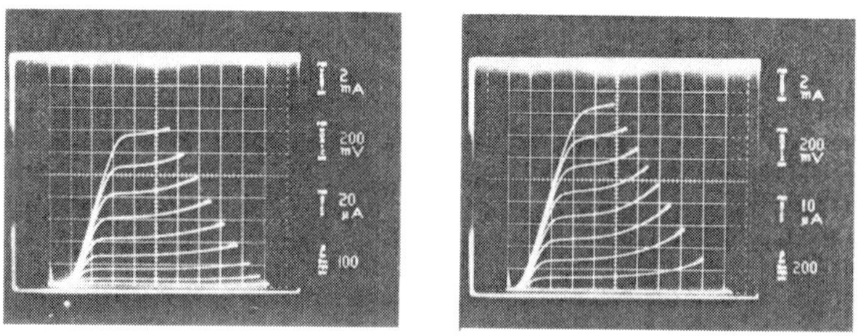

Figure 7. Common-emitter characteristics of the $Al_{0.48}In_{0.52}As/Ga_{0.47}In_{0.53}As$ heterojunction bipolar transistor with (a) abrupt emitter and (b) graded emitter at $300^{o}K$.

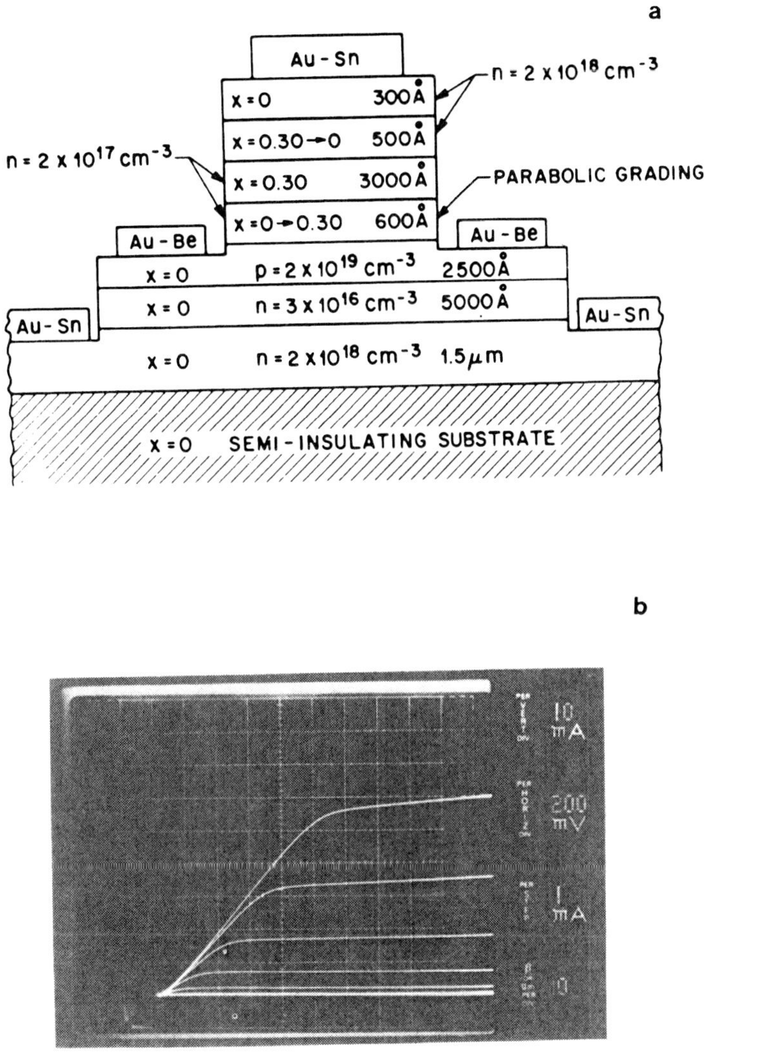

Figure 8a. Schematic diagram of bipolar transistor that has a parabolic grading width of 600Å at the emitter/base junction.

8b. Common emitter characteristics of the transistor shown in Figure 8a. Note the neglibible offset voltage.

It can be seen that the current gain β increases with increasing collector current and the maximum β obtained at a current level of about 15 mA is 140 for the abrubt emitter and 280 for the graded emitter. These high current gains are comparable to some of the best values obtained in (Al,Ga)As/GaAs HJBT's grown by MBE and demonstrates that this material and especially the $Al_{0.48}In_{0.52}As$ and graded $Ga_{0.47-x}Al_xIn_{0.53}As$ must be of high quality.

Note that a current gain increase by a factor of 2 is achieved through the use of the graded gap emitter which is attributed to a valence band barrier between the emitter and base under forward bias injection. The band-discontinuities at the $Al_{0.48}In_{0.53}As/Ga_{0.47}In_{0.53}As$ heterojunction are $\Delta E_c = 0.50$ eV and $\Delta E_v = 0.20$ eV.[10]

It is apparent from Fig. 7a and b that there is a relatively large collector-emitter offset voltage. This voltage is equal to the difference between the built-in potential for the emitter-base pn junction and that of the base-collector pn junction. No such offset is therefore present in homojunction Si bipolars.

It has recently been shown that by appropriately grading the emitter near the interface with the base such offset can be reducec and even totally eliminated[11]. The other advantage of grading the emitter is, of course, that the potential spike in the conduction band can be reduced, thus increasing the injection efficiency. The conduction band potential is the result of the sum of two potentials: the electrostatic potential Φ (The electrostatic potential is equal to V_{bi} (the built-in potential) - V_{be} the base/emitter voltage), which varies parabolically and the grading potential Φ . If linear grading is used there is always unwanted structure in the conduction band (spikes or notches). It now becomes obvious that any structure can be eliminated by grading with the complementary function of the electrostatic potential of a $Ga_{0.7}Al_{0.3}As$ P^+n junction $(1-\Phi_{es})$ over the depletion layer width at a forward bias equivalent to the base.

An HBT with such a parabolic grading has been favricated, using MBE with a $Ga_{0.7}Al_{0.3}As$ emitter and a GaAs base and collector. A schematic diagram of the transistor structure and the common emitter characteristic is shown in Figure 8a and 8b respectively. The emitter/base junction was graded from x=0 to x=0.3 on the emitter side over a distance of 600Å, the parabolic grading function being approximated by linear grading over nine regions. It can be seen from the characteristics shown in Figure 8b that the offset is very small about 0.03V. Virtually identical characteristics with offsets $\leq$0.03V were obtained for all devices on the wafer.

SUPERLATTICE AVALANCHE DETECTORS

We have demonstrated experimentally that in a superlattice (here $Al_{0.45}Ga_{0.55}As/GaAs$) the effective impact ionization rates for electrons and holes are very different (α (/β=8) even if they are the same in the basic bulk materials (in GaAs α (=β).[12]

To understand the superlattice APD consider the energy band diagram in Figure 9. Because of the very low doped material, the field is constant across the 2.5μm-long depletion layer. This consists of 50 alternationg GaAs (450Å) and AlGaAs (550Å) layers grown by MBE. For illustrative purposes, assume a field ε =2.7x10 V/cm; at this value a sizeable multiplication was observed.

Consider now, a hot electron accelerating in an AlGaAs barrier layer. When it enters the GaAs well, it abruptly gains an energy equal to the conduction band edge discontinuity ΔE_c=0.36eV. The effect is that the electron "sees" an ionization energy (E_{th} $\cong$1.6eV) reduced by ΔE_c with respect to the threshold energy in bulk GaAs (E_{th}=2.0eV). Since the impact ionization rate α increases exponentially with decreasing E_{th}, a large increase in the effective α with respect to bulk Gas is expected.

In contrast, the hole ionization rate β is not increased as much because the reduction in the hole ionization energy is smaller, i.e. ΔE_v=0.22e. The net results are in good agreement with the theory of this effect[13] and give α/β $\cong$7.

One potential problem in the superlattice APD is the trapping of electrons in the well. This can be minimized if the average carrier energy in the well is greater than ΔE_c. Both of these conditions are satisfied in the AlGaAs/Gas superlattice APD previously discussed. For large band edge discontinuities, this problem can be avoided by inserting a compositionally graded layer between the well and the barrier layers. The resulting bandgap grading smoothens out the barrier (Figure 10).

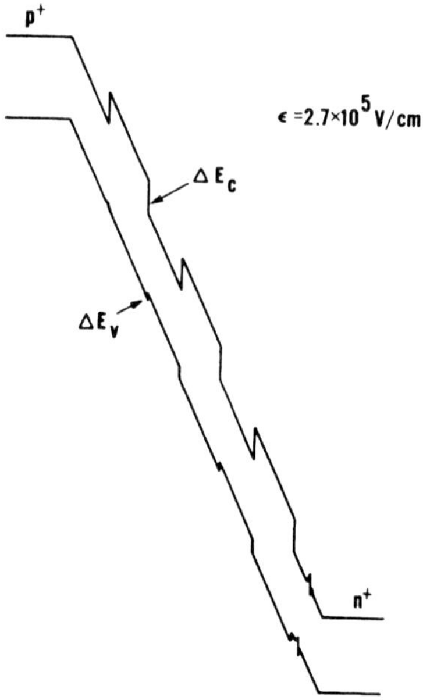

Figure 9. Energy band diagram of the $Al_{0.45}Ga_{0.55}$ As/ GaAs superlattice (ΔE_c=o.48eV, ΔE_v=0.08eV).

GRADED-GAP AVALANCHE DETECTOR

A novel device, the graded band-gap APD has been demonstrated.[14] The structure, grown by MBE, consists of a nominally intrinsic region, graded from GaAs to $Al_{0.45}Ga_{0.55}As$ (E_g =2.0eV) and sandwhiched between n^+ and p^+ regions. We have demonstrated that the effective α/β ratio of this structure is greatly increased over 1 (α/β 5-7.5), if the width of the graded layer is 0.5μm. The principle of the device is illustrated in Figure 11. It is seen that electrons experience a higher effective electric field and lower ionization energy than holes, since they are moving towards lower gap regions. The effect is a significant increase in the α/β ratio, when the grading exceeds 1 eV / μm. Another important property of this structure is the soft breakdown, which can lead to much greater gain stability compared to ungraded diodes.[14]

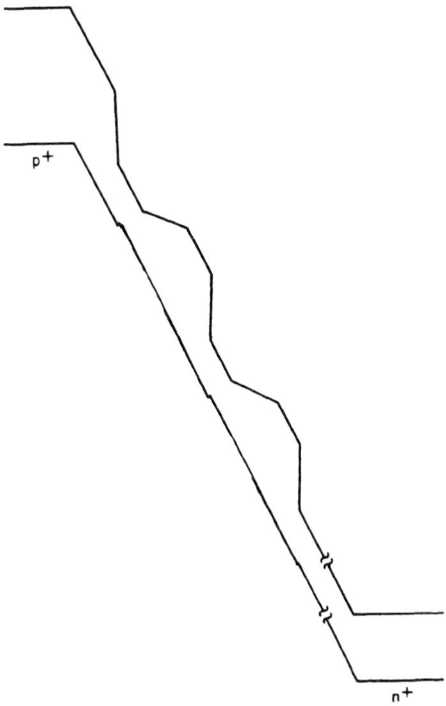

Figure 10. Energy band diagram of superlattice APD with graded sections to avoid electron trapping effects in the wells.

STAIRCASE STRUCTURES

Recently, Capasso, et al. introduced the concept of staircase potential. This innovative structure has several interesting applications. We shall concentrate on the staircase avalanche photodiode[15-19] and on the repeated velocity overshoot device.[20]

Figure 12a shows the band diagram of the graded-gap multilayer material (assumed intrinsic) at zero applied field. Each stage is linearly graded in composition from a low (E_{g1}) to a high (E_{g2}) band gap, with an abrupt step back to low band-gap material. The

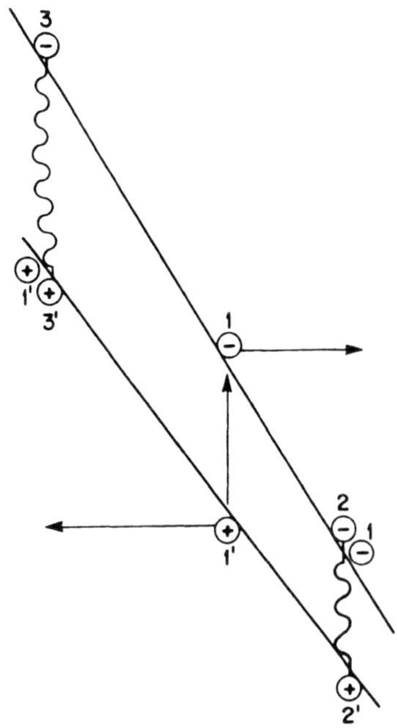

Figure 11. Schematic of impact ionization in a graded-gap semiconductor. The 1-1' e-h is the initial electron-hole pair. Note the lower electron ionization energy and the higher effective electric field seen by the electron.

conduction band discontinuity shown accounts for most of the bandgap difference, as is typical of many III-V heterojunctions. The materials are chosen for a conduction-band discontinuity comparable to or greater than the electron ionization energy E_{ie} in the low gap material following the step. The biased detector is shown in Figure 12b. Holes do not ionize since $\Delta E_v \cong 0$ and the field is too low to cause ionization. Consider a photoelectron generated near the p^+ contact. The electron does not impact ionize in the graded region before the conduction-band step because the effective field is too low. At the step, however, the electron ionizes since $\Delta E_v > E_{ie}$ and the process is repeated at every stage. Note that the steps correspond to the dynodes of a phototube. The gain is $M=(2-\sigma)^N$ where σ is the fraction of electrons that do not ionize per stage. The noise per unit bandwidth on the output signal, neglecting dark current, is given by $\langle i^2 \rangle = 2eI_{ph}M^2F$ where I_{ph} is the primary photocurrent and F the avalanche excess noise factor. A single calculation of F gives 16-18.

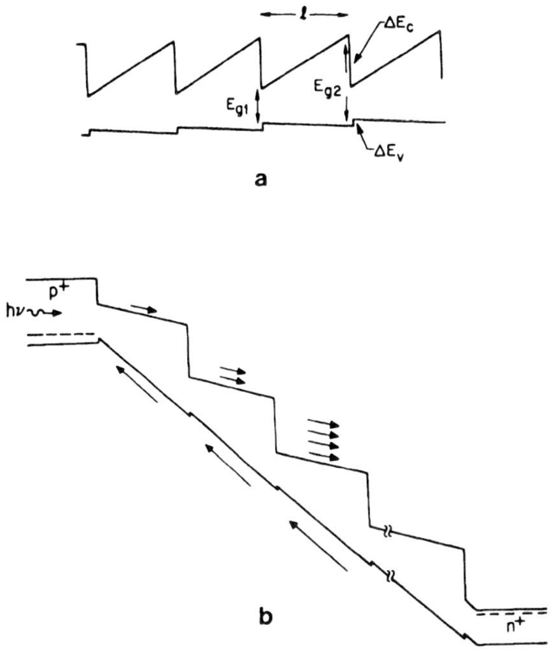

Figure 12. Band diagram of (a) unbiased-graded multilayer region and (b) the complete staircase detector under bias.

$$F = 1 + \frac{\sigma[1-(2-\sigma)^{-N}]}{(2-\sigma)} \tag{3}$$

Note that for small σ (≈ 0) $F \cong 1$ and is practically independent of the number of stages. Thus, the multiplication process is essentially noise free. It is interesting to note that the excess noise of this structure does not follow the McIntyre theory of a conventional APD[21]. In a conventional APD the minimum excess noise factor at high gain (>10) is 2 if one of the ionization rates is zero. The reason is that in the staircase APD the avalanche noise is lower than in the best conventional APD ($\alpha/\beta = \infty$) and can be understood as follows: in a conventional APD the avalanche is more random because carriers can ionize everywhere in the avalanche region, while in the staircase APD electrons ionize at well defined positions in space (i.e. the multiplication process is more deterministic). Note that, similarly, in a photomultiplier tube, the avalanche is essentially noise free ($F \cong 1$).

Finally, the low voltage operation of this device with respect to conventional APDs should be mentioned. For a five-stage detector and $\Delta E_c \cong E_{g1} \cong 1eV$, the applied voltage required to achieve a gain $\cong$ 32 is slightly greater than 5V. Possible material systems for the implementation of the device in the 1.3-1.6μm region are AlGaAsSb/GaSb and HgCdTe. In a practical structure one should always leave an ungraded layer immediately after the step having a thickness of the order of a few ionization mean free paths ($\lambda_i \cong$ 50-100Å) to ensure that most electrons ionize near the step. This modified staircase is shown in Figure 13.

Other interesting applications of staircase potentials have been proposed. We shall discuss here the repeated velocity overshoot device.[20] This structure offers the potential for achieving average drift velocities well in excess of the maximum steady state velocity over distances greater than 1μ. Figure 14a shows a general type of staircase potential structure. The corresponding electric field, shown in Figure 14b consists of a series of high-field regions of value E_1 and width d superimposed upon a background field E_o. To illustrate the electrical behavior and design considerations for a specific case, we consider electrons in the central valley of GaAs. The background field E_o is chosen so that the steady state electron energy distribution is not excessively broadened beyond its thermal equilibrium value, but at the same time the average drift velocity is still relatively high. For GaAs, an appropriate value would be around 2.5 kV/cm. At this field, the steady state drift velocity is 1.8×10^7cm/s and fewer than 2% of the electrons reside in the satellite valley. The electron distribution immediately downstream from the high-field region is shifted to higher energy by an amount $\Delta W = E_1 d$. (Note that while the distribution is shifted uniformly in energy, it is compressed in momentum in the direction of transport). We choose d so that the transit time across the high-field region is shorter than the mean phonon scattering time, which is about 0.13 ps in GaAs. The energy step ΔW is chosen to maximize the average velocity of the distribution after the step while still keeping most of the distribution below the threshold energy from transfer to the satellite valley. In GaAs, the intervalley separation is about 0.3 eV, so an appropriate value of ΔW would be about 0.2 eV, resulting in an average velocity of approximately 1×10^8cm/s immediately after the step. The momentum decays rapidly beyond the step due to scattering by polar optical phonons, with the result that the velocity decreases roughly linearly with distance as shown in Figure 14c. During this time, the distribution is broadened considerably in momentum. After the momentum (and velocity) have relaxed, the distribution requires additional time to relax to its original energy. Thus, the spacing L between high field regions must be large enough to allow sufficient cooling of the electron distribution before another overshoot can be attempted. This is necessary in order to avoid populating the high mass satellite valleys. The effect of the resulting repeated velocity overshoot shown in Figure 14c is the average drift velocities greater than the maximum steady state velocity can be maintained over very long distances.

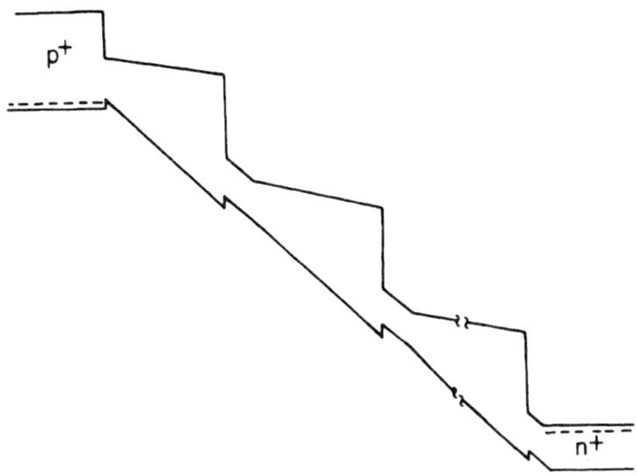

Figure 13. Band diagram of a modified staircase APD. The finite ionization mean free path above the ionization threshold is taken into account by introducing ungraded regions after the steps.

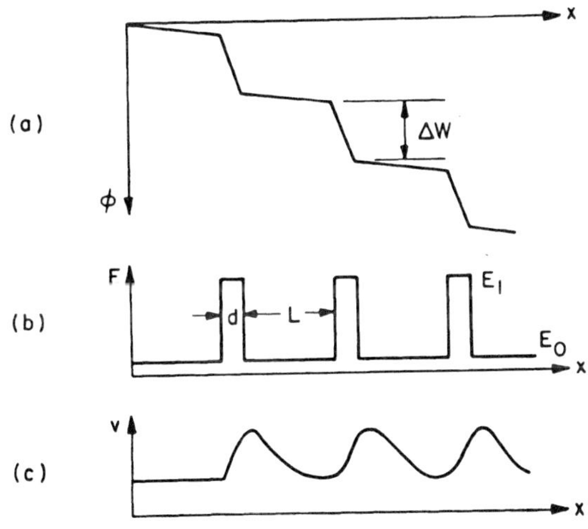

Figure 14. Principle of repeated velocity overshoot. Stairstep potential configuration (a) and the corresponding electric field (b). The ensemble velocity as a function of position is illustrated schematically in (c).

Two types of structures can be used to produce the desired potential profile. In the first class, the stair step in potential is obtained via the multistage graded-gap structure, used for the staircase detector. However, for this application, a much smaller conduction-band discontinuity should be used; one possibility is to grade repeatedly the composition from GaAs to $Al_{0.2}Ga_{0.8}As$, giving $\Delta E_c \cong 0.2eV$, every graded stage having approximately a thickness of 4000Å. The second class of structures makes use of alternating planar doped charge sheets.[20] However, in the case of the planar doped velocity overshoot device, the alternating charge sheets are placed in the drift region of PIN diode. For GaAs, we estimate that the spacing d between P- and N-doped sheets should be about 200Å, with the sheet thicknesses on the order of 20Å. The doping density should be about $6x10^{11}cm^{-2}$. Average velocities as high as $4x10^7cm/s$ have been estimated for this structure.[20]

CHANNELING DIODE

Figure 15 is a schematic of the structure which consists of alternated wide-gap p and low-gap n layers. A voltage is applied to the diode until all the layers are completely depleted. This new depletion scheme is illustrated in Figure 16. Figure 17 shows the band diagram and the operating principle of the structure[22,23]

Suppose that radiation of suitable wavelength is absorbed in the lower-gap layers thus creating electron-hole pairs. The two p-n heterojunctions formed at the interfaces between the relatively narrow band-gap and the surrounding higher band-gap layers serve to confine electrons to the narrow band-gap layers, while sweeping holes out into the contiguous wider band-gap p layers where they are confined by the potential. The parallel electric field ε causes electrons confined to the narrow band-gap layers to impact ionize. Holes generated in this way are swept out in the surrounding higher-gap layers before undergoing ionizing collisions in the narrower gap layers, since the layer thickness is made much smaller than the hole ionization distance $1/\beta$. In conclusion, electrons and holes impact ionize in spatially separated regions of different band-gap. The holes in the wider-gap layers impact ionize at a much smaller rate compared to the electron ionization rate in the relatively low-gap material, due to the exponential dependence of α,β can be made extremely large. Note that this structure has the unique advantage of providing a high (virtually infinite) α/β ratio at very high gains (>100) because electrons and holes avalanche in different materials. A typical design for 1.3-1.6μm detectors would have p and n layer thicknesses of 0.5-1.0μm and doping levels p-n $\cong 10^{16}/cm^3$. The p layers could be of InP or $Al_{0.48}In_{0.52}As$ and the n layers of $In_{0.53}Ga_{0.47}As$. These materials can be grown lattice matched to a semi-insulating Fe doped InP substrate. The estimated α/β ratio is $\cong$ 350 for a parallel field of $\cong 2x10^5V/cm$ at a gain of $\cong$ 150 for a layer length of $\cong$ 25μm.

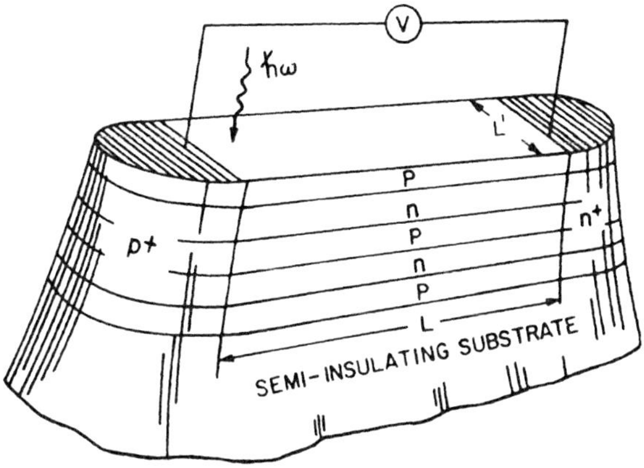

Figure 15. Schematic of the channeling APD. For simplicity, we assume equal acceptor and donor concentrations (n=p). The center layers have thickness d while the top and bottom layers have thickness d/2.

It is important to note that this structure has an interesting capacitance-voltage characteristic characterized by a conventional reverse biased diode behavior below the punch-through voltage V_{pth} followed by a large drop in capacitance for $V \cong V_{pth}$ and by a constant ultrasmall capacitance for $V>V_{pth}$ which is determined by the distance between the p^+ and n^+ contact regions.

This feature can be easily understood by considering Figure 16. It is very important to note that for a large change in capacitance complete depletion of all the layers is not required. Assume for example, that for the device in Figure 16, acceptor concentration is greater than the donor concentration; in this situation only the n layers will be completely depleted. The residual capacitance is that formed by the two undepleted sections of the p layers with the n^+ region. This capacitance is still much smaller than that before punchthrough.

This CV curve has recently been demonstrated in an AlGaAs/GaAs/AlGaAs pnp structure[24]. This device can also be used as an ultrasensitive photocapacitive detector[25].

The capacitance of these devices was measured at 1 MHz as a function of reverse bias, for different incident intensities (Figure 18). A 2mW He-Ne laser attenuated with neutral density filters was used as the light source. The top $Al_{0.45}Ga_{0.55}As$ is transparent to the $\lambda/6328\text{Å}$ radiation. Let us first discuss the dark C-V curve. The curve has three distinct regions. First, is a decrease of the capacitance with voltage, characteristic of a reverse biased diode, followed

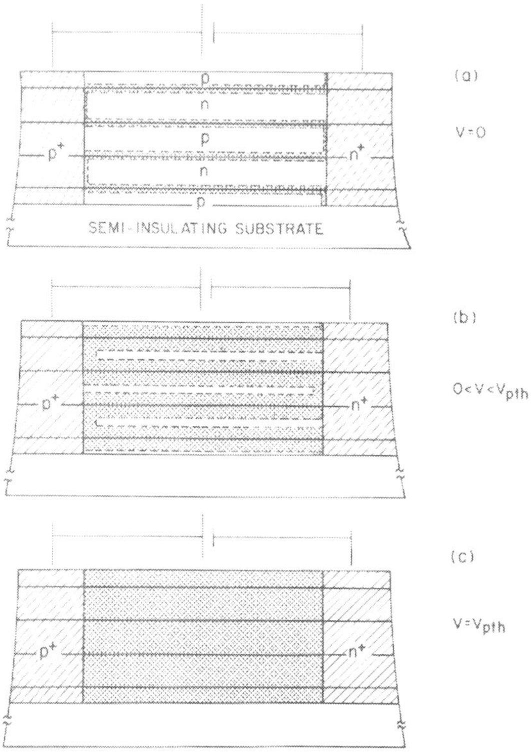

Figure 16. Cross section of the channeling diode at different reverse bias voltages V. The shaded portions of the layer represent the depleted volume. V_{pth} is the voltage at which the layers are completely depleted (c). Any further increase in reverse bias adds a constand field ε parallel to the device length. Equal doping levels (=n) ^{p} have been assumed.

by a one order of drop in capacitance over a small voltage range and a final region of nearly constant ultrasmall ($\leq$ 0.1pF) capacitance. The overall behavior agrees well with the previous predictions.

The other C-V curves in Figure 18 were obtained by varying the incident laser power over four orders of magnitude from 20 pW to 200 nW. Note the increase of the punch-through voltage with increasing power and the larger variations in capacitance (0.6 - 1.0 pF) (with respect to the "dark capacitance") produced by the low optical power levels used. It is clear that this device can be used as an ultrahigh sensitivity photocapacitive detector. The essential features of this novel photocapacitive phenomenon can be easily interpreted with the aid of Figure 16.

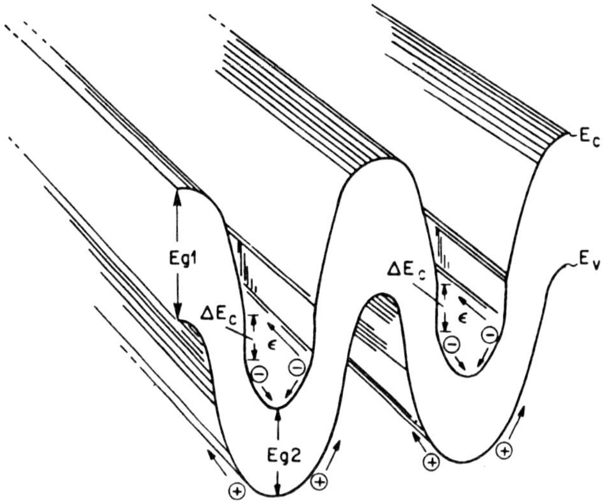

Figure 17. Band diagram of the channeling APD ($E_{g1}>E_{g2}$): ≺ is the parallel field causing carriers to ionize. Shown is also the electron-hole spatial separation mechanism. ΔE_v has been assumed negligible with respect to ΔE_c.

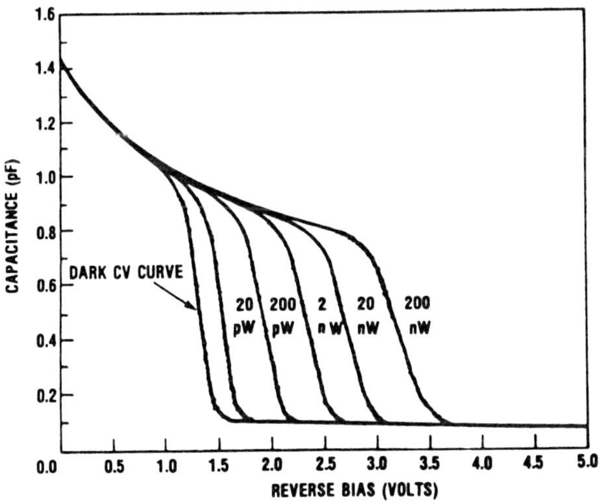

Figure 18. C-V characteristics of the channeling photodiode, for optical power levels in the range from 20 pW to 200 nW.

Assume that the device is biased at or slightly above the punch-through voltage so that the layers are completely depleted. When light shines on the device, the photogenerated electrons and holes are spatially separated and collected in the depleted n and p layers respectively, thereby partially neutralizing the ionized donor and acceptor space charge. The net effect is that the width of the depletion layer is reduced [Fig. 16b]; this produces a large change in capacitance. An additional voltage is thus required to completely deplete the layers. This explains the shift of the punch-through voltage with increasing optical power. The ultrahigh sensitivity of the structure is due to two factors. First, even a small (<10%) reduction of the depletion layer width produced by a very low incident power is sufficient to cause a large change in capacitance. Second, the spatial separation of optical generated electrons and holes greatly increases their recombination lifetime so that substantial quasistable excess densities of electrons and holes are present in the layers to compensate the ionized space charge. The increase in lifetime due to the spatial separation of electrons and holes was first discussed by Döhler in the contaxt of n-i-p-i superlattices[26].

It is important to stress that this detector can be used also as an ultralow capacitance, low-punch-through voltage p-i-n photodiode, provided it is biased at voltages such that the capacitance is not affected by the incident light. It is worth stressing the unique features and important differences of this structure with respect to conventional p-i-n diodes. The novel interdigitated p-n junction scheme allows one to achieve an ultrasmall capacitance, largely independent of the detector area between the p^+ and n^+ regions and of the layers doping. Thus, the sensitive area can be maintained reasonably large and the doping moderately high. Note that conventional pins require very low doping levels.

REFERENCES

1. F. Capasso, J. Vac. Sci. Technol. 1, 457 (1983).

2. H. Kroemer, RCA Rev. 18, 332 (1957).

3. F. Capasso, W.T. Tsang, C.G. Bethea, A.L. Hutchinson, and B.F. Levine, Appl. Phys. Lett. 42, 93 (1983).

4. B.F. Levine, C.G. Bethea, W.T. Tsang, F. Capasso, K.K. Thornber, R.C. Fulton and D.A. Kleinman, Appl. Phys. Lett. 43, 769 (1983).

5. J.R. Hayes, F. Capasso, A.C. Gossard, R.J. Malik and W. Wiegmann, Electron. Lett. 13, 410 (1983).

6. H. Kroemer, Proc. IEEE 70, 13 (1982).

7. F. Capasso, J.R. Hayes, R.J. Malik, A.C. Gossard, W. Wiegmann (unpublished).

8. D.L. Miller, P.M. Asbeck, R.J. Anderson, F.H. Eisen, Electron, Lett. 19, 367 (1983).

9. R.J. Malik, J.R. Hayes, F. Capasso, K. Alavi and A.Y. Cho, IEEE Electron, Dev. Lett., EDL-4, 383 (1983).

10. R. People, K.W. Wecht, K. Alavi and A.Y. Cho, Appl. Phys. Lett. 43, 118 (1983).

11. J.R. Hayes, F. Capasso, R.J. Malik, A.C. Gossard and W. Wiegmann, Appl. Phys. Lett. 43,949 (1983).

12. F. Capasso, W.T. Tsang, A.L. Hutchinson and G.F. Williams, Appl. Phys. Lett. 40, 38 (1982).

13. R. Chin, N. Holonyak, G.E. Stillman, J.Y. Tank and K. Hess, Electron, Lett. 16, 467 (1980).

14. F. Capasso, W.T. Tsang, A.L. Hutchinson and P.W. Foy, Inst. Phys. Conf. Ser. No. 67, 473 (1982).

15. F. Capasso, G.F. Williams and W.. Tsang, IEEE Spec. Conf. on Light Emitting Diodes and Photodetectors, Ottawa, 1982, Tech. Dig. p. 166 (1982).

16. F. Capasso, W.T. Tsang and G.F. Williams, IEEE Trans. Electron. Devices ED-30, 381 (1983).

17. F. Capasso and W.T. Tsang, Technical Digest of the International Electron Devices Meeting, San Francisco, 1982, p. 334 (1982).

18. F. Capasso, IEEE Trans. Nucl. Sci. Ns-30, 424 (1983).

19. G.F. Williams, F. Capasso and W.T. Tsang, IEEE Electron Device Lett. EDL-3, 71 (1982).

20. J.A. Cooper, F. Capasso and K.K. Thornber, IEEE Electron, Dev. Lett. EDL-3, 497 (1982).

21. R.J. McIntyre, IEEE Trans. Electron. Dev. ED-13, 164 (1966).

22. F. Capasso, Electron. Lett. 18, 12 (1982).

23. F. Capasso, IEEE Trans. Electron. Dev. ED-29, 1388, (1982).

24. F. Capasso, R.A. Logan and W.T. Tsang, Electron, Lett. 18, 760 (1982).

25. F. Capasso, R.A. Logan, W.T. Tsang and J.R. Hayes, Appl. Phys. Lett. 41, 944 (1982).

26. G. Dohler, Phys. Status Solid. B52, 533 (1972).

METAL-SEMICONDUCTOR INTERFACES

R.H. Williams

Physics Department
University College
P.O. Box 78
Cardiff CF1 1XL
U.K.

INTRODUCTION

The determination of those factors which influence the formation of Schottky barriers at metal-semiconductor interfaces has been an important goal for many decades. During the past few years considerable progress has been made as a result of the application of modern surface science techniques and parallel theoretical methods. However, it is true to say that the problem is still, at best, only partially understood. The formation of contacts, both ohmic and rectifying, has been an important technological topic even well before the transistor was invented and has in fact assumed increasing importance recently with the rapid development of very large scale integrated circuits.

One of the simplest possible contacts is shown in Figure 1(a) where a metal of large work function ϕ_m makes contact with a semiconductor of work function ϕ_s and electron affinity χ_s. In the absence of surface states on the semiconductor the Schottky model leads to a barrier ϕ_b given by

$$\phi_b = \phi_m - \chi_s \tag{1}$$

Clearly this barrier is highly dependent on the metal work function ϕ_m. In the presence of a large density of surface states on the semiconductor it is possible for the latter to be totally screened from the metal; ϕ_b may then be independent of ϕ_m. An example for acceptor states on an n-type semiconductor is given in Figure 1(b).

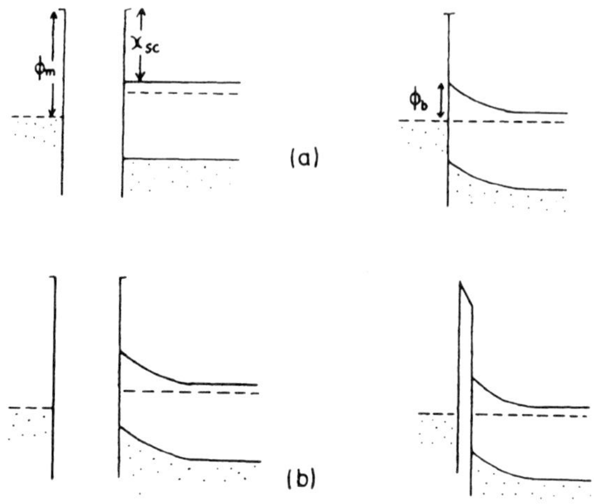

Figure 1a. Schottky barrier formation in the absence of Fermi level pinning.
1b. Form of the energy bands for a metal-semiconductor contact where the Fermi level is pinned by a high density of surface states.

This situation where ϕ_b is independent of ϕ_m is known as the Bardeen[1] limit and in practice it is found that many systems follow neither the Schottky nor Bardeen models. Many workers in the past, in their analysis of Schottky barrier height measurements for a range of metals on a given semiconductor, have assumed that ϕ_b varies linearly with ϕ_m i.e.

$$\phi_b = S(\phi_m - \chi_s) + C \quad (2)$$

where S and C are constants. Close examination of much of this work, however, reveals that the values of ϕ_m are often not known with enough precision for this 'linear model' to be verified.

The deposition of a metal onto a semiconductor surface of course leads to a change in the character of the free surface electron states and, as pointed out first by Heine[2], the above models assume a gross oversimplification of the real metal-semiconductor interface. More recently there have been numerous theories attempting to explain the relationship between Schottky barrier heights and the metals deposited, but most of them have been only partly successful. Some of these consider many body effects[3], the tails of metal wave functions[4], or new states generated by matching the metal wave functions to the semiconductor. In general it is assumed that the interface is abrupt and ordered, a situation which is not always realised in practice.

In this article we discuss in particular the role of imperfections at metal semiconductor interfaces and show that in many cases Schottky barrier formation can be dominated by complex chemical and metallurgical effects. This is so even for the most well controlled systems whereby atomically intimate interfaces are formed between metals and clean cleaved semiconductor surfaces under conditions of ultra high vacuum.

EXPERIMENTAL APPROACH

The existence of unknown contaminants on a semiconductor surface can lead to electron and hole states in the band gap which can complicate the interpretation of electrical barriers formed by the deposition of metals on such surfaces. Clearly, therefore, in order to study fundamental aspects of the problem it is desirable to study contacts to atomically clean surfaces. It is desirable to generate these surfaces by cleavage in vacuum rather than by heating or by argon sputtering followed by annealing. The latter two methods usually lead to a bulk doping profile which is nonuniform near the surface and often, for compound semiconductors, lead to surfaces that are non-stoichiometric chemically. This can lead to the generation of states in the band gap and in general give unwanted complications. The same is true for surfaces grown by molecular beam epitaxy as discussed briefly later.

The modern approach[5] involves the application of a range of 'surface science' techniques. It is customary to study simultaneously the structure of the semiconductor surface by low energy electron diffraction (LEED) or reflection high energy electron diffraction (RHEED), the chemistry by Auger electron spectroscopy (AES) or X-ray photoelectron spectroscopy (XPS), and the electronic structure by methods such as angle resolved photoelectron spectroscopy (ARUPS). The latter two techniques allow the formation of the Schottky barrier to be monitored in detail when quantities as small as one per cent of one atomic layer of a metal is deposited on the semiconductor. The growth of the metal film can then be progressively monitored using the above techniques and, when the film reaches a thickness of a few hundred Å, the Schottky barrier formed can then be measured, in situ, by current voltage and capacitance voltage techniques. The clean semiconductor surface can be exposed to a gas in a controlled way before deposition of the metal so that new states can be generated at the boundary essentially by 'doping' the interface.

SOME EXPERIMENTAL RESULTS

Surface States on InP

In order to illustrate the application of the above techniques to clean semiconductor surfaces we consider the case of surface states on the clean cleaved (110) surface of indium phosphide[6]. Detailed studies of Fermi level movement on n and p-type samples have established that, for high quality cleavages, there are no surface states in the band gap. Theoretical calculations show that this can only be so if the surface is relaxed in such a way that the In atoms move towards and the P atoms away from the bulk. The detailed surface structure has been measured by LEED[7] and calculation of surface states using this[7], or a similar structure due to Chadi[8], show that the surface states are removed from the gap so that vacant and full states overlap the conduction and valence band respectively. Figure 2(a) shows calculated states[6] superimposed on bulk states, for both unrelaxed and relaxed surfaces. Also shown is the measured surface states, close to the top of the valence band, established by ARUPS[6]. The distribution of charge associated with this state is shown in Figure 2(b); clearly this particular state is associated with a p-like distribution on the surface phosphorous atom, i.e. a phosphorous surface dangling bond.

With the possible exception of GaP it seems that high quality cleaved surfaces of all III-V compounds have surface states which overlap the valence and conduction bands rather than residing in the band gap. They therefore form highly ideal test beds to probe states generated in the band gap by deviations from perfection. Such defects are sketched in Figure 3; steps, vacancies, antisite defects, etc. can form can form gap states, as will be shown, and do lead to pinning of the Fermi level E_F. Note that 10^{12} states cm^{-2} can lead to considerable band bending at free surfaces, provided they have the right energy and the required donor or acceptor nature, and densities of around 10^{13} cm^{-2} can lead to appreciable Fermi level pinning at metal-semiconductor contacts. On a cleaved (110) GaAs surface the density of surface atoms is $\sim 8 \times 10^{14}$ cm^{-2}. Thus, if surface defects do form gap states, a very small density of them can have a large effect on cleaved (110) III-V surfaces, since their influence may not be masked by intrinsic surface states.

Stepped (110) Surfaces and Schottky Barriers on III-V's

It has been shown[9,10] that steps on the clean cleaved (110) surface of GaAs lead to acceptor states which pin the Fermi level in the band gap for n-type material. For a high state density this pinning energy is at $\sim$ 0.8 eV below the conduction band edge, as illustrated in Figure 3(a). Interestingly, a similar pinning is usually seen for GaAs surfaces grown by MBE[11] and for surfaces cleaned by sputtering and annealing[12]. For the latter it has been shown that a large density of atoms are at step edges[13] and usually the surface region is slightly deficient in the group V component. There have been attempts to account for these acceptor states theoretically in terms of the intersection of (110) cleavage planes[14], but this interpretation is not well established.

In Figure 3(b) we show the measured Schottky barriers for a large number of metals on n-type GaAs surfaces[15]. Interestingly they all appear, to first order, to yield pinning of the Fermi level at a

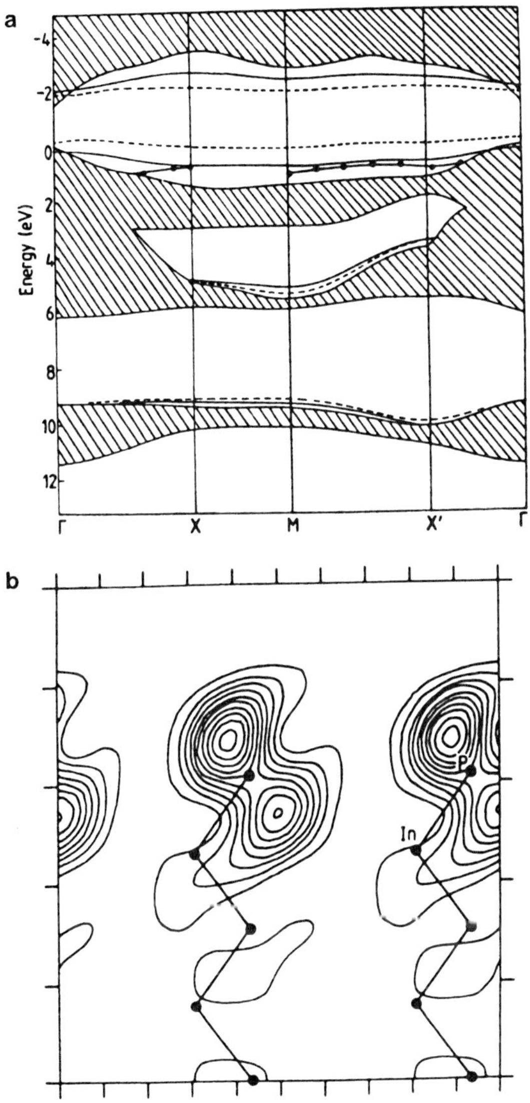

Figure 2a. Shows the tight-binding structure for the unrelaxed (broken curves) and relaxed (full curves) (110) surface of InP. Also plotted is the dispersion of the uppermost occupied surface state determined from the experimental results. The shaded region corresponds to projected bulk states.

2b. Electronic charge-density distribution at point X of the (110) surface of InP for the occupied dangling-bond band localised on the P atoms.

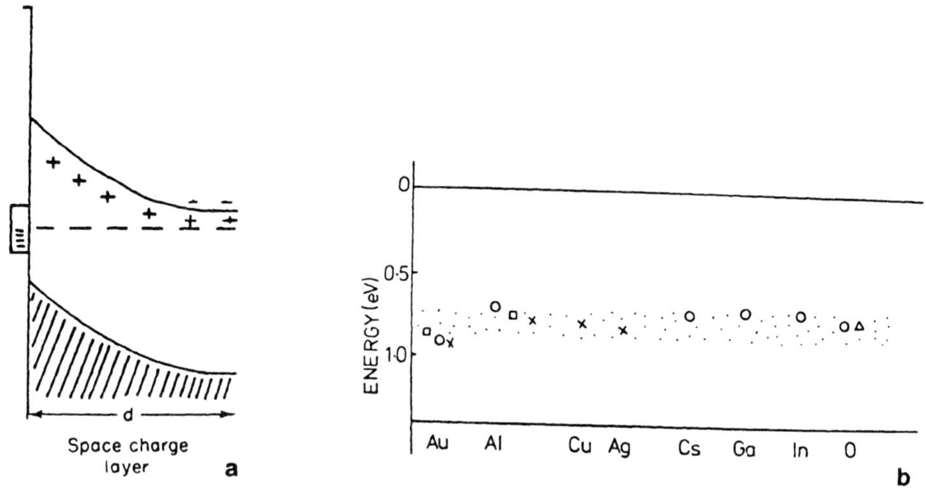

Figure 3a. Band bending associated with surface steps on n-GaAs (110) surfaces[9,10].
3b. Fermi level pinning energies for a range of metals on n-type GaAs (see reference 15).

fairly constant energy just below mid gap and close to the pinning energy observed for the free stepped surface. Detailed studies of this type have been carried out for a range of semiconductors, in particular GaSb[16] and InP[16,17]. For GaSb pinning levels close to the valence band are obtained for a range of metals, whereas for n-type InP the Fermi level pinning is usually in the upper half of the band gap. In the subsequent section we discuss the case of InP in more detail.

Metal Contacts to InP (110) Surfaces

In Figure 4 we show the Schottky barriers measured for a range of metals on clean cleaved (110) n-InP surfaces[17]. There is no simple relationship between ϕ_b and ϕ_m and the linear model outlined in Section 1 does not seem appropriate in this case. The horizontal scale in Figure 4 is the heat of reaction of each metal with P to form the most stable phosphide. It may be seen that the reactive metals in general give low barriers whereas the unreactive ones yield larger values. The detailed interaction of these metals with the InP surface has been studied and in Figure 5 and 6 we illustrate the major difference for unreactive and reactive metals for two selected examples, namely Cu and Ni. Figure 5 and 6 show the attenuation of the In4d and P2p core levels in photoemission as a result of increasing thicknesses of Ni and Cu respectively. For Cu the In4d and P2p emissions are attenuated at roughly the same rate consistent with a reasonable abrupt interface, and it is thought, a degree of island

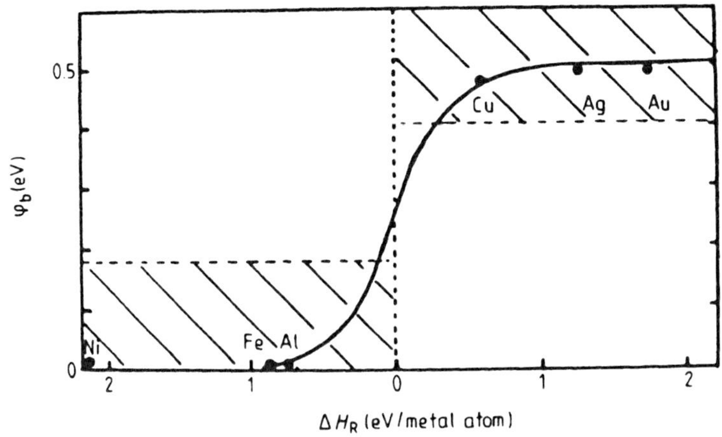

Figure 4. Plot of Schottky barrier height as a function of heat of reaction of the metal with phosphorus (after reference 17).

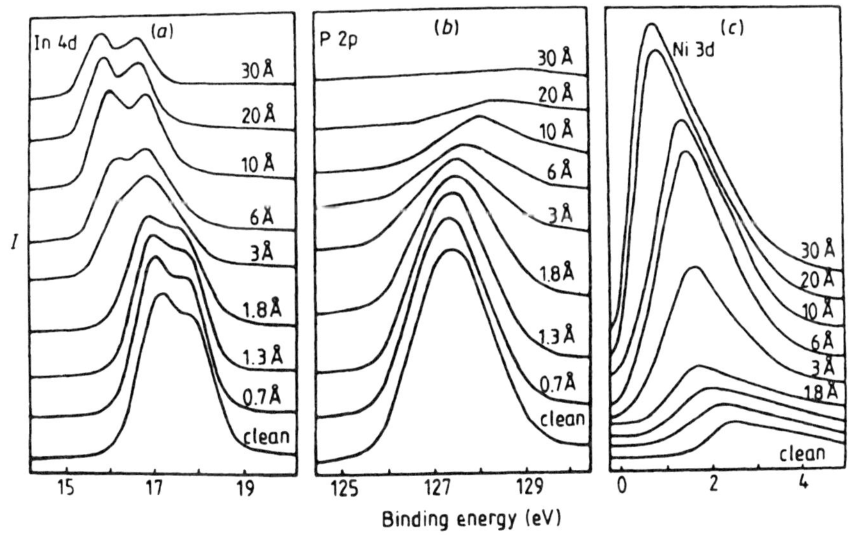

Figure 5. Photoemission intensity spectra for the In4d and P 2p core levels and the Ni3d band as a function of nickel thickness deposited on the surface. Photon energy 100 eV for the In4d peak and the Ni3d band; 170 eV for the P 2p peak.

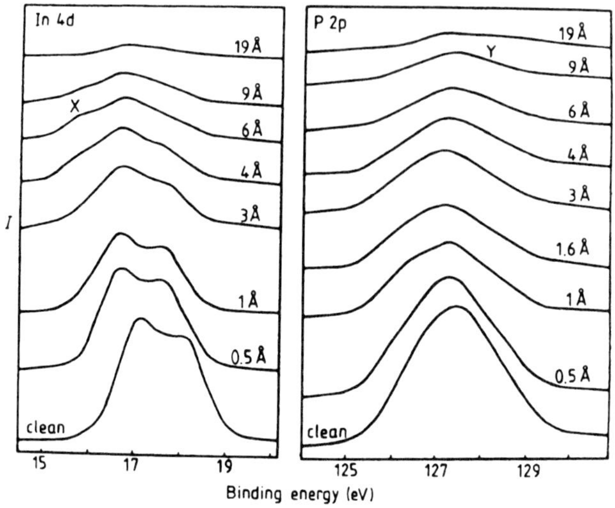

Figure 6. Photoemission intensity spectra for the In4d and P 2p core levels as a function of copper thickness deposited on the surface. Photon energy 100 eV for the In4d peak; 170 eV for the P 2p peak.

growth. For the case of Ni the situation is quite different; following the deposition of 20 Å the P-2p emission has largely disappeared whereas a chemically shifted emission from the In4d level is intense. Clearly, here the InP surface is dissociated, nickel phosphide is formed and In atoms are released and incorporated in the growing overlayer. The interface formed is nonabrupt and highly disordered. Figures 5 and 6 are characteristic of the unreactive and reactive groups and similar behaviour to Ni is also observed for Ga. To date the exact mechanisms which lead to the exchange reactions above are not fully understood but detailed total energy calculations for Al on (110) GaAs suggest that the Al atoms have considerable mobility and that the activation energy driving the interface reactions is derived from the heat generated when small Al clusters are formed on the GaAs surface. At any rate, for n-InP, strong chemical reactions and low Schottky barriers appear to be related.

Recent work has shown that this is to simple, e.g. Ga yields barriers of around 0.6eV on clean InP, though it is in the reactive category - see R.H. Williams, I.M. Dhavmadasa, M.H. Patterson, C. Maani and N.M. Forsyth, Surface Science, in press.

It is of interest to note that prolonged exposure of the clean n-InP surface to gases such as oxygen or chlorine lead to lower values of ϕ_b when Ag or Au contacts are deposited on these surfaces. The interaction with these gases leads to highly disordered nonstoichiometric surfaces also.

Models of Schottky Barrier Formation

Detailed studies such as those outlined above have been carried out for a number of intimate metal-III-V interfaces and for many of them the interfaces formed are not abrupt and ordered. Thus, models involving perfect interfaces are unlikely to be fully applicable. Several 'imperfect' models have been put forward to account for the various observations and just two of these will be mentioned here.

The first model, due to Freeouf and Woodall(18), assumes that cation outdiffusion into the metal contact leaves anion rich areas at the interface and that it is the work function of these anion rich regions that should be taken into account in the modelling of the barrier. The model accounts for a wide range of observations on a number of systems and may well be applicable particularly for systems where thick contacts have been subject to annealing. It is unlikely, however, to account for the strong Fermi level pinning often observed when the metal coverage on the clean surface is only around 25% of one atomic layer.

The second model we consider is the so-called "defect model' of Schottky barrier formation. Here, it is assumed that defects such as vacancies or antisite defects are formed at and just below the surface of the semiconductor when the deposited metal is incident on the surface. In this case, Fermi level pinning results from the defect levels which are characteristic of the semiconductor and not of the metal. At a disordered interface, such as Ni-InP, or Al-GaAs (110), the existence of such defects is inevitable and provided they form suitable levels in the band gap, they can certainly lead to strong pinning behaviour. Early calculations for P vacancies in InP(19) demonstrated that such a centre could form a donor/acceptor state just below the conduction band consistent with the low barriers observed for reactive systems. These calculations have been extended by Daw and Smith(20) and by Allen and Dow(21) who considered pinning by anion vacancies and antisite defects respectively in a range of III-V compound semiconductors. Some of the results of Daw and Smith(20) for the alloy system AlAs-GaAs-InAs are shown in Figure 7. Pinning by anion vacancies in the bulk and near the surface has been considered and the trends measured in the experiments are clearly reproduced. However, this trend is also reproduced by Allen and Dow(21) who considered antisite defects rather than vacancies. At this stage, however, it is not clear whether the tight binding calculations used are of sufficient accuracy to enable a great deal of weight to be placed on the above agreement between experiment and theory. However, it does seem certain that if defects are present at metal-III-V semiconductor interfaces in appreciable concentrations then strong pinning of the Fermi level may result. In many, perhaps most, cases the interfaces formed are neither abrupt nor ordered and the existence of defects cannot be ignored.

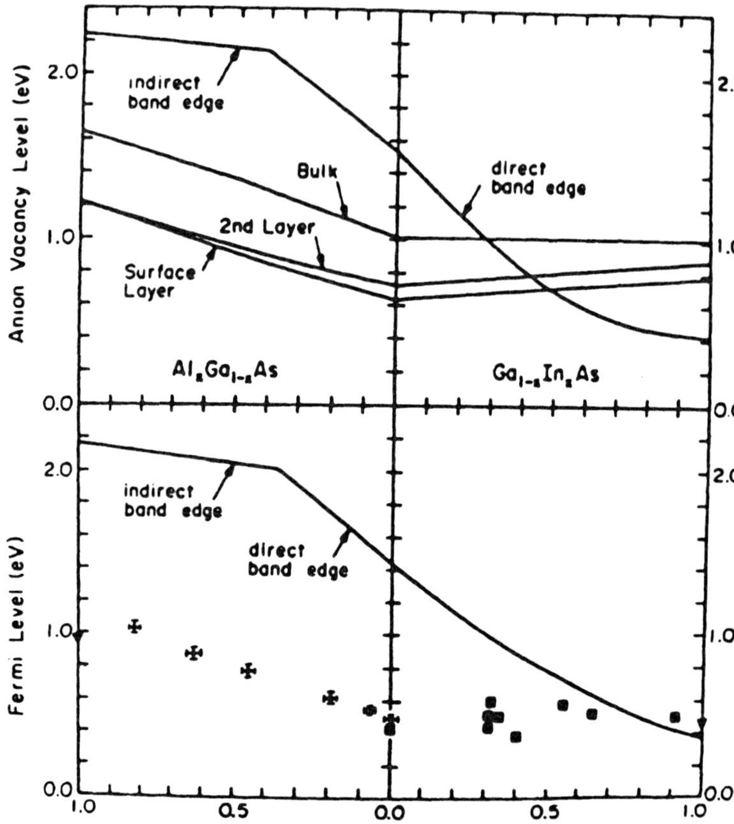

Figure 7. Lower: Fermi level pinning position relative to the valence band edge in alloys of AlAs, GaAs and InAs. After (33). Upper: Calculated anion vacancy level for a bulk defect and for surface defects in the same alloys. After (31).

Finally we mention the case of Sn contacts on n-InP. When deposited on the clean surface Sn often gives low barrier contacts, presumably as a result of In:Sn exchange leading to shallow donor levels and accompanying pinning of the Fermi level. Strangely, however, when the same experiment was carried out on clean n-GaAs large barriers were formed. This illustrates the many subtle differences that exist between cleaved GaAs and InP surfaces, many of which are not at present fully understood.

Case of Metals on GaSe

Our general conclusions in the previous section is that there are a whole range of processes which may contribute to Schottky barrier formation but in some instances one of these processes may dominate above all others. The case of reactive interfaces was considered and it is natural to ask the question whether the Schottky model will be followed for truly unreactive interfaces. The case we now consider is that for a series of metals on the layered semiconductor GaSe[22,23]. Because of its graphitic like structure GaSe cleaves to yield highly perfect, very inert surfaces and its band gap of ∿ 2 eV is large enough to allow large Fermi level shifts, in the absence of interface states.

Figure 8 shows typical Ga3d core level photoemission spectra for clean GaSe and following the deposition of two metals namely Au and Al. The attenuation of the core level is very slow when Au is deposited, due to island growth, and the observed shift of the emission line has been verified as a shift of the Fermi level from the equivalent shift of the Se3d emission. For Al, on the other hand, the behaviour is quite different. In this case, there is a small shift of the Ga3d emission associated with Fermi level movement but also there appears an additional chemically shifted component. With progressive deposition of Al the Se3d emission is attenuated whereas the chemically shifted Ga3d component remains strong. Clearly the Al is reacting with the surface to form a selenide and gallium is released. Detailed analysis of such photoemission

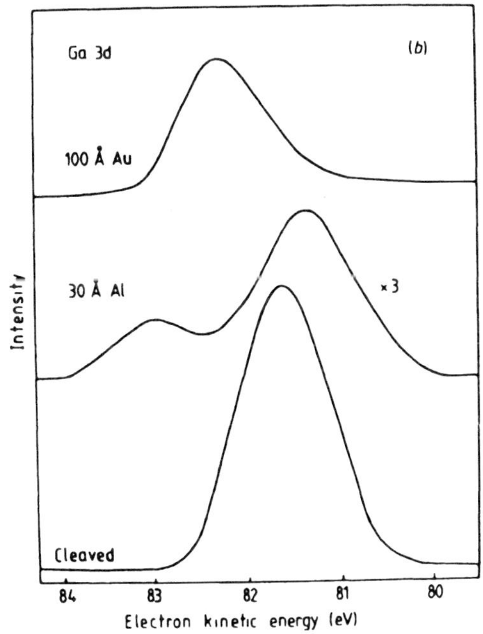

Figure 8. 3d gallium core level photemission for a cleaved surface and following deposition of Au and Al.

spectra for a range of metals enable metals which react strongly with the GaSe surface to be separated from those that interact only weakly.

In Figure 9 we show the shifts of the Fermi level at the surface for the range of metals studied. Those metals which react reasonably strongly with the surface show a roughly constant pinning energy whereas the unreactive metals Au and Ag show large shifts. Sn also belongs in this category. Interestingly, the differences in-Fermi level shifts for the two extremes of Au and Ag are close to those expected the the Schottky limit. Thus, in Figure 9 we see both Schottky-like and Bardeen-like behaviour on the same material, with the pinning energy levels for the latter being generated by chemical reactions between the GaSe surface and reactive metals i.e. by metal induced defect states.

Schottky Barriers and Adsorbates

In Section 3.3 the modification of Schottky barriers by absorbed gas layers at the metal semiconductor interface was referred to. In this section we consider this topic a little further. Modifications of GaAs-Ag[24] barriers by H_2S and Au-InP[25], also by H_2S, have been described in the literature. As an example of these effects, we consider the case of the Au-H_2O-InP system[26]. Figure 10, shows Auger electron spectra, form of the LEED patterns, and typical I-V characteristics for a gold front contact (with an ohmic back contact) when the gold is de-

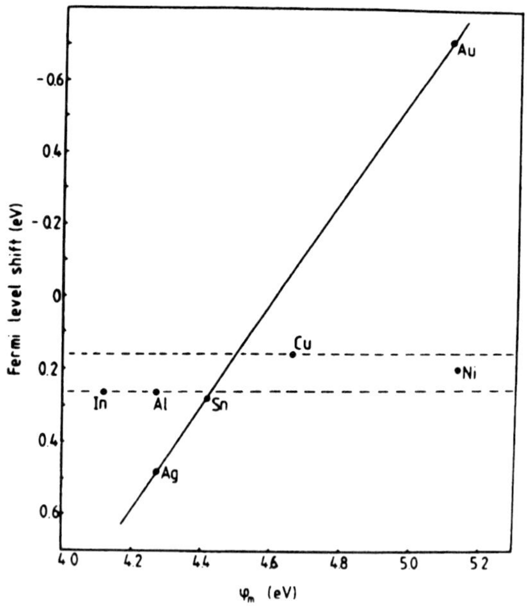

Figure 9. Shift of Fermi level at metal-GaSe interfaces. Negative shifts are from flat band towards the valence band.

posited onto a clean surface and on surfaces exposed to increasing amounts of water vapour. With increasing exposure the LEED pattern eventually disappears reflecting a disordered surface and the surface chemistry is drastically modified as various oxides are formed. The changing surface chemistry is paralleled by a reduction in the n-InP-Au Schottky barrier and the interpretation of this reduction has again been in terms of the defect model where shallow donor/acceptor levels are formed by the disorder and by the doping of the surface region by oxygen atoms.

Similar effects have been generated on cleaved surfaces of the II-VI semiconductors CdSe(27) and CdTe(28). On clean cleaved n-type CdTe crystals Au contacts yield values of $\phi_b \sim 0.9$ eV and Al $\phi_b \sim 0$ eV. On an air exposed surface both Au and Al yield $\phi_b \sim 0.9$ eV. By depositing very thin intermediate layers of Al on the clean surface followed by Au it is possible to generate a range of ϕ_b simply by varying the thickness of the Al layer. This is illustrated in Figure 11. The interpretation again assumes that shallow surface donor levels are generated by indiffusion of some Al atoms(28).

It is clear, therefore, that Schottky barriers can be changed and controlled in many situations by appropriate choice of gas-metal and metal-metal combinations. Barriers may be readily modified by the incorporation of very shallow dopants and indeed the situation in many ways resembles contacts formed where low energy ion implantation is used to control the effective Schottky barrier heights(29).

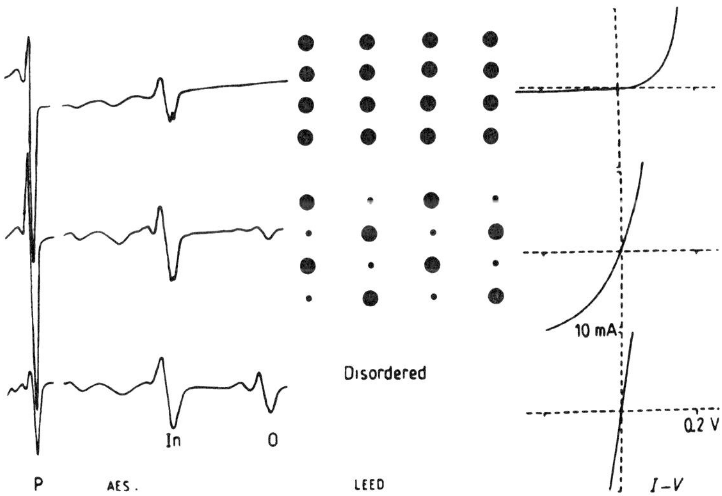

Figure 10. The current-voltage (I-V) characteristic for gold contacts to the clean InP (110) surface and to InP (110) surfaces exposed to progressively larger amounts of H_2O are shown, along with both LEED and Auger spectra for the respective surfaces.

Photoluminescence and Near Surface Defects

There is much evidence to suggest that near surface defects on InP crystal surfaces often lead to pinning of the Fermi level in the upper half of the gap and fairly close to the conduction band. The question arises as to whether such defects can be observed directly by techniques such as photoluminescence. We now describe experiments carried out by Street et al [30,31] for cleaved (110) InP surfaces using the photoluminescence technique.

Figure 12 shows photoluminescence spectra from (110) surfaces cleaved in ultra high vacuum so that the surfaces are atomically clean. Spectra are shown for smooth step free regions and also for highly stepped regions. A very distinct additional peak is seen for the stepped regions and this peak has been associated with near surface defects forming an energy level in the band gap close to the conduction band. Mechanical damage of the surface leads to an appreciable broadening of the emission associated with the additional peak, consistent with a series of relatively shallow energy levels. It is tempting to suggest that these levels are the same ones that lead to the pinning of the Fermi level for reactive metals on InP discussed in section 3.3 and to associate them with vacancies and/or antisite defects. Further work is required to probe these, however, and this is in progress.

Finally we consider the variation of intensity when the clean InP surface is exposed to oxygen. For the upper curve in Figure 12, i.e. associated with stepped regions, exposure to oxygen leads to very little difference in the peak intensities. For the smooth region, however, the intensity first reduces by an order of magnitude following small exposures, and then increases again for larger exposures, as illustrated in Figure 13.
If the oxygen is excited, e.g. by an ion gauge, the high intensity is recovered much more rapidly than for unexcited oxygen. The data suggests that in the clean situation the bands are flat up to the surface. On exposure to small amounts of oxygen acceptor levels are formed, and carriers are depleted from the bulk; the surface adsorbed species is probably O_2^-. The band bending generated separates the electrons and holes and decreases the luminescence intenity. Upon further exposure oxygen reacts with the surface, disorders it, and leads to shallow defect levels which pin the Fermi level close to the conduction band so that the bands are now flat and the corresponding photoluminescence intensity high. The variation of the luminescence intensity shown in Figure 13 mirrors the shift of the Fermi level measured by contact potential difference upon exposure of a clean n-InP surface to oxygen[32].

Clearly therefore the photoluminescence technique holds some promise in relation to the identification of near surface defects in semiconductors and in the investigation of the role of defects

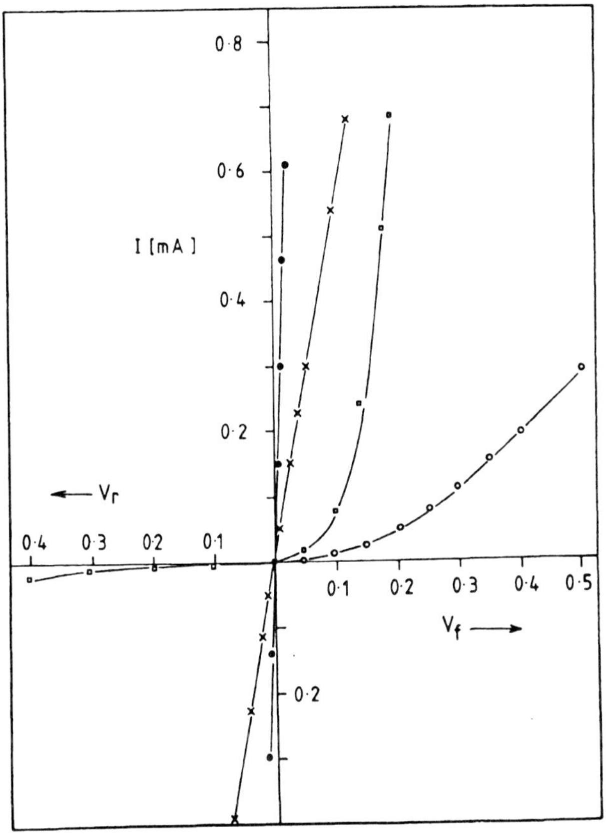

Figure 11. Current-voltage characteristics (dC) for (•) an Al-vacuum cleaved CdTe interface, (X) a 2 Å of Al/Au-vacuum cleaved CdTe interface, (□) a 0.8 Å of Al/Au-vacuum cleaved CdTe interface and (O) a Au/vacuum cleaved CdTe interface.

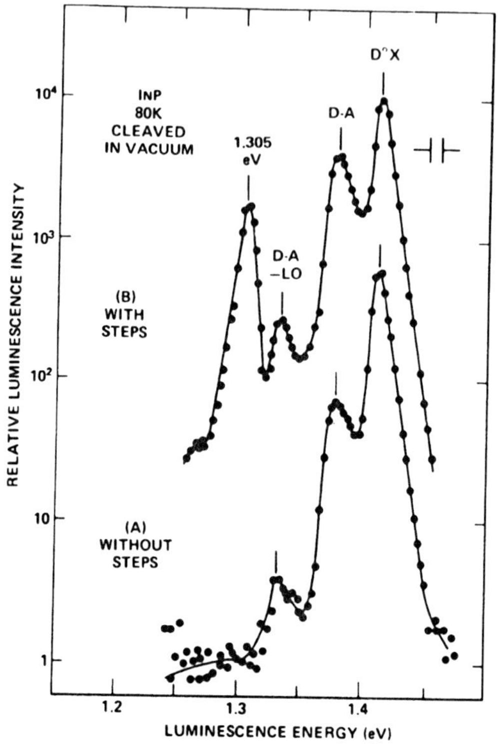

Figure 12. Photoluminescence spectra for cleaved (110) surfaces of InP both with and without steps. (After 30.)

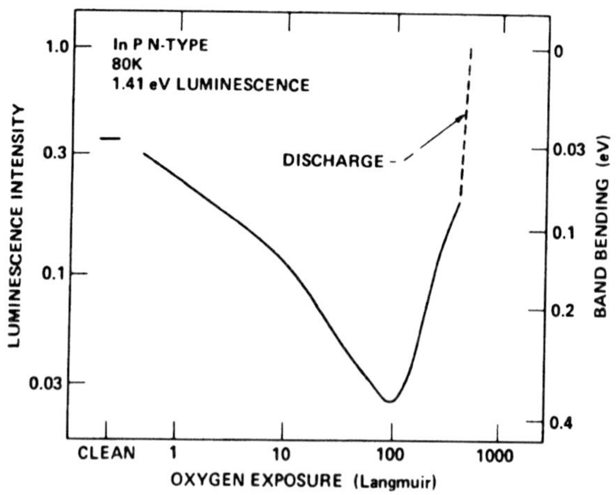

Figure 13. Variation of the photoluminescence intensity with exposure to oxygen for clean step free cleaved (110) InP surface. (After 30.)

in the formation of Schottky barriers. However, new and improved methods of characterising surface defects, particularly on a microscopic scale, are required if the progress made over the past few years is to be maintained.

CONCLUSIONS

It now appears that a number of different processes contribute the formation of Schottky barriers at metal-semiconductor interfaces. These include intrinsic surface states, metal induced interface states and metal induced defect states in the semiconductor. In some situations one process may dominate above all others; in some cases for metals on III-V semiconductor surfaces it appears that metal induced defects may dominate. In many instances of intimate interfaces, formed between metals and atomically clean semiconductor surfaces, models of the interface which assume an abrupt ordered situation are not useful. Chemical interactions and interdiffusion are commonly observed at such intimate interfaces. In the future, problems of contacts to small devices and to two dimensional systems and quantum well structures will be increasingly studied and there is a clear need to develop and apply microscopic experimental techniques which can explore the formation of barriers to these interesting systems.

REFERENCES

1. J. Bardeen, Phys. Rev. 71, 717 (1947).

2. V. Heine, Phys. Rev. A 138, 1689 (1965).

3. J.C. Inkson, J. Phys. C. 6, 1350 (1973)

4. S.G. Louie, J.R. Chelikowsky, and M.L. Cohen, Phys. Rev. B 315, 2154 (1977).

5. R.H. Williams, Contemp. Phys. 23 329 (1982).

6. G.P. Srivastava, I. Singh, V. Montgomery and R.H. Williams, J. Phys. C: Solid State Phys., 16 (1983) 3627-3640.

7. A. Kahn, E. So, P Mark and C.B. Duke J. Vac. Sci. Technol. 15 780 (1978).

8. D.J. Chadi Phys. Rev. B 18 (1800) (1978).

9. A. Huijser and J. Van Laar, Surface Sci. 52 (1975) 202.

10. W. Monch and J.H. Clemens, J. Vacuum Sci. Technol. 16 (1979) 1238.

11. J. Massies, Devoldere and N.T. Linh, J. Vacuum Sci. Technol. 16 (1979) 1244.

12. J.M. Palau, E. Testemale, A. Ismail and L. Lassabatere, Solid State Electron. 25 (1982) 285.

13. D.G. Welkie and M.G. Lagally, J. Vacuum Sci. Technol. 16 (1979) 784.

14. J.D. Joannopoulos and E.J. Mele, in Physics of Semiconductors 1978. Inst. Phys. Conf. Ser. 43 (Inst. of Physics, London, 1978) p. 1.

15. R.H. Williams, J. Vacuum Sci. Technol. 18 (1981) 929.

16. W.E. Spicer, I. Lindau, P.R. Skeath, C.Y. Su and P.W. Chye, Phys. Rev. Letters 44 (1980) 420.

17. See R.H. Williams Surface Science 132 (1983), 122.

18. J. Freeouf and J.M. Woodall, Appl. Phys. Letts. 39 (1981) 727.

19. G.P. Srivastava, in: Proc. Conf. on Semi-Insulating III-V Materials, Nottingham,(1980).

20. M.S. Daw and D.L. Smith, Appl. Phys. Letters 8 (1980) 690.

21 R.E. Allen and J.D. Dow, J. Vacuum Sci. Technol. 19 (1981) 383.

22. M.S. Daw and D.L. Smith, Solid State Commun. 37 (1981) 205.

23. G.J. Hughes, A. McKinley, R.H. Williams and I.T. McGovern, J. Phys. C, 15 (1982) L159.

24. J. Massies, J. Chaplart, M. Laviron and N.T. Linh, Appl. Phys. Lett. 38 (9) (1981).

25. V. Montgomery, R.H. Williams and G.P. Srivastava, J. Phys. C14 (1981) L191.

26. V. Montgomery and R.H. Williams, J. Phys. C, 15 (1982) 5887.

27. C.F. Brucker and L.J. Brillson, J. Vacuum Sci. Technol. 19 (1981) 617.

28. M.H. Patterson and R.H. Williams, J. Crystal Growth 59 (1982) 281-288.

29. J.M. Shannon, Appl. Physics Letts. 25 (1974) 75.

30. R.A. Street, R.H. Williams and R.S. Bauer, J. Vacuum Sci. Technol. 17 (1980) 1001.

31. R.A. Street and R.H. Williams, J. Appl. Phys. 52 (1981) 402.

32. P.W. Chye, C.Y. Su, I. Lindau, C.M. Garner, P. Pianetta and W.E. Spicer, Surface Sci. 88 (1979) 439.

NONEQUILIBRIUM PHONONS IN SEMICONDUCTORS: POWER DISSIPATION OF HIGHLY LASER- EXCITED ELECTRON-HOLE PLASMAS

P. Kocevar

Institut Für Theoretische Physik

Universität Graz, 8010 Graz, Austria

INTRODUCTION

Preparation of semiconductor surfaces by laser pulses of high intensity is one of the most promising techniques for microelectronics. The best studied applications are activation of implanted or surface-deposited dopants and recrystallization of amorphized (implantation damaged) surfaces. In spite of their technological importance the physical mechanisms underlying these highly non-equilibrium processes are still not understood. A widespread controversy exists about the way, in which the initially excited electron-hole plasma induces the annealing of the material.

The first detailed experimental and theoretical investigations (Auston et al. 1978, Baeri et al. 1979) were performed on crystalline as well as amorphized silicon surfaces. In particular the occurrence of a transient phase of high reflectivity corroborated the generally accepted thermal annealing model of an initially molten surface layer (of metallic reflectivity) and its refreezing after the end of the excitation pulse. The more surprising were the subsequent time-resolved reflectivity and Raman data of Lo and Compaan (1980), indicating the presence of a highly excited electron-hole plasma of very large density (of several $10^{22}cm^{-3}$, far above the intrinsic thermal density of $\approx 10^{19}cm^{-3}$) lasting for more than fifty nanoseconds. This long lifetime of the extremely hot and dense plasma implied a drastic reduction of the rate of energy transfer from the photoexcited free carriers to the lattice; these phonon emissions had been expected to take place on a picosecond time scale. The authors demonstrated the "cold" lattice by the existence of solid-phase Raman signals from the excited surface and by the fact that the ratio of Stokes to Antistokes intensities gave lattice temperatures only a few hundred degrees above room temperature. Such low temperatures would be in strong contrast to the melting temperatures of cyrstalline or amorphous silicon (e.g. T_m (c-Si)=1685°K). This was experimental support to an already existing alternative picture of the annealing process.

Based on an earlier theory of free-carrier induced softening of transverse acoustic modes (Heine, Van Vechten 1976), Van Vechten et al. (1979) had rejected the interpretation of Auston's data in terms of a thermal melting and proposed a nonthermal "plasma annealing" mechanism, in which sufficiently many valence-bonds are broken by an extremely dense carrier plasma to make the crystal fluid. This theory is supplemented by the assumption, that the high temperature gradients lead to local band-gap variations and corresponding gradients of the density and chemical potential of the carriers, resulting in negative diffusion coefficients and plasma self-confinement effects (Wautelet, Van Vechten 1981, Wautelet 1981). The remaining basic ingredient of this model, namely the assumption of a strongly reduced carrier-phonon interaction, was questioned in various theoretical papers (Yoffa 1980, Bok 1981, Combescot 1981), predicting maximal values of the actually present plasma densities much below the density-threshold for the plasma annealing mechanism.

In the meantime, a great number of experimental and theoretical investigations on pulsed laser annealing or spectroscopy gave further support to the thermal annealing model. Nathan et al. (1980) performed time-resolved reflectivity measurements in the plasma resonance regime, confirming Yoffa's (1980) theoretical results. Bentini et al. (1981) found no changes in the annealing characteristics after covering the silicon surface by a thin metallic layer to prevent photons from creating a carrier plasma in the semiconductor. Larsen et al. (1981), confirmed the melting mechanism in a time-resolved synchrotron X-ray analysis of the surface and Stritzker et al. (1981) again found surface temperatures $\approx T_m$ in time-of-flight measurements of surface evaporated atoms during the annealing process.

Nevertheless, more recent work (Aydinli et al. 1981, Compaan et al. 1982, Von der Linde 1982) still confirms the earlier Raman results with further refined techniques, so that Stokes/Antistokes ratios indeed give evidence for low lattice temperatures. In accordance with Van Vechten's theory, Eberhard et al. (1982) interpreted their photoemission data in terms of very high plasma densities, but this explanation was later challenged by experimental (Williams et al., 1982) and theoretical work (Car et al. 1983) invoking surface reconstruction effects. A similar controversy exists about measurements of the lattice constant during the annealing phase and their interpretation in terms of the temperature gap or of the plasma induced weakening of the valence-bonds (Van Vechten 1982, Car et al. 1983).

The most recent reflectivity measurements with picosecond and femtosecond laser pulses (Lowndes 1982, Kurz et al. 1982, Shank et al. 1983) support the thermal model, and a further sophistication of the earlier theories of Combescot and Bok predicts the possibility of a certain plasma instability at high temperatures, which is driven by the carrier-phonon interacton and might cause the melting of the lattic (Combescot and Bok 1982), with carrier concentrations around $10^{21}cm^{-3}$ at the transition temperature (see also Ambegaokar and Biswas 1983).

All these controversial experimental and theoretical results are still extensively debated, and it is obvious that more details of the dynamics of highly photoexcited carriers will have to be clarified before a consistent picture of laser annealing could evolve. One such aspect concerns the role of optical phonon amplification in reducing the cooling rates of the hot electron-hole plasma. This mechanism was proposed to explain an unexpectedly slow energy relaxation of carrier plasmas in gallium arsenide (GaAs) after picosecond or subpicosecond laser excitation (Shah 1981).

The following section presents a Boltzmann transport analysis of the coupled carrier-phonon system. It will be shown that in polar materials nonequilibrium phonon effects can indeed lead to a strongly reduced plasma-cooling and that this reduction does not depend on the details of the free carrier screening of the electron and hole phonon couplings. This latter finding corrects earlier attempts to explain the greater part of the reductions by the increased screening of the long-range couplings of the carriers to the longitudinal optical (LO) phonon modes (Shah 1981, Yoffa 1981). For elemental materials such as Si, however, the theory predicts small optical phonon disturbances, in agreement with recent experimental findings from time resolved Raman spectroscopy (Compaan 1982, Gobel 1982).

A TRANSPORT ANALYSIS OF THE PLASMA COOLING

Picosecond studies of the transmission (Von der Linde and Lambrich 1978, Leheny et al. 1979, Shah 1981) and of the luminescence (Gobel and Hildebrand 1978, Tanaka et al. 1980) of laser excited carriers in GaAs at low temperatures showed, that the cooling rates of the initially created hot and dense plasma were too low to be explained solely by the increased free carrier-screening of the couplings to the phonons. The properly screened power loss through emission of longitudinal optical (LO) phonons had to be artificially reduced by a further factor $^1/_2$ to get agreement with the experimentally determined plasma temperatures as functions of time after the excitation pulse (Leheny et al. 1979, Shah 1981). The existence of a plasma temperature T_c common to both electrons and holes was experimentally established by the spectral shape of the spectra.

The most reasonable explanation of this discrepancy was the presence of strong LO phonon heating during the pulses; in this way the efficiency of the phonon system as a cooling agent for the carriers should be strongly reduced. Later attempts to introduce this effect into the numerical analysis of the data used the concept of "phonon temperature", somewhere between the temperature of the lattice and the plasma, as adjustable parameter in the usual theoretical expressions for the energy relaxation of the carriers into an equilibrium phonon system (Van Driel 1979, Pugnet et al. 1981). At the same time the detailed analysis of Yoffa (1981) showed, that under conditions of phonon equilibrium free carrier screening would strongly reduce the carrier cooling in GaAs already for small plasma densities ($n_{e,h}>10^{17}cm^{-3}$), whereas this density threshold for Si should be much higher ($n_{e,h}>10^{21}cm^{-3}$), as consequence of the nonpolarity and many valley band structure of this material.

To eliminate the uncertainties of the phonon temperature concept and to obtain a parameter free theoretical estimate of the effects of nonequilibrium phonons a recent investigation (Kocevar et al. 1983, Potz and Kocevar 1983) extended earlier work on phonon disturbances in high field dc-transport (Kocevar 1972 and 1980) to the case of photo-excitation. The solution of the coupled time-dependent Boltzmann equations for electrons, holes, LO phonons, as well as TO (transverse optical) and acoustic phonons was achieved through the (experimentally justified) use of Fermi-distributions with time dependent T_c and chemical potentials for electrons and holes. Within this framework the authors performed a stepwise numerical time integration of the Phonon-Boltzmann equation and the energy balance for the carriers. During the pulse the carrier-generation rate was assumed to be constant, being determined by the experimentally given pair concentration n at the end of the laser-pulses.

Figure 1 shows the improvement of the theoretical carrier temperatures when phonon amplification is taken into account. Theoretical and experimental plasma temperatures are now within ≈35% (percent), a satisfactory result in view of the uncertainties in the analysis of the spectroscopic data (Shah 1981). The rather intricate interplay between LO and TO phonon emission and of the effects of free carrier screening can be inferred from **figure 2,** which shows T_c and the phonon-occupation numbers N_{LO} and N_{TO} of a strongly coupling LO and TO mode (of wavevectors q_{LO} and q_{TO} respectively) for the case $n=1.10^{18}cm^{-3}$ of figure 1, corresponding to excitation of GaAs (energy gap=1.50eV, lattice temperature T_L=10°K) with a 0.5 picosecond laser pulse of energy flux density $3.10^8 W/cm^2$ with photons of energy hν=1.65eV. $N_{c,LO}$ and $N_{c,TO}$ are Planck distributions heated to the carrier temperature T_c. They mark the borderline between predominant emisson (if $N<N_c$) or absorption (if $N>N_c$) of the phonons by the carriers.

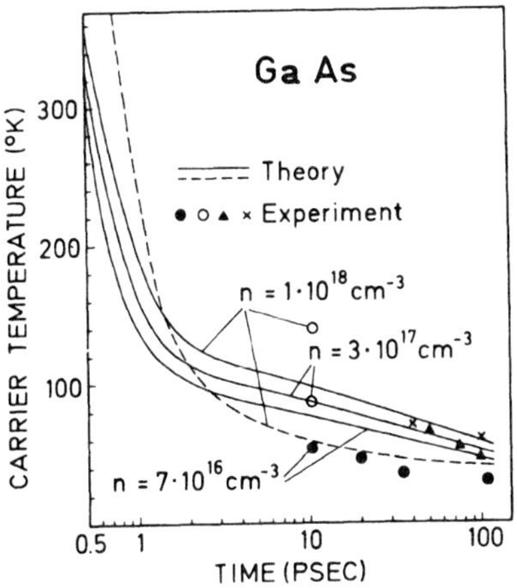

Figure 1. Carrier temperature as function of time (from Shah 1981). Experiment: full and open circles: Leheny et al. 1979; triangles: Von der Linde, 1978; crosses: Tanaka et al. 1980. Theory: full lines include optical phonon disturbances, dashed line for equilibrium phonons.

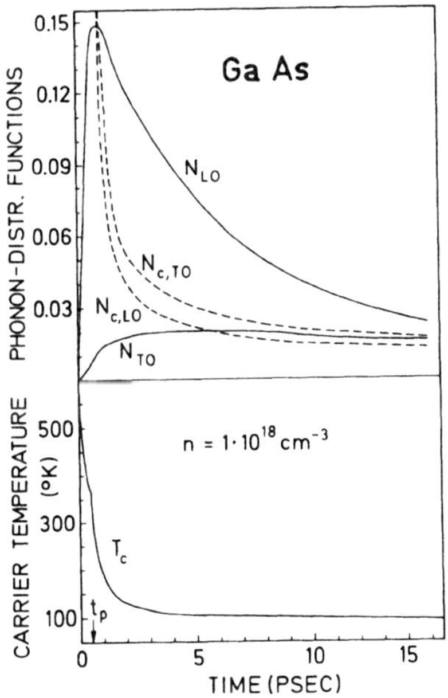

Figure 2. Theoretical phonon distributions and heated Planck distributions as functions of time for $n=1.10^{18}$ cm^{-3}; $q_{LO}=3.4.10^{6}$ cm^{-1}, $q_{TO}=5.5.10^{6}$ cm^{-1}; duration of laser pulse: $t_p=0.5$ picoseconds.

Since the polar coupling of electrons and holes to LO phonons of small wave vectors is much stronger than the short-range hole-TO phonon coupling, these LO modes grow much faster than the TO-modes during the excitation pulse. But as a consequence their effectiveness in cooling the carriers decreases rapidly, so that the TO phonon emission will soon dominate the energy relaxation of the carrier system. Because of the relatively long lattice-dynamical lifetime of optical phonons with respect to their decay into pairs of acoustic phonons (at T_L=10°K roughly 10 to 20 picoseconds), no losses of the excited LO phonons occur, until the further decreases of the carrier temperature T_c through TO phonon emission by the holes has diminished the instantaneous value of $N_{c,LO}$ below the actual value of the LO phonon distribution (see figure 2). From now on energy and momentum are fed back from the LO modes to the carriers by reabsorption processes. Obviously the interpretation of the apparent decay-time (≈5 picoseconds) of the LO phonon population must not be interpreted as the lattice dynamical LO phonon lifetime, but just as the manifestiation of the reabsorption of the initially excited phonons by the rapidly cooling carrier system. This aspect is of crucial importance for any "lifetime"-measurements of laser-excited optical phonons.

The energy relaxation of the carriers will now be driven by the emission of TO-phonons by holes through the optical deformation potential scattering. This interaction is of short range and, therefore, quite generally much less screened by free carriers than the long range polar coupling. For zone-center holes in III-V compounds the screening even vanishes by symmetry, so that in contrast to earlier expectations (Leheny et al. 1979, Yoffa 1981, Shah 1981) screening plays a minor role for the total power dissipation of the photoexcited plasma in gallium arsenide.

This is no longer true for the much higher carrier-densities under laser annealing conditions, with pulse durations of more than 10 nanoseconds. The much stronger screening of the polar couplings will then outweigh the proportionality of the phonon amplification to the carrier concentration, so that free carrier screening will be the decisive factor for the reduced energy transfer to the LO modes. The total power loss of the carrier plasma will again be dominated by the optical deformation potential coupling, moderately reduced by TO phonon amplification. Eventually, typically after 100 picoseconds, a steady state of the coupled system of carriers, LO and TO phonons will be established, during which the power input from the laser is balanced by the energy losses through carrier recombination and by the decays of optical phonons into pairs of acoustic phonons (Potz and Kocevar 1983).

For nonpolar materials such as silicon, the polar optical couplings are missing, so that even in the case of ultrashort excitation pulses no strong effects of phonon disturbances should occur, because of the lack of two competing energy relaxation channels for the carriers to different optical phonon modes. Moreover, the screening of the optical deformation potentials will now be allowed by symmetry and further reduce possible optical phonon disturbances. This agrees with recent time resolved Raman experiments, in which no optical phonon disturbances were detected (Compaan 1982, Gobel 1982), and justifies the assumption of equilibrium phonon distributions in Yoffa's (1981) estimate of the critical plasma densities for the onset of screening effects in laser excited silicon.

Since for silicon a quenching of the power transfer from the carrier system to the optical phonon modes has to be ruled out, the only remaining possibility of a nonequilibrium phonon effect as "bottle-neck" should be a trapping of the energy somewhere along the relaxation path of the stepwise decaying high frequency phonons. Since extremely long-lived zone boundary transverse acoustic modes have been recently found in polar semiconductors (Ulbrich 1981, Tsen et al. 1982), lifetime measurements of these modes in silicon might contribute to clarify the puzzling "cold lattice" -Raman spectra of highly laser-excited silicon surfaces.

ACKNOWLEDGEMENT

This work was supported by the Fonds zur Forderung der wissenschaftlichen Forschung in Osterreich, Projekt S22/07

REFERENCES

- Ambegaokar, V., and Biswas, R., 1983, Phys.Rev.Lett. 50; 285.
- Auston, D.H., Surko, C.M., Venkatesan, T.N.C., Slusher, R.E. and Golovchenko, J.A., 1978, Appl.Phys.Lett. 33: 437.
- Aydinli, A., Lo, H.W., Lee, M.C., and Compaan, A. 1981, Phys.Rev.Lett. 46: 1640.
- Baeri, P., Campisano, S.U., Foti, G., and Rimini, E., 1979, J.Appl.Phys. 50: 788.
- Bentini, G.G., Cohen, C., Desalvo, A., and Drigo, A.V., 1981, Phys.Rev.Lett. 46: 156.
- Bok, J., 1981, Phys.Lett. 84A: 448.
- Car, R., Selloni, A., Janik, J.F., and Pantelides, S.T., 1983, Physical 117B: 1007.

- Combescot, M., 1981, Phys.Lett. 85A: 308.
- Combescot, M., and Bok, J., 1982, Phys. Rev. Lett. 48:413.
- Compaan, A., Lo, H.W., Aydinli, A., and Lee, M.C., 1982, Proc. Mat. Res. Soc. Symp. (Boston 1982, to be published).
- Compaan, A., 1982, private communication.
- Eberhard, W., Brickmann, B., and Kaldor, A., 1982; Solid State Commun. 42: 169.
- Gobel, E.D., and Hildebrand, O., 1978, Phys. Status Solid (b) 88: 645.
- Gobel, E.O., 1982, private communication.
- Heine, V., and Van Vechten, J.A., 1976, Phys.Rev. B13: 1622.
- Kocevar, P., 1972, J.Phys.C; Solid State Phys.5: 3349.
- Kocevar, P., 1980, in Physics of Nonlinear Transport in Semiconductors, Eds. D.K. Ferry, J.R. Barker and C. Jacoboni (Plenum Press, New York 1980) p. 401.
- Kocevar, P., Potz, W., and Porod W., 1983, Physica 117B: 220.
- Kurz, H., Liu, J.M. and Bloembergen, N., 1983, Physical 117B: 1010.
- Larson, B.C., White, C.W., Noggle, T.S., and Mills, D., 1981, Phys.Rev.Lett. 48: 337.
- Leheny, R.F., Shah, J., Fork, R.F., Shank, C.V., and Migus, A., 1979, Solid State commun. 31: 809.
- Lo, H.W., an Compaan, A., 1980, Phys.Rev.Lett.44: 1604.
- Lowndes, D.H., 1982, Phys.Rev.Lett. 48: 267.
- Nathan, M.I., Hodgson, R.T., and Yoffa, E.J. 1980, Appl. Phys. Lett. 36: 512.
- Potz, W., and Kocevar, P., 1983, Phys.Rev. B28: 7040.
- Pugnet, M., Collet, J., and Cornet, A., 1981, Solid State Comm. 38: 531.
- Shah, J., 1981, J. Physique (Paris) 42 (C7): 445.
- Shank, C.V., Yen, Rl, and Hirlimann, C., 1983, Phys.Rev.Lett. 50: 454.

- Stritzker, S.. Pospieszcyk, Bl, and Tagle, J.A., 1981, Phys. Rev.Lett. 47: 356.

- Tanaka, S., Kobayashi, H., Saito, H. and Shionoya, H., 1980, J.Phys.Soc. (Japan) 49: 1051.

- Tsen, K.T., Abramsohn, D.A., and Bray, Rl, 1981, Phys.Rev. B 26: 4770.

- Ulbrich, R.G., 1981, J.Physique (Paris) 42 (C7): 423.

- Van Driel, H.M., 1979, Phys.Rev. B19: 5928.

- Van Vechten, J.A., and Tsu, R., 1979, Phys.Lett. 74A: 422.

- Van Bechten, J.A., 1982, Japan, J.Appl.Phys. 20: L125, and Solid State Commun. 43:807.

- Von der Linde, D., and Lambrich, R., 1978, Phys.Rev.Lett. 42: 1090.

- Wautelet, M., 1981: J.Physique (Paris) 42 (C7): 131.

- Wautelet, M., and Van Vechten, J.A., 1981, Phys.Rev.Lett. B23: 5551 and 5543.

- Williams, R.T., Kabler, M.N., Long, J.P., Rife, J.C., and Royt, T.R., 1982, in Laser and Electron Beam Interactions with Solids, Eds. B.R. Appleton and G.K. Celler (North Holland, Amsterdam 1982).

- Yoffa, E.J., 1980, in Laser and Electron Beam Processing of Materials, Eds. C.W. White and P.S. Peercy (Acad.Press, New York 1980).

- Yoffa, E.J., 1981, Phys.Rev. B23: 1909.

PICOSECOND MEASUREMENTS OF DEVICE AND CIRCUIT TRANSIENT RESPONSE WITH OPTOELECTRIC TECHNIQUES

Robert B. Hammond

Los Alamos National Laboratory
Los Alamos, New Mexico 87545

INTRODUCTION

In this seminar, I would like to give a tutorial introduction and overview of the recently developed optoelectronic techniques which now permit characterization of electronic device and circuit response with picosecond time resolution. It is appropriate that these techniques be discussed at this study institute, since they present the opportunity to make direct measurements of the response of submicron devices in the time regime where these devices operate.

These optoelectronic techniques rely on modelocked dye lasers which can produce optical pulses as short as 5.5×10^{-14} s. [1]. These pulses have been successfully compressed to as short as 3×10^{-14} s and amplified to powers greater than 10^9 W. [2]. Excite-and-probe (or pulse-and-probe) experiments using these extremely short laser pulses have permitted direct measurements of transient phenomena in physical [3], chemical and biological systems with femtosecond time resolution.

By using extensions of the optical excite-and-probe techniques, it is now possible with femtosecond laser pulses to make electrical measurements with picosecond time resolution. [4]. Short light pulses can be used to excite photoconductors to produce picosecond risetime electrical pulses; and can subsequently be used in conjunction with picosecond photoconductors or traveling wave Pockel cells to sample electrical transients with picosecond time resolution. The extremely high speed electrical sampling can also be performed with microvolt sensitivity.

In this paper, we will discuss the picosecond optoelectronic measurement techniques; and their promise for making direct measurements of transient response in the new classes of submicron semiconductor devices.

REQUIREMENTS FOR PICOSECOND ELECTRONIC CHARACTERIZATION

When one examines the problems inherent in making picosecond measurements of device and circuit response, one realizes that substantial improvements over conventional measurement techniques are necessary to make these measurements possible. Current sampling oscilloscope measurements are limited by at best 25 picosecond risetime and at best 10 picosecond trigger jitter. They are also limited by the necessity of transmitting the high bandwidth electrical transient signals with low dispersion to the sampling head. This becomes a particularly severe limitation when trying to make measurements on integrated circuits. The difficulty of getting high bandwidth signals off-chip are severe indeed. In fact, it is seldom possible to even take advantage of the intrinsic limits of the sampling scope when making this type of measurement.

The fundamental advantages of using an optical probe for making picosecond electrical measurements are considerable. First, the availability of extremely short optical pulses has already been mentioned. Also these very short optical pulses may be transmitted macroscopic distances without dispersion, either through optical fibers or through the atmosphere. Also, these pulses may be focused to micron dimensions and be transmitted to remote environments (for example low or high temperature regions) without dispersion. A further advantage is that two pulses can be brought to a single location with easily controllable temporal spacing and extremely low jitter ($\sim 10^{-15}$ s). Finally, the ready availability of 0.1 psec light pulses at $>10^8$ pulses per second permits high speed measurements with considerable signal averaging possible.

These capabilities of optical probes indicate that many orders of magnitude improvement over conventional techniques in electrical measurements may be possible with this approach. The real needs are to be able to generate clean, extremely fast risetime or short duration electrical pulses using short optical pulses, and second to be able to use the short optical pulses to sample electrical transients. In recent years, addressing these problems has been a growing area of research activity. In the following sections of this tutorial, we will discuss the latest results and how they may be used for extremely high-speed device response measurements.

OPTICAL EXCITE-AND-PROBE MEASUREMENTS

The optoelectronic measurement techniques which exploit short light pulses in order to make high-speed electronic measurements are direct outgrowths of measurement techniques developed by laser physicists. We shall first examine these "excite and probe" measurement techniques with an example before we look at the electronic measurement techniques themselves.

In order to investigate physical phenomena on the short time scales made possible by femtosecond lasers, purely optical "excite and probe" measurement techniques have been developed. An example of this type of measurement is the accurate measurement of the time duration of the laser pulses themselves. A commonly used method to measure femtosecond laser pulse length is Second Harmonic Generation (SHG) autocorrelation in potassium dihydrogen phosphate (KDP) crystals. This measurement is shown schematically in Figure 1.

A train of short laser pulses is produced by a modelocked laser. This pulse train is split into two trains in an optical beamsplitter. The first pulse train is used for excitation of the sample and the second pulse train is used as a probe of the sample. The optical path length of the probe beam is varied so that the sample may be probed at various times with respect to the optical excitation produced by the excitation beam. In the SHG autocorrelation measurement, the excitation pulse and the probe pulse pass through the same

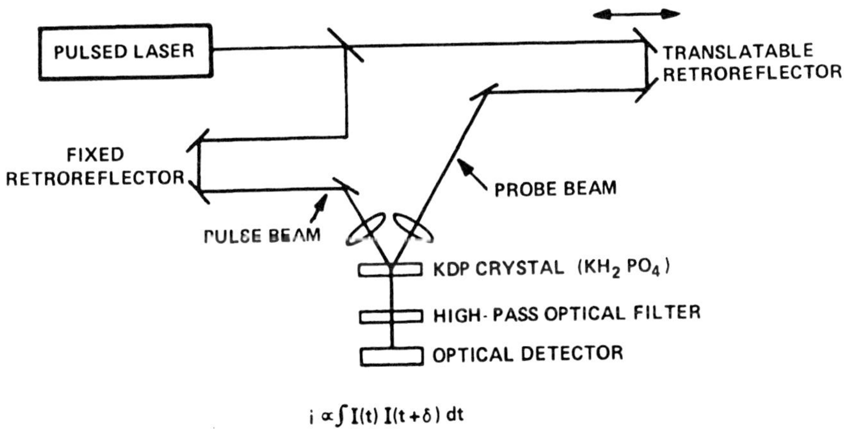

Figure 1. Schematic of the optical pulse length measurement by second harmonic generation autocorrelation in KDP. The train of light pulses from the laser is split into two beams which are subsequently brought back together at a focus in the KDP crystal. One of the two beams has a variable path length provided by the translatable retroflector. When pulses from the two beams simultaneously exist at the focus in the KDP, a second harmonic is produced which is detected by the optical detector.

focus in the KDP crystal. If the two pulses coincide in time at this focus, they can interact through the nonlinear refractive index of the crystal and produce an optical pulse having a momentum which is the vector sum of the momenta of the excitation and probe beams. Since the higher energy light pulses are produced only when the excitation and probe beams coexist in the KDP crystal, the time duration of the optical pulses may be determined by the variation in optical path length of the probe beam over which one sees the SHG signal. The intensity of the SHG light beam is proportional to the autocorrelation of the pulses produced by the modelocked laser:

$$i \propto \int I(t) \quad I(t+\delta)\, dt$$

In a general version of the excite and probe experiment, the excitation pulse produces a change in the sample under study. The probe pulse then samples that change by looking for changes in optical reflectivity, transmission, spectrum, etc.

Figure 2 shows a SHG autocorrelation of our modelocked laser at Los Alamos. The autocorrelation is displayed on an oscilloscope screen for convenience and the time scale is 5 ps/cm. The measurement indicates that the optical pulses are completed modelocked with a FWHM of <0.7 ps.

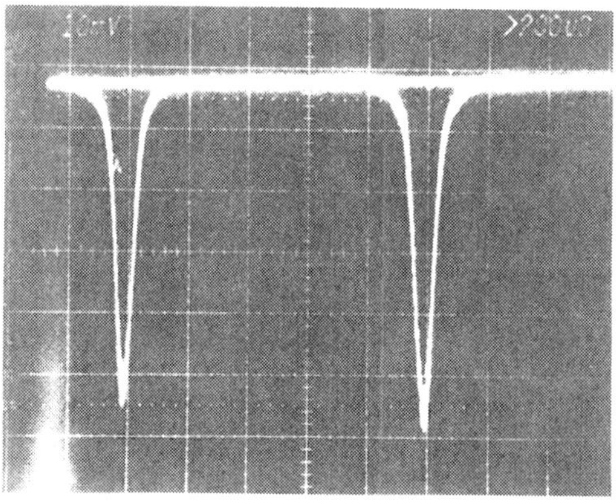

Figure 2. SHG autocorrelation of optical pulses from colliding pulse modelocked laser at Los Alamos. The time scale is 5 ps/cm. The autocorrelation was measured with a commercial optical autocorrelator manufactured by Spectra Physics. The autocorrelation shows no tails and shows an optical pulse full width at half maximum of 0.7 ps. This is actually the resolution of the instrument and is thus merely an upper bound to the actual optical pulse length. There are two separate autocorrelations on this oscilloscope trace which were taken with different time delays to determine the actual time scale.

PHOTOCONDUCTOR PULSERS AND SAMPLING GATES

Photoconductors offer the unique capability of converting short optical pulses into extremely fast risetime electrical pulses and with variable voltage amplitude from microvolts to kilovolts. (See Figure 3). Pulses with risetimes as short as 0.85 ps have been reported and amplitudes up to 130 kilovolts. This is possible due to the lateral, and thus low capacitance geometry of these devices; but mainly because carriers can be created uniformly and simultaneously throughout the entire active region in much less than one picosecond. Capacitances are usually in the tens of femtofarads for easily fabricated devices and can be much lower on special structures.

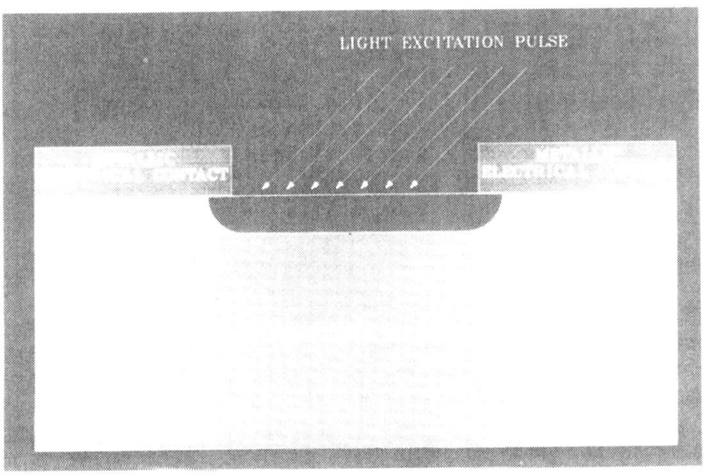

Figure 3. Photoconductor: Method of operation. The photoconductor is fabricated on a high resistivity material which preferably has high carrier mobility. The two electrical contacts are preferably ohmic and low resistance. The speed in turn-on of the photoconductor from an impulse light excitation is limited by the time taken to absorb the light and produce free carriers. There is no carrier transit time involved in turn-on because carriers are produced simultaneously across the entire active region between the contacts. There is also no transit time involved in the turn-off of the device. This is controlled by recombination of free carriers in the active region which occurs uniformly and simultaneously throughout. The fastest reported turn-on time has been reported by Valdmanis et al. at 0.63 ps and the fastest reported turn-off time has been reported by Auston et al. at 2.5 ps.

The current response of a photoconductor may be written:

$$I = \frac{(1 - R)\, q\, v_d\, E}{S\, w}$$

where R is the material reflectivity, q is the electronic charge, v_d is the carrier drift velocity, E is the total energy in the incident optical pulse, S is the photoconductor contact spacing, and w is the photon energy. One can see from this simple equation that with the 100 pJ optical pulse energy which is readily available from a colliding pulse modelocked laser, it is in principle possible to generate electrical pulses of several volts amplitude with subpicosecond risetime. Our results using semi-insulating, Fe-doped InP photoconductors, suggest that this should indeed be possible [5].

Besides their use as fast risetime, optically-activated, electrical pulsers; photoconductors may also be used as optically-activated, electrical sampling gates. In this case, the aperture time of the gate is determined by the bulk recombination of optically excited free carriers in the active region of the photoconductor. Aperture times as short as 2.5 ps have been reported [6].

Short lifetime may be obtained by a number of techniques in semiconductor photoconductive materials. These include using deep level impurity doping, radiation damage, polycrystalline material and amorphous material. Thus far, the most successful method of obtaining picosecond recombination lifetimes in material with moderate mobility has been radiation damage in III-V semiconductors [4]. Currently, research is ongoing at a number of laboratories attempting to develop higher mobility and shorter lifetime photoconductive materials.

Because photoconductors can both produce short pulses and sample them, it is possible to measure photoconductor response using an analog of the SHG autocorrelation technique for measuring short optical pulses. The experimental setup is shown in Figure 4. Once again, in this measurement one uses pulse and probe optical beams. Microstrip transmission lines are formed on the surface of the photoconductor material in the geometry shown. The photoconductors are the two gaps in the microstrip. One photoconductor acts as a voltage pulser and the other samples the voltage pulse produced. The average current measured is the autocorrelation of the photoconductor response, because the two photoconductors are identical. Figure 5 shows an actual photoconductor autocorrelation structure formed on a GaAs:Cr wafer. Figure 6 shows results of an optoelectronic autocorrelation measurement of the response of amorphous Si photoconductors [7].

In a similar measurement, it is possible to use a fast photodiode. Results of such a measurement are shown in figure 7. The photoconductor sampling time in this measurement was 12 ps. [8].

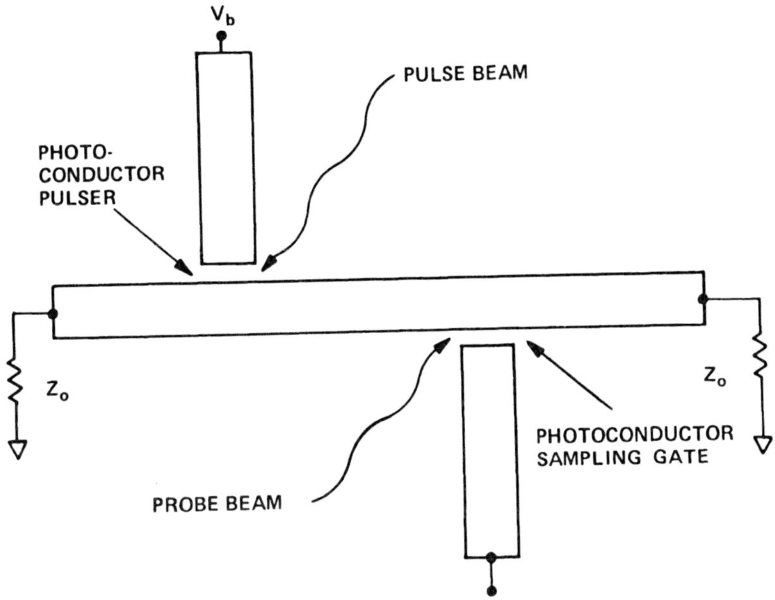

$$I_{avg} \propto \int i(t)\, i(t+\delta)\, dt$$

Figure 4. Experimental schematic of the optoelectronic autocorrelation measurement of photoconductor response.

Figure 5. This is an optoelectronic autocorrelation structure formed on a GaAs wafer. The metal lines form a 50 ohm mocrostrip transmission line. The gaps in the lines are Si photoconductors. The lines are 325 microns wide and the gap widths are 25 microns.

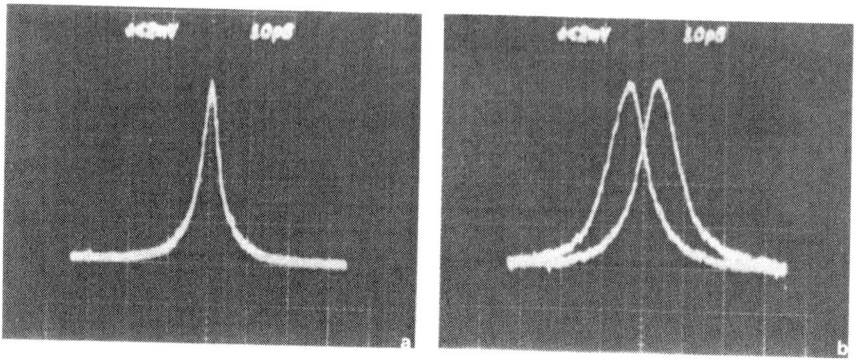

Figure 6. (a) SHG autocorrelation of optical pulses used for measurement in (b). (b) Optoelectronic autocorrelation of a Si film using circuit similar to that shown in figure 4. The two traces correspond to normal biasing scheme and time-reversed scheme with the bias and signal interchanged. (After D.H. Auston et al. [7]).

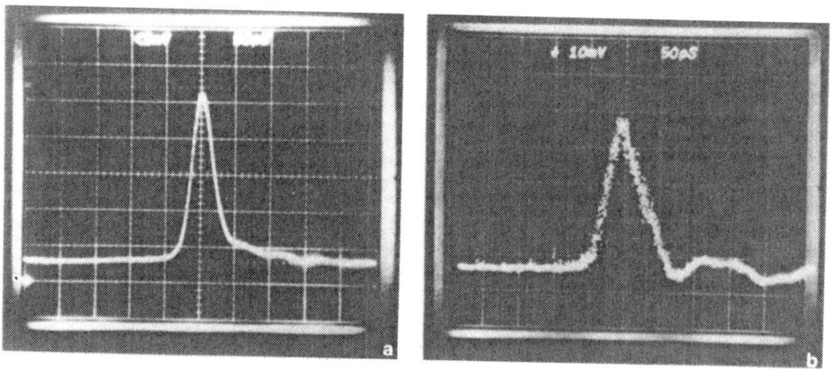

Figure 7. (a) Si photodiode response measured with a photoconductor sampling gate. The horizontal scale is 50 ps/cm. (b) Same photodiode measured on a sampling oscilloscope with a specified risetime of 25 ps. The horizontal scale is again 50 ps/cm. In contrast to the photoconducting sampling gate, the sampling oscilloscope introduces a considerable instrument broadening, has a higher noise level, and worse reflections. (After d.H. Auston and P.R. Smith [8]).

PICOSECOND DEVICE CHARACTERIZATION WITH PHOTOCONDUCTORS

With the combined capabilities of producing picosecond risetime or duration electrical pulses, and also sampling electrical transients with picosecond resolution; it is possible to fabricate a complete device or circuit characterization experiment using picosecond photoconductors in conjunction with a subpicosecond pulsed laser. The structure required for this measurement is shown schematically in Figure 8. A measurement of this type has been accomplished and reported in the literature by Smith and Auston [9].

The structure used by Smith and Auston is shown in Figure 9. In this structure there are two one-inch square Si-on-sapphire (SOS) wafers with 50 ohm microstrip transmission lines formed on them. Inbetween the two SOS wafers is a high-speed GaA FET (Avantek AT-8060). The pulsing photoconductor produced 15 ps FWHM voltage pulses with smplitude 0.3V. These pulses were transmitted to the gate of the FET along the microstrip. The sampling photoconductor had an aperture time of about 15ps and sampled the response of the FET to the voltage impluse provided by the pulsing photoconductor. The results of the measurement are shown in Figure 10. The measurement system had a FWHM response of 33 ps. This response was deteriorated considerably compared to the intrinsic response of the photoconductors by dispersion in the microstrip transmission lines. The very low jitter and large signal-to-noise in the measurement permits considerably more data analysis than would be possible with sampling scope data. This analysis permits an estimate of the fast component of the FET response as 23 ps. To improve the time resolution of this measurement, faster photoconductors and shorter transmission lengths are needed.

ELECTRO-OPTIC SAMPLING

A very promising technique for subpicosecond electrical sampling has recently been developed by J. Valdmanis [10] and coworkers at the University of Rochester. In their approach, a photoconductor is used to generate a fast rising electrical pulse which is then sampled optically in a traveling wave Pockel cell. The experimental setup is shown schematically in Figure 11. The time response of the sampling measurement is limited by the optical pulse length (0.2ps) and the time response of the Pockel effect (<<1 ps).

The traveling wave Pockel cell is shown in Figure 12. The photoconductor material (GaAs:Cr) and the Pockel cell material ($LiTaO_3$) are placed in close contact and have a continuous microstrip transmission line formed on them. The excitation beam causes the photoconductor to produce a fast rising electrical pulse which travels down the transmission line. The polarized probe beam passes through the $LiTaO_3$ crystal immediately under the center conductor of the transmission line. The electric fields produced by the electrical wave

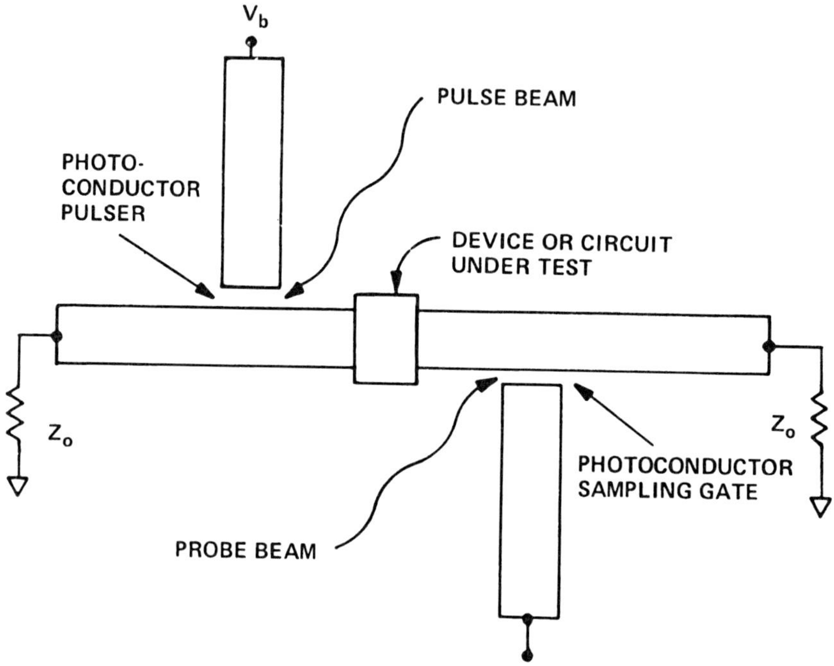

Figure 8. Schematic diagram of an optoelectronic characterization measurement. This diagram is identical to that shown in Figure 4 except that in between the pulsing and sampling photoconductors is placed a device or circuit to be tested. The pulsing photoconductor produces a fast risetime electrical pulse which goes to the device under test. The sampling photoconductor then samples the response of the device with high time resolution.

cause a change in polarization in the probe beam. The induced change in polarization is then a measure of the magnitude of the electrical pulse.

Figure 13 shows a photoconductor-generated, electrical step pulse which has been sampled using this electro-optic sampling technique [11]. The 10% to 90% risetime of this pulse is 0.85 ps. The voltage sensitivity of the measurement is less than 100 microvolts.

D.M. Bloom and coworkers [12] at Hewlett-Packard Research Lab have recently used the electro-optic sampling technique developed at the University of Rochester. They have assembled a 100 GHz characterization apparatus for measuring GaAs photodiode response.

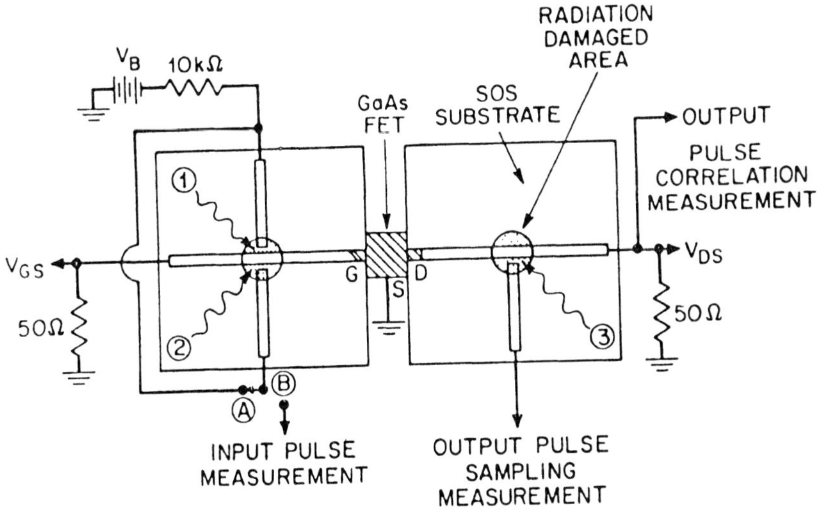

Figure 9. Schematic of optoelectronic circuit used to measure electronic impulse response of GaAs FET. Picosecond optical pulses indicated by wavey lines 1, 2 and 3 are focused on gaps in the microstrip transmission lines where very high-speed photoconductors have been made by radiation damage to the SOS wafers. (After P.R. Smith et al. [9]).

SUMMARY AND CONCLUSIONS

I have tried to show here how the intrinsic potential of short pulse lasers for making extremely high-speed device and circuit characterization measurements may be utilized. The experimental techniques are definitely in their early stage of development, but impressive results have already been achieved. Several approaches are available and probably more will be developed. Each has its advantages and disadvantages.

Electro-optic sampling has a significant advantage compared to photoconductor sampling. This is that extremely short lifetime photoconductors are not needed. The fast response resides in the intrinsic, and extremely high speed of the Pockel effect. The sensitivity of the measurement is in the microvolt range, sufficient for most practical applications. The disadvantage of this technieue compared to photoconductor sampling is the need for an electro-optic material, and the stricter requirements on the geometry of the measurement.

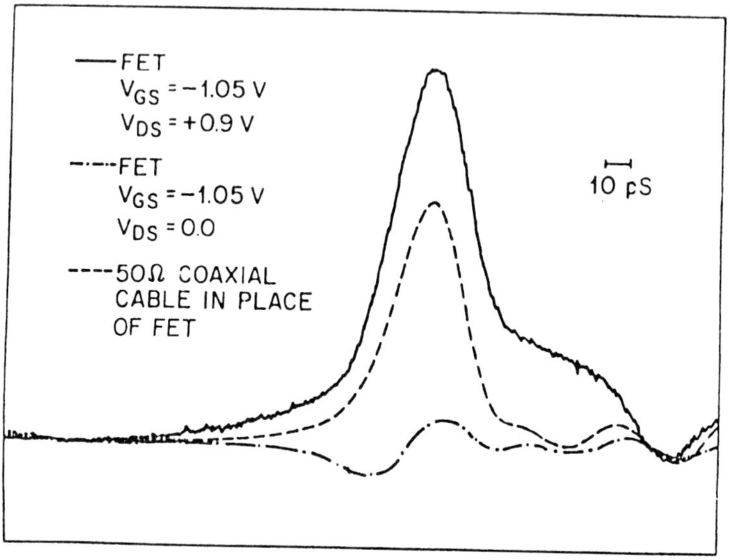

Figure 10. Electronic response of FET using circuit in Figure 9. Solid line is impulse response using photoconductor #1 as pulse generator and #3 as sampling gate under bias conditions for optimum gain (+3.7dB). Dashed curve is impulse response with FET replaced by 50 ohm transmission line. Dashed curve has been translated to align peak with solid curve. Dash-dot curve is impulse response of FET at no drain bias. (After P. Smith et al. [9]).

The electrical signal must be transmitted across a transition between transmission line materials, like the transition between GaAs and $LiTaO_3$ in the measurement described above. The photoconductor sampling has the advantage that the entire measurement can be formed on a single semiconductor substrate. Fast photoconductors have already been demonstrated in GaAs, InP and Si; so that the technologically important semiconductor devices and integrated circuits may be analyzed with this approach. This means that the very severe constraints imposed by signal dispersion in making picosecond device and circuit characterization measurements are considerably easier to deal with using photoconductor sampling.

With active research ongoing in this optoelectronic characterization area in a number of laboratories, we shall certainly see these techniques utilized extensively in the coming years to broaden our understanding of picosecond transport phenomena in semiconductors and picosecond device and circuit response.

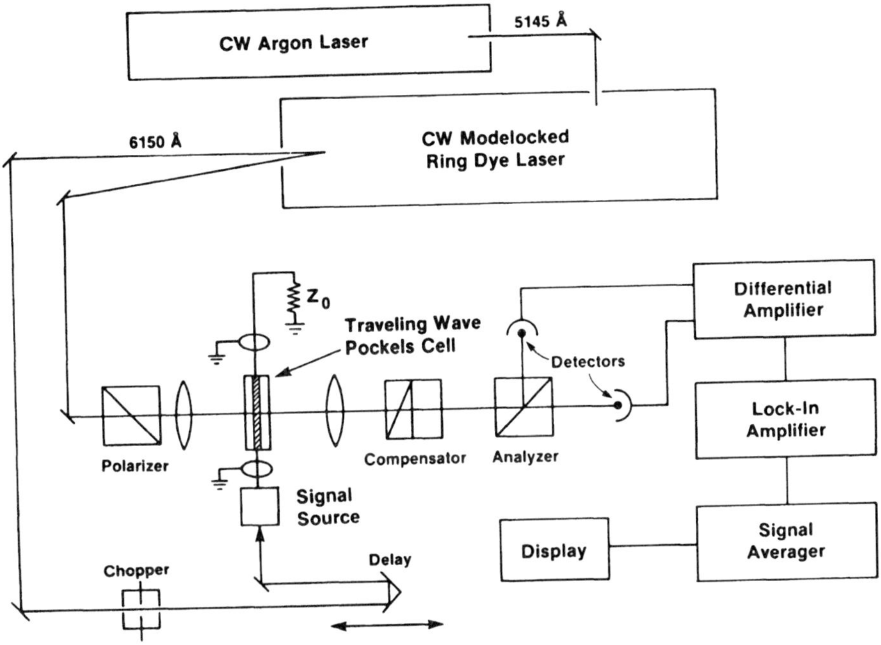

Figure 11. Experimental schematic of an electro-optic sampling measurement. A modelocked ring dye laser produces two beams of 0.12 ps pulses. One beam is chopped and used to excite a photoconductor pulser. The other beam passes through a $LiTaO_3$ traveling wave Pockel cell and samples the transmitted electrical signal. Two detectors measure the amplitudes in both polarizations of the transmitted beam and the difference of these signals indicates the electric field strength in the Pockel cell. (After Valdmanis et al. [10]).

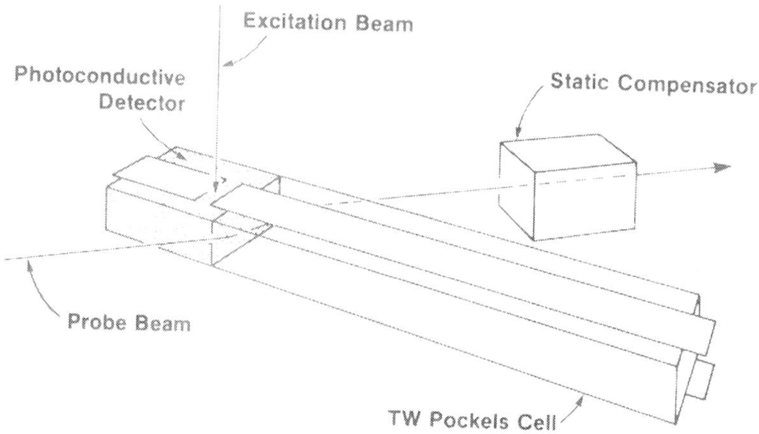

Figure 12. Traveling wave Pockel cell for electro-optic sampling measurement of Figure 11. The excitation beam causes the photoconductor to produce an electrical pulse which is transmitted down the transmission line and sampled by the probe pulse. (After Valdmanis et al. [11]).

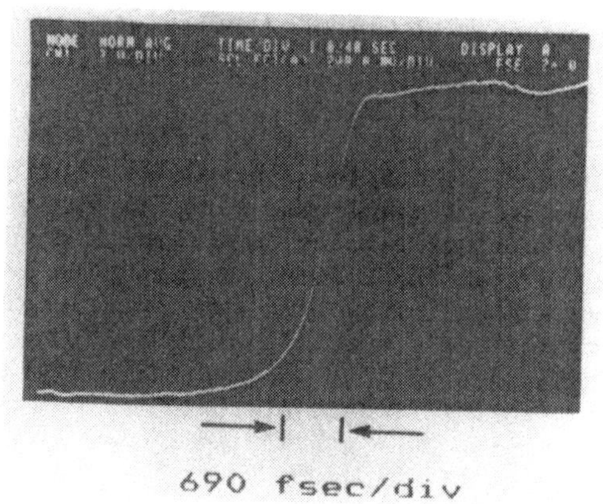

Figure 13. Photoconductor pulse risetime measured by electro-optic sampling. The Pockel cell was a 100-micron thick $LiTaO_3$ substrate and the photoconductor was a GaAs:Cr device. The risetime of the pulse is 0.85 ps. (After J.A. Valdmanis et al. [11]).

REFERENCES

1. J.M. Halbout and C.L. Tang. Autocorrelation Trace of a 55-fs Laser Pulse Train Obtained Without Pulse Compression. IEEE Journal of Quantum Electronics, Vol. QE-19, pp. 487-488, 1983.

2. C.V. Shank. Measurement of Ultrafast Phenomena in the Femtosecond Time Domain. Science, Vol. 219, pp. 1027-1031, 1983.

3. C.V. Shank, R.L. Fork, B.I. Greene, F.K. Feinhart and R.A. Logan. Picosecond Nonequilibrium Carrier Transport in GaAs. Applied Physics Letters, Vol. 38, pp. 104-105.

4. D.H. Auston and P.R. Smith. Picosecond Optical Electronics for High-Speed Instrumentation.. Laser Focus, April, 1982, pp. 89-93.

5. R.B. Hammond, N.G. Paulter, A.E. Iverson and R.C. Smith. Sub-100 ps Bulk-Recombination-Limited InP:Fe Photoconductive Detector. Technical Digest of the International Electron Devices Meeting, December, 1981, Washington, D.C., pp. 157-160.

6. R.B. Hammond, N.G. Paulter, R.S. Wagner, T.E. Springer and M. MacRoberts. InP:Fe Photoconductors as Photodetectors. IEEE Transactions on Electron Devices, Vol. ED-30, pp. 412-415, 1983.

7. D.H. Auston, Impulse Response of Photoconductors in Transmission Lines. IEEE Journal of Quantum Electronics, Vol. QE-19, pp. 639-648, 1983.

8. D.H. Auston, A.M. Johnson, P.R. Smith and J.C. Bean. Picosecond Optoelectronic Detection, Sampling and Correlation Measurements in Amorphous Semiconductors. Applied Physics Letters, Vol. 37, pp. 371-373, 1980.

9. D.H. Auston and P.R. Smith. Picosecond Optical Electronic Sampling: Characterization of High-Speed Photodetectors. Applied Physics Letters, Vol. 41, pp. 599-601, 1982.

10. P.R. Smith, D.H. Auston and W.M. Augustyniak. Measurement of GaAs Field-effect Transistor Electronic Impulse Response by Picosecond Optical Electronics. Applied Physics Letters, Vol. 39, pp. 739-741, 1981.

11. J.A. Valdmanis, G. Mourou and C.W. Gabel. Picosecond Electro-optic Sampling System. Applied Physics Letters, Vol. 41, pp. 211-212, 1982.

12. J. Valdmanis, G. Mourou and C. Gabel. Subpicosecond Electrical

Sampling. IEEE Journal of Quantum Electronics, Vol. QE-19, pp. 664-667, 1983.

13. S.Y. Wang, D.M. Bloom and D.M. Collins. 20 GHz Bandwidth GaAs Photodiode. Applied Physics Letters, Vol. 42, pp. 190-192, 1983.

14. B.H. Kolner, David M. Bloom and P.S. Cross. Characterization of High-Speed GaAs Photodiodes Using a 100 GHz EO Sampling System. Technical Digest of the Conference on Lasers and Electro-Optics, May, 1983, p. 170.

INDEX

CPSIA information can be obtained at www.ICGtesting.com
Printed in the USA
LVOW10s1910120114

369107LV00004B/57/P

9 781489 923844